metallurgy

Second Edition

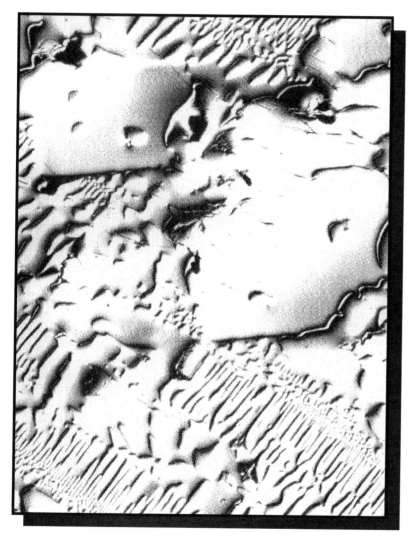

AMERICAN TECHNICAL PUBLISHERS, INC.
HOMEWOOD, ILLINOIS 60430

B. J. Moniz

Acknowledgments

The author and publisher are grateful to the following companies and organizations for providing technical information and assistance.

Aluminum Association
American Iron and Steel Institute
American Society for Metals
American Welding Society
Branson Ultrasonics Corporation
BUEHLER LTD.
Cianflone Scientific
CMI International
Copper Development Association Inc.
E.I. Du Pont De Nemours and Company
Equotip™ Associates

LeBlond Makino Machine Tool Company
LECO Corporation
National Association of Corrosion Engineers
Norton Company
Shore Instrument and Manufacturing Co.
Technicorp
Tempil, Big Three Industries, Inc.
Tinius Olsen Testing Machine Co., Inc.
TN Technologies, Inc.
Wilson® Instruments, Inc.

2 3 4 5 6 7 8 9 - 94 - 9 8 7 6 5 4

Printed in the United States of America

ISBN 0-8269-3509-5

Contents

Introduction

Metallurgy covers all aspects of metallurgical engineering, which include the three areas of extractive, mechanical, and physical metallurgy. The textbook includes both theory and application of metallurgical principles as applied to the conditioning, design, identification, selection, testing, and processing of metals and alloys. The textbook covers topics such as heat treatment, crystal structures, phase diagrams, materials standards, nondestructive and destructive testing, casting, forming, machining, and joining.

Measurements are generally shown in the metric system followed by the equivalent in the U.S. system in parentheses. The following formulas are used for temperature conversion:

$$°F = (1.8 \times °C) + 32$$

$$°C = \frac{(°F - 32)}{8.1}$$

nearest 5°?

Temperatures are not rounded for melting points and phase changes. They are rounded for ranges. For example, $140°C \times 1.8 + 32 = 284°F$ for melting points and phase changes and $285°F$ for ranges.

The formula for converting temperature differentials is $°F = °C \times 1.8$ or $°C = \dfrac{°F}{1.8}$. For example, a heating rate of 10°C per minute equals 20°F per minute.

$$°F = °C \times 1.8$$

$$°F = 10 \times 1.8$$

$$°F = 18 \text{ (round to 20)}$$

$$°F = \mathbf{20}$$

See page 219 for an example of temperature differentials.

Of particular importance are metallurgical principles that are illustrated through formulas and examples, which are presented in a step-by-step format. For example:

$$D = \frac{M}{V} \text{ where} \qquad D = \text{density} \qquad\qquad M = \text{mass of part (in g)} \qquad V = \text{volume of part (in cm}^3)$$

For example, a copper specimen has a mass of 89.6 g and a volume of 10 cm^3. What is the density of the copper?

$$D = \frac{M}{V} \qquad\qquad D = \frac{89.6}{10} \qquad\qquad D = \mathbf{8.96 \text{ g/cm}^3}$$

The comprehensive Appendix contains useful supplementary information. The Glossary provides an easy-to-find format of definitions of key terms. The Index is cross-referenced so that information can be found easily by using key words.

Metallurgy 1

Materials sciences are devoted to the study of engineering materials. There are five branches of materials sciences of which metallurgy is one. Metallurgy is the science and technology of metals. It is the oldest of the sciences devoted to the study of engineering materials. Metallurgy has evolved into three separate groups: extractive, mechanical, and physical. Extractive metallurgy is the study of the extraction and purification of metals from their ores. Mechanical metallurgy is the study of the techniques and mechanical forces (factors) that shape or make finished forms of metal. Physical metallurgy is the study of the effect of structure on the properties of metals. Metals and alloys may be identified in various ways, may exhibit distinct properties, and are processed in specific ways.

MATERIALS SCIENCES

Materials sciences consist of scientific and technological aspects of materials that are used to make engineering materials. Materials sciences consist of five branches, four of which are devoted to a particular class of engineering materials. These four branches are metallurgical engineering, ceramic engineering, polymer engineering, and composite engineering. Materials engineering is the final branch, and it compares the properties of the various classes of engineering materials. See Figure 1-1.

Metallurgical Engineering

Metallurgical engineering or *metallurgy* is the study of metals and is the oldest science devoted to the study of engineering materials. The growth of metallurgy during the past 150 years has led to its division into three well-defined groups: extractive, mechanical, and physical metallurgy. These three groups are usually pursued as separate areas of study.

Extractive Metallurgy. *Extractive metallurgy* is the study of the extraction and purification of metals

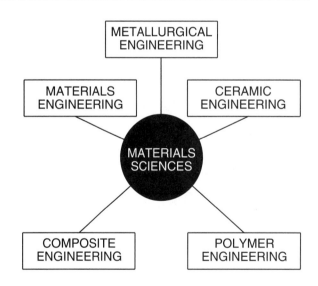

Figure 1-1. Materials sciences, which consist of the study of engineering materials, are divided into five branches.

from their ores. Extracting a metal from its ore is conducted in several process steps. Each process step increases the purity of the product by removing unwanted ingredients. For example, the extraction route from ore to refined metal includes any or all of the following process steps: benefication, mineral dressing, pyrometallurgy, hydrometallurgy, and electrometallurgy. See Figure 1-2.

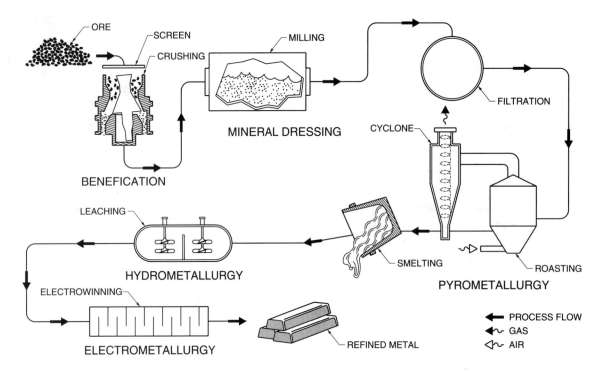

Figure 1-2. Extractive metallurgy is the study of the extraction and purification of metals from their ores.

Mechanical Metallurgy. *Mechanical metallurgy* is the study of the techniques and mechanical forces (factors) that shape or make finished forms of metal. Mechanical metallurgy studies the effects of stress, time, temperature, etc. on metal. For example, understanding these effects may result in improvements in surface quality and/or decreases in stresses that lead to distortion in finished forms of metal. See Figure 1-3.

SELECTED FACTORS STUDIED IN MECHANICAL METALLURGY
Stress
Time
Temperature
Rate of heating
Rate of cooling
Geometry of raw material
Geometry of desired finish product
Processing environment

Figure 1-3. Mechanical metallurgy is the study of the techniques and mechanical forces that shape finish forms of metals.

Physical Metallurgy. *Physical metallurgy* is the study of the effect of structure on the properties of metals. The two structures studied in physical metallurgy are the crystal structure and micro-structure. See Figure 1-4. The *crystal structure* is the arrangement of atoms in the metal. An *atom* is the smallest building block of matter that can exist alone or in combination. It cannot be divided without changing its basic character. The crystal structure is shown through modeling.

The *microstructure* is the microscopic arrangement of the components, or phases, within a metal. Microstructures are visible at high magnifications and may show the elements and grain structure present in a metal.

The properties of metals are intimately related to their structure. By modifying their structure, physical metallurgists tailor the properties of metals to properties that are more desirable and useful. For example, the structure of metals can be modified by chemical composition, alloying, and heat treatment.

When welded, certain stainless steels can lose their corrosion resistance due to a heat-induced metallurgical structure change. By modifying the chemical composition of susceptible stainless steels, the structure change is prevented and corrosion resistance is maintained.

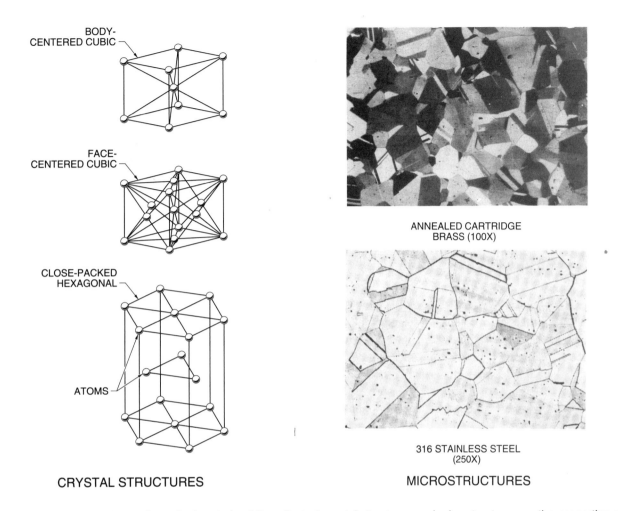

BODY-
CENTERED CUBIC

FACE-
CENTERED CUBIC

CLOSE-PACKED
HEXAGONAL

ATOMS

ANNEALED CARTRIDGE
BRASS (100X)

316 STAINLESS STEEL
(250X)

CRYSTAL STRUCTURES

MICROSTRUCTURES

Figure 1-4. Physical metallurgy is the study of the effect of crystal structures and microstructures on the properties of metals.

ductile-brittle transition

Carbon steels can become brittle at subzero temperatures. Specific alloying elements are added to carbon steels to help modify the crystal structure. These alloying elements increase the low-temperature toughness of carbon steels, which counteracts the tendency of carbon steels to become brittle.

The technology of heat treatment of steels is based on a specific crystal structure and microstructure change that occurs when steel is rapidly cooled from a high temperature. These changes lead to hardening and strengthening of steels.

Ceramic Engineering

Ceramic engineering, or *ceramics*, is the study of the development and production of products made from nonmetallic, inorganic materials by firing at high temperatures. Ceramic materials are divided into brick and advanced ceramics. Ceramics are used in applications such as porcelain, furnaces, engines, insulation, crucibles, and electronic components.

Polymer Engineering

Polymer engineering or *polymers* is the study of the development and production of synthetic organic materials. Polymers are divided into two groups: thermoplastics and thermosets. Polymers are used in applications such as adhesives, building products, fibers, sporting goods, and automotive and aerospace components.

Composite Engineering

Composite engineering, or *composites*, is the study of the applicability of combinations of materials. The combinations may be metallic, ceramic, or polymeric and show superior properties when compared to the individual ingredients. Composites are used to strengthen metals, ceramics, or polymers and improve their structural usefulness. Composites are used in applications such as high-temperature engine components, helicopter blades, high-pressure hoses, and aerospace structures.

Materials Engineering

Materials engineering, which crosses the boundaries of all the branches of materials sciences, is the study of the evaluation of the characteristic properties of all materials. This permits materials substitutions to be made in engineering components. The purpose of such substitutions is to improve the performance or lower the cost of the components.

No single property can be applied to distinguish all metals from all nonmetals. Some of the properties that distinguish metals from nonmetals are strength combined with toughness, electrical conductivity, and thermal conductivity. However, there is one dominant property that most often makes metals preferred over most nonmetals for structural applications, such as aircraft, bridges, and mechanical equipment. This dominant property is the ability of metals to yield in the presence of stress rather than break in a brittle manner.

Some polymers are similar to metals in that they yield rather than snap when a load is applied. These polymers usually lack stiffness and strength, particularly as the temperature is raised. Many ceramics are extremely strong, but do not yield when a load is applied. These ceramics tend to break easily in the presence of flaws. Composites often produce desirable combinations of properties by combining materials. Composites have limitations because of difficulties in maintenance and repair in the field. Metals yield in the presence of excessive stress and thus give warning of impending failure. This property is often why metals are preferred over nonmetals for structural components. See Figure 1-5.

SPECIMEN BEFORE TEST

DEFORMED SPECIMEN

BROKEN SPECIMEN

Figure 1-5. Metals yield in the presence of stress.

METAL IDENTIFICATION

Metal identification is performed by studying certain characteristics that metals exhibit. These characteristics relate to the description, specific properties, and process condition of metals. A metal is described as a pure metal or as an alloy and may be further divided into and identified as ferrous or nonferrous. Chemical analysis is used to determine if a metal is a pure metal or an alloy, if it is ferrous or nonferrous, and to identify the composition (chemical element makeup) of the metal. Metals and all other materials exhibit three types of properties that help identify the material. These properties are physical properties, mechanical properties, and chemical properties. The process condition is used to describe the condition that metals are supplied in, such as annealed, tempered, or cold worked.

Metals and Alloys

Although there is a difference between describing materials as metals or alloys, the two terms are used interchangeably. *Metals* refer strictly to pure metals, which are also chemical elements. For example, copper (Cu), iron (Fe), manganese (Mn), and zinc (Zn), are chemical elements and pure metals. Pure metals, which are usually soft and have low-strength, have extremely limited usage in engineering applications. *Alloys* are materials that have metallic properties and are composed of two or more chemical elements. At least one of the elements in an alloy is a metal. See Figure 1-6. Alloys are used when improved properties are required such as elevated temperature strength or high-damping capacity. Most materials used in engineering applications are alloys. For example, carbon steels are alloys of iron plus up to 1.2% carbon (C). Stainless steels are also alloys containing significant amounts of chromium (Cr) and often other elements such as nickel (Ni) and molybdenum (Mo).

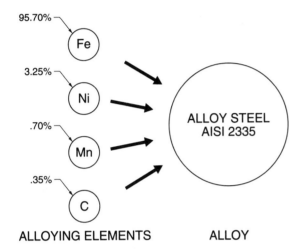

Figure 1-6. Alloys are made of two or more chemical elements, one of which is a metal.

Alloying profoundly affects the properties of pure metals. For example, very small amounts of oxygen (O) are added to commercially pure titanium (Ti). This addition doubles the concentration of oxygen from .2% O to .4% O. This results in an alloy that has double the strength of the pure titanium.

Ferrous and Nonferrous. Metals and alloys are also described as ferrous or nonferrous. *Ferrous metallurgy* encompasses alloys based on iron, where the major alloying element is iron. These alloys include the various families of carbon and low-alloy steels, tool steels, cast irons, and stainless steels. *Nonferrous metallurgy* encompasses all other pure metals and alloy systems.

Occasionally, the dividing line between ferrous and nonferrous alloys is indistinct. This is the case with some of the highly alloyed stainless steels. Many stainless steels are clearly ferrous; however, some high-alloy stainless steels have nickel as the major alloying element and are strictly nonferrous nickel-base alloys.

Chemical Analysis. Chemical analysis is the key to the identification of alloys and is used to determine the weight percentages of all the elements that make up alloys. The percentages are used to describe alloys and the sum must equal 100%. Each element has an allowable range or a minimum (min) or maximum (max) limit. There are coding systems that identify the alloys, their alloying elements, and the percentages of those alloying elements. For example, S30400 is the Unified Numbering System (UNS) identification code for a stainless steel. Each stainless steel contains specific quantities or ranges of certain alloying elements. See Figure 1-7.

COMPOSITION OF SELECTED STAINLESS STEELS								
UNS	**%C**	**%Mn**	**%Si**	**%Cr**	**%Ni**	**%P**	**%S**	**Others**
S30200	.15	2.00	1.00	17.0 to 19.0	8.0 to 10.0	.045	.03	—
S30215	.15	2.00	2.0 to 3.0	17.0 to 19.0	8.0 to 10.0	.045	.03	—
S30300	.15	2.00	1.00	17.0 to 19.0	8.0 to 10.0	.20	.15 min	.6 Mo (optional)
S30310	.15	2.5 to 4.5	1.00	17.0 to 19.0	7.0 to 10.0	.20	.25 min	—
S30323	.15	2.00	1.00	17.0 to 19.0	8.0 to 10.0	.20	.06	15 min Se
S30345	.15	2.00	1.00	17.0 to 19.0	8.0 to 10.0	.05	.11 to .16	.40 to .60 Mo; .60 to 1.00 Al
S30400	.08	2.00	1.00	18.0 to 20.0	8.0 to 10.5	.045	.03	.10 N
S30403	.03	2.00	1.00	18.0 to 20.0	8.0 to 12.0	.045	.03	.10 N
S30409	.04 to .10	2.00	1.00	18.0 to 20.0	8.0 to 10.5	.045	.03	—

Figure 1-7. Coding systems identify alloys, their alloying elements, and percentages of those alloying elements.

Properties

A *property* is a measurable or observable attribute of a material that is of a physical, mechanical, or chemical nature. See Figure 1-8. In order to identify and use metals reliably and cost-effectively, their properties must be measured and understood. Not taking these properties into consideration may lead to a costly failure.

Property determination is extremely important in alloy development and quality control. Designers rely on measured properties in order to select the size and shape of components. From these designs, metallurgists write specifications to indicate the alloy, product form, and quality level. Suppliers and purchasers then use the specifications as a basis for ordering materials.

Physical Properties. *Physical properties* are the characteristic responses of materials to forms of energy such as heat, light, electricity, and magnetism. Color, density, magnetic permeability, and weight of a material are physical properties. These properties can be measured without the application of force. Physical properties may be affected by alloying and/or through the manner that the material is worked or formed.

Mechanical Properties. *Mechanical properties* are the characteristic dimensional changes in response to applied external or internal mechanical forces. The shear strength, toughness, and stiffness of a metal are examples of mechanical properties. Measurement of mechanical properties can indicate

suitable metals for certain mechanical applications. Mechanical properties may be affected by alloying and/or through the manner that the material is worked or formed.

Chemical Properties. *Chemical properties* are the characteristic responses of materials in chemical environments. Corrosion resistance and resistance to acids and alkalies are examples of chemical properties. Understanding these properties can help determine how a metal will react in different service environments. Chemical properties may be affected by alloying and/or through the manner that the material is worked or formed.

Process Condition

The process condition of a metal is the identifiable state in which the metal is supplied. Metals may be supplied as cast, as wrought, or from powders. See Figure 1-9.

The properties of cast and wrought metals may be substantially and identifiably different. Cast metals are produced from molten metal solidifying in a mold cavity. Cast metals are metal objects produced by pouring molten metal into a mold cavity, which has the desired shape of the casting. The molten metal is allowed to solidify in the mold cavity and the casting is then demolded. Cast metal products obtain distinct and identifiable metallurgical structures through the process. The type of material used, the dimensions of the mold cavity, and the rate of cooling of the metal in the mold cavity have a distinct effect on the metallurgical structure of the

SELECTED PROPERTIES OF METALS		
Physical	**Mechanical**	**Chemical**
Coefficient of thermal expansion	Elongation	Chemical reactivity
Color	Fatigue limit	Corrosion resistance
Density	Hardness	Electrochemical potential
Electrical conductivity	Stiffness	Irradiation resistance
Lattice parameter	Shear strength	Resistance to acids
Magnetic permeability	Tensile strength	Resistance to alkalies
Weight	Toughness	Solubility

Figure 1-8. Property identification is an important part of failure analysis, alloy development, quality control programs, and design and purchasing specifications.

final product. Turbines, crankshafts, and gears are examples of cast metal products.

Wrought metals are worked into finished forms. They are worked using processes such as drawing, extruding, rolling, and pressing. Greater amounts of working have a greater effect on the properties of the finished form. The mechanical working performed on wrought metals alters the metallurgical structure. The direction at which a material is worked also has a distinct effect on the properties of the finished form. Wrought metals start as a cast ingot or billet, or metal powder billet. Finished forms include plate, bar, structural shapes, and sheet. Structural shapes include shapes such as beams and rails.

Metal powders are used when stringent composition controls are required. For example, metal powders are used in the production of superalloys. *Superalloys* are various high-strength, often complex alloys having resistance to elevated temperatures.

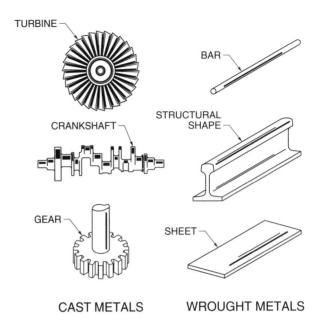

Figure 1-9. Metals may be supplied in cast or wrought process condition.

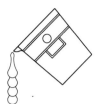

Physical Properties and Temperature Measurement

2

The physical properties of metals are their characteristic responses to forms of energy such as heat, electricity, and magnetism. Physical properties of metals are related to their atomic structure. These properties vary widely between different metals and alloys. Temperature measurement is critical in determining many physical properties. Accurate temperature measurement is also critical in many metallurgical operations such as heat treatment, forging, and casting. Many instruments (pyrometers and thermocouples) and techniques (temper and heat colors) are used for temperature measurement in these operations.

PHYSICAL PROPERTIES

Physical properties of metals are the characteristic responses of metals to forms of energy such as heat, electricity, and magnetism. Physical properties of metals are related to their atomic structure. The size and characteristics of the atoms that comprise a metal determine its physical properties. The physical properties of metals include density and specific gravity, melting point, thermal capacity and specific heat, thermal conductivity, electrical conductivity and resistivity, coefficient of thermal expansion and linear expansion, and magnetic susceptibility. These properties vary widely between different metals and alloys. See Figure 2-1.

Density

Density is the mass per unit volume of a material. It is measured in grams per cubic centimeter (g/cm^3) or pounds per cubic inch ($lb/in.^3$). The mass of the material is measured in grams or pounds and is divided by the volume of the material, which is measured in cubic centimeters or cubic inches. Density is found by applying the following formula:

$$D = \frac{M}{V}$$

where

D = density

M = mass of part (in g)

V = volume of part (in cm^3)

Example: Figuring density

A copper specimen has a mass of 89.6 g and a volume of 10 cm^3. What is the density of the copper?

$$D = \frac{M}{V}$$

$$D = \frac{89.6}{10}$$

$$D = \textbf{8.96 g/cm}^3$$

Castings are usually less dense than wrought materials of similar composition. Castings contain porosity from gas that is internally trapped during solidification. When a cast ingot is rolled or forged into a wrought product, the pores close. This produces a product with higher density.

SELECTED PHYSICAL PROPERTIES				
Metal or Alloy	Density (g/cm^3)	Coefficient of Thermal Expansion (μin/in.°C)	Electrical Resistivity ($\mu\Omega\cdot$cm)	Melting Point (°C)
Copper	8.96	16.5	1.673	1083 ± 0.1
Gold	19.3	14.2	2.35	1063 ± 0.1
Iron	7.87	11.76[a]	9.71	1535.5 ± 0.1
Zinc	7.13	39.7[b]	5.916[c]	419.505
Yellow brass	8.47	20.3	6.4	932.22
Low-silicon bronze	8.75	17.9	14.3	1060

[a] at 25°C
[b] from 20°C to 250°C for polycrystalline metal
[c] at 20°C

Figure 2-1. Physical properties vary widely between different metals and alloys.

Specific Gravity. *Specific gravity* is the ratio of the density of a material to the density of water. Specific gravity is found by dividing the density of the material by the density of water. It is not expressed as a unit of measure. Specific gravity is found by applying the following formula:

$$SG = \frac{D_M}{D_W}$$

where

SG = specific gravity

D_M = density of material

D_W = density of water

Example: Figuring specific gravity

Copper has a density of 8.96 g/cm^3 and water has a density of .9982 g/cm^3. What is the specific gravity of copper?

$$SG = \frac{D_M}{D_W}$$

$$SG = \frac{8.96}{.9982}$$

$$SG = \textbf{8.9761}$$

Low density, or low specific gravity, is an asset when lightness is important. The aircraft and aerospace industries spurred the development of low-density alloys, which also exhibit high strength to weight ratios. These low-density alloys are based on aluminum, magnesium (Mg), and titanium.

Melting Point

Melting point is the temperature at which a material passes from a solid to a liquid state. Pure metals and certain alloys possess a specific melting point. For example, aluminum has a melting point of 660° Celsius (660°C), or 1220° Fahrenheit (1220°F). Most alloys do not pass completely from the solid to the liquid state at a specified temperature, but melt over a range of temperatures. In this temperature range, the liquid state and solid state of the alloy coexist. For example, yellow brass, which is an alloy consisting of 65% Cu and 35% Zn, becomes soft when heated above 904°C (1660°F) and is completely molten at 932°C (1710°F).

The melting point depends on the strength of the atomic bonds. *Atomic bonds* are the forces that hold atoms together. When a metal melts, its atoms are no longer packed in the solid state, but move about freely with respect to one another. The higher the atomic bond strength, the less freely the atoms move. More thermal energy is required to break strong atomic bonds between the atoms. This results in a higher melting point.

As a metal approaches its melting point, its strength is significantly diminished. Most metals are not used in applications where temperatures approach their melting point. For certain nonstructural applications, the melting point or range is an important property. For example, when using alloys in applications such as fuses and solders, the relatively low melting point is an asset. See Figure 2-2.

Fuses are an electrical protective device with a metal part that is heated and melted by a passage of overcurrent. This passage of overcurrent breaks the metal part, which interrupts current flow. Solders are alloys that utilize low melting points to join two metallic surfaces. In PC board construction, solder is used to make an electrical connection between component leads and the foil of the PC board. The component lead is joined to the foil when the solder is melted and flows between the surfaces. The solder then solidifies, making the electrical connection.

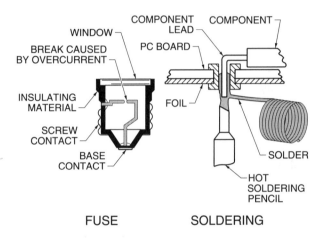

Figure 2-2. Specific melting points and ranges are an asset in certain applications.

Thermal Capacity

Thermal (heat) capacity is the amount of thermal energy required to raise a unit mass by one degree. This amount varies from metal to metal. Heat capacity is expressed in joules per kilogram degree Kelvin (J/kg°K) or British thermal units per pound degree Fahrenheit (Btu/lb°F). For example, the heat capacity for type 304 stainless steel is 500 J/kg°K and for pure copper is 380 J/kg°K. The heat capacity for the pure copper is approximately 25% less than the stainless steel.

Specific Heat. *Specific heat* is the ratio of the heat capacity of a material to the heat capacity of water. It is the common way of expressing heat capacity. Heat capacity of the material is divided by the heat capacity of water. Like specific gravity, specific heat is not expressed in any unit of measure. Specific heat is found by applying the following formula:

$$SH = \frac{HC_M}{HC_W}$$

where

SH = specific heat

HC_M = heat capacity of material

HC_W = heat capacity of water

Example: Figuring specific heat

Copper has a heat capacity of 380 J/kg°K and water has a heat capacity of 4183 J/kg°K. What is the specific heat of copper?

$$SH = \frac{HC_M}{HC_W}$$

$$SH = \frac{380}{4183}$$

$$SH = .091$$

Thermal Conductivity

Thermal conductivity is the rate at which thermal energy flows through a material. This is in contrast to specific heat, which is the property that controls the temperature increase in a metal. Thermal conductivity is measured in watts per meter degree Kelvin (W/m°K) or British thermal units per hour feet degree Fahrenheit (Btu/hr•ft•°F). Most metals are efficient conductors of heat, but the thermal conductivities vary widely from metal to metal.

Thermal conductivity is a significant factor in welding and casting operations. Copper is an efficient heat sink. A *heat sink* is a metal or nonmetal that has high thermal conductivity and may be used to rapidly conduct heat away from a location. Copper is used as a backing bar in welding or as a water-cooled mold in casting because the copper rapidly draws heat away from the solidifying metal. This improves the solidification pattern of the casting. However, its high thermal conductivity makes copper difficult to weld. Copper requires high heat input during welding because it rapidly conducts heat away from the heat source. In comparison, the 300 series austenitic stainless steels have a relatively low thermal conductivity. Their thermal conductivity is approximately one twenty-fifth that of copper. They require less heat input for welding because they retain the heat in the weld area.

Coefficient of Thermal Expansion

Coefficient of thermal expansion is the increase in a dimension of a metal per unit dimension per unit degree rise in temperature. Metals expand when heated and contract when cooled. Expansion occurs when a metal is heated. As the temperature of the metal increases, the atoms in the metal repel each other more and the distance between the atoms increases. The result is a dimensional change. Contraction occurs when a metal is cooled. As the temperature of the metal decreases, the atoms in the metal repel each other less and the distance between the atoms decreases. Again, the result is a dimensional change.

Dimensional change occurs in all three axes: length, width, and thickness. The length axis change, or linear dimensional change, is usually the greatest and is expressed as the coefficient of linear expansion.

Coefficient of Linear Expansion. *Coefficient of linear expansion* is the increase in length per unit length per degree temperature rise. It is expressed in microinch per inch degree Celsius (μin/in.°C) or microinch per inch degree Fahrenheit (μin/in.°F). The coefficient of linear expansion varies for the same metal with different temperature ranges. This is because the coefficient of expansion is not directly proportional to the temperature increase over the entire range of temperatures. Identifying the applicable temperature range when indicating the coefficient of thermal expansion for length, width, or thickness is essential.

In certain applications, the coefficient of linear expansion is a property of primary importance. An allowance must be made for expansion when designing equipment for application in an environment with elevated temperatures. For example, to allow for expansion in steam lines, loops are provided. Loops are offsets in the steam line that permit the thermal expansion of the line, which is caused by the steam. See Figure 2-3.

Electrical Conductivity

Electrical conductivity is the rate at which electrons move through atoms causing current to flow. The higher the electrical conductivity of a material, the easier the current flows through it. Electrical conductivity is one of the major ways of distinguishing metals from nonmetals, because nonmetals do not conduct electricity.

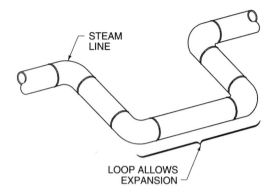

STEAM LINE

LOOP ALLOWS EXPANSION

Figure 2-3. Coefficient of linear expansion is a property of primary importance in many applications, such as loops in steam lines.

Electrical Resistivity. *Electrical resistivity* is the electrical resistance of a unit volume of a material and is measured in microhm per centimeter ($\mu\Omega$•cm). Resistivity is the reciprocal of electrical conductivity and it provides a convenient whole number. Resistivity is how conductivity is expressed. Pure metals possess the highest conductivity or lowest resistivity. Alloying of a metal increases the electrical resistivity of the metal. See Figure 2-4.

RESISTIVITY ($\mu\Omega$•cm) OF SELECTED METALS	
Silver	1.59
Copper	1.673[a]
Zinc	5.916[a]
Silicon	10[b]
Tin	11[c]
Yellow brass	6.4
Low-silicon bronze	14.3
Stainless steel (type 410)	57

[a] at 20°C
[b] at 0°C
[c] at 0°C for white tin

Figure 2-4. Alloying of a metal increases the electrical resistivity of the metal.

The two extremes of resistivity are metals and nonmetals. *Semiconductors* are materials that have a resistance to current flow that falls somewhere between the low resistance of metals (conductor) and the high resistance of nonmetals (insulator).

Superconductivity. *Superconductivity* is a complete reduction of electrical resistance in a material at temperatures near absolute zero (–273°C or –460°F). Superconductivity in some materials occurs at higher temperatures and does not require extensive cooling to exhibit this reduction of resistance. Superconductivity allows electricity to flow with little or no energy loss.

Magnetic Susceptibility

Magnetic properties of metals are utilized in the manufacture of magnets and in flaw detection techniques. Two examples are magnetic particle and eddy current inspection. In both, the magnetic properties of materials tested are used to explore the surface condition and internal quality. The magnetic effects of metals are the result of the distribution and spin of electrons in their shells.

Magnetic susceptibility is the intensity of magnetism produced in a material when it is placed in a magnetic field. The three types of magnetic susceptibility are ferromagnetic, paramagnetic, and diamagnetic. See Figure 2-5.

Ferromagnetic Metals. *Ferromagnetic metals* are metals that have high and variable magnetic susceptibility and are strongly attracted to a magnetic field. Iron, cobalt (Co), and nickel are examples of ferromagnetic metals.

Magnetic permeability is the measurement of the ease with which a ferromagnetic metal can be magnetized and demagnetized. Magnetic permeability depends on the strength of the magnetic field acting on the metal. A ferromagnetic metal becomes increasingly magnetized as the magnetic field, or magnetizing force, is increased. Induced magnetism increases rapidly at first and then levels off. At this point, large increases in the magnetizing force result in small increases in induced magnetism. As the magnetizing force is increased, the material eventually becomes magnetically saturated.

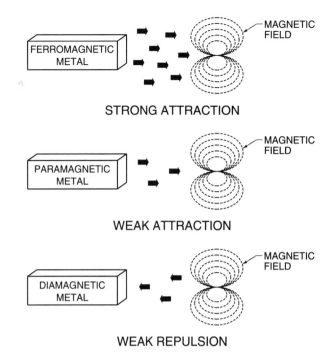

Figure 2-5. Ferromagnetic metals have high values of magnetic susceptibility, paramagnetic metals have low values, and diamagnetic metals have low negative values.

When the magnetizing force is removed, the induced magnetism falls back to zero or a residual level. Magnetically soft metals, or soft magnets, will not remain magnetized and the induced magnetism will fall to zero. Magnetically hard metals, or permanent or hard magnets, remain magnetized at some residual level of magnetism. To demagnetize a magnetically hard metal, a demagnetizing force must be applied. See Figure 2-6.

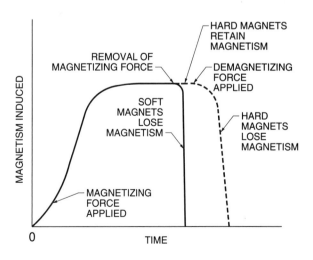

Figure 2-6. A magnetizing curve is a plot of the magnetism induced in a metal when a magnetizing force is applied.

Magnetostriction is the dimensional change that a ferromagnetic metal exhibits when magnetized. As the magnetizing force applied is varied, the dimensions of the metal also vary. This effect is used to convert electrical energy into mechanical energy. Transducers are magnetized using electrical energy. The frequency of the electrical energy causes the transducer to vibrate. If the vibrating transducer is placed in a fluid, it will transmit pulses through the fluid. Other applications of magnetostriction include echo-sounding (sonar) and ultrasonic cleaning.

Paramagnetic Metals. *Paramagnetic metals* are metals that have low values of magnetic susceptibility. These metals are weakly attracted to a magnetic field. Lithium (Li), sodium (Na), calcium (Ca), potassium (K), and most other metals are examples of paramagnetic metals.

Diamagnetic Metals. *Diamagnetic metals* are metals that have low negative values of magnetic susceptibility. These materials are weakly repelled by a magnetic field. Bismuth (Bi), copper, gold (Au), silver (Ag), and beryllium (Be) are examples of diamagnetic metals.

TEMPERATURE MEASUREMENT

Temperature measurement is an essential technique used to identify physical properties such as melting point, specific heat, and thermal conductivity. Temperature measurement is an important parameter in processes such as heat treating, welding, and forming. For example, solution annealing operations are sometimes performed relatively close to the melting point of the metal. Temperature readings lower than the actual value attained during solution annealing may result in localized melting of the metal. Localized melting leads to scrapping of the metal part.

Temperature-measuring devices are most often read in the Fahrenheit or Celsius scale and have a diversified range of complexity. In metallurgy most temperature measurements are conducted within the range of –100°C to 1370°C (–150°F to 2500°F).

Celsius and Fahrenheit Temperature Scales

Celsius and Fahrenheit are the two temperature scales used most often. See Figure 2-7. To convert a temperature from one scale to the other, the difference in the two bases of the scales and the ratio of the scales must be considered. The base on the

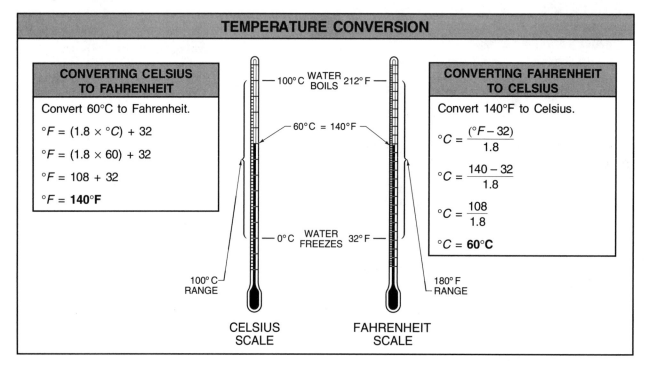

TEMPERATURE CONVERSION

CONVERTING CELSIUS TO FAHRENHEIT

Convert 60°C to Fahrenheit.

$°F = (1.8 × °C) + 32$

$°F = (1.8 × 60) + 32$

$°F = 108 + 32$

$°F = \mathbf{140°F}$

100°C — WATER BOILS — 212°F

60°C = 140°F

0°C — WATER FREEZES — 32°F

100°C RANGE

180°F RANGE

CELSIUS SCALE

FAHRENHEIT SCALE

CONVERTING FAHRENHEIT TO CELSIUS

Convert 140°F to Celsius.

$°C = \dfrac{(°F - 32)}{1.8}$

$°C = \dfrac{140 - 32}{1.8}$

$°C = \dfrac{108}{1.8}$

$°C = \mathbf{60°C}$

Figure 2-7. There is 1.8°F on the Fahrenheit scale for every 1.0°C on the Celsius scale.

Celsius scale is 0°C and the base on the Fahrenheit scale is 32°F. The 32° difference between the bases is used when converting temperatures. The ratio between the two is determined by the difference between freezing point (0°C or 32°F) and boiling point (100°C or 212°F) of water on the two scales. There is a range of 100°C between 0°C and 100°C on the Celsius scale, and a range of 180°F between 32°F and 212°F on the Fahrenheit scale. The ratio for the conversion is found by dividing 180 by 100, which is a 1.8 ratio. There is 1.0°C on the Celsius scale for every 1.8°F on the Fahrenheit scale.

To convert Celsius to Fahrenheit, multiply 1.8 by the Celsius reading and add 32. To convert Celsius to Fahrenheit, apply the following formula:

$$°F = (1.8 \times °C) + 32$$

where

$°F$ = degrees Fahrenheit

1.8 = ratio between bases

$°C$ = degrees Celsius

32 = difference between bases

Example: Converting Celsius to Fahrenheit

Convert 26°C to Fahrenheit.

$$°F = (1.8 \times °C) + 32$$

$$°F = (1.8 \times 26) + 32$$

$$°F = 46.8 + 32$$

$$°F = \textbf{78.8°F}$$

To convert Fahrenheit to Celsius, subtract 32 from the Fahrenheit reading and divide by the 1.8 ratio. To convert Fahrenheit to Celsius, apply the following formula:

$$°C = \frac{(°F - 32)}{1.8}$$

where

$°C$ = degrees Celsius

$°F$ = degrees Fahrenheit

32 = difference between bases

1.8 = ratio between bases

Example: Converting Fahrenheit to Celsius

Convert 70°F to Celsius.

$$°C = \frac{(°F - 32)}{1.8}$$

$$°C = \frac{(70 - 32)}{1.8}$$

$$°C = \frac{38}{1.8}$$

$$°C = \textbf{21.1}\overline{\textbf{1}}\textbf{°C}$$

Absolute Temperature Scale

Absolute temperature scale is measured from the absolute zero temperature. See Figure 2-8. *Absolute zero* is a theoretical condition at which no heat is present. The Kelvin (°K) and Rankine (°R) are two scales that use absolute zero as a common base. The Kelvin is the absolute temperature scale related to the Celsius scale. The Rankine scale is the absolute temperature scale related to the Fahrenheit scale.

On the Kelvin scale, absolute zero (0°K) is 273°C below 0°C. To convert Celsius to Kelvin, add 273 to the Celsius reading. To convert Celsius to Kelvin, apply the following formula:

$$°K = 273 + °C$$

where

$°K$ = degrees Kelvin

273 = difference between bases

$°C$ = degrees Celsius

Example: Converting Celsius to Kelvin

Convert 34°C to Kelvin.

$$°K = 273 + °C$$

$$°K = 273 + 34$$

$$°K = \textbf{307°K}$$

On the Rankine scale, absolute zero (0°R) is 460°F below 0°F. To convert Fahrenheit to Rankine, add 460 to the Fahrenheit reading. To convert Fahrenheit to Rankine, apply the following formula:

$$°R = 460 + °F$$

where

$°R$ = degrees Rankine

460 = difference between bases

$°F$ = degrees Fahrenheit

Example: Converting Fahrenheit to Rankine

Convert 96°F to Rankine.

$°R = 460 + °F$

$°R = 460 + 96$

$°R = \textbf{556°R}$

Pyrometers

A *pyrometer* is an instrument used for measuring temperatures (beyond the range of mercury thermometers) by the increase of electrical resistance in a metal, generation of electrical current of a thermocouple, or increase in intensity of light radiated by an incandescent body. Pyrometers are classified as recording or controlling. *Recording,* or *autographic pyrometers,* automatically measure temperature and record the information on a chart or store it in a computer. *Controlling pyrometers* use temperature-sensing elements and electrical circuitry to maintain a preset temperature in a furnace or any operation in which control is required, such as during a field stress-relieving heat treatment performed on welded piping.

Temperature is determined by measuring physical properties that change with temperature. These properties include the coefficient of thermal expansion of metals, the electrical resistance of a wire, the thermoelectric force of a bimetallic couple, or the intensity of radiant energy (radiation) emanating from a hot surface.

Applicability of the various types of pyrometers is based on the technique used, accuracy required, physical limitations, and convenience. Pyrometer types include the bimetallic coil, liquid expansion, gas or vapor pressure, resistance, thermoelectric, radiation, and optical.

Bimetallic Coil Pyrometer. A *bimetallic coil pyrometer* is a pyrometer with two strips of metal

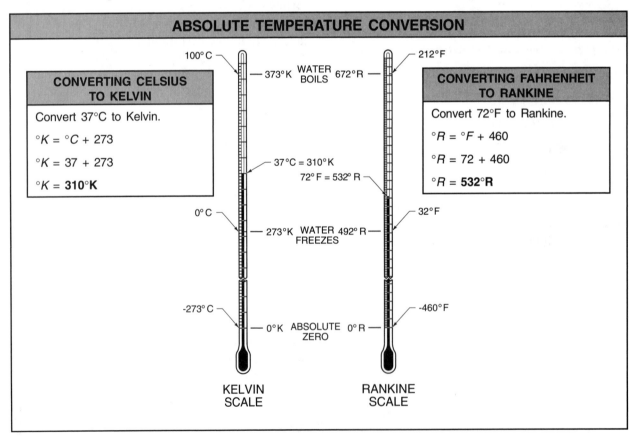

Figure 2-8. The Rankin scale is used for absolute temperature calculations in relation to the Fahrenheit scale. The Kelvin scale is used for absolute temperature calculations in relation to the Celsius scale.

bonded together. One strip has a high coefficient of thermal expansion and the other has a low coefficient of thermal expansion. The bimetal strip is wound into a helical coil. An increase in temperature causes the coil to unwind, and a decrease in temperature causes the coil to wind tighter. Rotary motion of a pointer (fixed to the moving coil) is produced and registers a temperature reading on a scale. Bimetallic coil pyrometers are rugged and can be used in the range of –73°C to 538°C (–100°F to 1000°F). The accuracy of the bimetallic coil pyrometer is low and the speed of response is slow because the coil must be enclosed in a protective tube. See Figure 2-9.

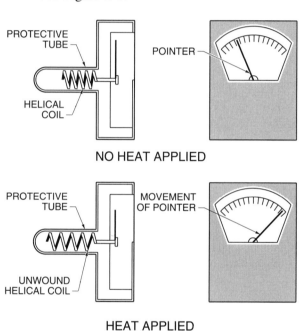

Figure 2-9. When heat is applied, the helical coil in the bimetallic coil pyrometer unwinds causing rotary motion of a pointer and producing a temperature reading.

Liquid Expansion Pyrometer. A *liquid expansion pyrometer* is a pyrometer with a bulb containing a liquid, such as mercury or alcohol, that is exposed to the metal to be measured. Temperature change causes the liquid in the bulb to expand or contract. The expansion or contraction causes a Bourdon tube, which is connected to the bulb, to expand or contract. This expansion or contraction of the Bourdon tube moves a temperature indicator.

The temperature range of a liquid expansion pyrometer depends on the type of liquid used and var-

ies from –37°C to 510°C (–35°F to 950°F). Liquid expansion pyrometers are fragile and have low-accuracy. The liquid expansion pyrometer is identical to the gas or vapor pressure pyrometer except for the expansion medium used in the bulb and Bourdon tube. See Figure 2-10.

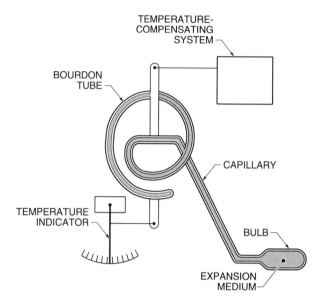

Figure 2-10. In a liquid expansion or gas or vapor pressure pyrometer, the expansion of the medium in the bulb produces movement in the Bourdon tube, which moves the temperature indicator.

Gas or Vapor Pressure Pyrometer. A *gas or vapor pressure pyrometer* is a pyrometer with a bulb filled with a volatile liquid that is exposed to the metal to be measured. Temperature change causes corresponding pressure variations in the saturated vapor above the liquid surface in the bulb. The pressure variations cause a Bourdon tube, which is connected to the bulb, to expand or contract. This expansion or contraction of the Bourdon tube moves a temperature indicator.

The temperature range of a gas or vapor pressure pyrometer depends on the liquid contained in the bulb and varies from –51°C to 260°C (–60°F to 500°F). Gas or vapor pressure pyrometers have low-accuracy and are primarily used for low-temperature applications such as plating and cleaning baths, degreasers, cooling water or oil, and in the subzero heat treatments.

Resistance Pyrometer. A *resistance pyrometer* uses the change in electrical resistance of a metal such as copper or nickel with the change in temperature. Depending on the metal used, the temperature varies from –212°C to 593°C (–350°F to 1100°F). Resistance pyrometers are accurate, but have a slow response. The primary use of resistance pyrometers is to calibrate other types of pyrometers.

Thermoelectric Pyrometer. A *thermoelectric pyrometer* is a pyrometer with a thermocouple and a millivolt-meter. An electromotive force (EMF), or voltage, is measured in millivolts (mV) and is generated when the thermocouple is heated and a current will flow causing a deflection of the millivolt-meter. The deflection represents temperature on a calibrated temperature scale. Thermoelectric pyrometers are the most widely used temperature-measuring instruments for metallurgical applications.

A *thermocouple* is a measuring device that consists of two electrically connected dissimilar metal wires. They measure potential created at the junction of the wires, the hot junction. The hot junction is connected to the copper wire of the measuring instrument circuitry, or cold junction. Thermocouples are made by cutting off suitable lengths of the dissimilar metal wires and carefully twisting them together for one or two turns. The ends are fused to form a well-rounded head, or hot junction. To maximize sensitivity to changes in temperature, there is no unnecessary mass of metal at the hot junction. See Figure 2-11.

Thermocouple wires should be in electrical contact only at the hot junction, since contact at any other point usually results in a lower EMF and an incorrect temperature indication. To prevent accidental contact along the length, the thermocouple wires are insulated from one another by porcelain beads or ceramic tubes. The extension lead wires are insulated positive and negative electrical conductors. These are used for connecting the thermocouple wires to the temperature indicator. Extension leads are relatively inexpensive and conserve the cost of expensive thermocouple wires. These extension leads are more robust and flexible than the rigid thermocouple wires.

A *thermowell* is a protective sheath used to protect the thermocouple from the corrosive effects of a high-temperature environment and also from mechanical abuse. A thermowell significantly extends the life of the enclosed thermocouple.

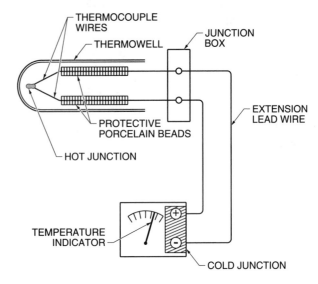

Figure 2-11. An EMF is produced in response to the temperature difference between the hot junction and cold junction of the thermoelectric pyrometer.

Certain cases do not require the ends of the thermocouple wires to be fused together because electrical contact is made through the object on which the thermocouple is placed or through the medium the thermocouple is immersed. See Figure 2-12. A contact pyrometer is used to read the temperature of a billet of hot metal before it is extruded in order to determine if it is at the appropriate temperature. An immersion pyrometer is used to read the molten material temperature when pouring metal castings.

Thermocouple metal combinations include four base metal combinations (J, K, T, and E) and three precious metal combinations (S, R, and B) and are identified by specific capital letters. This identification system was created by the American National Standards Institute (ANSI), American Society for Testing and Materials, and Instrument Society of America (ISA). The differences between the base metal and precious metal types are their compositions, resistivity, cost, service life, and precision.

The base metal combinations are nickel alloys and the precious metal combinations are platinium (Pt) and rhodium (Rh) alloys. The resistivity of the base metal combinations is four to eight times that of the precious metal combinations. Lower cost of the base metal combinations allows the thermocouple to be made of thicker wire, which lowers the resistance of the external circuit. A millivolt-meter is used for the temperature measurement in these instances. The precious metal combinations are made of thinner wire and require the use of a potentiometer, or high-resistance voltmeter, for temperature measurement. Precious metal combinations have longer service life at higher temperatures and are more homogeneous than the base metal combinations. The base metal types produce more accurate temperature measurements than the precious metal combinations.

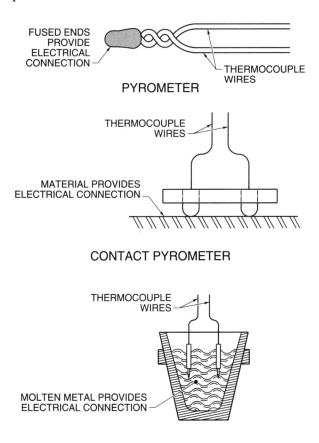

Figure 2-12. Thermocouple ends are fused together unless the medium to be measured provides the electrical connection.

The life of a thermocouple depends on several factors, which include the operating temperature, the time at temperature, the aggressiveness of the high-temperature environment, and number of temperature cycles. Thermocouple wires and extension

leads are supplied fully softened by a heat treatment process such as annealing. The wires are twisted, but not kinked or cold-worked, to form the hot junction. If strained by pulling, the extension leads may be damaged. Damages alter the electrical characteristics of the thermocouple and result in improper temperature indications. Homogeneity in the wires also causes improper temperature indications and can be tested by passing a small flame along the wires while the two ends of the thermocouple are connected to a galvanometer. There should be no deflection of the galvanometer, except when the flame passes the hot junction. See Figure 2-13.

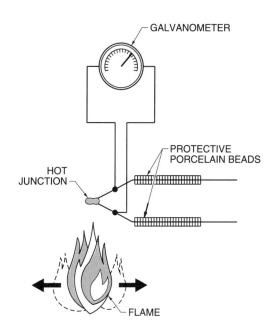

Figure 2-13. Homogeneity of the wires of a thermocouple can be tested by passing a small flame along the wires that are connected to a galvanometer.

Sensitive temperature responses require the thermocouple to be placed in direct contact with the metal part, which is accomplished by spot welding the thermocouple to the part. Other means may be used, such as using refractory brick to hold the thermocouple in contact with the part. In these cases, great care is taken to ensure no relative movement occurs.

In furnace temperature measurement, intimate contact is unnecessary and the thermocouple is enclosed in a thermowell. Thermowells are made from metals or nonmetals, depending on the required service temperature. Metal thermowells are more robust than the nonmetal types, but have lower temperature limitations. Metal thermowells are made from heavy wall tubing with a cap welded to one end or are drilled out as a one-piece unit from a bar.

The chief disadvantage of thermowells is a time lag introduced into the temperature measurement, which can be significant when temperatures are fluctuating. The time lag is caused by the time required for heat to be conducted through the wall thickness of the thermowell. See Figure 2-14.

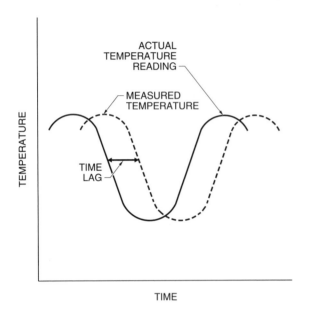

Figure 2-14. Thermowells introduce a time lag in temperature measurement when the temperature is fluctuating.

The EMF produced is a function of the difference in temperature between the hot and cold junctions and is measured by a potentiometer. To accurately measure the hot junction temperature, the cold junction temperature must be compensated for because the cold junction EMF is usually standardized at 0.00 mV at 0°C (32°F). The instrument contains a built-in cold junction compensator because the cold junction is usually above 0°C (32°F). This incorporates an allowance for the EMF difference between the cold junction temperature and 0°C (32°F). This feature permits accurate reading of the hot junction temperature directly from the scale on the measuring instrument. When the cold junction EMF of a chromel-alumel thermocouple is standardized at 0.00 mV at 0°C (32°F), the EMF difference between the cold junction and a hot junction at 205°C (401°F) is 8.31 mV.

Radiation and Optical Pyrometers. A *radiation pyrometer* measures, at a convenient distance from a hot source, the radiant energy emitted from the hot source. See Figure 2-15. Contact with the workpiece is unnecessary because the radiant energy from the hot source is focused on a thermopile. A *thermopile* is a group of thermocouples.

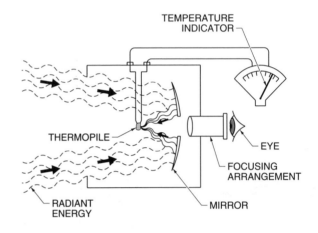

Figure 2-15. Radiation pyrometers focus the radiant energy on a thermopile, which generates an EMF that is translated into temperature.

The thermopile is contained within the pyrometer tube and is about the size of the eye of a needle. The rise in temperature of the thermopile is related to the amount of radiant energy it receives from the hot source. This energy is translated by the measuring circuitry in the pyrometer into the temperature of the hot source. The distance from the hot source is immaterial, but it is essential that the image received from the hot source is sufficiently large enough to completely cover the thermopile, otherwise an erroneous reading is obtained.

An *optical pyrometer* compares the intensity of light emitted from the hot source with the intensity of light emitted from a standard source, such as a lamp filament. The instrument does not contact the workpiece. The optical pyrometer consists of two major components, a telescope and a control box. The telescope contains a red glass filter that restricts the brightness of the hot source to one specific wavelength (that of red light) and a lamp with a calibrated filament. The filament current is adjusted so that its apparent brightness matches that of the target hot source. When the actual hot source reaches this point, the filament disappears from view. Measuring circuitry in the pyrometer converts the filament current value to a temperature reading on a temperature indicator. See Figure 2-16.

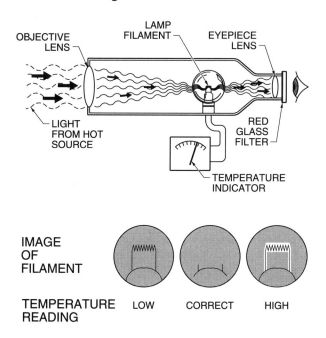

Figure 2-16. Optical pyrometers match the brightness of a lamp filament with the brightness of the hot source.

There are several distinct advantages in using radiation and optical pyrometers over other temperature-measuring devices. Radiation and optical pyrometers can focus and accurately measure temperatures of small areas or moving parts and can measure extremely high temperatures, ranging from 538°C to 5538°C (1000°F to 10,000°F). Radiation and optical pyrometers are not exposed directly to high temperature. This lack of exposure prolongs the life of the instrument.

A disadvantage in using radiation and optical pyrometers is that the hot sources do not necessarily emit all the radiation associated with their temperature. Emissivity controls the proportion of energy emitted, compared with the total available. *Emissivity* is a measurement of the extent to which a

surface deviates from the ideal radiative surface, which is a perfect emitter and absorber of thermal radiation. For example, a heated brick wall has an emissivity of 98%. This emissivity is high when compared with polished aluminum, which has an emissivity of 8%. When the emissivity becomes a matter of judgment, errors are introduced into the temperature measurement.

Another disadvantage in using radiation and optical pyrometers is when smoke or gas between the hot source and the pyrometer absorbs radiation to varying degrees and introduces errors into the measured temperature. With the optical pyrometer, additional errors may be introduced because the brightness match between the filament and the heat source is based on human judgement.

Temperature-indicating Crayons and Temper and Heat Colors

Temperature-indicating crayons and temper and heat colors are two simple, but relatively crude, temperature-measuring devices. *Temperature-indicating crayons* are used for monitoring the temperature of a surface that must meet some specified minimum or maximum value. Because they are a convenient and an approximate temperature guide, they are used for preheating temperature control during welding. The crayons are made of materials that melt at various temperatures. Temperature-indicating crayons are used by marking the workpiece with one or more of them, representing the needed range of temperatures. The crayons melt, indicating that the needed temperature range has been reached. An example of this type of crayon is the Tempilstik® produced by Tempil®. See Figure 2-17.

Temper and *heat colors* are an indication of the thickness of the oxide film that forms when steel is heated in air. As the temperature is raised, the oxide film thickens, causing an apparent color change. This provides an approximate visual method for estimating the temperature of steel when it is heat treated.

There are two distinct temperature ranges where temper and heat colors are applicable. The first temperature range is between 205°C and 370°C (400°F and 700°F). This temperature range is where some tool steels and high-carbon steels are tempered.

Tempil, Big Three Industries, Inc.

Figure 2-17. Melting characteristics of temperature-indicating crayons are used to measure the surface temperature of an object.

Temper colors range from straw to blue. The second temperature range is between 540°C and 1200°C (1000°F and 2200°F). This temperature range is where several kinds of heat treatments are performed on various types of steel. Heat colors range between faint red and yellow.

Many factors play a part in the visual appearance of steel when it is heat treated. Time and temperature is important because increasing time tends to thicken the film, which alters the observed color. The conditions of artificial or natural light, the composition of the oxide film, and the emissivity of the particular steel also affect the observed color. The method of estimating temperature by color is strongly dependent on operator experience and is used only as a guide.

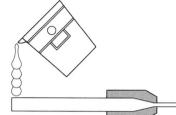

Mechanical Testing 3

Mechanical properties are obtained by mechanical testing. Mechanical testing is used for developing design data, maintaining quality control, assisting in alloy development programs, and providing data in failure analysis. Mechanical testing is usually destructive and requires test specimens of the material to be machined or cut to the specific shape required by the test method. The force involved in mechanical testing is static or dynamic and is applied once or repeatedly. Mechanical testing is conducted at various temperatures and in corrosive or noncorrosive environments. Hardness testing is a particular form of mechanical testing in which the applied force is static. Many types of mechanical tests have been developed and are used to compare the performance of materials.

MECHANICAL TESTING

A wide range of mechanical tests have been developed to measure the response of metals to mechanical forces in order to predict field performance as closely as possible. The stress configurations in the field are more complex than the simplified modes upon which mechanical tests are based. For example, a tensile test involves pulling a test bar until it fails. This test simulates the stretching of a component, such as a hanging beam. In the field, additional stress patterns (bending, twisting, vibration, etc.) usually complicate the stress configuration in this beam. Engineering judgement and experience must be used when the results of mechanical property tests are used in design.

Reasons for Mechanical Testing

Mechanical testing is used for developing design data and maintaining quality control. It is also used in assisting in alloy development programs and providing data in failure analysis.

Mechanical testing helps provide design data for engineering components. A safety factor is imposed by the designer on the test results in order to develop allowable stress data. *Allowable stress data* is the maximum allowable stress on a component under specific operating conditions. This data enables components to be sized and fabricated for individual service applications.

Mechanical testing is used in quality control programs to check or qualify products such as castings, forgings, or rolled plate. The testing is performed to standard procedures using standard test specimens removed from the product. The results indicate whether the product meets designated test requirements. Mechanical property requirements are often part of the purchase order.

Mechanical testing is used in alloy development programs as a comparison tool. The testing is used to compare new alloy compositions with established alloys. The results may be used to predict serviceability and indicate potential applications.

Mechanical testing is used in failure analysis to check on the quality of the material. Test specimens are usually obtained from regions adjacent to and away from the failure. Test specimens are taken from these regions to see if there is any difference between the failure and the rest of the material.

Applying Mechanical Forces

Mechanical force is applied using five methods: tension, compression, shear, torsion, and flexure. See Figure 3-1. *Tension* is a force that occurs when the load is applied axially (parallel to the axis) on the test specimen in a stretching manner. *Compression* is a force that occurs when the load is applied axially on the test specimen in a compressive manner.

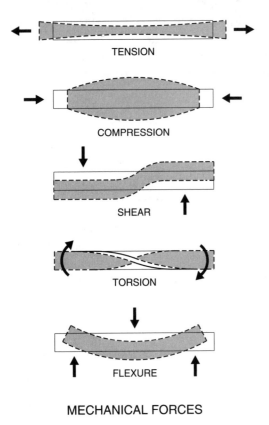

MECHANICAL FORCES

Figure 3-1. A mechanical force or load may be applied using five different methods.

Shear stress, or *shear,* is the stress caused by two equal and parallel forces acting upon an object from opposite directions. Shearing occurs when a force causes a material to separate along a plane parallel to the load. Pure shear is impossible to set up in a test without interference from other types of stresses in the devices that grip the test specimen. *Torsion,* a special case of shear, is an internal resisting force in which shear stresses occur by the twisting of the test specimen. The intensity of the shear varies from zero at the center of the test specimen to a maximum at the outside edge.

Flexure is a force that causes the bending of the test specimen. Bending introduces tension on the area being stretched and compression on the area being squeezed on the opposite side. All five methods of applying mechanical force induce a stress in the test specimen material and cause an accompanying strain.

Stress is the internal resistance of a material to an externally applied load. When a load is applied to a material, the atoms are displaced and exert a resisting force that attempts to return them to their original positions. Stress is accompanied by strain.

Strain is the accompanying change in dimensions when a load induces stress in a material. Strain is either elastic or plastic. Elastic strain occurs when the material is capable of returning to its original dimensions after removal of the load. For example, a spring with normal load returns to its original length when the load is removed. Plastic strain occurs when the material is permanently deformed by application of the load. For example, an overloaded spring will develop a permanent set or an increase in length. As the external force on a component is steadily increased, a point is reached where the strain changes from elastic to plastic.

Test Conditions

Mechanical testing is applied under a variety of conditions. These conditions include testing at high or low temperatures and in corrosive or noncorrosive environments. Test specimens must be selected with consideration of the location and orientation within the component. The preparation of the test specimen must ensure that reproducible tests are achieved.

Test Temperature. Mechanical testing is performed over three basic temperature ranges: low, ambient, and elevated. Test temperature is an important parameter because the behavior of a material may be significantly altered by the temperature at which it operates. For example, at low temperatures some steels become brittle.

Test Environment. The performance of materials under mechanical loads may be altered significantly by changing the environment. This is particularly important with fatigue (repeated) loads. Although most tests are conducted in air (considered noncorrosive for most purposes), it may be necessary to conduct the test in an applicable corrosive environment such as seawater.

Selection of Test Specimens. Selection of test specimens is a key consideration because metals often exhibit anisotropy (directional effects). This means mechanical properties vary in different directions (orientations). For example, a cold-rolled steel longitudinal test specimen is stronger than a transverse test specimen. The longitudinal test specimen is stronger because it is taken from the direction the steel is rolled. See Figure 3-2.

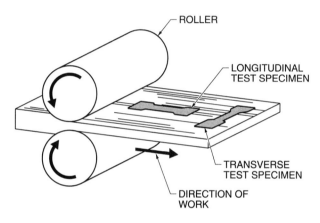

Figure 3-2. Longitudinal and transverse test specimens taken from cold-rolled plate material exhibit different mechanical properties.

The American Society for Testing and Materials (ASTM) has established guidelines for the orientation of test specimens. *ASTM A370 Mechanical Testing of Steel Products* is the publication containing these guidelines. The number of test specimens selected is usually based on experience. For example, tensile tests are performed on duplicate test specimens. Statistical analysis is used to determine the optimum number of test specimens.

Preparation of Test Specimens. In preparing a metal test specimen, a rough blank (outline of the test specimen) is first obtained by shearing, punching, flame cutting, or hacksawing. With production castings or forgings, the blank is in

the form of a separately cast bar or an appendage to the forging. The blank is machined into the desired shape of the test specimen. Sufficient excess metal is allowed for the removal of layers damaged by the severe mechanical work or heat involved in obtaining the blank. These layers are not representative of the true condition of the metal and give misleading results if incorporated into the test specimen.

The finished surface of the test specimen must be at least 3 mm ($\frac{1}{8}$ in.) from sheared faces and 6 mm ($\frac{1}{4}$ in.) from flame- or plasma-cut faces. The test specimen is finished by turning, planing, or milling. Finishing must be done without excessive heat buildup. The final surface finish must be fine and smooth so that it will not influence the way the test specimen breaks. For example, extremely coarse surfaces contain stress raisers (localized stresses) that cause localized failure. Sanding the surface with a fine abrasive paper provides a smoother surface and improves the reproducibility of the test results.

Rate of Application

The rate of application of mechanical forces for mechanical testing is divided into two main groups: dynamic and static. The two groups are based on the way the load is applied. With dynamic tests, the load is applied very rapidly and may also be applied continuously or repeatedly. The inertia of the test specimen and the rate of application of the load have a significant effect on the test results. Dynamic tests include damping capacity tests and the various types of toughness tests.

In most mechanical tests, the load is continuously applied until completion of the test. In certain dynamic tests, the load is applied repeatedly. The most common dynamic test is the fatigue test. In the fatigue test, the same load is applied, usually millions of times, until the test specimen fails.

In static tests the load is applied slowly enough so that the speed of testing has a negligible effect on the results. Static tests last from several minutes to several hours. Static tests include the tensile test, the compression test, special ductility tests, and the torsion test. Hardness testing is a type of static test in which the test specimen is not loaded to failure.

DYNAMIC MECHANICAL TESTS

In dynamic mechanical tests, the rate of application of the load or the repetition of the load exerts a significant effect on the properties of the material. These tests include the damping capacity test, fatigue testing, toughness testing, and fracture toughness testing.

Damping Capacity Test

Damping capacity is the rate at which a material dissipates energy of vibration (damps out vibrations). Chatter, noise, and fatigue failure are sometimes the result of vibrations in equipment. Damping capacity is related to internal friction, which is the resistance of the atoms in a metal to relative movement.

Metals with high internal friction transmit vibrations through the atoms and have low damping capacity. Metals with low internal friction have high damping capacity and absorb the energy of vibration by converting it into heat.

A *damping capacity test* is a dynamic mechanical test that measures the decrease in amplitude of the torsional vibrations of a twisted cylindrical bar. The bar is held vertically and clamped to a rigid base. An inertia bar, which is a piece of metal having relatively large mass in relation to the test bar, is clamped to the other end. The inertia bar is rotated by using magnets at a specified angle to achieve a desired stress. It is then released and the cylinder is set in torsional vibration. The rate of decrease in amplitude of the vibrations is measured and this measured value is the specific damping capacity.

Gray cast iron has the highest damping capacity of any common alloy and is used for equipment casings and supports. Similarly, the 400 series stainless steels are used in turbine blades because they damp resonant vibrations. Resonant vibrations are significant increases in the vibration of a part when its frequency of vibration matches that of another part exposed to it.

Some metals are used for their low damping capacity. For example, high-tin bronze is used for bells, and hard drawn high-carbon steel is used for piano or guitar strings.

Fatigue Testing

Fatigue is the failure of a material or component under alternating (cyclic) stresses, which has a maximum value less than the static tensile strength of the material. *Static tensile strength* is the tensile strength of a material before it is subjected to stresses. Fatigue is a problem that affects any component that moves or is subject to vibration. Automobiles, airplanes, ships, rotating or reciprocating parts, and piping connected to pumps are all subject to fatigue forces. The majority of failures in machines and engineering components are attributable to fatigue.

Except for ferrous materials, fatigue strength bears no fixed relationship to tensile strength or any of the other mechanical properties, such as percent elongation or hardness. Although there is no standard test, all fatigue tests involve repeated stress application of the test specimen. Test specimens range in size from tiny samples, which are studied under a microscope, to full-scale components, such as aircraft wings or marine propellers. Cyclic stresses are usually set up to simulate the type of stress experienced in service. For example, a reciprocating compressor experiences a push-pull alternating stress, while a rotating shaft experiences an alternate bending stress.

The stress ratio (R) of a cyclic stress is minimum stress divided by maximum stress ($\sigma_{min} \div \sigma_{max}$). The four classifications of cyclic stresses are low and high tensile stress (R > 0), zero stress and tensile stress (R = 0), unequal tensile stress and compression stress (R < 0), and equal tensile stress and compression stress (R = –1). See Figure 3-3. A *cycle* is each complete application of the stress. With reciprocating or rotating components, the time and stress magnitude of each cycle is usually equal (uniform loading). This is not the case with other components such as automobiles, airplanes, and ships that are randomly loaded.

The cyclic stress range ($\Delta\sigma$) is the difference between the maximum and minimum stress ($\sigma_{max} - \sigma_{min}$). The cyclic stress amplitude (σ_a) is the difference between the maximum and minimum stress divided by two ($[\sigma_{max} - \sigma_{min}] \div 2$). The mean stress ($\sigma_m$) is the sum of the maximum and minimum stress divided by two ($[\sigma_{max} + \sigma_{min}] \div 2$).

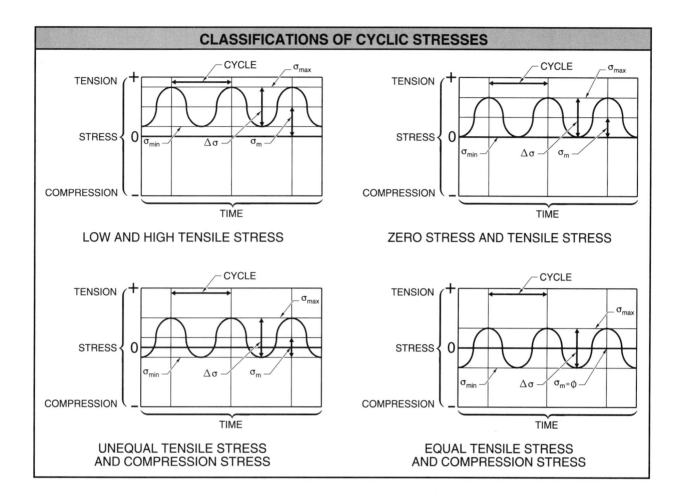

Figure 3-3. There are four classifications of cyclic stresses in fatigue. A cycle is each complete application of the stress.

Fatigue Strength. *Fatigue strength* is the stress at which a material fails by fatigue after a specific number of cycles. For ferrous alloys there is a limiting stress below which a repeating load may be applied an indefinite number of times without causing failure. This stress is the fatigue limit, or endurance limit. Nonferrous alloys do not exhibit a fatigue limit; therefore, one is specified for them. The fatigue limit is the stress corresponding to a specific number of cycles, usually 10 million (10^7). The magnitude of the fatigue limit depends on the mean stress. Most fatigue strength determinations are made using a fully reversed stress (R = –1) or a mean stress equal to zero.

From a series of fatigue tests an S-N curve is plotted. An *S-N curve* is a record of the amplitude of the cyclic stress (σ or S) plotted against the number of cycles to failure (N). See Figure 3-4. Two regions of fatigue, high-cycle and low-cycle fatigue, are indicated on the S-N curve. High-cycle fatigue occurs at low stresses and millions of cycles ($>10^4$). Low-cycle fatigue requires fewer cycles ($<10^4$) and occurs at high stresses.

An approximate correlation between fatigue strength and tensile strength exists for carbon and low-alloy steels. With carbon and low-alloy steels, the results of many fatigue tests indicate that under reversed bending conditions (equal maximum tensile stress and compression stress) the fatigue limit is approximately 50% the tensile strength.

Application of approximate correlations may be misleading because the size and surface finish of the component and the corrosiveness of the environment are significant in determining the fatigue limit. Fatigue test results made using polished test specimens are high when applied to full-scale components having machined finishes or surface irregularities caused by mechanical nicks or corrosion.

Toughness Testing

Toughness is the ability of a metal to absorb energy (high strain rates) and deform plastically before fracturing. A tough metal is more ductile and deforms rather than fracturing in a brittle manner, particularly in the presence of stress raisers such as cracks and notches. Since one of the most important requirements of structural metals is their ability to deform and give warning of impending failure, toughness is an important property to measure.

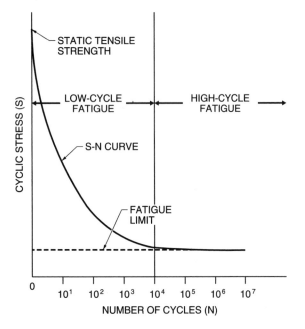

Figure 3-4. The magnitude of the fatigue limit depends on the stress repetition pattern, which is plotted on an S-N curve.

Mechanical properties are strongly affected by the rate of straining. A metal tested at a low strain rate may fracture with a large amount of strain (elongation), but a metal at a high strain rate may break with little or no elongation. A metal is tough and ductile at the low strain rate and is brittle at the high strain rate. See Figure 3-5.

Toughness is also affected by the test temperature and presence of stress raisers in the test specimen. The toughness of certain metals decreases significantly below a characteristic temperature.

Stress raisers such as a sharp change in section at the surface or internal inclusions may decrease toughness. Tests used to measure toughness include notched bar impact tests, nil ductility transition temperature tests, and fracture toughness tests.

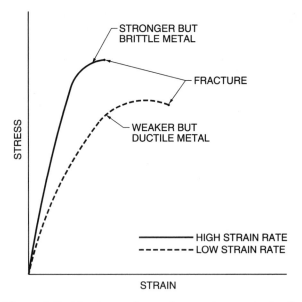

Figure 3-5. Metal tested at a low strain rate is ductile compared with the same metal tested at a high strain rate.

Notched Bar Impact Tests. A *notched bar impact test* is a test that measures the force (produced by a dynamic load) needed to break a small machine-notched test specimen. The two principal impact tests, the Charpy and Izod, are performed on a universal pendulum impact tester. See Figure 3-6. In both tests, the energy required to break the test specimen is measured. The resulting measurement is an indication of toughness.

Tinius Olsen Testing Machine Co., Inc.

Figure 3-6. A universal pendulum impact tester can perform both the Charpy and Izod impact tests.

The test specimen is a square-shaped bar containing a machined V-notch or keyhole notch (keyhole-shaped groove). A sawcut may also be used but is not recommended because the precision of reproducing the same cut is low. See Figure 3-7. These notches are characteristic of both the Charpy and Izod tests. The only difference is the location of the notch on the test specimen. The purpose of the notch in the test specimen is to facilitate fracture in a controlled location.

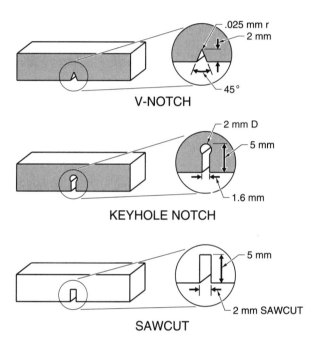

Figure 3-7. The V-notch is the most common Charpy and Izod impact test specimen.

During a Charpy test, the test specimen is placed horizontally against the two supports at the bottom of the tester. The pendulum is raised to a standard height, giving it a potential energy of 240 ft/lb (325 J). The pendulum is released and the test specimen is struck and broken by the hammer as it swings through its arc. The swing of the pendulum after it strikes the test specimen indicates the energy absorbed on impact and is measured in feet per pound or joules. When struck by the pendulum, tough materials absorb a significant amount of energy and brittle materials fracture with relatively little energy absorbed. Tough materials cause the pendulum to travel shorter distances after striking the test specimen. With brittle materials, the pendulum travels longer distances after impact. See Figure 3-8.

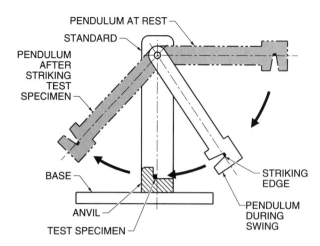

Figure 3-8. The swing of the pendulum after it strikes the test specimen indicates the energy absorbed on impact.

The Izod impact test operates on a similar principle to the Charpy. The main differences are in the position of the notch on and the method of support of the test specimen. See Figure 3-9. The notch is located toward one end of the test specimen, which is gripped vertically, instead of horizontally, in a vise.

Notched bar impact testing is widely used because of the conveniently small test specimen size and ability to correlate the results of many tests with service experience. The simple method of test specimen support is the main advantage of the Charpy over the Izod. This simple method of test specimen support allows Charpy testing to be performed over a range of test temperatures. Test specimens are heated in a furnace or cooled in a refrigerator to the test temperature and then they are rapidly tested with little or no change in temperature.

The small test specimen size required for notched bar impact testing is convenient because specimens are cut at various orientations within a component. Since the properties of metals may vary according to orientation, it is often necessary to check for properties in orientations that would exhibit the lowest quality in quality control programs. For example, with plate products a test specimen with a transverse orientation usually exhibits the lowest mechanical properties.

The behavior of metals in notched bar impact testing is extremely dependent on the rate of loading, test temperature, and type of notch. These variables make it difficult to translate the absorbed energy values into design criteria. However, the long history of notched bar impact testing, particularly the Charpy test, allows acceptance or rejection limits to be placed on lots (large quantities) of material. For example, some specifications require a minimum Charpy V-notch requirement for steel products of 15 ft/lb at the minimum expected service temperature. But this does not mean that a test specimen exhibiting 60 ft/lb is four times tougher than the minimum. The main value of notched bar impact testing is as a criterion of acceptance of material when reliable service behavior has been established.

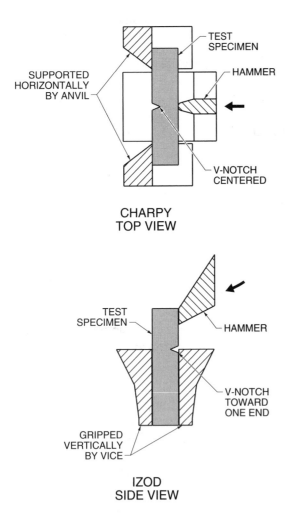

**CHARPY
TOP VIEW**

**IZOD
SIDE VIEW**

Figure 3-9. The main differences between the Charpy and Izod impact tests are the position of the notch and the method of support of the test specimen.

Nil Ductility Transition Temperature Testing. *Nil ductility transition (NDT) temperature* is the temperature at which the impact behavior of a metal

changes from ductile to brittle in the presence of a stress raiser. It is sometimes referred to as the ductile-to-brittle transition temperature (DBTT). Some metals, notably carbon and low-alloy steels, show a sharp transition in toughness when temperature decreases. This may become the controlling factor in determining their serviceability. For example, large steel storage tanks have failed catastrophically in cold weather because the NDT temperature of the plate material was higher than the atmospheric temperature at the time of failure. There were also critical stress raisers present in the design that caused initial fracture.

The Charpy test and drop weight test are used to determine the NDT temperature. The Charpy test is used to determine the NDT temperature by testing triplicate sets of test specimens over a range of temperatures. The results are plotted as impact strength against test temperature. See Figure 3-10. There is a transition from low to high impact strength. Depending on the type of steel, this transition may be gradual or sharp. The sharper the inflection of the curve, the easier the estimation of the NDT temperature.

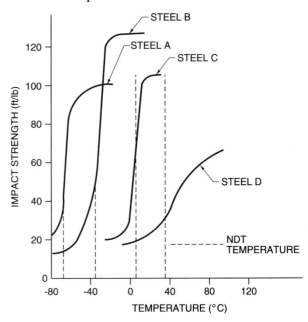

Figure 3-10. The sharper the inflection of the curve, the easier the estimation of the NDT temperature.

The drop weight test is a more reliable method than the Charpy. The test specimen is a slab or plate that is up to $\frac{5}{8}$ in. thick. A weld bead made from a brittle alloy is laid down the center of the plate. The plate is brought to the test temperature and placed in the test fixture. It is supported along both ends parallel to the weld, with the weld side facing down. A weight located vertically above the center of the plate is allowed to drop on it, causing the plate to bend. Cracking of the weld bead is initiated at 3° of bend. After that point the weld bead continues to crack, which initiates a fracture. To ensure the strain induced in the plate is elastic, a stop is placed below the weld bead. The stop limits the amount of deflection of the plate to 5° of bend. See Figure 3-11.

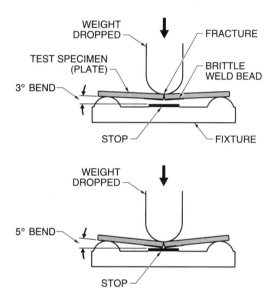

Figure 3-11. The drop weight test is more reliable than the Charpy when determining ductility.

If the temperature of the plate is below the NDT temperature, the crack will run and the plate will break into two pieces. At any temperature above the NDT temperature, the crack will come to a halt before it spreads out through the plate. The NDT temperature is the lowest temperature at which the plate will not break into two pieces. The drop weight test is described in *ASTM E 208 Drop Weight Test to Determine Nil Ductility Transition Temperature of Ferritic Steels.*

Fracture Toughness Testing

Fracture mechanics is the study of fracture toughness. *Fracture toughness* is the resistance of metals to brittle fracture propagation (spreading) in the presence of stress raisers. Flaws inside metals, such as inclusions and cracks, act in a similar manner to surface stress raisers. The high-stress concentrations at the tips of these internal flaws may produce a running (brittle) crack in some metals.

The fracture toughness (K_{1C}) of a metal at a given temperature is proportional to the stress level, which is measured in thousand pounds per square inch (ksi) or megapascals (MPa), and the square root of the crack length, which is measured in inches or meters. The unit of measure for fracture toughness is measured in ksi√in. or MPa√m using the plane-strain fracture toughness test.

Plane-strain Fracture Toughness Test. The plane-strain fracture toughness test is the most common method of measuring fracture toughness. Various types and sizes of test specimens are used. The compact tension test specimen is a block containing a machined notch. The test specimen is put into a fatigue testing machine to produce a small fatigue crack at the tip of the machined notch. The tip of the fatigue crack is a localized region of high stress intensity.

The test specimen is pulled to failure in a testing machine and the load is plotted against the opening of the notch. The load and crack extension at the sudden failure of the test specimen are measured and used to calculate the fracture toughness of the material. The test method is described in *ASTM E 399 Plane-strain Fracture Toughness Testing of Metallic Materials.* Fracture toughness testing is used to determine the critical stress intensity. This is a measure of the resistance of a metal to brittle fracture propagation in the presence of flaws and cracks. Pressure vessels, storage tanks, airplanes, and ships are examples of structures that are designed and manufactured in accordance with fracture mechanics principles.

STATIC MECHANICAL TESTS

Static mechanical tests include the tensile test, special tests for measuring ductility, compression test, and torsion test. Of these, the tensile test is the most widely used.

Tensile Test

The tensile test is a static test that measures the effects of a tensile force on a material. The data from this test includes tensile strength, yield point and yield strength, percent elongation and reduction in area, and modulus of elasticity. Tensile test results are used to indicate strength, ductility, stiffness, and proper parameters for heat treatment or processing.

Tensile Test Machine. A tensile test machine has two major components that are the means of applying the load to the test specimen and measuring the applied load. Some testing machines are designed for one type of test only, such as tension testing machines for testing chain and wire. Universal testing machines test specimens in tension or compression. See Figure 3-12.

With a universal testing machine, the load is applied mechanically to the test specimen by a screw and gears, or it is applied hydraulically by a hydraulic jack. The applied load is measured by a dynamometer (load cell) for mechanically driven machines. A load cell is a device that uses the elastic deformation of a spring or diaphragm that is calibrated to indicate the mechanical load applied to the test specimen. A Bourdon tube is used in hy-

draulically driven machines. The tube straightens out as the internal pressure on the fluid is increased. The motion of the tube is used to rotate a pointer over a scale that is calibrated to read the hydraulic load applied to the test specimen.

Tensile Test Specimens. Tensile test specimens are usually shaped like a dog bone, meaning that the central portion of the specimen is smaller in cross section than the two ends. This shape is used because it causes the test specimen to fail in the narrower central portion rather than at the ends, where the gripping devices affect the stress configuration.

The transition from the ends of the tensile test specimen to the reduced section is shouldered, or made with a fillet. The shoulder minimizes stress concentrations. See Figure 3-13. This is particularly important for brittle materials. The longitudinal axis of the test specimen should be symmetrical to avoid the introduction of bending loads during the test. The test specimen grips also must be symmetrical along the longitudinal axis of support to avoid the introduction of bending loads during the test.

The gauge length (distance over which the elongation measurement is made) is always less than the distance between the shoulders, but there is no

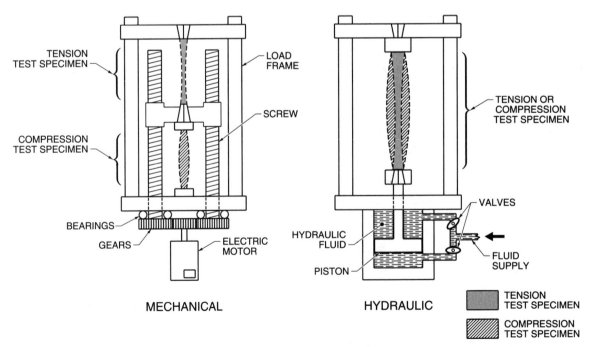

Figure 3-12. A universal testing machine can be mechanical or hydraulic.

specific guideline. The gauge marks are always an equal distance from the center of the length of the reduced section.

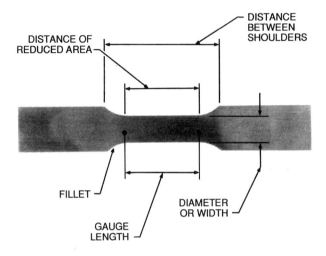

Figure 3-13. A fillet is used on the tensile test specimen to minimize stress concentrations, and the gauge marks are always an equal distance from the center of the length of the reduced section.

Tensile test specimens may be round or rectangular, depending on the stock from which they are obtained. For example, rectangular test specimens are obtained from plate or sheet, and round test specimens are taken from forgings or cast test bars. The shape of the ends of the test specimen is determined by the specimen gripping device that is used. See Figure 3-14.

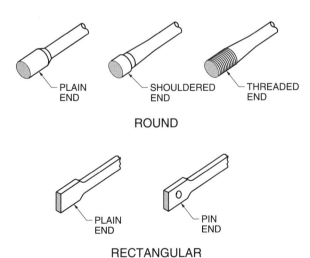

Figure 3-14. A variety of tensile test specimen ends are used to ensure secure and uniform gripping by the test machine.

The ends of round test specimens are either plain, shouldered, or threaded. Rectangular test specimens are generally made with plain ends, but sometimes they are pin ends. Pin ends are rectangular test specimens that contain a hole for a pin bearing.

Tensile Test Procedure. The tensile test procedure is conducted by fixing the test specimen firmly in the grips of the testing machine. An extensometer, a device for measuring the extension or elongation of the test specimen, is fitted to the specimen across its gauge length. See Figure 3-15. An axial load is applied and the test specimen is stretched. As the test specimen is stretched, a load-extension (stress-strain) curve is plotted. The extensometer is removed before the test specimen breaks. The tensile test procedure is described in *ASTM E 8 Tensile Testing of Metallic Materials*.

Figure 3-15. An extensometer measures the extension of elongation of the tensile test specimen.

The load-extension curve shows load and extension limits for metals. See Figure 3-16. Point A is the proportional limit. *Proportional limit* is the maximum stress at which stress is directly proportional to strain. Beyond point A, stress is no longer proportional to strain. Between points A and B, the line starts to curve. Up to point B, the tensile test specimen will return to its original length if the load is removed. Point B is the elastic limit. *Elastic limit* is the maximum stress to which a material is subjected without any permanent strain remaining after stress is completely removed. Beyond point B, strain is permanent, or the strain in the test specimen is plastic. A *plastic strain* is strain that remains permanent after the stress is removed. From point B, the shape of the curve varies for different metals.

Low- and medium-carbon steels show a jog in their curve, which peaks at point C, or the yield point. *Yield point* is the point at which strain occurs without an increase in stress. Between points C and D, the curve falls indicating a plastic strain. The curve continues down to point E, the lower yield point. The curve eventually regains its upward movement and peaks at point F. Point F is the ultimate tensile strength (a measure of the tensile strength). Between points F and G, the test specimen begins to neck down, or develop a pronounced waist. Point G is the point of failure. *Point of failure* is when the fracture occurs.

Only low- and medium-carbon steels exhibit yield point behavior. With all materials, the slope of the load-extension curve decreases and peaks at point F, with failure occurring at point G. With brittle metals, fracture may occur while the load is increasing toward point F.

When the tensile test is completed, the broken test specimen is removed from the testing machine and fitted together. The new increased gauge length, and the reduced diameter at the narrowest point are measured. This is usually at the break or immediately adjacent to it. See Figure 3-17. These measurements allow the percent elongation and percent reduction in area to be calculated.

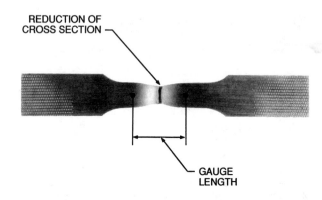

Figure 3-17. Increased gauge length and reduced diameter at the narrowest point are measured and used to calculate the percent elongation and percent reduction in area.

Tensile Strength. The tensile strength of the tensile test specimen is calculated by dividing the maximum

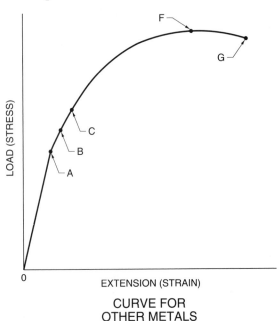

CURVE FOR
OTHER METALS

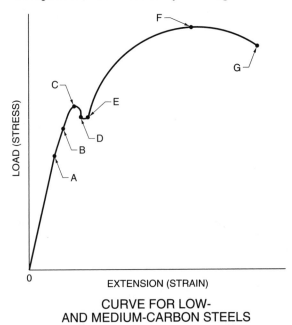

CURVE FOR LOW-
AND MEDIUM-CARBON STEELS

Figure 3-16. The load-extension curve shows load and extension limits for metals.

load applied by the original cross-sectional area of the test specimen. Tensile strength is not a precise value because the original cross-sectional area is not the same as the reduced cross-sectional area that actually exists at the maximum load. Tensile strength is measured in a thousand pounds per square inch or megapascals.

Yield Point and Yield Strength. Stress is the load divided by the original cross-sectional area. At the yield point, an increase in strain occurs without an increase in stress. Yield point behavior leads to Lüders bands (ripples) on the test specimen. Stretcher strains (elongated markings), a similar phenomenon, are observed in low-carbon steel pressings when deformed to the yield point.

For metals without a yield point, a yield strength (artificial value) is obtained from the load-extension curve. The yield strength, or .2% offset, is calculated by measuring the stress that causes a specific amount of permanent strain (usually .2%). See Figure 3-18.

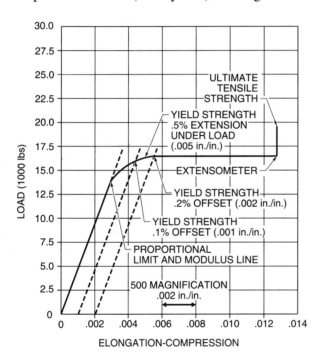

Figure 3-18. The yield strength, or .2% offset, is calculated by measuring the stress that causes a specific permanent strain (usually .2%).

Designers use tensile strength and yield point or yield strength values to develop maximum allowable stress data for materials. Designers incorporate a suitable safety factor with the tensile strength and yield point or yield strength values to develop maximum allowable stress data. For example, the American Society of Mechanical Engineers (ASME) indicates maximum allowable stresses in their *ASME Boiler and Pressure Vessel Code*. The maximum allowable stresses are based on one fourth of the tensile strength or two thirds of the yield point or yield strength, whichever value is lower.

Percent Elongation and Percent Reduction of Area. Percent elongation and percent reduction of area are measures of the ductility of a tensile test specimen. They indicate the amount of plastic deformation prior to fracture of the test specimen. Percent elongation of a tensile test specimen is found by applying the following formula:

$$\%E = \frac{L_f - L_g}{L_g} \times 100$$

where

$\%E$ = percent elongation

L_f = final length

L_g = gauge length

100 = constant

Example: Figuring percent elongation

What is the percent elongation for a tensile test specimen that has an initial gauge length of 2 in. and a final length of 2.45 in.?

$$\%E = \frac{L_f - L_g}{L_g} \times 100$$

$$\%E = \frac{2.45 - 2}{2} \times 100$$

$$\%E = \frac{.45}{2} \times 100$$

$$\%E = .225 \times 100$$

$$\%E = \mathbf{22.5\%}$$

Percent elongation is calculated from the gauge length. The longer the gauge length, the less the effect necking down of the test specimen has on the final length. This results in lower percent elongation for a given metal. See Figure 3-19. When the gauge length is made equal to k√A, where k is a constant equal to 4.47 and A is equal to the cross-sectional area of the test specimen, the percent elongation value remains practically constant for different gauge lengths. The most common gauge length in tensile testing is 2 in.

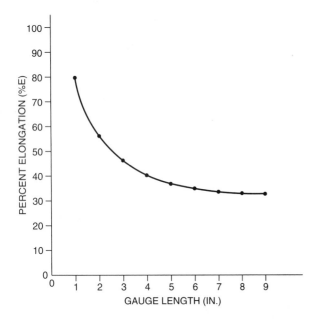

Figure 3-19. Percent elongation is calculated from the gauge length.

Percent reduction of area of a tensile test specimen is found by applying the following formula:

$$\%RA = \frac{D_o - D_f}{D_o} \times 100$$

where

$\%RA$ = percent reduction of area

D_o = original diameter

D_f = final diameter

100 = constant

Example: Figuring percent reduction of area

What is the percent reduction of area for a tensile test specimen with an original diameter of .505 in. and a reduced diameter of .350 in.?

$$\%RA = \frac{D_o - D_f}{D_o} \times 100$$

$$\%RA = \frac{.505 - .350}{.505} \times 100$$

$$\%RA = \mathbf{30.69\%}$$

Round tensile test specimens must be used to calculate percent reduction of area. Rectangular test specimens have significant rounding of their corners during the test, which makes measurement of the cross-sectional area less accurate.

Unlike the strength values that are used to develop allowable stresses for metals, ductility values obtained from the tensile test have limited significance for anticipating service performance. In service situations, the maximum amount of elongation usually tolerated is 1% to 2%. Unlike the tensile strength and yield point or yield strength, the ductility values are not used in the design of components.

It is important to measure ductility values because they are structure sensitive, or strongly influenced by the microstructure of the test specimen. Ductility values reveal undesirable microstructure, particularly in heat-treated steels that were exposed to improper heat treatment. Undesirable microstructure could lead to failure or an impaired service behavior.

Ductility values also have some relationship to the formability of metals. For example, the plastic strain capability of a metal is limited when the metal is drawn in a large amount of localized formations, such as corners.

Modulus of Elasticity. The *modulus of elasticity* (*Young's Modulus, E*) is the ratio of stress-strain in the region below the proportional limit on the stress-strain curve. The slope of the straight line is a measure of stiffness or springiness. The greater the stress required to produce a given strain, the stiffer the material. For example, steels are relatively stiff materials for the cost and have an approximate value of $E = 30 \times 10^6$ psi. To achieve an equivalent stiffness in materials such as aluminum or fiberglass-reinforced plastic, a thicker section is required. Modulus of elasticity is used in the design of cantilever and elliptical flat springs, which require high yield strength and low modulus of elasticity.

Specific Ductility Tests

Several qualitative or semiquantitative tests are used specifically to assess ductility. These specific ductility tests are used most often because they are simple and duplicate a particular processing operation where ductility is an important parameter. Specific ductility tests include the guided bend test and the various metal formability tests.

Guided Bend Test. The guided bend test consists of bending a rectangular piece of metal around a U-shaped die. This test is most commonly used to check the quality of welds. A welded test specimen is cut into a rectangular shape with the weld at the midpoint of the specimen. The weld is transverse to its axis. See Figure 3-20.

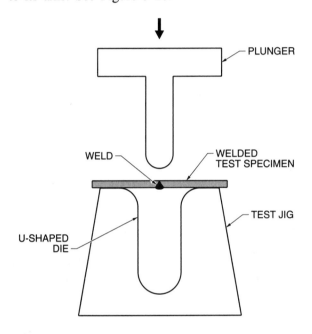

Figure 3-20. The guided bend test is an inexpensive and rapid method to check the quality of a weld.

The welded test specimen is placed in a test jig and bent by a plunger into a U-shaped die. The localized overstrain on the convex side of the U-shaped bend will quickly reveal the presence of weld defects such as lack of fusion. In some cases more costly techniques such as radiography might not detect these defects as easily. This type of testing is described in *ASTM E 190 Method for Guided Bend Test for Ductility of Welds.*

Formability Test. Formability tests measure the ductility of sheet metal used for deep drawing or stretching. In cupping tests, a metal sheet test specimen is stretched over an advancing punch with a rounded head to determine the fracture point. See Figure 3-21. Cupping tests are limited to predicting

gross differences in formability and used as an inspection tool. For example, cupping tests provide a rapid indication of the ductility of a stock of sheet. Cupping tests are described in *ASTM E 643 Method for Conducting a Ball Punch Deformation Test for Metallic Sheet Materials.*

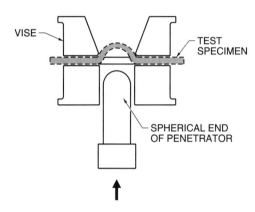

Figure 3-21. Cupping tests provide an indication of the formability of sheet metal.

Compression Test

The compression test, the opposite of the tensile test, is not often used for metals because of limitations in the test technique. The compression test is most often used as a quality control tool for nonmetallic materials such as concrete or refractory brick. For many alloys the stress-strain curve in tension and compression are similar in shape. The compression test also provides equivalent types of data to the tensile test. This data is useful in the analysis of structures exposed to compression stress or bending stresses in metalworking processes such as cold rolling or cold forging.

The compression test is limited because it is difficult to apply a true axial load to the test specimen, which results in the introduction of other stresses such as bending. To compensate, the test specimen is made squat (short) in order to stabilize the axial stress application. However, this makes the test specimen so short that it lessens the precision of the strain measurement. Another limitation of the compression test is the friction that results between the bare surfaces of the test specimen and the testing machine. This friction is produced from lateral (sideways) movement of the test specimen as it is compressed. Friction complicates the stress pattern and alters the test results.

Torsion Test

A torsion test is used to determine the shear resistance of a metal. The test is performed by applying torque (twist) to a cylindrical bar or tube-shaped test specimen in a specially designed torsion testing machine. See Figure 3-22. The torsion test is used to determine a metal's resistance to shear.

Tinius Olsen Testing Machine Co., Inc.

Figure 3-23. A tropometer is used to measure the degree of twist during a torsion test.

HARDNESS TESTING

The hardness of a material is its resistance to deformation (particularly permanent deformation), indentation, or scratching. Hardness testing has little practical application. For example, hardness testing could be used to predict the scratching or scuffing resistance of a material, but it is not used for this purpose. Hardness testing is actually the most widely used of all mechanical tests because it is a sensitive quality control technique. Hardness testing is not only rapid, but often nondestructive. The three main types of hardness tests are the scratching, rebound, and indentation hardness tests.

Tinius Olsen Testing Machine Co., Inc.

Figure 3-22. A torsion testing machine is used for determining a metal's resistance to shear.

The amount of torque, or the degree of twist, on the specimen is measured and recorded by a tropometer. A *tropometer* is a device used to measure the degree of twist and it operates like an extensometer. See Figure 3-23. A tropometer is mounted on the test specimen while it is in the testing machine and it measures the amount of torque produced.

Data obtained from torsion testing is similar to that obtained in a tensile test. Torsion testing is not common because resistance to shear is estimated from tensile test results. For example, the shear strength of medium-carbon steels is approximately one third to two thirds of their tensile strength. The modulus of rigidity (modulus of elasticity in shear) is approximately two thirds the modulus of elasticity in tension. The modulus of rigidity is the criterion for stiffness in the design of helical springs.

Scratching Hardness Tests

A scratching hardness test is a test that compares the hardness of one metal by scratching it with a material of a known hardness. Such tests are used as a relative guide to hardness. Scratching hardness tests include the Mohs' scale and the file hardness test.

Mohs' Scale. The Mohs' scale, the oldest hardness test, uses 10 minerals that are listed in order of increasing hardness. A mineral with a higher number scratches one with a lower number. For example, number ten (diamond) will scratch number nine (corundum). Corundum will scratch number two (gypsum) or any number between. The lowest number mineral that scratches a specific metal is its number on the Mohs' scale. The Mohs' scale is of little value in metallurgy because the actual hardness increments of the minerals numbered from one to ten are not uniform.

File Hardness Test. A *file hardness test* is a hardness test for metals that uses a file to rub against the surface of the metal and results in degree of bite, which indicates hardness. The technique is dependent on the experience of the test operator. On very hard materials, care must be taken not to damage the file. Damage is avoided by not putting excessive pressure between the file and the metal.

Another application for the file hardness test is as a positive indication that a part has been hard faced. *Hard facing* is the depositing of a surface filler metal to increase resistance properties. An example of hard facing is chromium-plated metals. If the file does not bite, then the specimen is chromium-plated.

Rebound Hardness Tests

Rebound hardness tests record the height of rebound of a hard object from the surface under calibrated conditions. The height of rebound increases with the hardness of the material. Rebound hardness tests include the scleroscope and Equotip tests.

Scleroscope Hardness Tester. A *scleroscope hardness tester* is an instrument that uses a test specimen that is freely supported horizontally and a glass tube that contains a diamond-tipped hammer positioned vertically over the specimen. The hammer is allowed to fall from a set height and the height of rebound is measured. See Figure 3-24. The test shows that the higher the rebound, the harder the specimen. The scleroscope hardness test is described in *ASTM E 448 Scleroscope Hardness Testing of Metallic Materials*.

Equotip Hardness Tester. The Equotip hardness tester, or Leeb hardness tester, is a commercial instrument with a tube containing a ball bearing. The tube is placed in contact with the test specimen in any one of five positions. The five positions are vertically down, vertically up, horizontally, 45° down, and 45° up.

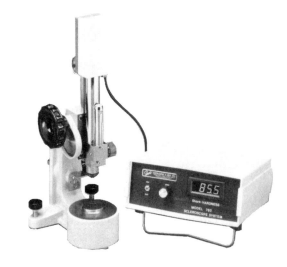

Shore Instrument and Manufacturing Co.

Figure 3-24. The scleroscope hardness tester uses the height of rebound of a diamond-tipped hammer from the test specimen surface to determine hardness.

A spring-loaded mechanism is triggered and propels the ball bearing toward the surface of the specimen. The rebound height is recorded electronically and displayed as a hardness number. See Figure 3-25.

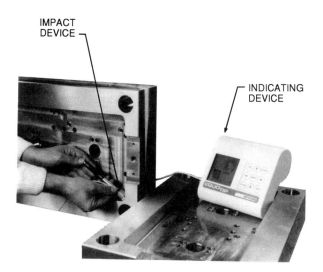

EQUOTIP™ Associates

Figure 3-25. The Equotip hardness tester, or Leeb hardness tester, can be used in five positions.

This hardness must be modified by subtracting a number that relates to the orientation of the tester with respect to the specimen. A graph is used to translate the modified number into one of the commonly used hardness numbers, such as Rockwell C.

Accuracy of Rebound Hardness Tests. A variety of factors influence the accuracy of rebound hardness tests. The tests must be performed on a firmly supported surface because vibration of the test specimen affects the height of rebound. This leads to an artificially low, or flat reading. To detect any chipping of the hammer or flattening of the ball bearing, instruments must be calibrated prior to each test using a standard block of known hardness. The surface of the test specimen must be free of oil, scale, and other contaminates and must not be rough. It may be necessary to make the surface smooth with a file prior to testing to remove any surface roughness.

Indentation Hardness Tests

Indentation hardness tests use the surface impression produced by a standardized shape indenter and standardized load to determine hardness. The depth or size of the impression is measured to obtain the hardness value of the test specimen. This hardness test is the most widely used. It is applied on soft or hard surfaces, on large components or to micro-constituents in metals, and with fixed laboratory testing equipment or portable field instrumentation.

Types of indentation hardness testing machines include the Brinell, Rockwell, Vickers, microhardness testers, and portable hardness testers. They all have similar general precautions that must be followed. Although each test has a particular hardness scale, it is possible to relate one scale to another. See Appendix.

Brinell Hardness Test. Brinell hardness tests use a machine to press a 10 mm diameter, hardened steel ball into the surface of the test specimen. See Figure 3-26. The machine applies a load for a specific period of time and causes an indentation that is used to calculate hardness.

Hardness is calculated by dividing the load by the area of the curved surface of the indentation. The Brinell hardness number is found by measuring

the diameter of the indentation and then finding the corresponding hardness number on a calibrated chart. The test is described in *ASTM E 10 Brinell Hardness Testing of Metallic Materials*.

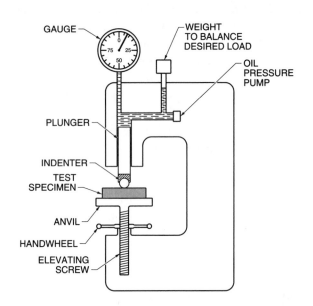

Figure 3-26. The Brinell hardness tester applies a load for a specific period of time and causes an indentation that is used to calculate hardness.

The load applied to the steel ball depends on the type of metal under test with the Brinell testing machine. With the Brinell testing machine, 500 kg is used for soft metals and thin stock, 1500 kg for aluminum castings, and 3000 kg for ferrous metals. The load is usually applied for 10 to 15 seconds. The diameter of the indentation is measured to ±.05 mm using a low-magnification portable microscope. With soft or hard metals, care must be taken to measure the exact diameter of the indentation and not the apparent diameter caused by edge effects that result in a ridge or depression encircling the true indentation. See Figure 3-27.

The Brinell hardness number followed by the symbol HB indicates a hardness value made under standard conditions using a 10 mm diameter, hardened steel ball; 3000 kg load; and an indentation time of 10 to 15 seconds.

A code is used for other test conditions. For example, 75 HB 10/500/30 indicates a Brinell hardness number of 75 was obtained in a test using a 10 mm diameter, hardened steel ball with a 500 kg load applied for 30 seconds. For extremely hard metals, a tungsten carbide ball is substituted for the steel ball, allowing readings as high as 650 HB.

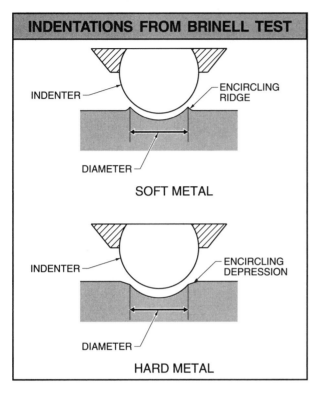

Figure 3-27. Soft or hard metals require careful measurement of their indentations in the Brinell test.

The Brinell ball makes the deepest and widest indentation of any hardness test, so that it indicates an average hardness value over many grains of the metal. Consequently, the Brinell hardness test is the least affected by surface irregularity or inhomogeneity. It is used on coarse stock, such as steel or aluminum castings, forgings, heavy plate, and heat-treated billets. Sometimes it is necessary to grind a flat spot on the surface to improve the diametrical measurement. The Brinell test is not suitable for very thin, case-hardened or hard-faced components.

Rockwell Hardness Test. The Rockwell hardness test is the most used and versatile hardness test. The testing machine has a variety of attachments that make it capable of measuring the hardness of a wide range of materials in many sizes and shapes.

A $\frac{1}{16}''$ diameter steel ball and a 120° diamond cone are the two types of indenters. The Rockwell hardness test uses two loads that are applied sequentially. See Figure 3-28. A minor load of 10 kg is applied that helps seat the indenter and remove the effect of surface irregularities. A major load, which varies from 60 kg to 150 kg, is then applied.

The amount of the major load determines the type of indenter used. For example, a steel ball is used with the 60 kg load and a diamond cone with the 150 kg load. The difference in depth of indentation between the major and minor loads provides the Rockwell hardness number. This number is taken directly from the dial on the machine. The Rockwell hardness test is described in *ASTM E 18 Rockwell*

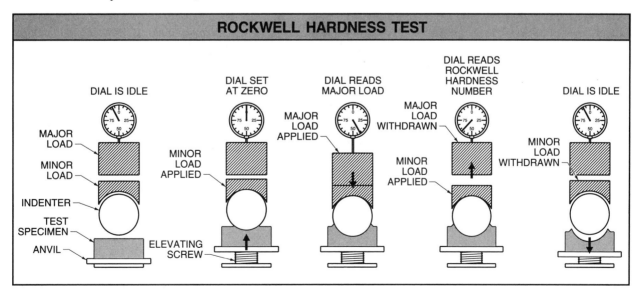

Figure 3-28. The Rockwell test uses two loads, a minor and a major, that are applied sequentially to determine hardness.

Hardness and Rockwell Superficial Hardness of Metallic Materials.

Several Rockwell hardness scales are used when measuring a variety of materials. See Figure 3-29. The designation system has a hardness number that is followed by HR, which is followed by another letter that indicates the specific Rockwell scale. For example, a test specimen exhibiting a hardness reading of 40 HRA has a Rockwell hardness reading of 40 on the A scale, the indenter is a diamond cone, and the black numbers on the dial are used for reading the hardness.

The two most common scales are Rockwell B (HRB) and Rockwell C (HRC). The Rockwell B scale uses a $\frac{1}{16}''$ diameter steel ball and a 100 kg load for relatively soft materials. For example, it is used on an annealed low-carbon steel, which may exhibit a hardness of approximately 85 HRB.

The Rockwell C scale uses the diamond cone and the 150 kg load for relatively hard materials. For example, a quenched and tempered low-alloy steel usually exhibits a hardness between 30 HRC and 45 HRC, depending on the tempering temperature.

Since Rockwell indentation is much smaller than the Brinell, test specimen preparation is significantly more important. Both sides of the test specimen must be clean, scale-free, dry, and parallel. Special jigs help support round or oversize test specimens to ensure immobility during the test.

ROCKWELL HARDNESS SCALES					
Group	Scale Symbol Prefix Letter	Indenter	Major Load (kg)	Dial Numbers	Typical Applications
Common Scales	B	$\frac{1}{16}''$ ball (1.6 mm)	100	Red	Copper alloys, soft steel, aluminum (Al) alloys, malleable iron.
Common Scales	C	Diamond cone	150	Black	Steel, hard cast iron, pearlitic malleable iron, deep case-hardened steel
General Scales	A	Diamond cone	60	Black	Cemented carbides, thin steel, shallow case-hardened steel
General Scales	D	Diamond cone	100	Black	Thin steel, medium case-hardened steel
General Scales	E	$\frac{1}{8}''$ ball (3.2 mm)	100	Red	Cast iron, aluminum and magnesium alloys, bearing metals
General Scales	F	$\frac{1}{16}''$ ball (1.6 mm)	60	Red	Annealed copper alloys, thin soft sheet metals
General Scales	G		150	Red	Phosphor bronze, beryllium, copper, malleable iron
General Scales	H	$\frac{1}{8}''$ ball (3.2 mm)	60	Red	Aluminum, lead (Pb), zinc
General Scales	K		150	Red	Aluminum, lead (Pb), zinc
Special Scales	L	$\frac{1}{4}''$ ball (6.4 mm)	60	Red	Bearing metals and other very soft or thin materials
Special Scales	M		100	Red	Bearing metals and other very soft or thin materials
Special Scales	P		150	Red	Bearing metals and other very soft or thin materials
Special Scales	R	$\frac{1}{2}''$ ball (12.7 mm)	60	Red	Bearing metals and other very soft or thin materials
Special Scales	S		100	Red	Bearing metals and other very soft or thin materials
Special Scales	V		150	Red	Bearing metals and other very soft or thin materials

Figure 3-29. The Rockwell designation system consists of the hardness number followed by HR, which is followed by the letter indicating the specific Rockwell scale.

Rockwell Superficial Hardness Test. The Rockwell superficial hardness test operates on the same principle as the regular Rockwell machine. It is used for thin strip or lightly carburized surfaces, small parts, or parts that might collapse under the conditions of the regular test. To achieve this, the minor load is reduced to 3 kg and the major load is 15 kg to 45 kg.

Two types of indenters, a $1/16''$ diameter steel ball and a 120° diamond cone, are used. The $1/16''$ diameter steel ball is given the designation T (meaning testing thin sheet). The 120° diamond cone is given the designation N (derived from its use for testing nitrided steel). For example, a superficial Rockwell reading of 30N-42 indicates a hardness value of 42 made with a 30 kg load and a 120° diamond cone. The Rockwell superficial test is described in *ASTM E 18 Rockwell Hardness and Rockwell Superficial Hardness of Metallic Materials.*

Vickers Hardness Test. The Vickers hardness test is similar in principle to the Brinell hardness test. In the Vickers hardness test, the hardness number is determined from the load divided by the surface area of the indentation. The major difference is that the indenter in the Vickers hardness test is a 136° square-base diamond cone. The load varies from 1 kg to 120 kg.

During a Vickers hardness test, the specimen is placed on an anvil and raised by a screw until it is close to the point of the indenter. The starting lever is tripped, allowing the load to be slowly applied to the indenter. The load is released, the anvil lowered, and a filar microscope is swung over to measure the diagonals of the square indentation to ±.001 mm. Diagonal measurements are averaged and the Vickers hardness number is followed by the letters HV. The Vickers hardness test is described in *ASTM E 92 Vickers Hardness Testing of Metallic Materials.*

Advantages of the Vickers hardness test are that extremely accurate readings can be taken and one type of indenter covers all types of metals and surface treatment. Test specimen preparation is important because a poor surface finish makes the measurement of the diagonals extremely difficult. A fine emery finish is the coarsest surface allowable.

Microhardness Testing. Microhardness testing is at the opposite end of the scale to the Brinell hardness test. A polished surface, coupled with light loads of less than 200 g, allow the hardness of individual grains of metal or other microconstituents to be measured. The two types of microhardness testers are based on the type of indenter. They are the 136° square-base pyramid and the elongated diamond pyramid with a 7:1 diagonal ratio (Knoop tester).

During microhardness testing, the test specimen is placed under the microscope of the microhardness tester. See Figure 3-30. The area of interest is focused at the intersection of the crosswires. The indenter is swung into place and the load applied for a set period of time. The load is then removed, the microscope swung back, and the length of the diagonals measured. From these measurements, the microhardness reading, either Vickers (HV) or Knoop (HK), is obtained from a chart. Microhardness testing is described in *ASTM E 384 Test Method for Microhardness of Metals.*

Microhardness testing is a useful tool for measuring the hardness of platings, coatings, foil, composition gradients in metals, bimetallic couples, very small parts, brittle materials, small diameter wire, powdered metals, and many other materials. The microhardness of the test specimen is always higher than the bulk surface hardness measured by any of the other indentation techniques. The surface hardening effect of the polishing operation and the extremely light load result in a relatively shallow impression and a higher apparent hardness.

Portable Testers. Portable indentation hardness testers are used on parts that are too unwieldy for the standard test machines, on fixed structures that cannot be moved, and/or in testing positions other than vertical. These portable testers use mechanisms other than dead weights to apply the load to the indenter.

Portable Brinell testers generally apply the load by means of a hydraulic cylinder equipped with a pressure gauge and a spring-loaded relief valve. The load is applied several times until pressure relief occurs. For a steel test specimen under a 300 kg load, three load applications are equivalent to 15 seconds holding time.

Wilson Instruments, Inc.

Figure 3-30. The microhardness measured by a microhardness tester is always higher than the bulk surface hardness.

Portable Rockwell testers generally apply the load by a screw connected to a calibrated spring and are equipped with two indicators. One indicator is a dial gauge that measures spring deflection, which indicates load, and the other is a dial gauge that indicates penetration. The minor load is first applied, the depth indicator index set, and the major load applied. The loading screw is turned back to release the major load and indicate the minor load reading. The hardness is read on the depth indicator as the difference in readings between the application of the minor and major loads.

There are several types of commercial portable indention hardness testers that do not closely follow the principles of the Brinell or Rockwell. Examples include the Telebrinneller®, Minibrinneller®, PTC®, and Barcol® portable indention hardness testers.

The Telebrinneller® and the Minibrinneller® portable hardness testers are two similar instruments. The Telebrinneller® is a larger version of the Minibrinneller®. The Telebrinneller® has a rubber anvil containing a 10 mm diameter Brinell ball that protrudes through the base of the anvil. A bar of known hardness is positioned on the back of the ball and the ball is placed over the area to be measured. The anvil is struck with the hammer, indenting the test

specimen and the test bar. The diameters of both indentations are measured with a portable microscope. A calibrated slide rule allows the unknown hardness of the test specimen to be calculated from the known hardness of the test bar and the two indentation diameters.

The PTC® portable hardness tester has a spring-loaded punch containing a hardened steel ball that is placed vertically over the test specimen and then released. The diameter of the indentation is read with a portable microscope that is calibrated to read the Rockwell C scale.

The Barcol® portable hardness tester is used on thermosetting plastics, but is applicable on soft metals such as aluminum. The instrument is placed on the test specimen in any position. Firm hand pressure is applied to force a needle-shaped indenter into the test specimen. The Barcol® hardness number is read from a dial fixed to the instrument.

General Precautions. General precautions in using indentation hardness tests include machine abuse, surface condition of the test specimen, specimen flatness, specimen thickness, spacing between and number of indentions, relationship of scales, and application of loads. Steel ball indenters may become flattened on very hard materials and mechanical abuse may chip diamond indenters. The quality of the indenter must be maintained for true, reproducible readings. The indenter must be checked regularly and replaced when necessary.

Surface preparation requirements become increasingly stringent as the size and depth of the indentation decrease. At the very minimum, it is necessary to remove rust or scale from the surface. When the sample is ground to prepare an area for the test, excessive heat must be avoided. Surface decarburization will lead to a lower hardness reading.

Test specimens must be supported to avoid rocking under the tester. It may be necessary to grind the back side of the test specimen to make it flat. The indenter should be perpendicular to the test specimen when making an impression. With a round test specimen such as bar, it is usually necessary to grind a small flat area to make a test.

The test specimen must be thick enough so that an anvil effect (bulge) does not appear on the opposite side when the indentation is made. For the

Rockwell and Brinell tests, the test specimen should be at least 10 times as thick as the depth of the impression. For the Vickers hardness test, the test specimen should be at least one and a half times as thick.

The minimum spacing depends on the type of test. If the indentations are too close there will be interaction between disturbed zones of metal. Indentations should be at least three diameters from the edge of the test specimen for the Brinell and Rockwell tests and two and a half diagonals for the Vickers. The minimum separation between indentations should be four diameters (center to center) for the Brinell and Rockwell and two and a half diagonals (center to center) for the Vickers.

More than one reading must be taken into account for surface irregularities and test specimen inhomogeneity. The minimum number of readings required for a specific test is determined by experience. For the Brinell, three readings are usually taken and averaged. For the Rockwell and Vickers, five readings are usually taken and averaged.

Indentation hardness readings are based on a combination of properties such as friction, elasticity, and viscosity of the indenter and the test specimen. These vary with the type of specimen and test. The distribution of plastic strain in the test specimen, which is caused by the particular type of indenter, is also an important factor. Consequently, care must be taken when converting hardness numbers.

Separate conversion tables are required for different families of metals. *ASTM E 140 Standard Hardness Conversion Tables for Metals (Relationship between Brinell Hardness, Vickers Hardness, Rockwell Hardness, Rockwell Superficial Hardness, and Knoop Hardness)* contains hardness conversion tables for several major families of alloys. Pocket-size conversion charts supplied by vendors are usually an extract for the steels portion of *ASTM E 140*. Regular conversion between different hardness scales should be avoided unless there is a large amount of experience and data available to justify making such correlations.

Applications of Indentation Hardness Tests. Indentation hardness testing has several major applications and is the most widely used quality control tool for metals. Indentation hardness testing

is invaluable during heat-treatment processes. A metal may be tested as raw stock, retested after heat treatment and rough machining, and tested again after finish machining. This is true provided that the test location does not affect tolerance or finish. The correlation between indentation hardness and properties obtained from heat treatment is usually good, especially for quenched and tempered steels.

The hardness value is a useful guide to the tensile strength of some metals because resistance to penetration in the hardness test is partly a measure of resistance to plastic flow. Resistance to plastic flow is a measure of stress versus strain. The relationship between hardness and tensile strength works best for carbon and low-alloy steels. The approximate relationship between hardness and tensile strength is found by applying the following formula:

$$TS = \frac{HB}{2}$$

where

TS = tensile strength (in ksi)

HB = Brinell hardness number

2 = constant

Note: At extremely high hardness values this relationship is no longer valid.

Example: Figuring relationship of hardness and tensile strength

What is the tensile strength of a steel with a hardness of 212 HB?

$$TS = \frac{HB}{2}$$

$$TS = \frac{212}{2}$$

$$TS = \textbf{106 ksi}$$

Hardness values provide an indication of the amount of cold work a metal has received. The hardness and springiness of alloys such as copper or austenitic stainless steels increase with increasing amounts of cold work. With small amounts of cold work, the hardness reading may be misleading because the increase in hardness is not uniform across the section but is greater toward the outside edges. Care must be taken when correlating hardness readings with the degree of cold work.

The Vickers, superficial Rockwell, and microhardness testers are used to verify the hardness of plating or case-hardened components that have a thin and hard surface layer. A microhardness tester is a useful tool for studying the hardness distribution within metals.

For example, microhardness testers are used to study base metal dilution on the hardness of single or multilayer hard-facing deposits, the hardness change in a weldment (area from the weld through the heat-affected zone to the parent metal). They are also used to study the presence of brittle constituents in a metal, such as grain boundary phases or inclusions, and the depth of hardening from case-hardening processes, such as nitriding or carburizing.

During indention hardness tests, the test specimen is sectioned, polished, and sometimes etched, which is a treatment of a chemical reagent to cause slight preferential dissolution. This is performed to highlight the items of interest. In most cases, metallographic mounting is necessary to preserve the edge of the specimen. For example, with a cast-hardened shaft, a slice is made perpendicular to the axis and the area of interest mounted. This section is polished and etched to reveal the microconstituents of interest for the microhardness survey.

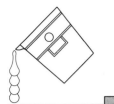

Structure of Metals 4

Each specific metal and nonmetal is known as a chemical element. The basic building blocks of chemical elements are atoms, which bond to each other to form physically visible solid materials. The atoms of metallic materials are arranged in a repeating pattern, or crystal structure. Most metals exhibit one of three crystal structure types. When different chemical elements are mixed to form an alloy, the resulting crystal structure depends on the proportion of the different chemical elements in the alloy and their crystal structures. The repeating pattern of atoms leads to three-dimensional crystallographic planes. Metals solidify from the molten state as small collections of atoms called nuclei. Atoms add to the nuclei to form growing solid crystals, which eventually form into aggregate crystals.

ATOMIC STRUCTURE

Atomic structure is the organization of atoms and their basic parts that comprise the chemical elements. *Chemical elements* are basic substances consisting of atoms of one type that alone or combined with other chemical elements constitute all matter. A chemical element is uniquely identified by its characteristic physical properties. Each specific metal and nonmetal is a chemical element. *Atomic bonding* is a process that occurs when atoms are bonded to each other, or held together by a force of attraction, to form physically visible solid materials. The four types of atomic bonding are metallic bonding, covalent bonding, ionic bonding, and Van der Waals bonding. The four types of atomic bonding strongly influence the physical properties of different material types.

Structure of Chemical Elements

A chemical element may be a solid, liquid, or gas. They are identified by characteristic physical properties, such as melting point and boiling point. Chemical elements are composed of atoms.

Atoms contain three types of particles: protons, neutrons, and electrons. Chemical elements are identified by their atomic number and weight. The *atomic number* of a chemical element is the number of protons in the nucleus of the chemical element's atom. The *atomic weight* of a chemical element is the sum of the protons and neutrons in a chemical element's atom. *Isotopes* are versions of the same chemical element with the same atomic number but different atomic weights.

Atoms. The structure (arrangement) of the particles of an atom is similar to the solar system. The central nucleus (core) is orbited by electrons in shells much like the sun is orbited by the planets. Protons and neutrons are in the nucleus of the atom and the electrons orbit the nucleus in shells. See Figure 4-1. Chemical elements have different atomic structures and this accounts for the different physical properties exhibited by the different chemical elements.

Protons are particles with a positive electrical charge. *Neutrons* are particles with a neutral electrical charge. *Electrons* are particles with a negative electrical charge and are equal in magnitude to the protons. The number of protons in an atom equals the number of electrons. This balance gives the atom a neutral electrical charge. Protons and neutrons are equal in mass. Electrons are particles $\frac{1}{1840}$ the mass of protons or neutrons and are contained in shells surrounding the nucleus of the atom.

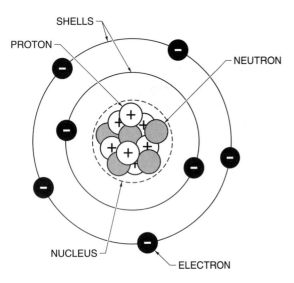

Figure 4-1. Protons and neutrons are contained in the nucleus (core) of the atom and the electrons orbit the nucleus in shells.

Each shell of electrons orbiting the nucleus contains a specific number of electrons. For example, the first shell (closest to the nucleus) may contain only 2 electrons, the second 8 electrons, the third 18 electrons, and the fourth 32 electrons. The shells fill from the first shell outward.

When the outermost shell contains its quota of electrons, the chemical element is completely stable and will not combine with other elements to form chemical compounds or molecules. For example, the inert gases argon and helium are used for inert gas welding because they do not react with metals to form undesirable intermetallic chemical compounds.

When the outermost shell is not completely filled with electrons, the chemical element can combine with other chemical elements to form chemical compounds or molecules. The electrons in an incompletely filled outermost shell are called valency electrons. When atoms combine to form chemical compounds or molecules, their valency electrons are shared. See Figure 4-2. The combined number of valency electrons equals the number required to completely fill a shell.

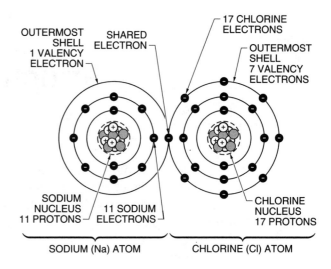

SODIUM CHLORIDE (NaCl)

28 PROTONS
28 ELECTRONS
8 ELECTRONS SHARED
IN OUTERMOST SHELL

Figure 4-2. Sodium and chlorine combine to form sodium chloride and produce a combined number of electrons in their shared outermost shell.

Atomic Weights and Numbers. Atomic weight is the sum of the protons and neutrons in the nucleus of a chemical element's atom. Atomic weight is used to compare chemical elements because atoms are extremely small and weigh very little. The system is based on the atomic weight of oxygen, which is equal to 16. The mass of the electrons in the atom is not used for calculating atomic weight because it is small and insignificant. The higher the atomic weight, the denser the chemical element. See Appendix.

Since protons and neutrons contribute equally to the atomic weight, oxygen, which has an atomic weight of 16, could be composed of 16 protons, 16 neutrons, or any combination of the two that add up to 16. Atomic weight does not distinguish between the number of protons and neutrons in the nucleus of the atom.

An atomic number is the number of protons in the nucleus of a chemical element. The atomic number increases in numerical sequence from the lightest to the heaviest chemical element. No two chemical elements have the same atomic number because no two chemical elements have the same number of protons in their nuclei.

Isotopes. Isotopes are forms of the same chemical element with different numbers of neutrons. For example, naturally occurring hydrogen (H) is a mixture of three isotopes. All three isotopes have one proton. See Figure 4-3. The hydrogen isotope (the predominant form) has a single proton and a single neutron in the nucleus. The hydrogen isotope deuterium (^{2}H) has one extra neutron. The hydrogen isotope tritium (^{3}H) has two extra neutrons. Naturally occurring hydrogen is a mixture of the three isotopes and has an average atomic weight of 1.008.

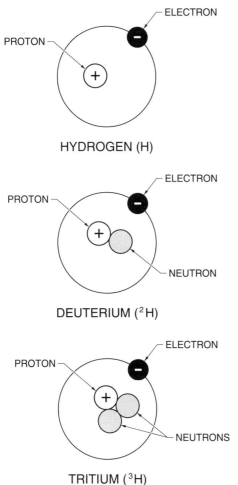

HYDROGEN (H)

DEUTERIUM (^{2}H)

TRITIUM (^{3}H)

Figure 4-3. Naturally occurring hydrogen is a mixture of three hydrogen isotopes and has an average atomic weight of 1.008.

The atomic weight of a chemical element is based on the average atomic weight in the mixture of isotopes that occurs naturally in the chemical element. Atomic weights are not necessarily whole numbers. Oxygen, which has an atomic weight of 16, is used as the basis for comparing atomic weights because it has no isotopes and is therefore a whole number.

Some isotopes such as the artificially prepared cobalt 60 (^{60}Co) and iridium 192 (^{192}Ir) are radioactive. These isotopes emit gamma rays as they spontaneously transform from one chemical element or isotope to another. Radioactive isotopes are used in radiographic inspection for defects in castings and welds.

Atomic Bonding

Atoms that make up materials are bonded by forces of attraction. The type of atomic bonding between atoms depends on the chemical element or elements involved and the physical state of existence. In the gaseous state, the atoms or molecules expand to fill the containment space and there is no atomic bonding between them. In the liquid state, there is some attraction and greater atomic bonding. In the solid state, there is much more attraction and the strongest atomic bonding. Solids exhibit four types of atomic bonding: metallic bonding, covalent bonding, ionic bonding, and Van der Waals bonding.

Metallic Bonding. *Metallic bonding* is a type of atomic bonding that occurs in a solid metal when the valency electrons leave individual atoms and are shared between all atoms in a free electron cloud. See Figure 4-4. Because of the abundance of protons, the atoms are positive-charged ions. An *ion* is an atom with a positive or negative electrical charge. The free electron cloud, or negative electron cloud, has a negative electrical charge. The free electrons are responsible for common metallic characteristics such as high levels of thermal conductivity and electrical conductivity.

To understand the behavior of metallic atoms, it helps to visualize them as hard spheres with a specific diameter. As the spheres move closer together, a force of attraction and a force of repulsion are produced. There is a balance between the force of attraction (negative-charged electrons for the positive-charged nuclei) and the force of repulsion (force between outer shells of the atoms). This balance of forces holds the atoms apart a certain distance. *Atomic diameter* is the closest distance two atoms approach each other in the solid state.

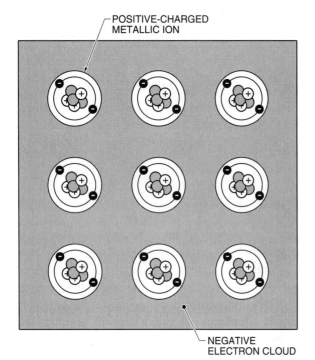

Figure 4-4. Metallic bonded atoms are positive-charged ions and the negative electron cloud is negative-charged.

Covalent Bonding. *Covalent bonding* is a type of atomic bonding that occurs when the valency electrons are shared between like atoms. Each like atom achieves a stable electron configuration. See Figure 4-5.

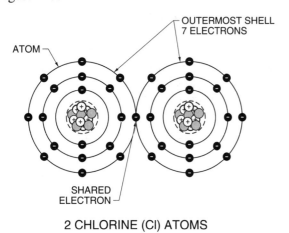

Figure 4-5. The sharing of valency electrons of unstable atoms produces a stable molecular configuration.

Covalent bonding leads to high levels of thermal resistivity (resistance to thermal conductivity) and electrical resistivity because there are no free elec-

trons. When atoms combine by covalent bonding, they form a molecule. For example, two chlorine atoms are much more stable when they combine to form a chlorine molecule.

Ionic Bonding. *Ionic bonding* is a type of atomic bonding that occurs when the valency electrons are exchanged between unlike atoms. Each atom gains or loses electrons to achieve a stable electron configuration. See Figure 4-6.

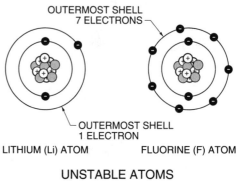

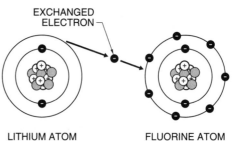

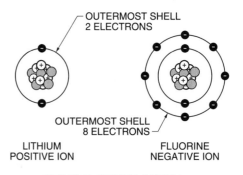

Figure 4-6. Valency electrons are exchanged between unlike atoms to achieve a stable electron configuration.

For example, lithium has an excess of one electron in the outermost shell and can combine with fluorine (F), which is deficient one electron in its outermost shell. The lithium atom gives up an electron in its outermost shell and the fluorine atom accepts that electron. The lithium acquires a net positive charge (+1) and becomes a positive ion. The fluorine acquires a net negative charge (−1) and becomes a negative ion. Ionic bonding leads to high levels of thermal resistivity and electrical resistivity.

Van der Waals Bonding. *Van der Waals bonding* is a type of atomic bonding that occurs when there is no exchanging or sharing of electrons. The centers of positive and negative charge in the atom do not coincide. See Figure 4-7. The atom behaves like a dipole. A *dipole* is an atom that has positive and negative centers of charge that are slightly separated. This separation causes weak bonding with other atoms or molecules. For example, the bonding between large molecules of thermoplastics such as polyvinyl-chloride (PVC) is an example of Van der Waals bonding. The bonding along each molecule is covalent. Van der Waals bonding provides cohesion between the molecules.

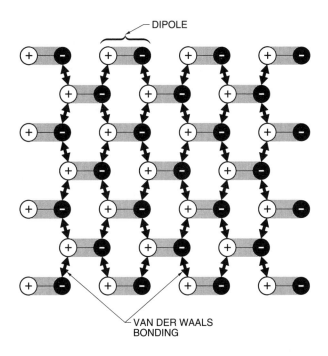

Figure 4-7. Van der Waals bonding is a type of atomic bonding that occurs when there is no exchanging or sharing of electrons.

CRYSTAL STRUCTURE

Most metals and most solids exhibit a crystal structure. Crystal structure is the configuration of atoms as they add to one another in an orderly and repeating three-dimensional pattern.

Amorphous solids are solids that do not exhibit a crystal structure. They possess the random arrangement of atoms that commonly occur in liquids. Examples of amorphous solids are glass, tar, and some polymers. Amorphous metals are metals without a crystal structure and are produced by special techniques. They have limited but specialized applications. Examples of amorphous metals include alloys of nickel-zirconium and molybdenum-rhenium.

Crystal structure may be present with any of the four types of atomic bonding. A space lattice and unit cell help to illustrate crystal structures. A *space lattice* is a regular array of points produced by lines connected through the points. A *unit cell* is the smallest arrangement of atoms that repeats itself through the space lattice. The space lattice or unit cell structure in alloys depends on the proportions of the various chemical elements, which leads to the formation of a solid solution or an intermediate phase.

The atoms in a crystal structure are arranged along crystallographic planes, which are designated by the Miller indices numbering system. The crystallographic planes and Miller indices are identified by X-ray diffraction.

Space Lattice and Unit Cell

A space lattice and unit cell are used to illustrate crystal structures. The repetition of atoms in three dimensions of a crystal structure may be shown by a three-dimensional array of points. The points represent the center of each atom or arrangement of atoms.

A two-dimensional example of a space lattice and unit cell is wallpaper. The wallpaper design represents the unit cell, which is repeated in a regular pattern. The regular pattern represents the space lattice. When there is one atom per space lattice point, the unit cell and space lattice become one and the same. As the number of atoms per lattice point increases, the unit cell becomes more complex. See Figure 4-8.

SPACE LATTICE AND UNIT CELL

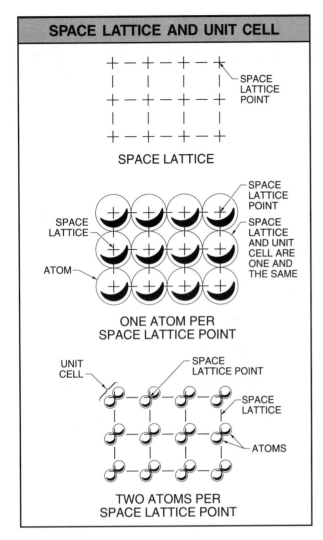

Figure 4-8. As the number of atoms per space lattice point increases, the more complex the space lattice becomes.

Although 14 types of unit cell types are possible, most metals exhibit one of three types. The three common types are body-centered cubic (BCC), face-centered cubic (FCC), and close-packed hexagonal (CPH). The atomic arrangements in the different unit cell types lead to significant differences in mechanical behavior. For example, FCC metals usually exhibit high levels of ductility.

The BCC unit cell is a cube with an atom at each corner and one in the center. Examples of BCC metals are alpha iron (αFe), delta iron (δFe), chromium, tungsten (W), molybdenum, and vanadium (V).

The FCC unit cell is a cube with an atom at each corner, one in the center, and one in the center of each side of the cube. Atoms in the FCC unit cell are more densely packed than in the BCC unit cell. Examples of FCC metals are gamma iron (γFe), aluminum, nickel, copper, gold, lead, and platinum.

The CPH unit cell is a hexagon with a reference atom that is surrounded by 12 atoms that are an equal distance from the reference atom. The reference atom lies in the basal plane surrounded by 6 of the 12 atoms. Two parallel planes on either side of the basal plane each contain three of the remaining atoms. The atoms in alternate planes in the CPH structure are in line with each other. It requires three unit cells to reveal the hexagonal shape of the CPH structure. Atoms in the CPH structure are as densely packed as in the FCC unit cell, but the difference between them is the stacking sequence (arrangement) of the atoms. Examples of CPH metals are magnesium, beryllium, zinc, cadmium (Cd), titanium, and zirconium (Zr).

Allotropic metals are metals that exhibit more than one unit cell structure. There are 15 metallic chemical elements that are allotropic. The most important allotropic engineering material is steel. Steel changes from FCC to BCC as it cools at a temperature between 871°C and 704°C (1600°F and 1300°F). The variance in temperature of the change depends on the carbon content. The opposite change occurs when the steel is heated. Allotropic changes make heat treatment of steel by quenching and tempering possible. Another example is titanium, which exhibits a CPH structure below 882°C (1620°F) and a BCC structure between 882°C and its melting point of 1727°C (1620°F and 3140°F).

Crystal Structure of Alloys. The crystal structure of alloys is determined by the proportions of alloying chemical elements present. Atomic mixing between chemical elements leads to the formation of a solid solution or an intermediate phase.

A solid solution has the space lattice or unit cell structure of the solvent (predominant) metal but incorporates the atoms of the solute (added) chemical elements. For example, red brass contains 70% Cu, which has a FCC structure. Red brass contains 30% Zn, which has a CPH structure. As a result, red brass has the FCC structure of the solvent (70% Cu) component of the alloy.

With an intermediate phase, the space lattice or unit cell structure differs from either component.

The intermediate phase usually forms at a fixed alloy composition. For example, iron carbide (Fe_3C), which is present in carbon steels and contains 6.67% C, has an orthorhombic crystal structure. An *orthorhombic crystal structure* is a crystal structure with crystals that have equal axes that are at right angles to each other. This differs from the BCC structure of iron or the CPH structure of the carbon.

Modeling of Atoms. Atoms in a space lattice or unit cell can be modeled using foam or table tennis balls. The model can have foam balls glued in contact with one another or table tennis balls connected to one another by wire links. See Figure 4-9. A model with foam balls glued together and in contact with one another correctly would illustrate the close packing of the atoms but might obscure the crystal planes. A model with table tennis balls separated from one another with wire links would illustrate the crystal planes but might also imply that the atoms are widely separated.

Crystallographic Planes

The repeating arrangement of atoms in a crystal leads to crystallographic planes in three dimensions. Crystallographic planes are the planes along which the atoms are arranged. Crystallographic planes are mathematically identified by Miller indices. Miller indices are used to describe the directions of crystallographic planes. X-ray diffraction is used to determine the crystallographic planes.

Miller Indices. Miller indices provide a method of defining a crystallographic plane in space. A corner of a unit cell is taken as the origin of three axes in space. See Figure 4-10. The crystallographic plane is defined by the reciprocal of the intercepts (intersections) it makes with the three axes. The lowest common denominator is found for the reciprocals and is used as a multiplier to produce fractions, which can be reduced. The reduced fractions are the Miller indices and are expressed in parentheses. For example, the plane *IJK* has a x-axis intercept of 3, y-axis intercept of 4, and z-axis intercept of 6. The reciprocals are $\frac{1}{3}$, $\frac{1}{4}$, and $\frac{1}{6}$ respectively. The lowest common denominator of these fractions is 12. To clear the fractions, they are multiplied by 12, which results in fractions of $\frac{12}{3}$, $\frac{12}{4}$, and $\frac{12}{6}$. The Miller index is found by reducing these fractions. The Miller index for the plane *IJK* is (4,3,2).

When a crystallographic plane is parallel to one of the axes, its intercept with the axis is at infinity. The resulting Miller index for the coordinate is zero because the reciprocal of infinity is zero. For example, plane *ABCD* is the face of a unit cube. The x-axis, y-axis, and z-axis have the arbitrarily chosen point of 0 as the origin. Since the plane *ABCD* is parallel to the y-axis, it has a y-axis intercept of infinity. It also is parallel to the z-axis, so it has a

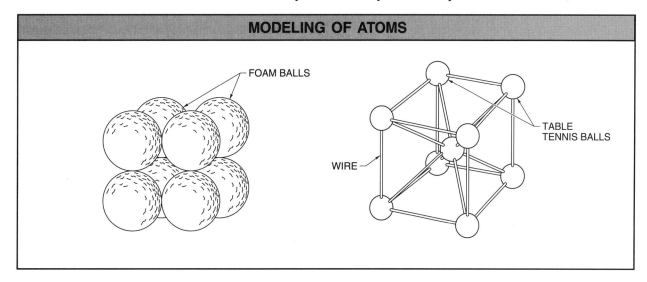

MODELING OF ATOMS

FOAM BALLS

TABLE TENNIS BALLS

WIRE

Figure 4-9. Atoms in a space lattice or unit cell can be modeled using foam or table tennis balls.

z-axis intercept of infinity. The plane intercepts the x-axis at +1. The Miller index for the plane *ABCD* is (100).

A negative intercept is expressed with a bar over the number. For example, since the origin is at 0 and the plane *EFGH* cannot be defined because it goes through it, the origin may be moved to point A. This results in an x-axis intercept of –1, y-axis intercept of infinity, and z-axis intercept of infinity. The Miller index for the plane *EFGH* is ($\bar{1}$00).

The direction of a plane is perpendicular to the plane and is expressed as the Miller index in brackets. To find the direction, the intercepts are traced from the origin to find a point in space. See Figure 4-11. Using the origin as a second point, a line is constructed through the point in space. For example, the plane *IJK* has an x-axis intercept of 3, y-axis intercept of 4, and z-axis intercept of 6. Plane *IJK* has a direction of [4,3,2] that is perpendicular to plane *IJK* and all planes parallel with a Miller index of (4,3,2).

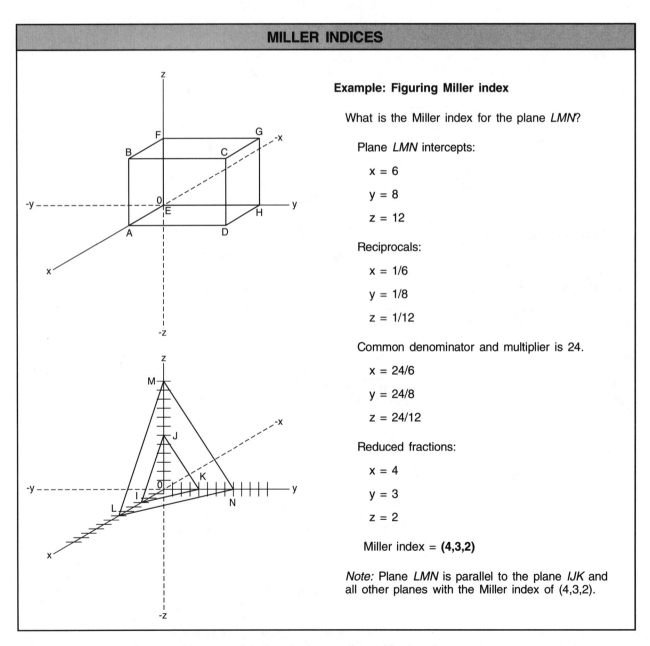

MILLER INDICES

Example: Figuring Miller index

What is the Miller index for the plane *LMN*?

Plane *LMN* intercepts:

x = 6

y = 8

z = 12

Reciprocals:

x = 1/6

y = 1/8

z = 1/12

Common denominator and multiplier is 24.

x = 24/6

y = 24/8

z = 24/12

Reduced fractions:

x = 4

y = 3

z = 2

Miller index = **(4,3,2)**

Note: Plane *LMN* is parallel to the plane *IJK* and all other planes with the Miller index of (4,3,2).

Figure 4-10. Miller indices provide a way of designating a crystallographic plane in space.

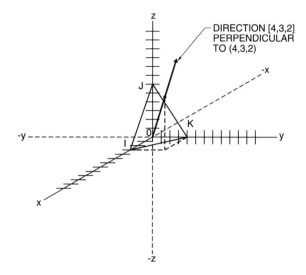

Figure 4-11. The direction of a plane is perpendicular to the plane and is expressed as the Miller index in square brackets.

X-ray Diffraction. X-ray diffraction is used to determine the atomic structure in crystals. The wavelength of X rays is approximately 1 Angstrom (Å), which is equal to 10^{-8} cm. The wavelength of the X ray is similar to the atomic spacing in crystals. X-ray diffraction causes a beam of incident X rays to be diffracted when passed through a crystal. The diffracted X rays produced are analyzed and used to identify the crystal structure of the metal. See Figure 4-12.

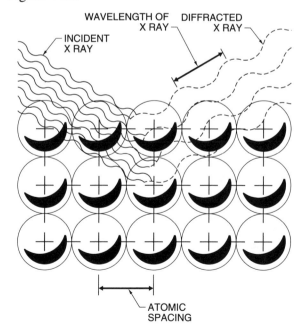

Figure 4-12. Diffracted X rays can be analyzed and used to identify the crystal structure of a metal.

X rays are used to determine complex atomic structures. For example, X-ray diffraction analysis revealed the structure of the human body's giant DNA molecule, which carries hereditary information to succeeding generations. In metallurgy, X-ray diffraction analysis has been most useful in identifying phases in alloys. The ability to identify the phases advanced the development of metallography.

GRAIN STRUCTURE

Metals do not exist as single crystals but as large numbers of grains. Grains are smaller crystals that have crystallographic planes in various orientations. A metal develops a grain structure during solidification from the molten state.

Boundaries between grains exert important effects on the mechanical properties of metals. Grain size is one of the most common microstructural measurements. Grain size can be manipulated with chemical additives to improve mechanical properties of alloys.

Metal Solidification

Metal solidification occurs in three stages. In the first stage, clusters of nuclei (solid atoms) form. Slight undercooling below the freezing point may be required to initiate the formation of nuclei in the molten metal. In the second stage, solid atoms add to the nuclei. This addition produces more unit cells. Crystal growth occurs in planes that correspond to the orientation of the crystallographic planes in the nuclei. In the third stage, dendritic growth begins. *Dendritic growth* is the solid three-dimensional growth that occurs in the melt (molten metal) in the form of shooting spikes. Eventually, the spikes of the dendrites begin to meet and interfere with the growth. The spaces between the dendrites fill with more branches until solidification is complete. Grains are the individual solidified dendrites. See Figure 4-13.

Most solids are polycrystalline materials. Polycrystalline materials are collections of crystals (grains) that have grown independently during solidification. They have the specific crystallographic orientations of the nuclei from which they formed.

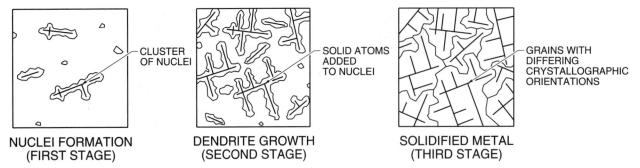

Figure 4-13. Solidification of a metal results in the formation of grains (many individual crystals).

Grain Boundaries

Grain boundaries are the interfaces between grains that affect the behavior of metals. Grain boundaries are regions of crystal misfit. The atoms adjacent to the grain boundaries are at a higher energy level than those in the bulk of the grain. See Figure 4-14.

AREAS AT
HIGHER
ENERGY LEVEL

BULK OF
GRAIN

GRAIN
BOUNDARY

Figure 4-14. Atoms on either side of the grain boundary are at a higher energy level than the bulk of the grain.

Because the grain boundary region is at a higher energy level, certain metallurgical effects occur more readily at grain boundaries. This is particularly the case with metallurgical effects associated with heat input. For example, the heat of welding causes intermetallic compounds to precipitate (develop) preferentially at grain boundaries in some stainless steels. The formation of these intermetallic compounds results in localized loss of corrosion resistance along the grain boundaries.

Grain Size

Grain size is a measure of the average dimensions of representative grains in a test specimen. The size of individual grains is determined by how much they can grow during solidification before meeting other growing grains. Grain size is directly related to the number of nuclei that originally formed in the molten metal. The greater the number of nuclei, the smaller the grain size and vice versa. Chemical compounds are sometimes added to molten metals just before casting to increase the number of nuclei and promote small grain size.

The grain size can exert a significant effect on the mechanical and physical properties of metals. For example, certain alloys are heat treated to develop coarse grain size. Coarse grain size improves an alloy's creep resistance. *Creep resistance* is the resistance to stress-induced strain. Fine grain size is a desirable characteristic for toughness in quenched and tempered low-alloy steels. Because it can be manipulated, grain size is a property of metals that can be specified when ordering metals.

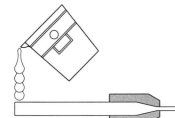

Metallography 5

Metallography is the microscopic examination of the microstructure (grain structure) of metals. Metallography, the most important tool in metallurgy, provides invaluable information on the processing history and properties of metals. Metallography is used as a quality control tool, in failure analysis, and for alloy development. To prepare a metallographic specimen, the component is usually sectioned or cut apart. Metallographic specimen preparation follows a rigid sequence of steps before the test specimen is ready for examination in a metallurgical microscope. Quantitative metallography is used to estimate the grain size or internal cleanliness of metals. Macroscopic examination is used to examine the overall structure of components with the naked eye or at low magnifications.

METALLOGRAPHIC EXAMINATION

Metallographic examination is the study of the microscopic features of material surfaces that have been specially prepared by cutting, grinding, polishing, and etching. The purpose of metallographic examination is to look for clues as to how a metal was made and/or how it performed. It may also give clues as to how a metal will perform in the future. Routine study, failure analysis, and alloy development are three areas of metallographic examination.

Most routine studies are conducted for the purpose of quality control. The standards for the examination and evaluation of the metallographic specimen are well-established. Those standards dictate the orientation of a specimen and preparation procedure used on a specimen. For example, there are standards for the examination and evaluation of measuring grain size. These standards dictate the orientation of the metallographic specimen and the preparation procedure used.

Failure analysis investigations are conducted to determine the cause of a failure. A metallographic specimen is taken from an area as close as possible to the failure initiation point (start of failure). It is compared to another metallographic specimen taken

from a representative region that has not failed. Close comparison of the two specimens may determine the cause and type of failure.

Alloy development programs correlate microstructure with mechanical properties. Metallographic specimens are often obtained from broken mechanical property test specimens of the alloys. The microstructure of the experimental alloys is compared with similar alloys, which have understood microstructures. They are compared to determine similar properties.

Metallographic examination is made at high magnification (microscopic) and requires small specimens that must be representative of the component. A sequence of steps is followed when preparing a specimen for metallographic examination. Steps include cutting and rough grinding, mounting and fine grinding, rough and final polishing, and etching and examination.

The metallographer (person preparing the metallographic samples) must skillfully prepare the specimen to avoid artifacts. *Artifacts* are false microstructural indications that do not correspond to the true microstructure and are caused when a metallographic mount is prepared. A *mount* is a device used for holding a specimen during metallographic preparation.

Cutting and Rough Grinding

Cutting and rough grinding is performed using selection techniques to obtain a representative metallographic specimen from the component. Special techniques may be required to preserve the specimen from damage so that essential information is not lost. Cutting is performed in steps to obtain a specimen that is a suitable size for rough grinding. Rough grinding removes coarse material and features resulting from the cutting process.

Specimen Selection. To obtain a representative specimen for examination, sometimes several specimens are required. For example, microstructural inhomogeneity in cast and wrought components may require more than one specimen taken from different locations of the component to verify the representative structure.

The most common specimen orientations are longitudinal (parallel to the axis) and transverse (perpendicular to the axis). Due to the orientation of the metal flow during working, longitudinal and transverse specimens removed from wrought components may show substantially different microstructures. See Figure 5-1.

Grain size may vary within specific components. Castings and welds are common components that exhibit varying grain size. Chills, or pieces of metal, are incorporated into a mold to locally increase metal solidification and alter the microstructure in castings. Exothermic feeding aids are incorporated into

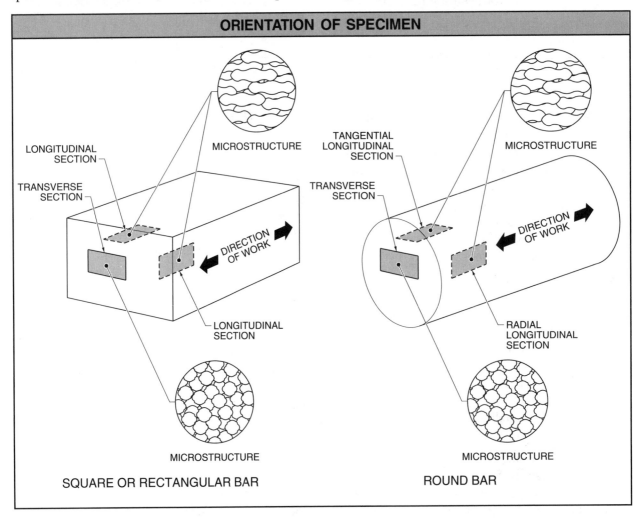

Figure 5-1. Longitudinal and transverse orientations of specimens from a worked metal exhibit different microstructures.

and alter the microstructure in the casting. See Figure 5-2. In welds, significant grain growth and refinement may occur in specific zones adjacent to the weld filler metal.

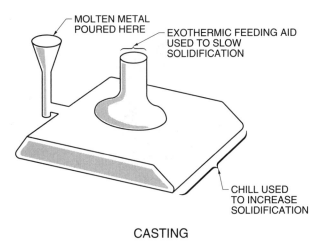

CASTING

Figure 5-2. Microstructures of specimens obtained from castings depend on the localized solidification in the casting.

When specimens are cut from coated, corroded, or surface-treated metals, care must be taken to preserve the surface. Special preparation techniques, such as surface plating, are required prior to cutting to protect the surface from the mechanical damage caused during specimen preparation. See Figure 5-3.

Fragile components, such as thin wall tubing or a porous material, require special pretreatment to avoid collapse during specimen preparation. For example, porous metal is impregnated with epoxy resin under vacuum to strengthen the specimen prior to preparation.

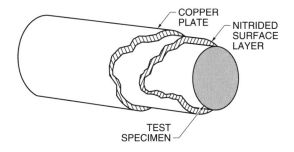

Figure 5-3. Surface plating is used for protection of the edge of coated, corroded, or surface-treated metals during specimen preparation.

Specimen Cutting. Cutting is the most common method of obtaining specimens from a component. Large specimens must be reduced in size using flame or plasma cutting. Subsequent cutting is accomplished by other means including a power hacksaw, band saw, mechanical saw, abrasive cutoff wheel, or diamond-tipped cutoff wheel. The power hacksaw, band saw, or mechanical saw are used on specimens that are too large or awkward to cut using an abrasive cutoff wheel. Abrasive cutoff wheels are used to obtain specimens that are or are close to the final size. Diamond-tipped cutoff wheels are used on small specimens where precision cuts are required and as a result the rough grinding steps are bypassed. See Figure 5-4.

BUEHLER LTD.

Figure 5-4. Diamond-tipped cutoff wheels are used on small specimens where precision cuts are required.

Overheating, which alters the microstructure, is damage caused by cutting. Flame or plasma cutting must be performed at least $\frac{1}{4}$ in. to $\frac{1}{2}$ in. from the area to be examined, so final cutting can be done with less damaging techniques. Cutoff wheels and saws use coolant at the cutting surface to prevent overheating. Materials with hardness above 35 HRC require an abrasive cutoff wheel or a diamond-tipped cutoff wheel for cutting operations.

Subsurface deformation is another form of specimen damage produced from cutting. Coarse cutting tools and heavy applications of force increase subsurface deformation. Unless the zone of subsurface deformation is removed, it can lead to artifacts.

Specimen Rough Grinding. Rough grinding prepares specimens for mounting by removing subsurface deformation, unnecessary roughness, flash, and scale caused by cutting operations. Specimens are ground flat on a wet abrasive belt sander using a 80 grit or 150 grit belt or are machined flat in a milling machine. When a diamond-tipped wheel is used to make the final cut, rough grinding is usually unnecessary.

Mounting and Fine Grinding

Specimens are mounted to prevent rounding of edges and because they are usually too small and awkward in shape to hold by hand during the polishing and etching stages. Selection of a mounting resin is based on a combination of factors, which must be considered so that the proper mounting resin is selected and examination results are correct. The metallographer selects the optimum resin for a specific mount. Coated or scaled specimens require special edge preservation techniques. Fine grinding prepares the mount for the final stages of specimen preparation.

Specimen Mounting. Most mounts are permanent, meaning that the specimen is permanently encased in resin. Some mounts have a temporary clamping device that holds the specimen flat and rigid during fine grinding. Before mounting, any burrs at the edges of the specimen caused by cutting or machining are carefully removed using a smooth file or coarse abrasive paper or cloth.

Mounting is usually performed in a mounting press, which encapsulates the specimen with a thermosetting resin under pressure and at an elevated temperature. See Figure 5-5. The specimen is placed face down in a vertical cylindrical mold in the mounting press. A predetermined quantity of thermosetting resin is poured into the mold and it is closed. The temperature is raised and pressure is maintained while the resin cures, making the resin hard and strong. After the mold cools, the mount is demolded.

A suitable mounting resin must cure at a temperature and pressure that does not alter the microstructure of the specimen. The mounting resin selected must resist chemical attack by the etchant, which is applied to the mount to reveal microstructural features.

Figure 5-5. Mounting presses use compression and heat to encapsulate the specimen in a plastic mounting resin.

The mounting resin must provide good adhesion to the edges of the specimen to prevent rounding of the edges and entry of lubricant or etchant during specimen preparation. Lubricant or etchant that enters will flow out after final preparation and drying. The lubricant or etchant will cause staining of the specimen. The mounting resin must fill pores and crevices on the exposed face of the specimen to prevent staining, and it must be electrically conductive if electrolytic polishing or etching is contemplated. Several types and forms of mounting resins are used. If side views of the specimen are required, the resin must be transparent. See Figure 5-6.

Cold mounting is performed when the specimen is too large for the mounting press or when the heat involved might alter the microstructure. Cold mounting is performed in room temperature and at atmospheric pressure using a thermoplastic resin. Cold mounting is also performed in a vacuum to remove air bubbles from the mount. For example, specimens from electronic components are often cold mounted in thermoplastic resins.

Mounting Coated and Scaled Specimens. Coated and scaled specimens require special mounting techniques because they are hard and brittle or soft and ductile. For example, nitrided or porcelain-enameled steel is hard and brittle and galvanized or nickel-plated steel is soft and ductile. Corrosion product scales are usually soft and friable (easily broken off). When transverse sections are prepared, the hard and brittle or soft and friable coatings tend to flake off, while the soft and ductile coatings tend to round off.

Coated and scaled specimens are mounted using a resin that abrades (erodes) at a lower rate than the specimen. Coated and scaled specimens may also be mounted by electroplating them prior to mounting. Electroplating provides a hard edge that resists mechanical abuse and rounding during grind-

ing and polishing. Whether the specimens are mounted using a resin that abrades at a lower rate than the specimen or the specimen is electroplated prior to mounting, the specimen must be polished as steadily as possible to avoid rocking and rounding of the edges.

Taper sectioning is a mounting technique that increases the magnification available when examining a specimen and is an effective technique for examining surface coatings, films, or layers. Specimens are mounted so that the axis of the coating, film, or layer makes a specific angle to the face of the mount. The coating, film, or layer is then magnified

MOUNTING RESINS

Plastic	Type	Molding Conditions			Heat-Distortion Temperature (°C)[a]	Coefficient of Thermal Expansion (in./in.°C)[b]	Transparency	Chemical Resistance
		Temperature (°C)	Pressure (psi)	Curing Time				
Phenolic molding powder	Thermosetting[c]	170	4000	5 min	140	$3.0–4.5 \times 10^{-5}$	Opaque	Not resistant to strong acids or alkalis
Acrylic (polymethyl methacrylate) molding powder	Thermoplastic	150	4000	none	65	$5–9 \times 10^{-5}$	Water white	Not resistant to strong acids
Epoxy casting resin	Thermosetting[d]	20–40	—	24 hr	60[e]	$4–7 \times 10^{-5}$	Clear but light brown in color	Fair resistance to most alkalis and acids; poor resistance to nitric and glacial acetic acids
Allyl molding compound	Thermosetting[f]	160	2500	6 min	150	$3–5 \times 10^{-5}$	Opaque	Not resistant to strong acids and alkalis
Formvar (polyvinyl formal) molding compound	Thermoplastic	220	4000	none	75	$6–8 \times 10^{-5}$	Clear but light brown in color	Not resistant to strong acids
Polyvinyl chloride molding compound	Thermoplastic[g]	160[h]	3000	none	60	$5–18 \times 10^{-5}$	Opaque	Highly resistant to most acids and alkalis

[a] as determined by the method described in *ASTM D 648-56,* at a fiber stress of 264 lb/in.^{2}
[b] coefficient of thermal expansion in most metals (in the range $1–3 \times 10^{-5}$ in./in./°C)
[c] wood-filled grade, preferably with low filler content
[d] liquid epoxy resin with an aliphatic amine hardener
[e] depends on curing schedule (can be as high as 110°C with heat curing)
[f] diallyl phthalate polymer with a mineral filler
[g] stabilized ridged PVC (for example, a mixture of 100 parts of paste-making grade of PVC, two parts dibasic lead phosphate, and two parts tribasic lead sulfate)
[h] must not exceed 200°C

Figure 5-6. Mounting resins must satisfy a variety of conditions.

by an amount dependent on the angle. Taper sectioning is used to examine surface layers such as electroplating, case hardening, or oxidation, which may be too thin to view by conventional means. See Figure 5-7.

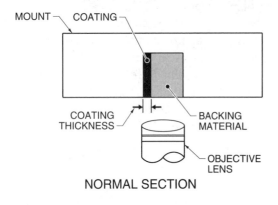

NORMAL SECTION

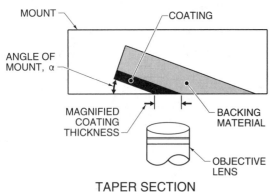

TAPER SECTION

Figure 5-7. In taper sectioning, the specimen is mounted at an angle that magnifies the coating thickness.

Mount Fine Grinding. Fine grinding prepares the mount for the final stages of specimen preparation by abrading the mount on a series of successively finer abrasive papers. Prior to fine grinding, any resin on the face of the specimen or any remaining burrs on the edges are removed by a 120 grit abrasive paper or cloth. Two types of fine grinding are four-stage manual grinding and four-stage belt grinding. During fine grinding using four-stage manual grinding or four-stage belt grinding, a successive series of water-lubricated papers (beginning with 240 grit and finishing with 600 grit) is used. In either type of fine grinding, the mount is lightly washed between abrasive papers or belts to prevent carryover of abrasive material.

Four-stage manual grinding uses an assembly of four strips of abrasive paper of increasing fineness.

The mounted specimen is moved up and down on each grade of paper, being careful not to rock the mount. Four-stage belt grinding uses a four-belt grinder to remove successively finer scratches by changing the abrasive paper between each stage. See Figure 5-8. The mounted specimen is abraded backward and forward without rotation until all grinding marks from the previous coarser abrasive paper have been eliminated.

BUEHLER LTD.

Figure 5-8. Belts used in four-stage belt grinding start with 240 grit and finish with 600 grit paper.

The direction of grinding is changed 90° with each change of abrasive paper, so that removal of the previous grinding marks is easily observed. See Figure 5-9. Four-stage belt grinding is also performed on a grinding wheel, but ample water lubrication must be used to prevent overheating.

In four-stage belt grinding, the amount of time spent on each abrasive paper is increased as finer grades of paper are used. Excessive grinding at any grade of abrasive paper is avoided because it may cause subsurface deformation that will not be eliminated by succeeding grades of abrasive paper and will lead to artifacts. The mount is thoroughly washed and dried after fine grinding is completed.

Rough and Final Polishing

Rough and final polishing procedures lead to a flat, scratch-free, mirror finish on the specimen. The specimen is polished using manual, mechanical, electrolytic, or chemical techniques. The surface must be free from pits (small, sharp depressions) and subsurface deformation effects that lead to artifacts when the specimen is etched. Pits are caused by the polishing operation that remove tiny non-metallic particles from the metal surface.

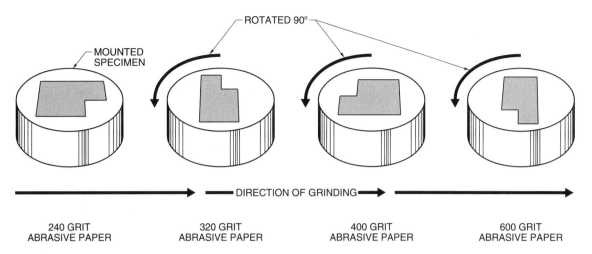

ROTATED 90°

MOUNTED
SPECIMEN

DIRECTION OF GRINDING

| 240 GRIT | 320 GRIT | 400 GRIT | 600 GRIT |
| ABRASIVE PAPER | ABRASIVE PAPER | ABRASIVE PAPER | ABRASIVE PAPER |

Figure 5-9. The mount is rotated 90° between successive papers and thoroughly washed between papers to prevent carryover of abrasive material.

Mount Rough Polishing. Rough polishing is a polishing process that is performed on a series of rotating wheels covered with a low-nap cloth (cloth containing a small amount of fiber). Successively finer grades of diamond rouge (polishing powders) are applied to each wheel, usually starting at 45 micron size. The grades usually decrease to 30 micron, 6 micron, and 1 micron. A small amount of lubricant is applied to the cloth to prevent overheating of the mount.

The mount is moved in an elliptical path against each wheel, using firm hand pressure that is decreased as the diamond rouge becomes finer. The mount is washed with liquid soap and water, alcohol, or acetone between each polishing wheel to prevent carryover of diamond rouge.

Final Polishing. In final polishing, a .3 micron to .05 micron alumina slurry in water is applied to a medium-nap cloth. If the specimen surface is to be subjected to microanalytical techniques such as energy dispersive X-ray analysis (EDXA) or electron probe microanalysis (EPM), alumina should not be used. Use of alumina in microanalytical techniques may lead to misinterpretation of the results of the analysis. In these cases diamond paste must be used.

Final polishing is similar to rough polishing, but during final polishing very light hand pressure is applied to the mount. After washing and drying in a current of warm air, the mount is examined under a metallurgical microscope for scratches. If the

mount is scratch-free, it is ready for etching and examination under a metallurgical microscope.

Automatic polishing is a polishing process that establishes a complex motion for the mount relative to the rotation of the polishing wheel. The rough and final polishing steps are performed in an automatic polishing machine. See Figure 5-10. The machine setting is determined from experience. Automatic polishing is used for large batches of repetitive work, for radioactive specimens, and for polishing techniques when corrosives are added to the wheel.

LECO Corporation

Figure 5-10. Automatic polishing in an automatic polishing machine establishes a complex motion for the mount relative to the rotation of the polishing wheel.

Electrolytic and Chemical Polishing. Electrolytic and chemical polishing are used in special circumstances and are methods of preparation that bypass the rough and final polishing stages. Electrolytic polishing is a polishing process in which the mount is the anode (connected to the positive terminal) in an electrolytic solution and current is passed from a metal cathode (connected to the negative terminal). See Figure 5-11. The current is passed through the electrolytic solution between the anode and the cathode. This current removes the rough peaks on the specimen surface. If the grain structure is homogeneous and single phase (consisting of one crystallographic component), a mirror-polished surface is obtained.

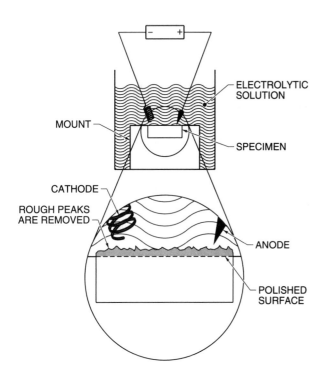

Figure 5-11. Electrolytic polishing removes rough peaks on a specimen with the flow of current between an anode and a cathode.

Chemical polishing is a polishing process that uses chemical reactions to remove the rough peaks on the specimen surface. Chemical polishing is similar to electrolytic polishing because the rough peaks on the specimen surface are removed. The mount is immersed in a specific chemical, which dissolves the high peaks on the rough specimen surface to produce a mirror-polished finish.

Etching and Examination

Etching and examination of the mount with a metallurgical microscope is the last stage of metallographic preparation. *Etching* is the selective attack by a chemical reagent that reveals the microstructural detail of the polished mount. The etched mount is examined under the reflected light in a metallurgical microscope. Microstructural characteristics are enhanced by various forms of illumination, which include brightfield illumination, darkfield illumination, polarized illumination, and Nomarski illumination.

Etching the Specimen. Etching (last stage before examination) is the controlled selective attack on a metal surface for the purpose of revealing microstructural detail. Prior to etching, the mount is examined with a metallurgical microscope in the as-polished condition. Besides revealing the minor scratches that must be removed, microstructural features such as inclusions (foreign material) and porosity (trapped air or gas) are easily observed at this stage.

The mount is then thoroughly degreased and dried and prepared for etching. Etchants selectively dissolve specific microstructural components, giving the as-polished surface a relief appearance. Etchants are selected to dissolve the microstructural components that will provide the best view of the microstructural features.

Etching is usually performed by immersion. The mount is immersed with the polished face upward in a small dish of etching solution, which is gently swirled. The mount is removed when a bloom appears. A *bloom* is a slight haze that appears and is evidence of the first appearance of the microstructure. See Figure 5-12. If necessary, further etching may be performed after examination under a microscope to strengthen any details. However, over-etching may cause loss of contrast. After etching, the mount is thoroughly rinsed in running water. Then acetone or alcohol is sprayed over the surface. The excess is allowed to run off against a cloth that is held at one side of the mount. The mount is then dried in a stream of hot air. The mount should be etched and fine polished at least twice to remove flowed metal from the surface.

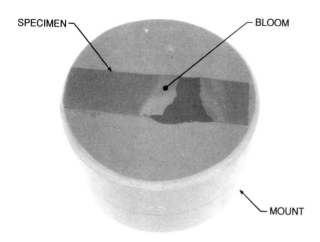

Figure 5-12. For optimum viewing of the microstructure, the mount is etched until a bloom appears on the surface.

Different etchants are available for revealing specific types of microstructural details. See Appendix. After the etching process, the specimen is ready for examination in a metallurgical microscope.

Metallurgical Microscopic Examination. A metallurgical microscope uses light reflected from the specimen surface to examine microstructural details. The surface of the specimen must be widely scanned to gain a representative view of the microstructure. Details are revealed because etching attacks the differently oriented crystallographic planes in the grains at different rates, which results in various shading effects. The proper amount of etching is required for optimum viewing of the microstructure. See Figure 5-13. Improper amounts of etching lead to overetching or underetching and result in false effects.

The etched specimen is placed in a metallurgical microscope and examined at low-power magnification of 25X or 50X to obtain an overall impression of the microstructure. It is then examined at increasing magnifications of 100X to 1000X to reveal fine detail. Higher magnifications up to 2500X cannot be achieved with an air space between the lens and the specimen. Higher magnifications require the use of water or oil immersion. A small amount of water or oil is daubed on the objective lens, which is raised to make contact with the specimen surface. If water or oil immersion is to be followed by lower magnification work, the water or oil is removed from the specimen and the mount may require repolishing and reetching.

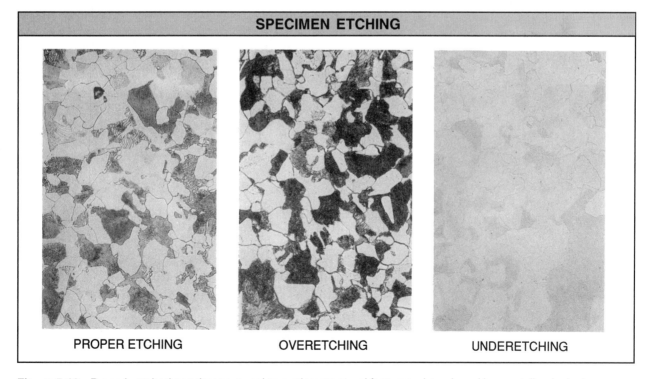

SPECIMEN ETCHING

PROPER ETCHING   OVERETCHING   UNDERETCHING

Figure 5-13. Properly etched specimens reveal true microstructural features when viewed in a metallurgical microscope.

When focussing the metallurgical microscope, contact with the object must be avoided to prevent surface damage. Focussing the microscope is accomplished in two steps. In the first step, the microscope stage is gradually moved toward the objective lens using the coarse adjustment. When the image appears, the second step is to complete the focussing using the fine adjustment. Some metallurgical microscopes are equipped with an automatic coarse adjustment feature.

Metallurgical microscopes vary from small benchtop units to larger units that have their own framework. See Figure 5-14. Some are equipped with a video camera and monitor that are used to view microstructures.

A *metallograph* is a metalurgical microscope equipped to photograph microstructures and produce photomicrographs. *Photomicrographs* are photographs of microstructures.

The four illumination forms for micrographs are brightfield, darkfield, polarized, and Nomarski. See Figure 5-15. Brightfield illumination is the most common form of illumination used to operate a metallurgical microscope. The surface of the specimen is placed perpendicular to the optical axis of the microscope and a white light is used. In brightfield illumination, surface features normal to the optical axis of the microscope appear the brightest.

In darkfield illumination, the specimen is illuminated at sufficient obliqueness (a narrow angle to the surface) so that the contrast is completely reversed from that obtained with brightfield illumination. Those areas that are bright in brightfield will be dark in darkfield and vice versa. Darkfield illumination is useful for highlighting microstructural features (inclusions, grain boundaries, and cracks), which are dark and difficult to distinguish under brightfield illumination.

Polarized illumination is used to reveal microstructural features in metals that are optically anisotropic. Optically anisotropic microstructural features have optical properties that vary with changes in viewing direction. The light is polarized by placing a polarizer in front of the condenser lens of the microscope and placing an analyzer behind the eyepiece. A *polarizer* is a device into which normal light passes and from which polarized light emerges.

Nomarski illumination uses polarized light that is separated into two beams by a biprism (two uniaxial double-refracting crystals). The beams are then reflected back through the biprism off the specimen surface. The biprism combines the beams into one beam, which is run through an analyzer and viewed through an eyepiece. Images produced are three-dimensional and vary in color. This variation in dimension and color is used to identify metals and their various phases.

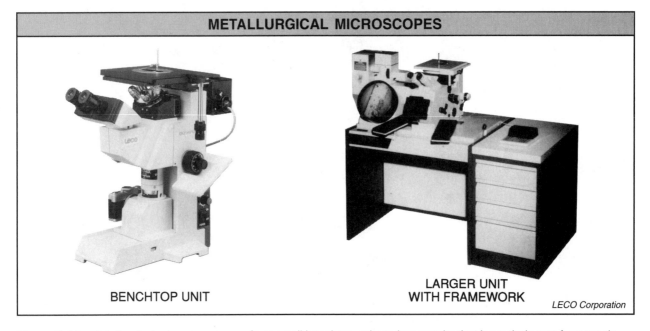

METALLURGICAL MICROSCOPES

BENCHTOP UNIT

LARGER UNIT
WITH FRAMEWORK

LECO Corporation

Figure 5-14. Metallurgical microscopes vary from small benchtop units to larger units that have their own framework.

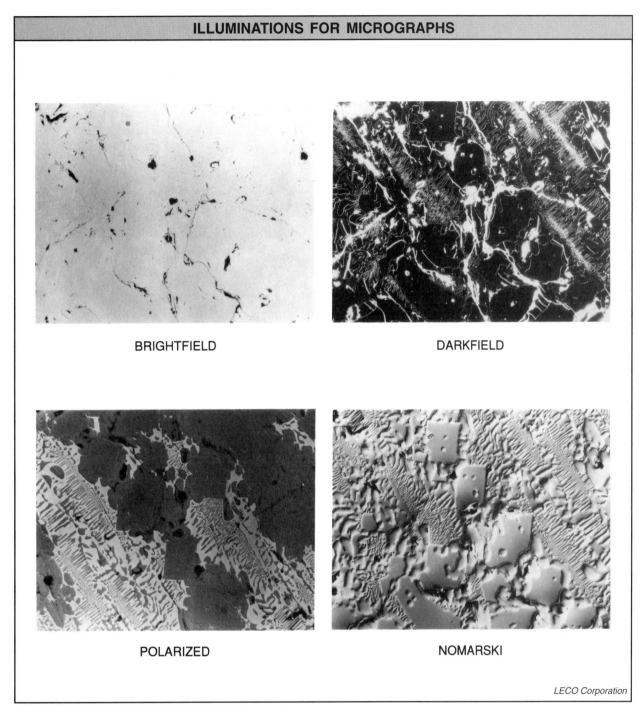

ILLUMINATIONS FOR MICROGRAPHS

BRIGHTFIELD

DARKFIELD

POLARIZED

NOMARSKI

LECO Corporation

Figure 5-15. The four illumination forms for micrographs are brightfield, darkfield, polarized, and Nomarski illumination.

Problems such as artifacts and surface films may appear before metallurgical microscopic examination. These problems may result in a poor examination of the microstructure. Artifacts are false microstructural indications that do not correspond to the true microstructure and are caused during met-allographic specimen preparation. Artifacts result from incomplete removal of a thin surface layer that has been affected by the specimen preparation process. For example, overheating during cutting may give the false impression that the specimen was heat-treated. See Figure 5-16.

PROPERLY PREPARED
SPECIMEN

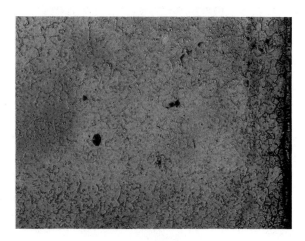

OVERHEATED
SPECIMEN

Figure 5-16. Heat treatment of a medium-carbon steel during cutting leads to artifacts in the microstructure.

Some specimens must be repolished and reetched several times to remove the affected surface layer and reveal their true structure. Alloys that form strong passive surface films, such as titanium and zirconium, require an etch-polishing technique. In the etch-pol- ishing technique, an etchant is sprayed on the polish- ing wheel to help chemically dissolve the films during the polishing process. To prevent artifacts, certain metals and alloys require different metallographic preparation procedures. See Figure 5-17.

METALLOGRAPHIC PREPARATION PROCEDURES		
Alloy Family	**Condition**	**Special Techniques**
Aluminum	Annealed (soft)	Use special final polishing methods to eliminate hard abrasive particles embedding
	Heat-treated	No particular problems
	Hard intermetallic phases	Requires care to prevent excessive relief between hard intermetallic phase and softer matrix
Copper	Copper is major component	No particular problems
	High-copper, single-phase	Requires care to prevent grinding and polishing artifacts
	Two-phase alloys	No particular problems
	Alloys with large amounts of lead	Requires techniques such as etch polishing
Austenitic stainless steel	All types	Requires care on cutting and grinding to prevent surface work hardening, such as adequate rough polishing to remove cold-worked surface layer
Carbon and low-alloy steel	All types	Requires care on cutting and grinding to eliminate local surface heat treatment Adequate rough polishing is necessary to eliminate local surface heat treatment
Magnesium and nickel	All types	No particular problems
Titanium	All types	Requires care to avoid surface work hardening during cutting and grinding Adequate rough polishing is necessary to eliminate surface work hardening Hot mounting may redistribute the hydride phase in alloys with high hydrogen Etch polishing is usually advantageous

Figure 5-17. To prevent artifacts, certain metals and alloys require different metallographic preparation procedures.

QUANTITATIVE METALLOGRAPHY

Quantitative metallography is the use of metallography to measure specific aspects of microstructures such as grain size and density of nonmetallic inclusions for quality control purposes. The two major quality control applications are grain size measurement and estimation of steel cleanliness.

Grain Size Measurement

Fine, or small, grain size adds to the strength and toughness of carbon steels and alloy steels. However, alloys that must be strong at high temperature are often heat-treated to achieve coarse, or large, grain size.

During solidification, individual grains in a metal grow independently of one another until they are impeded by other growing grains. This process leads to a variety of grain sizes and shapes in any microstructure. The spatial grain size, or actual grain size, is the average size of three-dimensional grains and may not be truly represented by a two-dimensional cross section of the material revealed by the etched metallurgical mount. For these reasons, grain size measurement is made on a statistically significant population of grains in the specimen examined at a fixed magnification.

The most common method of grain size classification is based on the ASTM grain size number. The comparison, planimetric, and intercept procedures are used to classify grain size according to the ASTM grain size number.

ASTM Grain Size Number. The ASTM grain size number (n) is obtained from the following formula: $N = 2^{n-1}$, where N is the number of grains per square inch. For example, a specimen with a grain size number of 8 has 128 grains per sq in. The higher the grain size number, the finer the grain size. See Figure 5-18. Grain size measurement is described in *ASTM Standard E 112 Determining Average Grain Size*.

When there is a mixed grain size, the situation should be recorded by estimating the area occupied by each grain size grouping. Materials that are heavily cold worked or partially recrystallized (fine grain size in a cold-worked material resulting from certain heat treatments) should not be measured for grain size. Two-phase (two crystallographic structured)

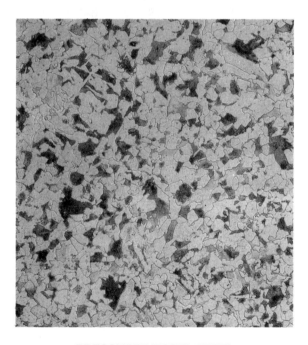

PRESSURE VESSEL STEEL

NICKEL-BASE ALLOY

Figure 5-18. A pressure vessel steel plate has a fine grain size for toughness. A nickel-base alloy used for high-temperature service has a coarse grain size for creep strength.

materials should be examined by special means outlined in *ASTM E 112*.

To measure the grain size, a polished specimen is first etched to outline the grain boundaries. One of the three procedures may be used to count the grains, which are examined at a fixed magnification of 100X.

Comparison Procedure. The comparison procedure uses standard charts, which have been developed for steels and copper-base alloys. A transparent outline of the applicable chart is superimposed over the image of the etched specimen in the metallurgical microscope at the correct magnification to allow direct comparison.

An advantage of the comparison procedure is that the standard chart is obtained from a real microstructure, which has a varied grain size within its standard specified range. This allows for rapid visual comparison. The comparison procedure is the most convenient, but it is applied only to equiaxed grain structures, which are grains having approximately the same dimensions in two perpendicular directions. The comparison procedure is accurate to plus or minus one whole grain size number, which is acceptable for most purposes.

Planimetric Procedure. The planimetric procedure uses a circle or rectangle of known area (usually 5000 mm^2) that is inscribed on a photomicrograph or on the ground-glass screen of a metallurgical microscope set at 100X. The inscribed area (field) should contain at least 50 grains and a minimum of three such fields are counted. The sum of all the grains included completely within the area, plus one half the grains intersected by the circumference or perimeter line of the area is the number of grains at the particular magnification. This number is multiplied by a factor that is dependent on the magnification and results in the number of grains per millimeter of length in the specimen. This is converted into the ASTM grain size number by a procedure described in *ASTM E 112*.

Intercept Procedure. The intercept procedure is based on counting the number of grains intercepted by one or more straight lines of sufficient total length to yield at least 50 intercepts. The intercept procedure is the most accurate. The grains are counted on a photomicrograph or on the ground-glass screen of the metallurgical microscope at a magnification of 100X. For equiaxed structures, the counts should be taken on at least three lines to assure a reasonable average. For structures that are not equiaxed, such as lightly cold-worked material, the measurements should be made both perpendicular and parallel to the longitudinal and transverse directions. Such measurements require two mounted and polished sections. The counts are converted to the ASTM grain size number by an equation in *ASTM E 112*. The intercept procedure is often used to resolve disputes because it is the most accurate procedure.

Image Analysis

Image analysis is a branch of quantitative metallography in which phases or constituents to be examined in the microstructure are detected and their relative amounts measured. A video camera is attached to a metallograph to record the microstructure. The video signal from the microstructure is fed to a digitizer, which breaks the image into a finite number of picture elements (pixels). To capture all of the video information, the resolution of the digitizer should be at least double that of the video signal from the metallograph.

Each pixel has an associated gray level, depending on the image and range from black to white. This is digitized by the instrument and is subjected to a variety of computerized enhancement techiques. High-resolution output monitors are used to view the processed image.

Specimen Preparation. Specimen preparation is a critical component of image analysis. Artifacts caused by improper specimen preparation must be removed, otherwise results of image analysis will be poor. To achieve consistent specimen preparation, an automatic polisher and a final polish using colloidal silica as the abrasive may be used. This method of specimen preparation minimizes surface damage during the final polishing.

Selective etching techniques that permit easier and more foolproof phase detection must be used.

Best results are obtained with etching procedures that maximize contrast. For example, when measuring grain size, the optimum image is one that fully reveals the grain boundary structure as black lines against a white unetched background.

Semiautomatic Image Analysis. Semiautomatic image analysis techniques use interactive devices through which the operator controls which microstructural features must be detected. The operator traces the outline of features that must be detected and edited on the monitor using a cursor or light pen. The instrument is programmed to measure the area, perimeter, diameter, or shape factor of similar features in the microstructural field being viewed.

An example of semiautomatic image analysis is the measurement of grain size. Special etching techniques are required to obtain complete grain boundary etching without etching of other features such as inclusions, which bias the results.

To obtain a good measurement of grain size, at least five fields are measured. The time required to detect the grain boundaries and edit the image is comparable to the time required for a manual count. However, an advantage is that the computer totals the results, makes the calculations, and stores the data as a permanent record.

Automatic Image Analysis. Automatic image analysis techniques use devices that automatically analyze large numbers of fields into clearly distinguishable features. Automatic image analysis techniques are used where a number of fields are required to obtain statistically sound data. For example, it is used to estimate percent inclusions, to estimate porosity or graphite in the microstructures, or to size particles.

Another application of automatic image analysis is the measurement of inclusions in steel. An as-polished specimen is used and the inclusion types (sulfides and oxides) can be separated by gray level differences in the microstructure. If the inclusions are irregularly distributed, it is necessary to measure approximately 100 fields per specimen and a substantial number of specimens per lot to obtain adequate statistics.

Inclusion Counting

Inclusion counting methods are used to estimate the internal cleanliness of metals. The density and shape of inclusions are the primary factors that determine the internal cleanliness of steels and other alloy systems. The internal cleanliness of low-alloy steels and other high-strength materials is an important consideration in their resistance to fracture. Depending on their shape and density, inclusions may present sites for fracture initiation. In other alloy systems, such as stainless steel, inclusions may be responsible for loss of corrosion resistance.

The *morphology* is the shape of an inclusion. For example, the shape may be block-shaped, round, or elongated. The most common method of inclusion counting is to use comparison charts, such as those developed in *ASTM E 45 Determining the Inclusion Content of Steel*. The specimen is polished to reveal the inclusions, and the morphology and density of the inclusions are compared with the standard charts so that a rating can be made.

Automatic inclusion counting is sometimes used for routine quality control in a production laboratory. Automatic inclusion counting requires standardized specimen preparation procedures. An image analyzer attached to the metallograph counts and distinguishes inclusions by the contrast in brightness between the inclusion and the background and also by the characteristic shape of the inclusion. Standardized specimen preparation procedures are described in *ASTM E 768 Preparing and Evaluating Specimens for Automatic Inclusion Assessment*.

MACROSCOPIC EXAMINATION

Macroscopic examination is used to reveal the general structure of large areas of a specimen because they might not be revealed under the higher magnifications used in metallographic examination. Macroscopic examination is performed with the naked eye or at magnifications up to 10X using a binocular microscope. Larger specimens are used for macroscopic examination than are used for metallographic examination. Specimens for macroscopic examination are often entire sections through components. A *macrograph* is a sketch of the etched surface of the specimen made from a macroscopic examination, which illustrates the key features. A *photomacrograph* is a photograph of the etched surface made from a macroscopic examination. The stages of macroscopic examination include specimen preparation, macroetching, and examination and preservation.

Specimen preparation is performed to prepare a specimen for macroscopic examination by removing (sawing or flame cutting) a slice in the plane to be examined. Macroscopic examination procedures are similar to those used in cutting, rough grinding, and fine grinding for metallographic specimen preparation. Fine grinding is performed to a final finish with a 240 grit abrasive paper and the specimen is not mounted.

Macroetching procedures, in contrast to metallographic etching, use macroetchants (deep etchants). The deep etchants are intended to develop gross features, such as the solidification structure of an ingot, the segregation of casting, the imperfections in a welded component, or the grain flow pattern in a forged bar. Macroetchants are designed to attack metals more deeply and faster than metallographic etchants. See Appendix.

Macroetching is usually performed by gently daubing the specimen with the macroetchant or by immersing small specimens in the macroetchant and gently swirling. Higher temperatures accelerate the etching rate. Prolonged etching is avoided because it leads to darkening of the specimen, which obscures detail. When the structural features are developed, the specimen is immediately rinsed in warm running water. During rinsing, the surface should be scrubbed with a soft bristle brush to remove deposits formed during macroetching. Deposits contain the macroetchant and lead to localized overetching if the macroetchant is not thoroughly scrubbed off. The washed specimen is dried by squirting it with alcohol or acetone, draining off excess by contact with a cloth at one edge, and drying in a current of warm air.

During the macroscopic examination and preservation procedures, the macroetched specimen is examined with the naked eye or under a binocular microscope and the specimen surface is immediately preserved with a clear lacquer coating. If the surface is not preserved, it will oxidize and darken with time and lose surface features.

Entire sections cut from welds, forgings, bars, and other components are macroscopically examined for defects that would result in rejection. Specified macroscopic examination requirements may be used as a quality control tool. For example, macroscopic examinations are used to expose the flow pattern in forged billets. Macroscopic examination is described in *ASTM E 381 Macroetch Testing, Inspection and Rating of Steel Products, Comprising Bars, Billets, Blooms, and Forgings.*

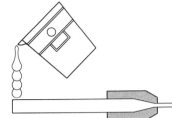

Phase Diagrams 6

When two or more chemical elements are combined to form an alloy, the resulting microstructure contains phases. Phase diagrams are graphic representations of the phases present in an alloy system at various temperatures and percentages of the alloying chemical elements. As the chemical composition is modified or the temperature changed, new phases form and existing phases disappear. Phase diagrams provide useful information about how an alloy will perform. Binary phase diagrams indicate the phases present in alloys composed of two chemical elements. Ternary phase diagrams indicate the phases present in alloys with three chemical elements. Phase changes occur as any specific composition is cooled from the molten state to room temperature and below. The same phase changes occur in reverse when a metal is heated to the melting point.

PHASE DIAGRAMS

Phase diagrams are graphic representations of the components (phases) present in an alloy system at various temperatures and percentages of the alloying chemical elements. A *phase* is any structure that is physically or crystallographically distinct and usually visible under a metallurgical microscope. As the chemical composition is altered or the temperature changed, new phases may form and existing phases may disappear.

An *alloy system* is a method of describing all the alloys that can be formed (all possible proportions) by the individual chemical elements of an alloy. Alloy systems contain three types of phases that are represented on phase diagrams. The three types are a pure metal component phase, a solid solution phase, and an intermediate phase. Specific experimental techniques are used to determine the temperature and composition limits of the phases.

Alloy Systems

Alloying modifies the crystal structure of a pure metal by forming a solid solution or a compound. Grain structure of a pure metal is also modified by

alloying. Phase describes a crystal structure that is present in an alloy and is evident in the grain structure. Phase changes may occur when the composition of an alloy is altered or the temperature is increased. At a specific temperature, any alloy composition begins to melt as it transforms from a solid phase to a liquid phase.

Theoretically, the number of compositions in an alloy system is infinite. A binary alloy system is composed of two chemical elements, a ternary alloy system of three chemical elements, a quaternary alloy system of four chemical elements, a quinary alloy system of five chemical elements, etc.

In most cases, only a handful of alloy compositions are commercially important. For example, the binary copper-zinc alloy system contains several commercially important alloys. This alloy system includes red brass (composed of 85% Cu and 15% Zn), cartridge brass (composed of 70% Cu and 30% Zn), and muntz metal (composed of 60% Cu and 40% Zn). Minor amounts of other chemical elements, such as tin, lead, arsenic (As), or antimony (Sb), are added to such binary alloys to improve their corrosion resistance, machinability, and/or mechanical properties.

Pure Metal Component Phase. The crystal structure of any phase based on one of the pure metal components in an alloy system is the same as the pure metal. This usually occurs when the addition of the other alloying chemical element(s) is not excessive, so that the crystal structure of the pure metal is not distorted by foreign atoms. Excessive amounts of alloying additions encourage the formation of a new phase.

For example, red brass exhibits the face-centered cubic (FCC) crystal structure of copper. The percentage of zinc in the red brass, which has a close-packed hexagonal (CPH) crystal structure, is not sufficient to alter the FCC crystal structure of red brass.

Solid Solution Phase. *Solid solutions* are products formed when the base metal (solvent metal) incorporates atoms of the other metal (solute metal) into its crystal structure. The two types of solid solutions are substitutional and interstitial. See Figure 6-1.

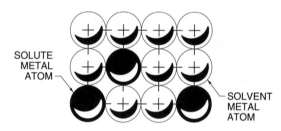

SUBSTITUTIONAL SOLID SOLUTION

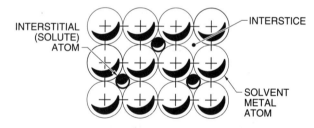

INTERSTITIAL SOLID SOLUTION

Figure 6-1. A substitutional solid solution is formed when atoms of a solute metal substitute for atoms of a solvent metal. Interstitial solid solution is formed when the interstitial atoms fit into the interstices of the solvent metal crystal structure.

Substitutional solid solutions are solid solutions formed when the solute metal atoms are substituted for the solvent metal atoms in the crystal structure. Substitutional solid solutions are formed under very specific conditions. For example, they are formed when both metals have similar crystal structures and atoms of relatively similar size.

The copper-nickel alloy system forms a single solid solution over the entire range of alloy compositions. Copper and nickel (both FCC) are mutually soluble in one another, whatever their respective amounts.

Interstitial solid solutions are solid solutions formed when the interstitial atoms (solute atoms) fit into the interstices (spaces) of the solvent metal crystal structure. The interstitial atoms must be very small compared with the solvent metal atoms. In many alloys the interstitial atoms are nonmetallic. For example, carbon forms an interstitial solid solution with a high-temperature version (allotrope) of iron (gamma iron, or γ-iron). Carbon atoms occupy the spaces of the FCC γ-iron crystal structure.

Intermediate Phase. *Intermediate phases* are structures usually formed between dissimilar chemical elements and often at a fixed composition or over a narrow composition range. Intermediate phases usually have covalent or ionic bonding and are nonmetallic in nature. For example, iron carbide (Fe_3C) is an intermediate phase that is formed between the low temperature BCC allotrope of iron (alpha iron, or α-iron) and carbon. Iron carbide is used to strengthen low-carbon steels.

Diagram Characteristics

Phase diagrams display the phases present in an alloy at any composition and temperature. Metallurgists use this information to predict service behavior of an alloy. For example, the temperature at which carbon steel is held prior to quenching (hardening) depends on the carbon content of the steel. This temperature is obtained from the iron-carbon phase diagram.

Percentages of the chemical elements are usually expressed in weight percent, which is the composition reported by chemical analysis. Impurities that enter the microstructure from processing operations are not usually indicated on phase diagrams. The two basic types of phase diagrams are equilibrium diagram (indicates the phases present under steady state conditions) and constitutional diagram (indicates phases at specific rates of heating and cooling).

Phase diagrams for binary alloy systems are plotted with temperature on the vertical axis (y-axis) and composition on the horizontal axis (x-axis). Phase diagrams for ternary alloys are plotted as triangular coordinates, pseudo-binary diagrams, and three-dimensional coordinates.

Weight Percent and Atomic Percent. Weight percent is the composition obtained by chemical analysis and is the one most frequently used on phase diagrams. The composition is sometimes given in atomic percent, rather than in weight percent. For a binary system with two chemical elements (A and B), weight percent is converted to atomic percent by applying the following formulas:

$$A\%_A = \frac{100W\%_A}{W\%_A + W\%_B(AW_A \div AW_B)}$$

where

$A\%_A$ = atomic percent of chemical element A

$W\%_A$ = weight percent of chemical element A

$W\%_B$ = weight percent of chemical element B

AW_A = atomic weight of chemical element A

AW_B = atomic weight of chemical element B

and

$$A\%_B = \frac{100W\%_B \times (AW_A \div AW_B)}{W\%_A + W\%_B(AW_A \div AW_B)}$$

where

$A\%_B$ = atomic percent of chemical element B

$W\%_B$ = weight percent of chemical element B

AW_A = atomic weight of chemical element A

AW_B = atomic weight of chemical element B

$W\%_A$ = weight percent of chemical element A

Note: In a binary system, the sum of the atomic percent of chemical element A and chemical element B is approximately 100%.

Example: Converting weight percent to atomic percent

For the binary system of aluminum (chemical element A) and copper (chemical element B), what are the corresponding atomic percents if aluminum has an atomic weight of 26.98 and a weight percent of 10, and if copper has an atomic weight of 63.54 and a weight percent of 90?

Converting weight percent to atomic percent for aluminum

$$A\%_A = \frac{100W\%_A}{W\%_A + W\%_B(AW_A \div AW_B)}$$

$$A\%_A = \frac{100(10)}{10 + 90(26.98 \div 63.54)}$$

$$A\%_A = \frac{1000}{10 + 90(.42)}$$

$$A\%_A = \frac{1000}{10 + 37.80}$$

$$A\%_A = \frac{1000}{47.80}$$

$$A\%_A = \mathbf{20.92\%}$$

Converting weight percent to atomic percent for copper

$$A\%_B = \frac{100W\%_B \times (AW_A \div AW_B)}{W\%_A + W\%_B(AW_A \div AW_B)}$$

$$A\%_B = \frac{100(90) \times (26.98 \div 63.54)}{10 + 90(26.98 \div 63.54)}$$

$$A\%_B = \frac{9000 \times (.42)}{10 + 90(.42)}$$

$$A\%_B = \frac{3780}{10 + 37.80}$$

$$A\%_B = \frac{3780}{47.80}$$

$$A\%_B = \mathbf{79.08\%}$$

Microstructure Impurities. Impurities are inclusions and/or porosity that enter the microstructure and are not shown on phase diagrams. *Inclusions* are usually nonmetallic substances, such as oxides, sulfides, and silicates, that are held or formed during solidification or subsequent reaction within the solidified metal. Inclusions most often enter metals from contamination of the molten metal with refractory mold material.

Porosity is pockets of gas inside a metal and may be macroscopic or microscopic. Porosity is caused when a metal shrinks during solidification and insufficient molten metal is available to take its place, or it is caused when gas is produced during solidification. For example, gas may be produced from the presence of moisture. Porosity is rarely present in wrought metals because regions of macroscopic porosity are usually scrapped and regions of microscopic porosity usually weld up during hot-working operations. See Figure 6-2.

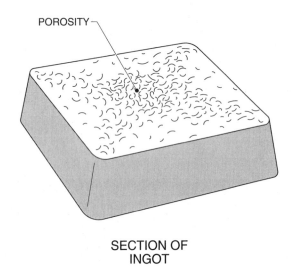

POROSITY

SECTION OF
INGOT

Figure 6-2. Porosity, which is caused by shrinkage at the center of a solidified ingot, is an impurity in the microstructure and is revealed by macroetching a slice taken through an ingot.

Equilibrium and Constitutional Diagrams. The two basic types of phase diagrams are equilibrium and constitutional. Equilibrium diagrams are more commonly available than constitutional diagrams. Equilibrium diagrams indicate the composition and temperature limits for phases under conditions of thermal equilibrium. In such cases, time is allowed for completion of phase changes at any temperature. Constitutional diagrams indicate the composition and temperature limits for phases under specific rates of heating and cooling. If the rate of heating or cooling is sufficiently high, completion of phase transformations may be prevented.

Diagram Determination

Phase diagrams are most often determined by measuring the temperatures at which phase changes occur. Identification of the phases involves the use of techniques such as thermal analysis, dilatometry, X-ray diffraction, and metallography.

A series of test bars, representing a range of alloy compositions in an alloy system, are produced. Specimens from the test bars are heated to a specific temperature and allowed to cool. As the specimen cools, phase transformations are noted by studying the appearance of the cooling curve, which is a plot of temperature against time. The two types of phase transformations are isothermal transformations and continuous cooling transformations.

Isothermal and Continuous Cooling Phase Transformations. Isothermal and continuous cooling phase transformations are the two types of phase transformations and are displayed on a cooling curve, which is a plot of temperature against time. An arrest (flattening) on the cooling curve is characteristic of an isothermal phase transformation. With a continuous cooling transformation, the gradient of the cooling curve changes, but the curve does not flatten. See Figure 6-3.

The energy required to initiate a phase change is reflected on the cooling curve by slight undercooling below the phase transformation temperature. Once the phase transformation begins, the temperature returns to the true value. Undercooling occurs because energy is required to create the surface between the new phase and the existing phase, which initiates the phase transformation.

Thermal Analysis. Thermal analysis is the plotting of the cooling curve of a molten specimen as it solidifies and cools to room temperature. With pure metals and certain alloy compositions, isothermal transformations correspond to the beginning and the end of solidification. For all other alloy compositions, a continuous cooling transformation corresponds to the beginning and end of solidification.

Thermal analysis is used for liquid-solid phase transformations or for solidification and melting. Thermal analysis is relatively insensitive and requires a large amount of heat evolution, such as occurs with a liquid-solid phase transformation.

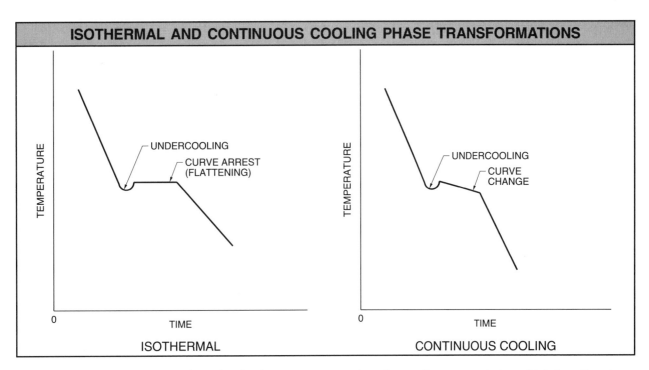

ISOTHERMAL AND CONTINUOUS COOLING PHASE TRANSFORMATIONS

Figure 6-3. An isothermal transformation is shown as an arrest on the cooling curve, during which time the phase transformation occurs. A continuous cooling transformation does not show an arrest, but the gradient of the cooling curve changes during the phase change.

Dilatometry. In dilatometry (dilatometric analysis), the length of a specimen is continuously monitored during cooling or heating. There is a measurable and abrupt change in length when a phase transformation occurs. See Figure 6-4. Dilatometry is used for phase changes in the solid state and is described in *ASTM E 80, Dilatometric Analysis of Metallic Materials*.

X-ray Diffraction. In X-ray diffraction, the heating or cooling cycle is performed while a specimen is subjected to an X-ray beam in a powder camera. A phase change is indicated by a change in lattice dimensions or by a change in crystal structure. The X-ray diffraction technique is precise and is used in determining phase changes in the solid state.

Metallography. In metallography, a specimen is heated to the required temperature and held for a sufficient time to allow the phase change reaction to complete. The specimen is cooled rapidly by immersion in water (quenched) and then polished and etched to reveal the microstructure present prior to quenching. Metallography is used to confirm parts

of the phase diagram identified by the other techniques. However, metallography does not work for alloys that undergo additional phase changes when quenched.

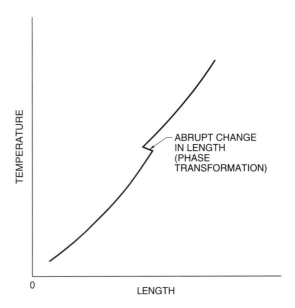

Figure 6-4. The length of the specimen, which changes abruptly during a phase transformation, is continuously monitored as it is heated or cooled to determine the temperature of the phase transformation.

BINARY PHASE DIAGRAMS

Binary phase diagrams are equilibrium or constitutional diagrams that indicate the phases present in binary alloy systems (alloys with two chemical elements). The alloy system of copper and nickel is an example of a binary system. Temperature is plotted on the y-axis and composition of the alloy is plotted on the x-axis. See Figure 6-5.

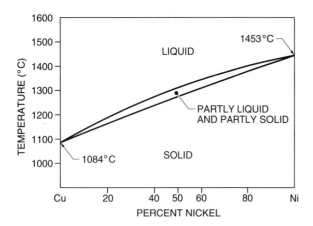

Figure 6-5. A binary phase diagram plots temperature on the y-axis and alloy composition on the x-axis.

The left and right extremes of the composition scale represent the pure metal components, which are 100% Cu on the left and 100% Ni on the right. Moving from left to right increases the percentage of nickel and decreases the percentage of copper, and vice versa. At any specific composition, the sum of the percentages of the two alloying chemical elements equals 100%. The phases that are stable at different alloy compositions and temperatures are indicated on the phase diagram.

The phase change reactions described for binary systems, such as eutectic, peritectic, etc., also occur in more complex alloy systems containing three or more major alloying chemical elements. The interpretation of alloy systems with three or more major alloy chemical elements is increasingly complex.

Liquid-state Phase Transformations

The phases formed by solidifying alloys include solid solutions, eutectics, intermediate phases, and peritectics. Phase transformations may also occur between molten phases that do not mix, such as the monotectic reaction. Pure metals only form solid solutions.

Pure Metal Solidification. The solidification of a pure metal is the isothermal transformation that takes place at the melting point. The resulting microstructure is single-phase and contains grains of the pure metal. See Figure 6-6. The solidification behavior of pure metal components of alloys is shown at the extreme ends of a phase diagram.

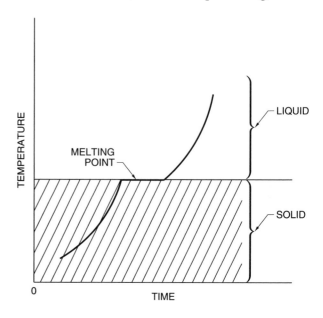

Figure 6-6. The melting point of a pure metal is indicated by an isothermal transformation on the cooling curve.

Solid Solution Alloy Solidification. Solid solution alloys solidify as a single phase. The chemical compositions of alloy phases is calculated by drawing tie lines on a phase diagram and using the lever rule. The lever rule states that solidifying alloys develop compositional inhomogeneity such as coring or segregation.

Solid solutions are formed between metals that have the same crystal structure and similar atomic size. The copper-nickel alloy system forms a single solid solution for all alloy compositions, because both copper and nickel have FCC crystal structures and atoms of similar size.

The cooling curves for pure copper and nickel display isothermal transformations at their melting points. The melting point for pure copper is 1084°C (1983°F), and the melting point for pure nickel is

1453°C (2647.4°F). The cooling curves for all copper-nickel alloys indicate continuous cooling transformations that begin when the first solid nuclei form in the molten metal and end when the last aggregates of molten metal have solidified. To construct a copper-nickel phase diagram, the first line (liquidus) is drawn linking the start temperatures of the continuous cooling transformations for the various alloys in the system. A second line (solidus) is drawn linking the finish temperatures of these transformations. See Figure 6-7.

A *liquidus* is the locus (connection) of points on a phase diagram representing the temperatures at which each alloy in the system begins to solidify during cooling or completes melting during heating. A *solidus* is the locus of points on a phase diagram representing the temperatures at which each alloy in the system completes solidification during cooling or begins to melt during heating.

The temperature at which an alloy crosses the liquidus on cooling marks the formation of the first solid nuclei in the melt. As cooling continues, increasing amounts of solid precipitate on the nuclei. The solidus marks the temperature at which solidification is complete.

Greek letters are used to indicate the different phases. The solid solution phase is alpha (α), or alpha solid solution. In binary alloy systems with

more than one solid solution, the succeeding letters of the Greek alphabet are used to describe them from left to right on a phase diagram. Examples are beta (β), gamma (γ), delta (δ), etc. The liquid phase is indicated by the letter *L*.

Solid solution alloys usually have improved mechanical or physical properties in comparison with component pure metals. For example, in the copper-nickel alloy system, maximum strength occurs at 66% Ni and 34% Cu. This is the composition of an alloy widely used for corrosion resistant applications. This alloy is alloy 400 (Monel). Other commercially important alloys in the copper-nickel alloy system are the cupronickels. *Cupronickels* are copper-nickel alloys that contain between 90% Cu and 30% Cu, and nickel as the remaining balance. Cupronickels are used extensively for tubing for heat exchangers and condensers.

Tie lines are horizontal lines drawn on a phase diagram that represent constant temperature and that connect the compositions of a pair of coexisting phases. Tie lines are used to figure the compositions and proportions of coexisting phases (conjugate phases) at a specific temperature. Alloy 400 has a composition of liquid and solid phases that coexist at 1343°C (2449°F) and is obtained by constructing a tie line that connects the liquidus

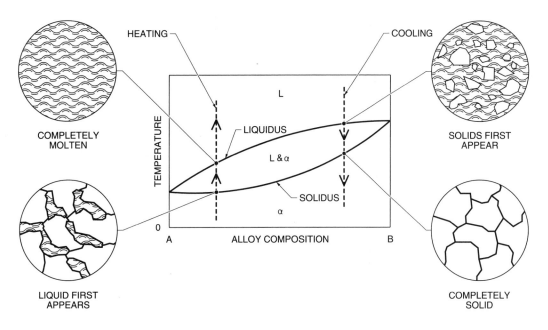

Figure 6-7. The liquidus and solidus are the upper and lower boundaries between completely liquid and completely solid alloys.

and solidus at that temperature. The liquid phase is 60% Ni and 40% Cu, and the solid phase is 70% Ni and 30% Cu.

The lever rule is used to figure the proportions of coexisting phases at a specific temperature. The lever rule states that the relative proportions of coexisting phases in an alloy, at a particular temperature, are obtained by constructing a tie line and obtaining a mechanical balance between the weights of each phase. The base alloy composition is used as a fulcrum.

With alloy 400 at 1343°C (2449°F), the lever rule indicates that the proportions of liquid and solid phase present under steady state conditions are 40% liquid phase and 60% alpha phase. See Figure 6-8. The sum of the percentages of the phases present at any temperature must total 100.

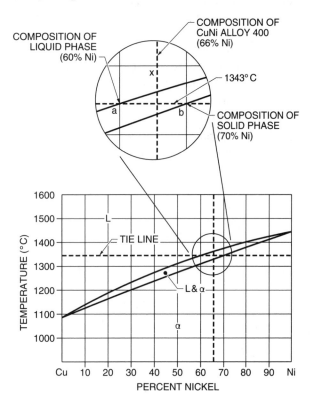

Figure 6-8. A tie line indicates the compositions of phases in an alloy coexisting at a specific temperature.

Coring and segregation are terms that describe chemical composition gradients in alloys caused by rapid cooling rates from the molten state. *Coring* is a condition of variable composition between the center and the surface of a unit of microstructure (such as a dendrite, grain, or inclusion). Coring re-

sults from nonequilibrium growth, which occurs over a range of temperatures and compositions. *Segregation,* a more general term, is any concentration of alloying chemical elements in a specific region of a metallic object.

The cooling rate of a casting may vary significantly depending on the type of mold. For example, components cast in a metal mold cool more rapidly than those cast in a sand mold. Most phase diagrams are equilibrium diagrams because they are developed under steady state conditions. In practice, however, most metals cool too rapidly for equilibrium to occur, and coring and segregation are inevitable in cast components.

A series of tie lines constructed over the solidification range of an alloy indicates that the solid phase composition changes along the solidus as solidification increases. See Figure 6-9. Each layer of solidifying metal deposited on the growing dendrites has a slightly different composition than the previous layers. The solidifying dendrites develop a chemical composition gradient, which causes the solidifying metal to become inhomogeneous.

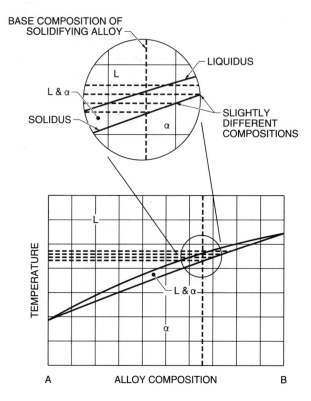

Figure 6-9. As an alloy solidifies, each layer of solidifying metal deposited on the growing dendrites has a slightly different composition.

The slower the cooling rate, the greater the time for atoms to diffuse (move through the crystal lattice). This atomic diffusion reduces inhomogeneity. The faster the cooling rate, the less time for atomic diffusion (greater inhomogeneity). The wider the temperature range between the liquidus and the solidus, the greater the opportunity for inhomogeneity.

Cored structures are most commonly found in as-cast metals. The last solid to form along the grain boundaries and between the interdendritic spaces is richest in the lower melting point component. Nonmetallic materials such as slag, sand, or inclusions are sometimes forced into these regions, so that cored structures may contain planes of mechanical weakness. The wider the temperature difference between the liquidus and the solidus, the greater the opportunity for coring.

Coring and segregation are removed by a homogenization type heat treatment. The cored alloy is heated to a temperature below the solidus to increase the rate of atomic diffusion. See Figure 6-10. The solidus temperature must not be exceeded because localized melting will occur. The homog-

enization temperature is maintained at a minimum of 55°C (100°F) below the solidus to avoid localized melting.

Mechanical working at elevated temperatures speeds atomic diffusion. Homogenization without the benefit of mechanical working requires a relatively long period of time (4 hours or more). A sufficient amount of time is necessary so that atomic diffusion will occur.

Eutectic Alloys Solidification. *Eutectic reaction* is an isothermal transformation in which a liquid transforms into two solid phases. Terminal solid solutions that are characteristic of eutectic and other reactions have limited composition ranges based on one of the components of the alloy system.

The lead-tin phase diagram displays a eutectic reaction. This phase diagram is characterized by a liquidus that falls to a minimum at 61.9% Sn and 38.1% Pb. See Figure 6-11. Alloys containing less than 19.2% Sn solidify as completely alpha solid solution at temperatures below the solidus. Similarly, alloys with more than 97.5% Sn solidify as

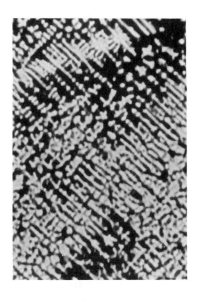

CAST MONEL METAL
REVEALING VARIABLE
COMPOSITION (25X)

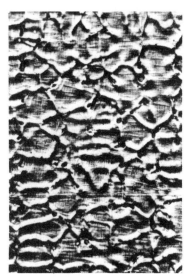

CAST MONEL METAL
REVEALING ROUGH CORED
STRUCTURE (100X)

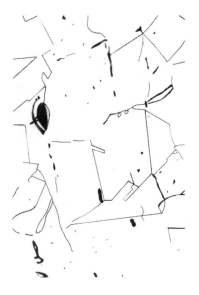

ROLLED AND ANNEALED
CAST MONEL METAL
REVEALING HOMOGENIZED
STRUCTURE (100X)

Figure 6-10. Coring in cast Monel metal is removed by hot working and annealing, which homogenizes the structure. Etching reveals coring that is caused by the variable composition gradient in the grains of the alloy.

completely beta solid solution at temperatures below the solidus.

Alloys containing between 19.2% Sn and 97.5% Sn solidify in a different manner. These alloys fall into two groups: alloys with <61.9% Sn and alloys with >61.9% Sn. Those with <61.9% Sn solidify by forming alpha solid solution, and those with >61.9% Sn solidify by forming beta solid solution. Solidification for both groups continues until the temperature reaches the eutectic reaction temperature, which is 183°C (361°F).

The percent of alpha solid solution and percent of liquid phase for alloys with <61.9% Sn is found by applying the following formulas:

$$\%\alpha = \frac{61.9 - \%Sn}{61.9 - 19.2} \times 100$$

where

$\%\alpha$ = percent alpha solid solution

61.9 = constant

$\%Sn$ = percent tin

19.2 = constant

100 = constant

and

$$\%L = \frac{\%Sn - 19.2}{61.9 - 19.2} \times 100$$

where

$\%L$ = percent liquid phase

$\%Sn$ = percent tin

19.2 = constant

61.9 = constant

100 = constant

Example: Figuring percent alpha solid solution

What is the percent alpha solid solution of a lead-tin alloy that contains 30% Sn?

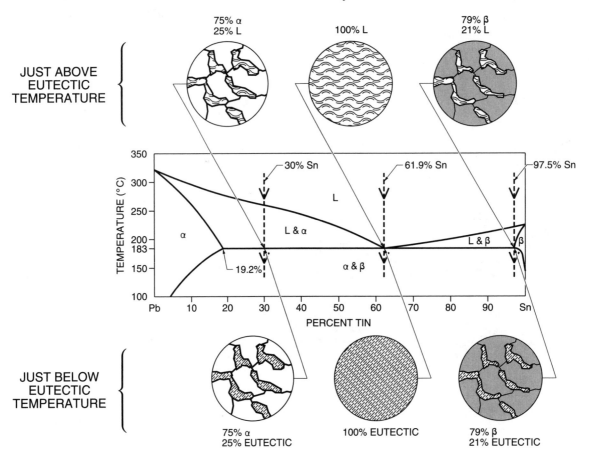

Figure 6-11. The lever rule is used to figure the proportion of liquid phase transforming to eutectic phase, which occurs on crossing the eutectic reaction temperature.

$$\%\alpha = \frac{61.9 - \%Sn}{61.9 - 19.2} \times 100$$

$$\%\alpha = \frac{61.9 - 30}{61.9 - 19.2} \times 100$$

$$\%\alpha = \frac{31.9}{42.7} \times 100$$

$$\%\alpha = .747 \times 100$$

$$\%\alpha = \mathbf{74.7\%}$$

Example: Figuring percent liquid phase

What is the percent liquid phase of a lead-tin alloy that contains 30% Sn?

$$\%L = \frac{\%Sn - 19.2}{61.9 - 19.2} \times 100$$

$$\%L = \frac{30 - 19.2}{61.9 - 19.2} \times 100$$

$$\%L = \frac{10.8}{42.7} \times 100$$

$$\%L = .253 \times 100$$

$$\%L = \mathbf{25.3\%}$$

When crossing the eutectic reaction temperature, all the liquid phase transforms isothermally to a mixture of alpha and beta. This mixture is the eutectic phase.

The percent of beta solid solution and percent liquid phase for alloys with more than 61.9% Sn is found by using the following formulas:

$$\%\beta = \frac{\%Sn - 61.9}{97.5 - 61.9} \times 100$$

where

$\%\beta$ = percent beta solid solution

$\%Sn$ = percent tin

61.9 = constant

97.5 = constant

100 = constant

and

$$\%L = \frac{97.5 - \%Sn}{97.5 - 61.9} \times 100$$

where

$\%L$ = percent liquid phase

97.5 = constant

$\%Sn$ = percent tin

61.9 = constant

100 = constant

Example: Figuring percent beta solid solution

What is the percent beta solid solution of a lead-tin alloy that contains 90% Sn?

$$\%\beta = \frac{\%Sn - 61.9}{97.5 - 61.9} \times 100$$

$$\%\beta = \frac{90 - 61.9}{97.5 - 61.9} \times 100$$

$$\%\beta = \frac{28.1}{35.6} \times 100$$

$$\%\beta = .789 \times 100$$

$$\%\beta = \mathbf{78.9\%}$$

Example: Figuring percent liquid phase

What is the percent liquid phase of a lead-tin alloy that contains 90% Sn?

$$\%L = \frac{97.5 - \%Sn}{97.5 - 61.9} \times 100$$

$$\%L = \frac{97.5 - 90}{97.5 - 61.9} \times 100$$

$$\%L = \frac{7.5}{35.6} \times 100$$

$$\%L = .211 \times 100$$

$$\%L = \mathbf{21.1\%}$$

When crossing the eutectic reaction temperature, all the liquid phase transforms isothermally to the eutectic phase. An alloy with 61.9% Sn is the eutectic composition. This alloy transforms isothermally to 100% eutectic phase at the eutectic reaction temperature. The eutectic composition is similar to the pure metal components of the alloy system. Eutectic composition has a fixed melting point. The closer an alloy is to the eutectic composition, the greater the proportion of eutectic phase in the microstructure of the alloy.

When cooling, the eutectic reaction is expressed as liquid yields alpha plus beta (L $\rightarrow \alpha + \beta$) at the eutectic reaction temperature. The microstructure of the eutectic phase is different from the alpha and beta solid solution of which it is composed. When the liquid phase transforms at the eutectic temperature, the alpha and beta phases grow upon each other in a finely divided and interspersed manner. This microstructure is often a source of strengthening in alloys.

In rare cases, the eutectic structure is not interspersed. In these rare cases, the eutectic structure is divided into two separate phases (divorced eutectic structure).

Alloy compositions to the left of the eutectic point are hypoeutectic alloys and those to the right are hypereutectic alloys. See Figure 6-12. Component chemical elements in a phase diagram are usually arranged so that commercially important alloys in the alloy system lie to the left of the eutectic composition.

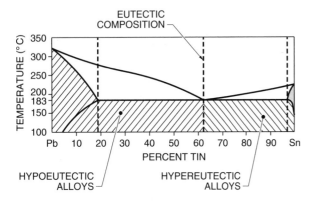

Figure 6-12. Hypoeutectic alloy compositions are located to the left of the eutectic composition, and hypereutectic alloy compositions are located to the right of the eutectic composition.

The formation of the eutectic phase lowers the melting point of an alloy. Depending on the application, lowering an alloy's melting point may be beneficial or detrimental. A lowered melting point is beneficial in lead-tin alloys because solders that are easier to work with are produced. By altering the alloy composition, the freezing range of solders can be manipulated. For example, solders that wipe easily remain soft over a wide temperature range and have a wider liquidus-solidus gap.

In the sulfur contamination of nickel alloys, a lowered melting point is detrimental. Small additions of sulfur to nickel and its alloys drastically lower the melting temperature (liquidus), which forms a eutectic compound at 25% S. Fuels con-

taining sulfur can lead to catastrophic failure of nickel-base alloy components operating at high temperature, such as gas turbines. Nickel-base alloys must be degreased before welding because grease and oil contain sulfur and cause lowering of the melting point. This condition causes welds to crack from the thermal stresses of solidification.

Terminal solid solutions are any solid phases of limited compositional range that are based on one of the components of the alloy system, pure metal, intermetallic compound, or interstitial compound. For example, in the lead-tin alloy system, the alpha and beta terminal solid solutions have the crystal structures of tin and lead.

The boundaries of terminal solid solutions are represented by the solidus and the solvus. The *solvus* is the locus of points representing the limits of solubility of a solid phase. See Figure 6-13. When an alloy composition crosses the solvus during heating or cooling, it undergoes a continuous cooling transformation.

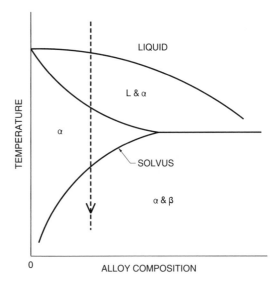

Figure 6-13. When an alloy crosses the solvus during cooling, it undergoes a continuous cooling transformation. The solvus represents the limits of solubility for a solid phase.

The composition range of a terminal solid solution depends on the temperature and the solubility of the components for each other. If the solubility is extremely low, the composition range is extremely narrow. If there is no solubility at all, the terminal solid solution disappears, leaving the pure metal component.

Intermediate Phases. Intermediate phases are distinguishable homogenous substances that have a composition range of existence that does not extend to any of the pure metal components of the phase diagram. Although they do not behave like metals or alloys, intermediate phases are similar to pure metals and eutectic alloys in that they have a fixed melting point. See Figure 6-14. Intermediate phases are identified by their chemical formula. The three types of intermediate phases are the interstitial compounds, the intermetallic compounds, and the electron compounds.

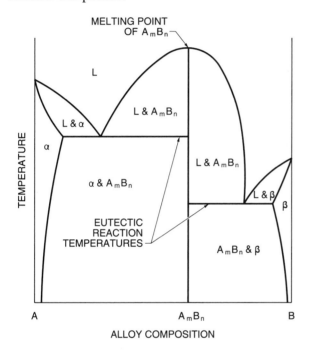

Figure 6-14. The intermediate phase divides a phase diagram into two parts, each of which contains a eutectic reaction.

Interstitial compounds are chemical compounds formed between metals, such as titanium, tungsten, tantalum (Ta), or iron, and nonmetals, such as hydrogen, oxygen, carbon, boron, and nitrogen (N). The component chemical elements of interstitial compounds bear a fixed ratio to one another. Examples of interstitial compounds are carbides, such as iron carbide, titanium carbide (Ti_4C), and tungsten carbide (W_2C). Carbides are developed in tool steels to increase hardness.

Intermetallic compounds are chemical compounds formed between metallic chemical elements that are nonmetallic in behavior. They are non-metallic in behavior because they have ionic or covalent atomic bonding, which gives them low ductility and low electrical conductivity. Examples of intermetallic compounds are materials such as the compounds of Mg_2Sn and Ca_2Se.

Electron compounds are compounds formed between metals such as copper and aluminum or silver and cadmium. Electron compounds have a wider range of solid solubility than interstitial and intermetallic compounds. Electron compounds are represented by Greek letters on the phase diagram. For example, the copper-rich end of the copper-aluminum phase diagram exhibits many electron compounds. See Figure 6-15.

Peritectic Alloys Solidification. A *peritectic reaction* is an isothermal reaction in which a solid phase reacts with the liquid from which it is solidifying to yield a second solid phase. Depending on the proportions of solid and liquid phases at the beginning of the peritectic reaction, the existing solid phase may be completely or partly consumed during the formation of the new phase.

If there is an insufficient amount of the existing solid phase, it is completely transformed. If it is in excess, what is not transformed is set in an envelope (matrix) of the new phase. The peritectic reaction is expressed as alpha plus liquid yields beta ($\alpha + L \rightarrow \beta$) when cooling at the peritectic reaction temperature.

For example, the silver-platinum alloy system exhibits a peritectic reaction. The cooling of a 70% Pt and 30% Ag alloy lies to the left of the peritectic reaction point. Alpha phase begins to solidify at 1590°C (2894°F). Just above the peritectic reaction temperature of 1186°C (2167°F), the mixture contains 43% alpha phase and 57% liquid phase. At the peritectic reaction temperature, the remaining liquid transforms to beta.

The liquid phase contains 33.7% Pt, and the beta phase contains 57.6% Pt. These compositions are obtained from the extremes of the tie line. For the liquid to transform to beta, it must enrich itself in platinum. The transformation of liquid to beta is achieved by the liquid phase interacting with the right amount of alpha. The amount of alpha must have 89.5% Pt for it to adjust its composition to that of the beta phase.

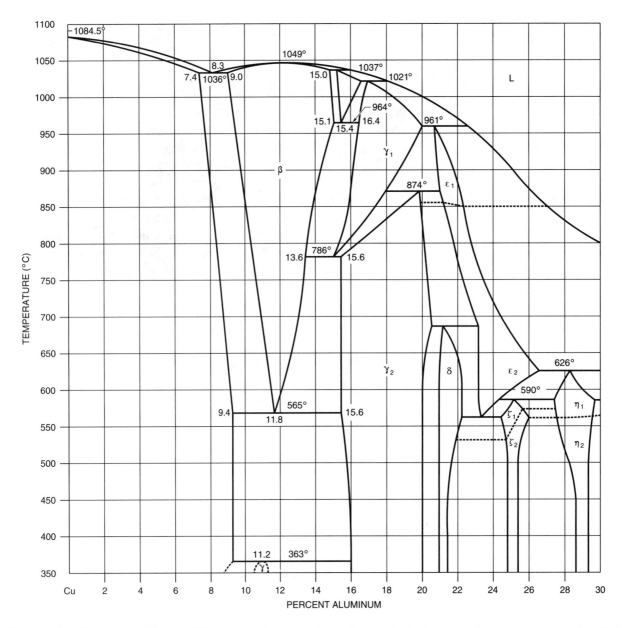

Figure 6-15. The copper-rich end of the copper-aluminum phase diagram indicates several electron compounds, which are illustrated by Greek letters.

The 57% liquid phase contains 19.2 units of platinum ([%L × 33.7] ÷ 100 = *units of Pt*, or [57 × 33.7] ÷ 100 = 19.2). See Figure 6-16. To transform liquid to beta, the liquid phase must contain 32.8 units of platinum ([%L × 57.6] ÷ 100 = *units of Pt*, or [57 × 57.6] ÷ 100 = 32.8). The liquid phase enriches itself by 13.6 units of platinum (32.8 − 19.2 = 13.6). The unknown value of the percent alpha phase with 89.5% Pt that recombines with the liquid phase to enrich itself in platinum is found by applying the following formula:

$$\frac{\%\alpha \times 89.5}{100} = 13.6 \text{ or } \%\alpha = \frac{13.6 \times 100}{89.5}$$

where

%α = percent alpha phase

89.5 = constant percent platinum

100 = constant

13.6 = units of platinum

Example: Figuring the amount of recombining alpha phase

$$\%\alpha = \frac{13.6 \times 100}{89.5}$$

$$\%\alpha = \frac{1360}{89.5}$$

$$\%\alpha = \mathbf{15.2\%}$$

As an alloy with 60% Ag and 40% Pt (which lies to the right of the peritectic reaction point) cools, it begins to solidify as alpha at 1316°C (2400°F). Just above the peritectic reaction temperature, the alloy contains 89% liquid and 11% alpha (66.3% Pt). The formula for finding percent liquid above

the peritectic reaction is ([%Ag − 10.5] ÷ [%Pt − 10.5]) × 100 = %L, or ([60 − 10.5] ÷ [66.3 − 10.5]) × 100 = 89%. At the peritectic reaction temperature, the alpha transforms to beta. After completion of the peritectic reaction, the alloy contains 74% liquid and 26% beta phase (57.6% Pt). The formula for finding percent liquid after completion of the peritectic reaction is ([%Ag − 42.4] ÷ [%Pt − 42.4]) × 100 = %L, or ([60 − 42.4] ÷ [66.3 − 42.4]) × 100 = 74%.

As the alloy continues to cool, the remaining liquid transforms to beta. Since the beta phase forms an envelope around the transforming alpha, it is difficult to achieve the peritectic transformation described.

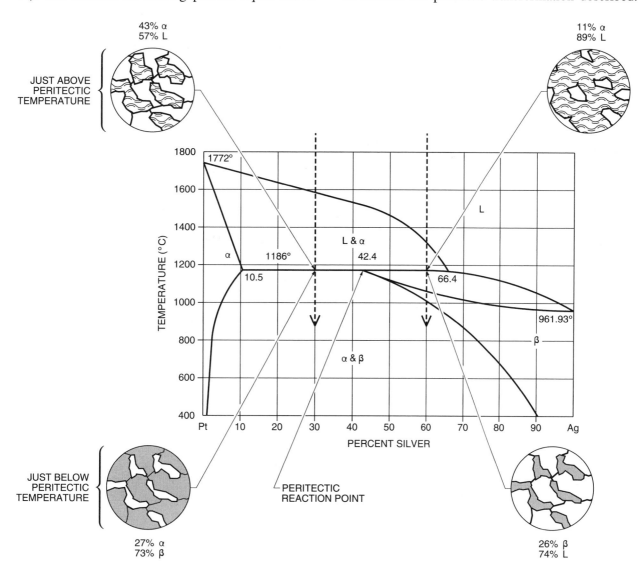

Figure 6-16. The silver-platinum alloy system exhibits a peritectic reaction.

The growing beta phase acts as a barrier to diffusion. The barrier stifles the reaction between the alpha and liquid to form beta and causes a peritectic reaction that is uncompleted. See Figure 6-17.

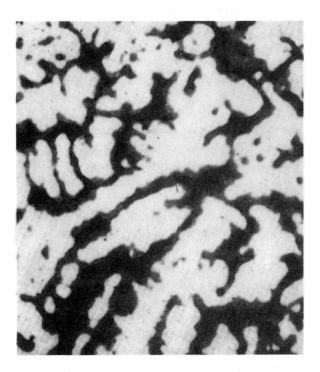

Figure 6-17. The room temperature microstructure of a 60% Ag and 40% Pt alloy exhibits a peritectic reaction that is uncompleted.

Monotectic Reaction. A *monotectic reaction* is one of several phase change reactions that exhibit immiscible (nonmixing) liquid phases over part of an alloy composition range. Although similar to a eutectic reaction, monotectic reactions have a liquid phase substituted for one of the solid phases. A monotectic reaction is expressed as liquid 1 yields liquid 2 plus alpha solid solution ($L_1 \rightarrow \alpha + L_2$), when cooling at the monotectic temperature. For example, the copper-lead phase diagram exhibits a monotectic reaction. See Figure 6-18.

A molten alloy containing 64% Cu and 36% Pb is made up of a single liquid phase until the alloy reaches a temperature of 993°C (1819°F). At this temperature, the monotectic reaction takes place isothermally and the original liquid phase transforms to a mixture of two phases. The two phases are alpha solid solution (pure copper) and a second liquid phase with 87% Pb.

As the alloy further cools, a eutectic reaction takes place at 326°C (619°F), with the eutectic point at 99.94% Pb. The second liquid phase present at the eutectic temperature is 36%, or ([36 – 0] ÷ [99.94 – 0]) × 100 = 36%. This second liquid phase transforms to a eutectic mixture of pure copper and pure lead.

Any alloy containing between 36% Pb and 87% Pb passes through the two-phase mixture of two liquids when cooling. These liquids, like oil and water, do not mix.

Lead has little solubility for many engineering alloys, such as copper alloys or steel. Lead is added to improve machinability of the alloys. The insoluble lead phase allows the tool to break down the leaded alloys into discrete (separate) chips.

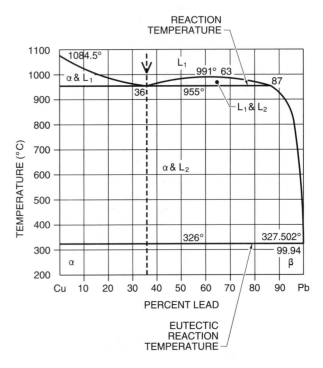

Figure 6-18. A monotectic reaction occurs in the copper-lead phase diagram. L_1 and L_2 are two immiscible liquid phases.

Solid-state Phase Transformations

Many alloys develop microstructures from phase transformations that take place at temperatures below the solidus (in the solid state). These solid-state phase transformations are strongly dependent on the cooling rate because diffusion of atoms is the driving force behind all phase transformations. The faster the cooling rate, the less time for atomic diffusion and for a predicted phase transformation to occur. Slow cooling rates allow time for diffusion, so predicted phase transformations are more likely to occur. The continuous cooling boundary line corresponding to the solidus or the liquidus is the solvus. The various solid-state phase transformations are allotropy, order-disorder transformations, second-phase precipitation, the eutectoid reaction, and the peritectoid reaction.

Allotropy. *Allotropy* is a reversible phase transformation exhibited by some pure metals that changes from one unit cell structure to another at a specific temperature. Iron, tin, cobalt, titanium, and manganese are examples of allotropic metals.

For example, pure titanium is CPH up to 882°C (1620°F). Above this temperature, pure titanium changes to BCC, and the reverse occurs on cooling. The low- and high-temperature solid solutions based on these allotropic forms of titanium are identified as alpha (low-temperature solid solution) and beta (high-temperature solid solution). See Figure 6-19.

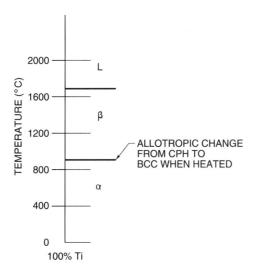

Figure 6-19. The pure titanium phase diagram exhibits an allotropic change at 883°C. The low- and high-temperature solid solutions based on the allotropic forms of titanium are alpha and beta.

Order-Disorder Transformation. *Order-disorder transformation* (ordering) is a reversible phase change between two solid solutions that have the same unit cell structure. The structure contains

atoms of the ordered phase that occur in a fixed sequence in the space lattice and atoms of the disordered phase that are randomly distributed.

Disordered phases are more common than ordered, but many solid solutions that are normally disordered may become ordered if the temperature is lowered sufficiently or if other specific conditions are satisfied. To illustrate that these solid solutions share a common crystal structure, the boundary line between ordered and disordered is dotted on a phase diagram. The ordered phase is identified with a Greek letter followed by a prime sign. For example, alpha prime (α′) is the notation for the ordered phase of the alpha solid solution. See Figure 6-20.

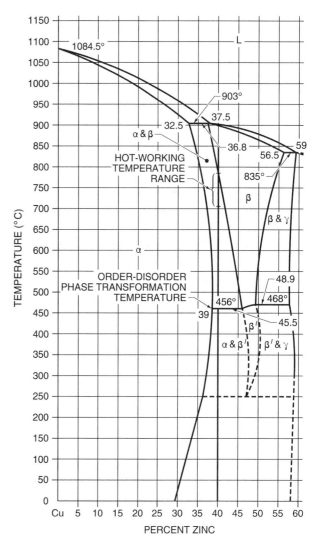

Figure 6-20. The order-disorder transformation for a copper with 40% Zn alloy occurs at 241°C (465°F).

Ordering occurs in the copper-zinc phase diagram at 60% Cu and 40% Zn (hot-worked brass). In the hot-working temperature range of 704°C to 788°C (1300°F to 1450°F), the alloy consists of alpha-plus-beta (α + β) phase. The beta (β) phase is disordered, resulting in an alloy with low strength and high ductility. The low strength and high ductility makes the alloy easy to hot work. As the alloy cools below 465°C (870°F), the ordered beta prime (β′) phase forms, resulting in an alloy that has relatively high strength and low ductility. Alpha-beta brasses, such as the 40% Zn alloy, are stronger and cheaper than single-phase alloys (alpha alloys), which contain less zinc. Because of this order-disorder transformation, alpha-beta brasses, such as the low-zinc alloys, are preferably hot worked rather than cold worked.

Second-phase Precipitation. A *precipitation reaction* is the formation of a second phase within the grains of an original phase. Precipitation occurs by Widmanstätten precipitation or precipitation (age) hardening (delayed precipitation).

When some alloys are cooled from a single-phase to a two-phase region, Widmanstätten precipitation occurs. The second phase precipitates at energetically favored sites, which are the grain boundaries of the transforming phase and along crystallographic planes within the grains. The precipitate has a geometrical appearance (Widmanstätten structure).

The widely used, high-strength titanium alloy Ti-6Al-4V, which has 6% Al and 4% vanadium (V), develops a Widmanstätten structure when cooling. The relevant phase diagram is a binary section (isopleth) drawn at 6% Al. This phase diagram indicates the effect of variations of titanium and vanadium. See Figure 6-21.

When Ti-6Al-4V is cooled from a temperature in the single-phase beta region, it crosses the solvus at about 982°C (1800°F) and enters the two-phase alpha-plus-beta region. The alpha phase precipitates in increasing quantities from the grain boundaries along the (100) planes, which are the energetically favored planes for precipitation. The resulting microstructure shows the new phase outlining the original phase as geometric interlacing needles.

Widmanstätten structures are fairly common, especially after hot working and cooling or after rapid

cooling. For example, the region of base metal immediately adjacent to a weld may exhibit Widmanstätten precipitation because this region is subject to extremely rapid cooling.

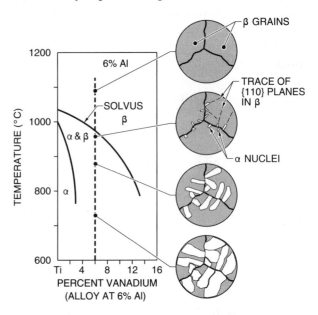

Figure 6-21. Widmanstätten precipitation occurs in the Ti-6Al-4V alloy as it cools from the beta-phase region into the alpha-plus-beta-phase region. A typical Widmanstätten structure has the microstructural appearance of geometrically interlacing needles.

Precipitation (age) hardening is a delayed precipitation reaction consisting of the precipitation of finely dispersed particles of a second phase in a supersaturated solid solution, or one containing a second phase in excess of its solubility limits.

Precipitation hardening is used as a strengthening mechanism in several commercially important alloys such as copper-aluminum and certain families of stainless steels. Precipitation hardening is achieved by rapidly cooling the alloy from a single-phase region into a two-phase region. The second phase is then released slowly from the supersaturated original phase.

The copper-aluminum (Cu-Al) alloy with 4% Al is an example of precipitation hardening. See Figure 6-22. The alloy is first solution annealed, or held at a suitable temperature in the single-phase solid solution region of the phase diagram. Solution annealing helps homogenize the structure.

The Cu-Al alloy is then quenched, or cooled from a heated state. The metal is quenched to room temperature at a rate fast enough to suppress the precipitation of the beta phase, which would normally occur with slow cooling below the solvus temperature of 500°C (932°F). Quenching is usually done by water or oil immersion.

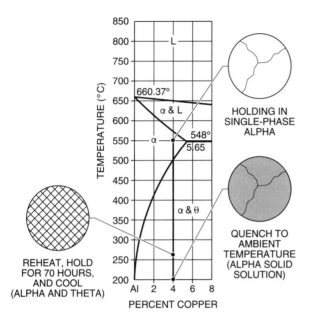

Figure 6-22. Precipitation hardening is a form of delayed precipitation. In a copper-aluminum alloy with 4% Al, precipitation hardening is achieved by quenching from the alpha-phase region and reheating in the alpha-plus-theta-phase region.

During the final stage of precipitation hardening, the beta phase is released in a controlled manner by heating the alloy to a temperature below the solvus. With some alloys, heating is unnecessary and precipitation occurs over a period of time at room temperature.

The hardness and strength of precipitation-hardened alloys rises and peaks with increasing combinations of temperature and time. *Overaging* is heat treating a metal for a longer time or at higher temperature and results in reducing hardness and strength. Under optimum precipitation-hardening conditions (optimum strengthening), the second phase is too fine to see (resolve) under an optical microscope. Overaging usually develops second-phase precipitates to a point where the precipitates are visible (resolvable).

Eutectoid Reaction. A *eutectoid reaction* is the isothermal transformation of a higher temperature solid phase into two new solid phases. A eutectoid reaction is expressed as alpha yields beta plus gamma ($\alpha \rightarrow \beta + \gamma$), when cooling at the eutectoid reaction temperature.

A eutectoid reaction is similar to a eutectic reaction, which involves the liquid phase. The tendency for solid-state reactions to occur is much lower than for liquid-state reactions and requires undercooling below the transformation temperature to drive the reaction. The speed of a eutectoid reaction increases with the increase of the amount of undercooling.

Hypoeutectoid alloys are to the left of a eutectoid composition, and hypereutectoid alloys are to the right of a eutectoid composition. A typical eutectoid structure is lamellar (laminations of the two phases). The lamellar structure is sometimes called pearlitic because it resembles the pearlite structure in carbon steels.

The finer the grain size of the transforming phase, the faster the transformation. The fine grain structure presents a greater number of grain boundary sites for the nucleation (initiation) of new phase transformations. The eutectoid structure grows in small nodules from each nuclei until the whole of the structure available for transformation consists completely of laminations.

The most important eutectoid reaction occurs in the iron-iron carbide alloy system and is illustrated on the iron-carbon diagram. The iron-carbon diagram is the basic phase diagram for carbon steel. Iron and carbon form the intermediate phase cementite (Fe_3C) at 6.7% C. See Figure 6-23. A steel containing .77% C transforms isothermally at 730°C (1346°F) from gamma phase to a lamellar structure of alpha plus Fe_3C ($\alpha + Fe_3C$).

A hypoeutectoid carbon steel containing .4% C is a single phase (γ-phase) between 788°C and 1454°C (1450°F and 2650°F). When cooling below the solvus, the alpha separates. At the eutectoid reaction temperature of 727°C (1340°F), the amount of alpha is 49%. The formula for finding percent alpha is ($[.77 - \%C] \div [.77 - .02]$) $\times$ 100 = %α, or ($[.77 - .4] \div [.77 - .02]$) $\times$ 100 = 49%. The remaining 51% gamma transforms isothermally to a lamellar structure of alpha plus Fe_3C ($\alpha + Fe_3C$).

A hypereutectoid steel containing 1.2% C is solely gamma between 871°C and 1343°C (1600°F and 2450°F). When cooling below the solvus, the Fe_3C

separates. At the eutectoid reaction temperature, the amount of Fe₃C is 7%. The formula for finding percent Fe₃C is ($[\%C - .77] \div [6.67 - .77]) \times 100 = \%Fe_3C$, or ($[1.2 - .77] \div [6.67 - .77]) \times 100 = 7\%$. The remaining 93% gamma transforms isothermally to a lamellar structure of alpha plus Fe₃C ($\alpha + Fe_3C$). It is possible to estimate the approximate carbon content of carbon steels by viewing microstructures of the metals. The carbon content is approximated from the proportions of alpha, eutectoid, and Fe₃C in the microstructure.

Proeutectoid phase is the phase that forms between the solvus and eutectoid temperatures. This formation range distinguishes it from its lamellar counterpart, which forms below the eutectoid temperature. For example, a metal with large grain size that has been subjected to rapid cooling rates, such as carbon steel welds, will have a proeutectoid phase. This metal will likely exhibit a Widmanstätten structure.

Peritectoid Reaction. A *peritectoid reaction* is the isothermal transformation of two solid phases to one solid phase when a metal is cooled. Peritectoid reaction is expressed as alpha plus beta yields gamma ($\alpha + \beta \rightarrow \gamma$) when cooling at the peritectoid reaction temperature.

A peritectoid reaction is similar to a liquid-phase peritectic reaction. In addition to other types of transformation products, peritectoid reactions may lead to the formation of intermediate phases or terminal solid solutions.

TERNARY PHASE DIAGRAMS

Ternary phase diagrams are equilibrium or constitutional diagrams that indicate the phases present in ternary alloy systems, or alloys consisting of three components (chemical elements). A ternary alloy system is a complete series of compositions produced by mixing three components (metallic or nonmetallic) in all proportions. Ternary alloy systems undergo similar phase transformations to binary systems, such as the eutectic, peritectic, eutectoid, peritectoid, etc. Since three components are involved, these phase transformations produce up to three phases instead of the two phases produced by two components. Ternary phase diagrams are illustrated on triangular coordinates, or isothermal sections (isotherms); on three-dimensional coordinates, or space diagrams; and on binary coordinates, or pseudo-binary diagrams (isopleths). See Figure 6-24.

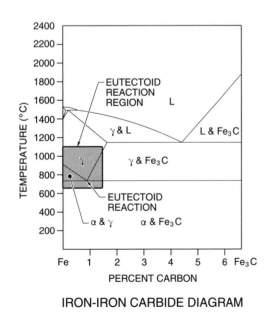

IRON-IRON CARBIDE DIAGRAM

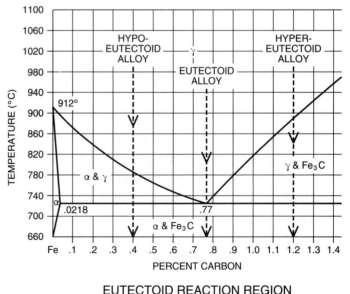

EUTECTOID REACTION REGION

Figure 6-23. The eutectoid reaction is the most important region of the iron-carbon diagram. This region is the iron-iron carbide diagram.

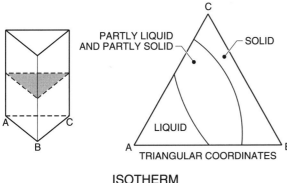

ISOTHERM

(TRIANGULAR
COORDINATES AT A
FIXED TEMPERATURE)

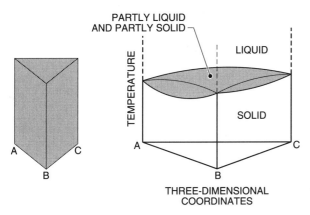

SPACE DIAGRAM

(THREE DIMENSIONS
SHOWING ALL COMPONENTS
AND TEMPERATURES)

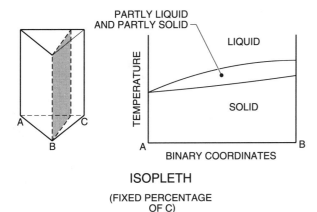

ISOPLETH

(FIXED PERCENTAGE
OF C)

Figure 6-24. Three ways to depict ternary phase diagrams are on triangular coordinates, on binary coordinates, and on three-dimensional coordinates.

Isotherms and isopleths are better visualized by relating them to the relevant three-dimensional space diagram. This diagram provides a complete picture

of a ternary alloy system, but it is difficult to use for practical applications. The space diagram is used to help understand the information displayed in isotherms and isopleths.

Isotherms

An *isotherm* (isothermal sections) is a section through a phase diagram that depicts all phases in equilibrium for an alloy composition at one temperature. An isotherm is the most widely used method of depicting ternary alloy systems. An equilateral triangle is drawn, and each side of the triangle represents the alloy systems of three binary phase diagrams. A triangle is constructed for each temperature to be studied. Isotherms are horizontal slices through the three-dimensional space diagram.

Ternary alloy compositions are obtained by drawing a series of lines parallel to the sides of the triangle through the alloy composition point. See Figure 6-25. The percentage of any component is given by the line opposite the apex of the triangle corresponding to 100% of the component. The intersection of this line with the sides of the triangle gives the percentage. The percentages of the three components must add up to 100%.

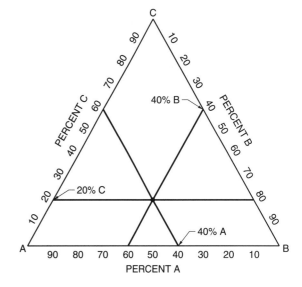

Figure 6-25. The percentages of each component of a ternary alloy are obtained by drawing lines through the alloy composition point parallel to the sides of the isotherm. The sum of the percentages must equal 100%.

Ternary Solid Solution Alloys. *Ternary solid solution alloys* solidify in a similar manner to the binary solid solutions. See Figure 6-26. At a temperature represented by point 1, alloy P is molten (the liquid phase field). Point 2 represents the liquidus surface where solid nuclei of alpha phase begin to form.

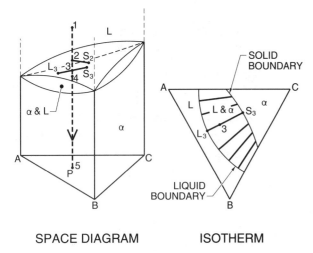

SPACE DIAGRAM   ISOTHERM

Figure 6-26. A space diagram illustrates the appearance of the component isotherms of a ternary solid solution alloy. Isotherms indicate a specific temperature during the cooling of the alloy when liquid phase and alpha phase are in equilibrium.

Point 3 represents a lower temperature where the alloy consists of a mixture of alpha plus liquid phases. An isothermal section at this temperature consists of three regions: liquid, liquid plus alpha, and alpha. The compositions of the liquid phase and solid phase in equilibrium at the temperature represented by point 3 are L_3 and S_3, respectively. Proportions of the phases are obtained from tie lines on the diagram.

To figure the percent of liquid phase, the distance between composition at P and the composition at S_3 is divided by the distance between composition at L_3 and the composition at S_3. The answer is then multiplied by 100. To figure the percent of solid phase, the distance between composition at P and the composition at L_3 is divided by the distance between composition at L_3 and the composition at S_3. The answer is then multiplied by 100.

Other tie lines are shown on the isothermal diagram, but they are not parallel and do not necessarily intersect the apex of alloy C. In ternary diagrams,

the composition of phases in equilibrium with each other and their tie lines must be determined experimentally. If the tie lines are not indicated on the diagram, the orientation must be estimated.

At point 4, alloy P is completely solidified. Point 5 represents alloy P at ambient temperature. The structure of the alloy is single-phase alpha.

Space Diagrams

The space diagram for a ternary eutectic alloy system consists of three binary eutectic systems. See Figure 6-27. The eutectic reaction temperatures are different in each system. Isothermal sections representing various points during the cooling of any alloy in the system from the liquid phase are also shown.

The cooling of alloy Q begins with the solidification of the gamma phase, which begins at the temperature of intersection with the liquidus surface (point 1). There are three, two-phase regions at the corners of the triangle and a central, single-phase liquidus region.

At a lower temperature represented by point 2, the liquid phase has a composition L_2 and the solid phase a composition of 100% alloy C. An experimentally determined tie line is unnecessary when a pure component of the alloy is in equilibrium with any phase. The tie line must connect with the apex of the triangle, which is the composition of the pure alloy C. The proportions of liquid and solid (gamma) are referenced from point 2.

To figure the percent of liquid phase, the distance between the composition at Q and the composition at C is divided by the distance between the composition at L_2 and the composition at C. The answer is then multiplied by 100. To figure the percent of solid phase, the distance between the composition at Q and the composition at L_2 is divided by the distance between the composition at L_2 and the composition at C. The answer is then multiplied by 100.

The temperature of the isothermal diagram representing point 2 is below the eutectic temperature of the B-C binary alloy system. Consequently, this isotherm exhibits a three-phase, liquid-plus-beta-plus-gamma $(L + \beta + \gamma)$ region. Three-phase regions are always triangular in shape.

At a temperature represented by point 3, which is below the eutectic temperature of the A-C binary alloy system, a binary eutectic of alpha and gamma begins to form from the liquid phase.

At a temperature represented by point 4, which is below the eutectic temperature for the A-B binary alloy system, the compositions of the three phases in equilibrium are given by the corners of the triangle, which surrounds alloy Q.

As the alloy cools further, the composition of the two solid constituents remains constant (as pure A and pure C). The composition of the liquid phase changes with the temperature. The crosses on the space diagram indicate how the liquid-phase composition changes with cooling. The locus of the composition of the liquid phase follows the valley formed between the two liquidus surfaces surrounding components A and C.

At the ternary temperature, represented by point 5 (TE on the space diagram), the remaining liquid precipitates as ternary eutectic (composed of alpha plus beta plus gamma). The composition of the remaining liquid at the ternary eutectic temperature is L. Composition L is at the junction of three valleys starting from three binary eutectic points.

No more phase changes occur when the alloy is further cooled. At ambient temperature, represented by point 6, alloy Q consists of a mixture of primary gamma (γ), binary eutectic alpha plus gamma ($\alpha + \gamma$), and ternary eutectic alpha plus beta plus gamma ($\alpha + \beta + \gamma$).

Plan Diagrams. The information from several isotherms may be combined in a plan diagram. A plan diagram indicates the valleys (boundaries) that separate primary phase fields in ternary or more complex alloy systems. See Figure 6-28.

Plan diagrams differ from isotherms because the boundary lines are not at constant temperature. The primary phase fields indicated on the plan diagram refer to the phases that first form on solidification. Isothermal lines drawn on the plan diagram indicate the boundaries of the liquidus surfaces at various temperatures. The extremes of the isothermal lines indicate the composition points of the eutectics, which are the valleys between the liquidus boundaries (eutectic composition). To simplify interpretation, arrows are drawn along the valleys to indicate

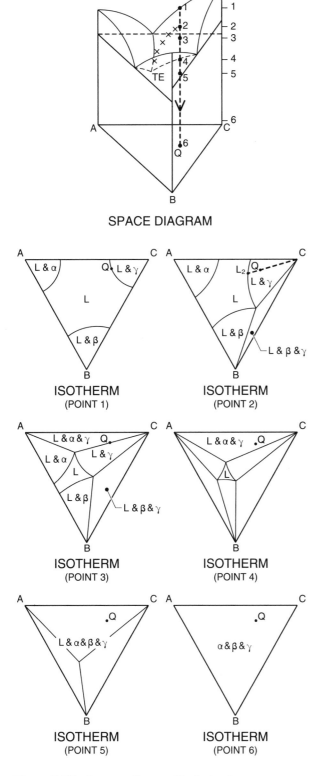

SPACE DIAGRAM

ISOTHERM
(POINT 1)

ISOTHERM
(POINT 2)

ISOTHERM
(POINT 3)

ISOTHERM
(POINT 4)

ISOTHERM
(POINT 5)

ISOTHERM
(POINT 6)

Figure 6-27. A space diagram illustrates the appearance of the component isotherms of a ternary eutectic alloy. Isotherms represent points during the cooling of alloys in the system.

the direction of decreasing temperature and the path of solidification when cooling.

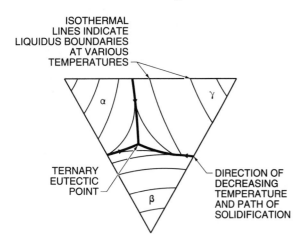

Figure 6-28. Isothermal lines indicate the boundaries of the liquidus surface at different temperatures.

Isopleths

Isopleths (pseudo-binary diagrams) are vertical sections through a space diagram, and they simplify ternary and more complex diagrams by indicating (freezing) the composition (ratio) of one or more alloying components. The freezing of the composition allows the behavior of any two alloying components to be displayed on binary temperature-composition axes.

An isopleth can represent a constant ratio of two alloying components and is illustrated by a tie line drawn through the apex of one of the components. An isopleth can also represent a constant proportion of one component, which would be a vertical section through a tie line drawn parallel to the side of the triangle. See Figure 6-29.

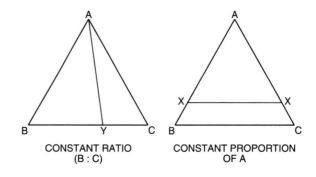

Figure 6-29. The two methods of constructing isopleths are constant ratio and constant proportion.

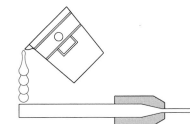

Effects of Plastic Deformation and Heat

7

Plastic deformation is an alteration of shape that remains permanent after removal of the applied load that caused the alteration. Metals deform by slip or mechanical twinning. The atomic arrangement in metals is not perfect because they contain lattice defects. Defects in the crystal lattice account for the difference between the theoretical and actual strengths of metals. Cold working (plastic deformation) exerts a significant effect on mechanical properties of metals. These effects can be removed by heat-treatment processes. Hot working (plastic deformation) also exerts a significant effect on mechanical properties of metals. Hot working is performed at higher temperatures than those used in cold working.

PLASTIC DEFORMATION AND STRUCTURE

Plastic deformation is an alteration of shape that remains permanent after removal of the applied load that caused the alteration. Plastic deformations may be in the form of slip or mechanical twinning. They are affected by different types and amounts of lattice defects present in the metal being deformed.

Slip

Slip is a process of plastic deformation in which one part of a metal crystal (grain) undergoes a shear displacement relative to another. When a metal is plastically deformed, atoms are displaced by slip and may be observed in polished specimens. The close-packed planes of atoms and slip bands in the crystal structure are the most favorable for slip.

Slip is dependent on the critical resolved shear stress (force bonding planes of atoms) and the number of slip systems available in the specific crystal structure. Slip is produced in a manner that preserves the crystal structure of the metal. See Figure 7-1.

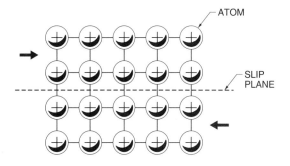

BEFORE SLIP

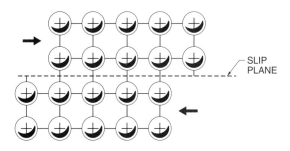

AFTER SLIP

Figure 7-1. During slip, one part of a metal crystal undergoes a shear displacement relative to another that preserves the crystal structure of the metal.

Slip occurs when the critical resolved shear stress of a metal is exceeded. Each type of crystal structure exhibits specific crystallographic planes (slip systems) that are optimum for slip. Metals with a low critical resolved shear stress and a large number of slip systems are relatively easy to deform plastically. When stressed, single crystals tend to rotate to make the crystallographic planes suitably oriented for slip. In polycrystalline metals (metals consisting of many crystals), the crystals with the most suitably orientated crystallographic planes are the first to exhibit slip.

Slip Bands. *Slip bands* are groups of closely spaced, parallel slip displacements that appear as single lines when observed under the optical microscope. As a metal is plastically deformed, certain parallel crystallographic planes undergo slip, much like the shearing (shuffling) of a pack of playing cards. The bulk of the material between the planes retains its original crystallographic orientation. See Figure 7-2.

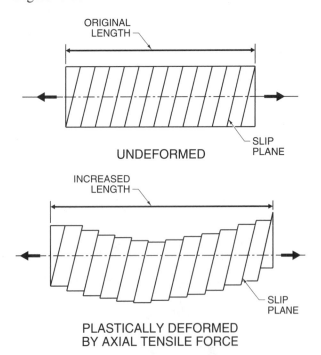

Figure 7-2. When a metal undergoes plastic deformation, certain parallel crystallographic planes undergo slip. The bulk of the material between the planes retains its original crystallographic orientation.

Slip bands change direction at the grain boundary because each grain has a different crystallographic

orientation. A polished metal surface subjected to plastic deformation exhibits a series of parallel lines, which are clusters of steps where the slip of parallel crystallographic planes has occurred. These clusters of steps are so close together that they appear as single lines in the structure. See Figure 7-3.

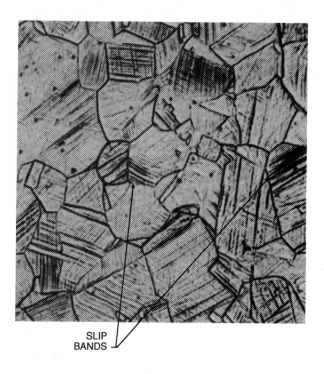

Figure 7-3. Slip bands change direction at the grain boundary because each grain has a different crystallographic orientation.

Critical Resolved Shear Stress. *Critical resolved shear stress* is the resolved shear stress required to cause slip in a designated slip direction on a given slip plane. Slip is induced by a tensile or compressive force on the crystal. Slip occurs when the resolved shear stress on a particular set of crystallographic planes reaches the critical resolved shear stress. Each metal has its own critical resolved shear stress (constant). For example, the critical resolved shear stress of copper is 71 psi and aluminum is 114 psi. The higher the critical resolved shear stress, the greater the resistance to plastic deformation.

To find the resolved shear stress acting on a set of crystallographic planes, the applied force is resolved (broken down) into two components (parallel and perpendicular to the slip planes). See Figure 7-4.

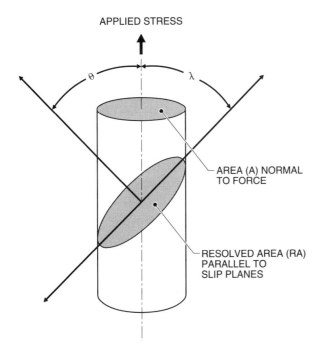

APPLIED STRESS

AREA (A) NORMAL TO FORCE

RESOLVED AREA (RA) PARALLEL TO SLIP PLANES

Figure 7-4. To figure the resolved shear stress acting on a set of crystallographic planes, the force applied is resolved into two components (parallel and perpendicular to the slip planes).

Angle lambda ($\angle\lambda$) is the resolved component that is parallel to the slip (shear) planes and equal to the angle of the slip planes. Angle theta ($\angle\theta$) is the resolved component that is perpendicular to the slip planes and equal to 90° minus $\angle\lambda$. Resolved shear stress is the force applied multiplied by the cosine of angle lambda ($cos\angle\lambda$) divided by the resolved area. Resolved area is the area (A) normal to the force applied divided by the cosine of angle theta ($cos\angle\theta$). The resolved area is parallel to the slip planes and $\angle\lambda$. Resolved area is found by applying the following formula:

$$RA = \frac{A}{cos\angle\theta}$$

where

RA = resolved area

A = area normal to force applied

$cos\angle\theta$ = cosine angle theta ($\angle\theta = 90° - \angle\lambda$)

Example: Figuring resolved area

What is the resolved area of a specimen that has an area 1 mm² and slip planes of 30° to the force applied ($\angle\lambda$)?

Figuring $\angle\theta$

$\angle\theta = 90° - \angle\lambda$

$\angle\theta = 90 - 30$

$\angle\theta = \mathbf{60°}$

Figuring resolved area

$$RA = \frac{A}{cos\angle\theta}$$

$$RA = \frac{1}{cos60}$$

$$RA = \frac{1}{.500}$$

$RA = \mathbf{2}$

Resolved shear stress is found by applying the following formula:

$$RSS = \frac{F \times cos\angle\lambda}{R}$$

where

RSS = resolved shear stress

F = force

$cos\angle\lambda$ = cosine angle lambda

RA = resolved area

Example: Figuring resolved shear stress

A specimen has an area 1 mm² and slip planes of 30° to the force applied ($\angle\lambda$) of 10 g/mm². What is the resolved shear stress?

Figuring $\angle\theta$

$\angle\theta = 90° - \angle\lambda$

$\angle\theta = 90 - 30$

$\angle\theta = \mathbf{60°}$

Figuring resolved area

$$RA = \frac{A}{cos\angle\theta}$$

$$RA = \frac{1}{cos60}$$

$$RA = \frac{1}{.500}$$

$RA = \mathbf{2}$

Figuring resolved shear stress

$$RSS = \frac{F \times cos\angle\lambda}{RA}$$

$$RSS = \frac{10 \times cos30}{2}$$

$$RSS = \frac{10 \times .866}{2}$$

$$RSS = \frac{8.66}{2}$$

$$RSS = \textbf{4.330 g/mm}^2$$

Slip Systems. Slip systems are the combination of slip planes and directions in the crystal lattice in which plastic deformation by slip occurs most favorably. Compared with any other planes, the most favorable planes for slip in any crystal structure are the closest (planes of atoms that are the closest together) of the close-packed planes. Close-packed planes are the most separated from similar parallel planes, so the interatomic bonding forces between close-packed planes are the weakest in the crystal lattice. The close-packed planes have the lowest critical resolved shear stress.

Slip systems for face-centered cubic (FCC), body-centered cubic (BCC), and close-packed hexagonal (CPH) crystal structures are individually different. Slip systems are indicated by the Miller indices des-ignations, which indicate planes and direction of planes in the crystal.

FCC metals have a total of 12 slip systems, con-sisting of four sets of (111) slip planes multiplied by three close-packed ⟨110⟩ slip directions. See Fig-ure 7-5. Since these slip systems are well-distributed in space, it is almost impossible to strain an FCC crystal without having at least one (111) plane in a favorable orientation for slip. The critical resolved shear stresses for FCC metals are low and these metals are easy to deform plastically. Silver, gold, copper, and aluminum are examples of FCC metals.

BCC metals also have 12 slip systems, consisting of six sets of (110) slip planes multiplied by two ⟨111⟩ slip directions. BCC metals do not have a truly close-packed plane where slip is relatively easy, so they actually have fewer slip systems and do not show high plasticity. Molybdenum, alpha iron, and tungsten are examples of BCC metals.

CPH metals have only three slip systems, con-sisting of the (0001) plane multiplied by three close-packed ⟨1120⟩ directions. Consequently, the critical resolved shear stress is relatively high. Other deformation mechanisms such as mechanical twinning also take place, so that CPH metals are more plastic than they would be if slip were the only available mechanism. Cadmium, magnesium, cobalt, and titanium are examples of CPH metals.

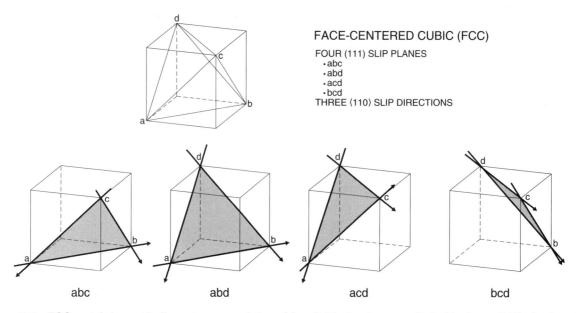

FACE-CENTERED CUBIC (FCC)

FOUR (111) SLIP PLANES
- abc
- abd
- acd
- bcd

THREE ⟨110⟩ SLIP DIRECTIONS

abc   abd   acd   bcd

Figure 7-5. FCC metals have 12 slip systems, consisting of four (111) slip planes multiplied by three ⟨110⟩ slip directions.

Crystal Rotation. *Crystal rotation* is the movement (rotation) of individual crystals that are under an applied force, so that the crystallographic planes move into the most favorable orientation for slip. Crystal rotation will facilitate plastic deformation in the metal. See Figure 7-6.

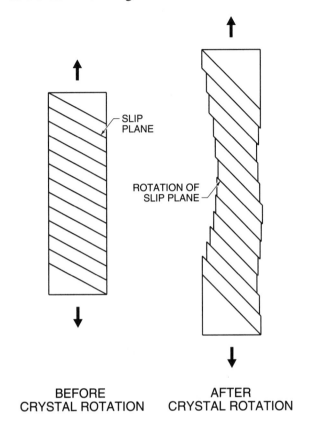

BEFORE CRYSTAL ROTATION   **AFTER CRYSTAL ROTATION**

Figure 7-6. Individual crystals tend to rotate under an applied force, so that the crystallographic planes move into the most favorable orientation for slip.

If the slip planes are parallel or perpendicular to the stress axis, slip will not occur. Materials will either deform by some other means (such as twinning) or fracture with no deformation. When the close-packed planes are at 45° to the stress axis, slip occurs easily because the critical resolved shear stress is equal to the applied force. In practice, neither of these extremes (deformation or fracture) is satisfied, and the close-packed planes are somewhere between the two extremes.

Polycrystalline Metals. *Polycrystalline metals* are metals that have the orientation of the crystallographic planes that vary from grain to grain. When a polycrystalline metal is stressed, slip originates in the grains with slip systems that are the most favorably oriented (45°) to the direction of applied force. Crystal rotation also brings more slip systems into favorable orientation. See Figure 7-7.

With small amounts of plastic deformation, a polycrystalline metal exhibits slip bands that change direction at grain boundaries. With increasing amounts of plastic deformation, the grains begin to elongate in the direction of stress.

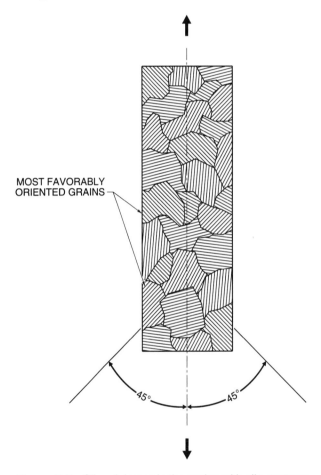

Figure 7-7. Slip originates in the grains with slip systems most favorably oriented (45°) to the direction of the applied force.

Mechanical Twinning

Mechanical twinning is the movement of planes of atoms in a lattice so that the two parts are mirror images of each other across the twinning plane (twin boundary). Mechanical twinning is a process of plastic deformation that assists slip and occurs in metals that do not have a large number of slip systems.

In mechanical twinning, a change in crystal shape occurs that is similar to the one in slip, but the crystal structure of the metal is preserved. See Figure 7-8. Mechanical twinning may accomplish plastic deformation or help to bring potential crystallographic planes into a favorable position for slip. Mechanical twinning is a major means of deformation in CPH metals that have relatively few slip systems. The noise (cry) produced when bending a tin sheet is a product of mechanical twinning.

The amount of atomic movement involved and the difference in microscopic appearance are the two major differences between slip and mechanical twinning. In slip, atoms move in whole blocks or whole numbers of atomic spaces. In mechanical twinning, atoms move in fractions of atomic spaces and the amount of space is proportional to the distance from the twinning plane.

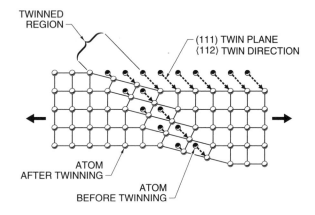

Figure 7-8. The crystal structure of a metal is preserved during mechanical twinning.

Under microscopic examination, slip appears as slip bands within the grains and mechanical twinning appears as distinct bands that are widely separated. See Figure 7-9. Parallel slip lines can be removed by polishing. Reetching does not bring them back into view. Mechanical twinning bands can also be removed by polishing, but reetching brings them back into view.

Lattice Defects

The atomic arrangement in metallic crystals is not perfect because they contain lattice defects. Lattice defects in metallic crystals give them relatively low strength and decrease their ability to harden with

plastic deformation. When there is significant relative motion of atoms, as in plastic deformation, defects in the crystal lattice play an important role in facilitating atomic movement. These defects are divided into point defects (vacancies and interstitial atoms) and line defects (dislocations and stacking faults).

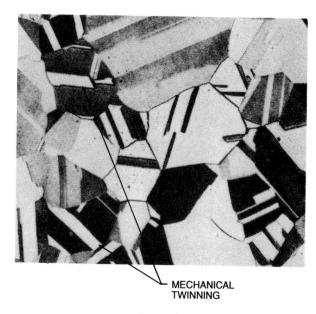

Figure 7-9. Slip appears as slip bands and mechanical twinning appears as bands in the microstructure.

Point defects are defects associated with discrete points in the crystal lattice and include vacancies and interstitial atoms. *Vacancies* are unoccupied lattice points. *Interstitial atoms* are smaller atoms of another element occupying a space between the atoms of the metal in the lattice. See Figure 7-10.

Line defects are defects associated with planes of atoms in the lattice and include dislocations and stacking faults. The major source of lattice defects are the two types of dislocations, which are edge and screw dislocations.

Theory of Dislocations. *Dislocations* are linear imperfections in a crystal structure. The theory of dislocation helps account for the way slip occurs in metals. The theory of dislocation states that if slip were to take place over a whole plane of atoms simultaneously, the critical resolved shear stress and the ultimate tensile strengths for metals would be 100 to 1000 times greater than found in practice.

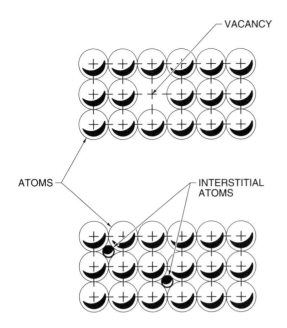

Figure 7-10. Point defects include vacancies and interstitial atoms.

Slip does not actually take place with complete planes of atoms sliding over one another. Slip instead takes place as each atom moves one place at a time. The movement of each atom during slip can be illustrated by describing the resisting force associated with sliding a heavy rug across a floor. If the rug is moved in one piece, the resistance to movement is relatively great. The rug, however, is moved with relative ease if wrinkles are made in the rug, and it is slid across the floor a little at a time by pushing each wrinkle along. Each wrinkle moves one place at a time like each atom move one place at a time during slip. See Figure 7-11.

The theory of dislocations accounts for the actual relatively low strength of metals compared with the theoretically high strengths. Edge dislocation and screw dislocation are the two main types of dislocations. Dislocations are often evident as combinations of both types.

A dislocation is specified by the orientation of the fault in the crystal caused by the dislocation line, together with the magnitude and direction of the associated shear. *Burgers vector* (*b*) is the displacement of the material above the slip plane relative to the material below. See Figure 7-12. The size of the Burgers vector is obtained from the atomic mismatch when the dislocation is enclosed along an atom-by-atom path. Burgers vector is the distance between where the atom-by-atom path originates to where it finishes. If no dislocation were present, the path would finish where it originated. The presence of the dislocation results in an additional atomic step being required for enclosure. If a dislocation were present, the path would not finish where it originated.

Edge dislocations are indicated by rows of mismatched atoms along a straight edge and are formed by extra partial planes of atoms within the body of a crystal. Edge dislocations have an extra half row of atoms inserted in the crystal. If the row of atoms were inserted above the slip plane, the atoms in the crystal would be squeezed together above the slip plane and spread out below it. This is represented by the perpendicular symbol (⊥). Alternatively, there may be an extra row of atoms below the slip plane, and this is represented by the inverted perpendicular symbol (⊤).

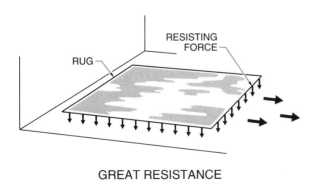

GREAT RESISTANCE

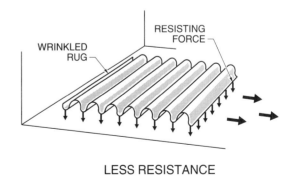

LESS RESISTANCE

Figure 7-11. The movement of each atom during slip can be illustrated by describing the resisting force associated with sliding a heavy rug across a floor.

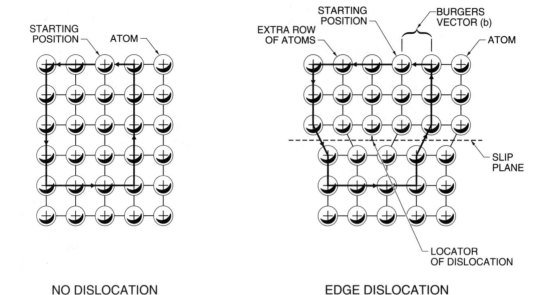

Figure 7-12. Edge dislocation is characterized by the Burgers vector (b), which is obtained by enclosing the dislocation in an atom-by-atom path and measuring the change in distance and direction from the starting position.

When a crystal containing an edge dislocation is mechanically loaded, the dislocation moves one Burgers vector (one atomic spacing). See Figure 7-13. The movement of the dislocation continues until it exits at a free surface. At this point, the crystal has deformed by one atomic spacing. The movement of edge dislocations across the slip plane explains the relatively low stress required for slip and fracture of metals as compared to the theoretical values of stress required.

A *screw dislocation* corresponds to the distorted lattice adjacent to the axis of a spiral structure in a crystal. The path taken around the dislocation axis is a helix (screw shape). See Figure 7-14.

Both a screw dislocation and an edge dislocation represent the boundary between slip regions and regions without slip in a crystal. When dislocations travel, there is slip of the atoms on the opposite sides of the plane of travel. The boundary between the slip region and the region without slip is perpendicular to the slip direction in an edge dislocation. Movement of the edge dislocation causes relative shear displacement in a direction perpendicular to itself. The boundary between the slip region and the region without slip is parallel to the slip direction in a screw dislocation. Movement of the screw dislocation causes a relative shear displacement in a direction parallel to itself.

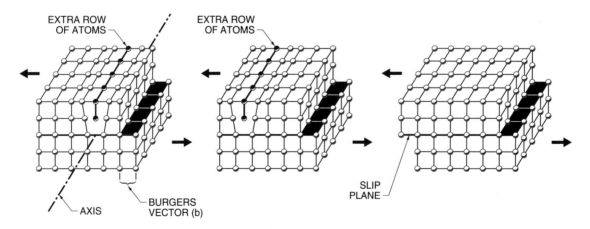

Figure 7-13. The mechanism of slip caused by movement of an edge dislocation greatly lowers the stress required for slip.

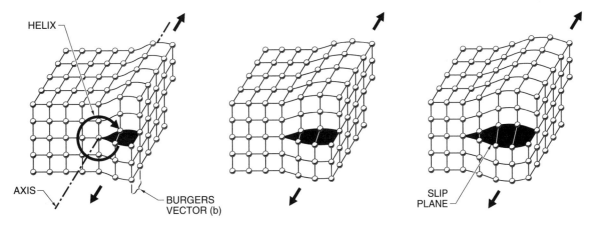

Figure 7-14. The path taken around the axis of the screw dislocation is a helix.

Dislocations are usually a combination of edge and screw types. See Figure 7-15. Enormous numbers of dislocations occur in materials and are measured as the dislocation density, which is the number of dislocations intersecting a random cross-sectional area of the material. Dislocations are first formed during the production of the metal. For example, a metal crystal that is newly formed by vapor deposition may contain 10^5 dislocations per mm^2. Plastic deformation greatly increases the dislocation density, so that a heavily cold-worked metal may contain more than 10^{11} dislocations per mm^2.

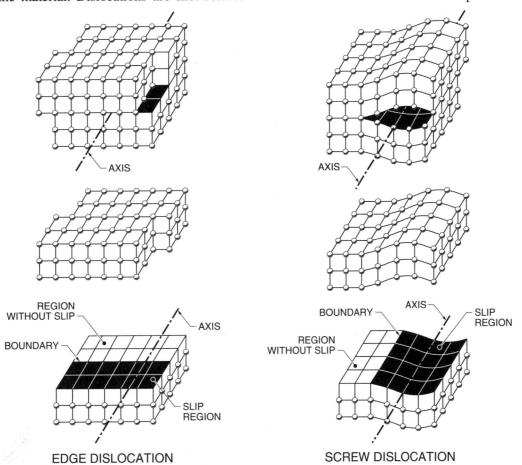

Figure 7-15. The boundary between the slip region and the region without slip is perpendicular to the axis of an edge dislocation and parallel to the axis of a screw dislocation.

Stacking Faults. *Stacking faults* are two-dimensional deviations from the normal stacking sequence of atoms in a crystal. They may be formed during the growth of a crystal or may also result from partial dislocations. Stacking faults are most common in close-packed planes of atoms. For example, the stacking sequence of close-packed layers in an FCC structure may contain a stacking fault that is the equivalent to a layer (two atoms thick) of CPH inserted into the FCC structure. See Figure 7-16.

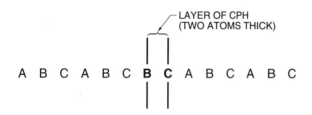

LAYER OF CPH
(TWO ATOMS THICK)

A B C A B C **B C** A B C A B C

STACKING FAULT IN
FCC STRUCTURE

Figure 7-16. Stacking faults are most common in close-packed planes of atoms.

PLASTIC DEFORMATION AND MECHANICAL PROPERTIES

Cold working is plastic deformation by controlled mechanical operations that is performed below the recrystallization temperature for the purpose of shaping a product. Plastic deformation performed below the recrystallization temperature has a distinct effect on the mechanical properties of a metal. Plastic deformation increases the strength of a metal and reduces its ductility by a process known as strain hardening, or work hardening. Strain hardening leads to anistropy (the development of directional properties). Cold-worked metals contain residual stresses. Residual stresses may present problems, such as distortion, during further processing or during operation.

Strain Hardening

Strain hardening is the increase in strength and hardness of a metal and the corresponding decrease in ductility due to plastic deformation by cold working. When a metal is cold-worked, slip occurs in preferentially oriented crystals and extends to other crystals, which have rotated or mechanically twinned to bring the slip systems into favorable orientation.

With increased cold working, the grains elongate in the direction of working. The greater the cold working (greater reduction of the cross-sectional area), the more elongated the grains become. See Figure 7-17. The dislocation density increases and the dislocations begin to intersect and tangle. These factors increase the resistance of the metal to plastic deformation, cause strain hardening, and are in sharp contrast to the facilitating role of dislocations in the early stages of plastic deformation.

Cold working causes strain hardening. Extreme cold working leaves few crystallographic planes unaffected by slip. As cold working increases, the hardness, yield strength, and tensile strength are increased. However, as cold working increases, the ductility of the metal decreases (becomes extremely brittle). See Figure 7-18. These properties are not only affected by the amount of cold working but may vary in relation to the direction of the cold working. Cold working makes metals anisotropic.

Anisotropy. *Anisotropy* is a characteristic of metal that exhibits different properties when measured in different directions in relation to the direction of cold working. The greater the amount of cold work, the greater the anisotropy. Mechanical properties become distinctly different in the directions parallel and perpendicular to the cold-working direction.

The metal parallel to the direction of the cold working exhibits an increase in tensile strength, yield strength, and hardness. The percent elongation, percent reduction in area, and notch toughness are reduced. The opposite occurs for mechanical properties measured in the perpendicular (transverse) direction.

Anisotropy is desirable only if a cold-worked metal is loaded (subject to an applied force) in a way that uses the increased strength developed in the cold-working direction. In applications that require uniform properties in all directions (isotropy), a metal may require an annealing heat treatment to fully soften it and restore uniformity. This is the case with deep drawing. With deep drawing, anisotropic properties may lead to the development of local thin spots and failure.

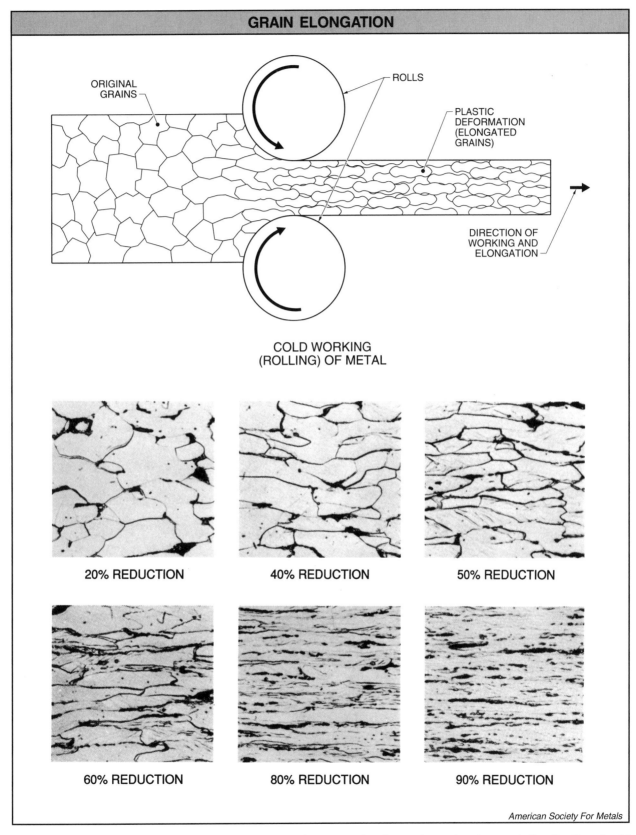

Figure 7-17. With increased cold working or reduction of cross-sectional area, the grains elongate in the direction of cold working.

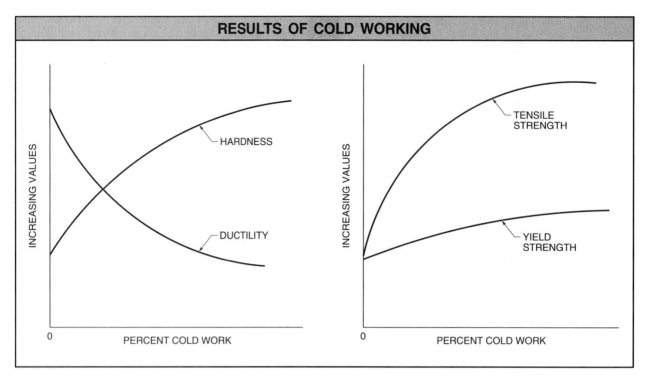

RESULTS OF COLD WORKING

Figure 7-18. The increase in strain hardening is the result of the increase of cold working.

Anisotropy developed from the elongation of the grains during cold work can be eliminated by an annealing heat treatment. *Fiber structure* is the elongation of inclusions and impurities in the direction of cold work. This anisotropy development is irreversible and subsequent annealing does not alter the morphology. Anosotropy may be revealed by sectioning the component in the longitudinal direction. The specimen is then ground and macroetched.

Residual Stresses. *Residual stresses* are stresses that remain within a metal as a result of plastic deformation. They may also be caused by other processing operations, such as casting or welding, and are generally undesirable. The surface layers of the metal are frequently in tension as a result of plastic deformation. When the surface layers are removed by machining, the compressed subsurface layers expand and raise the dimensional instability.

Components in a state of residual stress may crack (stress corrosion cracking) if the component is left in a corrosive environment for an extended period of time. Residual stress may be reduced or eliminated by a stress-relieving heat treatment.

HEAT, STRUCTURE, AND MECHANICAL PROPERTIES

Specific heat treatment processes are applied to cold-worked metals to change the structure and mechanical properties of the metal. These processes are collectively referred to as annealing. *Annealing* is a heat treatment process used to soften a metal. Annealing consists of heating a component to a suitable temperature, holding at temperature, and then cooling at a suitable rate. During annealing, the metal reverts to a softer condition and three major changes occur in the crystal and grain structure (recovery, recrystallization, and grain growth). *Hot working* is plastic deformation by controlled mechanical operations performed above the recrystallization temperature for the purpose of shaping a product. Hot working is performed at higher temperatures and does not cause strain hardening of metals, which is characteristic of cold working.

Annealing

A variety of heat treatment processes are referred to as annealing. See Figure 7-19. Annealing is performed for purposes such as reducing hardness;

improving machinability; facilitating cold working; producing a desired microstructure; or obtaining desired physical, mechanical, or other properties.

Annealing increases the strength and hardness of metals, but in some cases it is undesirable. For example, the increase in residual stresses from cold working enhances susceptibility to stress corrosion cracking or dimensional instability. Additionally, the strain hardening induced by cold working often hinders mechanical reduction operations by making a metal too hard and too brittle to undergo further working. This type of problem occurs in severe cold-working operations such as wire drawing and tube drawing.

The annealing process should be indicated in any written procedure to qualify the specific process used. When used without qualification, annealing implies a heat treatment process designed to soften a structure that has been hardened by cold working. As the annealing temperature is raised, the metal reverts to a softer condition and three major changes occur in the crystal and grain structure. These major changes are recovery, recrystallization, and grain growth.

Recovery. *Recovery* (stress relieving) is the reduction of residual stresses in a cold-worked, a fabricated, or a cast component that is affected by holding the component at an elevated temperature. Recovery is the first noticeable effect as the annealing temperature is raised. Recovery occurs without any apparent change in the mechanical or physical properties except for electrical conductivity, which usually rises rapidly. The effect of recovery is observed using X-ray diffraction. As recovery proceeds, the elongated spots on the film (indicative of residual stress in cold-worked metal) turn into rounded, smaller spots.

During recovery there is some *polygonization* (reduction in dislocation density), which is caused by the formation of minute subgrains (about 1 micron across) within the stressed grains. Recovery occurs over a specific temperature range for each metal or

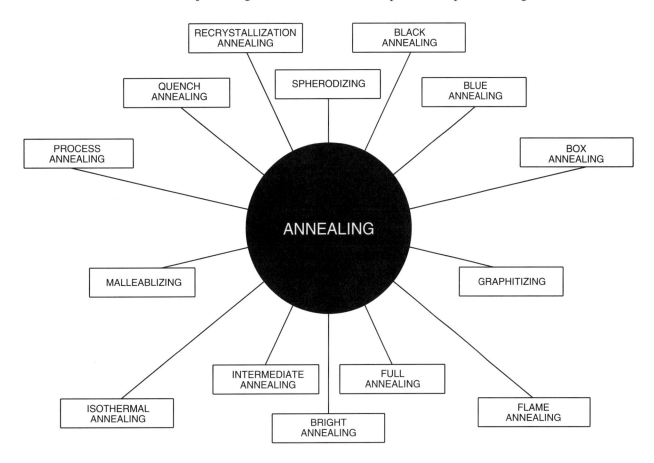

Figure 7-19. Annealing refers to a variety of specific heat treatment processes.

alloy family. The principal effect of recovery is the relief of internal stresses.

Stress-relieving operations consist of heating a component to a suitable temperature, holding it long enough to reduce residual stresses, and cooling it slowly enough to minimize the development of new residual stresses. Stress relieving is a common form of heat treatment and is performed on welded structures, castings, rough-machined parts, heat-treated components, etc. See Figure 7-20.

SELECTED STRESS-RELIEVING TEMPERATURE RANGES		
Alloy	**°C**	**°F**
Gray cast iron	540 to 565	1000 to 1050
High-carbon steels (wire for prestressed concrete)	315 to 425	600 to 800
Austenitic stainless steels	900	1650
Titanium alloys	480 to 595	900 to 1100
Aluminum alloys	345 to 400	650 to 750
Copper alloys	190 to 290	375 to 550

Figure 7-20. Stress-relieving temperature ranges vary for different types of alloys.

The longer the holding time at a particular stress-relieving temperature, the greater the percent of residual stress removed. Stress-relieving specifications always indicate a required holding time at temperature. See Figure 7-21.

Recrystallization. *Recrystallization* is the formation (usually accomplished by heating) of a new strain-free grain structure from an existing grain structure in a cold-worked metal. New grains first appear at the most severely deformed regions of the existing grains, which are usually the grain boundaries or slip planes.

Recrystallization leads to a sharp drop in strength and a corresponding rise in ductility. Although recrystallization occurs above the temperature range that is required for recovery, the upper recovery temperature overlaps the lower recrystallization temperature.

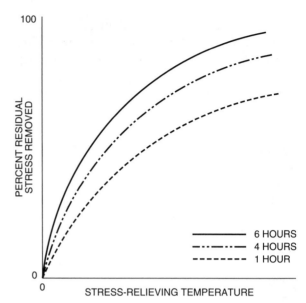

Figure 7-21. The percent residual stress removed that is achieved is related to the stress-relieving temperature and the time at temperature.

It is possible to remove residual stresses without significantly altering the mechanical properties. The altering of mechanical properties occurs during recrystallization. Residual stress relief is possible providing that recrystallization is not substantial. See Figure 7-22.

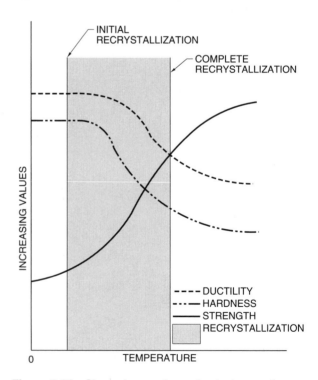

Figure 7-22. Sharp changes in mechanical properties are exhibited during recrystallization.

Unlike recovery, which starts almost immediately when heated, an incubation period is required to initiate recrystallization. Following the incubation period, recrystallization starts slowly and gains momentum. See Figure 7-23.

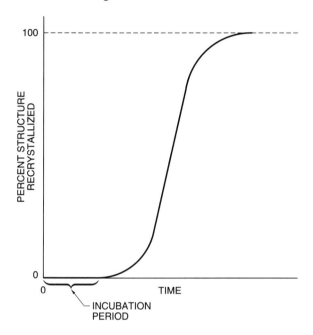

Figure 7-23. Recrystallization requires an incubation period to initiate the process.

Time at temperature, percent of cold working, purity of the material, rate of cold working, and temperature of cold working are factors that influence recrystallization. If the time at temperature, percent of cold working, purity of the material, or rate of cold work is increased, the temperature of cold working (recrystallization temperature) will decrease because of these factors.

The recrystallization temperature cannot be fixed. To provide a standard, recrystallization temperature is defined as the approximate minimum temperature at which full recrystallization occurs in 1 hour for a specified set of mechanical and metallurgical conditions. With increasing amounts of cold working, the recrystallization temperature and recrystallized grain size fall to minimum values. See Figure 7-24.

Metal purity has a profound effect on the recrystallization temperature. This effect is illustrated using the various levels of purity of aluminum. The recrystallization temperature of very pure aluminum ($\geq$99.9999% Al) is below room temperature. Super purity aluminum ($\geq$99.999% Al) has a recrystallization temperature of 80°C (176°F). For commercially pure aluminum ($\geq$99.0% Al), the temperature is 288°C (550°F).

For some metals, the recrystallization temperature is close to or below ambient temperature, so that the metals cannot be strengthened by cold working at ambient temperature. If attempted, the metal will recrystallize spontaneously, forming a strain-free grain structure.

Process annealing (heat treatment term for recrystallization) is used in the ferrous and nonferrous sheet and wire industries. Process annealing restores the ductility to cold-worked materials and permits

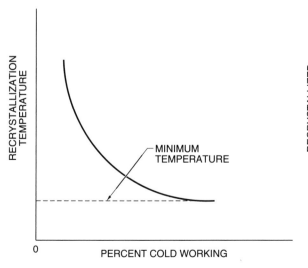
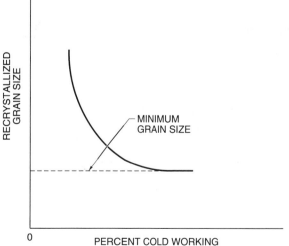

Figure 7-24. Recrystallization temperature and recrystallized grain size fall to minimum values with increasing amounts of cold working.

further cold working to achieve the required deformation or reduction.

Annealing twins are twin bands (mirror images) formed in the grain structure during recrystallization of certain cold-worked FCC metals. See Figure 7-25. Annealing such metals, notably copper-base alloys and the 300 series austenitic stainless steels, produces stacking faults, which are crystallographically equivalent to twins.

Annealing twins are evident in the microstructure after polishing and etching and have parallel bands across the grains (similar to mechanical twins). Annealing twins are evidence of prior cold work. For example, annealing twins are not present in an annealed casting.

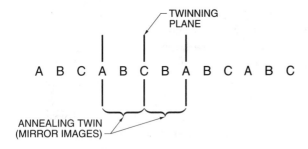

A B C A B C A B C A B C

SEQUENCE OF CLOSE-PACKED PLANES
(FCC STRUCTURE)

TWINNING
PLANE

A B C A B C B A B C A B C

ANNEALING TWIN
(MIRROR IMAGES)

SEQUENCE OF CLOSE-PACKED PLANES
WITH ANNEALING TWIN

Figure 7-25. A stacking fault formed after annealing of certain cold-worked FCC metals leads to the formation of an annealing twin (mirror images).

Grain Growth. *Grain growth* (grain coarsening) is an undesirable increase in grain size that occurs as a result of heating a metal to elevated temperatures. Grain growth occurs if the temperature is raised sufficiently above the recrystallization temperature, at which point the recrystallized grains absorb one another. This absorption causes individual grains to enlarge significantly.

As the temperature is raised and the time at temperature is increased, the grains continue to grow. Grain growth eventually stabilizes and attains a lim-

iting value for any specific time-temperature combination. Grain growth does not continue until the entire structure consists of one enormous grain. See Figure 7-26.

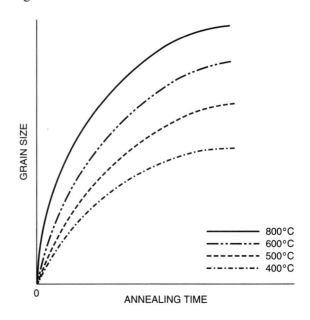

Figure 7-26. Amounts of grain growth increase with increasing temperature and time at temperature. As both increase, grain growth will eventually stabilize.

A thermodynamic driving force for grain growth is the reduction of the energy of the system. Grain boundaries are high-energy regions because of the atomic mismatch. Energy of the system is reduced each time a grain is consumed because there is one less grain boundary.

Unlike recovery and recrystallization, prior cold working is not required to initiate grain growth. A specific amount of cold working will cause an abnormally high grain growth, which is much higher than would be obtained under any other conditions. See Figure 7-27. The specific amount of critical strain (cold working) is relatively small and varies for different materials. For example, the critical strain value for iron is 10%. At 10% cold working, abnormal grain growth is present in iron. Other critical strain values are 2% for aluminum and 1% for lead.

In most engineering applications, grain growth is undesirable because coarse grain size leads to loss of toughness in a material. Annealing temperatures are kept below the value that causes grain growth. In some specific applications, particularly in those requiring high-temperature strength (creep resistance), large grain size is advantageous. In these cases, the material is annealed at a temperature high enough to produce a large grain size (ASTM grain size 4 or coarser). An example is the high-temperature annealing of alloy 800H. Alloy 800H is used in this condition in a wide variety of high-temperature applications, such as hot gas piping.

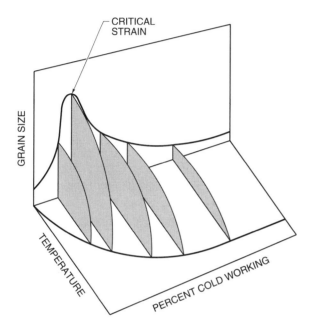

Figure 7-27. Critical strain produces abnormal grain growth in a material.

Hot Working

Hot working is plastic deformation by controlled mechanical operations performed above the recrystallization temperature for the purpose of shaping a product. When a material is plastically deformed, two opposing effects take place and their magnitude depends on the working temperature. The first is the hardening effect, which is caused by plastic deformation. The second is the softening effect, which is caused by recrystallization. Cold working takes place below the recrystallization temperature of a material, so there is no softening effect. Hot working takes place above the recrystallization temperature, so the hardening effect of plastic deformation is offset by the stress relaxation effect of recrystallization. If the hot-working temperature is high enough, the recrystallized grains also grow. See Figure 7-28.

Hot working and cold working have no relationship to the actual processing temperature. Lead and tin, which have recrystallization temperatures below ambient, may be hot worked at ambient temperature. However, steel may be cold worked above 538°C (1000°F) because the recrystallization temperature is above this value.

The recrystallization temperature increases with the rate of plastic deformation. If copper is slowly deformed, as in a tensile test, its recrystallization temperature is 399°C (750°F). However, rapid deformation, as in drop forging, raises the recrystallization temperature to 802°C (1476°F).

Hot working is performed just above the recrystallization temperature, but in most cases hot working is done at a significantly higher temperature than this value. A higher temperature is used to decrease the mechanical stresses required to perform the necessary plastic deformation, which in turn reduces the cost of the equipment.

High hot-working temperatures promote grain growth, which is usually undesirable in finished products. The final stages of reduction in hot-working operations must be conducted close to the recrystallization temperature to encourage the formation of fine grains.

Compared with cold working, hot working has a number of advantages. Advantages include the reduction in the power required for plastic deformation, break down and elimination of the undesirable cast structure of an ingot or billet, and redistribution of brittle films and constituents. Hot working welds up porosity in ingots or billets and improves the mechanical properties, especially strength. Hot working is widely used to reduce the size of ingots and to produce primary shapes, such as billets suitable for further processing.

A disadvantage of hot working is the development of directional properties from the alignment of inclusions, which is caused by plastic deformation. Another disadvantage is that a hot-worked metal may also contain internal laminations caused by the alignment of inclusions and slag. Hot-worked metals may also develop a thick oxide coating (scale), which is caused by a reaction of the metal with the air at high temperature. Scale hinders subsequent forming or machining operations and must be removed by grinding, blasting, or pickling (acid treatment).

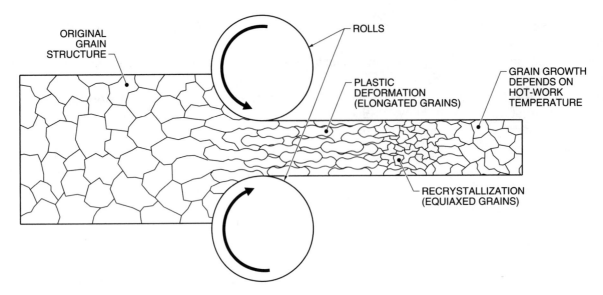

Figure 7-28. Hot working is plastic deformation by controlled mechanical operations performed above the recrystallization temperature of a material.

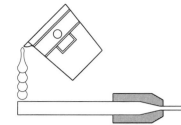

Fractography and Failure Analysis 8

Metals often fail by fracturing (developing cracks), which usually causes them to separate into two or more pieces. Fractography, the descriptive study of fracture surfaces, is of extreme importance in failure analysis investigations. Fractography reveals clues as to the causes of failures. Fractography is performed at macroscopic and microscopic levels of magnification. Failure analysis is the study of the causes of a failure of components and structures by fracture, distortion, seizure, corrosion, and/or wear. A failure analysis investigation is conducted because it is difficult to implement a solution to the problem without information as to the cause of the failure.

FRACTOGRAPHY

Fractography is the descriptive study of fracture surfaces using photographs, sketches, and text. In fractography the fractured surface of a failed component is examined for characteristic features (fingerprints), which reveal the mechanical loading and environmental conditions acting on the metal. Fractography is performed at macroscopic and microscopic levels of magnification. Macroscopic examinations are performed with the naked eye, magnifying glass, or light microscope. Macroscopic examinations reveal the fracture surface orientation. Microscopic examinations are performed with a scanning electron microscope and reveal the fracture surface morphology. See Figure 8-1.

Fracture Surface Orientation

Fracture surface orientation is the angular relationship between a fracture and the direction of the applied stress. Fracture surface orientation is dependent on the type of stress and the ductility of a metal. Fracture surface orientation is revealed by macroscopic examination. Different types of stress produce characteristic fracture orientations in ductile and brittle metals.

FRACTOGRAPHY		
Magnification Level	**Examination Method**	**Magnification Range**
Macroscopic	Naked eye	1X
Macroscopic	Magnifying glass	10X to 20X
Macroscopic	Light microscope	3X to 50X
Microscopic	Scanning electron microscope	15X to 15,000X

Figure 8-1. Fractography is performed at macroscopic and microscopic levels of magnification.

A metal subjected to an external mechanical force develops a stress, which is its internal resistance to the external mechanical force. A metal fractures when the breaking stress is exceeded. Tension, compression, and torsion are the three most common types of force. These forces produce characteristic fracture orientations in ductile and brittle metals.

Tension. Ductile metals loaded in tension (tensile) tend to be stretched in the direction of the force. When enough force is applied, a ductile metal deforms plastically by permanently changing its shape to accommodate the stress. See Figure 8-2.

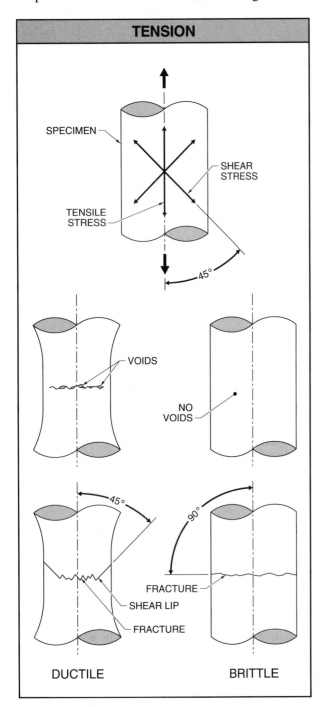

TENSION

SPECIMEN

SHEAR STRESS

TENSILE STRESS

45°

VOIDS

NO VOIDS

45°

90°

FRACTURE

SHEAR LIP

FRACTURE

DUCTILE

BRITTLE

Figure 8-2. Ductile metals loaded under tension will fracture at a 45° angle to the tensile stress. Brittle metals loaded under tension will fracture perpendicular to the tensile stress.

Plastic deformation occurs chiefly by shear. Shear is the movement of many millions of planes within the metal in relation to each other. Movement of these shear planes occurs at an angle of 45° to the tensile force.

When the shear stress exceeds the shear strength of a metal, a fracture occurs along the shear planes. For example, a ductile shaft of low-carbon steel loaded under tension will deform plastically along the shear planes. At a point during this process, the shear strength of the metal begins to be exceeded in tiny, localized regions. This deformation occurs easiest toward the center of the specimen. Tiny fractures form along the shear planes and create small voids in the metal. Voids link up to create a jagged crack that fans out toward the circumference of the specimen in a direction perpendicular to the tensile force. As the growing crack approaches the circumference, the remaining unaffected cross section of metal becomes incapable of supporting the load. Final fracture occurs rapidly at a 45° angle to the tensile force. Macroshear (shear lip) is the 45° final fracture. A shear lip is commonly observed in failures of ductile metals in the region of final separation.

Brittle metals loaded under tension cannot deform plastically by slip (like ductile metals), and they undergo little or no plastic deformation prior to fracture. When the tensile stress exceeds the force (cohesive strength) that holds the atoms together, a fracture is produced. For example, a brittle shaft of gray cast iron loaded under tension will fracture perpendicular to the tensile force with no shear lip present.

At the point where fracture first occurs in a ductile metal, the onset of void formation coincides with necking down. Necking down is the reduction of the cross-sectional area of a test specimen. As the elongation of the specimen proceeds toward failure, the necking down becomes more pronounced. Voids and necking down give warning of failure in a ductile metal under tension. A brittle metal gives little or no warning of failure. Necking down and a shear lip are not present in brittle metals. See Figure 8-3.

Compression. Ductile metals loaded under compression become shorter and wider as shear occurs along shear planes at a 45° angle to the direction of the tensile stress. For example, a ductile block of copper loaded under compression will bulge but does not usually fracture. Most mechanically worked metals are formed using compressive forces because metals can be subjected to much greater plastic deformation under compressive forces than under tensile forces.

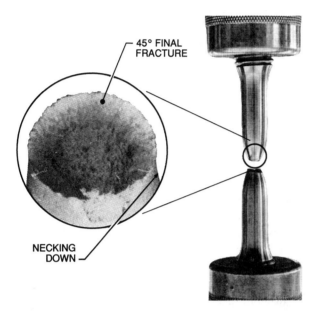

Figure 8-3. Necking down is the reduction of the cross-sectional area of a test specimen. Shear lip is the 45° final fracture of a test specimen.

Brittle metals loaded under compression exhibit a tensile stress in a direction perpendicular to the compressive force. See Figure 8-4. For example, a brittle block of gray cast iron that is loaded under compression and that has tensile stress exceeding the cohesive strength of the metal will fracture along planes perpendicular to the tensile stress (parallel to the compressive force). This method explains how brittle materials such as rocks are crushed.

Torsion. Ductile metals loaded under torsion (twisting) exhibit shear stress parallel to the direction of the torsional force. Shear occurs along the shear planes and plastic deformation occurs by twisting. When the shear stress exceeds the shear strength, the metal fractures in a direction parallel to the torsional stress. For example, a ductile shaft made of low-carbon steel that is loaded under torsion will fracture perpendicular to the axis of the shaft.

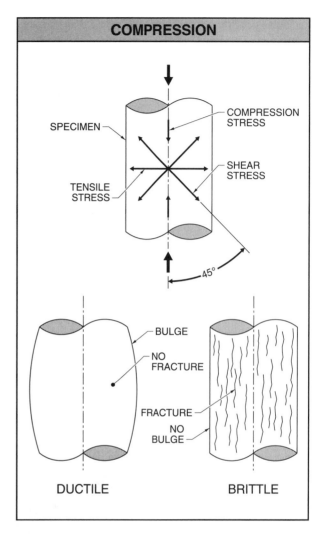

Figure 8-4. Ductile metals loaded under compression will bulge but do not usually fracture. Brittle metals loaded under compression will fracture along planes parallel to the compression stress but will not bulge.

Brittle metals loaded under torsion exhibit tensile stress at a 45° angle to the axis of the torsional force. See Figure 8-5. There is little or no plastic deformation by torsion before it fractures. When the tensile stress exceeds the cohesive strength of the metal, fracture occurs in a spiral at a 45° angle to the axis of the metal. For example, a brittle shaft of gray cast iron loaded under torsion exhibits a 45° angle spiral fracture. A stick of chalk exhibits the same fracture if carefully twisted.

Fracture Surface Morphology

Fracture surface morphology is the texture (topography) of the fracture surface. Unlike the fracture surface orientation, which is easily revealed by macroscopic examination, the morphology often requires microscopic examination for complete interpretation. The scanning electron microscope is most commonly used for interpretation of fracture surface morphology. The three most common types of fractures (ductile, brittle, and fatigue) exhibit distinctively different morphologies.

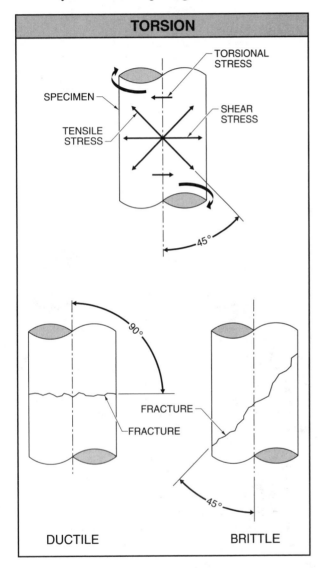

TORSION

Figure 8-5. Ductile metals loaded under torsion will fracture perpendicular to the axis of the metal (parallel to the torsional stress). Brittle metals loaded under torsion will fracture in a spiral at a 45° angle to the axis of the metal (parallel to the tensile stress).

Ridges or other markings in the morphology may be used to locate the origin(s) of the fracture. The roughness of the morphology indicates the relative speed of the fracture. The rougher the morphology, the greater the speed of the fracture.

A complete fracture surface may exhibit several different orientations and morphologies because the driving force for the propagation (growth of a fracture) may change as the fracture progresses. This change causes the orientation and morphology to change. There are usually three distinct regions (zones) on a completely fractured surface. The three regions are the initiation zone, propagation zone, and final failure zone.

Damage done to a fracture surface obscures the morphology and hinders fracture interpretation. Damage may occur during the course of the failure and/or during the failure analysis investigation.

Ductile Fracture. Ductile fractures begin on a microscopic scale at small imperfections (inclusions) in a metal, which produce weak areas that are prone to shear. *Microvoids* are tiny voids (cavities) formed as a metal separates in these weak areas prone to shear. The cross section of unaffected metal immediately adjacent to microvoids is reduced and subjected to higher shear stress, which increases the susceptibility of these regions to fracture. As these local regions fracture, the microvoids coalesce (join up).

Microvoid coalescence is the mechanism of ductile fractures that permits a fracture to zigzag across a metal surface. Microvoid coalescence is characterized by tiny voids (half cups) in either part of a fracture surface. *Ductile dimples* are a mass of tiny voids that make up the morphology of a ductile fracture surface. See Figure 8-6.

Brittle Fracture. Brittle fractures occur when the tensile stress resulting from the force applied to a metal exceeds the cohesive strength of the metal. Transgranular cleavage and intergranular separation are the two distinct brittle fracture morphologies that metals exhibit.

Transgranular cleavage is a brittle fracture morphology that exhibits a fracture that propagates within the grains along specific crystallographic planes. These specific crystallographic planes are planes along which atoms of a metal are aligned to create a crystal structure.

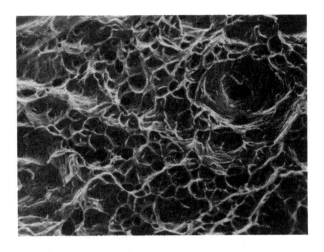

DUCTILE FRACTURE REVEALING
DUCTILE DIMPLES

Figure 8-6. Ductile dimples caused by microvoid coalescence are characteristic of ductile fractures.

Intergranular separation is a brittle fracture morphology that exhibits a fracture that propagates along the grain boundaries of a metal. The grain boundaries are formed by growing crystals within the metal during solidification or as modified by heat treatment and mechanical processing. See Figure 8-7. The brittle fracture morphology is dependent on whether the specific crystallographic planes or the grain boundaries are weaker.

Brittle fractures (catastrophic fractures) usually propagate rapidly with little warning. Brittle fractures are expected to occur in metals with low ductility or low toughness but are not expected to occur in normally ductile metals. Brittle fracture susceptibility in normally ductile metals may result from a number of causes such as improper heat treatment, exposure to a specific chemical environment, or operation below a specific temperature.

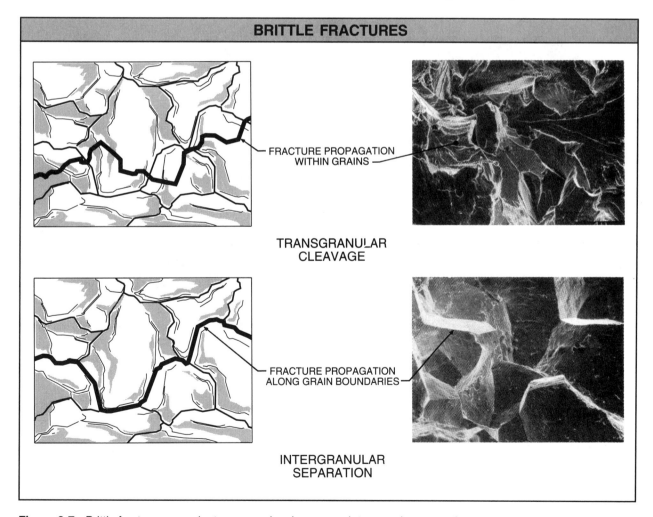

Figure 8-7. Brittle fractures occur by transgranular cleavage or intergranular separation.

Fatigue Fracture. Fatigue fractures occur under repeated or fluctuating stresses, which have a maximum value less than the tensile strength of the metal. A fatigue fracture propagates along a narrow crack front, which grows in stages under the action of the stresses.

Many fatigue fractures are progressive fractures and display arrest lines when examined macroscopically. *Progressive fractures* are fractures that grow in stages and have rest periods when no growth occurs (the component is idle or out-of-service). *Arrest lines* are imprints of the temporary positions of the fracture front during the growth of a progressive fracture. See Figure 8-8.

FRACTURE ORIGIN

ARREST LINES

Figure 8-8. Arrest lines are an indication of a progressive fracture.

Arrest lines are formed during the rest periods because the exposed face of a fracture tarnishes or corrodes up to the fracture front in the operating environment. Whether the component is idle or operating is not a factor because it will tarnish or corrode in either state (idle or operating).

The result of the series of rest periods is a series of arrest lines. The arrest lines point out the stoppages of a progressive fracture front. Beachmarks, clamshell marks, and conchoidal marks are terms used to describe these arrest lines.

Fatigue is the most common form of fracture in engineering components. Arrest lines are usually interpreted as a clear indication of fatigue, but they are actually a sign of any type of progressive fracture. Arrest lines are not confined to fatigue, and the absence of arrest lines on a fracture surface does not eliminate fatigue as a cause of failure. For example, the absence of arrest lines in a fatigue fracture may indicate that the component was run continuously (no rest periods) to failure.

Fatigue fractures begin at one or more initiation points, which are generally at the surface of the component. The initiation point is usually an area of localized stress concentration, such as a sharp change in section, a mechanical nick, or a corrosion pit. The metal locally hardens under the action of the repeated or fluctuating stress, which is aided by the localized stress concentration. The metal loses its ability to accommodate the stress by plastic deformation. Therefore, a small crack develops at the initiation point and grows under the influence of the fatigue stress. The fracture initiation region is minute compared to the bulk of the fracture surface. The morphology of the initiation region of a fatigue crack is difficult to identify, but it is relatively easy to locate the initiation point(s). See Figure 8-9.

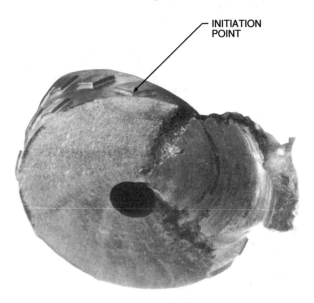

INITIATION POINT

Figure 8-9. Mechanical nicks or corrosion pits on the surface of a component can sometimes be used to locate the initiation points of a fatigue fracture.

A fatigue fracture grows in minute steps under the action of the fatigue stress. Using the extremely high magnification available in a scanning electron microscope, it is possible to identify fine parallel ridges (striations) on the fracture surface. *Striations* are minute steps that indicate the growth of a fatigue crack. See Figure 8-10. Striations propagate in a direction perpendicular to the stress axis. Each striation may represent one cycle of stress, but this is not always the case. Since striations are fine, they may be rubbed out as the mating fracture faces wear on one another. Striations may also be eliminated by corrosion or any other form of surface damage. The absence of striations does not eliminate fatigue from consideration.

FATIGUE FAILURE
REVEALING FATIGUE STRIATIONS

Figure 8-10. At high magnification, fatigue crack propagation is revealed as striations.

When examined macroscopically, the orientation of a fatigue fracture and the orientation of a brittle metal under a tensile load are similar. To indicate whether fatigue is involved, two pieces of evidence are considered. The orientation of a fatigue fracture surface is considered because it often exhibits a shear lip (characteristic of ductile metals) in the region of final fracture. The shear lip is formed when the remaining unaffected area of metal becomes too small to support the load. Also, when examined under a microscope, the morphology of a fatigue fracture is considered because it will exhibit microcracks that a brittle metal does not exhibit. *Microcracks* are small cracks that are the result of localized hardening that develops in a metal from the fatigue stress. Microcracks follow the shear planes of the

metal at a 45° angle to the stress axis. See Figure 8-11. Like striations, microcracks may be rubbed out during fracture propagation.

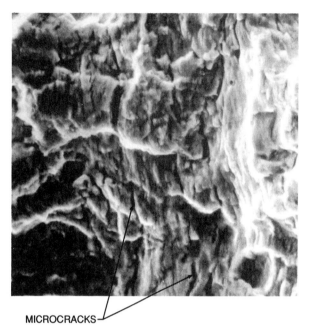

MICROCRACKS

Figure 8-11. Microcracks in the morphology are further evidence of fatigue.

Fractures with Single Origin. The origin of a fracture is the point where the fracture initiated. The identification of the origin helps determine the cause of a failure. In many cases the origin is a localized abnormality in the surface profile, metallurgical structure, or stress state of a metal. The origin is usually located by macroscopic examination. Visible markings on the fracture face are used to trace back the path of fracture propagation. The three most common markings used to locate the origin of a fracture are radial lines, chevron patterns, and arrest lines. See Figure 8-12.

Radial lines are markings that look like continuous rough peaks that appear on a fracture surface and point back toward the origin of the fracture. Radial lines are formed when a fracture propagates from the origin as a series of narrow bands, which grow sideways to meet one another. In order to completely separate the two halves of the fracturing metal, the last ligaments (narrow strips of metal between the bands) tear apart. These ligaments appear on the fracture surface as radial lines.

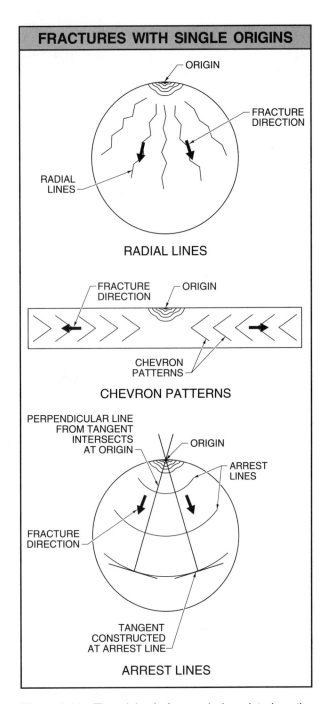

FRACTURES WITH SINGLE ORIGINS

RADIAL LINES

CHEVRON PATTERNS

ARREST LINES

Figure 8-12. The origin of a fracture is the point where the fracture is initiated.

Chevron patterns are V-shaped markings with apexes that point toward the origin of a fracture. Chevron patterns are similar to radial lines but are confined to high-speed fractures (brittle fractures). The V-shape indicates that the fracture is propagating faster along the edge of the component. Chevron patterns are used to locate the origin in a catastrophic

fracture of metals used for plate, piping, pressure vessels, and tank cars.

Arrest lines are markings that indicate rest periods of fracture propagation in a progressive fracture. They may be used to determine the fracture origin by constructing tangents to the arrest lines and drawing perpendicular lines from the tangents on the concave side of the arrest lines. The perpendicular lines meet and indicate at the fracture origin.

Fractures with Multiple Origins. Sometimes fractures propagate from multiple fracture origins. If the origins are in different planes, fractures propagate in the separate planes, but they eventually link up to form a continuous fracture plane. For this to occur, the ligaments between the planes separate and produce high and low points, giving the surface of the metal a ratchet appearance. *Rachet marks* are the high and low points on a fracture surface and are a sign of multiple fracture origins. See Figure 8-13.

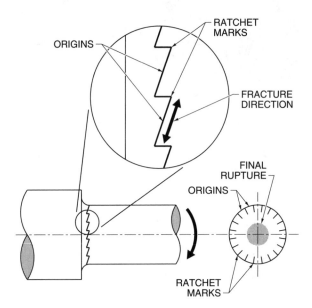

Figure 8-13. Ratchet marks are an indication of a fracture with multiple origins.

Failure Zones. The morphology of a fracture face will change as the force causing the fracture changes. The initiation zone, propagation zone, and final failure zone are the three characteristic zones on a fracture surface. The three zones are often revealed by macroscopic examination, although microscopic examination is usually required to confirm the fracture type present.

For example, all three zones are present on a rotating shaft with a mechanical nick on the surface. See Figure 8-14. An initiation zone occurs when the stress-concentrating effect of a mechanical nick and the fluctuating stresses in a rotating shaft develop into a crack. A propagation zone occurs when the rotation of the shaft imposes a fluctuating stress that causes a fatigue crack to grow from the initial crack (initiation zone). The fatigue crack continues to propagate until the remaining unaffected cross section of the shaft is incapable of supporting the stresses. A final failure zone occurs when the shaft fails rapidly due to a tensile overload. The relative speed of fracture growth in the propagation zone may be gauged by macroscopic examination of the fracture surface. The rougher the surface, the faster the speed of propagation.

Fracture Surface Damage. Damage to a fracture surface hinders interpretation of the morphology. Such damage may occur during the failure itself and also during a failure analysis investigation. Damage during a failure occurs when the two halves of the fracture surface rub against one another during the propagation of the fracture, when the two halves separate into two pieces during final failure, when the two halves corrode because the broken pieces are exposed to the environment, or when the two halves suffer abuse because the broken pieces are exposed to operations after the failure.

Damage occurring during a failure is unavoidable, but damage occurring during a failure analysis investigation is preventable. The two types of damage occur most frequently when attempting to fit the halves of the fracture together or when surface-cleaning procedures are used to remove unaffected metal. The halves of a fracture should never be fitted together. Surface-cleaning procedures are conducted to remove corrosion, such as rust, or surface contamination, such as grease. Any cleaning process that physically removes metal from the surface should not be used.

FAILURE ANALYSIS

A failure analysis investigation is conducted because it is difficult to implement a solution to the problem without information as to the cause of the failure. Four steps are completed in a failure analysis investigation. During the four-step investigation, the evidence is collected from the field, fractography

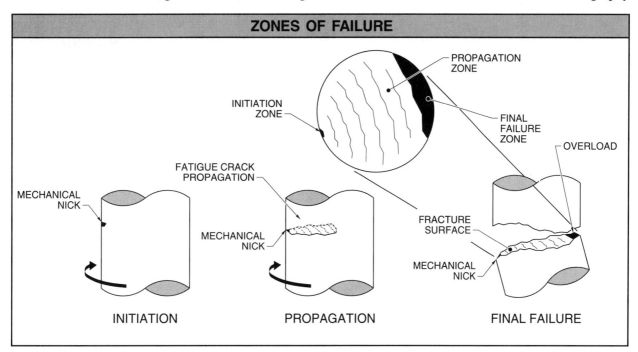

Figure 8-14. The morphology of a fracture face changes as the force for the fracture changes.

is performed, a destructive examination is conducted, and a failure analysis report is written. Completing the four steps requires skills such as interviewing those involved with the failure, researching events leading to the failure, deciding the scope of investigation, and ensuring that proper techniques and sequences are used to prevent loss of information.

When a failure occurs, not only the component or structure loses its usefulness, but there is also a loss in production and a possibility of injury or loss of life. Losses due to failure can rapidly increase costs. Even if the failure does not involve injury or loss of life, the importance of the failure analysis investigation is not diminished.

Field Evidence Collection

Three tasks performed during field evidence collection include developing a history of the operating conditions leading to the failure, documenting materials of construction, and selecting and preserving specimens for laboratory examination. To obtain the greatest benefit from field evidence collection, the failure analyst and the people who provide information about the failure must communicate. Good communication ensures that the correct specimens are removed for examination and that they are properly preserved and packaged for shipment.

The failure analyst may visit the failure site to collect evidence and to supervise the removal and preservation of specimens for examination. If shipment of the entire failure to the laboratory is impossible, a decision must be made about which parts or sections are to be removed for shipment. Selection and protection of specimens taken from the failure is an important aspect of the preliminary work because most failures are examined in the laboratory. The techniques used for the removal, preservation, and shipment of specimens have an important bearing on the quality of the laboratory examination.

A history of the operating conditions leading to the failure helps the failure analyst visualize the operation of the equipment at the time of failure. A checklist is invaluable in revealing whether the stress, temperature, and/or operating conditions were normal or excessive at the time of failure. See Figure 8-15.

TIME OF FAILURE CHECKLIST	
Response	**Question**
STRESS	
	Was loading static?
	Was loading cyclic?
	Was loading intermittent?
	Were vibration and resonance involved?
	What were design stresses?
	What proof tests were performed?
	What residual stresses were present?
TEMPERATURE	
	What was temperature at time of failure?
	Were nonuniform thermal stresses involved during operation?
	Were nonuniform thermal stresses involved on shut-down?
	Were nonuniform thermal stresses involved on start-up?
	Were nonuniform thermal stresses involved between materials with wide differences in coefficients of thermal expansion or conductivity?
OPERATING CONDITIONS	
	Was component properly installed?
	Was component properly aligned?
	Was component properly operated?
	Did operating conditions conform to those used as basis for design?
	Were operating conditions reasonably constant?
	Were operating conditions fluctuating widely?
	Was component operated continuously?
	Was component operated part-time?
	What was service life of component?
	What was service life of similar components?
	Were modifications made to component?
	Was component subject to operational abuse?
	Was component properly maintained?
	Was component properly lubricated?
	Was component properly cleaned?
	Was component properly inspected?

Figure 8-15. A checklist may be used to review the stress, temperature, and operating conditions at the time of failure.

Information about the materials of construction reveals whether the specifications were met and what preservice and in-service quality control procedures were conducted. Several sources of information may be required because records are rarely complete. For example, information may be required from the manufacturer or repair shop to complete information that was unrecorded or not kept up-to-date. A checklist may be used to indicate material characteristics and material information that is unrecorded or not kept up-to-date. See Figure 8-16.

MATERIAL CHECKLIST	
Response	**Question**
	What is material type?
	Under what standards (such as ASTM) was material purchased?
	What heat treatment and fabrication processes were used?
	Does material meet specified type?
	Does material meet heat-treatment specifications?
	Does material meet fabrication process specifications?
	What inspections were made?
	What tests were performed?
	Were repairs made during fabrication?
	Were repairs made during service?
	Were repairs properly conducted?
	Were repairs inspected?
	Were repairs tested?

Figure 8-16. A checklist may be used to indicate material characteristics.

Selecting and preserving specimens for laboratory examination is an essential part of the field evidence collection process. When possible, all parts of a failure are shipped to the laboratory for examination. All fracture faces are preserved and cushioned to avoid damage. When shipping components contaminated with regulated substances, care must be taken to ensure that the required documentation is produced, the proper authorities are notified, and shipping is conducted in a way that complies with regulations.

When it is impossible to ship an entire failure to the laboratory (size restrictions), examination of the failure helps the analyst determine regions to be examined (most important to investigation). It also helps them develop an overview of the problem. See Figure 8-17. The failure is carefully photographed before being cut into suitable pieces. Videotaping is recommended when the removal of specimens is a complex procedure. For example, videotaping is used during the removal of sections of turbine blades. If a heat-producing instrument, such as a torch, is used to remove specimens, the cuts must be made at a suitable distance from the fracture faces and other areas of examination to avoid damage and heat-induced metallurgical changes. Cuts should be made no closer than 12 mm (.5 in.) to the area of examination.

FAILURE OVERVIEW CHECKLIST	
Response	**Question**
	What was overall failure pattern?
	Where did failure start?
	How did failure start?
	Why did failure start?
	How did failure progress?
	Where did failure stop?
	Were cracks or defects other than main failure present?
	Were welds present within failure path?
	If welds were present, what was surface quality?
	Were welds located at points of maximum stress?
	Were welds located where there was a sharp transition in size?
	Was wear a factor?
	Was seizing a factor?
	Was galling a factor?
	Was fretting a factor?
	Were there signs of localized overheating?
	Were unusual deposits or material present?
	Were stress-concentrators such as gouges present?
	Were stress-concentrators such as nicks present?
	Were stress-concentrators such as sharp corners present?
	Were stress-concentrators such as fillets with poor radii present?
	Were dimensions of component same as those given in specifications?

Figure 8-17. A failure overview can pinpoint areas that require more detailed investigation.

The fracture surface must be preserved to protect it from corrosion or damage between the time of failure and the examination. Any preservatives used must prevent corrosion, be unreactive, and be easily and completely removable. Axle grease is a suitable short-term preservative. Lacquers are used for longer term storage but may not completely cover a rough fracture surface. Lacquers are more difficult to remove with solvents.

Replication tape is an acetate film used to make a replica of the fracture surface. When the shipping or removal of specimens is a problem, replication of the fracture face is an alternative method. *Replication* is a technique used to make impressions of some or all of the fracture face. Replication is achieved by pressing (using finger pressure) a small piece of solvent-softened replication tape on the area to be examined. Once the tape has fully hardened by solvent evaporation (approximately 30 minutes), it is carefully stripped off. The hardened face of the replica is a mirror image of the fracture face, and it is easily transported to the laboratory for examination of the fracture morphology. Key regions of the fracture face may be replicated, such as the initiation, propagation, and final failure zones. If the fracture surface is contaminated with corrosion, several replicas may be necessary to clean the surface before a satisfactory replica is made. Because the corrosion products that are collected on the contaminated replicas may be used to determine the reason for the fracture, they should not be discarded.

Replication tape is also a suitable preservative and covers rough surfaces more completely than grease or lacquers. The most reliable preservatives are the solvent-thinned (solvent cutback) type and the bitumens (tar) type. Solvent cutback types may be removed with organic solvents such as toluene or naphtha. Adhesive tape should never be used as a preservative because it does not intimately contact the fracture surface and may become corrosive if moisture is present.

Fractography

Fractography encompasses macroscopic and microscopic examination of a cleaned fracture surface. Fracture surface cleaning is conducted to ensure that the fracture surface is uncontaminated. Macroscopic examination is performed with the light microscope and with photography to document the evidence. Microscopic examination is performed with a scanning electron microscope. To indicate the chemical composition of areas examined, an energy dispersive X-ray analysis (EDXA) and electron probe microanalysis may be used with the scanning electron microscope.

Fracture Surface Cleaning. The technique used to clean the fracture surface depends on the amount and type of deposit materials. For example, loose material may be removed with a soft, organic fiber brush. Hard brushes may scratch the surface and should not be used.

Ultrasonic cleaning is a cleaning method that uses an organic solvent to remove lightly adhering grease, deposits, or liquid penetrant residues. An ultrasonic cleaning bath may be used to loosen surface contamination from a fracture surface. See Figure 8-18. The bath is vibrated at a high frequency for up to an hour. For more stubborn deposits, water-base detergents are used instead of organic solvents. When using water-base detergents, the immersion time must be reduced significantly (20 minutes or less) to prevent corrosion of the metal.

Branson Ultrasonics Corporation

Figure 8-18. An ultrasonic cleaning bath of organic solvent is used to remove light amounts of grease, deposits, or residues.

Replication cleaning is a cleaning method that uses replication tape to strip adherent deposits, such as corrosion products, from the fracture surface. Several applications of replication tape may be required. For example, it may take as many as six consecutive replicas to strip a heavily corroded surface. Replicas are retained in case the adherent deposits require analysis.

Cathodic cleaning is a cleaning method that uses the flow of electrical current to strip adherent deposits from a fracture surface. See Figure 8-19. The specimen is placed in a bath containing sodium cyanide, sodium hydroxide, or sodium carbonate and is connected to the negative terminal of a 12 V DC power source.

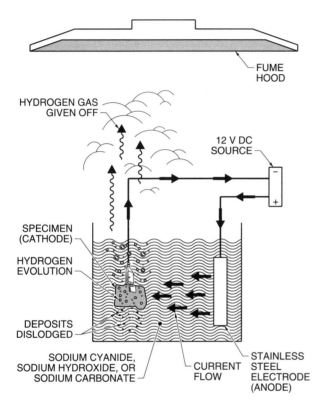

Figure 8-19. Cathodic cleaning is used to remove adherent deposits from a metal surface.

A stainless steel electrode is immersed in the bath and connected to the positive terminal of the power source. The passage of current from a stainless steel electrode (anode) to the specimen (cathode) causes hydrogen evolution on the specimen surface, which helps to dislodge deposits. Cathodic cleaning should not be used on metals susceptible to hydrogen embrittlement. For example, high-strength steels may suffer embrittlement in these solutions. Cathodic cleaning must be performed in a suitable fume hood because hydrogen is evolved during the passage of current. If sodium cyanide is used, there is the pos-

sibility of hydrogen cyanide evolution if acidic residues are present on the fracture surface.

Acid cleaning is a cleaning method that uses special corrosive chemicals to strip extremely stubborn deposits from the fracture surface. For example, hydrochloric acid containing a corrosion inhibitor is used to clean steel. A *corrosion inhibitor* is a chemical compound that when added to a corrosive chemical prevents attack of the metal, but it allows dissolution of scales and deposits to continue. To prevent corrosion, the correct corrosion inhibitor must be selected for specific metal-acid combinations. Acid cleaning requires a trial run duplicating the cleaning cycle on a coupon. A *coupon* is a specimen weighed and measured before cleaning that is made of the same metal as the failure specimen. After the trial run, the coupon is reweighed (to calculate the corrosion rate) and carefully examined for evidence of corrosion.

Macroscopic Examination. Light microscopes are used to macroscopically examine cleaned fracture surfaces, revealing the fracture orientation and other features in greater detail than can be seen with the naked eye. Specific checklist questions should be answered to help determine these features. See Figure 8-20.

The most convenient type of light microscope is the binocular microscope. The binocular microscope provides a three-dimensional view of the fracture surface. The magnification of the light microscope is limited by the required depth of field. *Depth of field* of a lens is the total depth of the image that can be maintained in focus. The rougher the fracture surface, the lower the useful magnification. The light microscope is limited to a magnification of 30X to 50X for most work.

Photography is an essential technique for documenting the orientation and morphology of fracture surfaces and component geometry. The overhead bellows camera, the most commonly used laboratory camera, provides instant 3 in. × 4 in. or 4 in. × 5 in. prints. The magnification of an overhead bellows camera approximates that of a light microscope. This makes the camera an ideal method of documenting macroscopic features observed in a light microscope. The magnification in an overhead bellows camera is found by applying the following formula:

$$M = \frac{B + f}{f}$$

where

M = magnification

B = bellows draw (distance from lens to object)

f = lens focal length

Example: Figuring magnification

What is the magnification of an overhead bellows camera that has a 40 mm focal length lens and a 36 cm bellows draw?

$$M = \frac{B + f}{f}$$

$$M = \frac{360 + 40}{40}$$

$$M = \frac{400}{40}$$

$$M = \mathbf{10X}$$

MACROSCOPIC FEATURES CHECKLIST	
Response	**Question**
	Did fracture occur suddenly in a single cycle of stress?
	Did fracture occur in a progressive manner?
	Did fracture originate at surface?
	Did fracture originate at subsurface?
	Was there more than one fracture origin?
	Does origin(s) coincide with stress raisers?
	Is there an obvious defect in metal?
	Does fracture surface suggest ductile behavior?
	Does fracture surface suggest brittle behavior?
	Does fracture surface suggest high-mean stress?
	Does fracture surface suggest low-mean stress?
	Does direction and plane of fracture relate to direction and magnitude of applied or residual stress in component?
	Was fracture surface clean?
	Was fracture surface heat-tinted?
	Was fracture surface corroded?
	Was fracture surface oxidized?

Figure 8-20. A checklist may be used to determine which features to look for during macroscopic examination.

To increase the magnification, the bellows draw must be increased for a given lens focal length, or a lens with a shorter focal length must be used. The shorter the focal length of the lens, the closer the lens must be to the specimen surface. A closer lens reduces the illumination available on the specimen surface. It is customary to work with a lens with the highest possible focal length for a given magnification. ASTM recommends focal lengths for particular magnifications. See Figure 8-21. To indicate the magnification, a scale may be placed alongside the specimen being photographed.

ASTM FOCAL LENGTH RECOMMENDATIONS	
Magnification	**Focal Length**
1X to 3X	152 mm
3X to 10X	72 mm
10X to 20X	40 mm
20X to 30X	32 mm

Figure 8-21. ASTM recommends focal lengths for particular magnifications.

Peaks and valleys of a fracture face must be in focus to achieve adequate resolution. Resolution is controlled by the depth of field of the lens. Depth of field of a lens is the total depth of the image that can be maintained in focus. The rougher the fracture surface, the greater the required depth of field. Depth of field is controlled by three factors: the lens focal length, the aperture (area) of the lens, and the distance from the fracture surface to the lens. Depth of field varies as the inverse square of the focal length. For example, if the focal length is reduced by one half, the depth of field increases by a factor of four.

Depth of field doubles as the aperture setting (f-stop number of the lens) doubles. The f-stop indicates the aperture size of the lens. Most lenses have depth of field scales inscribed on them that indicate the depth of field for any f-stop. Depth of field is proportional to the square of the distance of the fracture face from the lens. For example, if this distance is increased by a factor of three, the depth of field increases by a factor of nine.

Photographic lighting has the greatest overall effect on the appearance of a fracture face. Proper use of lighting sources and lighting methods permits key features of the orientation and morphology to be revealed. The four types of lighting sources include spotlight, diffused light, reflected light, and flashlight. See Figure 8-22.

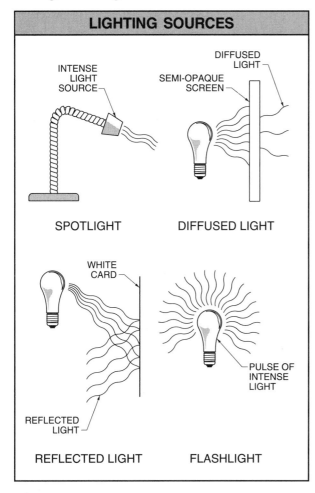

LIGHTING SOURCES

INTENSE LIGHT SOURCE

DIFFUSED LIGHT

SEMI-OPAQUE SCREEN

SPOTLIGHT

DIFFUSED LIGHT

WHITE CARD

REFLECTED LIGHT

PULSE OF INTENSE LIGHT

REFLECTED LIGHT

FLASHLIGHT

Figure 8-22. Next to sunlight, flashlight is the best light source for color photography of fractured surfaces.

Spotlight is an intense lighting source obtained from a single bulb in a reflector. *Diffused light* is a lighting source that uses a semi-opaque screen (such as ground glass) to diffuse the light source, reduce glare, and soften harsh details. *Reflected light* is a lighting source that bounces light off a white card, wall, or ceiling. The effect produced is similar to the effect produced by diffused light. *Flashlight* is a lighting source that provides a pulse of very intense light. Flashlight is the best light source (next to direct sunlight) for color photography of fracture surfaces.

The four types of lighting methods include main lighting, fill lighting, backlighting, and build-up lighting. These lighting methods use combinations of the four types of light sources to achieve the desired lighting effect. See Figure 8-23. *Main lighting* is a primary lighting method that has a light source at an angle to the subject that is within a 40° to 60° range. Any angle within this range is generally the angle that subjects are lighted in sunlight. *Fill lighting* is a lighting method that uses a small region of brighter light to increase detail on a dark area of a subject. The light source for fill lighting may be spotlight, diffused light, or reflected light. *Backlighting* is a lighting method that uses a diffused light source to eliminate or soften shadow detail. A light box (lighted ground-glass screen) behind the specimen is the most common diffused light source for backlighting. *Build-up lighting* is a lighting method that combines (adding or deleting) light sources to achieve the desired lighting effect.

Microscopic Examination. In microscopic examination a scanning electron microscope is used to examine and photograph cleaned fracture surfaces. Additional instrumentation may be used with the scanning electron microscope to conduct chemical analysis of the metal surface and surface deposits.

A diameter of about 1 in. and a thickness of $\frac{1}{2}$ in. or smaller is the optimum specimen size for a scanning electron microscope examination. This size allows the specimen to be rotated freely inside the specimen chamber for a wide viewing area. Larger sizes are permitted if limited rotation is not needed for complete examination of the component.

Specimens for a scanning electron microscope examination must be electrically conductive. Metals present no problem, but acetate replicas, deposit specimens, and metallurgical mounts require treatment with an electrically conductive coating.

Nonconductive specimens are usually coated in a vacuum with a thin layer. The thin layer is usually a coating of carbon or gold-palladium alloy that is less than 200 angstroms (Å) thick. An *angstrom* is a unit of measure equal to one ten-billionth of a meter. Metallurgical mounts are coated with carbon or a conductive bridge of graphite-containing paint. The paint is applied between the metal specimen and the specimen holder.

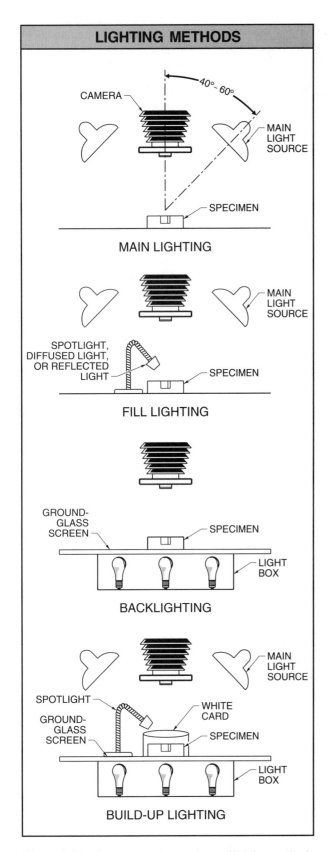

LIGHTING METHODS

40° ~ 60°

CAMERA

MAIN LIGHT SOURCE

SPECIMEN

MAIN LIGHTING

MAIN LIGHT SOURCE

SPOTLIGHT, DIFFUSED LIGHT, OR REFLECTED LIGHT

SPECIMEN

FILL LIGHTING

GROUND-GLASS SCREEN

SPECIMEN

LIGHT BOX

BACKLIGHTING

MAIN LIGHT SOURCE

SPOTLIGHT

GROUND-GLASS SCREEN

WHITE CARD

SPECIMEN

LIGHT BOX

BUILD-UP LIGHTING

Figure 8-23. Proper selection and use of lighting methods permits key features on a fracture surface to be revealed.

The specimen is placed in the chamber of the scanning electron microscope and air is pumped out to produce a vacuum of 10^{-6} to 10^{-9} torr. A *torr* is a unit of measure equal to $\frac{1}{760}$ of atmospheric pressure. The specimen is then bombarded with electrons from an electron gun positioned above it. This beam of electrons is focused into a spot (less than 100 Å in diameter) by a series of electromagnetic lenses. A system of scanning coils contained in the lower objective lens makes the electron beam scan in a raster pattern (side to side in lines from top to bottom) over the area to be examined on the specimen surface.

The electron beam interaction with the specimen causes various types of emissions from the specimen. See Figure 8-24. The backscattered electrons and secondary electrons are used to form an image of the specimen surface. *Backscattered electrons* are those electrons from the electron beam that are reflected back after interaction with the specimen surface. *Secondary electrons* are those electrons emitted from the specimen surface as a result of the electron beam interaction. The backscattered electrons and/or the secondary electrons are collected and amplified to display an image of the specimen surface on a cathode-ray tube. The image may be photographed to provide a permanent record. The magnification of the scanning electron microscope is controlled by the size of the area scanned by the electron beam. The larger the area, the lower the magnification. The smaller the area, the higher the magnification.

Greater magnification and depth of field are two advantages of the scanning electron microscope over the light microscope. Light microscopes are limited to about 50X, and scanning electron microscopes are capable of 15,000X or greater. The depth of field in scanning electron microscopes is one-hundred times that obtainable with light microscopes at comparative magnifications.

Scanning electron microscopes reveal details of fracture surface morphology that cannot be observed in light microscopes. If the surface is suitably cleaned and surface damage is not excessive, features such as fatigue striations, ductile dimples, transgranular cleavage, and intergranular separation are revealed. The morphology can be related to specific zones (initiation, propagation, and final failure zones) on the fracture surface revealed by macroscopic examination. Scanning electron microscopes are not limited to the study of the morphology of fractures. They are used for other important applications such as the study of surface finish, corrosion, and wear.

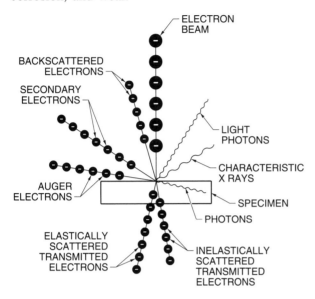

Figure 8-24. The electron beam interaction with a specimen causes various types of emissions. Two types, backscattered electrons and secondary electrons, can be used to form an image of the surface.

EDXA is a quantitative method of metal identification and is the chemical analysis of the surface area of a specimen scanned by an electron beam. EDXA is used in conjunction with a scanning electron microscope to determine the chemical composition of a specimen. EDXA uses fluorescent X rays (X-ray fluorescence) that are emitted by the interaction of the metal with the electron beam to determine the chemical composition of a specimen. See Figure 8-25. These fluorescent X rays have characteristic energies that correspond to the individual chemical elements present in the specimen.

EDXA has a semiconductor (solid-state detector) that analyzes the characteristic X rays and sorts them according to energies. Since the analysis is done simultaneously for all X-ray energies present, EDXA is an extremely rapid method of detecting the chemical elements present in the region examined. The quantity of each chemical element is also measured and is proportional to the intensity (peak height) of each characteristic X-ray energy detected. Concentrations of chemical elements as low as a few tenths of a percent are detectable using EDXA.

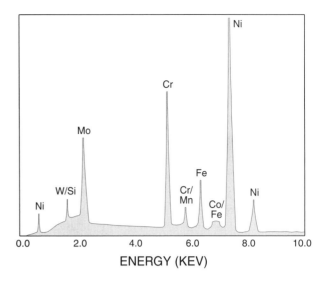

Figure 8-25. EDXA identifies and measures the concentration of chemical elements present in a specimen.

The cross-sectional area of the specimen volume excited by the electron beam is significantly greater than the beam area. The characteristic X rays emanate from a volume of the specimen in the shape of a teardrop. For an electron beam diameter of 100 Å, the widest diameter of the resultant teardrop is about 1 micron. See Figure 8-26.

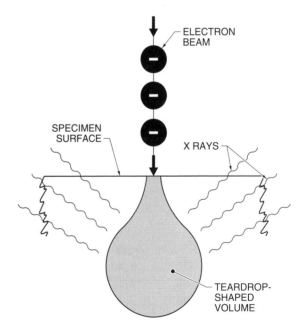

Figure 8-26. X rays are emitted from a teardrop-shaped volume of a specimen.

EDXA is limited to the detection of chemical elements with an atomic number of 11 or greater. For example, EDXA can detect sodium with an atomic number of 11 ($_{11}$Na) and sulfur with an atomic number of 16 ($_{16}$S). EDXA cannot detect important chemical elements with low atomic numbers. Chemical elements such as hydrogen ($_1$H), carbon ($_6$C), nitrogen ($_7$N), and oxygen ($_8$O) are undetectable by EDXA. These chemical elements are often present in metals, scales, and surface contaminants. Routine semiquantitative analysis provided by EDXA is normalized, meaning that the chemical elements analyzed are factored to 100%. If other chemical elements are present that EDXA cannot detect, the percentages of the elements analyzed are artificially high. This is often the case with corrosion products, scales, and inclusions. EDXA is a useful technique for identifying alloys because the normalized value can be compared to a known specimen.

Electron probe microanalysis is also used in conjunction with a scanning electron microscope to determine the chemical composition of a specimen. Electron probe microanalysis consists of an electron beam that scans the surface of a polished or lightly etched metallographic mount to provide an analysis of the constituents (such as chemical elements and grain structure) in the microstructure of a metal. Electron probe microanalysis is similar to EDXA because backscattered electrons are used to provide an image of a specimen surface. The characteristic X rays emitted are used to analyze the chemical elements present. Electron probe microanalysis uses wavelength dispersive X-ray analysis (WDXA) instead of EDXA to identify the chemical elements present.

In WDXA, the emitted characteristic X rays are diffracted (bent) by a diffracting crystal that separates them according to specific wavelengths. The characteristic X rays are identified by the wavelengths to determine the chemical elements present in the metal. A proportional counter detector is used to measure the intensity of the characteristic X rays and indicate the quantity of each chemical element present. The alignment of WDXA equipment (specimen, crystal, and proportional counter detector) must be on the circumference of one circle. See Figure 8-27. Compared to EDXA, WDXA requires critical alignment of the detection equipment with respect to the specimen surface.

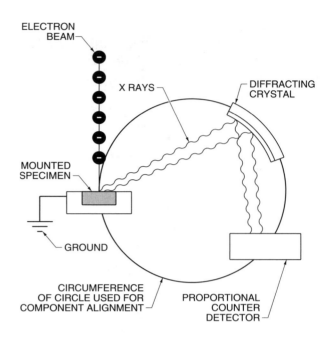

Figure 8-27. Electron probe microanalyzers use WDXA to analyze chemical elements present in micron-size areas of a specimen.

WDXA requires more time than EDXA. WDXA analyzes each chemical element one at a time, rather than all of them simultaneously. WDXA detects chemical elements with atomic numbers of 5 to 92. For example, WDXA can detect boron (B) with an atomic number of 5 ($_5$B), uranium with an atomic number of 92 ($_{92}$U) and all chemical elements with atomic numbers between these two. This additional detection capability is important because elements such as carbon, nitrogen, and oxygen are detectable. If the characteristic X-ray energies of one or more elements are too close to be separated (elemental material overlaps) using EDXA, they are often readily separated by wavelengths using WDXA. The most important of these elemental material overlaps is sulfur and molybdenum. Sulfur is present in oil, grease, and the atmosphere, as well as in inclusions in steel. Molybdenum is an alloying element in stainless steels and other alloys.

WDXA results can be expressed as X-ray maps or graphically. *X-ray mapping* is a qualitative method of displaying the results of WDXA by graphically highlighting the chemical elements to be detected. An *X-ray map* is an image of the distribution of the characteristic X rays emitted for a specific chemical element on a selected area of a specimen surface.

X-ray maps are particularly useful for documenting inclusions and other inhomogeneities in a specimen. X-ray maps for individual chemical elements may be compared with the electron image of a specimen surface to reveal what elements are associated with the specific features. See Figure 8-28.

Destructive Examination

Destructive examination is an examination of a specimen that requires the specimen be cut, machined, broken, drilled, melted, or dissolved. Destructive examination is conducted after the nondestructive work, such as fractography, is complete. In some cases, half of the fracture is used for fractography and the other half of the fracture is used for destructive examination.

Destructive examination is used to develop specific information about the mechanical properties, metallurgical structure, and chemical analysis of a specimen. Destructive examination techniques include mechanical property testing, metallography, and bulk chemical analysis.

ELEMENTAL X-RAY MAPPING

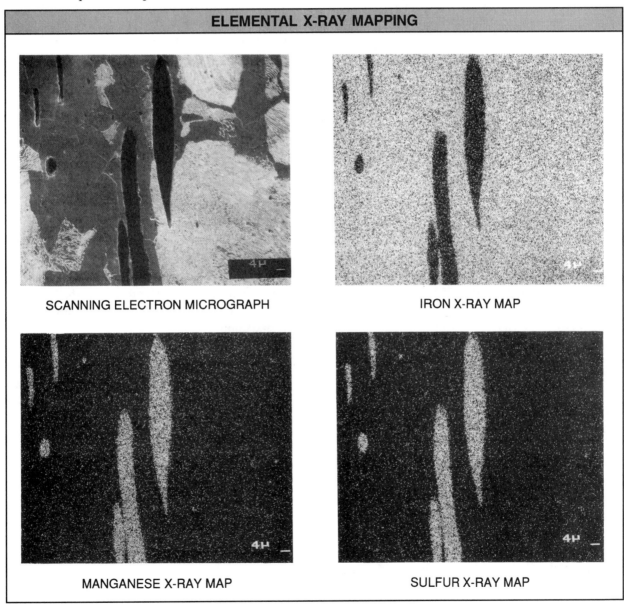

SCANNING ELECTRON MICROGRAPH

IRON X-RAY MAP

MANGANESE X-RAY MAP

SULFUR X-RAY MAP

Figure 8-28. An X-ray map can be compared to an electron micrograph of the fracture surface to reveal the chemical elements in a specimen.

Mechanical Property Testing. Mechanical property testing is used to develop mechanical property data and to provide standard fracture surfaces formed under specific mechanical loading conditions. These standard surfaces may be compared with the actual fracture surface. The information obtained from mechanical property testing is compared with certified test data for the material to check for discrepancies.

The most convenient and common mechanical property test is the hardness test. The hardness test reveals useful information about the mechanical work, heat treatment, and strength of a specimen without the need for elaborate specimen preparation. For many metals, strength increases with increasing hardness, while ductility and toughness decrease.

Microhardness testing is a technique for measuring the hardness of the constituents in the microstructure of a metal. Microhardness testing is performed on metallographically mounted, polished, and etched specimens and is a supplement of metallography. For example, microhardness testing may be used to detect decarburization (loss of carbon) at the surface of a carbon steel component resulting from heat treatment. Decarburization may lead to a decrease in fatigue strength and premature failure.

Tensile testing and impact testing are two other common mechanical property tests. Tensile tests use test specimens loaded under tension to determine the mechanical properties. Impact tests use test specimens that are impacted by a load to determine the mechanical properties. Test specimens must be machined from the component. Although specimen size and orientation limitations may prevent it, test specimens and test conditions should meet ASTM specifications when possible. The orientation of test specimens in relationship to the component must be carefully selected and recorded because of anisotropy.

If failure occurs at a temperature other than ambient, it may be necessary to test the specimen at the operating temperature. Testing is particularly important when the operating temperature is below ambient. Impact testing at the operating temperature may reveal important clues about the toughness of a material under operating conditions.

Mechanical property testing also provides standard fracture surfaces. Although fracture morphologies are consistent with the general rules for ductile, brittle, and fatigue fractures, the actual fracture surface morphology may be complex because of combinations of the loads or stress modes. The microstructure of a specimen may obscure the morphology. For example, the layered pearlitic microstructure in carbon steel may give the false impression of fatigue striations.

It is useful to obtain additional evidence in the form of fractures obtained under specific loading conditions. For example, a small part of the specimen may be pulled slowly to failure in a tensile test machine, while the other part is notched with a saw cut, tightened in a vise, and fractured with a hammer blow. The two fracture faces obtained under these loading conditions represent two vastly different rates of loading. The two differently loaded parts of the specimen are then compared with the actual fracture face. This comparison may help determine how the failure occurred.

Metallography. *Metallography* is a technique used to study the grain structure (microstructure) of a metal. In metallography, a specimen is cut from the fracture surface and mounted in a thermosetting resin to create a metallurgical mount. The metallurgical mount is polished to a mirror finish and then etched. The etched metallurgical mount is placed in a metallurgical microscope, which reveals the microstructure of the metal. The microstructure is then examined and studied.

The microstructure of a metal provides important evidence about the mechanical-working and heat treatment (thermomechanical) history of the metal. For example, the microstructure reveals whether the metal was cold worked or annealed, and a microstructure taken through the fracture may reveal whether the fracture propagation was transgranular or intergranular. The microstructure also reveals whether a component is in the specified metallurgical condition.

The orientation and locations of the specimens removed for metallography should be carefully selected. At least two specimens must be removed and mounted, one parallel (longitudinal mount) and the other perpendicular (transverse mount) to the axis of the failed component. See Figure 8-29. The two specimens provide information about the mechanical and thermomechanical history of the component. The location of the specimens reveals information about the metallurgical and physical condition of the metal at the origin of the fracture, compared to the bulk of the metal. To conduct a complete examination, specimens must be removed from the fracture origin and from the bulk of the metal.

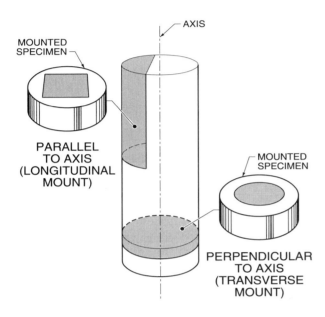

Figure 8-29. A minimum of two metallurgical mounts are produced from specimens removed from the bulk of the metal.

Care must be taken when cutting a specimen that was removed from the fracture origin. For example, cuts may be made with a diamond-tipped cutoff wheel. Cuts should be made so that the fracture origin is at one edge of a specimen. See Figure 8-30. The second cut that is made should be approximately $\frac{1}{16}$ in. from the fracture origin. The second cut is mounted facing upward. The mount is then ground and polished using standard procedures. The proper amount of grinding brings the fracture origin to the surface of the mount. However, excessive grinding to the mount causes the origin to be lost.

Bulk Chemical Analysis. Bulk chemical analysis is conducted on a bulk specimen to identify an unknown metal or to check that the metal meets the specified chemical analysis. Bulk chemical analysis is at the opposite end of the scale from

microanalytical techniques such as EDXA or WDXA. Microanalytical techniques examine small, precise areas of a specimen. Bulk chemical analysis is conducted using standardized techniques, such as those specified in ASTM standards.

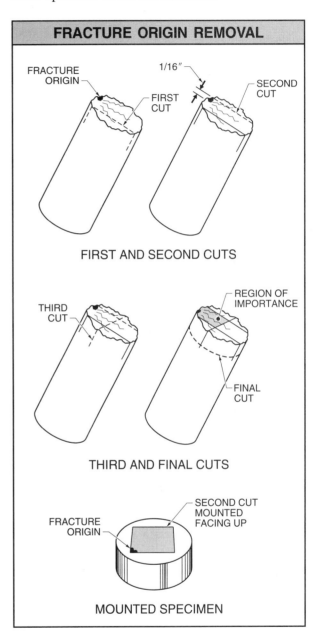

Figure 8-30. The specimen from the fracture face must be carefully cut, mounted, ground, and polished to reveal the fracture origin.

All bulk chemical analysis techniques are destructive and use a small specimen of the metal (from 1 g to 5 g) to conduct the analysis. After the specimen is removed by a process such as drilling, it is then dissolved, burnt, or sparked in order to make the analysis.

Quantitative field identification techniques such as X-ray fluorescence are also used for bulk chemical analysis. Quantitative field identification techniques are effective if the chemical elements required to confirm the identification are within the limitations of the instrumentation and if the specimen size is sufficient.

Failure Analysis Report

A *failure analysis report* is a document that describes the cause(s) of the failure and offers recommendations about preventing future failures. This report summarizes the evidence obtained in the three segments of the investigation (field, fractography, and destructive testing).

The three sections of a failure analysis report are the conclusions, discussion, and supporting evidence. See Figure 8-31. The conclusions are concise and clearly direct the reader to other sections. The three sections are written at increasing levels of depth so that readers may select what to read according to the reader level of interest in the subject.

The conclusions section consists of a brief history of the failure, description of the proposed cause(s), and recommendations to prevent the recurrence of the failure. The conclusions are placed at the beginning of a failure analysis report so the reader can decide whether the complete report requires study or whether reading the conclusions section is sufficient. The conclusions section lets the wide variety of people that read failure analysis reports gauge the need for reading the entire report.

The discussion section documents the reasons for the failure (revealed by the evidence obtained). The discussion section makes reference to the evidence found, which includes literature references and information obtained during the course of the failure analysis investigation. The discussion section does not describe the evidence in detail. Detailed information

about the evidence is included in the supporting evidence section of the failure analysis report.

FAILURE ANALYSIS REPORT
Conclusions
• Brief history
• Cause(s) of failure
• Recommendations
Discussion
• Detailed support of failure mechanism
• Detailed support of proposed solutions
• Referencing test work
Support Evidence
• Macroscopic examination results
• Photography
• Mechanical test results
• Metallographic examination results
• Scanning electron microscope examination results
• Microhardness test results
• Bulk chemical analysis results
• Microanalysis results (EDXA, X-ray mapping)

Figure 8-31. Failure analysis reports are documents that describe the cause(s) of a failure and offer recommendations about prevention of future failures.

The supporting evidence section consists of a description of the evidence obtained from the field, fractography, and destructive examination. When standardized tests are used, only the results are documented and the standard is referenced (such as the ASTM number). When special or unique tests are used or developed, the report should contain an adequate description of the test to provide the reader with an understanding about the methods used to obtain results.

Photographs and sketches are used to highlight the text and to eliminate unnecessary text. Photographs should indicate the magnification, orientation, and details (such as etchant in the case of a photomicrograph). Photographs and sketches are marked to indicate features described in supporting captions (words that describe the photograph).

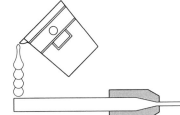

Field Identification of Metals   9

Field identification of metals consists of techniques that are used to rapidly check metals before they are fabricated into components, placed into service, or stored for later use. The three methods of field identification techniques are the manufacturer's identification, qualitative field identification, and quantitative field identification. The manufacturer's identification method includes documentation and physical identification markings. The qualitative field identification method includes analysis other than chemical analysis. The quantitative field identification method is chemical analysis.

FIELD IDENTIFICATION

A significant number of metal components and product forms are purchased for later use. With time, the certificate of analysis or physical identification markings (manufacturer's identification methods) may be lost or disappear due to environmental wear. Physical identification markings include color coding, stenciling, and stamping and embossing. In addition to manufacturer's identification methods, metals can be identified by qualitative field analysis (means other than chemical analysis) and quantitative field identification (chemical analysis). These methods of identification are used because it may become difficult to select the correct metal for an application from a mixed group of metals. As a result, the wrong metal can be substituted for a particular application.

For example, chromium-molybdenum (chrome-moly) low-alloy steels have a key use in critical applications, such as piping for handling high-temperature steam or hydrogen. Chrome-moly steels are similar in appearance to carbon steel, are magnetic like carbon steel, and rust like carbon steel if stored outdoors unprotected. A carbon steel should not be mistakenly substituted for a chrome-moly steel in

a critical application. It is easy to substitute one for the other because of their similarities. However, substitution may result in catastrophic failure because, in a critical application, carbon steel is likely to fail before chrome-moly fails.

Metal components and product forms are usually purchased from the manufacturer or vendor with accompanying certification of alloy type and chemical composition. Items for critical service applications require supplementary checks on the alloy to confirm identification. Supplementary checks are conducted in case an incorrect metal substitution has been made. For example, supplementary field identification programs are mandatory for critical items such as aircraft components or chemical equipment operating in hazardous service. The cost of supplementary field identification programs is minute compared to the cost of a failure resulting from an improper metal substitution. See Figure 9-1.

The scrap metal industry and plants that recycle metal are also dependent on field identification techniques. In the scrap metal industry, there is insufficient time, insufficient money, and/or no justification for complete chemical analysis of incoming material. Foundries and other plants that recycle metal for remelting into new products require field identification techniques to sort metals. When necessary, these plants rely on in-house chemical analysis laboratories to complete the identification.

Most field identification techniques are nondestructive, so physical removal of a specimen is unnecessary. To make an identification, however, it is usually necessary to prepare a small area of the surface. The inability to measure carbon content is the greatest limitation of all these techniques. Measurement of carbon content is essential for accurate separation of many grades of carbon, low-alloy steels, stainless steels, and some nickel-base alloys. Measurement of carbon content requires the removal of a small specimen that is subjected to a chemical analysis.

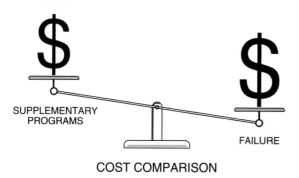

COST COMPARISON

Figure 9-1. Failure due to improper metal substitution outweighs in cost the implementation of supplementary field identification programs.

Manufacturer's Identification

The first step in field identification programs is to check the manufacturer's identifications. Record-keeping practices for the metals must be adequate so the metals can be separated and easily identified. Certificates of analysis and physical identification markings on the component are two forms of identification provided by product manufacturers.

Certificates of Analysis. *Certificates of analysis* are notarized statements of chemical analysis that are legally binding. The two types are the certificate of heat analysis and certificate of product analysis. They should be retained for the life of the product.

A *certificate of heat analysis,* or mill cert (mill certificate), is a statement of the chemical analysis and weight percent of the chemical elements present in an ingot or a billet. An ingot and a billet are the customary shapes into which a molten metal is cast.

These shapes are the starting points for the manufacture of wrought shapes by metal-forming processes such as rolling, drawing, forging, or extrusion. A certificate of heat analysis references a specific heat number and is the basic analysis provided by the manufacturer of a metal. A heat number is an identification code developed and maintained by the manufacturer and provides a record of melting practices and mechanical working of a metal.

A *certificate of product analysis* (check analysis) is a statement of the chemical analysis in weight percent of the end product (plate, wire, forgings, tubings, etc.) manufactured from an ingot or a billet. The specifications given in a certificate of product analysis allows a wider range of chemical composition for each chemical element analyzed than in the corresponding specifications given in a certificate of heat analysis. This wider range accounts for localized deviations in chemical composition (inhomogeneity) that are associated with the as-cast structure of an ingot or a billet.

Certificates of product analysis are not generally supplied to the purchaser unless they are specifically requested, usually at an additional cost. Certificates of product analysis provide more reliable identification than certificates of heat analysis. Certificates of product analysis uncover possible material substitutions between ingot or billet castings and manufacture of the finished forms.

Physical Identification Methods. Physical identification methods include the use of symbols and/or characters that are clearly marked and provide a code to the material identification of a metal component or product form. For example, the markings may be embossed, stamped, or stenciled on the component or product form.

Foundry marks are identification markings embossed on the exterior of castings. Foundry marks are incorporated into the pattern from which the mold is made. The markings are easily visible on the finished casting. The ASTM grade number, foundry name or logo, heat number, and foundry shorthand description for the alloy is the type of information included for identification. Additional information such as the design pressure and temperature rating of castings is sometimes included. See Figure 9-2.

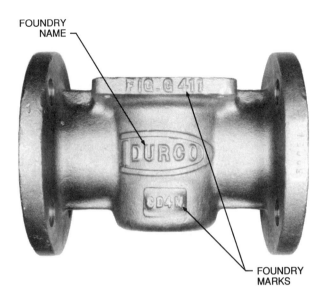

FOUNDRY
NAME

FOUNDRY
MARKS

The Duriron Company, Inc.

Figure 9-2. Foundry marks are identification markings that are embossed on the exterior of castings.

Wrought products (plate, sheet, bar, tubing, etc.) are identified by color coding or stencil markings. *Color coding* is an identification method consisting of colored stripes painted on one end of the product prior to storage at a metal service center or user's warehouse. See Figure 9-3. Color coding allows easy and rapid identification of metals. Color coding must clearly identify each metal in storage. Some color coding is set up to identify specific metals stored at a particular location, and for this reason there is no universal color coding system.

Stencil marking is an identification method that uses printed markings to describe alloy type relevant standard designations (ASTM designations) and dimensions of a product form. Stencil markings are repeated at regular intervals along the product form so identification is not lost when the form is sectioned. Stencil markings are not permanent and may degrade with outdoor storage or during service.

Certain chemical elements in some paints and working materials used for color coding or metal markings are potentially harmful. These chemical elements may cause catastrophic cracking in susceptible alloy systems, such as stainless steels and high-nickel alloys. Cracking is most likely to occur when the paint or marking material on the metal is exposed to heat (welding, high-temperature service, etc.) or to specific corrosive environments during service.

Chlorine (Cl), sulfur (S), and zinc are potentially harmful chemical elements. Marking materials that are used on susceptible alloys must contain low quantities (measured in parts per million, or ppm) of the harmful chemical elements. Quantities of 50 ppm to 100 ppm are allowable and lessen the potential for catastrophic cracking. Even if approved marking materials are used, it is good practice to remove them from areas that are to be welded, brazed, or soldered. An approved solvent should be used to remove marking materials. A marker with a fiber tip may be used to mark a metal because it does not leave amounts of solid residue that can cause cracking. Because markers leave no solid residue, a solvent is not needed for removal.

Stamping and embossing are identification methods that provide identification information that is stamped or embossed on metal forgings and fasteners. On forgings, the stamped impression is

COLOR CODING FOR SELECTED STEELS			
SAE Designation	**Color**	**SAE Designation**	**Color**
1010 Carbon steel	White	4640 Molybdenum steel	Green and pink
1025 Carbon steel	Red	3125 Nickel-chromium steel	Pink
1112 Free-cutting steel	Yellow	3325 Nickel-chromium steel	Orange and black
1120 Free-cutting steel	Yellow and brown	5120 Chromium steel	Black
2015 Nickel steel	Red and brown	6115 Chromium-vanadium steel	White and brown
2330 Nickel steel	Red and white	7260 Tungsten steel	Brown and aluminum
4130 Molybdenum steel	Green and white	9255 Silicon-manganese	Bronze

Figure 9-3. Color coding allows easy and rapid identification of metals.

produced with a metal die. The impression is usually located on the outside surface of the forging and consists of the ASTM type or other material type, the pressure and temperature rating of the forging, and the forge shop logo.

Fasteners are identified by an embossed or a stamped marking on the head or at the other end of a fastener. See Figure 9-4. Because space is restricted on fasteners, a code is used to identify the ASTM type or other material type and strength level. The name of the manufacturer may also be included. The Industrial Fasteners Institute publishes a list of manufacturers' logos used on fasteners.

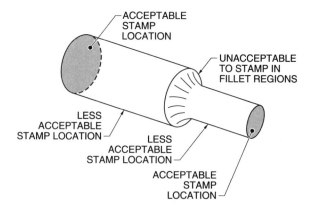

STAMPED IDENTIFICATION
LOCATIONS

Figure 9-5. Stamped identification locations on structural components must be carefully selected to avoid the introduction of surface mechanical stresses.

Welding filler metals (wire and covered electrodes) are identified by the American Welding Society (AWS) designations. For a covered electrode (welding filler metal with a flux coating), the AWS designation is stenciled on the flux coating at one end of the electrode. For a bare wire electrode, the AWS designation is printed on a paper flag that is glued to one or both ends of the wire.

EMBOSSED MARKING ON
FASTENER HEAD

Figure 9-4. Fasteners are identified by an embossed or a stamped marking on the head or other end of the fastener.

Location of stamped identifications on structural components must be carefully selected so surface mechanical stresses introduced by the stamping do not cause the initiation of cracking during service. Locations such as fillets on shafts are particularly susceptible to cracking and are unacceptable stamp locations. See Figure 9-5. As a further precaution, low-stress stamping should be used. Dies used for low-stress stamping have broken (rounded) corners on the impression symbols to minimize the intensity of the localized stresses that is caused by the stamped impression.

Qualitative Field Identification

Qualitative field identification techniques separate and distinguish metals using methods other than chemical analysis. These methods use the physical properties of metals, are fast, and are relatively low in cost. Although they are nondestructive, most qualitative field identification techniques require a small amount of surface preparation. Qualitative field identification techniques must not be conducted on finished surfaces such as seal or bearing faces. Qualitative field identification techniques include sorting by metal color, magnetic behavior, density measurement, spark testing, chemical spot testing, triboelectric sorting, thermoelectric potential sorting, metallographic identification, and optical emission spectroscopy.

A sequence of qualitative field identification techniques can be used to identify an unknown metal by a process of elimination. For example, to distinguish type 410 stainless steel from type 316 stainless steel, a simple check is made with a magnet to separate the two. Type 410 stainless steel is strongly magnetic and type 316 stainless steel is nonmagnetic. To distinguish type 316 stainless steel from type 304 stainless steel, a chemical spot test for molybdenum is used. A magnet check is not helpful because both types are magnetic. The two types can be separated by the chemical spot test because type 316 stainless steel contains nominally 2.25% Mo and type 304 stainless steel has none. However, to separate type 316 from many stainless steels containing all three alloys, two steps are required. In step one, a check with a magnet removes the type 410 stainless steel. In step two, a chemical spot test for molybdenum removes type 304 stainless steel.

Metal Color. Identification by metal color is a technique of field identification that groups metals according to the characteristic colors that a metal exhibits. See Figure 9-6. Heat tints from heat-treating operations and surface scales and tarnishes from exposure to the environment may hide the true color. To be sure of the true color, a small area of the surface must be cleaned by filing or rubbing with coarse abrasive paper.

Magnetic Behavior. Identification by magnetic behavior uses the response of an unknown metal to a magnetizing force to identify the metal. The simplest method is to bring a permanent magnet close to the metal.

The magnetic response is categorized as strong attraction, weak attraction, or no attraction with a possible weak repulsion. The type of attraction allows the unknown metal to be placed into a specific identification grouping. See Figure 9-7.

The magnetic response of some metals may be altered by heat-treating temperature or cold work, which may lead to false identifications. For example, type 302 and type 304 stainless steel are nonmagnetic in the annealed condition but become increasingly magnetic with increasing amounts of cold work. The effect of temperature on magnetic properties is illustrated in alloy 400 (Monel 400), which is slightly magnetic at ambient temperature. Monel 400 is nonmagnetic if its temperature is raised above the boiling point of water.

A more sophisticated method of magnetic behavior identification is performed using a coating thickness gauge. Coating thickness gauges are normally used to measure the thickness of paint and other nonmagnetic coatings on steel. When used in its normal mode, the coating thickness is read directly from the scale on the gauge. Before taking a reading, the instrument is calibrated on a specimen of known coating thickness.

When used for metal identification, the coating thickness gauge is calibrated using a nonmagnetic metal, such as type 304 stainless steel or aluminum, and given a rating of zero on the scale. The coating thickness gauge is then calibrated on a strongly magnetic material, such as carbon steel.

CHARACTERISTIC COLOR GROUPINGS	
Color	**Metal**
Red or reddish	Copper, >85% copper alloys
Light brown or tan	90% to 10% cupronickle
Dark yellow	Bronzes and gold
Light yellow	Brasses
Bluish or dark gray	Lead, zinc, and zinc alloys
Silvery white with soft luster	Aluminum
Silvery white with bright luster	Stainless steels
Gray	Carbon and low-alloy steels, 70% to 30% cupronickle
White or light gray	Nearly all others

Figure 9-6. Metals can be identified and grouped by the characteristic colors that the metal exhibits.

MAGNETIC BEHAVIOR GROUPINGS

Strong Attraction

- Carbon steels
- Cast irons
    Gray
    Ductile
    Malleable
- Cobalt
- Iron-silicon alloys (.05% Si to 4.5% Si)
- Iron-cobalt alloys
- Iron-molybdenum alloys
- Low-alloy steels
- Nickel
- Stainless steels
    Ferritic
    Martensitic (400 series)
    Martensitic precipitation hardening
- Tool steels

Weak Attraction

- Stainless steels
    Cast 300 series
    Cold-worked 302
    Cold-worked 304
    308 weld metal
    309 weld metal
    329
- Monel 400
    (Becomes nonmagnetic in boiling water)

No Attraction

- Alloy 20 types
- Commercially pure nonferrous metals
    (Except nickel and cobalt)
- Cupronickles
- Hastelloys
- Incoloys
- Inconels
- Stainless steels
    Austenitic precipitation hardening
- Stellite

Figure 9-7. Metals can be identified and grouped by the magnetic behavior they exhibit.

The carbon steel is given a rating of 10 on the scale. Unknown metals are then rated (1 to 10) on the scale according to the response recorded on the scale. See Figure 9-8. The coating thickness gauge method is relatively crude and is used to separate metals that exhibit significant differences in magnetic response.

Density Measurement. Metals are also identified by measuring the density in grams per cubic centimeter (g/cm³). Density is measured by obtaining a small specimen of the metal (1 cm cube or $\frac{1}{2}$ in. cube), a length of fine wire, an analytical balance, a small bench to straddle the analytical balance pan, and a 250 ml beaker that is filled with approximately 167 ml (two-thirds full) of distilled water.

CMI International

Figure 9-8. Coating thickness gauges identify unknown metals by comparing readings of unknown metals to readings of known reference metals.

Dirt and foreign matter are thoroughly removed from the surface of the specimen. The specimen is then washed with acetone and allowed to dry for 2 to 3 minutes. The specimen is weighed on an analytical balance to ±.001 g. The fine wire is also weighed to ±.001 g. The beaker containing distilled water is placed on the small bench that straddles the balance pan. See Figure 9-9. One end of the fine wire is firmly tied around the metal specimen. The other end is attached to the balance hook, so that the metal specimen is suspended and totally immersed in the beaker of distilled water. The metal specimen is reweighed when it is totally immersed in the distilled water. The density of a metal is found by applying the following formula:

$$D = \frac{W_a}{W_a - (W_d - W_w)} \text{ per cm}^3$$

where

D = density (in g/cm³)

W_a = weight of specimen (in g)

W_d = weight of specimen in distilled water (in g)

W_w = weight of fine wire (in g)

per cm^3 = 1 cm^3 (1 cm cube specimen)

Example: Figuring density with an analytical balance

What is the density of a specimen (1 cm cube) of 304 stainless steel that weighs 18.102 g in air, and weighs 15.960 g in the distilled water of an analytical balance, and that has a fine wire that weighs .151 g?

$$D = \frac{W_a}{W_a - (W_d - W_w)} \text{ per } cm^3$$

$$D = \frac{18.102}{18.102 - (15.960 - .151)} \text{ per } cm^3$$

$$D = \frac{18.102}{18.102 - 15.809} \text{ per } cm^3$$

$$D = \frac{18.102}{2.293} \text{ per } cm^3$$

$$D = \textbf{7.89 g/cm}^3$$

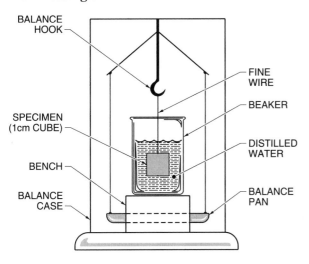

Figure 9-9. Metals can be identified by measuring the density of the metal with an analytical balance.

The four density identification groupings for metals are very high density, high density, average density, and low density. See Figure 9-10. From the figured density value, metals are placed in one of the groupings. Depending on the separation of the density values, metals inside the same grouping are distinguished from each other by checking the

figured densities against a table of known density values. See Appendix.

DENSITY IDENTIFICATION GROUPINGS		
Grouping	**Density Range (g/cm³)**	**Metals**
Very high density	12 to 22	Gold Rhodium Iridium (Ir) Ruthenium (Ru) Osmium (Os) Tantalum Palladium (Pd) Tungsten Platinum Uranium (depleted)
High density	9.8 to 11.9	Bismuth Molybdenum Lead Silver
Average density	6 to 9.7	Antimony Nickel alloys Cadmium Stainless steels Cast iron Steels Copper alloys Tin Nickel Zinc
Low density	1 to 5.9	Aluminum Magnesium Aluminum alloys Magnesium Beryllium alloys Beryllium alloys Titanium Titanium alloys

Figure 9-10. Figured density values can be used to place a metal in one of four groupings.

Spark Testing. *Spark testing* is a method of identifying metals such as carbon steels, low-alloy steels, and tool steels by visual examination of the spark stream (spark pattern). Spark stream characteristics are compared to spark charts to identify the unknown metal. See Figure 9-11. The spark stream is produced when the metal is held against a grinding wheel rotating at a high speed. The chemical composition of the steel influences the form of spark stream produced. Variations in content of carbon or other specific alloying chemical elements may be detected by the changes in the spark stream and used to identify an unknown metal.

The area of metal selected for spark testing must be free of scale and representative of the chemical composition of the metal. Decarburization is the chief problem and may lead to a false identification. The low-carbon decarburized surface layer exhibits a different spark pattern from the bulk of the steel. To obtain an accurate identification of a heat-treated bar, it may be necessary to saw, cut, or break a section through the bar. This ensures that the grinding wheel does not touch the decarburized rim of the bar.

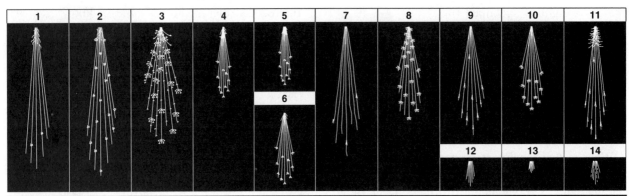

Metal	Stream Volume	Relative Length[a]	Color of Stream	Color of Streaks	Quantity of Spurts	Nature of Spurts
1. Wrought iron	Large	65	Straw	White	Very few	Forked
2. Machine steel (AISI 1020)	Large	70	White	White	Few	Forked
3. Carbon tool steel	Moderately large	55	White	White	Very many	Fine, repeating
4. Gray cast iron	Small	25	Red	Straw	Many	Fine, repeating
5. White cast iron	Very small	20	Red	Straw	Few	Fine, repeating
6. Annealed mall. iron	Moderate	30	Red	Straw	Many	Fine, repeating
7. High-speed steel (18-4-1)	Small	60	Red	Straw	Extremely few	Forked
8. Austenitic manganese steel	Moderately large	45	White	White	Many	Fine, repeating
9. Stainless steel (Type 410)	Moderate	50	Straw	White	Moderate	Forked
10. Tungsten-chromium die steel	Small	35	Red	Straw[b]	Many	Fine, repeating[b]
11. Nitrided Nitralloy	Large (curved)	55	White	White	Moderate	Forked
12. Stellite	Very small	10	Orange	Orange	None	
13. Cemented tungsten carbide	Extremely small	2	Light orange	Light orange	None	
14. Nickel	Very small[c]	10	Orange	Orange	None	
15. Copper, brass, and aluminum	None				None	

[a] actual length varies with grinding wheel, pressure, etc.   [b] blue-white spurts   [c] some wavy streaks   *Norton Company*

Figure 9-11. Spark charts are compared with spark stream characteristics to identify unknown metals.

Small portable grinders are most often used for spark testing, because they can be transported to test metals in the field. Stationary grinders may also be used if convenient. See Figure 9-12. To provide a satisfactory spark stream, the grinding wheel must rotate at high speeds (7500 fpm to 15,000 fpm) and must be hard (40 grain alumina wheel). For example, 16,000 rpm is suitable for a 2 in. diameter portable grinder with a 40 grain alumina wheel.

Before conducting a spark test, the grinding wheel is dressed clean with a diamond-wheel dresser to remove particles of metal from previous tests. If these particles were not removed, the spark stream of the specimen being examined would be contaminated by sparks from the previous tests.

The pressure between the grinding wheel and the specimen must be sufficient to maintain steady contact and to allow the spark stream to be given off approximately 1 ft horizontally and at right angles to the line of vision. There must be no objects obstructing the full-length view of the spark stream emanating from the grinding wheel.

Warning: Protective goggles and protective clothing must be worn when spark testing.

Conditions for spark testing must be standardized and testing should be conducted in diffused daylight, not bright sunlight or darkness. The spark stream should be protected from heavy drafts of air that may cause the tail of the spark stream to hook, which leads to an erroneous interpretation.

The entire spark stream is closely examined for its characteristic features. Characteristic features include carrier lines, forks, bursts, and arrowheads. Carrier lines are incandescent (glowing) streaks that trace the trajectories (paths) of each particle (spark).

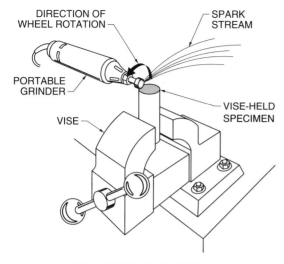

PORTABLE GRINDING

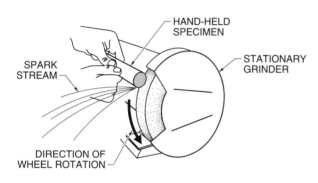

STATIONARY GRINDING

Figure 9-12. Spark testing is most often performed on portable grinders, but stationary grinders may also be used.

Forks are simple branchings of the carrier lines. Bursts are complex branchings of the carrier lines. Arrowheads are terminations of the carrier lines in the shape of arrowheads. See Figure 9-13.

A set of known specimens (generally 1 in. diameter by 4 in. length) should be maintained with the identification stamped and documented on the specimen. These specimens may be used to train inexperienced operators in identifying spark streams. Spark testing heat treats the surface layer of the metal, leading to localized hardening and possible cracking. Stock is discarded that is any closer than $\frac{1}{4}$ in. from the area of contact with the grinding wheel because of possible failure.

Chemical Spot Testing. Chemical spot testing is a method used to identify metals by the color changes that occur when a metal is contacted with specific chemical reagents that are applied to a metal or to a metal solution. The solution is produced by dissolving a small amount of the metal. Many chemical reagents are used to detect specific chemical elements present in a metal. Comprehensive, step-by-step identification techniques using chemical reagents have been developed to identify specific metals.

The application of a chemical reagent to a solution containing small amounts of the metal is the most common method of chemical spot testing. The solution is often produced using electric current (electrographic technique) to dissolve small amounts of the unknown metal in a chemical reagent. When the solution reacts with the chemical reagent, a color change occurs, which is used to identify the unknown metal.

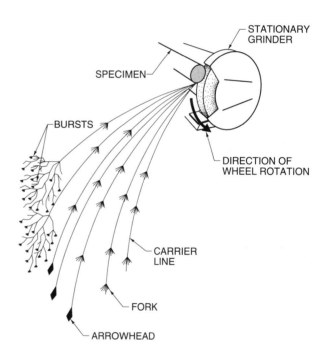

Figure 9-13. Characteristic features of spark streams include carrier lines, forks, bursts, and arrowheads.

During an electrographic chemical spot test, a metal surface is dressed with a file or emery paper, which removes scale or unnecessary roughness, and then the metal surface is degreased. A filter paper wetted with measured drops of a chemical reagent is placed on the metal surface. The unknown metal (anode) is electrically connected to the positive terminal of a 6 V DC battery. An aluminum cathode, connected to the negative terminal of the battery, is pressed against the wet filter paper. This connection completes the electrical circuit and allows current to flow until the cathode is removed and the circuit is disconnected. See Figure 9-14.

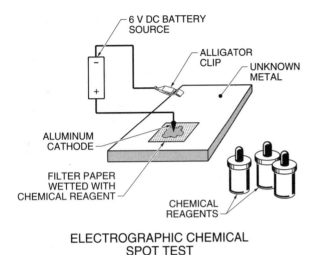

ELECTROGRAPHIC CHEMICAL
SPOT TEST

Figure 9-14. The electrographic chemical spot test is the most common chemical spot test.

The filter paper contains a small amount of the metal solution. Measured drops of a second chemical reagent are applied to the wet filter paper, causing a color change. The color change on the filter paper is used to identify the metal. Supplementary reagents may be applied to the filter paper that lead to additional color changes that narrow the metal selection. When the test is complete, the metal surface is thoroughly cleaned to remove the excess chemical reagent.

Commercial electrographic chemical spot test kits contain specific chemical reagents, filter papers, a battery, and instructions for various metal identifications. Instructions are strictly observed so that the correct inferences from the color changes are made. To familiarize operators with the color changes, a set of known standard specimens are used.

Warning: Some reagents in chemical spot testing kits are strong acids or alkalis and must be handled properly.

Triboelectric Sorting. Triboelectric sorting is a metal identification method that uses the measurement of the electrical potential (EMF) generated when two different metals are rubbed together.

In triboelectric sorting, the unknown metal and a probe (small file) are electrically connected to the terminals of a millivolt-meter. See Figure 9-15. The probe is rubbed across the surface of an unknown metal. The rubbing produces electron exchange that creates an EMF which is registered on the millivolt-meter. This EMF is used to identify the unknown metal by comparing it to EMFs produced by known metals.

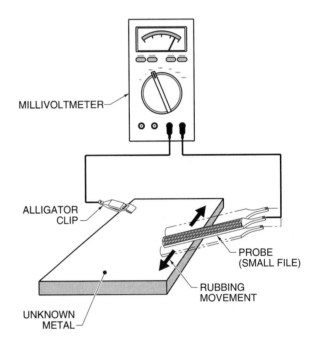

Figure 9-15. Triboelectric sorting uses the measurement of the EMF, which is generated when two different metals are rubbed together, to identify unknown metals.

Triboelectric sorting is limited because of the low EMF generated. This leads to an overlap between the recorded EMF values for many different metals. It is the reason for the insensitivity of the technique.

Thermoelectric Potential Sorting. Thermoelectric potential sorting is a metal identification method that uses measurement of the electrical potential (EMF) generated when two different metals are heated. This phenomenon is also used in pyrometry (science of temperatrue measurement).

The common and reproducible method of thermoelectric potential sorting is to standardize on the voltage generated by the heated junction of two dissimilar metals. This method allows a significantly greater amount of heat to be generated, which increases sensitivity.

The unknown metal is put in contact with the heated probe. The thermoelectric potential generated is recorded on a digital or analog readout. This value is compared with values of known specimens to make an identification.

The null point method is also used with thermoelectric potential sorting. The null point method is used for identifying an unknown metal or distinguishing it from other metals. A known standard specimen and a probe are electrically connected. The resulting potential is calibrated to read zero on the millivolt-meter. An unknown metal that reads the same as the known specimen will produce no deflection of the millivolt-meter. If the unknown metal is different from the known specimen, the millivolt-meter will deflect either side of zero. Thermoelectric potential sorting is described in ASTM E977, "Standard Practice for Thermoelectric Metal Sorting."

Metallographic Identification. Metallographic identification is a metal identification method that uses the interpretation of unknown metal microstructures. The microstructure is obtained by grinding and polishing an area on the unknown metal surface (size of a postage stamp) to a mirrored finish. The area is etched with a suitable etchant to reveal the microstructure. Using a portable metallurgical microscope, the microstructure is examined at magnifications of 100X to 500X to identify the unknown metal. See Figure 9-16.

Metallographic identification is used when the microstructure clearly identifies a metal or its condition. For example, metallographic identification distinguishes gray cast iron from nodular iron by the shape of the graphite constituents. The constituents are spidery in gray cast iron and globular in nodular iron.

A limitation of metallographic identification is the tendency to form artifacts, which are false microstructural indications resulting from the grinding and polishing operations. For example, excessive pressure or heating during grinding and polishing may heat treat a metal surface. This obscures the true microstructure of the metal and the surface microstructure may be unrepresentative of the bulk microstructure. Careful surface preparation is critical.

Where the location or orientation of the component makes it physically impossible to examine the microstructure in place, a replica (impression) of the polished and etched surface is made and examined in the laboratory. Replicas are small pieces of acetate tape (replication tape), sufficient in size to cover the polished and etched area. The acetate tape is applied on one side with a solution of acetate in a solvent to soften it. The softened side of the replication tape is pressed firmly against the polished and etched surface for about 30 seconds.

The replication tape is then allowed to harden for 15 to 20 minutes. When the edges of the tape break free from the metal surface, the replication tape is set and is carefully peeled off. The replication tape is then fixed to a glass slide with a silver backing and examined under a metallurgical microscope to reveal the microstructure of the metal. See Figure 9-17.

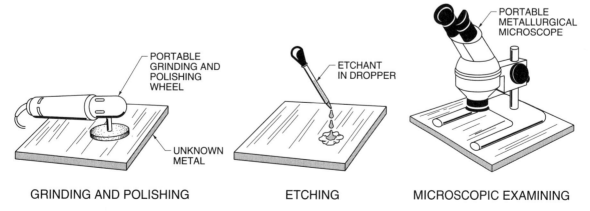

GRINDING AND POLISHING   ETCHING   MICROSCOPIC EXAMINING

Figure 9-16. Metallographic identification uses grinding and polishing, etching, and microscopic examination to interpret the microstructures of unknown metals.

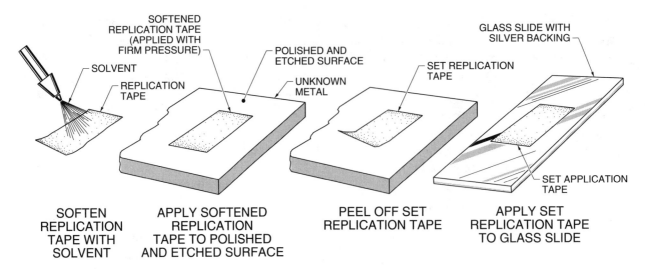

Figure 9-17. Replication tape is used to produce a replica when it is impossible to examine an unknown metal.

Optical Emission Spectroscopy. *Optical emission spectroscopy* (OES) is a metal identification method that uses light emitted from an unknown metal surface when it is arced (caused to spark) by an electric current. When an electric arc is struck on a metal surface, the light emitted is composed of various wavelengths. The chemical elements in the metal determine the component wavelengths produced. The intensity of each component wavelength is proportional to the concentration of its corresponding chemical element.

All light emitted from the arc is collected and passed through a glass prism, which diffracts it into the component wavelengths. The separated wavelengths are viewed as a series of lines of varying intensity and color. See Figure 9-18. The chemical elements detectable by OES are limited to those elements that have observable light wavelengths after diffraction and that are not volatilized (vaporized) in the heat of the arc. Low percentages of chemical elements may be undetected if the line obtained by diffraction is faint. OES can detect nickel, chromium, molybdenum, titanium, manganese, vanadium, copper, zinc, tungsten, magnesium, cobalt, lead, and niobium (Nb).

An optical emission spectrometer is an instrument used for OES that is placed on the surface of an unknown metal. A small area of the surface is intermittently sparked by striking an arc between the surface and a tungsten electrode, using a power source of 25 V to 40 V. The light emitted from the arc is diffracted through a glass prism and separated into characteristic wavelengths. By means of a viewfinder (focusing device), the operator can view the diffracted light as a series of vertical colored lines. By moving the lines from side to side using a vernier (scale that slides along the viewfinder), the operator is able to scan the entire series of lines produced and make measurements.

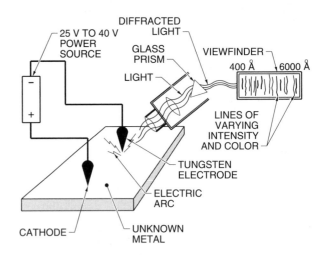

Figure 9-18. Optical emission spectroscopy (OES) uses the light emitted from unknown metal surfaces to identify the metal.

The lines can also be compared with those obtained from known standard metals. A camera that is connected to the eyepiece of the optical emission spectrometer permanently records the lines. The camera improves the sensitivity of the instrument because it records lines that are too faint for detection by the human eye. The camera also records lines from ultraviolet light.

Quantitative Field Identification

Quantitative field identification techniques separate and identify metals by chemical analysis of many chemical elements present in a metal. Although quantitative field identification techniques do not analyze for every chemical element that may be present, they are often comprehensive enough to identify many unknown metals with a high degree of accuracy. Quantitative field identification techniques have the advantage of being nondestructive, but the instrumentation used is higher in cost than qualitative field identification instrumentation.

Quantitative field identification techniques use X-ray fluorescence spectrography (XRF), which includes EDXA and WDXA. In this case, the techniques are used for field identifications of unknown metals. A gamma ray beam or X-ray beam is used to identify the unknown metal. The beam causes the atoms of specific chemical elements in the metal to fluoresce (exhibit fluorescent X rays). Fluorescent X rays have energy levels and wavelengths that are characteristic of the specific chemical elements in an unknown metal. The fluorescent characteristic X rays may be passed through one of two types of detectors that measure energy levels (EDXA) or wavelength intensities (WDXA). See Figure 9-19.

EDXA consists of placing the probe of the EDXA instrument over an unknown metal. A shutter in the probe is opened for a specific length of time to allow gamma rays from the source (such as iron 55 radioisotope) to be beamed on the unknown metal. A microprocessor built into the EDXA instrument displays the percentages of the chemical elements present in the unknown metal. The microprocessor is also programmed to display the names of specific alloys when the analysis matches the chemical composition of the unknown metal. EDXA instruments are portable and easily moved to the component to be analyzed. Within minutes, EDXA instruments can provide quantitative analysis of the indicated chemical elements.

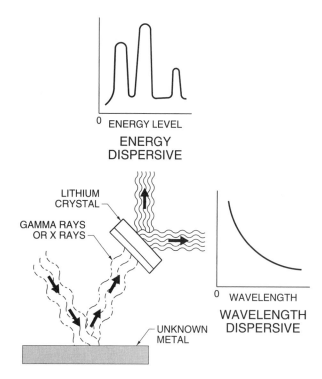

Figure 9-19. X-ray fluorescence spectrography (XRF) uses a detector that separates and identifies energy levels or energy wavelengths.

WDXA consists of placing the probe of the instrument over an unknown metal. A goniometer (detector system) is used to identify the chemical elements present in the unknown metal. The goniometer uses a lithium crystal to diffract the fluorescent X rays, which are supplied by an X-ray generator, into characteristic wavelengths. These characteristic wavelengths are used to identify the unknown metal. The X-ray generator used for WDXA must be water-cooled, which makes the instrument bulky and less mobile when compared to the EDXA instrumentation.

If the surface on which the probe is placed is not flat, a compensation factor must be applied. The factor is incorporated because curvature or gross surface irregularities cause the X-ray beam to miss part of the surface, resulting in lower than expected X-ray fluorescence and erroneous readings. Small sections of metal, such as weld wire, must be cut into pieces and banded together to provide an adequate cross-sectional area for the probe.

EDXA instruments such as the Alloy Analyzer® and the Metallurgist® are extremely portable and are used for quantitative field identification. See

Figure 9-20. WDXA instrumentation such as the Portaspec® are less portable because of the size of the X-ray generator and its cooling system. See Figure 9-21. Because WDXA instrumentation has less mobility, the unknown metal specimen is often brought out of the field to be identified.

The chief advantage of EDXA is that it can provide quantitative analysis of the indicated chemical elements within a few minutes. WDXA is significantly more time-consuming than EDXA because it analyzes for each chemical element one at a time, rather than all of them simultaneously. WDXA requires an additional 30 minutes to make a metal identification. The chief advantage of WDXA over EDXA is the significantly greater number of chemical elements that are detectable.

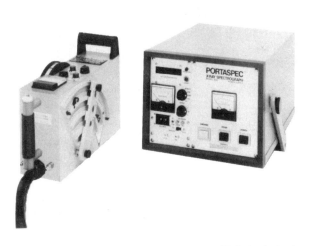

TN Technologies, Inc.

Figure 9-20. EDXA instrumentation, such as the Metallurgist®, are extremely portable for quantitative field indentification.

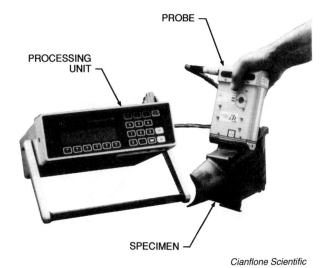

Cianflone Scientific

Figure 9-21. WDXA instrumentation, such as the Portaspec®, is less portable because of the size of the X-ray generator and its cooling system.

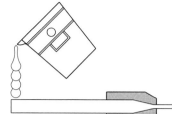

Materials Standards 10

Materials standards are developed by governmental, industrial, international, and national standards-developing organizations. A standard is a document (agreement) that serves as a model in the measurement of a property or the establishment of a procedure. Standards are created to ensure that materials meet the design specifications, which include mechanical, physical, and chemical property requirements. A material is acceptable if it meets the specified requirements that have been adopted by the various organizations. Written standards are used by suppliers and buyers as a basis for ordering materials.

STANDARDS DEVELOPMENT

A *standard* is a document (agreement) that serves as a model in the measurement of a property or the establishment of a procedure. The latest issue of any standard is always used because it supersedes all other issues of the standard. This rule applies to all standards. Standards are usually developed by committees of qualified personnel.

The first step of developing a standard is the production of a draft document that is reviewed by a committee that refines and improves it. Each subsequent draft is reviewed in a similar manner and the changes are balloted (voted on). The developing process may take several years, but the final document represents a consensus of committee opinion and indicates current industrial practices. A standard must be reviewed regularly (minimum of once every five years) to determine whether it will be reaffirmed or revised. If it is determined that the standard no longer contains relevant information, it is canceled.

Standards

The three classes of standards are specifications, test methods, and recommended practices. A code is any one of the three classes of standards and is

legally binding. See Figure 10-1. Standards are developed by committees having appropriate technical qualifications in the area addressed by the document.

Specification. A *specification* (material requirement) is a statement of technical requirements and commercial requirements that a product must meet. For example, alloy-steel and stainless steel bolting material for high temperature service are covered by *American Society for Testing and Materials A 193 (ASTM A 193)*.

Test Method. A *test method* is a set of instructions for the identification, measurement, or evaluation of the properties of a material. For example, a test method for notched bar impact testing of metallic materials is covered by *ASTM E 23*.

Recomended Practice. A *recommended practice* is a set of instructions for performing one or more operations or functions other than the identification, measurement, or evaluation of a material. For example, a recommended practice for the surface preparation of steel and other hard materials by water blasting prior to coating or recoating is covered by *National Association of Corrosion Engineers RP-01-72* (*NACE RP-01-72*).

Code. A *code* is a standard or set of applicable regulations that a jurisdictional (law enforcing) body has lawfully adopted. For example, the *American Society of Mechanical Engineers* (*ASME*) *Boiler and Pressure Vessel Code* contains regulations for the post-weld heat treatment of pressure vessels based on material type and thickness. The applicable heat treatment must be performed on welded pressure vessels in specific geographical locations where the *Boiler and Pressure Vessel Code* is a law.

STANDARDS

ASTM A 193
SPECIFICATION

FOR ALLOY STEEL AND STAINLESS STEEL BOLTING MATERIAL FOR HIGH-TEMPERATURE SERVICE

ASTM E 23
TEST METHOD

NOTCHED BAR IMPACT TESTING OF METALLIC MATERIALS

NACE RP-10-72
RECOMMENDED PRACTICE

SURFACE PREPARATION OF STEEL AND OTHER HARD MATERIALS

ASME
Boiler and Pressure Vessel Code
CODE

SECTION VIII, DIVISION 1, PARAGRAPH UCS-56 POST-WELD HEAT TREATMENT OF CARBON STEEL PRESSURE VESSELS

Figure 10-1. Standards are developed by committees that have appropriate technical qualifications in the area addressed by the document.

STANDARDS-DEVELOPING ORGANIZATIONS

Standards are written by committees of organizations that represent the producers and purchasers of materials. See Figure 10-2. These organizations include trade associations, technical societies, the American Society for Testing and Materials (ASTM), national and international standards organizations, United States government departments, and private organizations.

Trade Associations

Trade associations are organizations that represent the producers of metals or specific types of products. Metal manufacturers' trade associations were often the first to develop organized methods of designating the various alloys specific to their products and industry. Some of the other trade associations develop metal product standards to facilitate commerce within their business specialties.

The American Iron and Steel Institute (AISI), an organization that developed a series of standard designations for steel products, also developed the series *Steel Products Manuals*. The AISI designation system for carbon steel and low-alloy steel is a four digit system. For example, 1030 is a designation for a low-carbon steel containing between .28% C and .34% C. The AISI designation system is closely related to the system produced by the Society of Automotive Engineers (SAE). These designations are referred to as AISI-SAE designations.

For stainless steels, the designation system consists of three digits and is sometimes followed by a letter. For example, 304L stainless steel is a low-carbon version of 302 stainless steel. Steels in the 300 series are austenitic stainless steels. For tool steels, the designation system consists of a letter followed by a number. For example, D2 belongs to a family of high-carbon, high-chromium tool steels. All high-carbon, high-chromium tool steel family members have the prefix D.

Steel Products Manuals indicate steelmaking processes, chemical analysis ranges, available shapes, and tolerances for steel, stainless steel, and tool steel products (bar, sheet, and plate). *Steel Products Manuals* facilitate selection of steel and steel products based on accepted manufacturing conditions.

The Copper Development Association (CDA) is an organization that designates copper and copper alloys by means of a three digit numbering system. For example, copper alloy 836 is a cast leaded red brass containing specified amounts of zinc, lead, and tin. The CDA also publishes the series *Copper Standards Handbooks*, which provides data on manufacturing methods and available forms for copper alloy products.

The Aluminum Association (AA) is an organization that uses a four digit numbering system for wrought aluminum alloys, such as alloy 6011, and a three digit system for cast alloys, such as alloy 356. A temper designation (code) is used to indicate the mechanically worked or heat-treated condition of any alloy. The AA also publishes the series *Aluminum Alloy Data Books*. This series indicates the available forms of aluminum alloys and tolerances for aluminum alloys.

The Alloy Castings Institute (ACI) of the Steelfounders Society of America is an organization that uses a designation system for corrosion-resistant and heat-resistant cast alloys. The letter C indicates the corrosion-resistant series, and the letter H indicates the heat-resistant series. For example, CF-8 is a corrosion-resistant stainless steel, and HK-40 is a heat-resistant stainless steel.

The Manufacturers Standardization Society (MSS) of the Valve and Fittings Industry is an organization that publishes standards on steel castings for valves, flanges, fittings, and other piping components. For example, *MSS SP-55 Quality Standard for Steel Castings for Valves Flanges and Fittings and other Piping Components* is one of the standards developed by MSS.

The American Petroleum Institute (API) is an organization that publishes many standards for the production and handling of petroleum and petroleum products. For example, *API Std 650 Standard for Welded Steel Tanks for Oil Storage* is an API standard. The API also publishes and administers the pipeline welding code *API Std 1104 Standard for Welding Pipelines and Related Facilities.*

The American Water Works Association (AWWA) is an organization that publishes standards on components for handling municipal water supplies, such as steel tanks and piping. For example, *AWWA C106 Gray Iron Pipe Centrifugally Cast in Metal Molds for Water and Other Liquids* is one standard that the AWWA has developed.

SELECTED STANDARDS DEVELOPING ORGANIZATIONS

National Association of Corrosion Engineers

COPPER DEVELOPMENT ASSOCIATION INC.

ALUMINUM ASSOCIATION

AMERICAN IRON AND STEEL INSTITUTE

AMERICAN WELDING SOCIETY

Figure 10-2. Standards are written by committees of organizations that represent the producers and purchasers of materials.

Technical Societies

Technical societies are organizations that are comprised of groups of engineers and scientists united by a common professional interest. Several of these technical societies develop metals and other materials standards.

The American Society of Mechanical Engineers is an organization that administers the *Boiler and Pressure Vessel Code*. The *Boiler and Pressure Vessel Code* is published in a series of volumes (*The Code*) and is divided into 11 sections. See Figure 10-3. The volumes contain materials and fabrication guidelines for the building and repair of boilers, unfired pressure vessels, and nuclear plant components. ASME-approved materials (ferrous, nonferrous, and welding filler metals) have the prefix S added to the ASTM or AWS designation. For example, ASTM A 36 becomes ASME SA-36, and AWS A5.1 becomes ASME SA-5.1.

BOILER AND PRESSURE VESSEL CODE	
Section	**Title**
I	Power Boilers
II	Materials Specifications A–Ferrous B–Nonferrous C–Welding Rod, Electrodes, and filler metals
III	Nuclear Power Plant Components
IV	Heating Boilers
V	Nondestructive Examination
VI	Recommended Rules for Care and Operation of Heating Boilers
VII	Recommended Rules for Care of Power Boilers
VIII	Pressure Vessels
IX	Welding and Brazing Qualifications
X	Fiberglass-Reinforced Plastic Pressure Vessels
XI	Rules for Inservice Inspection of Nuclear Power Plant

Figure 10-3. The *Boiler and Pressure Vessel Code* is published in a series of volumes (*The Code*) and is divided into 11 sections.

Companies that insure boilers, pressure vessels, and power plants employ inspectors whose duty it is to confirm compliance with the requirements of *The Code* when equipment is fabricated or repaired. Strict enforcement of *The Code* during fabrication and repair has resulted in a significant reduction in the number of such items that have failed catastrophically, with attendant death, attendant injury, and property damage. ASME also administers the various codes for pressure piping, such as *B31.3 Chemical Plant and Petroleum Refinery Piping*.

The American Welding Society (AWS) is an organization that publishes standards on welding, brazing, and soldering of metals. For example, AWS Specifications A5.0 and A5.28 cover welding filler metals. AWS also administers *AWS D1.1 Structural Welding Code*, which provides rules for the construction of bridges, buildings, structures, and practically any other welded structure to which the *Boiler and Pressure Vessel Code* or other codes do not apply.

The Society for Automotive Engineers is an organization that develops and administers specifications for highway and off-road vehicles. SAE specifications are published annually in the *SAE Handbook*. The SAE also administers the approximately 2000 *Aerospace Materials Specifications* (*AMS*). These include procurement requirements for high quality materials for aerospace applications. For example, *AMS 4906 6A1-4V Continuously Rolled and Annealed Sheet and Strip* is one of the AMS administered by SAE.

The National Association of Corrosion Engineers is an organization that publishes standards for materials operating in corrosive environments. For example, *NACE Material Requirement MR-01-75 Sulfide Stress Cracking Resistant Metallic Material for Oil Field Equipment* is a standard for materials used for oil field equipment.

The American Society for Testing and Materials Standards

The American Society for Testing and Materials is a technical society, which is the world's largest source of voluntary consensus standards. ASTM develops standards on the characteristics and performance of materials, products, systems (such as magnetic particle examination), and services (such as assessment of test laboratories). ASTM standards are written by committees representing a balance of producers, end users, and general interest factions.

ASTM Book of Standards is published yearly and consists of approximately 68 volumes, containing approximately 8000 standards. The volumes are divided by subject matter, so the users can purchase the volumes appropriate for and related to their business interests. For example, there are five volumes dedicated to steel products.

The designation system for an ASTM standard consists of a letter (indicating the general classification of the standard), a space, a one to four digit serial number (the standard number), a dash, a two digit number (indicating the year of adoption or latest revision), and the title of the standard. See Figure 10-4. The metric version of a standard includes the letter M, which follows the standard number. Metric versions are identical to the other versions, except that all dimensional data is indicated in metric units. A tentative standard (issued on a trial basis) has the letter T following the year. If a standard is revised more than once in a year, the year of revision followed by the letter a is used for the first revision, b is used for the second revision, and so on.

For example, *ASTM B 408-87* is a specification for nickel-iron-chromium rod or bar published or last revised in 1987. At one time this standard existed as *ASTM B 408-77T*, meaning it was a tentative standard, and published or last revised in 1977. *ASTM A 36-81a* was a specification for structural steel that was published or last revised in 1981, and revised a second time that year. It now has a designation of *ASTM A 36/A 36M-89*, which means the standard is shown in both U.S. and metric systems of measurement and was published or last revised in 1989. Metric standards may also have a separate designation. For example, *ASTM B 133-89* is a specification for copper rod, bar, and shapes, whereas *ASTM B 133M-89* is the metric equivalent of the same standard.

In any reference to ASTM standards, the two digit year designation is left out. The latest issue of any standard is always used because it supersedes all other previous issues of the standard. This rule applies to all standards, not just ASTM standards.

Many ASTM standards contain supplementary requirements that are incorporated at the end of a standard and are identified by the letter S followed by a number. Supplementary requirements allow the user to select additional quality control requirements over and above those indicated in the general body of the standard. For example, a user may require a product analysis for a specific component over and above the normal heat analysis provided by the vendor. This requirement would be indicated

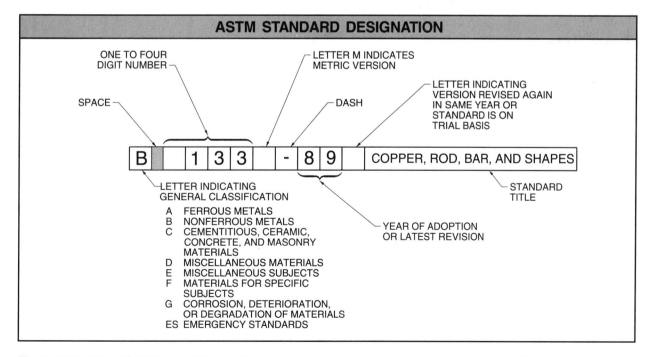

Figure 10-4. The *ASTM Book of Standards* is published yearly and consists of approximately 68 volumes, containing approximately 8000 standards.

by referencing the relevant supplementary requirement. Supplementary requirements are selected with care because they are performed at the expense of the purchaser.

Unified Numbering System (UNS). To satisfy the need for a common designation system for all families of alloys, ASTM and SAE jointly developed the UNS for metals and alloys. UNS is a common designation system that unifies all families of metals and alloys. UNS does not establish any requirement for form, condition, properties, or quality. It uniquely identifies that the chemical composition of alloys have been fixed by other specifications, such as one of the metal manufacturers' organizations. If the alloy is proprietary (produced by a limited number of suppliers), the chemical composition is established by the producer.

There are 17 series of UNS designations. See Figure 10-5. Each series consists of a capital letter followed by five numbers. The letter identifies the alloy family, and where possible, the five numbers are related to the pre-UNS designation of the alloy. For example, copper alloy C61400 was formerly CDA 614 (7% aluminum bronze), and S30400 was formerly AISI 304 stainless steel. Although the UNS has superseded many of the established designation systems such as those of the CDA, many standards indicate both designations.

Although many of the ASTM standards reference alloys by the numerical designations, such as UNS G10300 (AISI 1030) carbon steel, an equally large number does not reference any standard alloy designation. These ASTM standards are oriented toward achieving particular mechanical properties in the product and allow the chemical composition of the alloy to be adjusted to achieve the property requirements. For example, the tensile requirement for ASTM A 36 structural steel is achieved by substituting AISI 1030 steel. No alloy designation is indicated in the ASTM standard, which simply states the mechanical properties required for the specific product as structural steel shapes.

Some ASTM standards utilize internally developed designations for alloys. For example, grade B7 is a fastener equivalent to UNS G41420 (AISI 4142 low-alloy steel). Titanium grade 2 is a commercially pure form of titanium containing a specified amount of oxygen and is equivalent to UNS R50400. Many of these internal ASTM designations were developed when there was no alternative generic designation for a specific alloy.

UNS DESIGNATION	
Series	**Metals and Alloys**
Axxxxx	Aluminum and aluminum alloys
Cxxxxx	Copper and copper alloys
Exxxxx	Rare earth and similar metals and alloys
Fxxxxx	Cast irons
Gxxxxx	AISI and SAE carbon and alloy steels
Hxxxxx	AISI and SAE H-steels
Jxxxxx	Cast steels (except tool steels)
Kxxxxx	Miscellaneous steels and ferrous alloys
Lxxxxx	Low melting metals and alloys
Mxxxxx	Miscellaneous nonferrous metals and alloys
Nxxxxx	Nickel and nickel alloys
Pxxxxx	Precious metals and alloys
Rxxxxx	Reactive and refractory metals and alloys
Sxxxxx	Heat and corrosion resistant steels (including stainless steel), valve steels, and iron-base superalloys
Txxxxx	Tool steels, wrought and cast
Wxxxxx	Welding filler metals
Zxxxxx	Zinc and zinc alloys

Figure 10-5. UNS designations consist of a letter identifying the metal and alloy family and five numbers relating to the pre-UNS designation of the alloy.

National and International Standards Organizations

National and international standards organizations are sometimes affiliated with governmental organizations. National and international standards organizations include the American National Standards Institute (ANSI), the Canadian Government Standards Board (CGSB), foreign national standards organizations, European Community standards, and the International Organization for Standardization (ISO).

The American National Standards Institute is the coordinator of the American National Standards system. ANSI is a standard developing organization that adopts standards that are written and approved by member organizations. ANSI branches out and connects its member organizations by unifying their adopted standards. See Figure 10-6. ANSI also manages United States participation in international standards activities. ANSI is a source for many foreign standards.

ANSI approves the standards written by its member organizations as American National Standards when its member organizations agree that approval should be given. See Appendix. An approved standard retains the sponsor organization designation, but additionally carries on the title page a descriptor indicating it is an American National Standard. For example, ASTM A 36 is also an ANSI standard and is officially designated as ANSI/ASTM A 36.

Although ANSI generally coordinates standards written by other organizations, it also publishes some standards under its own name. These coordinated standards include standards for business data exchange, such as ANSI X-12.

Canadian Standards are administered by the Canadian Government Standardization Board (CGSB). These standards are identified by the letters CSA followed by a number.

Foreign national standards are developed by standards setting organizations in many countries and are often in coordination with government agencies. Foreign national standards are identified by their prefix letters. For example, the JIS prefix identifies the Japanese national standards and, the B.S. prefix identifies the Great Britain standards. See Figure 10-7.

European Community standards are developed by the European Community for Standardization and are known as Euronorms. Euronorms are identified by the letters EN followed by a number.

The International Organization for Standardization is a nongovernmental international organization that is comprised of national standards institutions of 91 countries (one per country). ANSI is the ISO representative for the United States. The main function of ISO is to provide a worldwide forum for the standards developing process.

United States Government Departments

Among the departments of the United States government that develop materials related standards, the work of the Department of Defense (DOD) is the most significant. DOD standards are known as *United States Government Military Standards* (*MIL Standards*). MIL Standards cover the specifications of many materials used by the armed forces, but are not to be restricted to them. For example, *MIL-C-267074B Requirements for Electroless Nickel Coatings* is equally applicable in civilian usage.

Other United States governmental organizations that develop materials related standards are the Department of Commerce (DOC), Department of Energy (DOE), Department of Transportation (DOT), Environmental Protection Agency (EPA), and Department of Agriculture (USDA). An effort

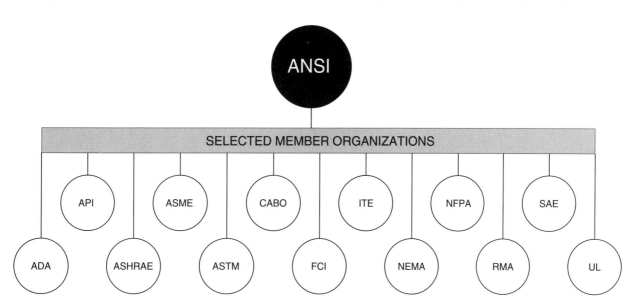

Figure 10-6. ANSI branches and connects its member organizations by unifying their adopted standards.

to merge the United States government standards into ASTM and other standards is being made to simplify the standards system.

STANDARDS SETTING COUNTRIES	
Country	**Prefix Letters**
Austria	ÖNORM
Belgium	NBN
Bulgaria	BDS
Canada	CSA
Czechoslovakia	CSN
France	AFNOR
Germany	DIN
Great Britian	B.S.
Hungary	MSZ
Italy	UNI
Japan	JIS
Poland	PN
Romania	STAS
Spain	UNE
Sweden	SS
Yugoslavia	JUS

Figure 10-7. Foreign national standards are identified by the prefix letters of the standard setting countries from which they originate.

Private Organizations

Private organizations create standards that are an accumulation of an organization's knowledge and experience with materials, methods, and practices. These standards are modified from other standards to meet the needs of the organization. For example, the material grade for heat exchanger tubing in a critical application may reference a specific standard, but the private organization set more strenuous standards to satisfy additional needs. These needs can include higher quality control requirements.

MATERIALS PROCUREMENT SPECIFICATION

A materials procurement specification must be written within the context of the intended application, which will dictate the quality requirements and the amount of quality control needed. For example, vacuum arc remelted aircraft quality forging stock is a key requirement for aircraft landing gear, because the manufacturing procedure improves internal metal purity and fiber structure, which are absolutely necessary for this critical component. The same requirement is overqualified and imposes unnecessary costs if applied to general purpose bolting material.

Whenever possible, a materials procurement specification should reference applicable ASTM standards because ASTM standards are the most widely used and generally the most accepted. References to ASTM standards will ensure that the material will be purchased in accordance to commonly used manufacturing practices and quality control procedures. When necessary, other standards should be referenced, such as in the case of special manufacturing procedures that are not described in ASTM standards. Where no reference standards exist, the particular requirements needed must be indicated. The format for a typical materials standard includes sections on scope, chemical composition, quality, test results, and special requirements.

Scope includes the size range, shape, and manufacturing method for the material. For example, the manufacturing method may be dictated by the end use and where possible, standard specifications should be referenced and documented.

Chemical composition references the UNS and other designations of the material and indicates the allowable composition range for elements in the material. The chemical composition may be indicated by heat analysis or product analysis.

Quality requirements indicate the testing procedures required to check that the material meets the specification. ASTM standards frequently include supplementary requirements, which may be specified by the user to incorporate higher quality tests.

Test results include the results expected from any mechanical and physical property tests performed using standard test methods. These methods can include hardness, impact, and tensile tests.

Special requirements include any special tolerances, surface finish, identification, packaging, shipping, inspection, or certification papers used. These may be required over and above what is described in any referenced standards.

Iron and Steel 11

Iron and steel manufacturing consists of four stages, which are ironmaking, steelmaking, production of semifinished forms, and production of finished forms. In ironmaking, iron ore is converted in a blast furnace to pig iron, an impure form of iron. Steelmaking is the refinement of raw materials to produce carbon steel and various alloy steels. The steelmaking furnace is the heart of the production of steel. The types of iron products manufactured are very limited. By contrast, steel products are very diverse. Steel is identified by composition designations and product specifications.

IRONMAKING

Ironmaking is the first stage in the manufacture of steel. Iron is made by reducing iron ore by the process of converting iron oxides to iron. This reduction is performed by blast furnace reduction or direct reduction. The types of iron products manufactured are wrought iron, ingot iron, and enamelling iron.

Blast Furnace Reduction

Blast furnace reduction is the reduction of iron ore to pig iron in a blast furnace. *Pig iron* is an impure form of iron. *Reduction* is the removal of the chemically combined oxygen in the iron ore. Reduction is performed in a blast furnace, which is a complex refractory-lined vertical structure more than 45 m (150 ft) high. See Figure 11-1.

Iron ore, metallurgical grade coke, and limestone are fed in at the top of the blast furnace. At the bottom of the furnace is the hearth where the molten metal is contained. A continuous supply of air, sometimes enriched with oxygen, is blasted in through tuyeres at the bottom of the furnace. The oxygen in the air combines with the carbon to form carbon monoxide, as shown in the following chemical equation:

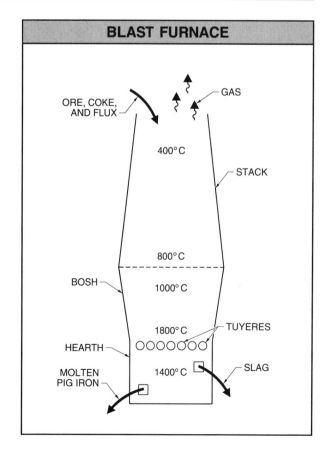

Figure 11-1. The blast furnace is used to convert iron ore into pig iron.

$$O_2 + 2C \rightarrow 2CO$$

The carbon monoxide reduces the iron ore to pig iron, as shown in the following chemical equation:

$$3CO + Fe_2O_3 \rightarrow 2Fe + 3CO_2 \uparrow$$

The heat of the chemical reactions melts the pig iron, which is tapped from the blast furnace every 3 to 4 hours. The purpose of the limestone is to flux (remove) some of the impurities from the molten pig iron. The impurities (slag) float to the top of the molten metal and are tapped through a separate outlet.

Pig iron is very impure and unsuitable for any practical purpose. It contains excessive amounts of carbon, phosphorus, and sulfur. These elements, and also manganese and silicon, must be refined to allowable amounts so the iron can be used in steelmaking. See Figure 11-2.

PIG IRON COMPOSITION	
Element	**Allowable Amounts**
Carbon	3.5% to 4.5%
Phosphorus	.05% to 2.00%
Sulfur	.01% to .10%
Manganese	.5% to 2.0%
Silicon	.3% to 2.0%

Figure 11-2. Elements present in pig iron must be refined to allowable amounts to meet steelmaking specifications.

The molten pig iron is usually poured into special refractory-lined railroad cars and transported to steelmaking furnaces. Molten pig iron also can be cast into small lengths on a pig casting machine. These castings are used for remelting and later usage, such as steelmaking.

Direct Reduction

Direct reduction (DR) is several reduction processes that produce metallic iron (direct-reduced iron, or DRI) from ores by removing most of the oxygen at temperatures below the melting points of the materials in the process. *Direct-reduced iron* is an iron product made by using methods other than blast furnace reduction. DR is performed in a variety of furnaces, which include a fluidized bed, a moving-bed shaft, a fixed-bed retort, and a rotary kiln. DR is not widely used because blast furnace reduction is a more energy efficient process, or because competitively priced scrap steel is available.

DRI is produced in varying carbon contents. The range is from the very low levels present in pure iron to the high levels in pig iron. It is usually produced in sponge or pellet form and is sent to steelmaking furnaces for conversion into steel.

Iron Products

The principal commercial iron products are wrought iron, ingot iron, and enamelling iron. Except for the names, there are no industry-wide specifications or designations for these products. Commercial iron products contain extremely low levels of carbon, which is the principal strengthening component in steel. As a result, commercial iron products are soft, ductile, and malleable. Depending on the carbon content and the impurity level, iron products may also exhibit magnetic and other desirable properties.

Wrought Iron. *Wrought iron* is pure iron containing silicate slag elongated in the hot-working direction. Wrought iron is made by several processes, the most significant being the Aston Process, or Byers Process. The first step in this process is to make steel, which is poured from the steelmaking furnace into a ladle containing a molten silicate slag. The temperature of the slag is maintained at several hundred degrees below the freezing point of the steel. This makes it solidify rapidly.

The rapid solidification liberates dissolved gases with enough force to shatter the molten metal into small fragments that settle at the bottom of the ladle. Because of the high temperature and the fluxing (cleaning) action of the slag, the fragments weld together to produce a ball (sponge-like mass). Immediately after pouring the steel into the ladle, it is transferred to another area so that the excess slag may be poured off. The sponge ball is dumped from the ladle and pressed into a bloom or billet suitable for rolling into various types of finished forms.

Wrought iron contains little carbon (<.1% C) and manganese (<.07% Mn), and a relatively high amount of silicate slag (approximately 2% Si). The microstructure consists of elongated silicate slag stringers in a matrix of almost pure iron. *Stringers* are the elongated configuration of foreign material aligned in the direction of working. The principal qualities of wrought iron, such as ease of welding and corrosion resistance, are dependent on the silicate slag. See Figure 11-3.

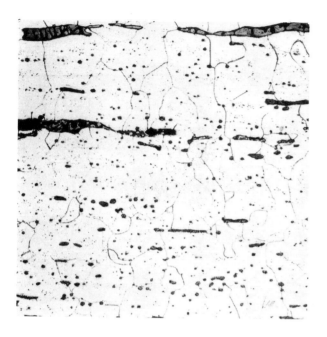

Figure 11-3. The microstructure of wrought iron consists of ferrite grains incorporating a large quantity of silicate slag elongated in the working direction.

Wrought iron is available as blooms or billets and in all types of hammered bars. Wrought iron is used chiefly in architectural applications, such as gates and fences.

Ingot Iron. *Ingot iron* is extremely pure iron that is soft and easily magnetized and demagnetized. It is used in soft magnets and in gaskets. Elements such as carbon, sulfur, nitrogen, and oxygen are reduced to very low values by refining operations. For example, the carbon content of ingot iron is <.003% C. Ingot iron is also known by commercial names, such as Armco Iron.

Enamelling Iron. *Enamelling iron* is a sheet product made by decarburizing sheet steel. *Decarburizing* is an annealing process in which the carbon is removed from the steel by diffusion. The sum of all the elements other than iron is <.03%. Enamelling iron is sometimes preferred over regular enamelling grades of steel for porcelain enamelled products. For example, it is used in refrigerators and washing machines because of its superior enamelling and forming properties.

STEELMAKING

Steelmaking consists of refining pig iron, directly reduced iron, and scrap steel into carbon steel and various alloys of steel. Steelmaking plants vary in size and complexity. This depends on the types of products made in the plants and the types of raw materials they process. The steelmaking furnace is the most important part of the steelmaking plant and steelmaking process.

Steelmaking is performed in two stages that consist of oxidation and deoxidation. *Oxidation* is a high-temperature reaction in which the metal forms an oxide. *Deoxidation* is the removal of oxygen from molten metal. To achieve steel with greater internal cleanliness, specific ladle refining processes are performed following the deoxidation stage. The production of semifinished forms is done using one of two processes, which are ingot casting and continuous casting.

The production of finished forms is achieved by rolling, forging, or drawing semifinished forms into a wide variety of items suitable for end use, such as structural shapes, wire rod, wire, tubing, and flat shapes. Carbon and alloy steel products are available in an extremely large number of compositions and forms. Steel products are specified by their composition, manufacturing method, and product form.

Steelmaking Plants

The two basic types of steelmaking plants are integrated steelworks and merchant mills. *Integrated steelworks* are large and complex operations consisting of all the necessary production units to manufacture a wide range of semifinished forms in carbon steels and alloy steels. They use iron ore and scrap steel as starting materials. Integrated steelworks offer economy of scale and a diversity of finished steel product forms.

Merchant mills (minimills) are smaller operations that produce a limited range of carbon steel finished forms. For example, they produce reinforcing steel bars, hot-rolled bars, light structural products, or wire rod. Merchant mills use scrap steel. As a result, there is no need for a blast furnace and its supporting equipment, such as a coke oven to manufacture metallurgical grade coke from coal to make pig iron. The scrap steel is melted in an electric furnace and then continuously cast into blooms, billets, slabs, or rounds. These semifinished forms are then rolled into the finished forms that comprise the merchant mill product line.

Steelmaking Furnaces

The steelmaking furnace is the heart of the production of steel. It is where raw materials are converted into molten steel. There are three types of steelmaking furnaces, which are the basic oxygen furnace, the open hearth furnace, and the electric furnace.

Basic Oxygen Furnace. The basic oxygen furnace is a large pear-shaped steel barrel lined with refractory. See Figure 11-4. The furnace is tilted so that the charge of molten pig iron and scrap steel can be poured into its mouth from the scrap-charging car. Depending on the furnace size, the charge weight can be from 80 to 350 tons, with 200 tons being the most common. An oxygen lance (pipe) is lowered into the furnace about six feet above the charge, and oxygen is blown in at very high velocities. This makes the oxygen ignite. The submerged injection (Q-BOP) process uses the basic oxygen furnace where the oxygen is blown through the bottom of the furnace.

Various fluxes are added in the hopper shortly after ignition. Fluxes combine with some of the undesirable constituents to form a slag. Other undesirable constituents escape in the hood (top of the vessel) as gases. When the steelmaking reaction is complete, the furnace is tilted and the refined steel is tapped into a ladle. A *ladle* is a refractory-lined receptacle, which is used for transferring and pouring the molten steel. Ladles usually have a capacity less than that of the furnace.

The basic oxygen furnace takes about an hour to produce a heat of steel. Since a large proportion of the charge is molten pig iron, the basic oxygen furnace is usually found in an integrated steelworks.

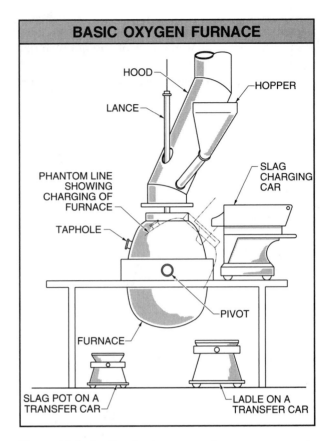

Figure 11-4. In the basic oxygen furnace, oxygen is lanced into the molten pig iron and scrap steel.

Open Hearth Furnace. The open hearth furnace is a large, refractory-lined shallow basin with an arched roof. See Figure 11-5. The furnace is charged with a mixture of liquid pig iron, solid pig iron, scrap steel, and fluxes. As much as 600 tons of metal may be charged in a single heat.

Liquid or gaseous fuel is burned with air or oxygen to melt the solid scrap and pig iron. After the burning gases pass over the molten mass in the center of the hearth, they are collected on the other side of the furnace and routed through a checker chamber. A *checker chamber* is a chamber containing a complex arrangement of heat absorbent bricks. The passage of the hot gases through the checker chamber heats the bricks. Incoming air to the furnace is then passed over these same hot bricks. This preheats the air and helps to improve the efficiency of the furnace.

After 15 to 20 minutes, the burning flame reverses direction and enters the hearth from the opposite side. The hot gases then heat the bricks on the other side as they pass over a second checker chamber.

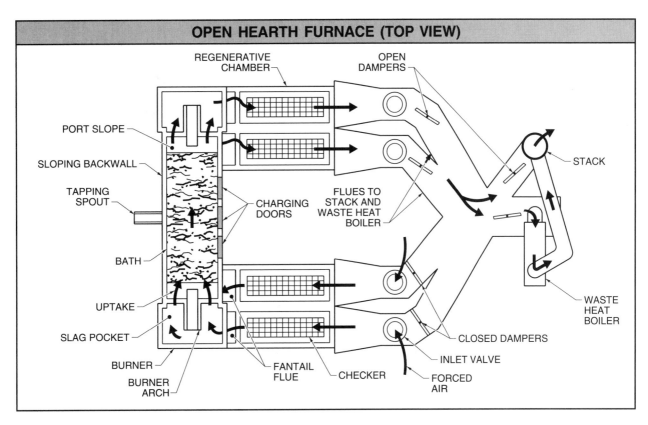

Figure 11-5. The open hearth furnace can make large amounts of steel in one heat.

This operation is reversed every 15 to 20 minutes to maintain the efficiency of the furnace. When the steelmaking reaction is complete, the furnace is tapped and the refined steel is poured into a ladle.

Open hearth furnaces are large and take 6 to 10 hours to make a heat of steel. These furnaces are situated inside integrated steelworks. Open hearth furnaces are becoming more rare. They are being replaced by the more efficient furnaces, such as the basic oxygen or electric furnaces.

Electric Furnace. The electric furnace differs from the basic oxygen and open hearth furnaces in that it uses electricity to supply the heat rather than gaseous or liquid fuel. There are several types of electric furnaces. The most common for steelmaking is the electric-arc furnace. See Figure 11-6. It resembles a giant, refractory-lined steel kettle with a spout at one end. The roof pivots and swings to one side so that the charge and flux can be loaded. Three large retractable electrodes extend up through the roof of the furnace.

The electrodes are lowered through the retractable roof into the furnace to contact the charge. A three-phase alternating current is passed through the electrodes. The current melts the charge. Although the capacity of most electric furnaces is usually much lower than the others, as much as 300 tons of steel may be manufactured in a single heat. Other reagents, such as flux, are added to refine the impure steel. It takes from 3 to 7 hours to make a heat of steel in an electric furnace. The furnace is tipped to pour the molten steel into a ladle. The slag is poured into a slag pot.

Electric furnaces are extremely versatile and may be operated in air or vacuum. They consume scrap, solid pig iron, and virgin (pure) materials. The virgin materials are supplied mainly in two forms, which are ingot iron and ferroalloys. Ingot iron is commercially pure iron. *Ferroalloys* are specific alloys of iron and other chemical elements that are used to introduce alloying elements into the steel.

Because of their versatility and the wide range of sizes available, electric furnaces are not restricted to the manufacture of carbon steel. They are used to make every kind of steel, including stainless steels and tool steels. This was the primary use of electric furnaces until the 1960s. Technological advances made the use of electric furnaces for the production of carbon steels competitive.

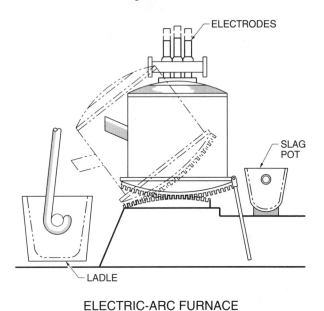

ELECTRIC-ARC FURNACE

Figure 11-6. The electric-arc furnace is the most common furnace used in making many alloys of steel.

Stages of Steelmaking

There are three stages of steelmaking. The first stage is oxidation, the second stage is deoxidation, and the final stage is ladle refining.

Oxidation. Each steelmaking process is designed so that controlled amounts of oxygen can be supplied to the molten charge to combine with unwanted elements in the impure metal. As oxidation proceeds, carbon and other elements that can be oxidized in preference to iron combine with the oxygen. These form compounds that are evolved as gases or separate as solid phases in the melt and can be removed by the flux, as shown in the following chemical equations:

$$2C + O_2 \rightarrow CO \text{ (gas)},$$

$$Si + O_2 \rightarrow SiO_2 \text{ (solid)},$$

and

$$Mn + 2FeO \rightarrow MnO_2 \text{ (solid)} + 2Fe \text{ (solid)}$$

Deoxidation. During deoxidation, a controlled amount of oxygen is removed from the molten steel. The deoxidation practice determines the amount of deoxidation performed and the basic steel type that is produced in the steelmaking operation. There are four levels of deoxidation practice. They influence the macrostructure of the ingot and result in the production of rimmed steel, capped steel, semikilled steel, and killed steel. See Figure 11-7.

Rimmed steel is steel with little or no deoxidation. The molten metal briskly bubbles as oxygen is evolved from it when it is poured into the mold and begins to solidify. The evolving oxygen reacts with the carbon at the boundary between the solidified metal adjacent to the mold and the remaining molten metal, forming carbon monoxide gas. This reaction causes the outer rim of the solidified metal to be very low in carbon and, consequently, very ductile. The resulting product is rimmed steel, which may be rolled to produce a very sound surface. Rimmed steels are used for sheet products, such as automobile bodies.

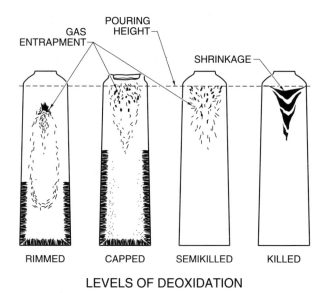

LEVELS OF DEOXIDATION

Figure 11-7. Each level of deoxidation practice exhibits a significant influence on the macrostructure of the ingot.

Capped steel is steel that is produced by a variation of the rimmed steel practice. The rimming (bubbling) action is allowed to begin, but it is then terminated after approximately 1 minute by sealing the mold with a cast iron cap. The surface condition of capped steel is very much like that of rimmed steel, but other characteristics are intermediate, between that of rimmed steel and semikilled steel.

Semikilled steel is steel in which a deoxidizer such as aluminum or silicon is added to the molten metal. These deoxidizers partially kill (suppress) the oxygen-carbon reaction, which leads to the bubbling in rimmed steel. Semikilled steel is more uniform in composition throughout the cross section of the ingot and is suitable for applications involving hot forging, cold extrusion, carburizing, or heat treatment.

Killed steel is steel that is completely deoxidized by a deoxidizing agent. The strongest and most often used is aluminum, but silicon is also used. A sufficient quantity of aluminum or silicon powder is added to the molten metal to completely kill the oxygen-carbon reaction. Aluminum killed steels exhibit fine grain size. The aluminum increases the number of solid nuclei that first form in the molten metal when it is poured into the mold. The fine grain size makes aluminum killed steels inherently tougher than the other types. Killed steels are used where improved toughness is important. For example, they are used in pressure vessels and high-strength components.

Ladle Refining. *Ladle refining* is the various secondary techniques of purifying molten steel prior to solidification. The purpose of ladle refining is to reduce detrimental internal characteristics in steel products such as dissolved gases, sulfur, or nonmetallic inclusions. Ladle refining is not done in the steelmaking furnace, but is performed in a ladle into which the molten steel is poured from the steelmaking furnace. Following the ladle refining process, the molten steel is poured into a tundish or mold, depending on the route to final product. A *tundish*, a refractory-lined receptacle much like a hot tub, is a reservoir at the top of the mold used to aid in the pouring of molten steel.

Specific ladle refining techniques are designed to achieve the desired quality level in the end product. These techniques include inclusion shape control, gas stirring, vacuum degassing, and protection of the pouring stream. Ladle refining processes usually result in killed steels.

Inclusion shape control is a refining technique that is used to alter the morphology of nonmetallic inclusions in steel. Inclusions consist of metal oxides, sulfides, and silicates that are formed during solidification or enter the molten metal by contamination from the refractory lining. Manganese sulfide inclusions, which are common, are formed by chemical combination of the manganese and sulfur in the steel.

Manganese sulfide inclusions become elongated when steel is mechanically worked to produce semifinished and finished forms. This leads to stringers. They are largely responsible for anisotropy of the steel. By adding calcium or rare earth elements into the ladle, it is possible to modify the composition of the manganese sulfide inclusions so that they do not become elongated during mechanical working. Inclusion shape control leads to a significant reduction of anisotropy.

Gas stirring is a refining technique that consists of bubbling gases, such as nitrogen or argon through the ladle, that promotes the floatation and removal of nonmetallic materials. Gas stirring also homogenizes the temperature and composition of the molten metal. Gas stirring helps produce a more uniform final product.

Vacuum degassing is a refining technique where molten steel is subjected to low pressure to help remove hydrogen and oxygen. Hydrogen is a major cause of embrittlement and many other defects in steel. Control of the oxygen content allows for better deoxidation. This control also minimizes the formation of solid oxides present in steels deoxidized by compounds containing elements such as silicon or aluminum. Vacuum degassing is performed by a variety of methods, such as pouring the molten steel into a closed ladle connected to a vacuum pump.

Protection of the pouring stream is a refining technique performed once ladle refining operations are complete. The stream of molten steel poured from the ladle is protected from air. This minimizes the oxygen that is picked up by the molten steel. One way of achieving this is to employ a shroud between the ladle and the tundish and between the tundish and the mold. The shrouds exclude air entry while the molten steel is being poured. See Figure 11-8.

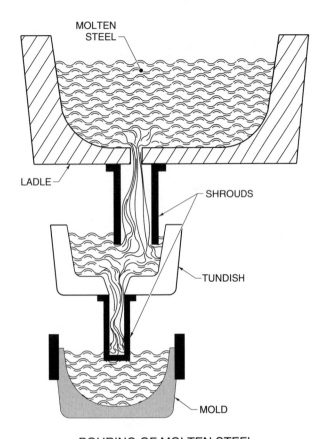

MOLTEN
STEEL

LADLE

SHROUDS

TUNDISH

MOLD

POURING OF MOLTEN STEEL

Figure 11-8. A shroud is used to protect the molten steel from picking up atmospheric gases as it is poured from the ladle into the mold.

Semifinished Forms

Semifinished forms are one of four basic shapes of which all finished steel products are produced, and include blooms, billets, slabs, and rounds. The semifinished forms are manufactured by ingot casting and rolling or continuous casting. Blooms, billets, slabs, and rounds are arbitrarily defined by their shape and dimensions.

A *bloom* is a square-shaped, semifinished form that is greater than 20 cm × 20 cm (8 in. × 8 in.). Where a bloom is too large for rolling into the finished product form, billets are used. A *billet* is a square-shaped, semifinished form that is less than 20 cm × 20 cm (8 in. × 8 in.). A *slab* is a rectangular-shaped, semifinished form that has a width-to-thickness ratio of 2:1 or greater. Slabs are usually 5 cm (2 in.) thick or greater. A *round* is a semifinished form that is a circular section of any diameter. See Figure 11-9.

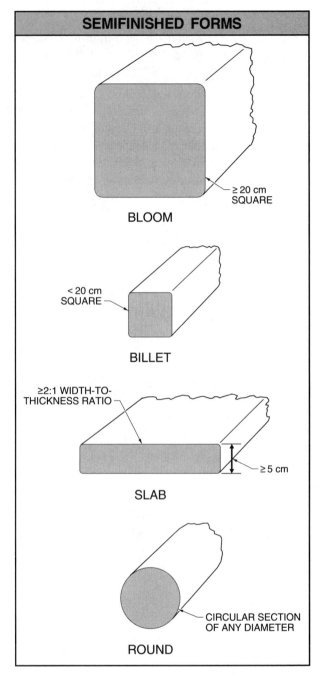

SEMIFINISHED FORMS

≥ 20 cm
SQUARE

BLOOM

< 20 cm
SQUARE

BILLET

≥2:1 WIDTH-TO-
THICKNESS RATIO

≥ 5 cm

SLAB

CIRCULAR SECTION
OF ANY DIAMETER

ROUND

Figure 11-9. The four types of semifinished forms (blooms, billets, slabs, and rounds) are defined by shape and dimensions.

Ingot Casting and Rolling. After all of the steel refining steps are completed, the ladle of molten steel is moved by an overhead crane to a pouring platform. The molten steel is teemed (poured) into a series of ingot molds. The steel solidifies in each ingot mold to form a casted ingot.

An ingot mold is shaped like a box. It is a tall container made of cast iron and weighs as much as one and a half times the weight of the ingots cast in it. The ingot mold is usually tapered to facilitate stripping (removal) of the ingot. The two basic types of molds are big-end-down and big-end-up. There are two variations of the big-end-down mold and three of the big-end-up mold. See Figure 11-10. The mold is set on a stool, and the molten steel is poured into the top of the mold.

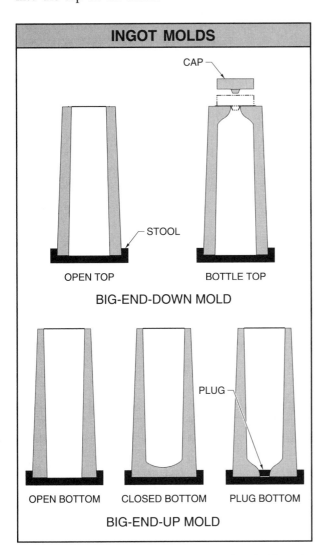

Figure 11-10. The ingot mold is tapered to facilitate its removal from the steel ingot after solidification.

During ingot solidification the molten metal to first contact the mold wall freezes rapidly and consists of small equiaxed (equal dimensions in all directions) grains in a band about 1.25 cm ($\frac{1}{2}$ in.)

wide. Columnar structures then appear. Columnar structures are large dendrites that grow inward principally along the longitudinal axis perpendicular to the mold wall. See Figure 11-11. The ingot is chemically inhomogeneous and is extremely segregated, with inclusions located between the dendrites and near the top of the ingot.

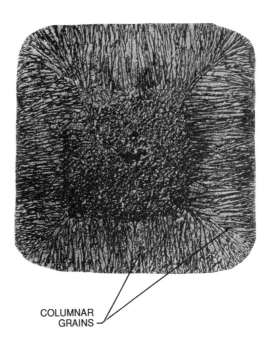

Figure 11-11. The macrostructure of a cast ingot consists of a chill zone of fine equiaxed grains adjacent to the mold wall, columnar grains that grow perpendicular to the wall, and equiaxed grains toward the center.

After solidification, the mold is stripped from the ingot, which is placed in a furnace known as a soaking pit. The burners soak the ingots, or supply heat so that the temperature equalizes through their cross sections. The ingots are removed from the soaking pit and hot rolled in a primary rolling mill (primary mill). A rolling mill consists of a series of parallel rolls through which products are passed in order to reduce their cross section. The first stage of ingot reduction is performed in a primary mill. A primary mill produces semifinished forms of convenient cross section, breaks down and homogenizes the cast structure of the ingot, and seals internal voids in the ingots. It may take a number of passes to achieve the wanted cross section. See Figure 11-12.

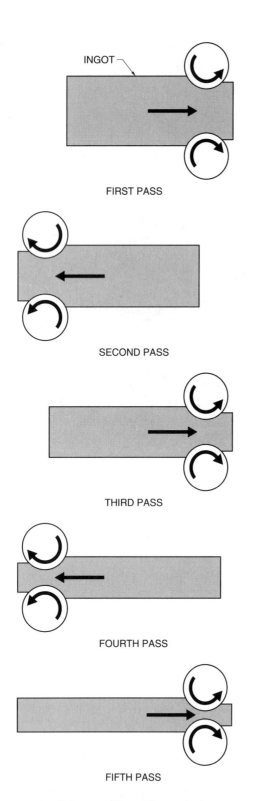

INGOT

FIRST PASS

SECOND PASS

THIRD PASS

FOURTH PASS

FIFTH PASS

Figure 11-12. Primary rolling reduces the ingot to semi-finished forms of the wanted cross section.

Continuous Casting. Continuous casting is a single line operation that replaces the sequential operations

of ingot casting, soaking, and primary rolling for the production of semifinished forms. It is performed using a continuous casting machine. A constant flow of molten steel is supplied to the continuous casting machine by pouring the molten steel from the ladle into a tundish. The tundish discharges into one or more vertical water-cooled copper molds. This causes the skin of the steel to solidify. The steel passes from the molds through a series of water spray, pinch rolls, and bending rolls. This series increases solidification and also changes the direction of the steel from the horizontal to the vertical.

The solidified steel emerges from the bending rolls as a bloom, billet, slab, or round. This depends on the shape and size of the water-cooled copper mold. The solidified product then passes through a straightener. The straightened product is then cut into convenient lengths for further processing. Three variations of continuous casting are vertical, curve mold, and bent strand. See Figure 11-13.

Compared with ingot casting, soaking, and primary rolling, continuous casting leads to a better yield of product per heat of steel. Energy savings, less pollution, reduction of capital and operating costs, and greater freedom from specific discontinuities characteristic of ingot cast products are also characteristic of continuous casting.

Discontinuities. Ingots may contain various types of discontinuities, which include blowholes, segregation, pipe, bursts, and scabs. Continuously cast steel is also susceptible to specific discontinuities. These include blowholes, segregation, cracks, scabs, and deformed product cross section.

Blowholes are holes formed by gas entrapped during solidification. They generally do not extend to the surface of the ingot, are generally clean, and weld shut during primary rolling. If they do extend to the surface of the ingot, they become oxidized by the atmosphere and produce a seam in the primary rolled product. Killed steels are usually free of blowholes.

Blowholes in continuous castings are also similar to those found in ingots. They are controlled by the deoxidation practice and shrouding of the tundish stream with a refractory tube.

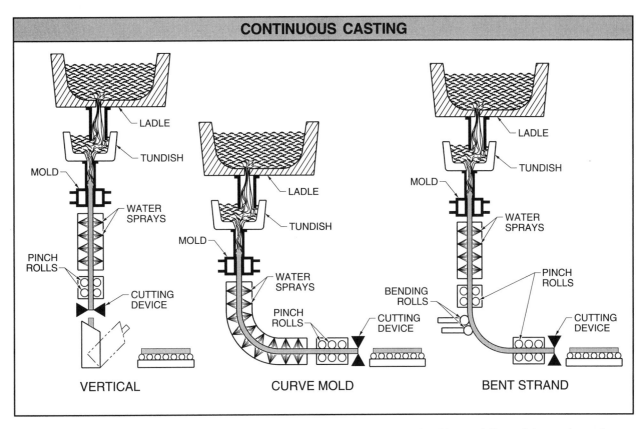

CONTINUOUS CASTING

VERTICAL   CURVE MOLD   BENT STRAND

Figure 11-13. Continuous casting is a direct and continuous method of producing blooms, billets, slabs, and rounds.

Segregation is any concentration of alloying chemical elements in a specific region of a metallic object. Segregation causes chemical inhomogeneity in the ingot. Generally, the first metal that solidifies at the mold wall has about the same chemical composition as the molten metal. However, as the rate of solidification decreases, dendrites of purer metal solidify first. These contain less carbon, manganese, phosphorus, sulfur, and other elements than the liquid from which they solidify. As a result, the remaining liquid metal is enriched in these elements and is less pure. Turbulence from gas evolution increases the tendency toward segregation. Consequently, killed steels are less segregated than semikilled, which in turn are less segregated than capped and rimmed steels.

Segregation in continuous casting is similar to that found in ingots, but there is one very important difference. Ingots have the benefit of soaking and primary rolling to break down the segregated structures and modify the grain structure. Segregation is a major disadvantage in continuously cast products.

Any subsequent reduction of the cross section to produce finished forms may not be sufficient to ensure adequate properties in specific applications.

Pipe is a central shrinkage cavity located in the upper portion of the ingot. It is formed during solidification of the ingot. See Figure 11-14. Pipe is most likely to be found in killed steels, where there is no vigorous gas evolution to offset the shrinkage. Hot tops are used to prevent pipe. A *hot top* is a refractory-lined container, which is placed on top of the ingot mold. It absorbs heat less rapidly than the ingot mold and therefore maintains a reservoir of molten steel. The reservoir feeds the solidifying metal below it. The hot top material is cropped off and discarded after primary rolling.

Bursts are internal fissures (narrow cracks) in ingots caused by internal tensile stresses resulting from heating and cooling. Bursts are welded shut if the ingot cross section is sufficiently reduced during primary rolling, unless they are severe enough to be open to the surface and become oxidized.

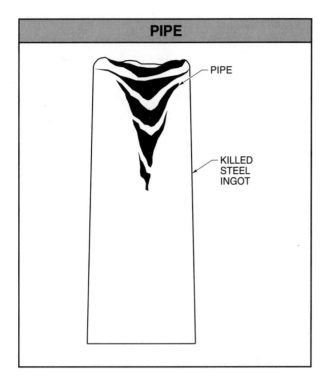

PIPE

PIPE

KILLED
STEEL
INGOT

Figure 11-14. Pipe is a shrinkage cavity located in the upper portion of the ingot.

Cracks in continuous castings are similar to the bursts that occur in ingots. They may be located at the surface or inside the metal. Cracks are eliminated by altering factors such as mold coatings, casting speed, casting temperature, machine alignment, etc.

Scabs in ingot casting are surface slivers caused by splashing and rapid solidification of the metal when it is first poured and strikes the mold wall. Scabs that are sufficiently thin are oxidized to iron oxide during soaking. Thicker scabs may be rolled into the surface of the steel during primary rolling, creating a surface discontinuity. Special mold coatings and pouring techniques help reduce or eliminate scabs.

Scabs in continuous castings are similar to those found in ingots. They are reduced or eliminated by proper control of the pouring stream between the tundish and the mold.

Deformed product cross section is a discontinuity common to continuous castings. It may be convex or concave and arise from shrinkage during the solidification process. Deformed product cross sections may be corrected by proper design of the mold and rolls.

Finished Forms

Finished forms for steel products are semifinished forms produced by ingot casting and rolling or by continuous casting that are further reduced and shaped. There are two principal processes for making finished forms, which are hot working and cold working. Both types of processes may lead to specific discontinuities in finished forms of steel. Steel products fall into several main groupings. These are structural shapes; bar, wire rod, and wire; pipe and tubing; and plate, sheet, and strip.

Hot Working. *Hot working* is plastic deformation processes performed above the recrystallization temperature so that work hardening of the steel does not occur. The principal methods of hot working are hot rolling, heavy press forging, and piercing and hot extrusion.

Hot rolling is the reduction of steel ingot size by rollers rotating in opposite directions and spaced at a distance less than the steel entering them. Hot rolling is widely used and is similar to primary rolling. The various semifinished forms are reduced in cross section to finished forms such as bar, rod, plate, sheet, and strip. Hot rolling usually begins at 1205°C (2200°F) and is completed in the range of 815°C to 955°C (1500°F to 1750°F).

Heavy press forging is plastically deforming metals between dies in presses at temperatures high enough to avoid strain hardening. It is used to directly reduce the cross section of ingots. Heavy press forging is performed in large hydraulic presses and is used when the very large starting size of the ingot must be reduced to produce the desired shape. See Figure 11-15.

Piercing is the cutting of holes into a metal. *Hot extrusion* is converting the ingot to uniform lengths and shapes by forcing the heated metal plastically through a die orifice. They are performed on billets, often as the starting point in the production of seamless pipe or tubing. The product (a long, hollow cylinder) is then drawn down to reduce the wall thickness and diameter to the required dimensions.

Cold Working. *Cold working* is plastic deformation processes performed below the recrystallization temperature, which lead to work hardening. Cold working is used to improve mechanical properties and also machinability, dimensional accuracy, and surface finish. Cold working allows the production of thinner gauges of sheet than are possible with hot working, and also narrower dimensional tolerances. Prior to any cold-working operation, the starting stock must be cleaned of scale by pickling or shot blasting to prevent surface defects from being worked in. Cold-worked forms include bar, wire, pipe, tube, sheet, and strip. The principal methods of cold working are cold rolling, cold drawing, and cold extrusion.

Figure 11-15. Heavy press forging is performed in a large hydraulic press in order to reduce the size of the ingot to a semifinished form larger than a standard bloom.

Cold rolling is the passing of unheated and previously hot-rolled bars, sheet, or strip through rollers at a distance from each other of less than the size of the steel entering them. This is repeated until the final size is obtained. Cold rolling is performed at ambient temperature, although frictional effects cause the temperature of the product to be raised to 120°C to 230°C (250°F to 450°F).

Cold drawing consists of pulling the steel through a series of tapered dies in order to reduce the cross section to the required size. Shapes produced by cold drawing are bar, rod, wire, or tube.

Cold extrusion is converting the ingot to uniform lengths and shapes by forcing the steel plastically through a die orifice. Cold extrusion is similar to hot extrusion, except that the process is performed at ambient temperature. In cold extrusion, the steel surface is specially treated in order to reduce frictional effects.

Shapes. Shapes that are produced from steel include structural shapes, bar, wire rod, wire, tubular shapes, and flat shapes. Each of these shapes has distinctive dimensional features and processes by which they are produced.

Structural shapes are hot-rolled, flanged shapes having at least one dimension of the cross section 3 in. or greater. They are used for bridges, buildings, ships, and for numerous other construction purposes. They are produced by passing billets or blooms through a series of grooved rolls. A portion of the shape is represented by the grooves in one roll, and the remaining portion is represented by the grooves in the opposite roll. Hot rolling of structural shapes is performed in three broad stages, which are roughing (rough forming), intermediate rolling, and finishing. See Figure 11-16.

Bar is an elongated hot-worked or cold-worked steel product that is relatively thick and narrow. Most bars have simple, uniform cross sections that are rectangular, square, round, oval, or hexagonal. Bar is produced in bar mills by rolling billets to the required dimensions, and is classified as hot rolled or cold finished.

Hot-rolled bar is available in two basic quality categories, which are merchant quality and special quality. These categories differ in characteristics such as internal cleanliness, relative freedom from surface imperfections, etc.

Cold-finished bar is available in various quality categories. Cold-finished bar is produced from hot-rolled bar using specific cold-finishing processes in order to improve surface finish, dimensional accuracy, alignment, or machinability. Cold-finishing processes include cold drawing, cold rolling, turning, grinding, polishing, or straightening. These processes may be used singly or in combinations with one another. When special properties are required, heat treatments such as annealing, normalizing, quenching and tempering, or stress relieving are performed.

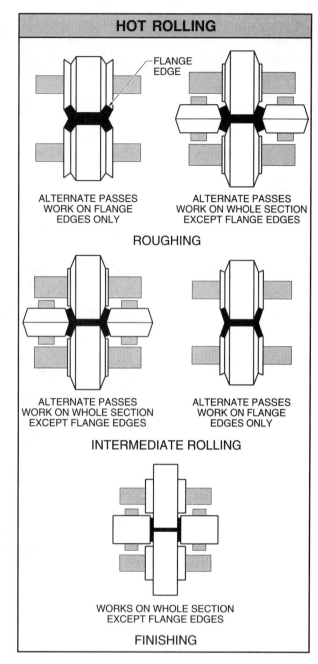

HOT ROLLING

FLANGE EDGE

ALTERNATE PASSES
WORK ON FLANGE
EDGES ONLY

ALTERNATE PASSES
WORK ON WHOLE SECTION
EXCEPT FLANGE EDGES

ROUGHING

ALTERNATE PASSES
WORK ON WHOLE SECTION
EXCEPT FLANGE EDGES

ALTERNATE PASSES
WORK ON FLANGE
EDGES ONLY

INTERMEDIATE ROLLING

WORKS ON WHOLE SECTION
EXCEPT FLANGE EDGES

FINISHING

Figure 11-16. Structural shapes are formed in several rolling sequences, known as roughing, intermediate rolling, and finishing.

Wire rod is a steel product rolled from billet in a rod mill and is used primarily in the manufacture of wire. It is also used for other items such as concrete reinforcing rod. Most wire rod is round in cross section. As wire rod comes off the rolling mill, it is wound into coils weighing from 300 lb to 4000 lb.

Wire is steel which is thin, flexible, continuous, and usually circular in cross section. Wire has an enormous number of uses, which include pins and needles, springs, screw stock, steel wool, and welding wire.

Wire is produced in several stages. The wire rod used to make wire is first descaled. It is then drawn through the tapered hole of one or several wire drawing dies. The number of dies and the number of drafts depend on the final size required. A *draft* is a single drawing step. To begin the drawing process, one end of the wire rod is pointed, inserted in the die, and then pulled by a powerful block through the die. Between drafts, the wire product is coiled. See Figure 11-17. After the final draft, the wire is often given a metallic coating for decoration and/or protection. Coatings include zinc (galvanizing), tin, and aluminum.

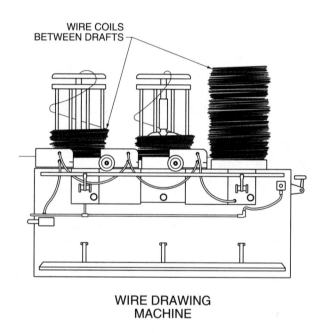

WIRE COILS
BETWEEN DRAFTS

WIRE DRAWING
MACHINE

Figure 11-17. Wire is drawn in several stages, known as drafts, on a wire drawing machine. After each draft the wire is coiled.

Various heat treatments are used to facilitate drawing or to improve the mechanical properties of wire products. These include process annealing, patenting, and oil tempering. Process annealing is used on low-carbon steel, and patenting is used on wire with >.4% C. The purpose of process annealing and patenting is to soften the wire, which becomes successively harder and more brittle with each draft. If a softening heat treatment were not performed, a point would be reached when the wire would break in the die. Oil tempering is a strengthening heat treatment performed on finished wire intended for specific end purposes such as overhead door springs and valve springs.

Tubular shapes are products that are hollow. Their two most common uses are in conveying fluids and in structural members. Tubular products are divided into two groups, which are tubing and pipe. These are further subdivided into various categories based on their end use.

Tubular products may be seamless or welded. See Figure 11-18. Seamless tubular products are made from forged cylindrical billets, which are pierced through the center to make them hollow and then rolled or extruded and drawn to size. Welded tubular products are made from skelp. *Skelp* is the starting stock for producing tubing or pipe and is usually a coil of flat steel sheet. The skelp is first uncoiled and then curved about its longitudinal axis into a circular configuration. The longitudinal edges of the skelp are welded together to form pipe or tubing. The skelp may then be reduced in size by cold drawing. Various welding process may be used to fuse the skelp. For example, electric resistance welding, with no welding filler metal, or tungsten inert gas welding, using welding filler metal, can be used.

Tubing is hollow product of round, square, or other cross section having a continuous perimeter. Tubing is designated in terms of its outside diameter (OD), inside diameter (ID), wall thickness (schedule), or weight per foot, depending on product shape. These designations refer to the actual values or nominal (named) values.

Pipe is a hollow product of circular cross section made to standard combinations of OD and wall thickness. Compared with tubing, there are fewer sizes of pipe due to the standard sizing, but a larger quantity of the standard sizes are produced.

Pipes sizes are classified by their nominal pipe size (NPS), which is a value based on the pipe diameter. Up to and including 12 in. NPS, there are wide differences between the NPS and the OD. For example, $\frac{1}{2}$ in. NPS standard weight pipe has an actual OD of .840 in. and a theoretical ID of .622 in. When different weights per foot or wall thicknesses are indicated for a particular NPS, the outside diameter of the pipe is not changed; only the inside diameter of the pipe is altered. In pipe sizes 14 in. NPS and greater, the actual OD of the pipe corresponds to the NPS.

Flat shapes or flat-rolled products include plate, sheet, and strip. The difference between them is the shape and size of their cross sections. Plate is a flat-rolled product that is 8 in. to 48 in. wide and is ≥.23 in. thick, or a flat-rolled product that is >48 in. wide and is ≥.18 in. thick. Sheet is thinner than plate and has a width-to-thickness ratio of >50 (50:1). For example, a steel sheet has a width of 5 in. and has a thickness of .10 in. if the width-to-thickness ratio is 50. Strip is narrower than sheet (material up to 12 in. wide) and has a width-to-thickness ratio similar to that of sheet.

Plate is used in the construction of buildings, bridges, ships, railroad cars, storage tanks, pressure vessels, etc. Sheet and strip are used in a wide variety of consumer and industrial goods. Where good surface finish and formability is required, rimmed or capped steel is sometimes specified.

Precoated sheet and strip can also be produced. Precoating consists of applying various types of coatings to the sheet or strip at the mill for the purpose of improving paint adhesion or corrosion resistance. Examples of these coatings systems are zinc, aluminum, tin, iron phosphate, and terne. *Terne* is an alloy of lead containing 3% Sn to 15% Sn. The coatings may be applied by dipping, spraying, or electrodeposition.

Plate, sheet, and strip are either hot rolled or cold rolled from slabs to their final dimensions. Prior to rolling, mill scale must be removed from the surface of the slab by pickling or shot blasting. Hot-rolled products are classified as finished or semifinished. For example, hot-rolled flat bar (a semifinished product) is sometimes made for further processing into cold-rolled strip. Cold-rolled products are always classified as finished.

Discontinuities. *Discontinuities* are interruptions in the normal physical structure of a component. Discontinuities in finished forms are categorized as surface discontinuities or internal discontinuities. They are generally similar to those found in semifinished forms.

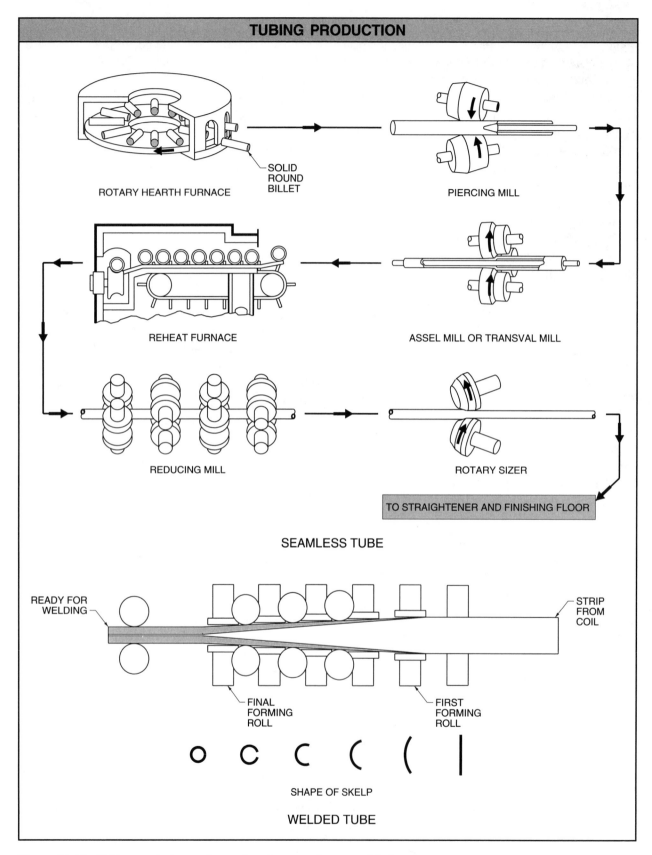

TUBING PRODUCTION

ROTARY HEARTH FURNACE

SOLID ROUND BILLET

PIERCING MILL

REHEAT FURNACE

ASSEL MILL OR TRANSVAL MILL

REDUCING MILL

ROTARY SIZER

TO STRAIGHTENER AND FINISHING FLOOR

SEAMLESS TUBE

READY FOR WELDING

STRIP FROM COIL

FINAL FORMING ROLL

FIRST FORMING ROLL

SHAPE OF SKELP

WELDED TUBE

Figure 11-18. Tubing products may be produced as seamless or welded.

Surface discontinuities are of particular concern when a specific surface finish is required. For example, the surface must be clean of discontinuities in components that are deep drawn or enamelled, or when a piece is to be upset forged. *Upset forging* is the increase of cross section due to an applied pressure during fabrication.

Internal discontinuities are of particular concern when the application of the product places severe restrictions on internal cleanliness. Examples of such restrictions include restrictions for tire cord or aerospace components. Ladle refining operations are often used to reduce internal discontinuities to acceptable levels.

Inclusions consist of metallic oxides, sulfides, and silicates that are held mechanically in the metal. They are formed during solidification or enter the molten metal from contamination with the refractory lining. Metallic oxide inclusions, such as alumina and calcium aluminates, are usually the result of deoxidizing additions. Sulfide inclusions are the manganese sulfide type. Sulfide inclusions are formed by chemical combination between the manganese and sulfur in the steel. Silicate inclusions are essentially silicate glass containing calcium, manganese, iron, and aluminum.

In as-cast steel, the inclusions are located at grain boundaries or between dendrites. Hot or cold working elongates the inclusions in the working direction if they are plastic at the working temperature. This leads to anisotropy. See Figure 11-19.

Specifications

In contrast to iron products, many specifications are available for steel products. The two specifications for steels fall into two categories, which are composition designations and product specifications. The composition designations were developed jointly by the American Iron and Steel Institute and the Society of Automotive Engineers. This numbering system is incorporated into the Unified Numbering System for Metals and Alloys. Product specifications are written by a wide range of societies that represent specific industry interests. The industry-wide system and most commonly used product specifications are those of the American Society for Testing and Materials.

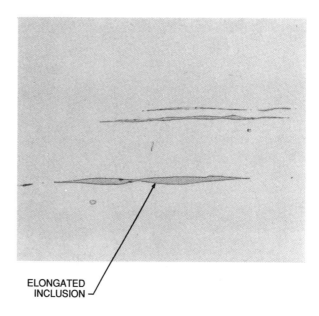

ELONGATED
INCLUSION

Figure 11-19. Inclusions become elongated in the working direction of the steel and lead to directional mechanical properties, known as anisotropy.

Steel Composition Designations. Steels are broadly classified as carbon steels or alloy steels. They are also individually designated by their chemical compositions. In the United States, the designation system is based on a four digit numbering system devised jointly by the AISI and the SAE. This numbering system is incorporated into the UNS, which consists of a letter followed by five digits.

The AISI and the SAE assign designations for steel products based on their compositions. Although these are two separate designation systems, they are nearly identical.

The *AISI-SAE designation system* consists of a four digit classification that is partially descriptive of the composition. See Figure 11-20. The first digit indicates the family to which the steel belongs. Thus, 1 indicates a carbon steel, 2 a nickel steel, 3 a nickel-chromium steel, and so on. In the case of simple alloy steels, the second digit indicates the percentage of the principal alloying element. Usually, the second, third, and fourth digits indicate the average carbon content in points (hundredths of a percent). For example, 2340 is a nickel steel with approximately 3% Ni and .4% C.

In addition to joint AISI-SAE designations, there are also specific AISI steels and SAE steels. To qualify for a separate AISI designation, the grade of steel must be made in production tonnages. Separately designated SAE steels qualify if they have specific engineering characteristics.

The AISI designations are published in the *AISI Steel Products Manuals*. These indicate the chemical composition ranges for the various designations, standard manufacturing practices, and related information. *AISI Steel Products Manuals* are not standards, but do indicate restrictions and tolerances applicable to various product forms based on accepted manufacturing practices.

SAE designations are published in the annual *SAE Handbook* under specific SAE standards. The *SAE Handbook* indicates the various chemical composition ranges for steels and other alloys.

In certain cases, the permissible chemical composition ranges vary with the product form. For example, the AISI-SAE composition ranges for steel plates are slightly wider than for bar products because the large rectangular ingots used for making plates are more susceptible to segregation than the smaller square or round ingots used for making bars. Specific amounts of residual alloying elements are

AISI-SAE DESIGNATION SYSTEM		
Numbers and Digits — **Type of steel and/or nominal alloy content**	**Numbers and Digits** — **Type of steel and/or nominal alloy content**	
Carbon steels	93xx — 3.25% Ni; 1.20% Cr; .12% Mo	
10xx — Plain carbon (1% Mn max)	94xx — .45% Ni; .40% Cr; .12% Mo	
11xx — Resulfurized	97xx — .55% Ni; .20% Cr; .20% Mo	
12xx — Resulfurized and rephosphorized	98xx — 1.00% Ni; .80% Cr; .25% Mo	
15xx — Plain carbon (1.00% Mn to 1.65% Mn max)	**Nickel-molybdenum steels**	
Manganese steels	46xx — .85% Ni and 1.82% Ni; .20% Mo and .25% Mo	
13xx — 1.75% Mn	48xx — 3.50% Ni; .25% Mo	
Nickel steels	**Chromium steels**	
23xx — 3.5% Ni	50xx — .27% Cr, .40% Cr, .50% Cr, and .65% Cr	
25xx — 5% Ni	51xx — .80% Cr, .87% Cr, .92% Cr, .95% Cr, 1.00% Cr, and 1.05% Cr	
Nickel-chromium steels	**Chromium steels**	
31xx — 1.25% Ni; .65% Cr and .80% Cr	50xxx — .50% Cr	
32xx — 1.75% Ni; 1.07% Cr	51xxx — 1.02% Cr } C 1.00% min	
33xx — 3.50% Ni; 1.50% Cr and 1.57% Cr	52xxx — 1.45% Cr	
34xx — 3.00% Ni; .77% Cr	**Chromium-vanadium steels**	
Molybdenum steels	61xx — .60% Cr, .80% Cr, and .95% Cr; .10% V and .15% V min	
40xx — .20% Mo and .25% Mo	**Tungsten-chromium steel**	
44xx — .40% Mo and .52% Mo	72xx — 1.75% W; 0.75% Cr	
Chromium-molybdenum steels	**Silicon-manganese steels**	
41xx — .50% Cr, .80% Cr, and .95% Cr; .12% Mo, .20% Mo, .25% Mo, and .30% Mo	92xx — 1.40% Si and 2.00% Si; .65% Mn, .82% Mn, and .85% Mn; 0% Cr and .65% Cr	
Nickel-chromium-molybdenum steels	**High-strength low-alloy steels**	
43xx — 1.82% Ni; .50% Cr and .80% Cr; .25% Mo	9xx — Various SAE grades	
43BVxx — 1.82% Ni; .50% Cr; .12% Mo and .25% Mo; .03% V min	**Boron steels**	
47xx — 1.05% Ni; .45% Cr; .20% Mo and .35% Mo	xxBxx — B denotes boron steel	
81xx — .30% Ni; .40% Cr; .12% Mo	**Leaded steels**	
86xx — .55% Ni; .50% Cr; .20% Mo	xxLxx — L denotes leaded steel	
87xx — .55% Ni; .50% Cr; .25% Mo		
88xx — .55% Ni; .50% Cr; .35% Mo		

Figure 11-20. The AISI-SAE designation system indicates the alloying elements and the partial chemical composition.

usually allowed in carbon steel and do not usually cause significant property changes.

UNS designations are a system of identification for steels and other alloys that consists of an uppercase letter followed by five numbers. For steel products, it consists of G, H, J, or K followed by the five numbers. The letters stand for specific groupings of steel. The five numbers following the letter indicate the type of steel with fixed compositional requirements, which are taken from standard AISI, SAE, or ASTM compositions.

AISI and SAE steels have the letter G assigned to them. Most other steels have the letter H (steels with specific hardenability requirements), J (cast steels), and K (miscellaneous steels). For example, ASTM A203 grade B, a nickel-containing pressure vessel steel has the UNS designation K22103. Not all commercial steels have a UNS designation, so the system is not universal. It does encompass most commercially available American produced steels.

Carbon steels are alloys of iron with carbon, manganese, and silicon, specifically containing up to 1.6% Mn and .6% Si, plus smaller amounts of sulfur and phosphorus. Carbon steels are designated by the 1xxx group in the AISI-SAE system.

The 1xxx group is subdivided in several smaller alloy systems. The 10xx carbon steels are plain carbon steels. The 11xx group is resulfurized, and the 12xx group is resulfurized and rephosphorized. The 12xx steels have sulfur and/or phosphorus added to improve machinability. The 13xx and 15xx groups have increased manganese contents to increase hardenability, which is the capacity of a steel to be strengthened in heavy sections by heat treatment.

The 1xxx group is divided into the three categories of low-, medium-, and high-carbon steels. Low-carbon steels contain less than approximately .20% C and are the weakest, most ductile, and most easily welded group. Medium-carbon steels contain approximately .20% C to .45% C and are stronger and less ductile. The approximate upper carbon limit for weldability without the need for preheating and post-weld heat treatment is .35% C. High-carbon steels contain greater than approximately .45% C and are significantly stronger and less ductile. High-carbon steels are weldable with careful attention to preheating, interpass temperature, post-weld heat treatment, and cooling rate from the welding temperature. The applications of low-, medium-, and high-carbon steels are strongly dependent upon the carbon content. See Figure 11-21.

Specific elements which are always present in carbon steels are carbon, manganese, silicon, sulfur, and phosphorus. *Residual elements* are additional chemical elements that are introduced by the steelmaking process and are not detrimental if present in suitably low amounts. The three main sources of residual elements are the steelmaking environment, scrap used to make steel, and the steel deoxidation practice.

The steelmaking environment may introduce nitrogen, hydrogen, and oxygen into the steel. Nitrogen assists in the strain ageing effect in steel, which is hardening introduced by cold working, and also has a marked strengthening effect. It is sometimes deliberately added to sheet steels to provide higher strength and moderately good formability at lower costs. Hydrogen has an embrittling effect and also plays an important role in flaking. *Flaking* is internal cracks or bursts usually occurring during cooling, after rolling or forging. For these reasons, hydrogen should be maintained below .0005% H. Oxygen greatly influences the consumption of deoxidizers used in steelmaking. Therefore, the oxygen content is controlled to enable the desired chemical composition and solidification structure to be met. For example, in rimmed steel, oxygen removal improves the cleanliness of the steel.

Scrap added to the steelmaking furnace may introduce a variety of metallic elements. For example, nickel, chromium, molybdenum, and copper increase the hardenability. The quantity of these elements is usually so low that the effect is not significant, unless ductility is an important characteristic, such as for deep drawing steels. In this respect, tin introduced from scrap is also harmful, but the amount in which it is present usually makes the effect negligible. In some cases, the presence of these residual elements that increase hardenability may cause cracking problems when the steel is welded. Preventing the problems may require alteration of the welding procedure, such as in raising the preheating temperature.

The steel deoxidation practice may also introduce a variety of metallic elements. Aluminum is generally desirable as a grain refining agent, and it also decreases susceptibility to strain ageing. However, aluminum tends to promote graphitization. Therefore, aluminum is undesirable in steels used for high-temperature applications. The other deoxidation elements, titanium, vanadium, and zirconium, are ordinarily present in such small amounts that they have no effect.

Alloy steel is a steel that contains specified quantities of alloying elements other than carbon and the common amounts of manganese, copper, silicon, sulfur, and phosphorous. Alloy steels have enhanced properties over carbon steels due to the presence of the one or more alloying elements, or due to larger proportions of the alloys ordinarily present in carbon steels. Like carbon steels, alloy steels are classified according to composition by the AISI-SAE system and the UNS designations.

The purpose of the alloying additions is to enhance mechanical properties, machinability, weldability, hardenability, abrasion resistance, corrosion resistance, and magnetic properties. In the AISI-SAE system, alloy steels are divided into groups (for example, 2xxx is nickel steels) and subgroups (for example, 23xx is a subgroup of nickel steels).

The 23xx and 25xx subgroups are nickel steels and have nominal nickel contents. The 23xx contains 3.5% Ni, and the 25xx contains 5% Ni. Nickel additons improve the strength and toughness of steels at low temperatures.

The 3xxx group consists of four subgroups whose principal alloying elements are nickel and chromium. These are the 31xx, 32xx, 33xx, and 34xx subgroups. The percentages of nickel and chromium are varied, but the ratio of nickel to chromium is maintained at approximately 2.5:1. The nickel and chromium addition increases toughness, ductility, hardness, and wear resistance.

Alloys in the 4xxx group contain molybdenum alone or combined with nickel and chromium. Molybdenum increases hardenability, high-temperature hardness, and strength. Nickel and chromium further improve these properties and also contribute toward toughness, weldability, and ease of case hardening. The 40xx and 44xx subgroups are molybdenum steels. The 41xx subgroup is chromium-molybdenum steels. The 46xx and 48xx subgroups are nickel-molybdenum steels. The 43xx and 47xx subgroups are nickel-chromium-molybdenum steels.

USES OF CARBON STEEL	
% C	**Applications**
.05 to .12	Chain, stampings, rivets, nails, wire, pipe, welding stock, where very soft, plastic steel is needed
.10 to .20	Very soft, tough steel; structural steels, machine parts; for case-hardened machine parts, screws
.20 to .30	Better grade of machine and structural steel; gears, shafting, bars, bases, levers, etc.
.30 to .40	Responds to heat treatment; connecting rods, shafting, crane hooks, machine parts, axles
.40 to .50	Crankshafts, gears, axles, shafts, and heat-treated machine parts
.60 to .70	Low-carbon tool steel, used where a keen edge is not necessary, but where shock strength is wanted; drop hammer dies, set screws, locomotive wheels, screw drivers
.70 to .80	Tough and hard steel; anvil faces, band saws, hammers, wrenches, cable wires, etc.
.80 to .90	Punches for metal, rock drills, shear blades, cold chisels, rivet sets, and many hand tools
.90 to 1.00	Used for hardness and high tensile strength, springs, high tensile wire, knives, axes, dies for all purposes
1.00 to 1.10	Drills, taps, milling cutters, knives, etc.
1.10 to 1.20	Used for all tools where hardness is a prime consideration; for example, ball bearings, cold-cutting dies, drills, wood-working tools, lathe tools, etc.
1.20 to 1.30	Files, reamers, knives, tools for cutting brass and wood
1.25 to 1.40	Used where a keen cutting edge is necessary; razors, saws, instruments, and machine parts where maximum resistance to wear is needed; boring and finishing tools

Figure 11-21. The end use of carbon steel is strongly influenced by the carbon content.

Nickel, chromium, and molybdenum are also present in other alloy steel subgroups. The combinations of these elements are adjusted in order to develop specific properties at an optimum cost. The nickel-chromium-molybdenum steels include the 81xx, 86xx, 87xx, 88xx, 93xx, 94xx, 97xx, and 98xx subgroups.

The 50xx, 51xx, and 52xx subgroups are chromium steels. Chromium improves wear resistance, high-temperature strength, and magnetic properties in specific cases. The 61xx subgroup consists of chromium-vanadium steels. The vanadium addition promotes fine grain size, improves hardenability, and increases hardness at elevated temperatures. The 72xx subgroup consists of tungsten-chromium steels. These are not used extensively for general engineering applications because relatively large amounts of tungsten are required to achieve improvements that may be obtained with smaller amounts of other alloying elements, such as molybdenum. The 92xx subgroup consists of silicon-manganese steels. Silicon and manganese increase strength, toughness, and hardenability at a relatively low cost.

Alloy steel of the xxBxx group contains boron, which is added to improve hardenability. Where specific hardenability is a requirement, the chemical composition of the regular grades may be slightly adjusted. Steels that meet specific hardenability requirements are designated Hxxxx. Steels produced by the electric furnace are designated Exxxx.

Steel Product Specifications. Steel product specifications contain information such as finished form, product size, deoxidation practice, finishing method, quality level, etc. Composition designations alone cannot be used to procure steel products. In order to procure material in specific shapes and forms, product specifications must be used. Steel products are specified by engineering societies and institutes whose members make, design, or purchase steel. The specifications are prepared by specialized organizations that are limited to their fields. Industry-wide specifications are published by ASTM.

ASTM specifications are the most widely used for the procurement of steels. ASTM specifications

have the prefix letter A followed by a number. For example, ASTM A 106 is a specification for seamless carbon steel pipe that is often specified for boilers. Specifications for steel products are written at two levels. The first level consists of generic requirements for a family of products, and the second level consists of individual specifications for particular product forms. The generic specifications reduce the length of individual specifications by collecting requirements in one place that are common to a family of products. See Figure 11-22.

EXAMPLES OF GENERIC ASTM SPECIFICATIONS	
Specifications	Applications
A6	Rolled steel structural plate, shapes, sheet piling, and bars, generic
A20	Steel plate for pressure vessels, generic
A29	Carbon and alloy steel bars, hot rolled and cold finished, generic
A505	Alloy steel sheet and strip, hot rolled and cold rolled, generic
A510	Carbon steel wire rod and coarse round wire, generic
A568	Carbon and high strength, low-alloy, hot rolled and cold rolled steel sheet and hot rolled strip, generic
A646	Premium-quality alloy steel blooms and billets for aircraft and aerospace forgings
A711	Carbon and alloy steel blooms, billets, and slabs for forging

Figure 11-22. Generic ASTM specifications indicate common characteristics of families of product forms and prevent excessive repetition in the individual product specifications.

ASTM specifications contain the items necessary to procure specific steel products. These include the product dimensions, steelmaking practice, chemical composition, mechanical test requirements, quality control tests, and marking and packaging information. In many cases, product dimensional tolerance restrictions are the same as the corresponding items in the *AISI Steel Products Manuals*.

Each ASTM specification usually encompasses several steels, which are identified by the terms "grade," "type," or "class." These terms are used interchangeably in the various specifications to denote items such as strength level, finishing process, deoxidation practice, surface quality, chemical composition, etc. For example, ASTM A 516 grade 70 is fully killed carbon steel pressure vessel plate having a tensile strength of 70 ksi.

The title of the ASTM specification usually includes a quality descriptor. This indicates the characteristics that make the product particularly suited to a specific application or subsequent fabrication operation. An example of a quality descriptor is specification *ASTM A 646 Premium Quality Alloy Steel Blooms and Billets for Aircraft and Aerospace Forgings*. The quality descriptor is the term "Premium Quality" that describes the quality level of the blooms and billets. See Figure 11-23.

Many ASTM specifications contain supplementary requirements, which are add-ons that may be selected at the option of the purchaser. Supplementary requirements have the letter S followed by a number. They allow the purchaser to request additional quality checks at additional cost. For example, a product analysis is usually a supplementary requirement because most standard specifications only provide heat analysis.

The American Society for Mechanical Engineers writes specifications for the steels used in boilers, pressure vessels, and nuclear plant components. Design rules for these items are published in the *ASME Boiler and Pressure Vessel Code*.

In order to develop its specifications, the ASME utilizes relevant ASTM specifications. In some cases, minor changes or deletions are made. ASME steel specifications are identified by the original ASTM number, with the addition of the prefix S and a hyphen. For example, ASTM A 516 grade 70 pressure vessel steel plate is also designated as ASME SA-516 grade 70. ASME specifications for steel products are published in Section IIA of the *ASME Boiler and Pressure Vessel Code*.

The American Petroleum Institute develops standards related to petroleum refining, storage, and transportation, especially in areas where particular ASTM standards are not considered relevant. An example of an API standard is the standard for steel line pipe used in the transportation of hydrocarbon products (*Spec 5L for line pipe*).

Aerospace Materials Specifications (*AMS*) are published by the SAE. The *AMS* are used for the procurement of materials intended for aerospace applications. These specifications often contain significantly higher quality requirements than those for grades of steel having similar compositions, but intended for other applications. For example, stringent cleanliness requirements, such as consumable electrode remelting care (*AMS 6415 Consumable Electrode Remelted Normalized and Tempered 4340 Low-Alloy Steel*), are common in the *AMS*.

SELECTED QUALITY DESCRIPTORS IN ASTM STANDARDS	
Alloy Steels	**Carbon Steels**
Drawing quality	Structural quality
Pressure vessel quality	Cold drawing quality
Aircraft quality	Cold pressing quality
Structural quality	Cold flanging quality
Bearing quality	Special hardenability
Special surface quality	Forging quality
Cold heading quality	Marine quality
Pressure tubing	Merchant quality
Mechanical tubing	

Figure 11-23. Quality descriptors in ASTM standards are used to distinguish products well suited to specific applications.

Carbon Steels 12

Carbon steels are alloys of iron, carbon, and manganese. The small amount of carbon that carbon steels contain significantly affects the crystal structure and phase changes that occur. Most carbon steels have ferrite-pearlite microstructures. Some have microstructures of pearlite or pearlite-cementite. The percentages of the phases present in carbon steels (ferrite, pearlite, or cementite) are obtained from the iron-carbon diagram, which describes alloys of iron and iron carbide. Mechanical properties of carbon steels depend on the carbon content, the manganese content, and the microstructure that the steel exhibits.

CARBON STEELS

Carbon steels consist principally of iron, but the small amount of carbon that they contain significantly affects the crystal structure and phase changes that occur in the steel. Carbon steels are divided according to carbon content. Mechanical properties of carbon steels are additionally affected by other chemical elements present and by the microstructure that the steel exhibits.

Crystal Structure and Phases

Iron and steel undergo crystal structure changes when heated and cooled. Carbon steels and corresponding phase changes are plotted on the iron-carbon diagram. The eutectoid reaction is the most important phase change reaction on the iron-carbon diagram because it indicates the transformation of austenite (a carbon solid solution) to ferrite (a carbon solid solution), cementite (iron carbide or Fe_3C), and pearlite (an aggregate of ferrite and cementite).

Crystal Structure of Iron. Iron is an allotropic metal. Allotropic metals exhibit more than one type

of crystal structure, depending on their temperature. Iron undergoes three phase changes when cooled from the melting point to the ambient temperature. See Figure 12-1.

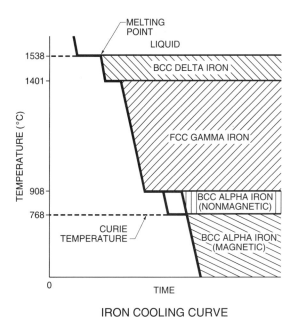

Figure 12-1. The cooling curve for pure iron exhibits three crystal structure changes that occur when cooled from the melting point to the ambient temperature.

181

Iron solidifies at 1538°C (2800°F) into the BCC crystal structure of delta (δ) iron. With further cooling, at 1401°C (2554°F), the second change occurs in which the atoms rearrange into the FCC crystal structure of gamma (γ) iron. When the temperature reaches 908°C (1666°F), the third change occurs, and the atoms revert to a BCC crystal structure of alpha (α) iron. The three changes occur in reverse order at equivalent temperatures when iron is heated. Although there is no accompanying crystal structure change, alpha iron becomes magnetic when cooled below the Curie temperature. *Curie temperature* is the temperature of magnetic transformation above which a metal is nonmagnetic, and below which it is magnetic. The Curie temperature for alpha iron is 768°C (1414°F).

Crystal Structure of Iron-Carbon Alloys. The basis for carbon steel is carbon interstitially dissolved in iron. The carbon atoms occupy the spaces between the iron atoms in the iron crystal structure. See Figure 12-2. Carbon dissolves in the BCC and FCC forms of iron. The result is ferrite and austenite respectively. Austenite is capable of dissolving significantly more carbon than ferrite. Although the BCC lattice has more spaces than the FCC lattice, the spaces are not large enough to accommodate a large number of carbon atoms.

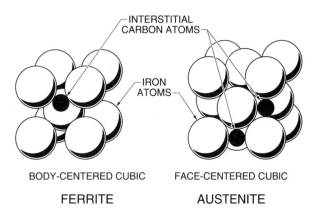

Figure 12-2. Carbon dissolves in the body-centered cubic (BCC) and face-centered cubic (FCC) forms of iron to produce ferrite and austenite, respectively.

Phase changes similar to those in iron occur in iron-carbon alloys. The temperatures at which the changes occur vary in relation to the amount of carbon dissolved in the iron. Allotropic forms of iron-carbon have different names than their pure iron counterparts. Delta iron (BCC) is referenced as delta ferrite, gamma iron (FCC) is referenced as austenite, and alpha iron (BCC) is referenced as alpha ferrite. Because the high-temperature delta ferrite is rarely encountered in engineering applications, alpha ferrite is referred to as ferrite. Ferrite and austenite in iron-carbon alloys are of commercial importance.

Iron-Carbon Diagram. The iron-carbon diagram is an equilibrium (phase) diagram that indicates the phases present in alloys of iron and carbon. It is used to predict the microstructure of steels, especially carbon steels. The iron-carbon diagram is actually a section of the phase diagram between pure iron and an interstitial compound (cementite), which contains 6.67% C.

The iron-carbon diagram is used to predict the phases present in carbon steels under conditions of equilibrium or relatively slow heating and cooling (as occur in normalizing or annealing heat treatments). The iron-carbon diagram exhibits three phase transformation reactions, which are the peritectic, the eutectic, and the eutectoid. Of these, the eutectoid reaction is the most important. See Figure 12-3. Most commercially important carbon steels are represented by the region of the iron-carbon diagram extending from 0% C to 1.2% C.

The five solid phases in iron-carbon diagrams are ferrite (alpha, α), austenite (gamma, γ), cementite, delta ferrite (delta, δ), and pearlite (ferrite and cementite). With the exception of delta ferrite, which is present at temperatures much higher than used in heat treating and metalworking operations, these phases play an important part in the structure and properties of carbon steels.

Phase Descriptions. The four most important phases in carbon steels and those involved in the eutectiod reaction are austenite, ferrite, cementite, and pearlite. *Austenite* is a gamma solid solution of one or more elements in FCC iron. When it is applied to carbon steels, it is defined as an interstitial solid solution of carbon in gamma iron. Austenite can dissolve up to 2% C at 1129°C (2065°F). Austenite is not stable below 723°C (1333°F).

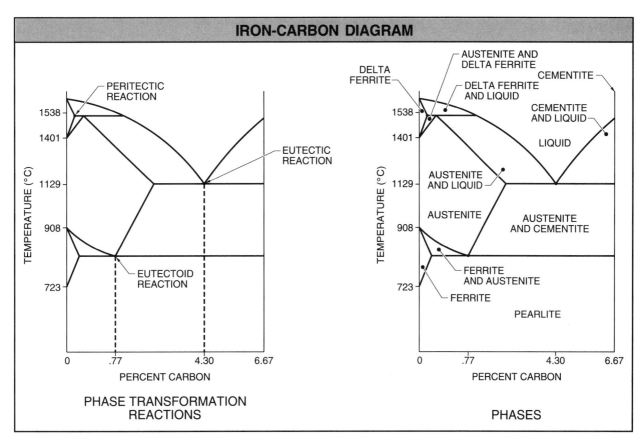

Figure 12-3. The iron-carbon diagram contains peritectic, eutectic, and eutectoid phase transformation reactions and ferrite (α), austenite (γ), cementite (Fe_3C), delta ferrite (δ), and pearlite (ferrite and pearlite) solid phases.

Ferrite is an alpha solid solution of one or more elements in BCC iron. When applied to carbon steels, it is defined as an interstitial solid solution of carbon in alpha iron. Ferrite dissolves considerably less carbon than austenite, with the maximum amount being .025% C at 723°C (1333°F).

Cementite is a compound of iron and carbon referred to as iron carbide. Cementite is an intermetallic compound that has a chemical composition that is slightly altered by the presence of other carbide forming elements in carbon steels. Cementite, however, is still considered to be iron carbide. Cementite contains 6.67% C and the remainder is iron.

Pearlite is a lamellar aggregate of ferrite and cementite formed from the eutectoid decomposition of austenite during slow cooling. This leads to a diffusion controlled phase transformation. When it is applied to carbon steels, pearlite consists of a lamellar aggregate of ferrite and cementite and has the microstructural appearance of fingerprints. See Figure 12-4.

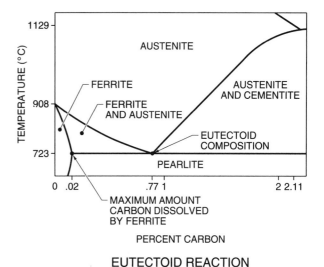

EUTECTOID REACTION
REGION

Figure 12-4. The eutectoid reaction region is the most important region of the iron-carbon diagram.

Pearlite is present in the room temperature microstructure of slow-cooled carbon steels. The phase is actually metastable, which means it is not thermodynamically stable. Under the right conditions, pearlite decomposes to iron and graphite.

Decomposition does not occur at ambient temperature because the thermodynamic driving force (force causing decomposition) is too low. Decomposition occurs at elevated temperatures over prolonged time. Carbon steels are not used for applications that are subject to prolonged temperatures above 425°C (800°F) because this can produce graphitization. *Graphitization* is the slow transformation of pearlite to graphite, which results in rendering the steel brittle.

Classifications

Carbon steels are classified as hypoeutectoid if ferrite-pearlite is formed when cooled, eutectoid if pearlite is formed, or hypereutectoid if pearlite-cementite is formed. Critical temperatures are used to define the temperatures of the key phase changes under equilibrium conditions or at specific rates of heating or cooling.

Hypoeutectoid Steels. *Hypoeutectoid steels* are carbon steels with a carbon content that lies to the left of the eutectoid point on the iron-carbon diagram and contain less than approximately .8% C. Hypoeutectoid steels encompass the largest group of carbon steel compositions and provide product designers with a range of strength and fabrication characteristics to work with. See Figure 12-5.

Hypoeutectoid steel with .2% C has good weldability and formability. It is widely used in components such as boiler tubing and structural shapes. After solidifying, the microstructure of the .2% C steel consists of austenite solid solution. On crossing the solvus (line GJ), the allotropic phase transformation to ferrite begins. *Solvus* is the temperatures at which various compositions of the solid phases coexist with other solid phases. This ferrite (proeutectoid ferrite) forms initially at the austenite grain boundaries and grows inward. Since ferrite dissolves little carbon compared with austenite, the carbon rejected by the newly formed ferrite is absorbed by the remaining austenite. The austenite becomes progressively richer in carbon.

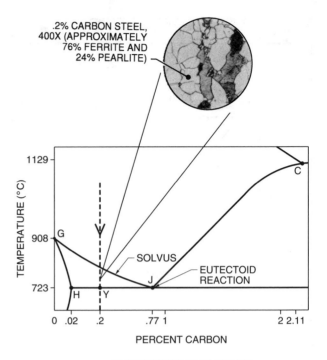

Figure 12-5. Hypoeutectoid steels contain <.8% C and exhibit a slow-cooled microstructure consisting of ferrite and pearlite. A hypoeutectoid steel with .2% C consists of approximately 76% ferrite and 24% pearlite.

At the eutectoid temperature (line HJ) of 723°C (1333°F), the amount of ferrite and austenite is given by the tie line relationship as follows:

$$\%\alpha = \frac{JY}{HJ} \times 100$$

where

$\%\alpha$ = percent ferrite

JY = value of line between points J and Y

HJ = value of line between points H and J

and

$$\%\gamma = \frac{HY}{HJ} \times 100$$

where

$\%\gamma$ = percent austenite

HY = value of line between points H and Y

HJ = value of line between points H and J

Example: Figuring %α

$$\%\alpha = \frac{JY}{HJ} \times 100$$

$$\%\alpha = \frac{.77 - .2}{.77 - .02} \times 100$$

$$\%\alpha = \mathbf{76\%}$$

Example: Figuring %γ

$$\%\gamma = \frac{HY}{HJ} \times 100$$

$$\%\gamma = \frac{.2 - .02}{.77 - .02} \times 100$$

$$\%\gamma = \mathbf{24\%}$$

At the eutectoid temperature, the amount of carbon dissolved in the remaining austenite rises to .77% C. The remaining austenite undergoes the eutectoid reaction by transforming to pearlite. No additional phase transformations occur when further cooled to ambient temperature. The resulting microstructure of a .2% C hypoeutectoid steel consists of approximately 76% proeutectoid ferrite and 24% pearlite.

Eutectoid Steel. *Eutectoid steel* is a carbon steel that has a carbon content of approximately .8% C (typically used for railroad rails). This composition is selected for its combination of strength and wear resistance. After solidifying, the microstructure of a eutectic steel consists entirely of austenite solid solution. See Figure 12-6.

When the temperature reaches 723°C (1333°F), the austenite transforms isothermally to pearlite. The microstructure of the eutectic steel at room temperature consists of 100% pearlite.

Transformation of austenite to pearlite takes place by atomic diffusion. The first stage is the precipitation of cementite in localized regions at the austenite grain boundaries. To facilitate precipitation of this high-carbon phase, the austenite immediately adjacent to the cementite is depleted of carbon. Atoms locally rearrange into the BCC crystal structure to form ferrite. This results in a layer of ferrite forming on either side of the cementite. The process then reverses in the austenite that is immediately adjacent to the layers of ferrite, which has become enriched in carbon.

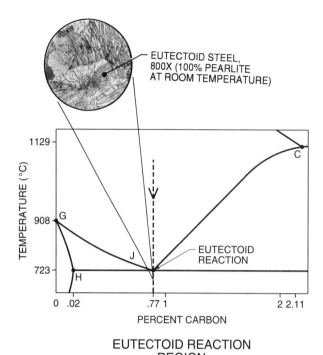

EUTECTOID REACTION REGION

Figure 12-6. Eutectoid steel contains approximately .8% C and exhibits a slow-cooled, 100% pearlite microstructure consisting of ferrite and cementite.

This process forms a layer of cementite on either side of the ferrite. The alternating layers of ferrite and cementite grow into the austenite grains until all the austenite is completely transformed to pearlite. See Figure 12-7.

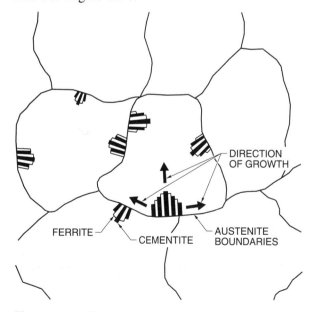

Figure 12-7. The formation of pearlite from austenite is diffusion controlled and leads to a layered microstructure of ferrite and cementite.

Hypereutectoid Steels. *Hypereutectoid steels* are carbon steels with a carbon content that lies to the right of the eutectoid point on the iron-carbon diagram and contain more than approximately .8% C. These steels exhibit hardness and high tensile strength and are used in components such as axes and chisels.

After solidifying, the microstructure of a hypereutectoid steel containing 1% C consists entirely of austenite. See Figure 12-8. When the temperature reaches the solvus (line CJ), the austenite begins to transform to cementite (proeutectoid cementite), which precipitates along the grain boundaries. To accomplish this, the adjacent austenite is depleted in carbon and the concentration of carbon moves along the line CJ.

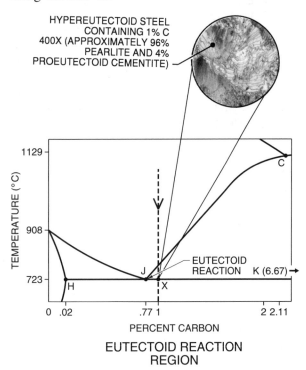

Figure 12-8. Hypereutectoid steels contain >.8% C and exhibit a slow cooled microstructure consisting of cementite and pearlite. A hypereutectoid steel with 1% C consists of approximately 96% pearlite with approximately 4% cementite outlining the grain boundaries.

At the eutectoid temperature of 723°C (1333°F), the amounts of cementite and austenite are given by the tie line relationships as follows:

$$\%Fe_3C = \frac{JK - XK}{JK} \times 100$$

where

%Fe₃C = percent cementite

JK = value of line between points J and K

XK = value of line between points X and K

and

$$\%\gamma = \frac{JK - JX}{JK} \times 100$$

where

%γ = percent austenite

JK = value of line between points J and K

JX = value of line between points J and X

Example: Figuring %Fe₃C

$$\%Fe_3C = \frac{JK - XK}{JK} \times 100$$

$$\%Fe_3C = \frac{5.90 - 5.67}{5.90} \times 100$$

$$\%Fe_3C = \mathbf{3.9\%}$$

Example: Figuring %γ

$$\%\gamma = \frac{JK - JX}{JK} \times 100$$

$$\%\gamma = \frac{5.90 - .23}{5.90} \times 100$$

$$\%\gamma = \mathbf{96.1\%}$$

At the eutectoid temperature, the remaining austenite transforms isothermally to pearlite. The room temperature microstructure of a slow-cooled 1% C steel consists of approximately 4% proeutectoid cementite network surrounding grains of pearlite.

Critical Temperatures. *Critical temperatures* are temperatures in any specific steel composition at which the austenite phase change begins or is completed (for a specific rate of heating or cooling). Critical temperatures are obtained from temperature-time heating or cooling curves that are plotted for specific steel compositions. The critical temperatures are indicated by arrests on these curves, which represent points on the solvus or eutectic lines on the iron-carbon diagram.

Critical temperatures are designated by the uppercase letter A (for arrest) followed by a subscript. A₁ is the boundary between the pearlite phase field and the austenite phase field, which is the eutectic transformation line. A_3 is the boundary between the ferrite-austenite phase field and the austenite phase field. A_{cm} is the boundary between the cementite-austenite phase field and the austenite phase field. See Figure 12-9.

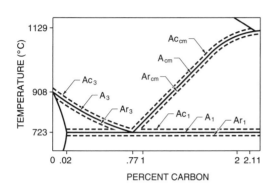

EUTECTOID REACTION REGION

A_1 = Critical temperature between pearlite phase field and austenite phase field (eutectic transformation line)

Ar_1 = Critical temperature between pearlite phase field and austenite phase field on cooling

Ac_1 = Critical temperature between pearlite phase field and austenite phase field on heating

A_3 = Critical temperature between ferrite-austenite phase field and austenite phase field

Ar_3 = Critical temperature between ferrite-austenite phase field and austenite phase field on cooling

Ac_3 = Critical temperature between ferrite-austenite phase field and austenite phase field on heating

A_{cm} = Critical temperature between cementite-austenite phase field and austenite phase field

Ar_{cm} = Critical temperature between cementite-austenite phase field and austenite phase field on cooling

Ac_{cm} = Critical temperature between cementite-austenite phase field and austenite phase field on heating

Figure 12-9. Critical temperatures are temperatures at which phase transformations begin or are completed in the eutectoid region of the iron-carbon diagram.

Under equilibrium conditions, the critical temperatures are sometimes designated by the lowercase letter e (for example, Ae_1, Ae_3, and Ae_{cm}). With faster heating or cooling rates, there is less time for diffusion of the atoms, so the critical temperatures are displaced from the equilibrium values. Critical temperatures are displaced downward with cooling and displaced upward with heating.

To differentiate these new critical temperatures from the equilibrium values, a letter r (from the French word refroidissant) is used to designate cooling, and a letter c (from the French word chauffant) is used to designate heating. The displacement of critical temperatures from the equilibrium values increases with the rate of heating or cooling.

For complete transformation to austenite, a steel must be heated above its Ac_3 or Ac_{cm} temperature (depending on its carbon content). For complete transformation from austenite to ferrite, pearlite, or cementite, a steel must be cooled below its Ar_1 temperature. For example, when a hypoeutectoid steel is austenitized prior to quenching and tempering, it is heated to a temperature above the Ac_3 to ensure that the entire microstructure has transformed to austenite. Complete transformation does not occur if the steel is not heated above its Ac_1.

Mechanical Properties

The mechanical properties of carbon steels are determined by the carbon and manganese contents, pearlite interlamellar spacing, and grain size. Grain size is refined (decreased) through the steel deoxidation practice or by cycling the steel through the critical temperature range. *Grain growth* is an undesirable phenomenon caused by heating steel above the grain-coarsening temperature.

Effects of Carbon and Manganese. Carbon and manganese are the principal alloying elements in carbon steels and contribute to the enhancement of mechanical properties. Carbon steels usually contain from .05% C to 1.2% C and .25% Mn to 1.7% Mn. Increasing the carbon content increases the percentage of pearlite or cementite. This leads to a major strengthening effect consisting of strength and hardness increases, and toughness and ductility decreases. See Figure 12-10.

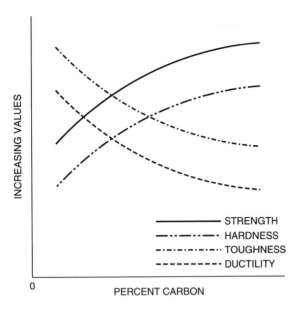

EFFECTS OF CARBON CONTENT

Figure 12-10. Increased carbon content of carbon steels increases the strength and hardness, but decreases the toughness and ductility of the steel.

Manganese helps increase the strength of carbon steels, but, unlike carbon, the effect occurs through solid solution hardening. Manganese atoms displace iron atoms in the crystal lattice during solid solution hardening. The effects of carbon and manganese on the mechanical properties of steels are additive. Carbon makes the greater contribution by the effect it has on pearlite formation. See Figure 12-11. Carb-

on and manganese increase hardenability, which is a property in steel that determines the depth to which it hardens when quenched.

Pearlite Interlamellar Spacing. *Pearlite interlamellar spacing* is the distance between the ferrite and cementite lamellae. Decreasing the pearlite interlamellar spacing increases hardness and toughness. *Normalizing* is a heat treatment that decreases the pearlite interlamellar spacing and refines the grain size. Normalized steel is heated in a furnace into the austenite region (above the Ac3), held for a specified period of time, removed from the furnace, and allowed to cool in still air. Structural steel are supplied in the normalized condition to optimize the mechanical properties.

Effects of Grain Size. Grain size is an important property in all steels. Steel may be coarse-grain (large grain size) or fine-grain (small grain size). See Figure 12-12. Grain size is usually reported as the austenite grain size in terms of the ASTM grain size number. To measure grain size, standardized techniques of specimen preparation and examination are employed.

The steel deoxidation practice is the most significant factor in controlling the grain size. The austenite grain size in solidifying steel dictates the grain

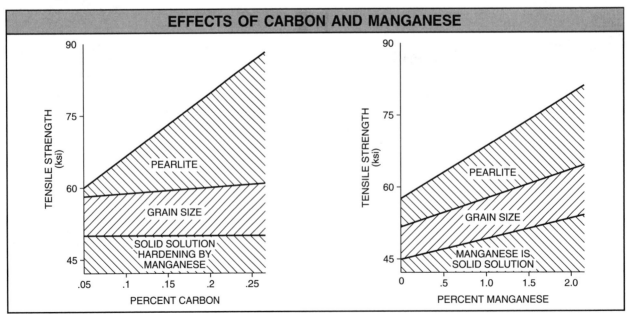

Figure 12-11. Carbon steels are strengthened by the addition of carbon and manganese.

size of the ferrite and pearlite, which forms from the transformation of the austenite. The larger the austenite grain size, the coarser the ferrite-pearlite grain size. Coarse-grain steel is generally deoxidized with silicon, and fine-grain steel is deoxidized with aluminum (producing a fully killed steel). Fine-grain size is generally beneficial in carbon steels because it increases ductility and toughness.

GRAIN SIZE AND PROPERTIES OF CARBON STEELS		
Property	Coarse-grain	Fine-grain
Quenched and tempered steel		
Hardenability	Deeper	Shallower
Toughness	Less	Greater
Distortion	More	Less
Quench cracking	More likely	Less likely
Internal stress	Higher	Lower
Annealed or normalized steel		
Rough machinability	Better	Worse
Fine machinability	Worse	Better

Figure 12-12. Grain size can have either a detrimental or a beneficial effect on properties of carbon steels.

Grain refinement is a method used to induce finer-than-normal grain size. Grain refinement is achieved by thermally cycling a steel through the critical temperature range. See Figure 12-13. At point G, which is the Ac_1 (approximately 732°C or 1350°F), the pearlite recrystallizes to form fine-grain austenite. The ferrite does not transform at this temperature, but as the temperature is raised, increasing amounts of ferrite recrystallize to form fine-grain austenite. When the temperature reaches point H, which is the Ac_3 (approximately 843°C or 1550°F), the entire microstructure consists of fine-grain austenite. If the steel was cooled at this point, the fine-grain austenite would transform to fine-grain ferrite and pearlite. Thermal cycling (heating and cooling) through the upper and lower critical temperatures is a method that is sometimes used to refine a coarse-grain steel.

Grain refinement by thermal cycling also occurs during the multiple-pass welding of steel. The as-deposited weld beads are coarse-grain, but each bead is subjected to grain refinement by the next weld pass due to the temperature cycling through the lower and upper critical temperatures. Consequently, multiple-pass welds in carbon steel are tougher than single-pass welds.

Grain growth is an undesirable increase in grain size. It occurs as a result of heating the steel to

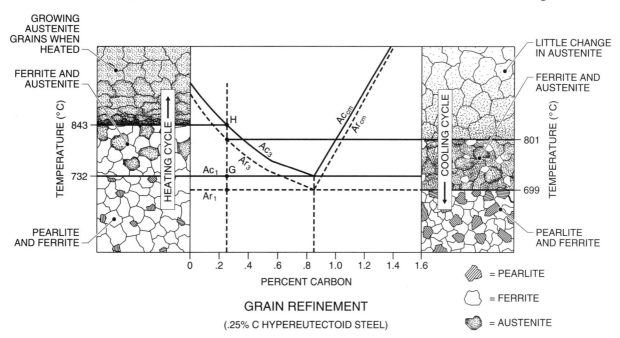

Figure 12-13. Grain size of a carbon steel can be refined by cycling it through its critical temperatures.

elevated temperatures above the upper critical temperature. Coarse-grain steel is more susceptible to grain growth. Grain growth in coarse-grain steel occurs at lower temperatures than in fine-grain (fully killed) steel.

In coarse-grain steel, the grain size increases gradually and consistently as the temperature is increased above the A_1. In fine-grain steel, the grain size increases slightly, if at all, as temperature is increased until a specific temperature is reached. This inhibition of the grain growth is caused by aluminum nitride particles in the steel. Aluminum nitride particles are formed during the steel deoxidation process. The particles help pin (restrain) the austenite grain boundaries and prevent them from enlarging. At a specific temperature (the grain-coarsening temperature), an abrupt increase in grain size begins. See Figure 12-14.

The grain-coarsening temperature of fully killed steel is 925°C to 980°C (1700°F to 1800°F) and is dependent on a variety of factors. For example, it is dependent on the amount the steel has been cold worked. One advantage of fully killed steel is the resistance to grain coarsening during high-temperature heat treating operations (for example, carburizing). These operations are performed at temperatures as high as 925°C to 955°C (1700°F to 1750°F).

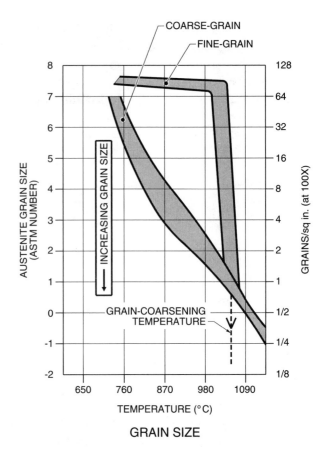

GRAIN SIZE

Figure 12-14. The grain size of coarse-grain steels gradually increases with increasing temperature, whereas fine-grain (fully killed) steels maintain fine-grain size until the grain-coarsening temperature is exceeded.

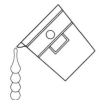

Cooling Rate and Hardenability of Steels   13

The austenite transformation products that form when carbon steel is cooled are significantly altered by the cooling rate. With fast cooling, austenite transforms to martensite or bainite rather than the pearlitic products produced by slow cooling. Isothermal transformation (I-T) and continuous cooling transformation (C-T) diagrams are used to predict austenite transformation products under fast cooling conditions. Hardenability is a property of steel that determines the depth of the martensite formed by quenching. Hardenability is influenced by the chemical composition and the microstructure of steel.

COOLING RATE

Cooling rate is the amount of degrees a metal cools per unit of time. It is usually measured in degrees Celsius per second (°C/sec) or degrees Fahrenheit per second (°F/sec). Martensite and bainite are transformation products that form when carbon steel is quenched from the austenitizing temperature. Martensite forms at extremely fast cooling rates by a diffusionless (shear-type) mechanism. Bainite formation requires carbon diffusion and occurs at intermediate cooling rates. *Carbon diffusion* is the spontaneous movement of carbon atoms within a material. For either product, the cooling rates involved are significantly faster than those leading to pearlitic transformation products.

Iron-carbon diagrams do not predict the transformation products when austenite is quenched rather than slowly cooled. Under these conditions, isothermal transformation (I-T) diagrams and continuous cooling transformation (C-T) diagrams are used to indicate the transformation products obtained at specific cooling rates. Specific I-T diagrams and C-T diagrams must be developed for each steel composition.

Austenite Transformation Products

Austenite transformation products are transformation products produced when steel is slowly cooled from the austenitizing temperature and include the formation of ferrite, pearlite, or cementite, depending on the carbon content of the steel. Pearlitic products are formed from austenite as a result of carbon diffusion. As the cooling rate is increased, a point is reached where carbon diffusion has no time to occur. The carbon is trapped in the austenite lattice. Since austenite is not thermodynamically stable at low temperatures, some type of transformation must occur. When diffusion is not possible, shear is a more favorable method of transformation. This leads to the formation of martensite. At intermediate cooling rates, when limited diffusion is possible, bainite is formed.

Hardness of the various austenite transformation products increases progressively from the pearlitic products, to bainite, and then to martensite. Bainite or martensite formation is a key step in the heat treatment of steel, but martensite formation becomes a problem when welding high-carbon and low-alloy steels.

Pearlitic Products. *Pearlitic products* are austenite transformation products that consist of ferrite, cementite, and pearlite, which are formed when the cooling rate is slow enough to allow carbon to diffuse out of the austenite. Ferrite is low in carbon, cementite is high in carbon, and pearlite is an aggregate of ferrite and cementite. The iron-carbon diagram predicts the types and quantities of the phases that form during slow cooling or under equilibrium conditions, which allow maximum carbon diffusion.

Martensite. *Martensite* is an austenite transformation product that forms when austenite is cooled rapidly, resulting in no time for the carbon to diffuse and form pearlitic products. All the carbon is trapped in the austenite, resulting in exactly the same carbon content in the martensite as in the austenite from which it is formed.

Martensite is formed by a diffusionless, shear-type mechanism. The face-centered cubic crystal structure of austenite is transformed by shear into the body-centered tetragonal structure of martensite. See Figure 13-1. This transformation mechanism allows cooperative movement of atoms from one crystal structure to the other. The new structure is equivalent to a supersaturated solution of carbon in alpha iron (ferrite).

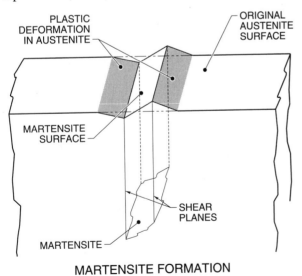

MARTENSITE FORMATION

Figure 13-1. The face-centered cubic crystal structure of austenite transforms by shear into the body-centered tetragonal cubic crystal structure of martensite.

Trapped carbon atoms develop considerable strain in martensite, making it extremely strong and hard, but it is also brittle and unsuitable for many structural applications. Martensite must be toughened by a tempering process to make it useful and significantly stronger and tougher than any of the slow-cooled pearlitic products.

The hardness and strength of martensite increases with increasing carbon content of the steel. See Figure 13-2. Low-carbon martensite in low-carbon steel is soft and does not necessarily require tempering to improve its toughness. For example, carbon content is the principal factor in determining whether fusion-welded steel components require preheating and post-weld heat treatment to maintain toughness of the weld heat-affected zone.

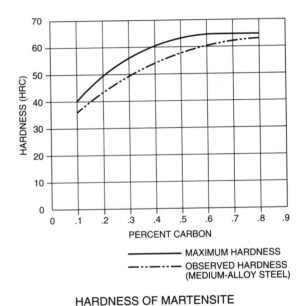

HARDNESS OF MARTENSITE

Figure 13-2. The hardness of martensite is a function of the carbon content of the steel.

Martensitic transformations also occur in some nonferrous alloy systems, such as copper-aluminum and gold-cadmium, but in engineering metallurgy, the term "martensite" is reserved for steels. Martensite formation begins at a specific temperature (M_s) and the transformation is completed at some lower temperature (M_f). The values of M_s and M_f depend on the composition of the steel. M_f can be as high as 480°C (900°F), and for some steel compositions it is well below room temperature. For all the austenite to transform to martensite, the steel must pass through the M_s and M_f. If it does not, the transformation to martensite is partial, and other phases are present in the final product. For example, if a steel is quenched in a molten salt bath with a temperature between the M_s and M_f, it partially transforms to martensite, and the remainder consists of a mixture of other transformation products. The same result occurs if the M_f is below room temperature and the steel is quenched to ambient.

The morphology of martensite may be either lath martensite or plate martensite, depending on the carbon content of the steel. Lath martensite forms in steels with <.5% C (the majority of engineering steels) and has the appearance of bundles of laths (narrow strips). Plate martensite forms in steels with >1% C and has the appearance of lens-shaped needles packed in different orientations. Steels between .5% C and 1% C exhibit mixed structures of lath and plate martensite. Both types of martensite exhibit an indistinct grain structure because the individual grains of laths or plates tend to merge into one another at the grain boundaries. The detail within the laths and plates is hard to resolve in the optical microscope, even at extremely high magnification. An electron microscope must be used to determine the morphology type. See Figure 13-3.

Bainite. *Bainite* is an austenite transformation product that is formed under continuous cooling or isothermal transformation conditions. These conditions are between those that produce pearlitic or martensitic structures. Bainite resembles pearlite because both are composed of ferrite and cementite and require carbon diffusion. Unlike pearlite, in bainite, the ferrite and cementite is not as layered. Bainite occurs as elongated particles of cementite in a matrix of ferrite.

The morphology of bainite consists of upper bainite and lower bainite. Upper bainite forms closer to the temperature range for pearlitic products and has a feathery structure. Lower bainite forms closer to the martensite and has a needle-shaped structure. See Figure 13-4.

Bainite is generally harder and tougher than pearlitic products and softer and tougher than martensite. Lower bainite is generally stronger and tougher than upper bainite. The toughness and strength of lower bainite compares favorably with tempered martensite of similar hardness. Certain heat treatment operations, such as austempering, are specifically designed to produce lower bainite rather than martensite. This eliminates the tempering step that follows quenching.

LATH MARTENSITE

PLATE MARTENSITE

Figure 13-3. Lath martensite and plate martensite require the use of an electron microscope for complete resolution.

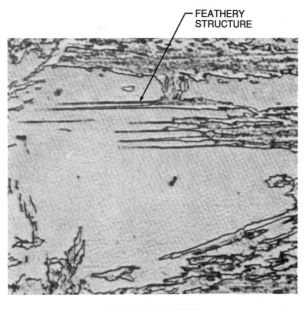

UPPER BAINITE   LOWER BAINITE

Figure 13-4. The morphology of upper bainite consists of a feathery structure, and that of lower bainite consists of a needle-shaped structure.

Austenite Transformation Diagrams

The three main diagrams used to depict the transformation products of austenite are the iron-carbon diagram, I-T diagrams, and C-T diagrams. They take into account the cooling rate and the method of transformation (isothermal versus continuous cooling). The iron-carbon diagram is applicable to all carbon steel and indicates the transformation products of austenite when it is cooled slowly enough to form pearlitic products. I-T diagrams are specific to each steel composition and indicate all transformation products of austenite under isothermal conditions. C-T diagrams are also specific to each steel composition and indicate all transformation products of austenite under continuous cooling conditions.

Iron-Carbon Diagram. The iron-carbon diagram indicates the phase changes that occur when carbon steel is slowly cooled from the austenitizing temperature. The phase transformation occurs by the diffusion of carbon out of austenite to form ferrite, pearlite, or cementite. Hypoeutectoid steels, eutectoid steel, and hypereutectoid steels are the three main groupings of steels identified on the iron-carbon diagram. See Figure 13-5.

Hypoeutectoid steels contain less than approximately .8% C and consist of ferrite and pearlite. Eutectoid steel contains approximately .8% C and consists entirely of pearlite. Hypereutectoid steels contain more than approximately .8% C and consist of pearlite and cementite.

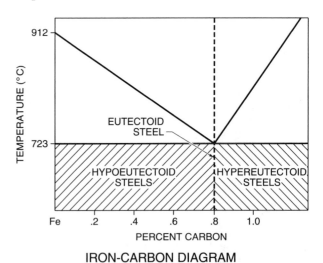

IRON-CARBON DIAGRAM

Figure 13-5. Hypoeutectoid steels, eutectoid steel, and hypereutectoid steels are the three main groupings of steels identified on the iron-carbon diagram.

As the cooling rate increases, diffusion of carbon has less time to occur, which results in a slight change in shape of the iron-carbon diagram. See Figure 13-6. The principal changes are the reduction of the upper and lower critical temperatures for cooling and the movement in the eutectoid composition. The eutectoid composition shifts to the left for hypoeutectoid steels and to the right for hypereutectoid steels. Faster cooling rates lead to finer pearlite grain size, which is harder and tougher than the coarse pearlite formed at slower cooling rates and higher transformation temperatures.

The iron-carbon diagram is used to predict the structure of carbon steels from heat treatments such as annealing. It also can be used to estimate critical temperature, stress-relieving temperature, or other key heat treatment temperatures.

Isothermal Transformation Diagrams. *Isothermal transformation diagrams* are plots of temperature against log time that indicate austenite transformation products for specific steels under isothermal conditions. I-T diagrams are also referred to as time-transformation-temperature diagrams, or T-T-T diagrams.

I-T diagrams are developed for each specific steel composition. The shapes of I-T diagrams vary with the steel composition, but they typically exhibit a nose, a bay, and/or a flat bottom shape. See Figure

13-7. The experimental method involves studying the transformation products obtained in small steel samples. The samples are austenitized and then held for set time periods at various temperatures below the lower critical temperature.

Small test samples, which are 6.5 mm in diameter by 1.5 mm in thickness (.25 in. in diameter by .0625 in. in thickness), are used in order to improve response to temperature changes. For example, five samples of eutectoid carbon steel are austenitized and then immersed in a molten salt bath at 620°C (1150°F), which is below the lower critical temperature. The samples are removed from the molten salt bath at increasing set time intervals and water quenched to room temperature. They are polished and etched to reveal the microstructures. The microstructures are indicative of the transformation products obtained in the eutectoid steel held at increasing set time intervals under isothermal conditions.

The sample that is air cooled to room temperature and the sample that is held in the molten salt bath for *a* seconds and water quenched develop a microstructure composed of 100% martensite. The sample held for *b* seconds forms a small amount of pearlite, with the bulk of the microstructure being 85% martensite. As the holding time is increased to *c* seconds, the pearlite in the microstructure increases to 80% and the martensite decreases to 20%.

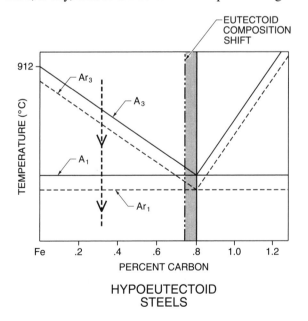

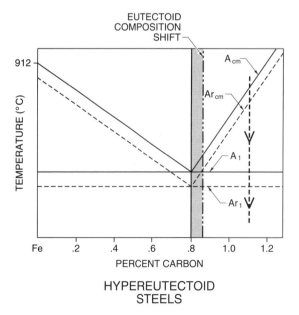

Figure 13-6. As the cooling rate increases, diffusion of carbon has less time to occur, which results in a slight change in the shape of the iron-carbon diagram. The eutectoid composition is shifted to the left for hypoeutectoid steels and to the right for hypereutectoid steels.

The sample held for *d* seconds consists of 100% pearlite with no traces of martensite.

A graph is plotted with temperature on the y-axis and log time on the x-axis. The isothermal conditions *a* to *d* indicate the percentages of martensite and pearlite formed at specific time intervals at a temperature of 620°C (1150°F).

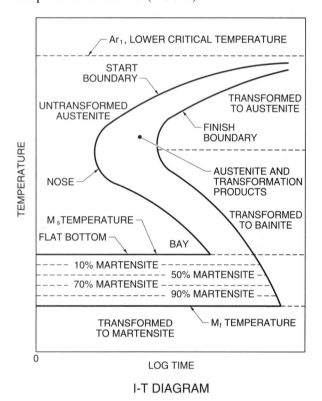

Figure 13-7. Isothermal transformation (I-T) diagrams typically exhibit three distinctive regions and the temperature and time boundaries for the transformation of austenite.

From the microstructure, two points representing the longest time interval in which 100% martensite is formed and the shortest time interval in which 100% pearlite or bainite is formed are estimated and plotted on the graph. These two points indicate the start and the finish of the isothermal transformation. This experiment is repeated at different temperatures below the lower critical temperature in order to develop an I-T diagram for the eutectoid steel. All of the start points and finish points of transformation are joined up and result in two staggered S-shaped curves. The curves indicate the boundaries for 100% martensite and 100% pearlite or bainite formations at specific temperatures.

Mixed structures composed of various percentages of martensite, pearlite, or bainite are formed between these boundaries.

The area to the left of the start boundary and above and below the nose consists of austenite that has not transformed. The area between the boundaries consists of austenite plus transformation products. The area to the right of the finish boundary consists of pearlite above the nose and bainite below it. The M_s and M_f temperatures are indicated on the flat bottom below the bay and are extensions of the start and finish boundaries. The area between the M_s and M_f lines is composed of a series of parallel curves indicating the progress of martensite transformation from 0% to 100%.

A variety of microstructures are obtained from the isothermal transformation of a eutectoid steel. The microstructure formed depends on the rate that the steel cools to the isothermal transformation temperature. See Figure 13-8. If the steel is quenched rapidly enough to avoid the nose of the start boundary line to a temperature below M_f, then 100% martensite is formed. If the steel is cooled to a temperature above the nose and is allowed to transform isothermally through the start and finish lines, then 100% pearlite is formed. If it is quenched rapidly enough to miss the nose of the first boundary and is allowed to transform isothermally at a temperature above M_s, then 100% bainite is formed.

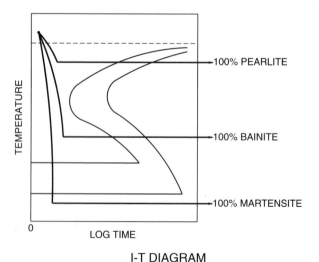

Figure 13-8. The microstructure of steel depends on the rate that it cools to the isothermal transformation temperature.

I-T diagrams for hypoeutectoid and hypereutectoid steels contain an additional region not present on the eutectoid diagram. This region is located above the nose and between the start and finish boundaries. See Figure 13-9. For a hypoeutectoid steel, the region consists of austenite plus proeutectoid ferrite. For a hypereutectoid steel, the region consists of austenite plus proeutectoid cementite.

Continuous Cooling Transformation Diagrams. *Continuous cooling transformation diagrams* are plots of temperature against log time and indicate the austenite transformation products for specific steels when continuously cooled. Many heat treatment operations rely on quenching a steel from the austenitizing temperature and are more accurately represented by C-T diagrams. C-T diagrams are also referred to as C-C-T diagrams.

Despite the general similarity of shape between C-T diagrams and I-T diagrams for identical steels, the data is presented differently. On C-T diagrams, phase changes are recorded within the start and finish boundaries, whereas on I-T diagrams, this region indicates the transforming phases. On C-T diagrams

the products of transformation appear at the bottom of the diagram, whereas on I-T diagrams, they are indicated on the right-hand side of the finish boundary. See Figure 13-10.

Although similar in shape to I-T diagrams, the nose of C-T diagrams is shifted down and to the right, indicating that more time is available for martensite formation than shown on I-T diagrams. I-T diagrams, which are more readily available for different grades of steel, actually err by indicating a faster cooling rate than necessary to form 100% martensite when steel is quenched. The error is usually on the safe side because the goal of many heat treatment operations is to produce 100% martensite. Producing 100% martensite optimizes the available strength obtained by tempering.

C-T diagrams are developed for each specific steel composition using various experimental methods. For example, a Jominy end-quench specimen (cylindrical steel bar), is machined to standardized dimensions of 25 mm in diameter by 102 mm in length (1 in. in diameter by 4 in. in length). Thermocouples are attached along the length of the bar, which is heated in a furnace to the austenitizing temperature range.

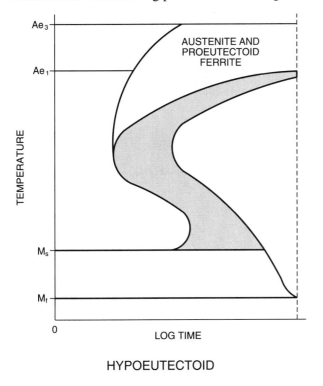

HYPOEUTECTOID

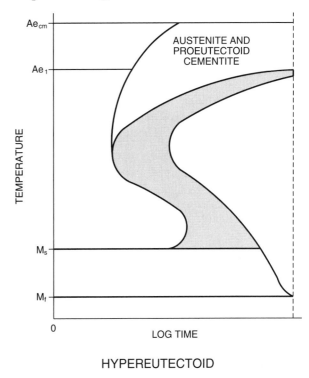

HYPEREUTECTOID

Figure 13-9. I-T diagrams for hypoeutectoid steels include a region for proeutectoid ferrite. I-T diagrams for hypereutectoid steels include a region for proeutectoid cementite.

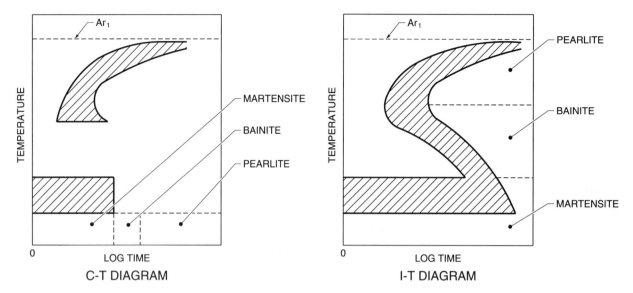

Figure 13-10. The products of transformation are indicated along the bottom of C-T diagrams and on the right-hand side of I-T diagrams.

The bar is quickly removed from the furnace, suspended vertically in a holding fixture, put in a testing rig, and rapidly cooled with a jet of water. The water is sprayed against the lower face of the bar. The cooling rate is measured by the thermocouples attached to the bar. The cooling rate decreases progressively along the bar from the quenched end.

After it has cooled to room temperature, the bar is sectioned, polished, and etched to examine the microstructures at specific locations along the length. The microstructures are correlated with the cooling rates calculated from the thermocouple readings along the bar. A C-T diagram that indicates the transformation products obtained at various cooling rates is developed from the information. With a eutectoid steel, the microstructure changes from 100% martensite to 100% pearlite as the cooling rate decreases.

HARDENABILITY

Hardenability is a measure of the depth of hardening obtained when a metal is quenched. It is a measure of how a steel responds to heat treatment and is used to predict the mechanical properties attainable by heat treatment for steel having specific composition requirements and various section sizes.

Hardenability and hardness are both measurements, but they measure essentially different properties of a steel. Since hardenability is a measure of the response of a steel to quenching from the austenitizing temperature, cooling rate is a critical factor of hardenability. Various tests and methods of presenting the data are used to measure hardenability. Hardenability allows the most cost-effective (least alloyed) steel to be selected for a given application. The chemical composition and the microstructure of a steel affect the level of hardenability that the steel exhibits.

Hardenability Concepts

Steel with a high level of hardenability is needed where high levels of strength and high levels of toughness are required in large section sizes. Hardenability is a property of steel that determines the depth and distribution of hardness induced by quenching. The deeper a steel can be hardened, the deeper it can be toughened when tempered. This should not be confused with hardness, which is a measurement obtained from the surface of a component after it is quenched.

For the practical application of heat treatment, it is not sufficient to know whether a steel forms martensite or upper bainite when quenched. It is important to know to what depth these products form in the steel at various cooling rates because the cooling rate decreases from the outside to the center of a section. To obtain the required strength in a component, it may be necessary to form martensite through the entire section.

Hardenability and Hardness. Hardenability is a measure of the depth of hardening obtained on quenching, whereas hardness is a measure of the hardness obtained at the surface of the steel. Hardness is a function of the carbon content of a steel and hardenability is a function of several factors. These factors are expressed by their influence on the shape of I-T diagrams and C-T diagrams and their positions on the time axis.

For example, the difference between hardenability and hardness can be illustrated by heat treating bars of 1080 carbon steel (.8% C) and 4140 low-alloy steel (.4% C plus small amounts of chromium and molybdenum). The bars are approximately 50 mm in diameter by 100 mm in length (2 in. in diameter by 4 in. in length).

The bars are austenitized, quenched, and then sectioned diametrically (across the diameter of the bar) at their midpoints. A slice from each is carefully wet ground to avoid overheating. Diametrical hardness traverses are then made across the ground face of each slice using a Rockwell hardness tester. See Figure 13-11.

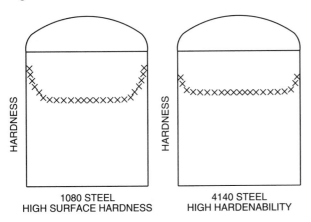

1080 STEEL
HIGH SURFACE HARDNESS

4140 STEEL
HIGH HARDENABILITY

HARDENABILITY
(× = HARDNESS VALUES)

Figure 13-11. A 1080 steel develops higher surface hardness when quenched, but a 4140 steel has higher hardenability because it retains hardness across the section thickness.

The traverses reveal that the low-hardenability 1080 steel exhibits a high level of surface hardness, which rapidly falls and then levels off to a low value across the bulk of the section. The high-hardenability 4140 steel exhibits a lower surface hardness, but it is maintained across the section with a slight drop at the center.

Critical Cooling Rate. *Critical cooling rate* is the slowest continuous cooling rate that produces 100% martensite in a steel. Cooling at less than the critical cooling rate leads to the formation of other transformation products.

In any section thickness of steel, the cooling rate decreases from the surface towards the center. To achieve 100% martensite throughout the section, the critical cooling rate must be exceeded throughout the section. On the I-T diagrams and C-T diagrams, the cooling rate throughout the section must miss the nose of the curve. See Figure 13-12.

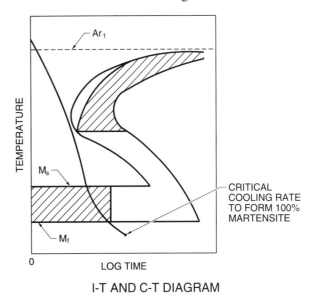

I-T AND C-T DIAGRAM

Figure 13-12. The critical cooling rate is the slowest cooling rate that misses the nose of the I-T or C-T diagram.

Hardenability Measurement

Hardenability is measured by the Jominy end-quench test, which is used to develop end-quench hardenability curves for specific steels. Hardenability bands are the scatter bands that define the boundaries for the minimum and maximum end-quench hardenability curves for standard steels. The cooling rates along the Jominy end-quench test specimen may be correlated with the corresponding C-T diagram for the steel to indicate the microstructures obtained.

Jominy End-Quench Test. The *Jominy end-quench test* is a laboratory procedure for determining the hardenability of steels. The standards for the test procedure are found in *ASTM A255* and *SAE J406, End-Quench Test for Hardenability of Steel*. See Figure 13-13.

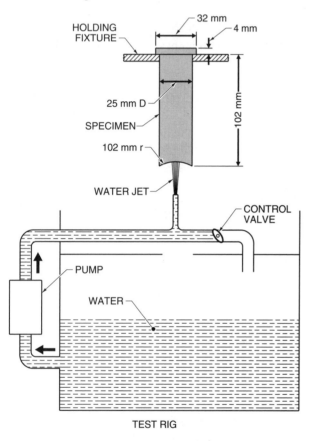

JOMINY END-QUENCH TEST

Figure 13-13. The Jominy end-quench specimen is austenitized and quenched under standardized conditions.

To conduct the end-quench test, a 25 mm in diameter by 102 mm in length (1 in. in diameter by 4 in. in length) specimen is first normalized. This consists of heating above the upper critical temperature and air cooling to eliminate the variable of inhomogeneous microstructure. It is then heated uniformly to a standard austenitizing temperature. The specimen is removed from the furnace, sus-

pended vertically in a holding fixture, placed in a testing rig, and immediately end-quenched by a jet of water. The water contacts the end face of the specimen without wetting the sides. The quenching is continued until the entire specimen has cooled.

Two longitudinal flat surfaces are ground on opposite sides of the quenched specimen, and hardness readings (HRC) are taken at 1.5 mm (.0625 in.) intervals for the first inch from the quenched end, and at greater intervals beyond that point until a hardness level of 20 HRC or a distance of 50 mm (2 in.) from the quenched end is reached. An end-quench hardenability curve is created to determine the hardenability of the specimen. An *end-quench hardenability curve* is a plot of hardness readings on the y-axis versus the distance from the quenched end on the x-axis. See Figure 13-14.

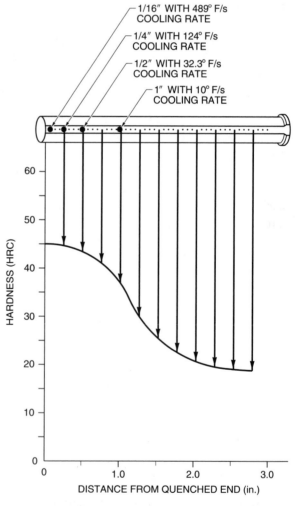

Figure 13-14. On an end-quench hardenability curve, hardness is plotted against distance from the quenched end of the Jominy bar.

Since the quenching conditions are standardized, each location on the Jominy end-quench test specimen represents a certain cooling rate. Also, because the thermal conductivities of various steels are approximately the same, this cooling rate is the same at a given position on the test specimen regardless of the composition of the test steel.

Each test specimen is therefore subjected to a series of cooling rates that vary continuously from very rapid at the quenched end to very slow at the air-cooled end. The ASTM graph paper used for plotting the end-quench hardenability curve indicates the cooling rate as a function of distance from the quenched end. See Figure 13-15.

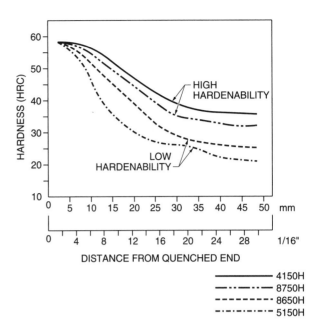

Figure 13-16. High-hardenability steels exhibit hardness that is maintained for greater distances from the quenched end of the Jominy bar than low-hardenability steels.

Hardenability Bands. *Hardenability bands* are bands that define the boundaries for the minimum and maximum end-quench hardenability curves for standard steels. They indicate the range of hardenability expected from such alloys. Hardenability bands have been established by the AISI and SAE by analysis of end-quench hardenability curves

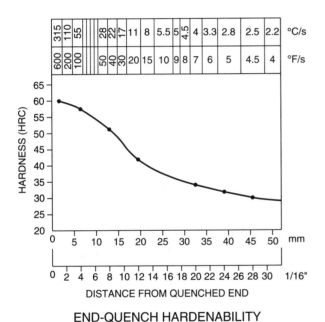

END-QUENCH HARDENABILITY
CURVE FOR AISI 8650

Figure 13-15. The ASTM graph paper used for plotting end-quench hardenability curves indicates the variation of cooling rate with distance from the quenched end of the Jominy bar.

Hardenability differences between specific grades of steel are readily compared if end-quench hardenability curves are available. High-hardenability steels maintain their as-quenched hardness values to greater distances along the Jominy specimen. See Figure 13-16.

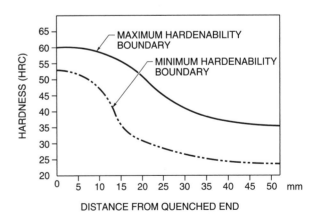

HARDENABILITY BAND
(AISI 4140H STEEL)

Figure 13-17. A hardenability band indicates the maximum and minimum hardenability boundaries for a given grade of steel.

collected from hundreds of heats (ingot) of standard steels. See Figure 13-17.

The selection of steel is often dependent on hardenability. Hardenability bands allow the user to anticipate the range of hardenability expected for a standard steel. This is based on the allowable composition range and minor heat treatment variables. Hardenability bands guarantee hardenability in purchased products.

Standard steels that fall within the limits of their hardenability bands are designated with an uppercase letter H. For example, a heat of 4140 low-alloy steel that falls within the limits of its hardenability band is designated 4140H.

End-quench Hardenability Curves and C-T Diagrams. C-T diagrams indicate the phases that form in a steel at different cooling rates. Correlating the end-quench hardenability curve with the C-T diagram for a particular steel makes it possible to identify the various phases expected at different locations along the Jominy end-quench specimen. See Figure 13-18.

For steel, 100% martensite is formed up to point A, and mixed structures containing martensite, ferrite, and bainite are formed at lower cooling rates. As the hardenability increases, 100% martensite is produced at increasing distances from the quenched end of the Jominy specimen. The higher the hardenability, the further the nose of the C-T diagram is along the time axis.

Hardenability Applications

End-quench hardenability curves are used to predict the hardness obtained at various depths in oil-quenched and water-quenched bars. The severity of the various types of quenching media used to harden steel is ranked numerically. End-quench hardenability curves are used to select steel to meet specific toughness requirements and indicate the largest bar

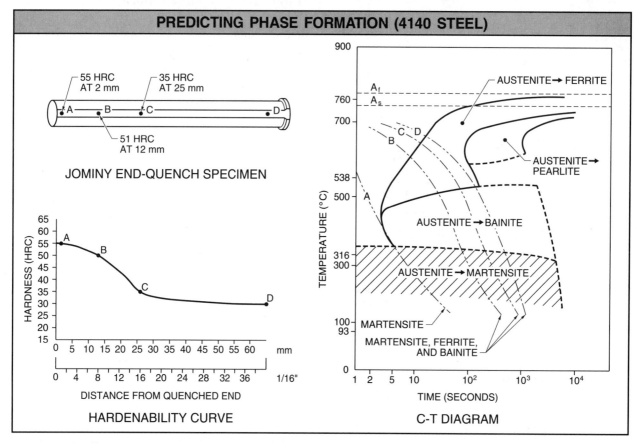

Figure 13-18. Correlation of the end-quench hardenability curve with the matching C-T diagram enables the phases formed at different locations along the Jominy end-quench specimen to be predicted.

diameter that will harden completely through its section when quenched.

Depth of Hardening. The most common criterion for hardenability is the distance along the end-quench specimen where the microstructure consists of 50% martensite. This value is selected because 50% martensite is easy to distinguish in the microstructure and is clearly indicated as the point of inflection on the end-quench hardenability curve. For example, with an 8650 alloy steel end-quench hardenability specimen, the point of inflection occurs at 16 mm (.625 in.) from the quenched end. See Figure 13-19.

Experiments confirm that the cooling rate at a given distance from the quenched end of the Jominy

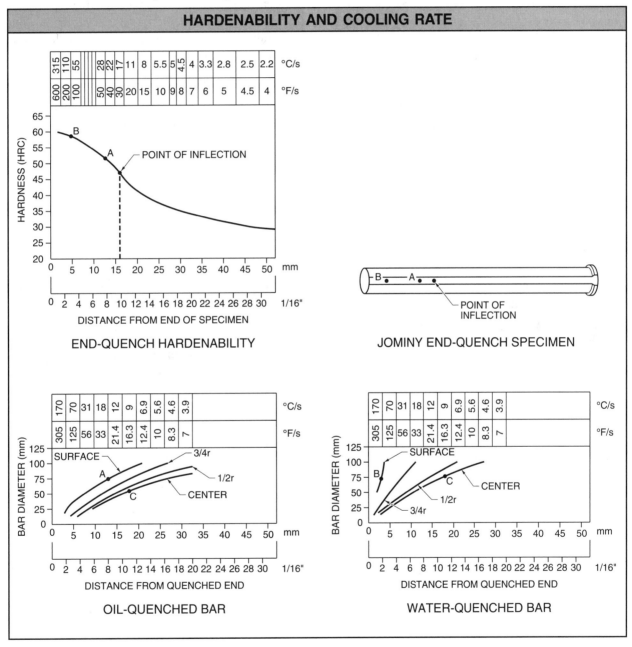

Figure 13-19. The most common criterion for hardenability on the end-quench hardenability curve is the point of inflection (50% martensite). Cooling rates at given distances from the quenched end of the Jominy bar can be correlated to the cooling rates at four different locations in the quenched specimen.

specimen may be correlated with the cooling rate at various depths in bars or thicknesses of plate. Curves are available that show the relationship between the cooling rates along Jominy end-quench specimens and the diameter of bars at four locations from the center to the surface for water quenching and oil quenching.

The surface hardness of a 75 mm (3 in.) diameter oil-quenched bar, which is point A on the oil-quenched curves, corresponds to the hardness obtained at a distance of 13 mm (.5 in.) from the quenched end of the Jominy bar, which is point A on the end-quench hardenability curve and has a hardness of 53 HRC. The surface hardness of a water-quenched bar of the same size, which is point B on the water-quenched curves, corresponds to the hardness obtained at a distance of 3 mm (.125 in.) from the quenched end, which is point B on the end-quench hardenability curve and has a hardness of 58 HRC. The difference in hardness indicates that water quenching is more severe than oil quenching.

Using the same curves for the point of inflection as for the requirement for hardenability, the maximum allowable thickness for through-hardened bars may be calculated for the various quenching media. With oil quenching, through hardening is achieved in a 57 mm (2.25 in.) diameter bar, which is point C on the oil-quenched curves. With water quenching, through hardening is achieved in a 75 mm (3 in.) diameter bar, which is point C on the water-quenched curves. Therefore, water quenching causes deeper hardening than oil quenching.

Severity of Quench. Severity of quench (H) is a quantitative measure of the cooling power of a quenching medium. See Figure 13-20. The higher the value of H, the more severe the quench. The order of increasing severity in common quenching media is as follows: air, oil, water, and brine. Increasing the circulation or agitation of any quenching medium increases its severity of quench.

Severity of quench is determined experimentally by quenching a series of round bars of a given steel. For example, various diameter bars of SAE 3140 alloy steel are austenitized. Some of the bars are quenched in oil and some in water. All the bars are then transversely sliced at their midpoints to examine the microstructures. Etching reveals a transition point between dark and light corresponding to approximately 50% martensite. As with end-quench hardenability curves, this location is taken as the criterion for depth of hardening.

The experiment indicates that the greater the bar diameter (D), the greater the unhardened diameter (D_u). A graph of D_u/D versus D is plotted for the oil-quenched and water-quenched samples. To obtain severity of quench, the oil-quench or water-quench curves are matched to one of the calculated curves of D_u/D versus HD. These are available for a wide range of quench severities. See Figure 13-21.

Matching is performed by finding the best correspondence of the measured D_u/D versus HD curves against the calculated D_u/D versus HD curve. The HD value obtained by the match is used to determine the severity of quench. Severity of quench is found using the following formula:

$$H = \frac{HD}{D}$$

where

H = severity of quench

SEVERITY OF QUENCH				
	Air	**Oil**	**Water**	**Brine**
No circulation of fluid or agitation of piece	.02	.25 to .30	.9 to 1.0	2
Mild circulation (or agitation)	—	.30 to .35	1.0 to 1.1	2 to 2.2
Moderate circulation	—	.35 to .40	1.2 to 1.3	—
Good circulation	—	.4 to .5	1.4 to 1.5	—
Strong circulation	.05	.5 to .8	1.6 to 2.0	—
Violent circulation	—	.8 to 1.1	4	5

Figure 13-20. Severity of quench increases from air, to oil, to water, to brine. The amount of agitation of the quenching medium also increases the severity of quench.

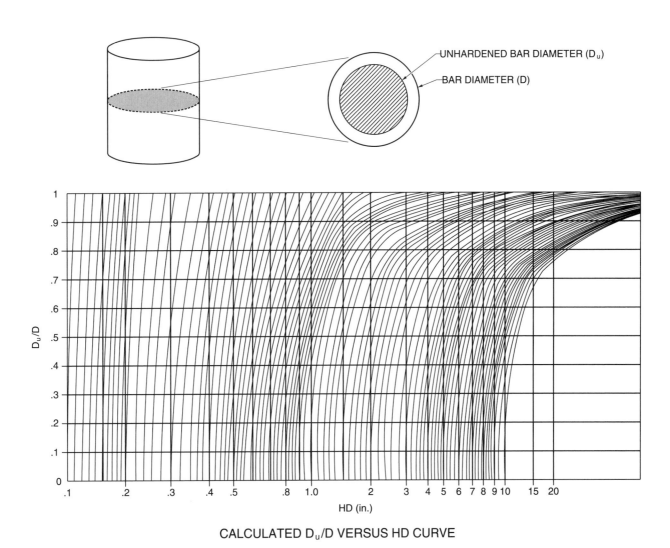

CALCULATED D_u/D VERSUS HD CURVE

Figure 13-21. Curves of D_u/D versus HD are used for estimating the severity of quench (H) of the quenching medium.

HD = HD value obtained by matching the curves

D = bar diameter

Example: Figuring severity of quench

When the curves are matched, a point on the water-quench curve corresponding to a bar diameter of 46.5 mm (1.83 in.) falls on an HD value of 66 mm (2.6 in.) on the calculated curve. What is the severity of quench?

$$H = \frac{HD}{D}$$

$$H = \frac{66}{46.5}$$

$$H = \textbf{1.42}$$

Ideal Critical Diameter. *Ideal critical diameter* (D_I) is the largest diameter of any specific steel bar that is hardened to 50% martensite by a perfect quench. *Perfect quench* is a theoretical quench in which the surface of the bar cools instantaneously from the austenitizing temperature to the temperature of the quenching medium. A perfect quench has a severity of quench value of infinity ($H = \infty$). Ideal critical diameter values allow quantitative ranking of the hardenability of steel. The higher the ideal critical diameter, the greater the hardenability of the steel. See Figure 13-22. Ideal critical diameters are expressed as a range of values. This allows for acceptable variations of chemical composition and grain size within the steel specification.

IDEAL CRITICAL DIAMETER VALUES (IN.)							
Steel	**D_I**	**Steel**	**D_I**	**Steel**	**D_I**	**Steel**	**D_I**
1045	.9 to 1.3	4047 H	1.7 to 2.4	5120 H	1.2 to 1.9	8640 H	2.7 to 3.7
1090	1.2 to 1.6	4053 H	2.1 to 2.9	5130 H	2.1 to 2.9	8641 H	2.7 to 3.7
1320 H	1.4 to 2.5	4063 H	2.2 to 3.5	5132 H	2.2 to 2.9	8642 H	2.8 to 3.9
1330 H	1.9 to 2.7	4068 H	2.3 to 3.6	5135 H	2.2 to 2.9	8645 H	3.1 to 4.1
1335 H	2.0 to 2.8	4130 H	1.8 to 2.6	5140 H	2.2 to 3.1	8647 H	3.0 to 4.1
1340 H	2.3 to 3.2	4132 H	1.8 to 2.5	5145 H	2.3 to 3.5	8650 H	3.3 to 4.5
2330 H	2.3 to 3.2	4135 H	2.5 to 3.3	5150 H	2.5 to 3.7	8720 H	1.8 to 2.4
2345	2.5 to 3.2	4140 H	3.1 to 4.7	5152 H	3.3 to 4.7	8735 H	2.7 to 3.6
2512 H	1.5 to 2.5	4317 H	1.7 to 2.4	5160 H	2.8 to 4.0	8740 H	2.7 to 3.7
2515 H	1.8 to 2.9	4320 H	1.8 to 2.6	6150 H	2.8 to 3.9	8742 H	3.0 to 4.0
2517 H	2.0 to 3.0	4340 H	4.6 to 6.0	8617 H	1.3 to 2.3	8745 H	3.2 to 4.3
3120 H	1.5 to 2.3	X4620 H	1.4 to 2.2	8620 H	1.6 to 2.3	8747 H	3.5 to 4.6
3130 H	2.0 to 2.8	4620 H	1.5 to 2.2	8622 H	1.6 to 2.3	8750 H	3.8 to 4.9
3135 H	2.2 to 3.1	4621 H	1.9 to 2.6	8625 H	1.6 to 2.4	9260 H	2.0 to 3.3
3140 H	2.6 to 3.4	4640 H	2.6 to 3.4	8627 H	1.7 to 2.7	9261 H	2.6 to 3.7
3340	8.0 to 10.0	4812 H	1.7 to 2.7	8630 H	2.1 to 2.8	9262 H	2.8 to 4.2
4032 H	1.6 to 2.2	4815 H	1.8 to 2.8	8632 H	2.2 to 2.9	9437 H	2.4 to 3.7
4037 H	1.7 to 2.4	4817 H	2.2 to 2.9	8635 H	2.4 to 3.4	9440 H	2.4 to 3.8
4042 H	1.7 to 2.4	4820 H	2.2 to 3.2	8637 H	2.6 to 3.6	9442 H	2.8 to 4.2
4047 H	1.8 to 2.7					9445 H	2.8 to 4.4

Figure 13-22. The hardenability of various steels is rated using the ideal critical diameter (D_I) values. The higher the D_I, the greater the hardenability.

Ideal critical diameter is related to the actual critical diameter (D) for any specific quench by the severity of quench. For example, for a perfect quench, the actual critical diameter is equal to the ideal critical diameter. See Figure 13-23.

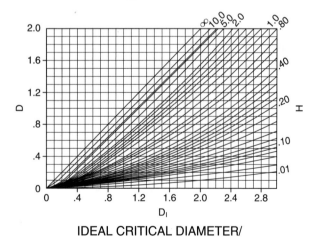

IDEAL CRITICAL DIAMETER/
ACTUAL CRITICAL DIAMETER

Figure 13-23. Ideal critical diameter (D_I) is related to the actual critical diameter (D) by the severity of quench (H). For a perfect quench (H = ∞), D_I and D are equal.

Chemical Composition and Microstructure

Hardenability is affected by the chemical composition and the microstructure of a steel. These factors influence the temperature and the time available to complete the transformation. C-T diagrams illustrate how the chemical composition and microstructure of steels influence hardenability.

Alloying Elements. Carbon is the principal alloying element that influences the hardenability of steel. Additions of carbon increase hardenability by increasing the time available for martensite to form. The effect of increasing carbon content is the movement of the nose of the C-T diagram along the time axis and additionally to alter the M_s and M_f temperatures. The C-T diagram is shifted to the right with the increase in carbon content. Carbon also causes significant lowering of M_s and M_f temperatures. For steels with greater than approximately .6% C, the M_f falls below room temperature. See Figure 13-24.

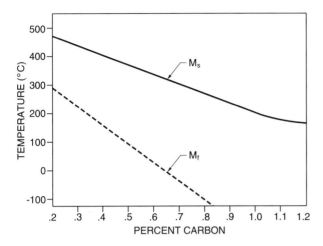

Figure 13-24. Increasing carbon content significantly lowers the M_s and M_f temperatures.

In most cases, increasing the alloy content in a steel delays the start of transformation and increases the time available for its completion. Like carbon, most alloying elements have a depressing effect on the M_s and M_f temperatures. The effect of the alloying elements is not as great as the effect produced by carbon. See Figure 13-25. A primary reason for alloying steel is to increase its hardenability at acceptably low carbon contents. This is equivalent to making martensite form more easily by delaying the start of transformation so that slower cooling rates can be used in heat treatment operations. This leads to fewer problems.

ALLOY ELEMENT EFFECT ON M_s TEMPERATURE		
Element	**Change in M_s Temperature**	
	°C	**°F**
Carbon	−285	−510
Manganese	− 33	− 60
Chromium	− 22	− 40
Nickel	− 17	− 30
Molybdenum	− 11	− 20
Tungsten	− 11	− 20
Silicon	− 11	− 20
Cobalt	+ 6	+ 10
Aluminum	+ 17	+ 30

Figure 13-25. Like carbon, most alloying elements have a depressing effect on the M_s temperature.

Microstructure. Austenite microstructure and austenitizing temperature influence hardenability by altering the position of the C-T diagram on the time axis. Higher austenitizing temperatures increase hardenability by causing more carbon to enter the austenite. This alters the M_s and M_f temperatures and shifts the C-T diagram to the right.

Austenite grain size and homogeneity prior to quenching have several effects on hardenability. Increasing the austenite grain size increases hardenability by delaying the start and completion of the transformation, because the larger the austenite grain size, the less grain boundary area available for transformation. This reduces the number of sites available for the diffusion of carbon to form pearlitic products and increases the time available to form martensite or bainite. The inhomogeneity of austenite reduces hardenability by speeding up the start of the transformation because those regions in the austenite that are lower in carbon content than the bulk composition will transform to produce undesirable pearlitic products.

Retained austenite is austenite that has survived a heat treatment cycle in which it would have been expected to transform to other products. Depending on the chemical composition of the steel and the austenitizing temperature, a significant amount of retained austenite can be produced. Retained austenite is undesirable because it reduces hardenability, and it must be removed as much as possible by modification of the heat treatment procedures.

At the M_f temperature, all the austenite should transform to martensite. In practice, not all of the austenite transforms. The amount of retained austenite is not significant in many steels. A small amount (5%) is tolerable, but as much as 30% to 40% can be present in high-carbon and high-alloy steels, which exhibit M_f temperatures below the ambient temperature. High austenitizing temperatures also increase the percentage of retained austenite by allowing extra amounts of carbon to enter the austenite.

The purpose of quenching is to form hard, strong martensite. The presence of soft, weak retained austenite prevents a quenched steel from achieving the expected mechanical properties. Also, retained austenite does not remain stable when tempered but transforms to cementite, bainite, or martensite. This transformation causes dimensional instability and may cause embrittlement.

Several methods are employed to eliminate or reduce retained austenite to acceptable levels. These include subzero quenching and double or triple tempering (repeated tempering operations). Subzero quenching, performed at temperatures as low as $-100°C$ ($-148°F$), cools the steel near or below its M_f, resulting in complete austenite transformation. Double or triple tempering helps transform the retained austenite to martensite, which is then tempered in the next tempering cycle. Double or triple tempering is often used on tool steels to optimize toughness and improve dimensional stability.

Retained austenite is usually difficult to detect in the microstructure but can sometimes be observed as white patches in a martensite structure. See Figure 13-26. The patches are difficult to resolve in the optical microscope. X-ray diffraction is used to measure the amount of retained austenite in steels.

COARSE PLATES OF MARTENSITE —   WHITE PATCHES OF RETAINED AUSTENITE

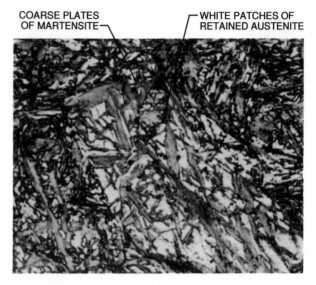

RETAINED AUSTENITE

American Society for Metals

Figure 13-26. Retained austenite is usually difficult to resolve in the optical microscope, but it is sometimes observed as white patches in a martensite structure.

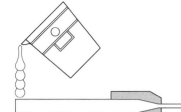

Heat Treatment of Steels 14

Heat treatment is the process of heating and cooling metals in order to obtain desired properties. Steels are heat treated for various reasons, which include improving their strength and toughness, softening them, and producing a hard and wear resistant surface layer. Heat treatment processes are divided into austenitizing, quenching, and tempering; interrupted quenching; annealing; and case hardening. Heat treatment may introduce physical problems in the steel. The component design and the fabrication process may also cause problems during heat treatment. If these problems are not corrected, they can lead to the premature failure of a component.

AUSTENITIZING, QUENCHING, AND TEMPERING

Heat treatment is the process of heating and cooling metals in order to obtain desired properties. Austenitizing, quenching, and tempering is the most common heat treatment process for steels. It allows steels to be produced in a wide variety of strength levels. There are several problems that may occur if this heat treatment process is not carefully performed. Interrupted quenching techniques are special heat treatment processes used to minimize distortion and cracking, which is sometimes associated with regular quenching and tempering.

Austenitizing

Austenitizing is the process of heating a steel to a temperature above the upper critical temperature to produce a microstructure of austenite. Austenitizing consists of preheating the metal and holding it at the austenitizing temperature. The furnace atmosphere has a direct effect on the results of the heat treatment process. Quenching and tempering are performed after the steel has been austenitized.

Preheating. Preheating to the austenitizing temperature is done at a rate that minimizes the development of high stresses in the component. This is particularly important with nonsymmetrical (complex) components and heavy sections because they absorb heat nonuniformly and develop differential expansion characteristics. This produces excessive mechanical stresses that cause distortion and cracking in the component or section.

Nonsymmetrical sections and heavy sections are preheated at a carefully controlled rate to minimize the development of stresses. This is accomplished by staged heating. *Staged heating* is heating at a controlled rate to a set temperature, holding the component until the temperature equalizes throughout the section, and then continuing the heating at a higher rate to the austenitizing temperature. Typical preheating rates for steel are 110°C/hr to 220°C/hr (200°F/hr to 400°F/hr) up to 315°C (600°F).

Holding at Austenitizing Temperature. Holding at the austenitizing temperature for sufficient time is done to produce uniform transformation of the microstructure to austenite. The holding time depends on the thickness of the component and is generally performed for 1 hour per inch (hr/in.) of cross section after the component has reached the austenitizing temperature. The minimum holding time is usually 30 minutes, except for some tool steels where very short holding times of a few minutes are used. For example, a component with a maximum cross section of 4 in. is held for 4 hours, and a component $\frac{1}{4}$ in. across is held for 30 minutes.

The selected austenitizing temperature must be high enough to dissolve carbides and produce an austenite microstructure containing enough carbon to achieve the required hardening when quenched. The temperature must not be so high that excessive grain growth occurs. See Figure 14-1.

For hypoeutectoid and eutectoid steels, the optimum austenitizing temperature is approximately 60°C (110°F) above the upper critical temperature (A_3). Hypereutectoid steels exhibit problems such as severe grain growth, decarburizing, and scaling if they are heated 60°C (110°F) above their upper critical temperature. *Decarburization* (decarb) is the annealing process in which the carbon is removed from the steel by diffusion. *Scaling* is the loss of metal from a metal surface by the formation of a scale. Hypereutectoid steels are therefore austenitized at 60°C (110°F) above their lower critical tem-

perature (A_1). Although this causes less carbon to be dissolved in the austenite, it is not a problem because hypereutectoid steels contain enough carbon to be dissolved in the austenite so that the required hardness is obtained when quenched.

Furnace Atmospheres. Furnace atmospheres are tailored to provide surface protection to steels during heat treatment and also to supply the necessary ingredients for case-hardening operations. Heated steel of any composition is quick to react with whatever is in contact with the surface. Oxygen reacts with the surface to form scale. Some gases react to add carbon to the steel, while others remove carbon from the surface. Consequently, the surface of the steel must be surrounded by a suitable atmosphere that minimizes undesirable changes in surface composition.

Two types of gaseous atmospheres commonly used for protecting steel at high temperatures are exothermic gas and endothermic gas. *Exothermic gas* is a gaseous atmosphere made by passing a partially burned gas-air mixture over a catalyst that is heated by the heat from the partial combustion. *Endothermic gas* is a gaseous atmosphere made by

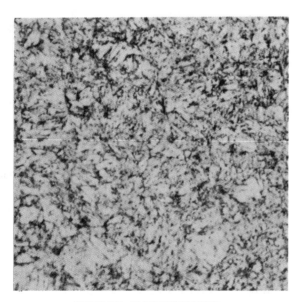

NORMAL AUSTENITIZING
TEMPERATURE

HIGH AUSTENITIZING
TEMPERATURE

Figure 14-1. Excessively high austenitizing temperatures lead to grain coarsening and loss of toughness.

passing a mixture of fuel gas (usually propane) and air over an externally heated catalyst.

Exothermic gas is considerably less costly than endothermic, but it is usually limited to very low-carbon steels since it tends to decarburize any steel with appreciable carbon content. Endothermic gas can be used safely over the entire range of varying carbon contents in steel because it can be varied in composition.

Carbon potential is the criterion used to evaluate an endothermic gas. *Carbon potential* is the percentage of carbon in steel that will neither be carburized nor decarburized by the gas. The carbon potential of an endothermic gas is a function of its dew point. *Dew point* is the temperature at which the gas is saturated with water vapor. Since the dew point of a gas is easily measured, it is used as an indicator of the carbon potential. Low dew points have high-carbon potential and are used to protect high-carbon steels. High dew points have low-carbon potential and are used to protect low-carbon steel.

Problems and Solutions. Austenitizing is designed to produce a homogeneous, fine-grain austenitic microstructure prior to quenching. The optimum austenitizing temperature is a compromise between achieving rapid dissolution of carbon and minimizing other problems, including excessive carbide dissolution, grain growth, scaling, decarburization, insufficient carbide dissolution, and burning.

Excessive carbide dissolution is the process in which an excessive amount of carbon dissolves in the steel. This occurs when hypereutectoid steels are austenitized at too high of a temperature. These steels should be austenitized just above their lower critical temperature. If austenitized above the upper critical temperature (as with hypoeutectoid steels), excessive carbon is dissolved and leads to a significant amount of retained austenite when quenched. This causes cracking and distortion.

Grain growth occurs when a steel is austenitized at too high of a temperature. Quenched steels with large austenite grains are susceptible to cracking during quenching and lack toughness. Fine-grain steels (aluminum deoxidized) do not exhibit grain growth up to approximately 980°C (1795°F). They retain a fine austenite grain size during austenitizing.

Coarse-grain steels (silicon deoxidized) may exhibit significant grain growth, especially if held in the austenitizing temperature range for a prolonged time. This may happen during carburizing, which is often performed for many hours at temperatures of approximately 930°C (1705°F).

Excessive scaling destroys surface finish or dimensions. Scaling is reduced substantially by using the correct endothermic or exothermic gas mixture. Special furnace atmospheres are used to avoid scaling. For example, scaling is avoided by placing the component in a vacuum furnace.

Decarburization leads to reduced surface hardening compared with the bulk of the section when the component is quenched. For example, decarburization causes substantial reduction of fatigue life in shafts. To reduce decarburization, the component is wrapped with paper. The paper burns off, which leaves behind a surface richer in carbon. This helps counteract the tendency for carbon to be lost from the surface. In shafts, where fatigue resistance is especially important, a machining allowance must be included so that all decarburization is removed.

Insufficient carbide dissolution is incomplete austentizing that occurs when the austenitizing temperature is too low. For example, it occurs when hypoeutectoid steels are treated below the upper critical temperature. If the steel is not completely austenitized, the ferrite or pearlite that is present does not harden when quenched. The steel is softer than specified.

Burning is permanent damage to a metal caused by austentizing too close to the melting temperature. This rarely happens unless the furnace is out of control. Burning leads to permanent damage caused by intergranular oxidation or incipient melting along the grain boundaries of the steel. See Figure 14-2.

Quenching

Quenching is the rapid cooling of a heated metal. This process is performed to obtain the desired transformation products. Quenching of a metal increases the strength and hardness and it decreases the toughness and ductility. The mechanism of quenching takes place in three steps.

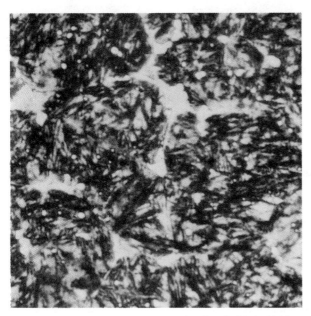

American Society for Metals

Figure 14-2. Burning leads to permanent damage caused by intergranular oxidation or incipient melting along the grain boundaries of the steel.

Quenching media are the materials that are used to quench a metal. Quenching media are selected on the basis of their cooling power, convenience in handling, and cost.

The slowest quenching rate that achieves the desired transformation is selected so that distortion and cracking are minimized. The power of any quenching medium can be increased by increasing the speed of agitation or by incorporating additives into a metal.

Mechanism of Quenching. The mechanism of quenching consists of three steps, which are vapor blanket, vapor transport, and liquid cooling. See Figure 14-3. The geometry of the component exerts a considerable influence on the way heat is abstracted during quenching. The quenching procedure must be designed so that vapors are not trapped anywhere.

Vapor blanket cooling occurs when the component is first immersed in the quenching medium. An unbroken vapor blanket develops, surrounding and insulating the component. Relatively slow cooling occurs by radiation through the blanket.

Vapor transport cooling occurs when the continuous vapor blanket collapses with the falling surface temperature of the component. This leads to violent boiling of the quenching medium and rapid heat removal, mostly as heat from vaporization of the

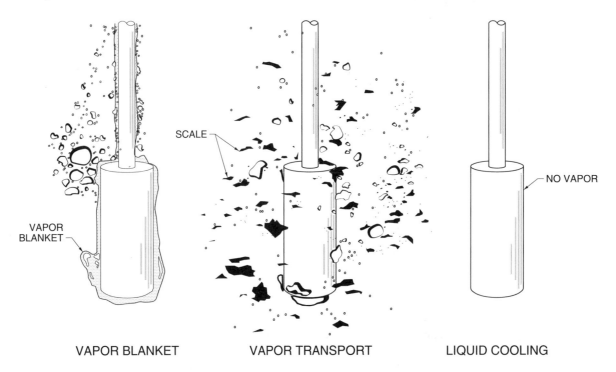

Figure 14-3. Quenching occurs in three steps, which are vapor blanket, vapor transport, and liquid cooling.

quenching medium. The boiling point of the quenching medium determines the duration of the vapor transport cooling step. The lower the boiling point, the longer vapor transport cooling lasts.

Liquid cooling occurs when the surface temperature is reduced below the boiling range of the quenching medium. Cooling in this stage is by conduction and convection. As in vapor blanket cooling, the rate of liquid cooling is relatively slow.

Quenching Media. The common quenching media are water, brine, oil, polymer, and air. Each medium has specific characteristics that result in a variety of properties that are exhibited after quenching.

Water quenching is inexpensive, readily available, without disposal problems, and used whenever practical. It provides a rapid quench up to the boiling point of water, but can cause complex components to crack. Vapor pockets of evolved steam also form and may become trapped in complex components, which leads to uneven as-quenched hardness. Water quenching is restricted to symmetrical components with low hardenability, such as carbon steel bars.

Brine quenching is performed in aqueous (water) solutions containing various percentages of salt, such as sodium chloride or calcium chloride. A common brine quench is 10% sodium chloride. Compared with water, brine quenching is faster, forms fewer vapor pockets, and the vapor blanket cooling step terminates at a higher temperature. Despite the faster quenching rate, cracking and distortion are less likely. Brines are corrosive, so expensive corrosion protection measures must be applied to the quenching equipment, and corrosion inhibitors must be added to the brines. Brine quenching is the fastest quenching medium and is used where water does not provide a sufficiently rapid quench, such as for components with low hardenability.

Oil quenching is performed in mineral oils that usually have a paraffin base, such as linseed oil. Oil quenching is significantly slower than water or brine quenching. The principal advantage of oil quenching is uniform heat extraction, which slows down toward the end of the quenching cycle. Oil quenching reduces the dangers of cracking that are associated with water and brine quenching. A disadvantage of oil quenching is that oil presents a flammability hazard. Oil quenching is used for com-plex components with moderate to high hardenability, such as low-alloy steels.

Polymer quenching is performed in aqueous solutions of organic polymers, such as polyalkene glycol. Polymer quenching rates are between those of water and oil quenching. It is advantageous because the process is free of occupational health hazards, corrosion hazards, and fire hazards.

Air quenching is performed in a fast moving stream of air. It is the slowest method and can only be used on steels with extremely high hardenability, such as tool steels. Air quenching minimizes the distortion and cracking in complex components.

Quenching Medium Selection. The selection of a quenching medium requires compromises between achieving the as-quenched hardness requirement and minimizing or avoiding distortion and cracking of the component. The optimum quenching medium allows the substitution of less expensive (lower alloyed) steels of lower hardenability. It also results in productivity improvement through reduced heat treatment cycle times, and it minimizes occupational health hazards and corrosion hazards.

The more severe the quenching medium and the less symmetrical the component, the greater the distortion from quenching. Severity of quench for a quenching medium must exceed the value required to achieve the minimum allowable hardenability in a steel. See Figure 14-4. The severity of quench of any quenching medium is increased by agitation.

Problems and Solutions. Quenching problems are generally associated with dimensional changes that occur as a steel is cooled. They include quench cracking, retained austenite, and slack quenching.

Quench cracking is the fracture of a steel during quenching. It is associated with the volume expansion that occurs when a steel transforms from austenite to martensite when quenched. The surface of the component is the first to transform, and its expansion is not restricted. When martensite begins to form inside the component, its expansion is restricted by the outer layers of martensite. This results in internal stresses that place the inner layers of the component in compression and the outer layers in tension. Quench cracking is avoided by reducing the severity of the quench, using an interrupted quenching technique, or modifying the design of the component to eliminate undesirable stress raisers. A *stress raiser* is a change in contour that increase localized stresses. Less severe quenching media usually require steels with higher hardenability. See Figure 14-5.

Retained austenite is the austenite that has survived a heat treatment cycle in which it would have been expected to transform to other products. A small amount of retained austenite is not damaging, but in steels with higher carbon content and alloy steels, the amount may become significant. Many methods are used to avoid large amounts of retained austenite and reduce them to acceptable levels. For example, subzero quenching may be used when the M_f is below room temperature.

Slack quenching is incomplete hardening of a steel caused by quenching from the austenitizing temperature at a rate slower than the critical cooling rate. This results in martensite and additional transformation products. To avoid slack quenching, a powerful enough quenching medium must be used. The components must be stacked in the quenching medium in a manner that leaves room between them for adequate circulation.

Tempering

Tempering (drawing) is the heating of a quenched steel in a furnace to a specified temperature below the lower critical temperature and then allowing it to cool at any desired rate. Tempering must be performed as soon as possible after quenching to minimize the risk of cracking. Components are usually removed from the quenching medium when they are warm to the touch and tempered immediately. The holding time at the tempering temperature is 1 hr/in. of cross section, plus the time allowed for the furnace temperature to steady out (return to required temperature) after the component is placed in it. Most steels are tempered in the range of 205°C to 595°C (400°F to 1100°F).

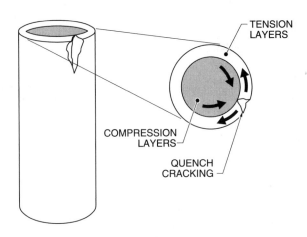

QUENCH CRACKING

Figure 14-5. The outer layers of the component transform first and are placed in tension, whereas the inner layers transform last and are placed in compression. With severe quenching, this leads to quench cracking.

Hardness is the key quality control parameter in tempering. Generally, ±20 Brinell hardness numbers (points) is permitted. This allows for a spread of 40 points. This translates to ±2 Rockwell C (a spread of 4 points). Alternatively, a minimum or maximum hardness may be specified, plus the allowable deviation. If the specified hardness range

	Flow Rate		Severity of Quenching Medium			
Agitation	**m/min**	**ft/min**	**Air**	**Mineral Oil**	**Water**	**Brine**
None	0	0	.02	.20/.30	.9/1.0	2.0
Mild	15	50		.20/.35	1.0/1.1	2.1
Moderate	30	100		.35/.40	1.2/1.3	
Good	60	200	.05	.40/.60	1.4/2.0	
Strong	230	750		.60/.80	1.6/2.0	4.0

TYPICAL SEVERITY OF QUENCH VALUES

Figure 14-4. The severity of quench for a quenching medium must exceed the value required to achieve the minimum allowable hardenability in a steel.

is too narrow, a second complete heat treatment is needed to achieve the hardness required.

The primary purpose of tempering is to increase the mechanical properties of the steel. Goals are to increase ductility and toughness while reducing hardness. Tempering relieves stresses and improves dimensional stability. Also, significant microstructural changes occur during tempering.

Mechanical Property Changes. Significant mechanical property changes occur during tempering. As-quenched steels are too brittle for most applications. They contain residual stresses and are dimensionally unstable. There are very few cases in which as-quenched steels are used in service. Tempering decreases strength and hardness, and it increases elongation and toughness. These changes in the steel become more pronounced with increasing tempering temperature.

Stress relief occurs during tempering. Residual stresses are introduced by fabrication processes, such as welding, or from distortion that occurs during quenching. At any tempering temperature, a specific amount of residual stress can be removed up to a maximum value. The higher the tempering temperature and the longer the time at that temperature, the greater the percentage of removal. See Figure 14-6. Most of the removal occurs within the first 2 hours at the tempering temperature.

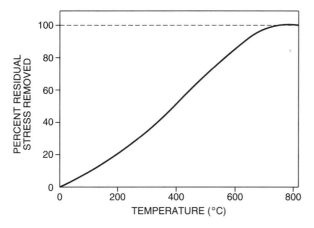

Figure 14-6. At any tempering temperature, a specific maximum amount of residual stress can be removed up to a maximum value.

The value of tempering in reducing susceptibility to brittle fracture in service increases with the carbon and alloy content of the steel. In applications where essentially no reduction of the as-quenched hardness can be tolerated, low-temperature tempering is performed at 150°C (300°F). This reduces the brittleness of the martensite without having a significant effect on hardness. For example, low-temperature tempering is performed with hardened tools made from carbon or low-alloy steels.

Microstructural Changes. The microstructural changes that occur after tempering are related to the tempering temperature. As the temperature is raised, several stages (changes) take place. The first stage is the formation of epsilon iron carbide (ε-Fe_3C) and lowering of the carbon content of the martensite. The second stage is the transformation of retained austenite to ε-Fe_3C and ferrite. The third stage is the transformation of ε-Fe_3C and low-carbon martensite to cementite and ferrite. See Figure 14-7.

The temperature ranges for the three stages overlap and depend on the tempering time. The first stage occurs between 100°C and 250°C (210°F and 480°F). In this stage, ε-Fe_3C begins to precipitate from the martensite, which is depleted in carbon content. Epsilon iron carbide is finely divided and etches up dark. This produces a darkening effect of the martensite in the microstructure. There is no appreciable loss of hardening at this stage because carbide precipitation and the accompanying lowering of carbon in the martensite tend to offset one another with respect to hardness.

The second stage occurs between 200°C and 300°C (390°F and 570°F). In this stage, retained austenite transforms to ferrite and cementite and begins after the ε-Fe_3C is well established.

The third stage occurs between 250°C and 350°C (480°F and 660°F). In this stage, the ε-Fe_3C transformation to ferrite and cementite takes place as required by the iron-carbon diagram. With some alloy steels and many tool steels, there is a fourth stage referred to as secondary hardening. *Secondary hardening* is the formation of alloy carbides, which occurs in the temperature range of 500°C to 650°C (930°F to 1200°F).

The purpose of tempering is to toughen a component by reducing the as-quenched hardness and residual stresses. Unless a subzero quench is being performed, the component is transferred to the tempering furnace when it reaches 90°C (195°F). In all cases, components should be tempered immediately after they have been quenched because stresses may cause cracking.

Problems and Solutions. Problems that occur during tempering of steel are retained austenite, blue brittleness, temper embrittlement, and tempered martensite embrittlement. Retained austenite does not remain stable on tempering but transforms to cementite, bainite, or martensite. This leads to dimensional instability and possible embrittlement. To avoid retained austenite, double or triple tempering is used for steels in which retained austenite is present in significant amounts, such as in tool steels.

Blue brittleness is brittleness that occurs in some carbon and alloy steels after being tempered between 200°C and 370°C (390°F and 700°F). This temperature range corresponds to the blue heat range. Steels that are prone to blue brittleness should not be used tempered or used in this temperature range.

Temper embrittlement is embrittlement that occurs mostly in low-alloy steels that are tempered and then slowly cooled between 575°C and 360°C (1065°F and 680°F). Temper embrittlement is not a problem in plain carbon steels containing <.3% Mn. Steels having a higher manganese content or

steels containing nickel or chromium, such as 4140 steel, are susceptible to temper embrittlement.

To avoid temper embrittlement, susceptible steels should not be slow cooled through the damaging temperature range or used in this range. Temper embrittlement may occur in heavy sections that are tempered at a higher temperature and then allowed to cool. Such components may require quenching through the susceptible temperature range.

Tempered martensite embrittlement is the formation of embrittling cementite from the decomposition of retained austenite during the second stage of tempering. It occurs in high-strength alloy steels when tempered between 205°C and 370°C (400°F and 700°F). To avoid tempered martensite embrittlement, susceptible steels, such as 4340 steel, should not be tempered in this range.

Interrupted Quenching Techniques

Interrupted quenching techniques are stepwise quenching processes that develop specific microstructures in steels and minimize distortion and cracking. These techniques involve quenching components in a molten salt bath just above the M_s, holding for sufficient time to permit temperature equalization through the section thickness, and then allowing transformation to proceed. They are specifically used

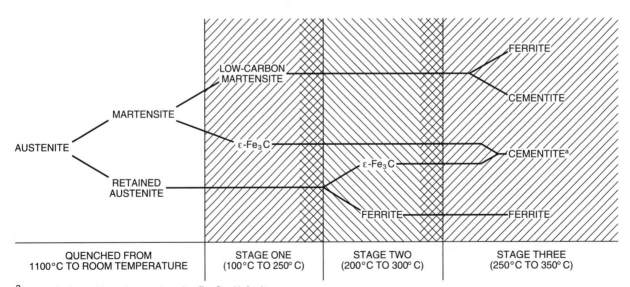

a cementite formed from the reaction of ε-Fe₃C with ferrite and/or low-carbon martensite

Figure 14-7. Tempering takes place in three stages that overlap one another as the temperature is raised.

to minimize distortion and cracking. The two interrupted quenching techniques are austempering and martempering (marquenching).

Austempering. *Austempering* is an interrupted quenching technique that consists of austenitizing followed by quenching in a medium maintained in the bainite transformation temperature range for the steel. This allows the temperature to equalize in the component and the austenite to transform isothermally to lower bainite. The cooling rate throughout the section thickness must be fast enough to miss the nose of the transformation curve. After transformation is complete, the component is allowed to cool to room temperature. The lower bainite formed has similar strength and superior ductility to tempered martensite of the same composition. See Figure 14-8.

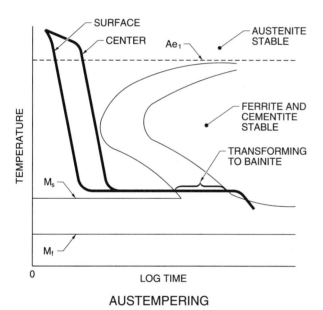

Figure 14-8. Austempering consists of austenitizing a steel, quenching it into a suitable medium maintained at a temperature in the bainite transformation range, and cooling it in air.

Molten salts are the most common quenching media and have a melting temperature of 205°C to 370°C (400°F to 700°F). This is the transformation temperature range for lower bainite in most steels. Because molten salts have a low rate of heat extraction compared with water or oil, austempering is restricted to components of small section size.

Austempering is used for the heat treatment of high-carbon steels in the form of sheet, wire, and strip.

Martempering. *Martempering* is an interrupted quenching technique that consists of quenching an austenitized component into molten salt or hot oil close to the M_s temperature, holding until temperature equalization has taken place, and cooling in air to a temperature below the M_f to complete the transformation to martensite. Martempering is performed to minimize the distortion that occurs during conventional quenching to martensite. With martempering, martensite formation occurs at a reasonably uniform rate throughout the component cross section. This reduces the residual stresses caused by conventional quenching.

Like austempering, the cooling rate from the austenitizing temperature must be rapid enough to miss the nose of the transformation curve. See Figure 14-9. Unlike austempering, the component must be conventionally tempered after martempering in order to restore toughness. Martempering is used for the heat treatment of tools, bearings, and dies in which quench cracking is encountered in conventional quenching and tempering processes.

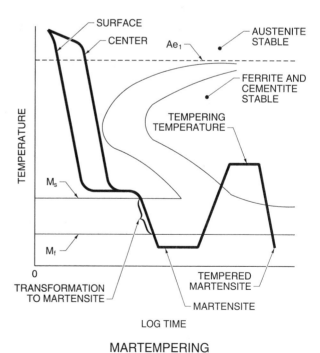

Figure 14-9. After martempering, the component must be conventionally tempered in order to restore toughness.

ANNEALING

Annealing processes are heat treatments that produce pearlitic microstructures (ferrite, pearlite, and cementite). They are performed to homogenize the microstructure, increase ductility, remove residual stresses, and improve machinability. Annealing processes consist of heating to temperature, holding at temperature, and cooling at a controlled rate. The soaking temperatures and rates of cooling may differ significantly from those used in quenching and tempering processes in which the objective is to develop martensitic products. The principal annealing processes are full annealing, normalizing, spheroidizing, process annealing, and stress relieving.

Full Annealing

Full annealing is a heat treatment in which a steel component is held in the austenitizing temperature range and then cooled inside the furnace. It produces a soft, coarse pearlitic microstructure.

The holding temperature is a function of the carbon content. For hypoeutectoid and eutectoid steels, the holding temperature is just above the upper critical temperature, and for hypereutectic steels, it is just above the lower critical temperature. Hypereutectoid steels must not be heated above the upper critical temperature. Overheating will cause a continuous band of proeutectoid cementite, which will form along the grain boundaries during slow cooling. This leads to embrittlement.

Components are held at temperature for 1 hr/in. of section. Slow cooling is achieved by switching off the furnace. When the temperature of the component has fallen below the lower critical temperature, the furnace door may be opened or the component removed to speed up the cooling rate. At this point, no further microstructural changes occur. The soft, coarse pearlitic structure obtained in full annealing is ideal for rough-machining operations.

Normalizing

Normalizing is a heat treatment that decreases pearlite interlamellar spacing and refines grain size. The steel component is held in the austenitizing temperature range and allowed to cool in still air. Only steels that form pearlitic microstructures when air cooled are normalized. Air-hardening steels such as tool steels cannot be normalized because they would form martensite.

The key differences between normalizing and full annealing are the holding temperatures and the cooling rates. See Figure 14-10. The soaking temperature in normalizing is higher, approximately 55°C (100°F) above the upper critical for both hypoeutectoid and hypereutectoid steels. Compared with the temperature for full annealing, this higher holding temperature achieves a greater uniformity of microstructure for relatively short holding times. Approximately 1 hr/in. of section is the standard holding time, but it is generous because of the faster homogenization obtained with normalizing.

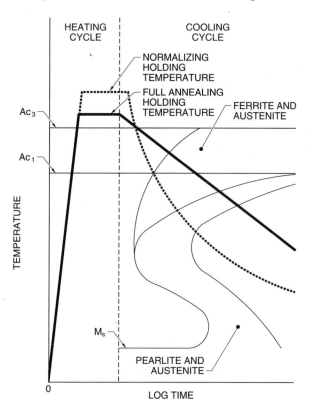

Figure 14-10. The key differences between annealing and normalizing are the holding temperatures and the rates of cooling.

The cooling rate for normalizing is faster than for full annealing. The faster cooling rate results in a uniform grain size that has a narrower and more closely spaced pearlite lamellae. This leads to greater strength and toughness and higher hardness than is obtained in full annealing.

When the component is a dull black color, which indicates it is below the lower critical temperature, it may be water quenched or oil quenched to save time. Thick sections should be allowed to air cool to even lower temperatures before quenching to ensure that all of the section is below the lower critical temperature. Forced air cooling with fans is sometimes used to accelerate the cooling rate from the austenitizing temperature to obtain more desirable properties. This can be done with heavy sections in which the internal cooling rate would otherwise be too slow.

The objectives of normalizing are to improve toughness and machinability or to prepare a steel for further heat treatment. Normalizing is often used for hot-worked or cast steel products that have a coarse, inhomogeneous structure. It is also used as a preparatory step in heat treatment operations, such as quenching and tempering, because it helps improve response to heat treatment.

Spheroidizing

Spheroidizing is a heat treatment that produces a spheroidal (globular) form of carbide in a steel product. It is primarily performed to produce maximum softness in a steel product.

Spheroidizing is extremely important for low-carbon and medium-carbon steels that are cold formed. The low level of hardness of spheroidized steels is also important for improving the machinability of high-carbon steels that are machined prior to hardening.

Spheroidizing can be performed through a variety of procedures. The most common are prolonged heating to a temperature below the lower critical temperature, heating and cooling alternately between temperatures just below and just above the lower critical temperature, heating above the lower critical temperature and then cooling very slowly in a furnace, and holding just below the lower critical temperature.

Process Annealing

Process annealing is a heat treatment performed below the lower critical temperature and is designed to restore ductility to cold-worked steel products. See Figure 14-11. It consists of soaking a cold-worked component at 10°C to 20°C (20°F to 40°F) below the lower critical temperature. Strict temperature control is necessary to prevent overheating above the lower critical temperature.

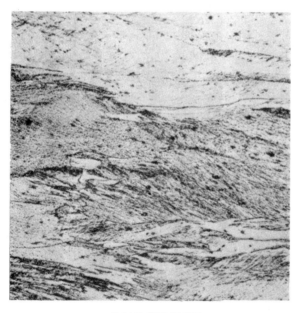

COLD-WORKED
STEEL

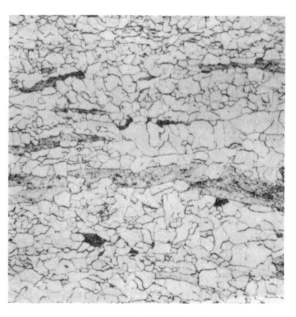

PROCESS ANNEALING
SHOWING RECRYSTALLIZATION

Figure 14-11. Process annealing of a cold-worked steel causes recrystallization and evolution of fine equiaxed grains.

When process annealing is employed to prepare a material for processes with minimal cold work, such as cold sawing or cold shearing, temperatures well below the lower critical temperature may be used. After soaking, the component may be air cooled or water quenched. Process annealing is used extensively in the sheet and wire industries to soften material that has been drawn or rolled prior to further cold reduction.

Stress Relieving

Stress relieving is a heat treatment in which a steel is heated to a suitable temperature below the lower critical temperature, held for long enough to reduce residual stresses induced by cold deformation or thermal treatments, and cooled slowly enough to minimize the development of new residual stresses. There are many sources of residual stresses in steel products. Residual stresses can develop during processing of the material from ingot to final product; during rolling, casting, or forging; during forming operations, such as shearing, bending, grinding and machining; and during fabrication processes, such as welding.

Stress relieving is performed by heating the component or a selected area of it to the stress relieving temperature and holding for approximately 1 hr/in. of section after the temperature has steadied out. Components must be properly supported during stress relieving to ensure that they do not distort under their own load. Cooling is performed at a rate, which is between 55°C/hr and 110°C/hr (100°F/hr and 200°F/hr), that minimizes the introduction of new stresses.

It is extremely important that the stress relieving process is monitored with thermocouples, which are attached to various parts of the component. This allows the heating and cooling rates to be monitored and checks that the actual stress relief temperature is within the specified range. To ensure adequate stress relief, thermocouples should be attached to the extremities of the component and at locations of different thicknesses. The percentage of stress relief achieved depends upon the time and temperature. Longer times at lower temperatures achieve the same percentage of stress relief as shorter times at higher temperatures. For example, holding com-

ponents at 595°C (1105°F) for 6 hours provides the same amount of stress relief as 650°C (1200°F) for 1 hour. See Figure 14-12.

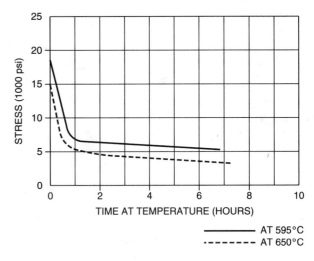

STRESS RELIEVING
(1025 STEEL)

Figure 14-12. The majority of the stress relief is obtained after approximately 2 hours at any temperature.

In the stress relieving temperature range, the material undergoes microscopic creep. Creep resistant materials designed for high-temperature service, such as chromium-molybdenum low-alloy steels, require higher stress relieving temperatures than carbon steels and low-alloy steels. For example, typical stress relieving temperatures for carbon and low-alloy steels are in the range of 595°C to 675°C (1105°F to 1245°F), and for chromium-molybdenum low-alloy steels, the range is 705°C to 730°C (1300°F to 1350°F).

CASE HARDENING

Case hardening is a group of heat treatment processes that develop a thin, hard surface layer on a component and leave the core (bulk of the section) relatively soft, strong, and tough. See Figure 14-13. Case hardening can often provide the optimum combination of properties. For many components, wear and the most damaging mechanical stresses act only on the surface. Therefore, components may be made of a low-carbon or medium-carbon steel, or low-alloy or medium-alloy steel, for toughness and ease of fabrication, and the surface case may be hardened to optimize its wear and fatigue resistance.

Case hardening also reduces distortion and eliminates cracking that might occur with through hardening, especially in large section sizes. Localized case hardening of selected areas may be performed by masking off the areas that must not be hardened. Masking off is accomplished by copper plating the nonhardened areas or coating them with special compounds. Case-hardening processes alter the surface composition or maintain the surface composition of the steel.

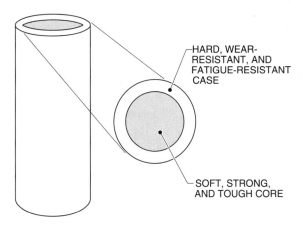

CASE HARDENING

Figure 14-13. Case-hardening processes develop a hard case with a soft, strong, and tough core.

Altering Processes

Case-hardening processes that alter the surface composition cause local increases in the carbon or nitrogen content. These processes consist of exposing the metal at high temperature to an environment rich in carbon, nitrogen, or both. This exposure leads to migration, which is inward diffusion of the carbon, and chemical modification of the surface layer of the steel. The altering processes include carburizing, carbonitriding (gas cyaniding), nitriding, and nitrocarburizing.

Carburizing. *Carburizing* is a case-hardening process for low-carbon steels that uses an environment with sufficient carbon potential and a temperature above the upper critical temperature. Carbon is absorbed into the surface and by diffusion creates a carbon concentration gradient between the surface and the interior of the steel.

The high-carbon surface is then hardened by austenitizing, quenching, and tempering. This produces a high-carbon martensitic case with good wear and fatigue resistance superimposed on a tough and low-carbon steel core. The principal carburizing reaction that introduces carbon into the surface of the steel is the following:

$$CO_2 + C \leftrightarrow 2CO\uparrow$$

Carburizing takes place when the reaction goes to the left. The maximum amount of carbon that dissolves in austenite at the carburizing temperature is indicated on the iron-carbon diagram by the A_3 line. For example, it is approximately 1.2% C for a .2% C steel at a carburizing temperature of 925°C (1700°F). The rate of carburizing increases as the temperature increases.

Carbon diffuses into the steel and develops a high-carbon surface layer. The carbon concentration falls steadily below the surface until it is equal to the carbon content of the steel. The higher the temperature or the longer the carburizing time, the deeper the carbon case. A carburizing temperature of 925°C (1700°F) is most common. It represents a compromise between an acceptable carburizing rate and useful life of furnace equipment.

The principal carburizing processes are gas, pack, and liquid carburizing. Gas carburizing uses methane and carbon monoxide as the carburizing medium, pack carburizing uses charcoal or coke, and liquid carburizing uses molten salt containing calcium cyanide. Each carburizing process produces different case depths depending on temperature and time. See Figure 14-14.

Gas carburizing is a case-hardening technique in which the component is placed in a furnace containing a gaseous carburizing environment. The environment is produced by combustion of natural gas or another hydrocarbon gas in exothermic or endothermic gas generators. The environment contains methane, carbon monoxide, hydrogen, carbon dioxide, water vapor, and nitrogen. Although many other reactions occur, the basic carburizing equation applies in gas carburizing.

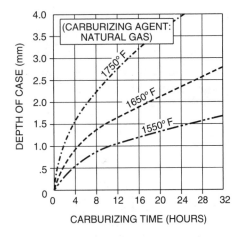

GAS CARBURIZING

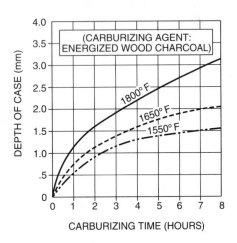

PACK CARBURIZING

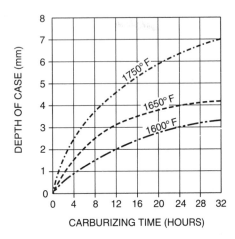

LIQUID CARBURIZING

Figure 14-14. Higher carburizing temperatures or longer holding times at the carburizing temperature result in a deeper carbon-enriched case.

Gas carburizing is performed in batch or continuous furnaces. In batch furnaces, the components are charged and discharged in batches (groups). In continuous furnaces, the components enter and leave the furnace in one continuous stream. Heating in both batch and continuous furnaces is accomplished by radiant tubes. Gas carburizing allows greater control of case depth compared with other methods of carburizing and can produce heavy cases up to 10 mm (.4 in.) deep.

Pack carburizing is a case-hardening technique in which carbon monoxide derived from a solid carbon-containing compound decomposes at the metal surface into nascent (newborn) carbon and carbon dioxide. The nascent carbon is absorbed into the steel, and the carbon dioxide reacts with carbon in the compound to produce more carbon monoxide. Pack carburizing is performed by packing components in heavy boxes containing charcoal or coke plus an energizer, such as barium or sodium carbonate, which accelerates the carburizing reaction.

The box may be built around the component, which is a big advantage for large or awkward shapes. The component may be slow cooled from the carburizing temperature. This is sometimes advantageous if finish machining is required before quenching and tempering.

Control of case depth is less accurate in pack carburizing. This is disadvantageous for thin cases or those requiring close tolerances. The component cannot be directly quenched after carburizing if control of case depth is required to obtain the desired mechanical properties. Also, more time is required for the cycle because the box and packing materials reduce heating and cooling rates of the component. Pack carburizing produces moderate cases of .65 mm to 6.5 mm (.025 in. to .25 in.) deep.

Liquid carburizing is a case-hardening technique in which the component is held in a molten salt bath to introduce carbon and sometimes nitrogen into the surface. The active ingredient in the salt bath is calcium cyanide. Liquid carburizing produces light to moderate cases, up to .75 mm (.03 in.) deep, with low-temperature baths operating between 845°C and 900°C (1555°F and 1650°F). Heavier cases, up to 3 mm (.12 in.) deep, are produced in molten salt baths operating between 900°C and 955°C (1650°F and 1750°F).

Fast heating and rapid penetration are characteristics of liquid carburizing. Compared with the other carburizing techniques, liquid carburizing has greater uniformity of case depth. There is a slight increase in surface hardness from the small amount of nitrogen absorbed. Also, there is freedom from high-temperature oxidation and soot formation, which occur with pack and gas carburizing. Disadvantages of liquid carburizing include environmental and occupational health hazards associated with cyanide-containing salts, problems with complex shapes that cannot be handled, and the need for regular checking of the bath composition to assure uniform case depths.

Carbonitriding. *Carbonitriding* is a case-hardening process for carbon and alloy steels that consists of holding them above the upper critical temperature to simultaneously absorb carbon and nitrogen. This process is followed by cooling to room temperature at a rate that produces the desired case and core properties. Carbonitriding is performed between 815°C and 870°C (1500°F and 1600°F), and light cases of up to .75 mm (.03 in.) deep are obtained.

Carbonitriding uses the carbon and nitrogen to harden in different ways. Carbon provides high case hardness, which occurs through martensite formation when the component is cooled. Nitrogen also contributes by increasing case hardness, but it also reduces the critical cooling rate. This allows the cooling to be done by oil or forced-air quenching.

An endothermic carrier gas is used for carbonitriding. Mixed with the gas is 5% to 10% natural gas, which supplies the carbon, and 3% to 10% anhydrous ammonia, which supplies the nitrogen. Ammonia dissociates into hydrogen and nitrogen. The nitrogen reacts with the surface of the steel. The bulk of the nitrogen reacts to form very hard nitride compounds, which contribute appreciably to the high hardness of the surface. The remaining nitrogen goes into solid solution to increase hardenability.

Nitriding. *Nitriding* is a subcritical case-hardening process that introduces nitrogen into the surface of a steel. Nitriding is performed in an atmosphere containing atomic nitrogen at a temperature between 510°C and 565°C (950°F and 1050°F). This is below

the temperature of the iron-nitrogen eutectic compound, which is 590°C (1095°F). Before nitriding, the component is austenitized, quenched, and tempered above the nitriding temperature. No other heat treatment is necessary.

Nitridable steels are generally medium-carbon with additions of strong nitride forming elements such as chromium, aluminum, vanadium, and molybdenum. Other nitridable alloys include stainless and tool steels. Several steels with high-aluminum content (Nitralloys) are made specifically for nitriding and give the best obtainable results.

Most nitriding operations produce a surface layer consisting of two zones. See Figure 14-15. The white layer is the zone at the surface and consists of a thin layer of an iron-nitrogen compound. This layer is brittle and undesirable, so the nitriding conditions are adjusted to minimize it. Below the white layer is the nitrogen-enriched diffusion zone, which may be a solid solution or contain precipitates of the alloying elements in the steel. The diffusion zone is in extreme compression and has enhanced fatigue resistance. A nitrided case is extremely hard, registering approximately 70 HRC.

WHITE LAYER

DIFFUSION ZONE

Figure 14-15. In conventional nitriding processes, the surface consists of a very thin, white layer and a deeper diffusion zone.

Nitriding takes from 60 to 90 hours, and the case depth obtainable is extremely light. It is usually from .13 mm to .25 mm (.005 in. to .01 in.) deep. Despite these disadvantages, nitriding is popular because the low temperature and light case depth results in components that can be heat treated and machine finished in advance. This reduces possible distortion. The case is exceptionally hard and has excellent wear, seizing (cohering of moving parts resulting from galling), and galling (localized welding) resistance under operating conditions where heat is generated by friction between moving components in contact. Nitriding consist of three divisions, which include gas, ion, and liquid nitriding.

Gas nitriding is a nitriding technique that is performed with ammonia (NH_3). Gas nitriding is the most common case-hardening process and The ammonia dissociates on the surface of the steel according to the following equation:

$$NH_3 \rightarrow 2N + 3H_2$$

The resulting nitrogen is absorbed at the surface of the steel.

The components are very carefully cleaned before nitriding. Nitriding is performed on fixtures in a gastight retort with an inlet for the ammonia and an outlet for the exhaust gas. The high temperature in the retort causes some of the ammonia to dissociate. The amount of dissociation is controlled to regulate the ammonia flow.

The standard nitriding cycle, which runs at 495°C (925°F) for the full time, is frequently modified to improve the characteristics of the case. For example, in double stage nitriding, the normal run at 495°C to 525°C (925°F to 975°F) is followed by a diffusion period in which the ammonia dissociation percentage is raised to approximately 80%. The diffusion step may be run in the normal temperature range or be raised to 550°C to 565°C (1020°F to 1050°F). The diffusion step is usually as long as, or longer than, the nitriding step of the cycle.

Ion nitriding is a nitriding technique performed in a vacuum where high-voltage electrical energy forms a plasma through which nitrogen ions accelerate to impinge on the component. The ion bombardment heats up the component and scours it clean. Ion nitriding provides excellent dimensional and surface finish control without a white layer. This eliminates the need for grinding. Alloys that benefit from ion nitriding are more limited than those that can be used in conventional nitriding operations.

Liquid nitriding is a nitriding technique performed in a molten salt bath containing a mixture of sodium or potassium chloride and sodium or potassium cyanide or cyanate. A small amount of sodium or potassium carbonate is also included to accelerate the process. The ratios of these ingredients are varied according to the type of job, but must be held within reasonably close limits. Liquid nitriding introduces problems in handling highly toxic compounds, so effective safety measures must be taken.

The bath operates in the range of 510°C to 565°C (950°F to 1050°F). Nitriding rates are about the same as for gas, but the process has the advantage of being able to nitride plain carbon steels as well as more expensive alloy steels.

Nitrocarburizing. *Nitrocarburizing* is a subcritical case-hardening process that consists of diffusing carbon and nitrogen into the surface of the component for 3 hours at a temperature below 680°C (1255°F). This produces a very light case. Nitrocarburizing, unlike carbonitriding, is performed below the lower critical temperature so that no further heat treatment is necessary.

The component is exposed to a mixture of endothermic gas and ammonia. The case consists of a thin white layer and an underlying diffusion zone. These combine to provide good scuffing resistance, antifrictional properties, and fatigue strength. The short exposure time is the primary reason for the very light case.

Maintaining Processes

Maintaining processes are case-hardening processes that maintain the surface composition of the steel, which consist of heating the component very rapidly so that only the surface layer is austenitized. This is followed immediately by quenching to transform the case to martensite. Before case hardening, the component is heat treated to obtain the desired bulk hardness. After case hardening, tempering is usually performed at a lower temperature to supply ductility to the case. The heat required to austenitize the surface layers of the component may be applied by induction or flame.

Induction Hardening. *Induction hardening* is a case-hardening process in which the surface layer of the component is heated by electromagnetic induction. The component is heated to a temperature above the upper critical temperature and immeditely quenched.

The depth of hardening (current penetration) depends chiefly on the frequency of the alternating current, and increases as frequency decreases. See Figure 14-16. By varying the frequency, the case depth may be altered to suit the application. Induction hardening frequencies are usually 1000 hertz (Hz) or higher. Many types of coils are used for induction hardening to achieve the required hardening pattern. Coils must be water-cooled to prevent them from overheating.

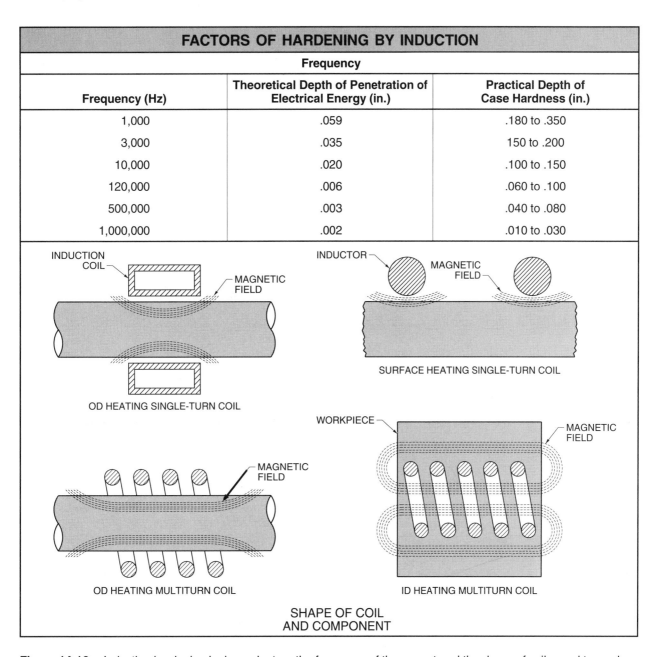

FACTORS OF HARDENING BY INDUCTION		
Frequency		
Frequency (Hz)	Theoretical Depth of Penetration of Electrical Energy (in.)	Practical Depth of Case Hardness (in.)
1,000	.059	.180 to .350
3,000	.035	150 to .200
10,000	.020	.100 to .150
120,000	.006	.060 to .100
500,000	.003	.040 to .080
1,000,000	.002	.010 to .030

SHAPE OF COIL
AND COMPONENT

Figure 14-16. Induction hardening is dependent on the frequency of the current and the shape of coils used to produce the current.

Electromagnetic induction is achieved by placing a metal coil close to the component. Alternating current passed through the coil establishes a highly concentrated, rapidly alternating magnetic field within it. The strength of the field depends on the magnitude of the current flowing in the coil. The magnetic field in the coil induces an electric potential in the component that causes a flow of current in the component. The electrical resistance of the component causes heat to be generated. The pattern of heating obtained by induction is determined by the shape of the coil, the number of turns in the coil, the frequency of the alternating current, the power input, and the shape of the component.

Medium-carbon and high-carbon steels are usually induction hardened to improve fatigue and wear resistance because they contain sufficient carbon to achieve high case hardness. Alloy steels may also be hardened where heavier cases are required. The duration of the high-frequency induction heating cycle is extremely short (a few seconds). Consequently, the time available for austenite formation is extremely limited, which is compensated for by increasing the austenitizing temperature. Due to this rapid heating rate, grain growth does not have time to occur.

The steel is immediately quenched. Water is the most common quenching medium and is usually incorporated into the induction hardening equipment so that it can be done automatically. See Figure 14-17. Induction hardening is used to produce widely varying case depths, but light to moderate case depths are most common. Typical induction-hardened components are piston rods, pump shafts, spur gears, and cams.

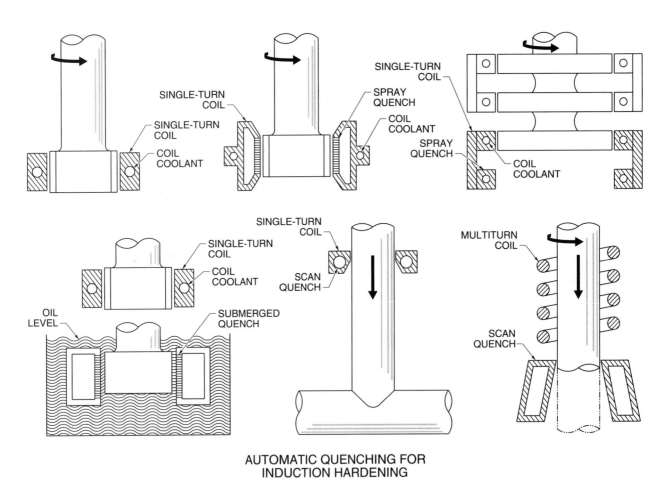

AUTOMATIC QUENCHING FOR
INDUCTION HARDENING

Figure 14-17. During induction hardening, the component is immediately quenched using special designs that incorporate quenching into the equipment.

Flame Hardening. *Flame hardening* is a case-hardening process in which an intense flame is used to heat the surface layer of the component to a temperature above the upper critical temperature. This is immediately followed by quenching. The high-temperature flame is obtained by combustion of a mixture of fuel gas with oxygen or air.

Flame hardening is extremely versatile because of the wide range of heating conditions obtainable with burners and torches. Various methods of flame hardening are used. See Figure 14-18. Localized or spot hardening may be performed by directing a stationary flame head toward an area of a stationary component. In progressive flame hardening, the torch travels over a stationary component or the component travels under a stationary torch. In spinning methods, the component is rotated within an array of torches.

The component is immediately spray quenched in water and then tempered to improve toughness and relieve stresses. Flame hardening is generally used to produce heavy cases of approximately 6.5 mm (.25 in.) deep. Light cases, as low as .75 mm (.03 in.) deep, may be obtained by increasing the speed of heating and quenching. Flame hardening is extremely versatile and may be used on components too large or complex to be hardened by other methods, such as in applications where only local areas require hardening. These applications include gear teeth, shafts, lathe ways, cams, and hand tools.

Process Selection

Selecting a case-hardening process requires detailed analysis of the application and the type of steel. All case-hardening processes develop hard and wear resistant surfaces. As a result, selection of the specific process is determined by a variety of other factors, such as load on the component, dimensional stability, type of steel, and mechanical properties required of the core.

Carburizing produces moderate case depths that have excellent capacity for contact loads, good bending fatigue strength, good resistance to seizure, and excellent freedom from quench cracking. Carburizing is applied to low-cost to medium-cost steels and generally requires a high capital investment.

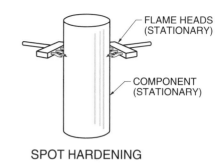

SPOT HARDENING

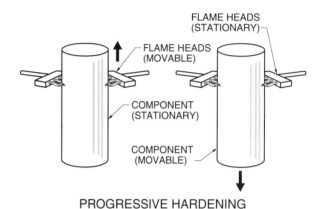

PROGRESSIVE HARDENING

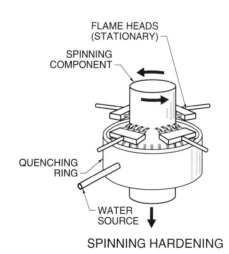

SPINNING HARDENING

Figure 14-18. During flame hardening, the component and the flame heads move relative to one another.

Carbonitriding and nitrocarburizing produce light case depths that have fair capacity for contact loads, good resistance to seizure, and excellent freedom from quench cracking. Carbonitriding and nitrocarburizing are applied to low-cost steels and require a medium capital investment.

Nitriding produces light case depths that have fair capacity for contact loads, good bending fatigue strength, excellent resistance to seizure, excellent dimensional control, and relative freedom from quench cracking. Nitriding is applied to medium-cost to high-cost steels and requires a medium capital investment.

Induction hardening produces heavy case depths that have good capacity for contact loads, good bending fatigue strength, fair resistance to seizure, fair dimensional control, and fair freedom from quench cracking. This process is applied to low-cost steels and requires a medium capital investment.

Flame hardening produces heavy case depths with good capacity for contact loads, good bending fatigue strength, fair dimensional control, and fair freedom from quench cracking. It is applied to low-cost steels, and it requires a low capital investment.

COMPONENT DESIGN AND FABRICATION

Problems in heat treatment may arise from the component design and the fabrication process. See Appendix. These are valid for carbon and low-alloy steels and increase with the alloy content of the steel. Tool steels, which are the most complex low-alloy steels, are the most sensitive to heat treatment difficulties. They leave little room for improper practices.

Components designed for heat treatment must contain generous fillets in the corners and be of uniform section as much as possible. Sharp corners or reentrant angles cause high concentrations of stress. Large changes of cross section lead to differential thermal expansion or contraction during heating or cooling. To avoid these stresses, the mass of a section may be balanced by boring out holes from bulky components and using generous radii.

Fabrication processes that may lead to heat treatment difficulties include cold-forming operations, cutting operations, identification markings, welding, and machining and grinding. Cold-forming operations, such as deep drawing or stretching, leave high-residual stresses that may locally alter the mechanical properties. This may result in warping of the component when heat treated. If the critical strain is exceeded, such as in a severely deformed section, recrystallization and local grain growth may occur during heat treatment. To avoid these problems, it may be necessary to employ stress relieving before heat treatment.

Cutting operations, such as shearing, blanking, and piercing, leave residual stresses. These stresses are detrimental to the component. Sharp or torn edges must be avoided and removed because these areas cause stresses.

Identification markings made by punching or electroetching may cause cracking, especially with high-hardness steels (>50 HRC). Identification markings should not be made in locations of high bending or torsional stresses. Marking these locations will cause failure in the component.

Welding leads to high levels of residual stresses that may aggravate cracking during heat treatment. Welded components should be stress relieved before heat treatment so that cracking does not occur.

Machining or grinding operations often leave residual stresses, which lead to cracking problems. See Figure 14-19. The heat of grinding may austenitize the surface of the component. This produces a martensitic structure that contains residual stresses. The martensitic structure will crack before or during heat treatment.

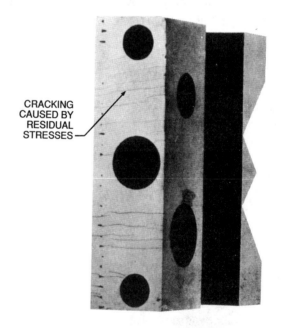

CRACKING CAUSED BY RESIDUAL STRESSES

Figure 14-19. Machining and grinding operations must be carefully controlled because residual stresses, which lead to cracking before or during heat treatment, will be produced.

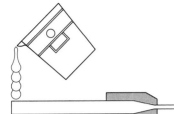

Tool Steels 15

Tool steels are a group of steels that generally have high carbon and high alloy contents. Tool steels exhibit high levels of hardness, wear resistance, and strength sometimes in conjunction with reasonable toughness. Carbon content, other alloying elements, and the manufacturing process affect the level of these properties. These factors and properties are used to designate tool steels. Designation systems for tool steels include the Unified Numbering System for metals and alloys, manufacturers' trade names, and the AISI system.

TOOL STEELS

Tool steels are a group of steels that generally have high carbon and high alloy contents. They are characterized by high hardness and wear resistance, which are sometimes accompanied by toughness and resistance to elevated temperature softening. Although tool steels are mostly used to make tools for cutting, shaping, or forming materials, they are not restricted to these applications. Tool steels are used in many applications where their hardness, high strength, and reasonable toughness are advantageous.

Tool steels typically have high carbon contents, which is partly responsible for their high hardness. Most tool steels contain between .6% C and 1.3% C, and the most commonly used tool steels contain between .8% C and 1.1% C. See Figure 15-1.

In some applications, tool steels are subjected to high loads that are often applied rapidly. They must be able to withstand these loads without breaking or deforming. In other applications, they must provide these capabilities under conditions that cause high temperatures to develop in them.

Tool steels contain significant amounts of other elements such as chromium, cobalt, manganese, molybdenum, nickel, silicon, tungsten, and vanadium to form various alloy carbides and help increase the hardness, wear resistance, and elevated-temperature softening resistance. No single alloy can provide all these properties, so the selection of a tool steel is based on trade-offs that achieve the optimum combination of properties. These properties are primarily a function of the chemical composition of the alloy. See Appendix.

Tool steels are heat treated by austenitizing, quenching, and tempering. Tool steels may also be austempered or case hardened. Great care is used in any heat treatment and fabrication process to avoid cracking or distortion. Designation systems have been created to identify the various tool steels, their chemical compositions, and in what form they are supplied.

Three common methods of identifying tool steels are the Unified Numbering System for metals and alloys, manufacturers' trade names, and the AISI system. Each is important both for the historical perspective and for the current requirements (for example, safety regulations).

In the UNS, all tool steels are designated by the uppercase letter T followed by five numbers. UNS designations have AISI equivalents. For example, the tool steel UNS T30402 is equivalent to the AISI designation D2.

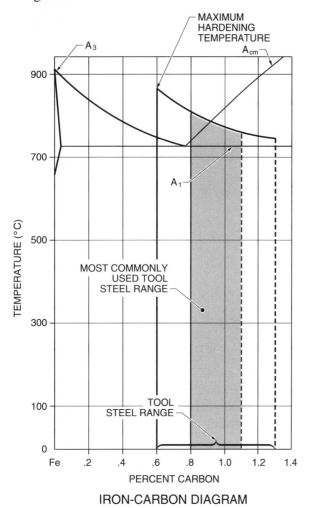

Figure 15-1. The portion of the iron-carbon diagram in which most tool steels are found is in the range of .6% C to 1.3% C.

Commercially, tool steels are identified by their manufacturers' trade names. See Figure 15-2. The equivalent UNS and AISI designations are provided by the manufacturer for the commercial alloys to aid in tool steel selection. Significant variations in melting and forming practices may exist between different manufacturers, so manufacturers' recommendations should always be followed when heat treating or fabricating tool steels. This is particularly important in critical applications where variation in the required properties must be held to a minimum.

The AISI designation system is the most commonly used designation system for tool steels. Tool steels are divided into seven divisions, which are water-hardening, cold work, shock-resisting, special-purpose, mold steel, hot work, and high-speed. These seven divisions are further divided into 11 distinct families.

TOOL STEEL IDENTIFICATION				
		Manufacturers and Trade Names		
AISI	**UNS**	**Crucible**	**Jessop**	**Teledyne Vasco**
D2	T30402	AIRDI 150	CNS-1	OHIO DIE
M1	T11301	REX TMO	MOGUL	8-N-2
A2	T30102	AIRKOOL	WINDSOR	AIR HARD
O1	T31501	KETOS	TRUFORM	COLONIAL NO. 6
H13	T20813	NU-DIE V	DICA B VANADIUM	HOT FORM V
S1	T41901	ATHA PNEU	—	PAR-EXC
T1	T12001	REX AA	—	NEATRO

Figure 15-2. Tool steels are identified by UNS designations, manufacturers' trade names, and AISI designations.

The families are identified by the individual uppercase letters W, O, A, D, S, L, F, P, H, T, and M. See Figure 15-3. The letter represents a key characteristic of the tool steel. The key characteristic may be the method of hardening, a significant mechanical feature, or a major alloying element. For example, the letter D identifies the family of high-carbon, high-chromium cold work tool steels. Each family contains specific alloys that are identified by one or two numbers following the letter. For example, D2 (1.5% C, 12% Co, 1% V, and 1% Mo cold work steel) is one of the most commonly used steels in this family due to its versatility.

Diversification of properties and characteristics influence the number and types of uses of tool steels. The uses range from chisels and punches to mold material and structural materials. Each group of tool steels has certain uses for which it is suited. See Figure 15-4.

Water-hardening Tool Steels

Water-hardening tool steels are hypereutectoid steels made according to relatively stringent melting practices. They are the least costly and have the most applications. They contain small amounts of alloying elements. Water-hardening tool steels consist of group W steels.

TOOL STEEL FAMILIES

AISI	UNS	Identifying Elements	AISI	UNS	Identifying Elements
W1	T72301	.6% C to 1.4% C	H14	T20814	.4% C, 5% Cr, and 5% W
W2	T72302	.6% C to .1.4% C and .25% V	H19	T20819	.4% C, 4.25% Cr, 2% V, 4.25% W, and 4.25% Co
W5	T72305	1.1% C and .05% Cr	H21	T20821	.35% C, 3.5% Cr, and 9% W
S1	T41901	.5% C, 1.5% Cr, and 2.5% W	H22	T20822	.35% C, 2% Cr, and 11% W
S2	T41902	.5% C, 1% Si, and .5% Mo	H23	T20823	.3% C, 12% Cr, and 12% W
S5	T41905	.55% C, .8% Mn, 2% Si, and .4% Mo	H24	T20824	.45% C, 3% Cr, and 15% W
S6	T41906	.45% C, 1.4% Mn, 2.25% Si, 1.5% Cr, and .4% Mo	H25	T20825	.25% C, 4% Cr, and 15% W
S7	T41907	.5% C, 3.25% Cr, and 1.4% Mo	H26	T20826	.5% C, 4% Cr, 1% V, and 18% W
O1	T31501	.9% C, 1% Mn, .5% Cr, and .5% W	H42	T20842	.6% C, 4% Cr, 2% V, 6% V, and 5% Mo
O2	T31502	.9% C and 1.6% Mn	T1	T12001	.75% C, 4% Cr, 1% V, and 18% W
O6	T31506	1.45% C, .8% Mn, 1% Si, and .25% Mo	T2	T12002	.8% C, 4% Cr, 2% V, and 18% W
O7	T31507	1.2% C, .75% Cr, and 1.75% W	T4	T12004	.75% C, 4% Cr, 1% V, 18% W, and 5% Co
A2	T30102	1% C, 5% Cr, and 1% Mo	T5	T12005	.8% C, 4% Cr, 2% V, 18% W, and 8% Co
A3	T30103	1.25% C, 5% Cr, 1%, and 1% Mo	T6	T12006	.8% C, 4.5% Cr, 1.5% V, 20% W, and 12% Co
A4	T30104	1% C, 2% Mn, 1% Cr, and 1% Mo	T8	T12008	.75% C, 4% Cr, 2% V, 14% W, and 5% Co
A6	T30106	.7% C, 2% Mn, 1% Cr, and 1.25% Mo	T15	T12015	1.5% C, 4% Cr, 5% V, 12% W, and 5% Co
A7	T30107	2.25% C, 5.25% Cr, 4.75% V, 1% W, and 1% Mo	M1	T11301	.8% C, 4% Cr, 1% V, 1.5% W, and 8% Mo
A8	T30108	.55% C, 5% Cr, 1.25% W, and 1.25% Mo	M2	T11302	.85% to 1% C, 4% Cr, 2% V, 6% W, and 8% Mo
A9	T30109	.5% C, 5% Cr, 1% V, 1.4% Mo, and 1.5% Ni	M3, 1	T11313	1.05% C, 4% Cr, 2.4% V, 6% W, and 5% Mo
A10	T30110	1.35% C, 1.8% Nn, 1.25% Si, 1.5% Mo, and 1.8% Ni	M3, 2	T11323	1.2% C, 4% Cr, 3% V, 6% W, and 5% Mo
D2	T30402	1.5% C, 12% Cr, 1% V, and 1% Mo	M4	T11304	1.3% C, 4% Cr, 4% V, 5.5% W, and 4.5% Mo
D3	T30402	2.25% C and 12% Cr	M6	T11306	.8% C, 4% Cr, 2% V, 4% W, 5% Mo, and 12% Co
D4	T30404	2.25% C, 12% Cr, and 1% Mo	M10	T11310	.85% to 1% C, 4% Cr, 2% V, and 8% Mo
D5	T30405	1.5% C, 12% Cr, 1% Mo, and 3% Co	M30	T11330	.8% C, 4% Cr, 1.25% V, 2% W, 8% Mo, and 5% Co
D7	T30407	2.35% C, 12% Cr, 4% V, and 1% Mo	M33	T11333	.9% C, 4% Cr, 1.15% V, 1.5% W, 9.5 Mo, and 8% Co
L2	T61202	.5% C to 1.1% C, 1% Cr, and .2% V	M34	T11334	.9% C, 4% Cr, 2% V, 2% W, 8% Mo, and 8% Co
L6	T61206	.7% C, .75% Cr, .25% Mo, and 1.5% Ni	M36	T11336	.8% C, 4% Cr, 2% V, 6% W, 5% Mo, and 8% Co
F2	T60602	45% C to 1% C, .8% Mn, .5% Si, .95% Cr, .2% V, and .25% Mo	M41	T11341	1.1% C, 4.25% Cr, 2% V, 6.75% W, 3.75% Mo, and 5% Co
P2	T51602	.07% C, 2% Cr, .2% Mo, and .5% Ni	M42	T11342	1.1% C, 3.75% Cr, 1.15% V, 1.5% W, 9.5% Mo, and 8% Co
P3	T51503	.1% C, .6% Cr, and 1.25% Ni	M43	T11343	1.2% C, 3.75% Cr, 1.6% V, 2.75% W, 8% Mo, and 8.25% Co
P4	T51604	.07% C, 5% Cr, and .75% Mo	M44	T11344	1.15% C, 4.25% Cr, 2% V, 5.25% W, 6.25% Mo, and 12% Co
P5	T51505	.1% C and 2.25% Cr	M46	T11346	1.25% C, 4% Cr, 3.2% V, 2% W, 8.25% Mo, and 8.25% Co
P6	T51506	.1% C, 1.5% Cr, and 3.5% Ni	M47	T11347	1.1% C, 3.75% Cr, 1.25% V, 1.5% W, 9.5% Mo, and 5% Co
P20	T51620	.35% C, 1.70% Cr, and .4% Mo			
P21	T51621	.2% C, 1.2% (Al), and 4% Ni			
H10	T20810	.4% C, 3.25% Cr, .4% V, and 2.5% Mo			
H11	T20811	.35% C, 5% Cr, .4% V, and 1.5% Mo			
H12	T20812	.35% C, 5%, Cr, .4% V, 1.5% W, and 1.5% Mo			
H13	T20813	.35% C, 5% Cr, .4% V, 1.5% W, and 1.5% Mo			

Figure 15-3. Tool steel families are designated by a letter that stands for a key characteristic of those steels, which may be the method of hardening, a significant mechanical feature, or a major alloying element.

TOOL STEELS		
Tool Steel Group	**Properties and Characteristics**	**Common Use**
Group W	Tough core and hard and wear resistant surface	Cutlery, forging dies, and hammers
Group O	Wear resistant to moderate temperatures	Dies and punches
Group A	Minimum distortion and cracking on quenching	Dies, punches, and forming rolls
Group D	High hardness and excellent wear resistance	Long run dies and brick molds
Group S	Excellent toughness and high-strength	Chisels, rivet sets, and structural applications
Group L	High toughness and good strength	Arbors, cams, and chucks
Group F	Tough core, hard surface, and gall resistant	Burnishing tools and tube-drawing
Group P	Low hardness and low resistance to work hardening	Dies and molds
Group H	Good resistance to softening at elevated temperatures and good toughness	High stressed components and high-temperature extrusion dies
Group T	High hardenability, high hardness	Cutting tools and high-temperature structural components
Group M	High hardenability, high hardness	Cutting tools

Figure 15-4. Diversification of properties and characteristics influence the number and types of uses of tool steels.

Group W Steels. Group W steels are water-hardening tool steels consisting of W1, W2, and W5. Group W steels contain approximately 1% C. Small amounts of chromium and vanadium are added to increase hardenability and maintain fine grain size, which improves toughness.

Group W steels are shallow hardening and generally limited to sections <13 mm ($\frac{1}{2}$ in.) if through hardening is required. These steels develop a hard case with a soft core when quenched. The hard case and soft core are advantageous because they provide tough core properties in combination with a hard, wear resistant surface.

To maximize the depth of hardness, group W steels are quenched as rapidly as possible. Quenching must be performed in water or 10% sodium chloride brine. This increases the chance of cracking, so tempering is performed immediately after hardening and preferably before the component reaches room temperature. Toughness initially increases with increases in tempering temperature to approximately 180°C (355°F) but falls rapidly to a minimum at approximatley 260°C (500°F). See Figure 15-5. Double tempering (subjected to two tem-

pering cycles) may be necessary to temper any martensite that forms from the retained austenite left after the first tempering cycle.

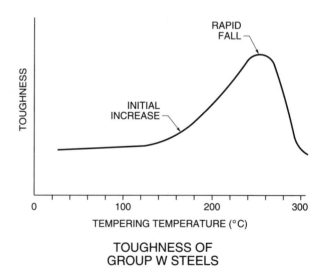

TOUGHNESS OF
GROUP W STEELS

Figure 15-5. Toughness of group W steels initially increases with tempering temperature but falls rapidly to a minimum.

Wear resistance of group W steels is lower compared to higher alloy tool steels that have similar carbon contents. This results in a shorter-than-normal service life. For example, they are used for cold header dies. In this application, the group W steels are superior to most other alloys.

Group W steels are used for items such as cold header dies, cutlery, embossing tools, forging dies, hammers, reamers, twist drills, and woodworking tools. Each group member is manufactured to specific quality levels, which are achieved by careful control of melting practices, alloy content, and heat treatment. The higher the quality level, the better the control of hardenability and hardness.

Cold Work Tool Steels

Cold work tool steels are steels that have alloy compositions designed to provide moderate-to-high hardenability and good dimensional stability during heat treatment. They have high wear resistance, but they begin to soften at temperature above 205°C to 260°C (400°F to 500°F). The toughness of cold work tool steels is generally poor to fair. Cold work tool steels are the most commonly used tool steels. The majority of tool applications can be served by one or more of the cold work tool steels. Cold work tool steels consist of group O, group A, and group D steels.

Group O Steels. Group O steels are oil-hardening cold work tool steels consisting of O1, O2, O6, and O7. They have high carbon contents and contain sufficient quantities of other alloying elements. This allows through hardening of moderate section sizes with oil quenching. Group O steels are usually normalized to homogenize the grain structure prior to austenitizing. They are quenched from temperatures of 790°C to 815°C (1450°F to 1500°F). Oil quenching is performed in agitated baths maintained at 50°C to 70°C (120°F to 160°F). Surface hardness of 56 HRC to 65 HRC is obtained by oil quenching followed by tempering of the steel at 175°C to 315°C (350°F to 600°F). See Figure 15-6.

The most important property of group O steels is wear resistance up to moderate temperatures. This is primarily due to the high carbon content. Graphite in the microstructure of O6 improves machinability in the annealed condition and helps reduce galling

and seizing when fully hardened. Although the wear resistance is satisfactory in many cases, it is lower than the other cold work tool steels (for example, groups A or D). Group O steels are extensively used in dies and punches for blanking, trimming, drawing, flanging, and forming. They are also used for machinery components, such as cams, bushings and guides, and gauges, where dimensional stability and good wear properties are needed.

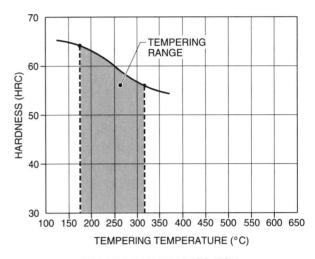

SURFACE HARDNESS FOR
GROUP O TOOL STEELS

Figure 15-6. The hardness of group O tool steels is progressively reduced as the tempering temperature is increased.

Group A Steels. Group A steels are air-hardening, medium-alloy cold work tool steels consisting of A2 (most commonly used), A3, A4, A6, A7, A8, A9, and A10. The carbon content of group A steels varies from approximately .5% C to 2% C. Group A steels also contain enough manganese, chromium, and molybdenum to achieve full hardness in sections at least 100 mm (4 in.) in diameter when air cooled. The air hardening characteristics of group A allow minimum distortion and cracking on quenching.

Group A steels consist of two subgroups. The subgroups are steels with high chromium content (approximately 5% Cr) and those with high manganese content (approximately 2% Mn to 3% Mn). High-chromium steels develop moderate resistance to softening at elevated temperatures from chromium carbide particles in the microstructure. They include A2, A3, A8, and A9. High-manganese steels have improved hardenability. For example, A7 through hardens in sections as large as a cube 175 mm (7 in.) wide. Other significant high-manganese steels are A4, A6, and A10.

Silicon is added to improve the toughness of A8. Silicon and nickel are added to A9 and A10 to improve their toughness. The high carbon content and high silicon content of A10 cause graphite to be formed in the microstructure. This gives the steel much better machinability in the annealed condition and much better galling and seizing resistance in the fully hardened condition.

Group A steels are usually supplied as-annealed. See Figure 15-7. A2 tool steels are annealed at 845°C (1550°F) and consist of massive carbides and fine spheroidal carbide particles in a matrix of ferrite. To minimize distortion, A2 steels are preheated before austenitizing. Austenitizing is performed in protective atmospheres, such as molten salt, to minimize oxidation and/or decarburization. A2 tool steels are austenitized at 955°C (1750°F), air cooled, and tempered at 150°C (300°F). Excessively high austenitizing temperatures are avoided because they promote retained austenite. All group A steels are air cooled.

Tempering begins when the steel reaches 50°C to 65°C (120°F to 150°F). Double or triple tempering is performed to maximize the transformation of retained austenite. If stabilizing at subzero temperatures is performed, it is performed immediately and is followed by tempering to avoid cracking of newly transformed martensite. A2 and A7 steels are sometimes nitrided to improve surface hardness in specific applications.

Group A steels have exceptional toughness, but they have moderate wear resistance and elevated temperature softening resistance. They have better toughness, machinability, and ability to be ground compared to group D steels, but they sacrifice in wear resistance. Group A steels are used in dies for blanking, forming, laminating, coining, drawing, thread rolling and trimming; for punches, forming rolls, slitting cutters, gauges, broaches, knurls, mandrels, burnishing tools, plastic molds, master hubs, and shear blades.

Group D Steels. Group D steels are high-carbon, high-chromium cold work tool steels consisting of D2 (most commonly used), D3, D4, D5, and D7.

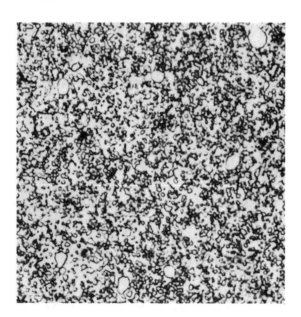

ANNEALED A2 TOOL STEEL

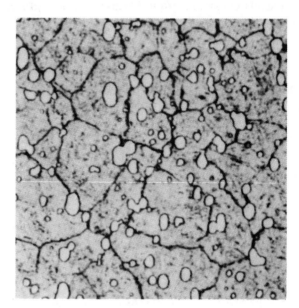

AUSTENITIZED, QUENCHED, AND TEMPERED
A2 TOOL STEEL

American Society for Metals

Figure 15-7. After annealing at 845°C (1550°F), the microstructure of A2 tool steel consists of massive carbide and fine spheroidal carbide particles in a ferrite matrix. After austenitizing at 955°C (1750°F), air cooling, and tempering at 150°C (300°F), the microstructure of A2 tool steel consists of spheroidal carbide particles in a matrix of tempered martensite.

D2, D4, D5, and D7 contain 1.5% C to 2.35% C, 12% Co, and 1% Mo. D3 contains 2.25% C and 12% Co, but does not contain molybdenum. D2 and D7 also contain vanadium, and D5 contains cobalt. All group D steels are air hardened, except for D3, which is oil quenched. They all have high hardness and excellent wear resistance, which is obtained from alloy carbides formed by the high carbon and high alloy content. D7, with the highest carbon and vanadium content, has the best wear resistance.

Group D steels are supplied as-annealed and are not normalized. After forging, group D steels should be re-annealed before hardening. Preheating before austenitizing reduces distortion in hardened components by minimizing non-uniform dimensional changes that may occur during austenitizing. To avoid decarburizing, austenitizing must be performed in a controlled atmosphere, such as a vacuum. Excessively high temperatures during austenitizing will promote retained austenite. The austenitizing time must be as long as the minimum time recommended in order to dissolve carbides.

Quenching of group D steels, with the exception of oil-quenched D3, is performed in air. Tempering is performed when the steel reaches 50°C to 65°C (120°F to 150°F). Double or triple tempering is performed to maximize the transformation of retained austenite. Low-temperature stabilizing treatments are sometimes performed after quenching to improve dimensional stability, which reduces the amount of retained austenite. Nitriding is performed to increase surface hardness. Group D steels show a slight increase in hardness (secondary hardening) when tempered in the range of 450°C to 550°C (840°F to 1020°F). This is caused by the precipitation of alloy carbides and helps to improve high-temperature softening resistance. See Figure 15-8.

Group D steels have excellent wear resistance, especially at moderate temperatures. They are used for long-run dies for blanking, forming, thread rolling, and deep drawing. They are also used for brick molds, gauges, burnishing tools, burnishing rolls, shear blades, and slitter cutters.

Shock-resisting Tool Steels

Shock-resisting tool steels are steels that have a relatively low carbon content (.4% C to .6% C) and contain manganese, silicon, tungsten, and molybdenum. They offer a combination of high strength, high toughness, high ductility, and low-to-medium wear resistance. Shock-resisting tool steels are designed for applications involving impact loading because of their combination of high strength and toughness under repeated shock and low-to-medium wear resistance. Shock-resisting tool steels consist of group S steels.

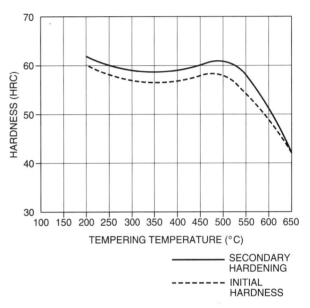

SECONDARY HARDENING
FOR GROUP D STEELS

Figure 15-8. Group D steels show a secondary hardening effect when tempered in the range of 450°C to 550°C (840°F to 1020°F).

Group S Steels. Group S steels are shock-resisting steels consisting of S1, S2, S5, S6, and S7. Group S steels vary from shallow hardening to deep hardening. S2 is water quenched, whereas S1, S5, and S6 are oil quenched. S7 is air cooled, except for large sections, which are oil quenched.

Group S steels are not normalized, and when annealed, care must be taken not to exceed the recommended annealing temperature. Special care is taken if the silicon content is high because the presence of silicon increases susceptibility to embrittlement from graphitization. Graphitization occurs at elevated temperatures above the recommended annealing temperature.

The microstructure of S1 consists of a dispersion of fine spheroidal particles of carbide in a matrix of ferrite. See Figure 15-9. After austenitizing at 1040°C (1900°F), oil quenching, and tempering at 220°C (425°F), the microstructure consists of tempered martensite and spheroidal carbide particles.

Except for extremely intricate components of widely varying section thickness, stress relieving before hardening is seldom required. When stress relieving is employed, the steel should be furnace cooled to 510°C (950°F) and then air cooled. Except for large sections, preheating prior to austenitizing to decrease time at the austenitizing temperature is not usually necessary. Controlled atmospheres are often employed during austenitizing to avoid decarburization. All shock-resisting steels are tempered immediately after quenching to prevent cracking. S1 is often carburized or carbonitrided to increase its surface hardness and wear resistance.

The excellent toughness at high-strength levels makes group S steels useful for machine components where toughness is of prime importance. Applications include chisels, rivet sets, punches, and driver bits. Where some heat resistance is required, such as in hot punching and shearing, S1 and S7 are sometimes used. Group S steels are often considered for non-tooling or structural applications.

Special-purpose Tool Steels

Special-purpose tool steels are steels that contain small amounts of chromium, vanadium, nickel, and molybdenum and are used in applications requiring good strength, good toughness, scratch-resistant surfaces, and gall resistance. Special-purpose tool steels consist of group L and group F steels.

Group L Steels. Group L steels are low-alloy, special-purpose tool steels consisting of L2 and L6. They contain between .5% C and 1% C. L2 contains chromium and vanadium for improved hardenability and fine grain size. L6 contains chromium and molybdenum, in addition to 1.5% Ni, for improved toughness. Group L steels are almost as tough as group S steels.

Group L steels are normalized after forging or hot working. Stress relieving prior to hardening is desirable to minimize distortion. Group L steels are oil-quenched, except for large sections of L2, which

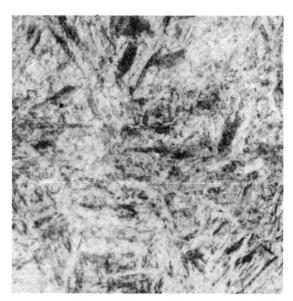

ANNEALED S1 TOOL STEEL

AUSTENITIZED, QUENCHED, AND TEMPERED
S1 TOOL STEEL

American Society for Metals

Figure 15-9. Microstructure of S1 tool steel annealed at 790°C (1450°F) consists of a dispersion of fine carbide particles in a ferrite matrix. After austenitizing at 1040°C (1900°F), oil quenching, and tempering at 425°C (800°F), the microstructure of S1 tool steel consists of tempered martensite and spheroidal carbide particles.

are often water-quenched. Tempering is performed immediately when the component reaches 50°C to 80°C (125°F to 180°F). Double tempering is used to eliminate the effects of retained austenite.

Group L steels are used for machine components, such as arbors, cams, chucks, and collets. Some special applications may require good strength and toughness characteristic of the group L steels.

Group F Steels. Group F steels are carbon-tungsten, special-purpose tool steels consisting of one principal member, which is F2. When hardened, F2 acquires a slippery surface that is difficult to scratch and does not gall as readily as other types of tool steels. F2 develops a hard case and tough core. Applications for F2 include burnishing tools, tube-drawing, wire-drawing, cold extrusion dies, and brass cutting tools.

Mold Steels

Mold steels are steels that contain chromium and nickel as the principal alloying elements. They have a low-to-medium carbon content and a total alloy content of 1.5% to 5%. Mold steels consist of group P steels.

Group P Steels. Group P steels are mold steels consisting of P2, P3, P4, P5, P6, P20, and P21. P2, P3, P4, P5, and P6 are low-carbon grades of steel that are surface hardened by carburizing. P20 and P21 are medium-carbon grades of steel that are carburized or nitrided.

P2, P3, P4, P5, and P6 exhibit low hardness and low resistance to work hardening in the annealed condition. This facilitates its use for cold-hobbing operations to create mold impressions. After the impression is formed, the mold is carburized, hardened, and tempered to a surface hardness of 50 HRC to 58 HRC.

The heat treatment of group P steels closely follows practices employed for carbon steels and low-alloy steels. It has little similarity to the methods used for most other groups of tool steels (especially those with a higher carbon content). P20 and P21 are normally supplied heat treated to 30 HRC to 36 HRC. In this condition, they can be machined into

intricate dies and molds. P20 and P21 are not hardened by quenching and tempering after machining, but can be carburized. P21, which contains aluminum, can also be nitrided.

Hot Work Tool Steels

Hot work tool steels are steels that have been developed to withstand combinations of heat, pressure, and abrasion associated with manufacturing operations performed at high temperatures from 480°C to 760°C (900°F to 1400°F). They resist wear up to temperatures of 540°C (1000°F). See Figure 15-10. Hot work tool steels have good hardenability and toughness. They are the group H steels and consist of three subgroups, which are chromium, tungsten, and molybdenum hot work tool steels. These represent the principal alloying element of the tool steel. They have medium carbon content (.35% C to .45% C) and chromium, tungsten, molybdenum, and vanadium contents in the range of 6% to 25%.

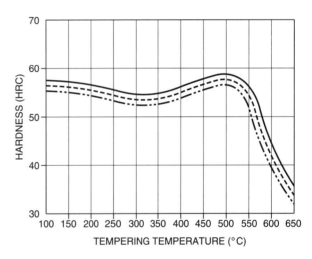

HARDNESS OF SELECTED
GROUP H STEELS

Figure 15-10. Group H steels resist softening during continued exposure to temperatures up to 540°C (1000°F).

Group H Steels. Chromium hot work steels are group H steels consisting of H10, H11, H12, H13, H14, and H19. The most commonly used are H11, H12, and H13. All chromium hot work steels have good resistance to softening at high temperatures because of the medium chromium contents and carbide forming elements, such as molybdenum, tungsten, and vanadium. The low carbon and low alloy contents promote toughness in a working hardness range of 40 HRC to 55 HRC. Higher tungsten and molybdenum contents increase hot strength but slightly reduce toughness. Vanadium is added to increase resistance to washing (erosive wear) at high temperatures. Increasing silicon improves resistance to oxidation at temperatures up to 800°C (1475°F).

Chromium hot work steels are fully or partially air hardened and should be annealed prior to heat treatment rather than normalized. Preheating rate for annealing should be slow and uniform to prevent cracking, especially when annealing hardened tools. These steels are extremely susceptible to carburizing or decarburizing. They must be protected against carburizing and decarburizing by packing, controlled furnace atmospheres, or vacuum.

Stress relieving, which minimizes distortion during hardening, is beneficial if performed after rough machining and before final machining. These steels should be preheated prior to austenitizing. Slow heating to the austenitizing temperature is preferred for all members except for H19. The time at austenitizing temperature for H19 should only be long enough to allow for complete heating.

H10, H11, H12, H13, H14, and H19 hot work steels are cooled in still air and tempered immediately. Multiple tempering transforms retained austenite and minimizes cracking due to hardening stresses. Steels with <.35% C are sometimes carburized or nitrided to achieve higher surface hardness.

The chromium hot work steels are especially adapted to hot die work of all kinds, particularly dies for the extrusion of aluminum or magnesium. They are also used as die casting dies, forging dies, mandrels, and hot shears. H11 and H13 are used to make certain highly stressed components, particularly in the aerospace industry. The material for these demanding applications is produced by vacuum arc remelting. One chief advantage of H11 and H13 over conventional high-strength steels is the ability to resist softening during continued exposure to temperatures up to 540°C (1000°F) while providing moderate ductility. Another chief advantage is the room temperature tensile strength of 250 ksi to 300 ksi (1720 MPa to 2070 MPa).

Tungsten hot work steels are group H steels consisting of H21, H22, H23, H24, H25, and H26. H21 is the most commonly used. They all are more highly alloyed than the chromium hot work steels. They contain at least 9% W and 2% Cr to 12% Cr. This improves the resistance to high-temperature softening but reduces shock resistance from the normal working hardness of 45 HRC to 55 HRC.

Heat treatment practices for the tungsten hot work steels are similar to those used for the chromium hot work steels, however, rapid heating to the austenitizing temperature is required for the tungsten hot work steels and the soaking time for them should be consistent with complete heating of the component. Tungsten hot work steels are used to make mandrels and extrusion dies for high temperature applications, such as extrusion of brass, nickel alloys, and steel. Tungsten hot work steels are also used for hot forging dies of rugged design.

Molybdenum hot work steel is a group H steel consisting of the principal alloy of H42. It contains 5% Mo, 4% Cr, 2% V, and 6% W. It is similar to the tungsten hot work steels and has almost identical characteristics and uses as they do.

H42 is more resistant to heat checking than the tungsten hot work steels. *Heat checking* is the cracking of metal due to alternating heating and cooling of the extreme surface metal. H42 requires greater care when heat treated, particularly with regard to decarburizing and control of the austenitizing temperature. The advantage of H42 over the tungsten hot work steels is the lower cost.

High-speed Tool Steels

High-speed tool steels are steels that are typically employed for high-speed cutting operations. They resist softening and maintain sharp cutting edges at high service temperature because of their excellent red hardness, which is the resistance to soften at red heat (temperatures of 540°C, or 1000°F). This property is attributed to the presence of complex alloy carbides in the microstructure of the steel. The toughness of high-speed tool steels is normally not high. High-speed tool steels consist of group T and group M steels.

Group T Steels. Group T steels are tungsten high-speed tool steels consisting of T2, T4, T5, T6, T8, and T15. The base composition is T1, also known as 18-4-1. This represents the nominal percentages of tungsten (18% W), chromium (4% Cr), and vanadium (1% V). Group T steels have medium carbon contents, and some may also contain cobalt. The hardenability of group T steels is very deep. Sections up to 75 mm (3 in.) across can be hardened to 65 HRC or greater by quenching in oil or molten salt.

Group T steels are usually fully annealed prior to austenitizing or after forging. They are slowly preheated in two stages prior to austenitizing to minimize stress concentrations, which could lead to stress cracking. After the second preheating stage, the steels are heated rapidly to the austenitizing temperature. Austenitizing is performed close to the melting point to facilitate the dissolution of complex alloy carbides, so accurate temperature control is absolutely necessary.

Group T steels may be quenched in air, oil, or molten salt. Group T steels are normally subjected to two separate tempering treatments. Refrigeration at −85°C (−120°F) is used after the first tempering treatment to transform retained austenite. The steel is then retempered. See Figure 15-11. These steels may also be nitrided to increase surface hardness and improve frictional properties.

Group T steels are used for cutting tools such as broaches, chasers, cutters, drills, reamers, and taps. They are also used for making dies, punches, and high-load, high-temperature structural components.

Group M Steels. Group M steels are molybdenum high-speed steels consisting of M1, M2, M3 class 1, M3 class 2, M4, M6, M7, M10, M30, M33, M34, M36, M41, M42, M43, M44, M46, and M47. They are similar to group T steels in performance but are more commonly used because of lower cost for equivalent grades. Group M steels contain carbon in the 1% C range in addition to various amounts of molybdenum, tungsten, chromium, vanadium, and/or cobalt. M1, M2, and M4 are the commonly used group M steels. M41, M42, M43, M44, M46, and M47 are the ultrahard high-speed tool steels and exhibit high levels of hardness.

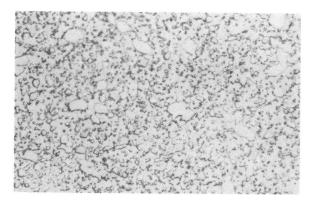

ANNEALED T1 STEEL

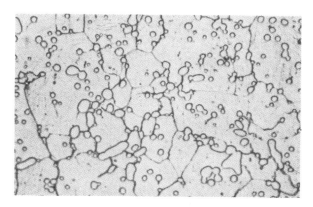

AUSTENITIZED, QUENCHED, AND TEMPERED T1 TOOL STEEL

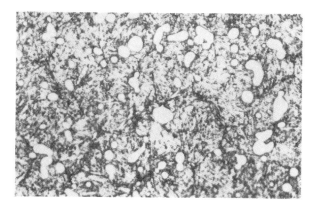

DOUBLE TEMPERED T1 STEEL

American Society for Metals

Figure 15-11. The microstructure of T1 steel annealed at 900°C (1650°F) consists of large and small spheroidal carbide particles in a ferrite matrix. After austenitizing at 1280°C (2335°F), salt quenching to 605°C (1125°F), and air cooling, the microstructure of T1 tool steel consists of undissolved carbide particles in an untempered martensite matrix. After tempering the same sample at 540°C (1000°F), the microstructure of T1 tool steel consists of undissolved carbide particles in an austenite matrix.

Heat treating procedures for these steels are similar to those used for group T steels. For example, with both groups it is important to temper on the high side of the tempering curve. See Figure 15-12. There are two tempering temperatures indicated that will achieve a hardness of 63 HRC. Temperature at point B, which is on the high side of the curve, is preferred over temperature at point A because it obtains the same hardness as well as increased toughness and dimensional stability. Increasing the carbon and vanadium contents of group M steels increases wear resistance. Increasing the cobalt content improves red hardness, but it also lowers toughness. As with group T steels and unlike group W steels and group D steels, resistance to softening at elevated temperatures is high. Group M steels are primarily used for cutting tools, such as drills, reamers, taps, end mills, milling cutters, hobs, shearing tools, slitting tools, blanking dies, piercing dies, and press forming dies.

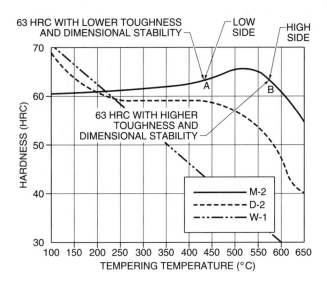

Figure 15-12. Group M steels have high resistance to softening at elevated temperatures compared to other tool steels. Like group T steels, group M steels are tempered on the high side of their tempering curve.

Cast Irons 16

The six divisions of cast irons are gray iron, white iron, malleable iron, ductile iron, compacted graphite iron, and alloy iron. Cast irons are designated by ASTM and other standards. The principal advantage of cast irons is the ease with which they can be cast into complex shapes. Cast irons are very fluid when molten, undergo only slight shrinkage on solidification, and do not form undesirable surface films when poured. The principal disadvantages of cast irons are relatively low levels of toughness and ductility. Poor weldability is also a disadvantage, but there is a need to weld cast irons, and considerations must be taken to do so.

CAST IRONS

Carbon and silicon are the principal alloying elements in cast irons. By varying the balance between carbon and silicon and alloying with various other elements, a wide variety of alloys can be cast. The carbon equivalent (CE) is used to determine the basic composition of all cast irons except alloy cast irons. *Carbon equivalent* for cast irons is a factor that indicates the total relationship of the percentage of carbon, silicon, and phosphorus in the chemical composition.

The term "cast" is usually omitted from the description of cast irons. For example, ductile cast iron is usually referred to as ductile iron. There are six divisions of cast irons, which are gray iron, white iron, malleable iron, ductile iron, compacted graphite iron, and alloy iron.

Some cast irons are sold under the name Meehanite. This is the name of a company that licenses cast iron foundry technology, including additives that are put into molten metal before it is cast. The various grades of Meehanite include members of different cast iron families. The mechanical properties of meehanite cast irons are similar to those of cast irons made to standard specifications.

Carbon Equivalent

Gray irons with a CE that is <4.3 are hypoeutectic, and those with a CE that is >4.3 are hypereutectic. CE is calculated using the following formula:

$$CE = \%C + \tfrac{1}{3}(\%Si + \%P)$$

where

CE = carbon equivalent

$\%C$ = percent carbon

$\%Si$ = percent silicon

$\%P$ = percent phosphorus

Example: Figuring carbon equivalent

What is the carbon equivalent of a gray iron containing 3.4% C, 2.4% Si, and .12% P.

$$CE = \%C + \tfrac{1}{3}(\%Si + \%P)$$

$$CE = 3.4 + \tfrac{1}{3}(2.4 + .12)$$

$$CE = 3.4 + \tfrac{1}{3}(2.52)$$

$$CE = 3.4 + .84$$

$$CE = \mathbf{4.24\%}$$

Note: This gray iron is hypereutectic.

When the CE is close to the eutectic value of 4.3%, the molten metal persists to a relatively low temperature, and solidification takes place over a narrow temperature range. These attributes are important in promoting uniformity of properties within a given iron casting.

The total carbon, silicon, and phosphorus contents of cast irons, as expressed in the CE, establish the solidification temperature range of the alloy and the resulting microstructure. However, two irons with the same CE but different carbon and silicon contents will have different casting properties. For example, carbon is more than twice as effective in preventing solidification shrinkage than silicon, and silicon is more effective in preventing thin sections from becoming hard.

Gray Iron

Gray iron is the most widely used cast iron. The microstructure of gray iron consists of a matrix of pearlite (iron carbide), ferrite, or martensite containing a distribution of graphite flakes. Gray iron can be hardened and tempered like steel. The graphite flakes in the microstructure make the gray iron extremely brittle.

Microstructure. Most gray irons are hypoeutectic alloys of iron, carbon, and silicon. A typical gray iron with an alloy composition of 3% C and 2% Si first solidifies by forming austenite. See Figure 16-1. At the eutectic temperature of approximately 1150°C (2100°F), the remaining liquid transforms to kish graphite (free graphite). The kish graphite grows into the austenite matrix in the shape of three-dimensional flakes. When the temperature falls below the eutectoid temperature, approximately 760°C (1400°F), the austenite transforms to either ferrite or pearlite. The transformation depends on the rate of cooling.

At room temperature, the microstructure consists of a matrix of ferrite or pearlite that contains a distribution of graphite flakes. Pearlitic gray iron contains a matrix of pearlite, which is the most common matrix for industrial applications. A martensitic matrix is produced by austenitizing and quenching pearlitic or ferritic gray iron.

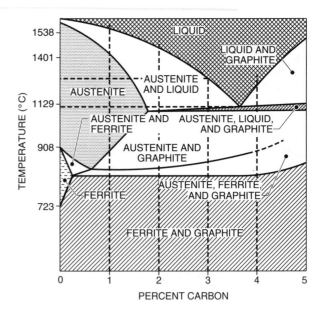

Figure 16-1. A section through the iron-carbon-silicon phase diagram at 2.5% Si illustrates the microstructure obtained on cooling a typical gray iron.

The cooling rate from the molten state has a significant influence on the microstructure of gray iron. With extremely rapid cooling, iron carbide, not graphite, is formed. Some martensite may also accompany the pearlite that is formed. Cooling rate also influences the size of the graphite flakes. Slow cooling rates produce coarse graphite flakes, and fast cooling rates produce fine graphite flakes.

Properties. The properties of gray iron are dominated by the flake graphite constituent. The three-dimensional flakes provide a path for brittle fracture propagation with little or no plastic strain. As a result, gray iron has little or no ductility. Although the ductility of gray iron is influenced by the graphite flakes, its strength depends almost entirely on whether the matrix is pearlitic, ferritic, or martensitic. Unlike the cementite phase in steels, the graphite flakes in gray iron lower hardness and strength. The hardness of gray iron is between that of the soft graphite and the matrix phase.

The graphite flakes provide gray iron with one of the highest damping capacities in any engineering material. The graphite flakes also result in improved machinability, high-temperature scaling resistance in air, and thermal shock resistance. They significantly reduce metal shrinkage during solidification and provide gray iron with excellent castability compared with cast steel.

The distribution of the flake graphite may take several different forms but is usually of no significance, providing the specified tensile properties are met. Sometimes the foundry worker aims for a specific graphite distribution pattern to optimize the properties for a particular end use. For example, a random graphite distribution is favored for internal combustion engine cylinders because of the properties it produces.

Designations. ASTM designations are most commonly used for gray iron. ASTM designates gray irons by their minimum tensile strength in ksi or MPa. The designations range from class 20, with a minimum tensile strength of 20 ksi (140 MPa) to class 60, with a minimum tensile strength of 60 ksi (410 MPa). Class 40 is the most commonly used gray iron. See Figure 16-2.

GRAY IRON DESIGNATIONS				
Class	UNS	Minimum Tensile Strength (ksi)	Hardness (HB)	Micro-structure
G1800		18	187 max	Ferrite-pearlite
20	F11401	20		
G2500		25	170 to 229	Pearlite-ferrite
25	F11701	25		
G3000		30	187 to 241	Pearlite
30	F12102	30		
G3500		35	207 to 255	Pearlite
35	F14201	35		
G4000		40	217 to 269	Pearlite
40	F12801	40		
45	F13101	45		
50	F13502	50		
55	F13801	55		
60	F14101	60		

Figure 16-2. Gray iron is designated by its grade number, which indicates its minimum tensile strength in ksi.

Ambient and elevated temperature strength, the ability to produce a fine machined finish, modulus of elasticity, and wear resistance are properties of gray iron that generally increase from grade 20 to grade 60. Machinability, resistance to thermal shock, damping capacity, and castability in thin sections are properties that generally decrease with increasing tensile strength from grade 20 to grade 60.

Gray iron is also specified in SAE standards. Many gray iron compositions are identified in the UNS with the uppercase letter F. Gray iron has many applications due to its castability and low cost. It is used for bases and supports for moving components to dampen vibrations, in pressure applications such as cylinder blocks, for wear-resistant and scuff-resistant materials used in cylinder sleeves and valve guides, and in general municipal applications such as manhole covers and hydrants.

White Iron

White iron is extremely hard cast iron. White iron is formed when the carbon does not precipitate as graphite during solidification but combines with the iron and any of the alloying elements to form carbides. This formation occurs during extremely fast cooling from the molten state and is achieved by the use of metal or graphite chills embedded in the mold. Chills rapidly extract heat from the molten metal, unlike the regular sand molds normally used for cast iron.

Fast cooling suppresses the formation of graphite and causes cementite to precipitate. Carbides are produced when alloying elements such as molybdenum, chromium, and vanadium are present. The cementite and the alloy carbides make white iron extremely hard and brittle. White iron is used chiefly for its exceptional hardness and wear resistance. When broken, it produces a white crystalline fracture face. This is where the term "white iron" is derived. See Figure 16-3.

Chilled Iron. For certain applications, a combination of white and gray iron microstructures is produced. Chills are placed in the mold wherever the finished casting must have a hard and abrasion-resistant surface. *Chilled iron* is an area of the casting that solidifies more slowly and has a readily machinable gray iron microstructure that contains graphite. *Mottled iron* is the mixed structure of iron carbides and graphite in the boundary zone between the white and gray regions in chilled iron. Alloy composition and foundry practice must be carefully controlled if the boundary between the chilled and unchilled components of the casting is to be held at the specified location in the design. Chilled iron is used in applications such as railway wheels, stamp shoes, and heavy-duty machinery components. See Figure 16-4.

Malleable Iron

Malleable iron is a ductile form of cast iron that is produced by heat treating white iron. Malleable iron has moderate strength and significantly greater ductility than gray iron.

Microstructure. The microstructure of malleable iron is produced by heat treating white iron. The white iron is heated in a controlled atmosphere furnace to a temperature above the eutectoid temperature, usually approximately 925°C (1700°F), and held for several hours. The heat treatment causes several microstructural changes.

First, the carbon leaves the cementite and dissolves in the austenite. It then migrates from the austenite to preferred locations within the microstructure and precipitates as temper carbon (irregularly shaped nodules). The casting is then allowed

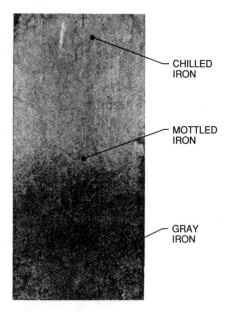

Figure 16-4. A fracture through a flanged gray iron specimen with a chilled (white) iron area shows the boundary region between the two kinds of microstructures as mottled iron.

to cool slowly in the furnace through the eutectoid transformation range. Remaining excess carbon in the austenite precipitates onto the existing temper carbon nodules. The austenite that is depleted of carbon transforms to ferrite. The resulting microstructure consists of a ferrite matrix containing temper carbon nodules. See Figure 16-5.

GRAY IRON
REVEALING GRAY APPEARANCE
OF GRAPHITE FLAKES

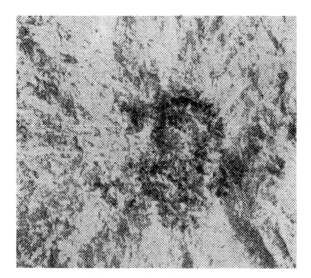

WHITE IRON
REVEALING WHITE
CRYSTALLINE STRUCTURE

Figure 16-3. Gray iron is the most widely used cast iron. White iron derives its name from its fracture appearance, which is white and crystalline.

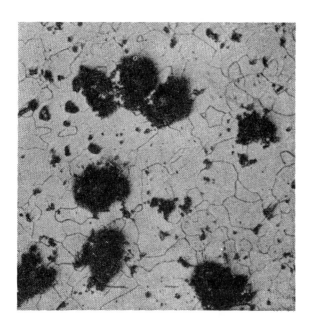

Figure 16-5. The microstructure of ferritic malleable iron consists of temper carbon nodules in a ferrite matrix.

If a pearlite or martensite matrix is required to improve strength, the cooling rate from the austenitizing temperature is increased. Also, alloying elements are added to increase hardenability.

Properties. The principal difference between malleable and gray iron is ductility. Malleable iron is significantly more ductile because the temper carbon nodules do not present a continuous fracture path like that presented by the graphite in gray iron. Malleable iron is also characterized by moderate strength, good toughness, castability, and machinability. See Figure 16-6.

Malleable iron has many industrial uses, which include axle and differential housings, camshafts, and crankshafts in automobiles; and gears, chain links, sprockets, and elevator brackets in conveying equipment. Other uses of malleable iron include rolls, pumps, nozzles, cams, and rocker arms in machine components; gun mounts, tank parts, and pistol parts in ordinance; and wrenches, hammers, clamps, and shears in small tooling.

Ductile Iron

Ductile iron contains similar amounts of carbon and silicon to gray iron, but differs in the shape of the graphite constituent. In ductile iron, the graphite is spheroidal. Ductile iron is also referred to as nodular iron or spheroidal graphite (SG) iron. It has equivalent strength to gray iron, but ductile iron has significantly greater ductility.

PROPERTIES OF MALLEABLE IRON CASTINGS				
Class or Grade	Minimum Tensile Strength (ksi)	Minimum Yield Strength (ksi)	Minimum Elongation in 2 in. (%)	Hardness Range (HB)
ASTM A47				
32510	50	32.5	10	145 max
35018	53	35	18	156 max
ASTM A220				
40010	60	40	10	149 to 197
45006	65	45	6	156 to 207
50005	70	50	5	179 to 229
60004	80	60	4	197 to 241
ASTM A602				
M4504	65	45	4	163 to 217
M5003	75	50	3	187 to 241
M5503	75	55	3	187 to 241

Figure 16-6. Malleable iron has greater ductility than gray iron.

Microstructure. The microstructure of ductile iron is achieved by spheroidization of the graphite. Spheroidization is accomplished when a controlled amount of magnesium or cerium (Ce) is added to the molten metal. The magnesium must be combined with other elements, such as nickel, iron, silicon, and calcium, to prevent it from vaporizing instantaneously. Various proprietary alloys of magnesium combined with one or more of the elements nickel, iron, silicon, and calcium are frequently used. A residual concentration of .035% Mg is necessary to produce full spheroidization of the graphite.

The microstructure of ductile iron most commonly consists of graphite nodules surrounded by an envelope of ferrite in a matrix of pearlite. The matrix may be completely transformed to ferrite by annealing and completely transformed to martensite by quenching. See Figure 16-7.

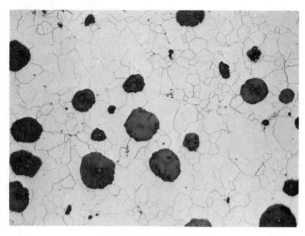

FERRITIC DUCTILE
CAST IRON

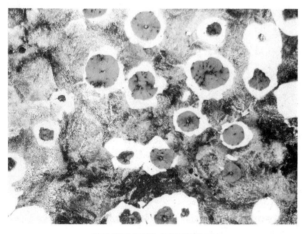

PEARLITIC DUCTILE
CAST IRON

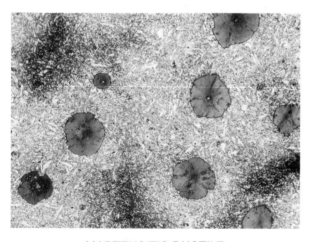

MARTENSITIC DUCTILE
CAST IRON

Figure 16-7. The microstructure of ductile iron consists of spheroidal graphite in a matrix of ferrite, pearlite, or martensite.

Properties. The properties of ductile iron are similar in strength to gray iron. See Figure 16-8. However, ductile iron is significantly tougher, more ductile, and its damping capacity is significantly less than that of gray iron. The improvement in toughness and ductility is caused by the morphology of the graphite constituent. Unlike flake graphite, spheroidal graphite does not present a continuous brittle fracture path. For example, tensile test bars of ferritic ductile iron exhibit an 18% minimum elongation at a minimum tensile strength of 60 ksi (410 MPa). This compares with about a .6% elongation for gray iron of similar strength.

PROPERTIES OF DUCTILE CAST IRON			
Grade	Minimum Tensile Strength (ksi)	Minimum Yield Strength (ksi)	Minimum Elongation in 2 in. (%)
60-40-18	60	40	18
65-45-12	65	45	12
80-55-06	80	55	6
100-70-03	100	70	3

Figure 16-8. Ductile iron exhibits considerable elongation compared with equivalent strength grades of gray iron.

Designations. Ductile iron is designated in the ASTM designation system using a sequence of three numbers separated by dashes. These numbers indicate the minimum tensile strength in ksi, the yield strength in ksi, and the percent elongation. These numbers are obtained from representative test bars. For example, 60-40-18 has minimum tensile strength of 60 ksi, yield strength of 40 ksi, and minimum elongation of 18%. Ductile iron is also specified by the SAE and is identified in the UNS by the uppercase letter F.

The cost of the elements added to the melt makes ductile iron more expensive, but because prolonged heat treatment is not required, it competes favorably with malleable iron. Ductile iron is used for many structural applications, particularly those requiring strength and toughness combined with good machinability at low cost. The automotive and the agricultural implement industries are the principal users of ductile iron castings. They are employed for items such as crankshafts, front wheel spindle supports, steering knuckles, and disk brake calipers. Piping such as culvert, sewer, and pressure pipe are other applications of ductile iron castings.

Compacted Graphite Iron

Compacted graphite iron is a form of cast iron that has a shape between the flake graphite of gray iron and the spheroidal graphite of ductile iron. They are high in ductility and thermal shock resistance.

Microstructure. Compacted graphite iron is produced by inoculating the molten metal to prevent the formation of flake graphite. This is similar to the manner in which ductile iron is produced. However, the graphite is in the form of interconnected flakes with blunted edges and a relatively short span. See Figure 16-9. Consequently, the surface area-to-volume ratio of the compacted graphite is much less than that of regular flake graphite.

Figure 16-9. The morphology of the graphite in compacted graphite iron is more blunted than the graphite in gray iron.

Properties. The intermediate morphology of the compacted graphite results in a combination of desirable properties. For example, compacted graphite iron shrinks less on solidification than ductile iron and has better thermal shock resistance. Compared to gray iron, compacted graphite iron has greater ductility. Compacted graphite iron is used for specific applications where it provides the optimum combination of properties, such as in disc brake rotors and diesel engine heads.

Alloy Iron

Alloy iron is a cast iron that contains one or more intentionally added alloying elements, up to a total of 30%, in order to enhance specific properties. Alloying has a great effect on the properties and the end use of the alloy iron. This effect is used to separate the alloy irons into the three groups of abrasion-resistant iron, corrosion-resistant iron, and heat-resistant iron. The heat-resistant irons are further divided into heat-resistant ductile iron and heat-resistant gray iron.

Effects of Alloying. The effects of alloying elements in gray iron are divided according to the effect they exert during solidification of the iron. Alloying elements are either graphitizers or carbide stabilizers. *Graphitizers* are alloying elements that promote graphite formation. *Carbide stabilizers* are alloying elements that promote cementite or alloy carbide formation. The major alloying elements in cast irons are silicon, nickel, chromium, molybdeum, copper, and vanadium, but other elements are used. See Figure 16-10.

Silicon is the strongest graphitizer. The silicon content in cast iron can be increased above its normal level to form alloy iron with excellent oxidation and corrosion-resistant properties. Nickel is also a graphitizer, but it is not nearly as powerful as silicon. Nickel decreases carbide stability while it increases the fineness and stability of the pearlite. This results in an increase in the strength of the iron. Aluminum is a strong graphitizer and is added primarily to some heat-resistant irons to improve scaling resistance and growth. It decreases ductility and castability. Copper is a mild graphitizer. Carbon additions increase the strength and hardness of iron by reducing the tendency of ferrite formation and finer pearlite formation.

Chromium is a strong carbide former. Chromium increases the tendency to produce white iron during solidification and increases the retention of a higher combined carbon in the solidified metal. This increases wear resistance and, at elevated temperatures, retards oxidation. Molybdenum is a mild carbide former. Molybdenum is useful in improving elevated temperature strength and creep resistance. Vanadium is a strong carbide former. Vanadium is used in small amounts to refine the graphite size in gray iron.

EFFECTS OF ALLOYING		
Element	During Solidification	During Eutectoid Reaction
Silicon	Strong graphitizer	Promotes ferrite and graphite formation
Nickel	Graphitizer	Mild pearlite promoter
Aluminum	Strong graphitizer	Promotes ferrite and graphite formation
Copper	Mild graphitizer	Promotes pearlite formation
Chromium	Strong carbide former Forms very stable, complex carbides	Strong pearlite former
Molybdenum	Mild carbide former	Strong pearlite former
Vanadium	Strong carbide former	Strong pearlite former

Figure 16-10. Alloy elements in alloy irons are either graphitizers or carbide stabilizers.

Abrasion-resistant Iron. Abrasion-resistant irons are alloys of white iron and consist of three types, which include low chromium-nickel, high-chromium, and high chromium-molybdenum. Low chromium-nickel abrasion-resistant irons are the Ni-Hard® irons. Ni-Hard® irons generally contain from 1% Cr to 4% Cr and 3% Ni to 5% Ni. In the Ni-Hard® irons, the alloying is varied from 2.5% C to 3.6% C, from 3.5% Ni to 5.0% Ni, and from 1.4% Cr to 3.5% Cr. Nickel suppresses the austenite to pearlite transformation, ensuring a martensitic matrix with a little retained austenite. Chromium ties up excess carbon as carbides, counteracting the graphitizing effect of nickel. Abrasion resistant irons have no graphite in the microstructure.

Ni-Hard® irons are used for handling abrasive materials. For example, they are used for grinding and pulverizing equipment in the coal mining and mineral processing industries. Other applications of Ni-Hard® include pipe elbows and slurry pumps.

High-chromium abrasion-resistant irons contain approximately 25% Cr. Chromium carbides in the microstructure provide the abrasion resistance. Martensite is the ideal matrix for abrasion resistance and is achieved in thin sections in sand molds by the natural cooling rate from the casting temperature. Thicker sections, which cool more slowly, develop a pearlite matrix and have substantially less abrasion resistance. Provided they are not too thick, they are austenitized, quenched, and tempered to develop a martensite matrix. Since the hardenability of high-chromium irons is relatively low, such heat treatment is only used on moderately thick sections. Consequently, the high-chromium abrasion-resistant irons are limited to applications such as pump volutes and impellers of moderate section thickness.

High chromium-molybdenum abrasion-resistant irons contain approximately 18% Cr and up to approximately 3% Mo. The high chromium content produces abrasion-resistant carbides. The molybdenum is added to increase hardenability. The high chromium-molybdenum abrasion-resistant irons are used for mining applications, such as grinding mill liners, impact hammers, and pump impellers. They are also used for well drilling applications, such as mud pump liners.

Corrosion-resistant Iron. Corrosion-resistant irons are based on gray or ductile iron microstructures. They consist of the high-nickel (austenitic) and the high-silicon types.

High-nickel corrosion-resistant irons have the trade name Ni-Resist®. They contain from 13.5% Ni to 36.0% Ni and 1.8% Cr to 6.0% Cr. One grade contains from 5.5% Cu to 7.5% Cu. The austenitic irons are produced as either gray or ductile iron and are identified by various specifications.

Gray austenitic iron contains high percentages of nickel, which ensures that graphite flakes form when cooled, even when the composition is relatively high in chromium. The microstructure of the gray austenitic iron consists of flake graphite in an austenite matrix. Copper is added to improve corrosion resistance. In addition to excellent corrosion resistance, gray austenitic iron has good high-temperature oxidation resistance. However, toughness and ductility are relatively low.

Ductile austenitic iron contains varying percentages of nickel and chromium, depending on the end use. Unlike the gray austenitic iron, ductile austenitic iron contains no copper because copper interferes with spheroidizing. The microstructure of the ductile austenitic iron consists of graphite nodules in an austenite matrix. These alloys have a wide range of corrosion resistance.

Gray and ductile austenitic irons are used in many corrosion-resistant applications, such as handling seawater, acids and alkalis, and sour gas. These irons are typically used for pump impellers, casings, and suction components.

The high-silicon corrosion-resistant irons are gray irons with 14.20% Si to 14.75% Si. The matrix consists of ferrite. Silicon assists in the development of a tough protective surface film, resulting in extremely high corrosion resistance, particularly to acids such as sulfuric and nitric. High-silicon irons are known commercially as Duriron®.

High-silicon irons have poor mechanical properties, especially to thermal and mechanical shock. They are used in the production of chemicals, fertilizers, textiles, and explosives. Other applications include pumps, agitators, mixing nozzles, and valves. They are also used as impressed current anodes in cathodic protection.

Heat-resistant Iron. Heat-resistant irons are gray or ductile irons containing alloying elements that are designed to improve high-temperature strength and oxidation resistance. Alloying elements added include silicon, chromium, nickel, molybdenum, and aluminum.

Growth occurs in some cast irons, and alloying with carbide stabilizers and aluminum help to reduce it. *Growth* is a permanent increase in volume as a result of prolonged exposure to elevated temperatures or repeated cycles of heating and cooling. It has an appearance of fine cracks on the surface.

Silicon and chromium increase resistance to oxidation (scaling) by forming a tight, impervious surface film. Both elements also reduce strength and toughness at elevated temperatures. Nickel increases elevated temperature strength and toughness. Molybdenum increases high-temperature strength. Aluminum reduces growth and scaling but also reduces room temperature mechanical properties.

Heat-resistant irons consist of austenitic irons, intermediate-silicon irons, and high-chromium white irons. Austenitic irons consist of gray or ductile irons alloyed with nickel to develop an austenite matrix, plus chromium for scaling resistance and molybdenum for high-temperature strength. High-nickel austenitic irons are highly resistant to growth.

Although the room temperature strength of gray austenitic iron is equivalent to that of gray iron, the elevated temperature strength is better. Gray austenitic irons are used in steam service from 400°C to 815°C (750°F to 1500°F) for applications such as turbine diaphragms, valves, and nozzle rings.

At equivalent temperatures, ductile austenitic irons have greater strength than gray austenitic irons. Ductile austenitic irons contain 5% Si to 6% Si for improved scaling resistance and are used for manifolds and valve guides for heavy-duty engines.

Intermediate-silicon heat-resistant irons are gray and ductile irons with 4.5% Si to 8.0% Si. The gray iron version is known commercially as Silal®. The ductile iron versions are more popular because of higher strength and ductility. Nicrosilal®, a nickel-containing ductile iron, is used in the range 315°C to 540°C (600°F to 1000°F) and has excellent scaling resistance up to 815°C (1500°F). Other intermediate-silicon irons contain small amounts of molybdenum to improve high-temperature strength.

High-chromium white irons consist of white cast irons with 15% Cr to 35% Cr. The high chromium content helps form a tight, protective chromium oxide-rich film on the surface, which improves scaling resistance, and precipitates chromium carbides, which minimizes growth. They are used up to 980°C (1800°F) for their excellent scaling resistance, but they have poor toughness and should not be used where there is severe impact or shock loading. High-chromium white irons are used for tray and breaker bars in sintering furnaces, burner nozzles, glass molds, and valve seats for engines.

Welding Considerations

Cast irons are welded or braze welded to repair casting defects, fabricate components, and salvage components that malfunction while in service. *Braze welding* is a method of welding that uses a welding filler metal with a liquidus above 450°C (840°F) and below the solidus of the base metal. Cast irons are difficult to weld, chiefly because of their high carbon content and low ductility. These factors, accompanied by the high-shrinkage stresses that occur in welding, promote cracking. With the exception of white iron, all families of cast irons may be welded. However, cast irons are considerably more difficult to weld than carbon steel, and special precautions must be taken to avoid cracking. Welding is avoided whenever possible, and if mechanical repair is possible, it should be performed.

Several considerations must be reviewed when cast irons are welded or braze welded. They include percentage of carbon and other elements, joint preparation, welding filler metal composition, and preheating and post-weld heat treatment.

Carbon and Other Element Content. The high carbon content of cast irons is the chief cause of poor weldability. See Figure 16-11. The heat-affected zone of base metal adjacent to the welding filler metal is heated into the austenite region. It cools so rapidly that high-carbon martensite forms, accompanied by the formation of brittle carbides, which, coupled with high stresses, are susceptible to failure.

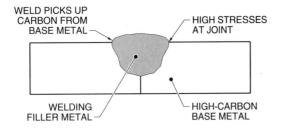

Figure 16-11. Difficulties in welding cast irons are caused chiefly by their high carbon content and low ductility.

As with steels, phosphorus in cast iron causes embrittlement during welding. For weldability, the phosphorus content must be held below .1% P. Repair welding of high-phosphorus castings requires special techniques. Other elements such as nickel, chromium, or molybdenum increase hardenbility and cracking susceptibility in the base metal.

Joint Preparation. Proper preparation of the surfaces to be joined is critical to ensure good wettability by the welding filler metal. *Wettability* is the ability of a liquid to adhere to a solid. All casting skin and foreign materials must be removed from the joint surface and adjacent areas. Castings that have been in service are often contaminated with oil and grease that can be removed with solvents or steam cleaning. Where possible, the casting is heated uniformly at approximately 370°C (700°F) for approximately 30 minutes, or for a shorter time at 540°C (1000°F) using a gas torch.

The graphite on the surface of gray iron can be oxidized by searing the surface with an oxidizing flame or heating the casting in a strongly decarburizing atmosphere. This is followed by wire brushing to remove the graphite.

Abrasive blasting with steel shot is not suitable for gray iron, but it is used on malleable or ductile iron. Other methods are also used to ensure wettability, such as molten salt bath cleaning. Before any cleaning procedure is used in production, wetting tests should be performed using the proposed welding filler metal and joining procedures. The welding filler metal should be applied to the cleaned, flat surface and then examined visually for satisfactory wetting. If the surface is not uniformly wetted, it has not been properly prepared.

Welding Filler Metal. The composition of the welding filler metal is varied to suit the requirements for the weld. The factors that influence the selection of a welding filler metal include the type of cast iron, mechanical properties desired in the joint, need for welding filler metal to deform plastically and relieve welding stresses, machinability of the joint, color matching between base and welding filler metal, allowable dilution (chemical composition change caused by melting and mixing of the welding filler metal and the base metal), and cost.

Cast irons can be welded with cast iron, steel, nickel alloy, and copper alloy electrodes. See Figure 16-12. Cast irons can be braze welded with copper alloy electrodes. Cast iron electrodes are most often used for oxyacetylene welding of gray and ductile iron. Extensive heat input is required before, during, and after welding to prevent cracking and maintain machinability.

WELDING FILLER METALS FOR CAST IRON		
Description	**Form**	**Process**[a]
Cast Iron		
Gray iron	Welding rod	OAW
Gray iron	Bare electrode	BMAW
Alloy gray iron	Welding rod	OAW
Ductile iron	Welding rod	OAW
Steel		
Carbon steel	Covered electrode	SMAW
Carbon steel	Bare electrode	GMAW
Nickel Alloys		
93% Ni	Bare electrode	GMAW
95% Ni	Covered electrode	SMAW
53 Ni-45 Fe	Covered electrode	SMAW
53 Ni-45 Fe	Flux cored electrode	FCAW
55 Ni-40 Cu-4 Fe	Covered electrode	SMAW
65 Ni-30 Cu-4 Fe	Covered electrode	SMAW
Copper Alloys		
Low-fuming brass	Welding rod	OAW
Nickel brass	Welding rod	OAW
Copper-tin	Covered electrode	SMAW
Copper-tin	Bare electrode	GMAW
Copper-aluminum	Covered electrode	SMAW
Copper-aluminum	Bare electrode	GMAW

[a] OAW = oxyacetylene welding
BMAW = bare metal arc welding
SMAW = shielded metal arc welding
GMAW = gas metal arc welding

Figure 16-12. Matching and nonmatching welding filler metals may be used to weld cast irons.

Carbon steel welding filler metal is prone to embrittlement by dilution with the high content of carbon in the cast iron, and by sulfur and phosphorus, which may also be present. This leads to a joint with low ductility and poor machinability. Carbon steel welding filler metal is used for joints that are not highly stressed, where machining is not a major concern, or where cast iron is joined to steel. The welding process is designed to minimize heat input, which keeps dilution at acceptable levels. High preheating and post-weld heat treatment temperatures are used to minimize stresses.

Certain nickel alloy compositions, such as the Ni-Rod® series, are specifically designed for welding cast irons. Deposited metal from these electrodes has a carbon content well above the solubility limit for carbon. The excess carbon is rejected as graphite during solidification, which increases the volume of the weld metal and minimizes shrinkage during solidification. As a result, this expansion reduces residual stresses. Nickel alloy welding filler metals remain ductile and are machinable after welding. Additionally, preheating may not be necessary, except in highly restrained sections.

When copper alloy electrodes are used for the braze welding of cast iron, the joint is soft and ductile when hot. The joint yields appreciably during cooling, which reduces welding stresses. The deposited metal is readily machinable, but it does not provide a color match.

Heat Treatments. Preheating and post-weld heat treatment help to minimize cracking and restore ductility. Preheating assists in softening the martensite that forms on welding in the base metal immediately adjacent to the welding filler metal. Post-weld heat treatment helps to relieve stresses resulting from the welding process.

A certain amount of preheating is essential for welding any cast iron. The preheating temperatures depend on the cast iron, its mass, welding process, and welding filler metal. See Figure 16-13. With large or complex castings, preheating must be slow and uniform in order to prevent cracking from unequal expansion.

When a casting is welded with an extremely high preheating temperature, such as 650°C (1200°F), post-weld heat treatment may not be necessary if the casting is allowed to cool slowly from the welding temperature. This is performed by covering it with an insulating blanket, vermiculite, or sand. For proper control, the preheating temperature should be monitored using contact pyrometers, temperature indicating crayons, or thermocouples.

Post-weld heat treatment is performed to stress relieve fully restrained welds, or welds that are intended for use in a severe environment. Normally, post-weld heat treatment is performed immediately after welding by increasing the temperature to 590°C to 620°C (1100°F to 1150°F), and holding the casting at temperature for about 1 hr/in. of thickness. The cooling rate should be 30°C/hr (50°F/hr) until the casting has cooled to 370°C (700°F).

Depending on the heat input during welding, malleable iron may require an additional heat treatment after welding. For this reason, brazing is sometimes preferred. During brazing, the temperature is low enough to prevent the iron from reverting to white iron in the area of the joint.

RECOMMENDED PREHEATING TEMPERATURES		
Cast Iron	**Temperature Range (°C)**	
	Arc Welding	**Oxyacetylene Welding**
Gray	20 to 320	430 to 650
Malleable (ferrite matrix)	20 to 150	430 to 650
Malleable (pearlite matrix)	20 to 320	430 to 650
Ductile (ferrite matrix)	20 to 150	200 to 650
Ductile (pearlite matrix)	20 to 320	200 to 650

Figure 16-13. The preheat temperature selected for welding cast irons depends on the restraint in the joint, the welding process, the type of welding filler metal, the cast iron microstructure, and whether or not post-weld heat treatment is to be performed.

Stainless Steels 17

Stainless steels are alloys of iron and carbon that contain between 12% Cr and 30% Cr plus other alloying elements. The chromium, which helps develop a passive surface oxide film, provides corrosion resistance in stainless steels. Stainless steels are wrought or cast and consist of several families that are named for their metallurgical structure. Stainless steels are fabricated and welded by all common processes and are used for their corrosion resistance, high-temperature strength, scaling resistance, and low-temperature toughness. These properties account for their extremely wide use in practically every industry.

IDENTIFICATION OF STAINLESS STEELS

Many designations are used to specify and separate the various stainless steels. Stainless steels are affected by the different alloying elements, which affect the predominant metallurgical structures that stainless steels exhibit. There are five families of wrought stainless steels, which include martensitic, ferritic, austenitic, precipitation-hardening, and duplex stainless steels. Cast stainless steels exhibit various types of metallurgical structures and are classified as a sixth family. See Figure 17-1.

The six families of stainless steels are the most widely used of all corrosion-resistant alloys. These steels are applied in architecture, transportation systems, furniture, power generation, laundry equipment, food processing and kitchen equipment, agriculture, textiles, hospital and surgical equipment, chemical and petrochemical plants, and aerospace equipment.

SELECTED APPLICATIONS FOR STAINLESS STEELS	
Martensitic	Valves, nuts and bolts, screws, cutlery, scissors, and turbine buckets and valves
Ferritic	Heat exchanger tubes, toasters, refrigerator trays, and automotive trim
Austenitic	Fuel lines, refrigerator cars, shovels, milk cans, chemical processing equipment, fasteners, architectural materials, food and beverage processing and handling equipment, and household items
Precipitation hardening	Aerospace components, high-strength fasteners, and valves
Duplex	Heat exchanger tubing
Cast	Petrochemical processing equipment

Figure 17-1. There are five families of wrought stainless steels, which include martensitic, ferritic, austenitic, precipitation-hardening, and duplex stainless steels. Cast stainless steels exhibit various types of metallurgical structures and are classified as a sixth family.

Stainless Steel Designation Systems

Several organizations produce designations for stainless steels. See Figure 17-2. Wrought stainless steels are American Iron and Steel Institute designated by three numbers sometimes followed by a letter, such as 304, 304L, and 410 stainless steels. The AISI numbering system has no rationale.

SELECTED DESIGNATIONS FOR STAINLESS STEELS				
	AISI	**UNS**	**SAE**	**ACI**
Martensitic	403	S40300	51403	—
	420	S42000	51420	—
Ferritic	405	S40500	51405	—
	442	S44200	51442	—
Austenitic	201	S20100	30201	—
	304	S30400	30304	—
Precipitation hardening	630	S17400	—	—
	600	S66286	—	—
Duplex	329	S32900	—	—
	—	S31803	—	—
Cast	—	J94224	—	HK
	—	J92600	—	CF-8

Figure 17-2. Several organizations produce designations for stainless steels.

The Society of Automotive Engineers designation system consists of five numbers of which the last three correspond to the AISI designation. For example, SAE 30304 corresponds to AISI 304. The Alloy Castings Institute designation system identifies nonproprietary cast stainless steels. It consists of two uppercase letters, a dash, and a series of numbers or letters. The first letter identifies the major group (H for heat resistant and C for corrosion resistant). The second letter provides clues to the alloying elements. The letters and numbers after the dash provide clues to the percentages of carbon and key alloying elements. For example, HK-40 is a heat-resistant stainless steel with 26% Cr, 20% Ni, and a nominal carbon content of .4% C. CF-8 is a corrosion-resistant stainless steel with 19% Cr, 9% Ni, and a maximum carbon content of .08% C.

The UNS combines all standard and many proprietary cast and wrought stainless steels into a common numbering system. The UNS consists of a letter followed by five numbers. Wrought stainless steels are identified by the uppercase letter S, and cast alloys are identified by the uppercase letter J. Ex-

amples of this designation system are UNS S30400 (AISI 304) and UNS J92600 (ACI CF-8).

Stainless steel welding filler metals are designated by the American Welding Society in standards AWS A 5.2 for covered electrodes and AWS A 5.9 for bare wire electrodes. Examples from these standards are E308 and ER308, respectively. The E symbolizes an electrode, and the ER symbolizes an electrode, rod, or wire.

Many organizations develop product standards for stainless steels using the designation systems. Product standards are used for specifications and purchasing, and include requirements for chemical compositions, manufacturing procedures, heat treatments, testing procedures, and mechanical properties. The ASTM is the major product standard writing organization. ASTM standards for stainless steels are identified by the uppercase letter A followed by a number. For example, ASTM A 240 is a specification for high-quality plate, sheet, and strip that is made to specific grades of stainless steels. Stainless steel product standards are also contained in the ASME specifications.

Alloying and Metallurgical Structure

The metallurgical structures of stainless steels are influenced by alloying elements. The changes in structures account for the widely differing properties and methods of heat treatment used for stainless steels. This leads to an enormous number of applications for stainless steels in a variety of industries.

Chromium is the principal element influencing the corrosion resistance of stainless steels. Chromium is a ferritizer (promotes ferrite) and influences the metallurgical structure. Nickel, carbon, and nitrogen are powerful austenitizers. Other elements, such as niobium, titanium, or molybdenum, are added to improve mechanical properties or corrosion resistance and are also classified as ferritizers, austenitizers, or carbide formers.

Effects of Chromium. Chromium alters the metallurgical structure in a unique way. When 12% Cr or more is added to iron or steel, they become stainless (immune to rusting). Chromium has a form of corrosion resistance known as passivity and is the key ingredient in the development of a passive surface film. A *passive surface film* is film on the outer surface layer of a metal that has superior corrosion resistance. The passive surface film primarily contains chromium and oxygen, is tenacious, and forms instantaneously on the surface.

If the passive surface film is broken or stripped, corrosion resistance is impaired. Additions of extra chromium and alloying elements, such as molybdenum, help strengthen the passive surface film. These additions may be necessary in situations where the absence of oxygen favors the breakdown of the passive surface film, leading to serious corrosion. Although increasing amounts of chromium improve the quality of the passive surface film, there are limits to chromium addition. Alloys containing >30% Cr are difficult to fabricate and are not of any practical importance.

When chromium is added to steel, the austenite phase field contracts. With an addition of >12% Cr, it becomes an island, forming a gamma loop on the phase diagram. The gamma loop is the isolated austenite phase field. The entire phase field to the right of the gamma loop consists of delta ferrite. See Figure 17-3. Stainless steels with compositions inside the gamma loop form martensite (martensitic stainless steels) when cooled down, and stainless steels with compositions to the right of the gamma loop form ferrite (ferritic stainless steels).

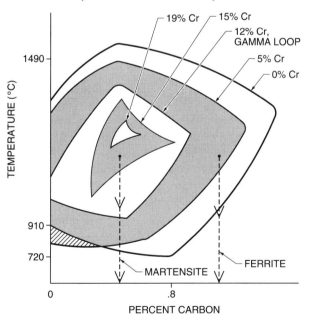

Figure 17-3. With >12% Cr addition, the austenite phase field becomes an island, forming a gamma loop on the phase diagram.

Effects of Nickel. Nickel is the second most important alloying element of stainless steels. It counteracts the metallurgical effect of chromium by promoting austenite formation. Nickel enlarges the gamma loop to the extent that austenite is stable at room temperature and below. An alloy with 18% Cr and 8% Ni is austenitic at room temperature and is the base composition for austenitic stainless steels. The base composition contains 18% Cr because greater or lesser amounts of chromium require more nickel to ensure a completely austenitic structure at room temperature. Since nickel is more expensive than chromium, this is not a cost effective approach of ensuring the complete austenitic structure. See Figure 17-4.

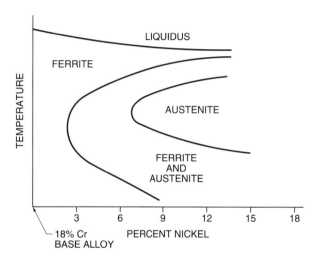

Figure 17-4. The base composition contains 18% Cr because greater or lesser amounts of chromium require more nickel to ensure a completely austenitic structure at room temperature.

Although the phase diagram for the 18% Cr, 8% Ni alloy indicates that ferrite should be present at room temperature, in practice there are two reasons that it is not. First, with nickel present, the transformation of austenite to ferrite or martensite is sluggish and does not occur at normal rates of cooling. Second, the phase diagram does not exhibit the effects of carbon and nitrogen. These are austenitizers and help enlarge the gamma loop.

Effects of Other Elements. Other elements in stainless steels are classified as ferritizers or austenitizers, depending on whether they act like chromium, which stabilizes ferrite, or like nickel, which stabilizes austenite. Some alloying elements are carbide formers and preferentially combine with carbon. Ferritizers, austenitizers, and carbide formers are added to stainless steels for their effects on properties. See Figure 17-5. Other elements are also added to stainless steels to improve specific properties. For example, sulfur, selenium, and phosphorus improve machinability.

To predict the structure when all the alloying elements are factored in, each alloying element is given a weighting number (n_W). The weighting number of each ferritizer is multiplied by its percent content. The chromium equivalent (CrE) is the total of all the sums (weighting numbers times percent content) of the ferritizers. The weighting number of each austenitizer is multiplied by its percent content. The nickel equivalent (NiE) is the total of all the sums (weighting numbers times percent content) of the austenitizers. To predict the metallurgical structure from the composition, the CrE is plotted against the NiE on the Schaeffler diagram. See Figure 17-6. The *Schaeffler diagram* is a geographical representation of the phases present by plotting nickel equivalent against chromium equivalent. The CrE and the NiE are found by applying the following formulas:

$$CrE = (Cr_W \times \%Cr) + (Si_W \times \%Si) +$$
$$(Mo_W \times \%Mo) + (V_W \times \%V) +$$
$$(Al_W \times \%Al) + (Nb_W \times \%Nb) +$$
$$(Ti_W \times \%Ti) + (W_W \times \%W)$$

where

CrE = chromium equivalent

Cr_W = chromium weighting number

$\%Cr$ = percent chromium

Si_W = silicon weighting number

$\%Si$ = percent silicon

Mo_W = molybdenum weighting number

$\%Mo$ = percent molybdenum

V_W = vanadium weighting number

$\%V$ = percent vanadium

Al_W = aluminum weighting number

$\%Al$ = percent aluminum

Nb_W = niobium weighting number

$\%Nb$ = percent niobium

Ti_W = titanium weighting number

$\%Ti$ = percent titanium

W_W = tungsten weighting number

$\%W$ = percent tungsten

and

$$NiE = (Ni_W \times \%Ni) + (C_W \times \%C) +$$
$$(Mn_W \times \%Mn) + (Co_W \times \%Co) +$$
$$(Cu_W \times \%Cu) + (N_W \times \%N)$$

where

NiE = nickel equivalent

EFFECTS OF ALLOYING ELEMENTS ON STAINLESS STEELS		
Element	**Maximum Quantity Added (%)**	**Properties Improved**
Molybdenum	6.50	Corrosion resistance and high-temperature strength
Manganese	16.00	Maintains austenitic structure and is substituted for nickel for cost
Copper	4.50	Corrosion resistance in some stainless steels and strength in others
Aluminum	1.50	Strength in some stainless steels
Titanium	1.00	Corrosion resistance in some stainless steels
Niobium plus Tantalum	.50	Corrosion resistance in some stainless steels
Sulfur or Selenium	.25	Machinability, but at loss of weldability
Silicon	3.00	High temperature scaling resistance when added above normally specified amount of 1%

Figure 17-5. Ferritizers, austenitizers, and carbide formers are added to stainless steels for their effects on properties.

Ni_W = nickel weighting number

$\%Ni$ = percent nickel

C_W = carbon weighting number

$\%C$ = percent carbon

Mn_W = manganese weighting number

$\%Mn$ = percent manganese

Co_W = cobalt weighting number

$\%Co$ = percent cobalt

Cu_W = copper weighting number

$\%Cu$ = percent copper

N_W = nitrogen weighting number

$\%N$ = percent nitrogen

Example: Figuring the *CrE* and the *NiE* (predicting metallurgical structure)

Predict the metallurgical structure of type 302 stainless steel when given:

$Cr_W = 1$	$\%Cr = 17.5$
$Si_W = 2$	$\%Si = .8$
$Mo_W = 1.5$	$\%Mo = 0$
$V_W = 5$	$\%V = 0$
$Al_W = 5.5$	$\%Al = 0$
$Nb_W = 1.75$	$\%Nb = 0$
$Ti_W = 1.5$	$\%Ti = 0$
$W_W = .75$	$\%W = 0$
$Ni_W = 1$	$\%Ni = 9.2$
$C_W = 30$	$\%C = .12$
$Mn_W = .5$	$\%Mn = 1.5$
$Co_W = 1$	$\%Co = 0$
$Cu_W = .3$	$\%Cu = 0$
$N_W = 25$	$\%N = 0$

Figuring *CrE*

$CrE = (Cr_W \times \%Cr) + (Si_W \times \%Si) +$
$(Mo_W \times \%Mo) + (V_W \times \%V) +$
$(Al_W \times \%Al) + (Nb_W \times \%Nb) +$
$(Ti_W \times \%Ti) + (W_W \times \%W)$

$CrE = (1 \times 17.5) + (2 \times .8) + (1.5 \times 0) +$
$(5 \times 0) + (5.5 \times 0) + (1.75 \times 0) +$
$(1.5 \times 0) + (.75 \times 0)$

$CrE = 17.5 + 1.6 + 0 + 0 + 0 + 0 + 0 + 0$

$CrE = $ **19.1**

Figuring *NiE*

$NiE = (Ni_W \times \%Ni) + (C_W \times \%C) +$
$(Mn_W \times \%Mn) + (Co_W \times \%Co) +$
$(Cu_W \times \%Cu) + (N_W \times \%N)$

$NiE = (1 \times 9.2) + (30 \times .12) + (.5 \times 1.5) +$
$(1 \times 0) + (.3 \times 0) + (25 \times 0)$

$NiE = 9.2 + 3.6 + .75 + 0 + 0 + 0$

$NiE = $ **13.55**

Note: The metallurgical structure of the type 302 stainless steel is completely austenitic.

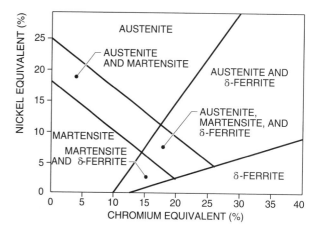

Figure 17-6. To predict the metallurgical structure from the composition, the *CrE* is plotted against the *NiE* on the Schaeffler diagram.

The principal reason for carbide formers is to combine preferentially with carbon. This has two important effects, depending on the type of alloy. In some alloys, it prevents chromium from combining with carbon, which is detrimental to corrosion resistance. In other alloys, carbides provide a high-temperature strengthening effect.

Martensitic Stainless Steels

Martensitic stainless steels contain the least amount of chromium and, therefore, are the least corrosion resistant. They are hardened by quenching and tempering to develop high levels of strength and abrasion resistance. Martensitic stainless steels are strongly ferromagnetic.

Metallurgical Structure. The metallurgical structure of martensitic stainless steels is influenced principally by the chromium and carbon content. The three groups of martensitic stainless steels are low-carbon, nickel-bearing low-carbon, and high-carbon. The basic martensitic stainless steel is type 410 (S41000), and it is closely related to other martensitic stainless steel family members. See Figure 17-7.

Low-carbon martensitic stainless steels contain <.15% C and 12% Cr to 14% Cr. Nickel-bearing low-carbon types contain some nickel to help improve toughness, plus carbide forming elements, such as vanadium. High-carbon martensitic stainless steels contain >.15% C, which expands the gamma loop to allow chromium additions up to 18% Cr. This provides increased strength and hardness. See Figure 17-8.

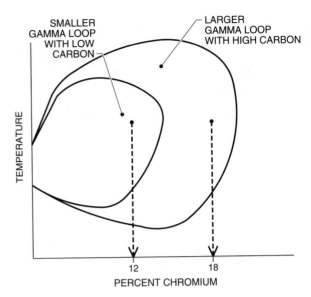

Figure 17-8. High-carbon martensitic stainless steels contain >.15% C, which expands the gamma loop to allow chromium additions up to 18%.

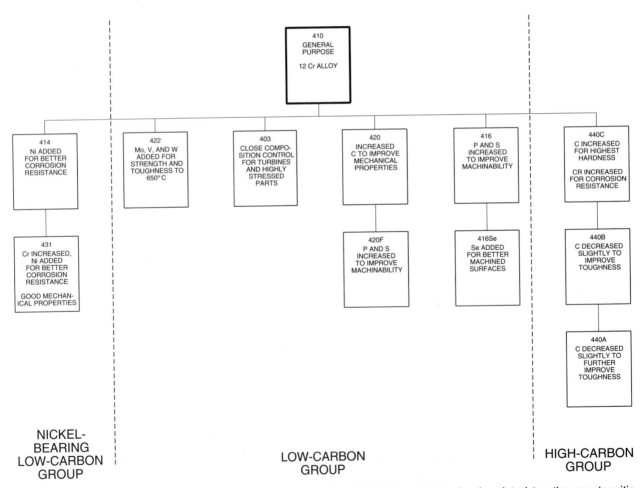

Figure 17-7. The basic martensitic stainless steel is type 410 (S41000), and it is closely related to other martensitic stainless steel family members.

Mechanical Properties and Heat Treatment. The martensitic stainless steels are softened by annealing and strengthened by austenitizing, quenching, and tempering. Two methods of annealing are full annealing and process annealing. Full annealing is annealing performed above the upper critical temperature, followed by very slow cooling. Full annealing develops a structure of carbide particles in a ferrite matrix.

Process annealing consists of heating below the lower critical temperature, followed by air, oil, or water quenching. Process annealing develops a martensitic structure with enough ductility and machinability so that some fabrication processes may be performed. To prevent cracking, the rate of cooling after any annealing or hot-working process must be extremely slow, between 15°C/hr and 25°C/hr (30°F/hr and 40°F/hr). More rapid cooling, such as air cooling, should be performed below 540°C (1000°F) to avoid temper embrittlement.

Austenitizing, quenching, and tempering improve strength and toughness. When quenched and tempered, the low-carbon types achieve a maximum hardness of approximately 45 HRC and the high-carbon types achieve a maximum hardness of approximately 60 HRC. Retained austenite can be a problem, but it is reduced by minimizing the austenitizing temperature and double tempering. Most martensitic stainless steels are air hardening, so they can be air quenched. The low-carbon types may be oil quenched.

The martensitic stainless steels have strong resistance to softening during tempering. They are tempered at higher temperatures than low-alloy steels to restore their ductility and toughness. Temperatures range from 595°C to 760°C (1100°F to 1400°F). Martensite stainless steels must not be tempered or slow cooled through the range 440°C to 540°C (825°F to 1000°F) to avoid temper embrittlement. Furnace or blanket cooling must not be performed through this temperature range, but air cooling is rapid enough to avoid the problem of temper embrittlement.

Tempering induces carbide precipitation, which further increases strength, hardness, and abrasion resistance. The higher the carbon content, the greater this secondary hardening effect. Thus, the high-carbon type 440C (S44004) is significantly harder and stronger than any other martensitic stainless steel. Carbide forming elements are added to some martensitic stainless steels to form alloy carbides, which also enhances the secondary hardening effect. Examples of these types of martensitic stainless steels are types 442 (no UNS) and 414 (S41400).

The nickel-bearing low-carbon martensitics, such as types 414 and 431 (S43100), have a lower austenite temperature boundary than the others. Consequently, they cannot be tempered at too high of a temperature. This results in low ductility, despite the high strength.

Ferritic Stainless Steels

Ferritic stainless steels contain more chromium than martensitic stainless steels. Ferritic stainless steels are relatively weak and cannot be hardened by quenching. They are used chiefly for their corrosion and scaling resistance and are strongly ferromagnetic. Ferritic stainless steels are divided into regular ferritics and low-interstitial ferritics. The base composition of ferritic stainless steels is type 430 (S43000), which is closely related to other members of the ferritic stainless steel family. See Figure 17-9.

Metallurgical Structure. The relatively high chromium content of ferritic stainless steels places them to the right of the gamma loop. Theoretically, their metallurgical structure is ferrite from the melting point to room temperature. However, in many ferritic stainless steels, the combination of alloying elements causes expansion of the austenite-ferrite phase field immediately adjacent to the gamma loop. Consequently, some austenite is present at high temperatures. This austenite transforms to martensite when cooled. The small amount of martensite is not usually a problem, although annealing is sometimes necessary to soften it.

The low-interstitial ferritic stainless steels contain very low amounts of carbon, nitrogen, and residual elements. They are produced by clean steelmaking processes, such as argon-oxygen decarburizing and vacuum degassing. With these processes, carbon and nitrogen are maintained below .015% (compared with .12% in type 430). This results in a noticeable improvement in toughness.

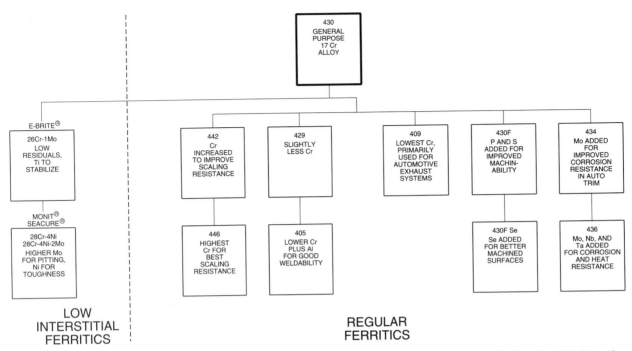

Figure 17-9. The base composition of the ferritic stainless steels is type 430 (S43000) and is closely related to other members of the ferritic stainless steel family.

Carbide forming elements, such as titanium, are also added to maintain corrosion resistance in the as-welded condition. Molybdenum is added to some alloys to improve corrosion resistance. The low-interstitial ferritic stainless steels are limited to sections of <.5 in. The size is limited because the toughness in thicker sections is unpredictable.

E Brite® (S44627) is the original low-interstitial ferritic stainless steel containing 26% Cr and 1% Mo. Higher molybdenum-containing alloys are used in seawater applications and contain a small amount of nickel to improve toughness. Examples of these are Monit® (S44635) and SeaCure® (S44660).

Ferritic stainless steels have lower carbon contents than the martensitic stainless steels and generally have higher chromium contents. They are weak, only slightly stronger than carbon steel, but have better corrosion resistance and better high-temperature scaling resistance than the martensitic stainless steels.

Mechanical Properties and Heat Treatment. Ferritic stainless steels cannot be strengthened by quenching and tempering and are not appreciably hardened by cold work. These features, coupled with low as-welded toughness, limit the use of ferritic stainless steels in structural applications.

Annealing is performed to relieve stresses that develop from welding or cold working. It also improves corrosion resistance by homogenizing the microstructure. Annealing causes martensite to transform to ferrite and causes carbides to spheroidize. Grain growth becomes excessive at approximately 930°C (1700°F). Depending on the alloy content, annealing is performed between 760°C and 815°C (1400°F and 1500°F).

Ferritic stainless steels are susceptible to embrittlement when exposed to high temperatures. This further limits structural applications of ferritic stainless steels. The 475°C (885°F) embrittlement and the sigma phase embrittlement are the two forms of high-temperature embrittlement that occur.

The 475°C (885°F) embrittlement occurs during slow heating or cooling through the range 400°C to 525°C (750°F to 975°F) and takes place most rapidly at 475°C (885°F). Ductility is restored by annealing above 595°C (1100°F), followed by fast cooling through the susceptible temperature range. The 475°C (885°F) embrittlement is attributed to the transformation of ferrite (α) into the brittle alpha prime (α′) phase.

Sigma phase embrittlement occurs during prolonged heating in the range 540°C to 870°C (1000°F to 1600°F). See Figure 17-10. Sigma phase embrittlement is not usually a problem during heat treatment operations, when the holding time at temperature is relatively short. Sigma phase is a brittle iron-chromium compound that has a formation that is favored by high-chromium content and ferritizers. Type 446 (S44600), with the highest chromium content, is the most susceptible alloy.

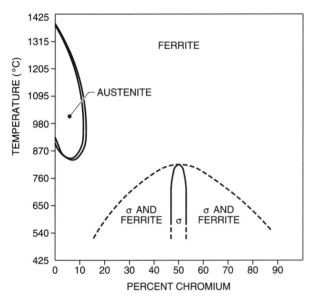

Figure 17-10. Sigma phase embrittlement occurs during prolonged heating in the range 540°C to 760°C (1000°F to 1400°F).

Austenitic Stainless Steels

Austenitic stainless steels are the largest and most widely used family of stainless steels. They have excellent corrosion resistance, weldability, high-temperature strength, and low-temperature toughness. Austenitic stainless steels are generally nonmagnetic. They are only hardened by cold work. Heat treatment operations are used to relieve stresses, restore softness, or improve corrosion resistance. Sensitization is a problem that occurs during certain heat treatment and welding operations.

Metallurgical Structure. The FCC structure of austenitic stainless steels accounts for their excellent high-temperature strength and low-temperature toughness. The structure of austenitic stainless steels

is based on the composition 18% Cr and 8% Ni. The austenitic structure is achieved by the presence of nickel. Other strong austenitizers that are present include nitrogen and manganese. Carbon, another strong austenitizer, is not used in large amounts because it reduces corrosion resistance. Carbon is deliberately reduced in some grades to improve corrosion resistance in the as-welded condition. Carbide stabilizers, such as niobium, tantalum, and titanium, are added for the same reason. Chromium and nickel are increased and molybdenum is added to improve corrosion resistance.

Standard grades of austenitic stainless steels are designated in the AISI 200 and 300 series. The basic austenitic stainless steel alloy is type 302 (S30200), and the majority of the grades belong to the 300 series. See Figure 17-11. Type 304 (S30400) is a lower carbon version of type 302. It is the most commonly used stainless steel. Types 316 (S31600) and 317 (S31700) contain increasing amounts of molybdenum for increased corrosion resistance. In these alloys, nickel is increased over the base alloy composition to counteract the ferritizing effect of molybdenum. Alloy 254SMO (S31254), a 20Cr-18Ni-6Mo low-carbon grade, has outstanding resistance to chloride pitting and stress cracking.

Types 304L (S30403), 316L (S31603), and 317L (S31703) are lower carbon versions of types 304, 316, and 317. These steels are designed to maintain their corrosion resistance in the as-welded condition. Certain types, such as 304/304L or 316/316L, are dual marked (have properties of two types of stainless steels). These alloys have an extra-low carbon content to meet the as-welded corrosion resistance of 304L or 316L, and they also contain additions of nitrogen to meet the slightly better strength rating of 304 or 316. Dual marking reduces suppliers' inventories because these steels can be used for a variety of products. Types 304N (S30451) and 316N (S31651) are regular carbon versions with additions of nitrogen to increase strength.

Sulfur or selenium (Se), which improves machinability, is added to the basic alloy to produce types 303 (S30300) and 303Se (S30323), respectively. These alloys are available in bar or rod and are used in mass production operations on screw machines. The disadvantage of these types of stainless steels are their low levels of corrosion resistance.

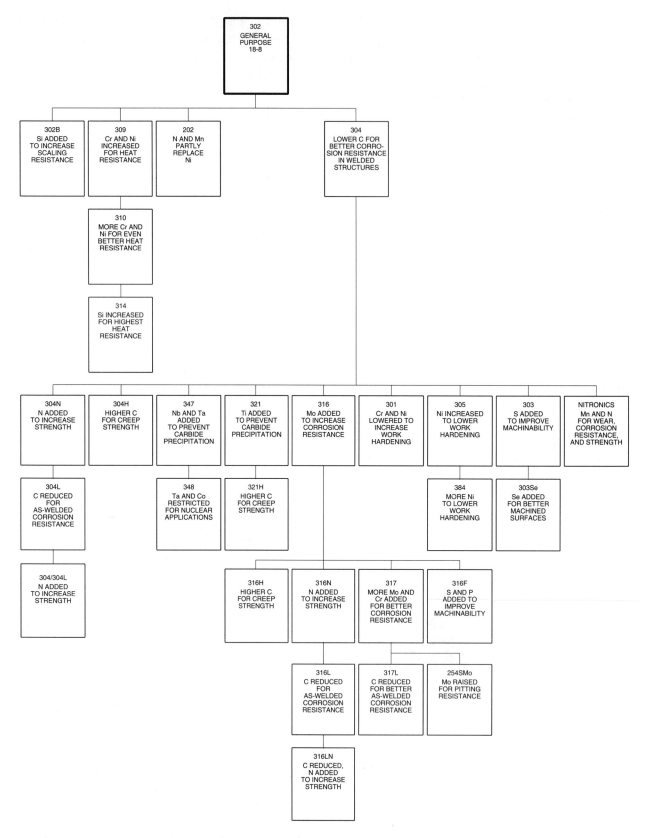

Figure 17-11. The basic austenitic stainless steel alloy is type 302 (S30200), and the majority of the grades belong to the 300 series.

The 200 series consists of a limited number of alloys in which manganese is substituted for half the nickel content. Manganese-substituted austenitics, such as types 201 (S20100) and 202 (S20200), were developed during periods such as a wartime nickel shortage. These alloys are similar to those in the 300 series.

In addition to the standard types, many proprietary austenitic stainless steels have been developed. These are developed for higher strength and corrosion resistance and generally contain increased amounts of nickel, molybdenum, or nitrogen. Another group contains large amounts of manganese substituted for nickel, in addition to large amounts nitrogen for strength. This group, which includes the Nitronic series, has substantially increased strength over the regular types in the 300 series. Many high-alloy stainless steels are nickel-base alloys because the major alloying element is nickel and not iron.

Mechanical Properties and Heat Treatment. The austenitic stainless steels cannot be strengthened by heat treatment. They are strengthened by cold work, known as work hardening, or strain hardening. Annealing after cold work leads to softening. Work-hardened alloys should not be welded if full strength is required in the heat-affected zone of the weld.

The low-nickel austenitic stainless steels that include types 301, 302, 201, and 202 form martensite when cold worked. This increases the strength and makes the steel slightly magnetic. The nickel-containing martensite that forms is not the brittle variety found in iron-chromium alloys and does not require tempering. Type 301 is the most work hardenable of these grades and is used extensively in sheet form for structural applications. See Figure 17-12.

Type 305 (S30500) is a higher nickel-containing grade that acts to stabilize austenite and prevent martensite formation. This results in a low rate of work hardening that is needed in severe forming operations. Type 384 contains more nickel to further reduce the rate of work hardening.

Austenitic stainless steels have high levels of toughness at extremely low temperatures. This makes austenitic stainless steels one of the most important groups of alloys used for cryogenic ap-

plications (applications at temperatures below the boiling point of liquified gases, –100°C or –150°F).

Solution annealing is an annealing process that is performed between 1040°C and 1175°C (1900°F and 2150°F) and is followed by rapid quenching in water. The rapid quenching prevents carbides from precipitating, which results in reduced corrosion resistance. Solution annealing is the most common heat treatment.

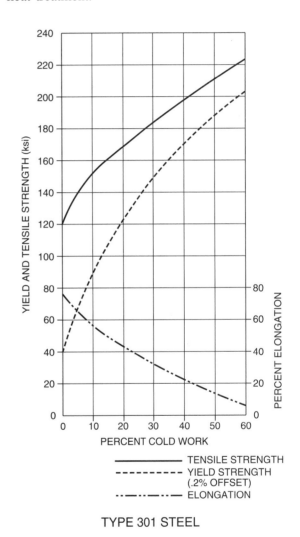

TYPE 301 STEEL

Figure 17-12. Type 301 is the most work hardenable of the low-nickel austenitic stainless steels and is used extensively in sheet form for structural applications.

Stress relieving of austenitic stainless steels must be performed at higher temperatures than for most steels because of their greater high-temperature strength. The optimum stress-relieving temperature may lead to carbide precipitation, a loss of corrosion resistance, and dimensional instability. The stress relieving procedure must be carefully selected to achieve the required result without compromising other properties. See Figure 17-13.

Sigma phase embrittlement is not usually a problem with austenitics stainless steels, except in alloys that contain small amounts of ferrite or large percentages of ferritizers. Examples include certain welding filler metals, and certain alloys with high chromium or high molybdenum contents. The ferrite transforms to sigma when held in the temperature range 540°C to 760°C (1000°F to 1400°F), leading to embrittlement on cooling to room temperature. Susceptible welding filler metals and alloys should not be held in the sigma phase temperature range for an extended time. Resistant materials, such as austenitic nickel-base welding filler metals, must be substituted for the susceptible welding filler metals and alloys.

SUGGESTED STRESS RELIEVING TEMPERATURES [a, b]			
(In Order of Decreasing Preference)			
Fabrication or Service Conditions	**Types 304L and 316L**	**Types 321 and 347**	**Types 304 and 316**
Stress relief between forming operations	A, B, C	B, A, C	C[c]
Stress relief after severe forming	A, C	A, C	C
Prevention of cracking from severe forming and high-service loading	A, C, B	A, C, B	C
Dimensional stability in machining or service	D	D	D

[a] A = hold at 1065°C to 1120°C (1950°F to 2050°F), slow cool
 B = hold at 900°C (1650°F), slow cool
 C = hold at 1065°C to 1120°C (1950°F to 2050°F), quench or cool rapidly
 D = hold at 205°C to 480°C (400°F to 900°F), for 4 hours per inch of section, slow cool
[b] to allow optimum use of stress relieving, stabilized or low-carbon grades should be used
[c] treatments A or B may also be used, if followed by treatment C when forming is completed

Figure 17-13. The stress relieving procedure must be carefully selected to achieve the required result without compromising other properties.

Sensitization. *Sensitization* is the precipitation of chromium carbide in austenitic stainless steels. It is a problem in other alloy systems, such as ferritic stainless steels, cast stainless steels, and nickel-base alloys, that contain chromium. The optimum temperature range for sensitization varies for these different alloys.

In austenitic stainless steels, sensitization occurs between 425°C and 815°C (800°F and 1500°F) and most rapidly at 650°C (1200°F). Sensitization most commonly occurs during operations such as welding or stress relieving. Although chromium carbide precipitation usually does not impair the mechanical properties of austenitic stainless steels, it can severely reduce corrosion resistance.

When heated into the sensitization temperature range, carbon and chromium combine to form discrete precipitates of chromium carbide. Precipitation is favored at the grain boundaries and leads to a localized zone that is depleted in chromium. This zone is extremely susceptible to corrosion (intergranular corrosion) in specific environments. See Figure 17-14. With severely sensitized microstructures, entire grains may drop out, leading to extremely high corrosion rates.

The various methods used to reduce sensitization include lowering the carbon content, specifying stabilized alloys, and employing solution-annealed material. Lowering the carbon content may minimize carbide precipitation during short-term heating operations, such as welding or stress relieving. Low-carbon austenitic grades are designated by the uppercase letter L. For example, types 304L and 316L are low-carbon grades. In these grades, the carbon content is reduced from the regular .08% C maximum to .03% C maximum.

Stabilized grades contain alloying elements that preferentially combine with carbon to form alloy carbides. This limits the chromium that combines with the carbon to form chromium carbide. Examples of standard stabilized grades include type 321 (S32100), which is stabilized with titanium, and type 347 (S34700), which is stabilized with niobium and tantalum. Type 321 is stabilized to optimize its corrosion resistance. Stabilizing of type 321 is performed by heat treating at 620°C to 900°C (1150°F to 1650°F) for 2 to 5 hours and air cooling to tie up the carbon as a titanium carbide precipitate.

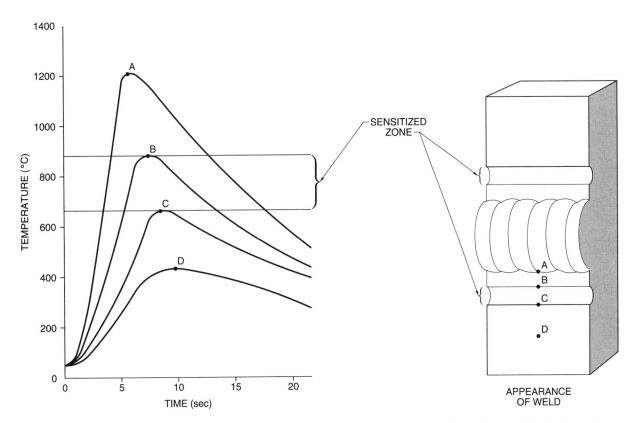

Figure 17-14. When heated into the sensitization temperature range, carbon and chromium combine to form discrete precipitates of chromium carbide.

Performance at High Temperatures. Austenitic stainless steels have excellent high-temperature strength and scaling resistance. Austenitic stainless steels are the strongest of all stainless steels used in services above 540°C (1000°F). Depending on the operating stress, they are used from 760°C to 870°C (1400°F to 1600°F). See Figure 17-15. Grades designated by the uppercase letter H have better and more predictable high-temperature strength. This is achieved by maintaining the carbon content on the high side of the specification range and solution annealing to increase the grain size. Examples of these grades include types 304H, 316H, and 321H. Low-carbon grades, such as types 304L and 316L, have inferior high-temperature strength and should not be considered for structural applications above 540°C (1000°F).

The scaling resistance of austenitic stainless steels is a result of the chromium content. Types 310 (S31000) and 309 (S30900) have higher chromium contents than the basic composition. The higher the

chromium content, the better the scaling resistance. The high nickel content of 310 also contributes excellent high-temperature strength. Silicon is sometimes added to some alloys, such as types 314 (S31400) and 302B (S30215), to improve scaling and carburizing resistance.

The coefficient of thermal expansion of austenitic stainless steels is higher than that of other steels and nickel-base alloys. This creates problems when austenitic stainless steels are used at high temperatures or when heat treated. Dimensional stability is poor during annealing or stress relieving. Allowances to prevent excessive stresses must be made for expansion and contraction when austenitics are joined to ferritic steels.

The high coefficient of thermal expansion and low thermal conductivity create problems in welding austenitic stainless steels. Heat dissipates from the joint at a relatively slowly rate, increasing the possibility of distortion.

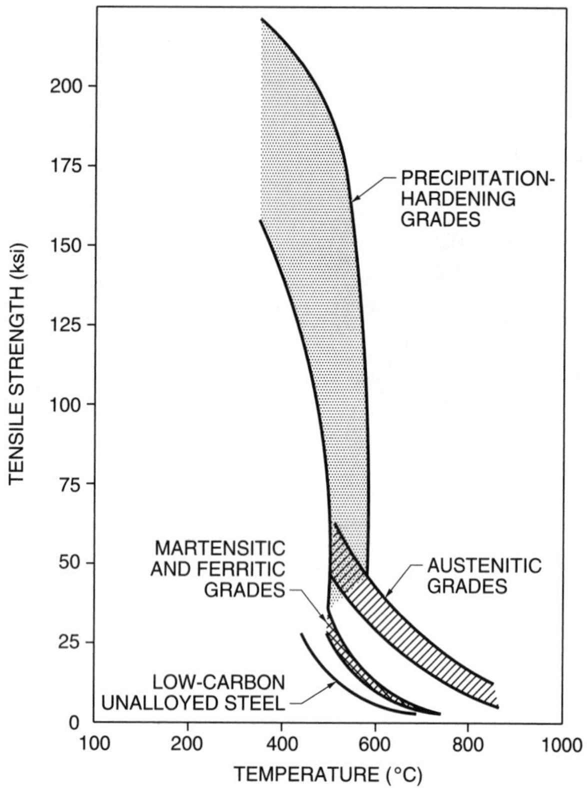

Figure 17-15. Austenitic stainless steels are the strongest of all stainless steels used in services above 540°C (1000°F). Depending on the operating stress, they are used from 760°C to 870°C (1400°F to 1600°F).

quenched from the solution-annealing temperature. After machining, working, or stamping to the desired shape, the steels are precipitation hardened at relatively low temperatures to achieve the desired strength and hardness, with very little distortion or scaling. Most precipitation-hardening stainless steels are designated by the AISI 600 series, but they are better know by trade designations. See Figure 17-16. The precipitation-hardening stainless steels consist of the martensitic, semi-austenitic, and austenitic groups.

Metallurgical Structure. Most precipitation-hardening stainless steels contain at least 4% Ni and small amounts of aluminum or titanium. The precipitates that lead to strengthening are usually nickel-aluminum or nickel-titanium compounds. Copper, niobium, and molybdenum may also be added to form precipitates.

The mechanical properties are varied by changing the aging temperature. Lower aging temperatures result in increased strength but reduced toughness. Higher aging temperatures result in lowered strength and improved toughness.

Many of these stainless steels are used in critical high-strength applications where resistance to fracture propagation is of extreme importance. They are produced by secondary melting techniques, such as electroslag remelting or vacuum arc remelting, to develop a cleaner microstructure that has greater resistance to brittle fracture.

Precipitation-hardening Stainless Steels

Precipitation-hardening stainless steels can be heat treated to higher strengths than any other stainless steels. They are relatively weak and soft when

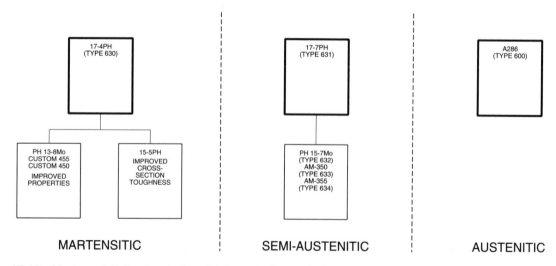

Figure 17-16. Most precipitation-hardening stainless steels are designated by the AISI 600 series, but they are better known by trade designations.

Martensitic Precipitation-hardening Stainless Steels. Martensitic precipitation-hardening stainless steels are the most widely used group. These steels are low-carbon martensitic stainless steels with 12% Cr to 15% Cr and .05% C. They are solution treated at 1040°C (1900°F) and quenched to form a soft, low-carbon martensite. After fabrication, they are aged between 480°C and 620°C (900°F and 1150°F). One method of designating the aged condition of some martensitic precipitation-hardening stainless steels is the uppercase letter H followed by the aging temperature in degrees Fahrenheit. For example, H1100 is a martensitic precipitation-hardening stainless steel with an aging temperature of 1100°F.

The basic alloy is type 630 (S17400), or 17-4PH, and the maximum strength is obtained in the H900 condition. Overaging at higher temperatures leads to a reduction in strength, which is offset by an improvement in ductility and toughness. With over-aging at increasingly higher temperatures, the precipitates become more and more visible in the microstructure. A development of 17-4PH is 15-5PH (S15500). This alloy is less prone to delta ferrite formation, which is undesirable because it reduces transverse toughness. The corrosion resistance of the martensitic precipitation-hardening stainless steels is slightly better than that of the regular martensitic stainless steels.

Semi-austenitic Precipitation-hardening Stainless Steels. Semi-austenitic precipitation-hardening stainless steels are similar in composition to the standard 18Cr-8Ni austenitic stainless steels, but they have a lower nickel content. The basic alloy is type 631 (S17700), or 17-7PH. Semi-austenitic precipitation-hardening stainless steels are available chiefly in sheet form.

Semi-austenitic precipitation-hardening stainless steels form austenite plus approximately 20% delta ferrite on quenching from the solution-annealing temperature. The austenite and delta ferrite formation results because the alloying elements depress the temperature at which martensite transformation begins (M_s) to below room temperature. Martensite does not form at this low temperature.

For semi-austenitic precipitation-hardening stainless steels, a twofold heat treatment is used. The alloys are first conditioned, or heated, between 730°C and 955°C (1350°F and 1750°F). This occupies some carbon as chromium carbide and raises the M_s. When the alloy is cooled, either to room temperature or below zero, soft low-carbon martensite forms. The higher the conditioning temperature, the lower the M_s. After conditioning, the alloys are fabricated and aged between 455°C and 565°C (850°F and 1050°F) to achieve the required strength.

Austenitic Precipitation-hardening Stainless Steels. Austenitic precipitation-hardening stainless steels retain austenitic structures at all temperatures. Quenching from the solution-annealing temperature produces austenite. These steels are fabricated and then aged up to 730°C (1350°F). The basic alloy is type 600 (S66286), or A286, and like other austenitic stainless steels, it is nonmagnetic and has good toughness at cryogenic temperatures.

Because of the higher aging temperatures, austenitic precipitation-hardening stainless steels can be used at higher temperatures than the other groups. The corrosion resistance of austenitic precipitation-hardening stainless steels is inferior to that of the regular austenitic stainless steels.

Duplex Stainless Steels

Duplex stainless steels have a duplex (two-phase) microstructure, consisting of ferrite and austenite. This is achieved by balancing the ferritizers and the austenitizers in the chemical composition. Duplex stainless steels possess certain desirable qualities that austenitic and ferritic stainless steels do not. For example, duplex stainless steels have better strength and chloride stress-corrosion cracking resistance than austenitic stainless steels. They also have a better ability to be fabricated and better toughness than ferritic stainless steels.

Metallurgical Structure. Most duplex stainless steels contain approximately 70% Fe, 20% Cr to 25% Cr, 4% Ni to 7% Ni, and 2% Mo to 4% Mo. See Figure 17-17. Nitrogen is added to ensure a balance between austenite and ferrite under fast cooling conditions, such as welding. The addition also improves tensile strength and corrosion resistance. Duplex stainless steels are stronger than austenitic stainless steels. Their coefficient of thermal expansion is midway between carbon steel and austenitic stainless steel.

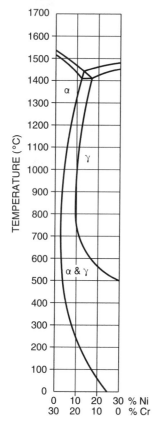

Fe-Cr-Ni ALLOY SYSTEM
(70% Fe)

Figure 17-17. Most duplex stainless steels contain approximately 70% Fe, 20% Cr to 25% Cr, 4% Ni to 7% Ni, and 2% Mo to 4% Mo.

Duplex stainless steels are hot worked or annealed in the range 1000°C to 1150°C (1830°F to 2100°F). Fast cooling is required to prevent the precipitation of undesirable phases leading to embrittlement. The room temperature microstructure of duplex stainless steels contains approximately equal amounts of ferrite and austenite. See Figure 17-18.

Duplex stainless steels exhibit high-temperature embrittlement similar to that of the ferritic stainless steels. With the exception of type 329 (S32900) and 2205 (S31803), most duplex stainless steels are proprietary alloys, such as Carpenter 7-Mo Plus® and Ferralium® alloy 255 (S32550).

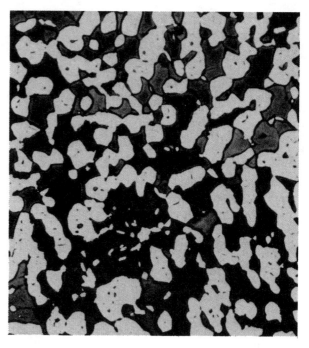

Figure 17-18. The room temperature microstructure of the duplex stainless steels contains approximately equal amounts of ferrite and austenite.

Cast Stainless Steels

Cast stainless steels are martensitic, ferritic, austenitic, or duplex in structure. Certain duplex types are precipitation hardening. They are divided into the corrosion-resistant group (C series) and heat-resistant group (H series).

Corrosion-resistant Castings. Corrosion-resistant castings are designated by the uppercase letter C followed by a letter that indicates the approximate alloy content. The higher the letter (A being the lowest and Z being the highest), the greater the alloy content. See Figure 17-19. Numbers and letters following a dash indicate the carbon content and the presence of important alloying elements. For example, CF-8 contains 19% Cr, 9% Ni, and .08% C maximum. CF-3 is a similar alloy in which the carbon content is maintained at .03% C maximum.

CA-15 (J91150) is martensitic and is hardened by quenching and tempering for improved wear resistance. CA-6NM (J91540) is also martensitic, but it has better toughness, better resistance to cavitation, and improved weldability. CB-30 (J91150) is ferritic and cannot be hardened by heat treatment. It contains significantly more chromium than the martensitic types and is more corrosion resistant.

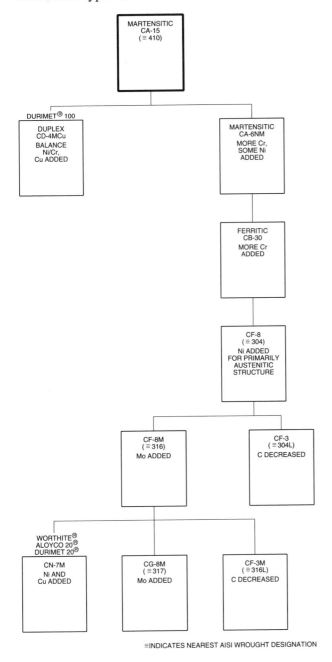

≡INDICATES NEAREST AISI WROUGHT DESIGNATION

CORROSION-RESISTANT SERIES

Figure 17-19. Corrosion-resistant castings are designated by the uppercase letter C followed by a letter that indicates the approximate alloy content.

Austenitic castings are solution annealed and quenched to ensure that all carbides are dissolved, which maximizes corrosion resistance. CF-8

(J92600) and CF-8M (J92900) are two of the most widely used austenitic castings and have similar corrosion resistance to the wrought types 304 and 316 stainless steels, respectively. CF-3 (J92700) and CF-3M (J92800) are the low-carbon equivalents in which carbon is maintained at <.03% C to decrease susceptibility to sensitization. CG-8M is a modification of CF-8M, with higher molybdenum content for improved corrosion resistance.

Although designated austenitic, CF-8, CF-8M, CF-3, CF-3M, and CF-8M are not completely austenitic. They usually contain from 5% ferrite to 20% ferrite, which is distributed throughout the matrix. The distribution is deliberately performed to improve castability. The optimum ferrite content is obtained by balancing the amount of ferritizers and austenitizers. Excessive amounts of ferrite result in an undesirable continuous network of ferrite in the matrix. For example, prolonged elevated temperature exposure would precipitate a continuous network of sigma or chi phase (phase similar to sigma that forms in higher molybdenum containing grades), leading to embrittlement or loss of corrosion resistance.

The Schaeffler diagram is used to estimate the allowable amount of ferrite in austenitic castings. The *ferrite number* is an arbitrary, standardized value indicating the ferrite content of an austenitic stainless steel casting or weld metal. See Figure 17-20. The actual ferrite number of a casting is obtained by comparing the magnetic behavior of samples representative of the casting with standard samples having known ferrite numbers. The measured ferrite number is then compared with the allowable ferrite number that is obtained from the Schaeffler diagram.

CN-7M (J95150) is a high-alloy austenitic casting in which nickel is the major alloying element. It is completely austenitic. The alloy was developed originally for sulfuric acid, and it has the best corrosion resistance of all the C series alloys. CN-7M is sometimes referred to as alloy 20, and has trade names such as Worthite® and Durimet®.

CD-4MCu (J93370) has a duplex structure of austenite and ferrite with corrosion resistance sometimes exceeding that of CF-8M. It also has outstanding erosion resistance. CD-4MCu is most often used in the solution-annealed condition to optimize corrosion resistance. It may also be precipitation hardened to high-strength levels.

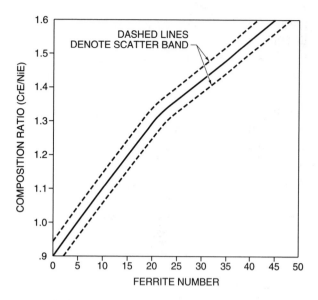

Figure 17-20. The Schaeffler diagram indicates the amount of ferrite in austenitic castings.

Heat-resistant Castings. Heat-resistant castings include a wide range of alloy compositions encompassing stainless steels and nickel-base alloys. Heat-resistant castings are designated by the uppercase letter H, followed by a letter that indicates the approximate alloy content. The higher the letter, the greater the percentage of alloying elements. See Figure 17-21. Heat-resistant castings are divided into three categories, which are straight chromium, chromium-nickel, and nickel-chromium.

Heat-resistant castings are used for service applications above 650°C (1200°F) for their creep strength, elevated-temperature corrosion resistance, scaling resistance, and dimensional stability. The completely austenitic grades are required when high-temperature strength or retention of room temperature toughness are needed. Ferritic grades are required where high-temperature scaling resistance is needed. Long exposures at 595°C to 870°C (1100°F to 1600°F) lead to the loss of room temperature ductility and toughness.

Straight chromium heat-resistant castings contain 9% Cr to 28% Cr, and some have minor amounts of nickel. HA (9Cr-1Mo) is martensitic and contains enough chromium to provide good scaling resistance up to 650°C (1200°F). HC (J92605), with 28% Cr, is ferritic and highly resistant to scaling, especially in sulfur bearing gases. It has low levels of ductility

and impact strength at room temperature. HD (J93005), with 29% Cr and 5% Ni, has a duplex structure and is similar to HC, except that the small amount of nickel slightly improves high-temperature strength in HD.

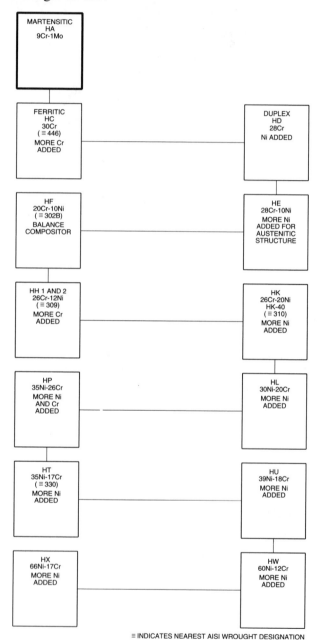

≡ INDICATES NEAREST AISI WROUGHT DESIGNATION

HEAT-RESISTANT SERIES

Figure 17-21. Heat-resistant castings are designated with the uppercase letter H followed by a letter that indicates the approximate alloy content.

Chromium-nickel heat-resistant castings are the most widely used. They contain from 20% Cr and

10% Ni to 30% Cr and 20% Ni. The chromium-nickel heat-resistant castings are almost completely austenitic because of the nickel content. HE (J93403), with 28% Cr and 10% Ni, has excellent high-temperature scaling resistance and corrosion resistance. It is not completely austenitic and is susceptible to sigma phase embrittlement when exposed in the temperature range 540°C to 870°C (1000°F to 1600°F).

HF (J92603), with 20% Cr and 10% Ni, is completely austenitic if the composition is balanced. HF has excellent corrosion resistance. If the composition is not properly balanced, some ferrite exists, which causes the alloy to become susceptible to sigma phase embrittlement.

HH (J93503), with 26% Cr and 12% Ni, is the most widely used heat-resistant casting. It comprises approximately one-third of all production of heat-resistant castings. It exhibits a combination of high-strength and oxidation resistance up to 1095°C (2000°F). The composition balance of HH alloy is critical to maintain a completely austenitic structure. Two versions of HH alloy (HH type 1 and HH type 2) are produced. HH type 1 contains a small amount of ferrite in the microstructure and is susceptible to sigma phase embrittlement. HH type 1 should only be used where the operating temperature is continuously above 870°C (1600°F). HH type 2 is completely austenitic and free from sigma phase embrittlement. HH type 2 is used where temperature cycling occurs below 870°C (1600°F).

HK (J94224), with 26% Cr and 20% Ni, is less oxidation resistant than the other alloys but is one of the strongest heat-resistant castings at temperatures above 1040°C (1900°F). It consists of an austenitic matrix containing a network of carbides. These carbides accumulate under continuous high-temperature service and contribute to high-temperature creep resistance. The creep resistance is also influenced by the carbon content. There are three grades of HK, each having narrower carbon ranges than indicated in the allowable composition range. They are identified by the midpoint of their carbon range, and are HK-30 (.3% C), HK-40 (.4% C) and HK-50 (.5% C). HK-40 is widely used in high-temperature processing equipment in the chemical and petrochemical industries.

Nickel-chromium heat-resistant castings are completely austenitic because of their high nickel contents. Nickel is the major alloying element. HL (N08604), with 30% Ni and 20% Cr, is similar to HK, except that its higher nickel content provides better resistance to hot gases.

HP, with 35% Ni and 26% Cr, consists of an austenitic matrix containing massive primary carbides. When HP is exposed to elevated temperatures, secondary carbides are precipitated. This increases creep resistance to much greater values than that of HK. Proprietary alloys, such as Manaurite 36X®, Kubota KHRC 35®, and Wiscalloy 25-35 Nb®, are based on HP and have even better creep resistance.

HT (N08605), with 35% Ni and 17% Cr, has good resistance to thermal shock. It is also resistant to high-temperature oxidation and carburization. HU, with 39% Ni and 18% Cr, is similar to HT, but the higher nickel and chromium contents provide greater resistance to high-temperature corrosion.

HW, with 60% Ni and 12% Cr, is suited for applications where wide or rapid temperature cycling occurs. In addition, it has excellent resistance to high-temperature oxidation and carburization. It is susceptible to attack by sulfur and sulfur compounds. HX, with 66% Ni and 17% Cr, is similar to HW but contains more nickel and chromium. This results in better high-temperature corrosion resistance for HX.

STAINLESS STEEL MANUFACTURING

Stainless steels are usually melted in an electric furnace and refined by argon-oxygen decarburizing (AOD). Ladle refining techniques, such as vacuum-argon decarburizing, are also used. The molten metal is cast and then rolled or forged into semifinished forms, which include blooms, billets, slabs, and tube rounds. Continuous casting is also used to produce semifinished forms.

The semifinished forms are hot or cold finished by rolling, drawing, extrusion, or forging into tubing, bar, plate, sheet, strip, or wire. Strict temperature control and avoidance of sulfur contamination is required during hot-working operations. Hot-finished and cold-finished forms may be annealed and pickled. *Pickling* is the removal of surface oxides from metals by chemical or electrochemical reaction. Cold-finished forms are also supplied as-finished.

Stainless Steel Fabrication

Stainless steels are joined by welding, brazing, or soldering. The key factor influencing joint quality is proper removal of the surface oxide film during the joining operation. Surface cleanliness is also important in avoiding cracking. With welding, the need for preheating and post-weld heat treatment depends on the metallurgical structure of the stainless steel.

Stainless steels are fabricated by machining, cold forming, hot forming, and cutting. In these operations, maintenance of the specified surface finish is extremely important. The chief problem with machining is the tendency for galling. During cold forming, problems arise from springback caused by work hardening. Special care must be taken when finishing stainless steels because the surface finish has a profound effect on the corrosion resistance.

Welding. Cleaning and joint preparation are critical in welding stainless steels. Contamination from grease and oil must be avoided so that corrosion resistance is not impaired through carbon pick up during welding. Carbon steel files and brushes must not be used because they leave fragments of high-carbon material adhering to the stainless steel. Stainless steel brushes and clean grinding disks must be used for cleaning. After welding, the job must be thoroughly cleaned to remove fluxes and weld spatter. A variety of welding filler metals are produced for welding the various stainless steels. See Figure 17-22.

Martensitic stainless steels are preheated before welding to offset the cracking tendency caused by martensite formation. Because martensitic stainless steels are air hardenable, they must also be post-weld heat treated to temper the martensite. If optimum mechanical properties are required, the welded component is reaustenitized, quenched, and tempered. If full strength is not required in the weld, austenitic welding filler metals such as E310 or ER310 may be used. If full strength is required, matching martensitic stainless steel welding filler metals must be employed. The preheating temperatures for marten-

sitic stainless steels are dictated by the carbon content. See Figure 17-23.

Ferritic stainless steels present a problem when welded because of grain growth and loss of toughness. Preheating between 150°C and 230°C (300°F and 450°F) helps to reduce residual stresses in highly restrained joints where martensite forms when cooled. Grades that do not contain carbide stabilizers, such as types 430, 442, and 446, are susceptible to sensitization when welded and require solution annealing.

Austenitic stainless steels are the most weldable stainless steels. They require less heat input and use less current than carbon steel because they have a lower melting point and have higher electrical resistivity. However, the relatively high coefficient of thermal expansion and relatively low thermal conductivity means that there is more chance for distortion and warping.

Completely austenitic welding filler metal is sensitive to hot cracking (hot shortness), which occurs as intergranular cracking when cooled from the molten state. It is prevented by adjusting the weld composition to produce small amounts (2% to 6%) of delta ferrite in the microstructure.

The composition of welding filler metal that produces the required percentage of delta ferrite may be predicted from the Schaeffler diagram. The Schaeffler diagram displays the amount of ferrite present in terms of its ferrite numbers (FN). The DeLong diagram, a modification of the Schaeffler diagram, is more commonly used. A ferrite content of 2-12FN is required for welding stainless steels. A ferrite content exceeding 12FN can lead to a continuous network of ferrite, which increases the risk of embrittlement from sigma phase formation. Excessively low ferrite increases the risk of hot cracking. Welding filler metal suppliers certify austenitic stainless steels with the FN. For example, type 308 stainless steel rod and wire, used for joining 304 stainless steel, is usually certified 2-12FN.

Austenitic stainless steels are susceptible to sensitization when welded. Sensitization is minimized through the use of low-carbon or carbide stabilized grades. Solution annealing and quenching after welding, as a means of eliminating sensitization, is not recommended because of the excessive distortion that results.

WELDING OF STAINLESS STEELS		
Type	**Final Service Condition**	**Welding Filler Metal[a]**
Austenitic Stainless Steels		
201 and 202	As-welded or fully annealed	308
301, 302, 304, 305, and 308	As-welded or fully annealed	308
302B	As-welded	309
304L	As-welded or stress-relieved	347 and 308L
303 and 303Se	As-welded or fully annealed	312
309 and 309S	As-welded	309
310 and 310S	As-welded	309, 310, and 316
316	As-welded or fully annealed	310
316L	As-welded or stress-relieved	316-Cb and 316L
317	As-welded or fully annealed	317
317L	As-welded or stress-relieved	317-Cb
321	As-welded or after stabilizing and stress-relieving heat treatment	321 and 347
347	As-welded or after stabilizing and stress-relieving heat treatment	347
348	As-welded or after stabilizing and stress-relieving heat treatment	347
Ferritic Stainless Steels		
405	Annealed	405-Cb and 430
	As-welded	309, 310, and 410-NiMo
430	Annealed	430
	As-welded	308, 309, and 310
430F	Annealed	430
430F Se	As-welded	308, 309, and 312
446	Annealed	446
	As-welded	308, 309, and 310
Martensitic Stainless Steels		
403 and 410	Annealed or hardened and stress-relieved	410
	As-welded	309, 310, and 410-NiMo
416 and 416Se	Annealed or hardened and stress-relieved	410
	As-welded	308, 309, and 312
420	Annealed or hardened and stress-relieved	420
431	Annealed or hardened and stress-relieved	410
	As-welded	308

[a] electrode or rod

Figure 17-22. Many welding filler metals are produced for welding the various stainless steels.

PREHEAT TEMPERATURES FOR MARTENSITIC STAINLESS STEELS		
%C	Preheat Temperature	Remarks
Less than .1	15°C min (60°F)	PWHT* optional
.1 to .2	205°C to 260°C (400°F to 500°F)	Cool slowly, PWHT optional
.2 to .5	260°C to 315°C (500°F to 600°F)	PWHT required
Over .5	260°C to 315°C (500°F to 600°F)	Weld with high heat input, PWHT required

* PWHT-Postweld Heat Treat

Figure 17-23. The preheat temperatures for martensitic stainless steels are dictated by the carbon content.

Quench welding is sometimes used to reduce sensitization. *Quench welding* is a joining technique where a small length of metal is welded and then quenched with a wet rag. This causes rapid cooling and avoids sensitization. The technique is repeated until the entire component is welded.

If optimum mechanical properties are required, precipitation-hardening stainless steels must be re-solution annealed and aged after welding. For martensitic types, aging alone is often sufficient. Welding filler metals match the parent metal composition. Aluminum tends to oxidize from the heat produced by welding. This is minimized by using the lowest practical heat input. Manufacturers' instructions should be followed when welding precipitation-hardening stainless steels.

Special problems are encountered when heat-resistant alloys that have been in service require welding. Precipitated carbides make them brittle and susceptible to cracking during welding. Before welding, heat-resistant alloys must be solution annealed to dissolve the carbides and restore ductility.

Brazing. All stainless steels can be brazed. As in welding, the weld joint area must be clean before brazing. Fluxes are not required for furnace brazing in strongly reducing, inert, or vacuum atmospheres, but are needed for induction or torch brazing. The brazing filler metals used are complex alloys based on silver, nickel, or copper.

Soldering. Soldering is similar to brazing, but the process temperature is lower. No problems of sensitization, hardening, or tempering occur at the lower temperatures of soldering. However, soldered joints are not as strong as brazed joints. Stainless steels are rarely soldered because the residues from soldering decrease the corrosion resistance of the joint.

Machining. The machinability of stainless steels varies widely between the types. As a group, stainless steels are more difficult to machine than carbon and low-alloy steels. Martensitics are usually machined in the annealed condition, except for finish machining, which is done after heat treatment. The machinability of martensitics is between the austenitics and ferritics. The ferritics are the easiest family to machine and the austenitics are the most difficult to machine because of their gumminess.

Significantly improved machinability is obtained by the addition of sulfur, selenium, or phosphorus. Free-machining stainless steels, such as types 416, 430F, and 303, have poor weldability and lower corrosion resistance than other stainless steels. For materials that tend to stick or gall during machining, the work and tool must be firmly held and the lubrication must be effective. Because of the work-hardening tendency of austenitic grades, a positive cut must be made at all times. If the tool is allowed to run idle against the surface of the job, hard spots, which are difficult to machine, develop. The correct geometry must be used on tools for machining stainless steels. See Figure 17-24.

Cold Forming. Stainless steels are cold formed by a wide variety of processes. Rod, bar, and wire are cold formed by cold heading, cold riveting, drawing, extrusion, and swaging. In these operations, ferritic stainless steels behave similarly to low-carbon steels. Ferritic type 430 tolerates the most severe deformation. Austenitic stainless steels work harden to a greater extent, tend to stick, and cause galling. They will suffer more deformation before they fracture than the ferritics will. Intermediate annealing may be required to soften austenitics in order to permit further deformation.

In forming ferritic stainless steel sheet, more power must be supplied for carbon steels because of their higher yield strength. With heavy sections, warming to 150°C (300°F) may be necessary to counteract the poor room temperature ductility. Austenitic stainless steel sheet has a higher rate of work hardening and requires greater working forces than the ferritics. The work-hardening effect is even more pronounced in grades that tend to form martensite when mechanically worked. Intermediate annealing may be necessary to soften the material and permit further cold working.

TOOL GEOMETRY FOR MACHINING STAINLESS STEELS	
Tool	**Geometry**
Drills	Included angle 130° to 140° Cutting-edge clearance 6° to 15° Webs thin
Reamers	Spiral fluted preferable Lead in chamfer 30° minimum with 2° to 3° on land below chamfer Clearance angle approximately 7° Land widths narrow
Taps	Chamfer 9° minimum Positive rake (hook) 15° Spiral flutes
Threading dies	Positive rake (hook) 15° for straight threads, up to 30° for circular and tangent chasers
Milling cutters	Positive rake 5° to 20° (a negative rake on carbide cutters) Clearance angle 5° to 10°
Saws	Approximately 8 teeth per inch for heavy stock, more for lighter sections Reciprocating saws must clear work on return stroke

Figure 17-24. The correct geometry must be used on tools for machining stainless steels.

Hot Forming. Stainless steels can be hot formed, but all processes, such as forging or rolling, become increasingly difficult as the alloy content increases. The temperature of hot forming varies according to the alloy type.

Martensitic stainless steels are heated in two stages. First, the steels are heated to 815°C (1500°F) to equalize the temperature and then to the forming temperature. After forming, they are removed from the forming temperature, held at 815°C (1500°F), and allowed to cool very slowly below the lower critical temperature to prevent air hardening.

Ferritic stainless steels are also heated to the forming temperature in the two stages. Subsequent cooling is not as critical as in the hardenable martensitic grades. Although grain growth occurs in the upper part of the forming temperature range, grain refinement can be accomplished by finishing the working process at a lower temperature of 760°C (1400°F). Stress relieving between 260°C and 315°C (500°F and 600°F) helps remove forging stresses.

Austenitic stainless steels are much stronger than the martensitics or ferritics at elevated temperatures. They are hot worked at correspondingly higher temperatures to reduce energy costs. Hot working must be completed above the sensitization temperature, and cooling must be as rapid as possible. Time at temperature must be minimized to reduce scaling. After hot working, solution annealing and quenching is required to fully restore corrosion resistance.

Cutting. Stainless steels are usually cut by shearing or blanking. They require greater power in shearing than used for carbon steels. With austenitic stainless steels, the shearing or blanking action must take the moving edge through the sheet thickness because the cut component does not drop away as with carbon steel. To avoid a poor edge from the effects of work hardening, blades and punches must be accurately set, well sharpened and clean, and with clearances running at approximately 5% of sheet thickness. Brittle ferritic grades of stainless steel are warmed before cutting.

Stainless steels are hot cut using plasma-arc cutting, which is a fast and accurate method of cutting. Air-carbon arc gouging is another cutting process used. With air-carbon arc gouging, the edges are contaminated and must be removed by grinding.

Finishing. Stainless steels are produced in a variety of surface finishes. Surface finishes are designated by a number that indicates the level of surface finish. Number 4, a general purpose finish, is the most popular. See Figure 17-25. A great amount of care must be taken during fabrication to avoid damaging the surface finish.

Organic materials, which include finger prints, must be removed before any heat treatment can be performed, or a pattern will be burned on the surface. Scale removal may be required after heat treatment. Salt bath descaling, sand blasting, or grit blasting is used to remove scale pickling. With pickling, the component is immersed in an acidic solution, such as 15% nitric acid plus 3% hydrofluoric acid at 60°C (140°F). Appropriate precautions must be taken when handling pickling solutions. With salt bath descaling, the component is immersed in molten salts, such as sodium hydride or sodium hydroxide. Sand blasting or grit blasting is suitable for heavy components and is followed by pickling. Carbon steel shot must not be used because it leaves small particles of iron on the surface of the stainless steel, which causes contamination.

SURFACE FINISHES FOR STAINLESS STEELS	
Finish Number	Description
1	Frosty white. Produced by hot rolling followed by annealing and descaling.
2D	Dull cold-rolled finish, similar to No. 1 but brighter. Produced by cold rolling, annealing, and descaling. The dull finish may result from descaling or may be developed by a final, light, cold-roll pass on dull rolls.
2B	Bright, dense. Produced the same as No. 2D, except that the annealed and descaled sheet receives a final, light, cold-roll pass on polished rolls.
3	Bright. A polished finish obtained with abrasives (100-mesh). Sheets may or may not be polished during fabrication.
4	Bright, good luster. Sheets are finished with 120-mesh to 150-mesh abrasive.
6	Dull satin finish, lower reflectivity than No. 4. Produced by Tampico brushing the No. 4 finish.
7	High reflectivity. Produced by buffing finely ground surface. Grit lines not removed.
8	Mirror finish. Obtained by polishing and buffing extensively. Surface is free of grit lines.

Figure 17-25. Surface finishes for stainless steels are designated with a number that indicates the degree of surface roughness.

CORROSION RESISTANCE

The corrosion resistance of stainless steels increases with the amount of alloying elements, particularly chromium, nickel, molybdenum, and nitrogen. The austenitic stainless steels have the broadest range of corrosion resistance.

Corrosion Types

Stainless steels generally fail because of localized forms of corrosion. The most common of these are pitting and crevice corrosion, intergranular corrosion, and stress-corrosion cracking. Pitting and crevice corrosion occur when the passive surface film is broken in discrete locations, leading to deep attack of small areas of the surface. Pitting of stainless steels is particularly associated with the chloride ion in aqueous environments, such as seawater or brackish (salty) water. Pitting can lead to complete perforation of a component. Crevice corrosion is similar to pitting, except that it occurs in crevices or shielded areas on the metal surface, such as under gaskets in pipe flanges or under deposits.

Pitting and crevice corrosion resistance are improved by agitation and vigorous aeration of the solution to increase the flow rate and oxygen content, which helps maintain the quality of the passive surface film. Operating at high alkalinity (pH) decreases pitting tendency. High-chromium, high-molybdenum alloys are more resistant to pitting and crevice corrosion. Inhibitors, such as nitrates, may be added to decrease pitting tendency. Operating at the lowest possible temperature also helps because both pitting and crevice corrosion are favored by high temperatures.

Intergranular corrosion is corrosion of a metal along its grain boundaries. When heated into the sensitization temperature range, carbon and chromium combine to form discrete precipitates of chromium carbide. Precipitation is favored at the grain boundaries and leads to a localized zone that is depleted in chromium. This zone is extremely susceptible to corrosion in specific environments. With severely sensitized microstructures, entire grains may drop out, leading to extremely high corrosion rates. Methods of preventing intergranular corrosion include using stabilized or low-carbon alloys and solution annealing followed by quenching. Lowering the carbon content may minimize carbide precipitation during short-term heating operations, such as welding or stress relieving.

Stress-corrosion cracking is crack formation in an alloy exposed to a specific corrosive, often intensified by the presence of tensile stresses. With austenitic stainless steels, the chloride ion is the most common ion associated with stress-corrosion cracking. Chloride stress-corrosion cracking most often occurs in situations where the chloride ion is allowed to concentrate to levels above 1500 ppm to 2000 ppm and the temperature is above 60°C (140°F). For example, the actual chloride content of most water is relatively low (approximately 100 ppm), but if the operating conditions favor the concentration of chlorides, cracking may eventually occur. Chloride stress-corrosion cracking can occur at tube-tubesheet joints in heat exchangers, under thermal insulation, or in any location where the chloride ion concentrates under the right temperature conditions. See Figure 17-26.

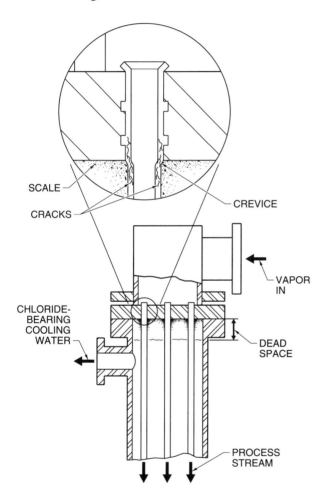

Figure 17-26. Chloride stress-corrosion cracking can occur at tube-tubesheet joints in heat exchangers or under thermal insulation, or in any location where the chloride ion concentrates under the right temperature conditions.

Ferritic and duplex stainless steels are practically immune to chloride stress-corrosion cracking. This is one of the reasons that they are sometimes pre-ferred over austenitic stainless steels as materials for construction.

Austenitic stainless steels are also susceptible to caustic stress-corrosion cracking, which is caused by the hydroxyl ion (OH⁻). As with chloride stress-corrosion cracking, caustic stress-corrosion cracking can occur in environments with extremely low hydroxyl ion contents where there is opportunity for concentration to occur, such as in steam condensation. Caustic stress-corrosion cracking of austenitic stainless steels occurs above 80°C (175°F).

Hardenable martensitic stainless steels are subject to sulfide stress-corrosion cracking in solutions containing acidic sulfide ions if their hardness exceeds 22 HRC. Extremely hard martensitic stainless steels, with hardness in excess of 48 HRC, may fail from hydrogen assisted stress-corrosion cracking in moist or humid air.

Resistance to Specific Corrosives

Stainless steels have wide ranging applications in many corrosives, including the atmosphere, waters, acids, salts, organics, and high-temperature environments. See Figure 17-27. Most stainless steels resist rusting and staining in the atmosphere. In marine environments that contain the chloride ion, pitting and staining may occur in low-alloy grades, such as the martensitics. In these environments, grades containing molybdenum (type 316) are superior.

Although stainless steels are resistant to natural waters, pitting can occur under stagnant conditions, especially where the chloride concentration is high or where microbe-containing organisms can settle out. Continuous flow is required to prevent this problem. In order to prevent pitting, stainless steel equipment that is hydrotested, or tested by filling it with water, should be completely drained and dried after testing.

Ferritic stainless steels are highly resistant to oxidizing acids, such as nitric acid, because of their high chromium content. Austenitic stainless steels possess good resistance to oxidizing acids. The grades containing molybdenum have moderate resistance to reducing acids, such as dilute sulfuric acid. A *reducing acid* is an acid that reduces (dissolves) the passive surface film. All stainless steels have very poor resistance to highly reducing acids, such as hydrochloric acid.

Stainless steels have good resistance to most alkalis. Hot concentrated sodium hydroxide will stress-corrosion crack austenitic stainless steels.

Stainless steels have good resistance to a broad range of organic compounds, such as acetic anhydride or formaldehyde. Hot organic acids and compounds contaminated with water or chlorides may cause pitting and intergranular corrosion of sensitized stainless steels. Pitting and intergranular corrosion occur because small amounts of hydrochloric acid are formed by degradation.

Stainless steels generally possess excellent high-temperature scaling resistance. Reducing sulfur compounds, such as hydrogen sulfide, and high-temperature atmospheres containing halogens, such as hydrogen chloride, reduce high-temperature scaling resistance of the stainless steels.

Some hot organic compounds, such as acetic acid and formic acid, may cause pitting and intergranular corrosion, especially if they are contaminated with chlorides and if the stainless steels are sensitized. Chlorinated hydrocarbons are extremely corrosive, especially at high temperatures where they degrade in the presence of water to form hydrochloric acid.

CORROSION RESISTANCE[a] OF SELECTED STAINLESS STEELS							
Solution	Type 302	Type 316	Type 430	Solution	Type 302	Type 316	Type 430
Acidic							
Arsenious		NA			NA		NA
Benzoic		NA			NA		NA
Chromic		A			A		A
Gallic		NA			NA		NA
Hydrochloric		A			A		A
Lactic		NA			NA		NA
Nitric		NA			NA		NA
Phosphoric		NA			NA		SA
Stearic		NA			NA		NA
Sulfuric		NA			NA		A
Saline							
Aluminum sulfate		NA			NA		—
Ammonium hydroxide		NA			NA		NA
Calcium carbonate		NA			NA		NA
Copper nitrate		NA			NA		NA
Ferric chloride		A			NA		A
Lead acetate		NA			NA		NA
Potassium nitrate		NA			NA		NA
Silver nitrate		NA			NA		NA
Sodium chloride		NA			NA		SA
Sulfur chloride		SA			—		—
Zinc chloride		A			SA		—

[a] NA = not affected
  A = affected
  SA = sometimes affected

Figure 17-27. Stainless steels have wide ranging applications in many corrosives.

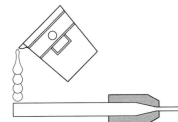

Copper 18

Copper is produced by smelting and refining concentrated copper ores and is worked into a variety of finished forms. The extraction and refining steps used for copper depend on the type of ore. Copper alloys are designated according to their compositions and their thermal and mechanical treatments. Copper is alloyed with many elements. Copper and copper alloys are easy to fabricate. The chief properties of copper and copper alloys are high thermal and electrical conductivity and good corrosion resistance. Some copper alloys are strengthened to high levels.

COPPER MANUFACTURING

Copper is produced by smelting and refining concentrated copper ores and is worked into a variety of finished forms. Copper alloys are strengthened by solid solution hardening, work hardening, or precipitation hardening.

Extraction and Production

The extraction and production steps used for copper depend on the type of ore. Sulfide ore is concentrated, melted, and refined. Nonsulfide ore is leached and refined. An additional major source of copper is scrap. See Figure 18-1.

The most common ores are copper sulfide containing minerals. These ores vary from very rich, with as much as 15% Cu, to very low grades with <.7% Cu. The ore is crushed, ground, and concentrated to 25% Cu. The concentrate is melted in a reverberatory furnace or flash smelter to produce a copper sulfide matte, which contains 60% Cu.

The copper sulfide matte is oxidized in a converter. The oxidation process transforms the iron sulfide to an iron oxide slag and reduces the copper sulfide to blister copper, containing at least 98.5% Cu. The blister copper is fire refined, which removes most of the oxygen and other impurities. The remaining product is fire-refined copper, which contains 99.5% Cu and is cast into anodes. Some fire-refined copper is used as is, but most is electrolytically refined to a purity of 99.95% Cu.

A small amount of copper is produced from nonsulfide ores, such as oxides, silicates, and carbonates. These ores are not concentrated like the sulfide ores. Instead, the copper is leached out of them and recovered by solvent extraction. The finished copper has the same purity as copper made by the electrolytic refining.

A large amount of copper is recovered from scrap, such as turnings from screw-machined rods, used electrical cable, and automobile radiators. Scrap is fed directly to the converters.

Production of Finished Forms. Finished forms (coils, billets, and blooms) of copper and copper alloys are produced by specific hot-rolling and cold-rolling operations. The coils, billets, and blooms are annealed and further mechanically worked into strip, sheet, plate, bar, extrusions, tube, pipe, wire, and cable. Most of these finished forms are further worked using specific fabrication processes before end use.

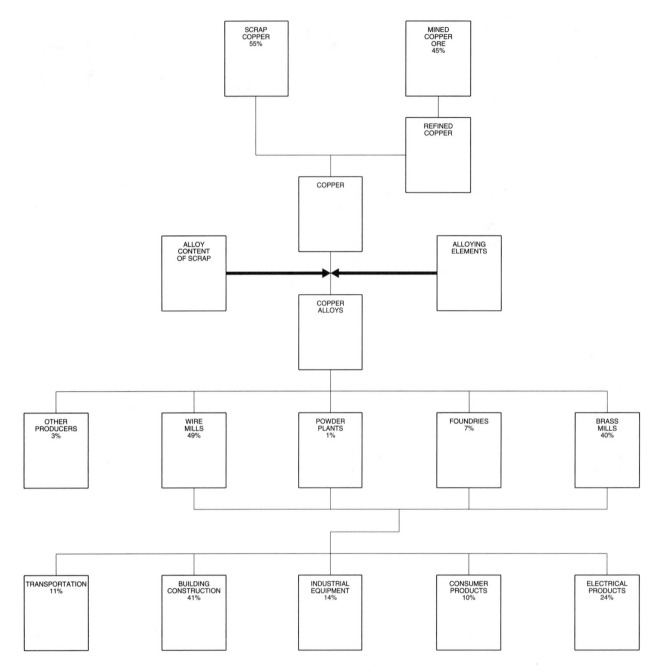

Figure 18-1. Copper is extracted chiefly from sulfide ores and processed into specific finished forms.

Many types of copper alloys are used to produce tube and pipe. Shells are first made by extruding or piercing billets. These shells are cold drawn to the required tube or pipe size.

Commercially pure coppers, modified coppers, and certain bronzes are used to produce wire and cable. Wire and cable are made from wire rod, which is successively drawn down in dies of decreasing diameter. Wire products are often covered with an insulating material.

Casting. Casting is used to make components such as plumbing goods, bearings, and cylinders, which are too complex to produce by working or machining alone. Foundries that cast copper alloys use prealloyed ingot, scrap, and virgin metal for raw materials.

Copper alloy castings have wider permissible chemical composition ranges than equivalent wrought copper alloys because the effects of composition on hot or cold workability are less important. For example, cast copper alloys may contain significantly greater amounts of lead than equivalent wrought alloys. Lead is added to improve machinability and galling resistance.

Sand casting is the most economical casting process. Sand casting has the greatest flexibility of casting size and shape. All copper alloys can be successfully sand cast.

Cast alloys are classified as high-shrinkage alloys or low-shrinkage alloys. High-shrinkage alloys are extremely fluid in the molten state and, with careful design, produce high-grade castings by sand, permanent mold, plaster, die, and centrifugal casting. High-shrinkage alloys include the manganese bronzes, aluminum bronzes, silicon bronzes, and some nickel silvers. Low-shrinkage alloys (less fluid) are usually limited to sand, permanent mold, plaster, and centrifugal casting. Low-shrinkage alloys include the yellow brasses. See Figure 18-2.

Casting brasses, tin bronzes, and leaded copper alloys have moderate strength, moderate hardness, and high elongation. The manganese bronzes, aluminum bronzes, silicon bronzes, and some nickel silvers are used for their greater strength. Most of the higher-strength alloys have better than average resistance to corrosion and wear.

The mechanical properties of cast copper alloys are measured from tensile test specimens that are machined from separately cast test bars. The mechanical properties of the castings themselves are almost always lower than those of the separately cast test bars. The mechanical properties of the cast copper alloys depend on the section size of the casting and the casting process.

FOUNDRY PROPERTIES FOR COPPER ALLOYS[a]						
UNS Number	Common Name	Shrinkage Allowance (%)	Approximate Liquidus Temperature °C	°F	Castability Rating[b]	Fluidity Rating[b]
C83600	Leaded red brass	5.7	1010	1850	2	6
C84400	Leaded semi-red brass	2.0	980	1795	2	6
C84800	Leaded semi-red brass	1.4	955	1750	2	6
C85400	Leaded yellow brass	1.5 to 1.8	940	1725	4	4
C85800	Yellow brass	2.0	925	1700	4	4
C86300	Manganese bronze	2.3	920	1690	6	2
C86500	Manganese bronze	1.9	880	1615	6	2
C87200	Silicon bronze	1.8 to 2.0			8	3
C87500	Silicon brass	1.9	915	1680	7	1
C90300	Tin bronze	1.5 to 2.0	980	1795	3	6
C92200	Leaded tin bronze	1.5	990	1815	3	6
C93700	High-lead tin bronze	2.0	930	1705	1	6
C94300	High-lead tin bronze	1.5	925	1700	1	6
C95300	Aluminum bronze	1.6	1045	1910	8	5
C95800	Aluminum bronze	1.6	1060	1940	8	5
C97600	Nickel silver	2.0	1145	2090	5	7
C97800	Nickel silver	1.6	1180	2160	5	7

[a] sand casted
[b] Relative rating for casting in sand molds. The alloys are ranked from 1 to 8 over-all castability and fluidity; 1 is the highest or best possible rating.

Figure 18-2. To a large extent, the foundry properties of copper alloys depend on whether they are low-shrinkage or high-shrinkage alloys.

Working and Heat Treatment

Copper and copper alloys are hot or cold worked. Copper and copper alloys are annealed to soften or homogenize them. They are then hardened by precipitation hardening, transformation hardening, or spinodal decomposition.

Hot and Cold Working. Hot working is used on copper alloys that remain ductile above their recrystallization temperatures. Compared to cold working, hot working permits more deformation. Hot working is very beneficial in refining the as-cast grain size and in softening the metal for cold finishing.

Cold working increases both tensile strength and yield strength, but the effect on yield strength is the greatest. In many cases, the tensile strength of the hardest tempered copper alloy (the alloy with the greatest degree of cold work) is approximately double the tensile strength of the alloy in the annealed condition. For the same tempered copper alloy, the yield strength is approximately five times greater than the annealed alloy. The hardness measurement has a poor correlation with the increasing strength achieved by cold work. Hardness is not a reliable indicator of the degree of cold work.

Annealing Heat Treatments. Annealing heat treatments are performed to homogenize, soften, and stress relieve the metal. Homogenizing is used to eliminate coring in cast metals before cold working. Homogenizing is performed at high temperatures for long time intervals.

Softening is performed on cold-worked material to make it acceptable for further cold working. Softening is accomplished by heating the alloy above its recrystallization temperature. The greater the amount of prior cold working, the lower the recrystallization temperature. Stress relieving is performed to reduce residual stresses from cold-forming operations, which might contribute toward stress-corrosion cracking or dimensional instability in cold-formed components. Stress relieving is performed at low temperatures. See Figure 18-3.

Hardening Heat Treatments. The most common hardening heat treatments for copper alloys consist

of precipitation hardening, transformation hardening, and spinodal decomposition. Precipitation (age) hardening is achieved in high-strength coppers (alloyed with zirconium, chromium, or nickel plus phosphorus) and beryllium coppers. These alloys are solution treated and quenched to produce a soft condition, fabricated, and then precipitation hardened at an intermediate temperature for a time usually not exceeding 3 hours. Various combinations of strength, hardness, conductivity, and toughness can be achieved by varying the alloy, the hardening temperature, and the degree of cold work before precipitation hardening.

ANNEALING TEMPERATURES	
UNS	**°C**
Wrought Coppers	
C11000	250 to 650
C12000	325 to 650
C12200	375 to 650
C10200, C14500, and C18700	425 to 650
C11300 to C11600 and C12700 to C1300	400 to 475
Wrought Copper Alloys	
C21000 and C22000	425 to 800
C22600, C2600, and C60600	425 to 750
C23000	425 to 725
C24000, C27000, and C35300	425 to 700
C31400, C35600, and C37000	425 to 650
C28000, C36500 to C38500, C44300 to C48500, C66700, C67400, C67500, and C68700	425 to 600
C51000 to C54400 and C65100	475 to 675
C50500	475 to 650
C71500	650 to 815
C70600 and C74500 to C78200	600 to 815
C65500	475 to 700
C61300 and C61400	815
C63800	Above 650
C63000 and C64200	575 to 650
C17000 to C17600	775 to 1050[a]

[a] solution treating temperature

Figure 18-3. Annealing heat treatments are used to homogenize, soften, or stress relieve copper alloys.

Transformation hardening may be performed on certain two-phase aluminum bronzes. These alloys are hardened in two stages, first by cooling rapidly from a high temperature to produce a martensitic type of structure, and then by tempering at a lower temperature to stabilize the structure and restore some ductility and toughness.

Spinodal decomposition may be performed on certain copper alloys containing a small amount of chromium, such as a 70Cu-30Ni alloy. 70Cu-30Ni is homogenized at a high temperature and then slowly cooled through a specific temperature range in which the diffusion rate of the copper and nickel atoms are such that compositionally different waves develop within the crystal structure, causing a hardening effect.

Copper Designation Systems

Copper alloys are designated according to their compositions and thermal and mechanical treatments.

Unified Numbering System designations are used to identify specific alloys. Temper designations are used to indicate the thermal and mechanical condition.

UNS Designations. UNS designations are used to identify copper alloys. They consist of the uppercase letter C followed by five numbers. The UNS designations for copper alloys are expansions of the original three number designation system developed by the Copper Development Association. For example, CDA copper alloy 377 for forging brass is equivalent to UNS C37700.

Each family of copper alloys occupies a specific series of UNS designations. Wrought alloys are assigned C10000 to C79999, and cast alloys are assigned C80000 to C99999. For example, wrought silicon bronzes occupy UNS designations C64700 to C66100. See Figure 18-4.

COPPER ALLOY FAMILIES		
UNS	**Description**	**Major Alloying Elements**
Wrought		
C10100 to C15500	Coppers	Cu
C16200 to C19500	High-copper alloys	Cu plus Cd, Be, and Cr or Fe
C20500 to C28200	Brasses	Cu and Zn
C31400 to C38600	Leaded brasses	Cu, Zn, and Pb
C40500 to C48500	Tin brasses	Cu, Zn, and Sn
C50100 to C52400	Tin bronzes	Cu and Sn
C53400 to C54800	Leaded tin bronzes	Cu, Sn, and Pb
C60600 to C64200	Aluminum bronzes	Cu and Al
C64700 to C66100	Silicon bronzes	Cu and Si
C66400 to C69800	Special brasses	Cu, Zn plus Mn, Si, Al, etc.
C70100 to C72500	Copper nickels	Cu and Ni
C73200 to C79900	Nickel silvers	Cu, Ni, and Zn
Cast		
C80100 to C81100	Coppers	Cu
C81300 to C82800	High-copper alloys	Cu plus Cr, Be, Co, Ni, and Si
C83300 to C83800	Red brasses, leaded red brasses	Cu, Sn, Zn, and Pb
C84200 to C84800	Semi-red brasses, leaded semi-red brasses	Cu, Sn, Zn, and Pb
C85200 to C85800	Yellow brasses, leaded yellow brasses	Cu, Sn, Zn, and Pb
C86100 to C86800	Manganese bronzes, leaded manganese bronzes	Cu, Zn, Al, Mn, and Pb
C87200 to C87900	Silicon bronzes and brasses	Cu, Zn, and Si
C90200 to C91700	Tin bronzes	Cu and Sn
C92200 to C92900	Leaded tin bronzes	Cu, Sn, and Pb
C93200 to C94500	High-leaded tin bronzes	Cu, Sn, and Pb
C94700 to C94900	Nickel-tin bronzes	Cu, Sn, and Ni
C95200 to C95800	Aluminum bronzes	Cu, Al, Fe, and Ni
C96200 to C96600	Copper nickels	Cu, Ni, and Fe
C97300 to C97800	Nickel silvers	Cu, Ni, and Zn
C98200 to C98800	Leaded coppers	Cu and Pb
C99300 to C99700	Special alloys	Cu, Ni, Fe, Al, and Zn

Figure 18-4. Copper alloys are designated in the Unified Numbering System for metals and alloys.

Temper Designations. Temper designations are codes that indicate the exact thermal and mechanical condition of any copper alloy. They are described in ASTM recommended practice *B601, Temper Designation System for Copper and Copper Alloys, Wrought and Cast.*

For example, the uppercase letter O indicates annealed tempers in which annealing is the principal heat treatment used to meet property requirements. See Figure 18-5. See Appendix. Numbers that follow the O narrow down the description of the alloy, such as O10 (cast and annealed) or O30 (hot extruded and annealed).

COPPER ALLOYS

There are several copper alloy systems, which include commercially pure coppers, modified coppers, beryllium coppers, brasses, casting brasses, tin bronzes, aluminum bronzes, silicon bronzes, copper-nickels, and nickel-silvers. Each system exhibits different properties that are characteristic of that system. See Figure 18-6.

Commercially Pure Coppers

Commercially pure coppers contain at least 99.9% Cu, plus extremely small amounts of other elements.

Commercially pure coppers are used primarily for their high electrical conductivity, and, to a lesser extent, thermal conductivity. Impurity elements decrease electrical and thermal conductivity by disturbing the periodicity (repetitiveness) of the crystal lattice. Impurity elements must be kept to permissible low levels.

Cast commercially pure coppers have lower electrical and thermal conductivity than nominally equivalent wrought alloys because the elements added to ensure a sound casting (silicon, tin, zinc, aluminum, and phosphorus) decrease the conductivity. Calcium, boron, or metallic lithium may be added to offset this effect and improve conductivity.

Commercially pure coppers are soft, weak, and very ductile. Commercially pure coppers are used in all product forms for architectural applications, electrical components, electrical wire, and gaskets. They consist of tough pitch, deoxidized, and oxygen-free coppers.

Tough Pitch Coppers. Tough pitch coppers are purified commercially pure coppers. The impurities are removed by oxidation, which leaves residual oxygen in the molten metal. The oxygen is detected as particles of cuprous oxide (Cu_2O) in the microstructure of the copper.

ASTM COPPER AND COPPER ALLOY TEMPER DESIGNATIONS		
Designation/ Temper Name or Condition	**Designation/ Temper Name or Condition**	**Designation/ Temper Name or Condition**
Cold-worked tempers	**As-manufactured tempers**	**Solution-treated and precipitation-hardened temper**
H00 ⅛ hard	M07 As-continuous cast	
H01 ¼ hard	M10 As-hot forged and air cooled	TF00 TB00 and precipitation hardened
H02 ½ hard	M11 As-forged and quenched	**Cold-worked and precipitation-hardened tempers**
H03 ¾ hard	M20 As-hot rolled	
H04 Hard	M30 As-hot extruded	TH01 TD01 and precipitation hardened
H06 Extra hard	M40 As-hot pierced	TH02 TD02 and precipitation hardened
H08 Spring	M40 As-hot pierced and rerolled	TH03 TD03 and precipitation hardened
H10 Extra spring	**Annealed tempers**	TH04 TD04 and precipitation hardened
H12 Special spring	O10 Cast and annealed (homogenized)	**Precipitation-hardened and cold-worked tempers**
H13 Ultra spring	O11 As-cast and precipitation heat treated	
H14 Super spring	O20 Hot forged and annealed	TL00 TF00 cold worked to ⅛ hard
H50 Extruded and drawn	O25 Hot rolled and annealed	TL01 TF00 cold worked to ¼ hard
H52 Pierced and drawn	O30 Hot extruded and annealed	TL02 TF00 cold worked to ½ hard
H55 Light drawn; light cold rolled	O31 Extruded and precipitation heat treated	TL04 TF00 cold worked to full hard
	O40 Hot pierced and annealed	
	O50 Light annealed	

Figure 18-5. Temper designations indicate mechanical working and heat treatment conditions of copper alloys.

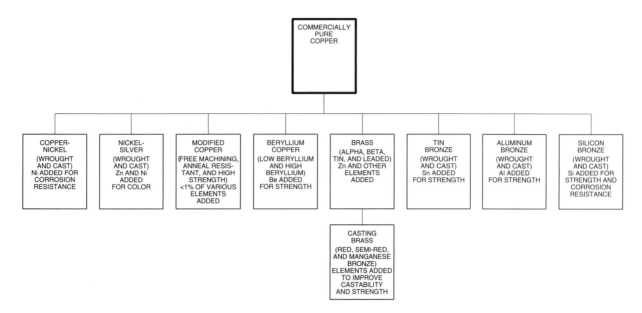

Figure 18-6. Each system exhibits different properties that are characteristic of that system.

C11000, electrolytic tough pitch (ETP) copper, contains .03% O_2 to .06% O_2. ETP copper is obtained from electrolytically refined or cathode copper and contains <50 ppm metallic impurities, which includes sulfur. It is the most common commercially pure copper.

C12500, fire-refined tough pitch copper, contains approximately the same oxygen as ETP copper, but it has greater amounts of other impurities. This increase in impurities reduces its electrical conductivity by approximately 7%. Fire-refined tough pitch copper is made by melting and deoxidizing blister copper anodes.

Both types of tough pitch copper become brittle when heated above 400°C (750°F) in a reducing gas atmosphere, such as during annealing or welding with a reducing flame. The embrittlement is caused by hydrogen in the atmosphere. The hydrogen diffuses into the copper, reducing the cuprous oxide particles to copper and producing steam, according to the following equation:

$$2H_2 + Cu_2O \rightarrow 2Cu + H_2O\uparrow \text{ (steam)}$$

The evolution of steam causes a network of porosity to develop, which drastically reduces strength and ductility. Embrittlement is recognized by thick boundaries or cracks that develop on bending. As a result, tough pitch coppers are unsuitable for welding. See Figure 18-7.

American Society for Metals

Figure 18-7. The microstructure of tough pitch copper heated in hydrogen for 2 minutes at 850°C (1560°F) shows black voids, which correspond to porosity created by pockets of water vapor (steam).

Deoxidized Coppers. Deoxidized coppers are commercially pure coppers that are treated in the molten state with phosphorus or another deoxidizer so that oxygen is completely removed from the metal.

Deoxidized coppers are used for tubing and in welding applications.

C12200, phosphorus deoxidized copper, contains .02% residual phosphorus. The phosphorus substantially lowers conductivity, but prevents embrittlement during annealing, brazing, or welding.

Oxygen-free Coppers. Oxygen-free coppers are commercially pure coppers that are melted and cast in a nonoxidizing atmosphere. Because of the nonoxidizing atmosphere, deoxidation is not required.

C10100 and C10200 are oxygen-free high-conductivity (OFHC) coppers. OFHC coppers have electrical conductivity equivalent to ETP copper. They are not susceptible to embrittlement and are suited to applications that require high conductivity and high ductility. C10100, which is preferred for electronic applications, is usually referred to as oxygen-free electronic copper.

Modified Coppers

Modified coppers are commercially pure coppers containing <1% of the alloying elements. This results in improved machinability, resistance to elevated temperature softening, and strength. Alloying elements have a minor effect on electrical conductivity. Modified coppers consist of free-machining coppers, anneal-resistant coppers, and high-strength coppers.

Free-machining Coppers. Free-machining coppers are modified coppers containing tellurium (C14500), lead (C14700), or sulfur (C18700). The alloying of copper with these elements results in significantly improved machinability. Free-machining coppers are used for electrical components that are machined at high production rates.

Anneal-resistant Coppers. Anneal-resistant coppers are modified coppers containing cadmium or silver, which increase resistance to elevated temperature. Cadmium coppers include C14300, C14310, and C16200. The anneal-resistant copper that contains silver is C11400. Anneal-resistant coppers are used for electrical conductors and components where the manufacturing or operating temperature is above 205°C (400°F), such as in the soldering of automotive radiators and in high-temperature conductors.

High-strength Coppers. High-strength coppers are modified coppers containing zirconium, chromium, or nickel and phosphorus, which develop strength by precipitation hardening. Zirconium coppers include C15000 and C15100. Zirconium coppers are solution treated at 900°C to 925°C (1650°F to 1700°F) and aged at 500°C to 550°C (930°F to 1020°F) for 1 to 4 hours. Strength is significantly increased by cold working before aging. Cold-worked material is aged at a lower temperature of 370°C to 480°C (700°F to 900°F). Aging increases electrical conductivity.

Chromium coppers include C18200, C18400, and C18500. Chromium coppers are deoxidized with silicon because phosphorus deoxidation severely reduces electrical conductivity. Chromium coppers are solution treated at 955°C to 1010°C (1750°F to 1850°F), and aged at 425°C to 480°C (800°F to 900°F) for 1 to 3 hours.

Nickel-phosphorus coppers include C19000. Nickel-phosphorus coppers are solution treated at 705°C to 760°C (1300°F to 1400°F) and aged at 425°C to 480°C (800°F to 900°F) for 1 to 3 hours.

High-strength coppers are used for a wide variety of components requiring strength, particularly at elevated temperature, such as spring clips, electrical connectors, circuit breakers, resistance welding tips and wheels, and fasteners.

Beryllium Coppers

Beryllium coppers are wrought and cast copper alloys containing small amounts of beryllium, which are precipitation hardened to extremely high levels of tensile and fatigue strengths comparable with low-alloy steels. Small amounts of cobalt or nickel may be added to aid in hardening and to refine the grain size.

The solubility of beryllium in alpha solid solution decreases from 2.1% at 1065°C (1950°F) to <.25% at room temperature. See Figure 18-8. Quenching from the solution annealing temperature of 790°C (1450°F) produces a single-phase structure. Aging at an intermediate temperature causes the precipitation of very fine (CuCo)Be berylide inclusions within the alpha matrix, which increases strength and hardness.

Figure 18-8. Aging develops (CoCu)Be berylide inclusions in the copper-rich matrix of beryllium coppers.

Low-beryllium coppers and high-beryllium coppers are the two groups of beryllium coppers. The low-beryllium coppers contain approximately .5% Be plus 1.5% Co or 2.5% Co. The low-beryllium coppers develop useful strength with relatively high electrical conductivity. The high-beryllium coppers contain approximately 2% Be, and develop very high strength with low electrical conductivity. See Figure 18-9. Beryllium coppers are used for Bourdon tubes, bellows, tubing, diaphragms, fasteners, lock washers, springs, switch components, relay components, electrical and electronic components, retaining rings, roll pins, valves, pump components, spline shafts, rolling mill components, welding equipment, instrument housings, and molds for plastic components.

Caution: To keep the airborne concentration of beryllium within the allowable limits, proper safety precautions must be taken when melting, welding, flame cutting, polishing, buffing, grinding, and machining beryllium coppers.

Heat Treatment. Heat treatment of beryllium coppers is performed in two stages, which are solution annealing and precipitation hardening. Beryllium coppers are usually supplied in the solution-annealed condition. In this condition, the material is suitable for fabricating into components. If solution annealing is performed in air, an excessively thick beryllium oxide layer develops. The beryllium oxide layer causes severe tool wear. Although chemical or mechanical cleaning may be used to remove the beryllium oxide layer, special reducing furnace atmospheres are employed to prevent its formation.

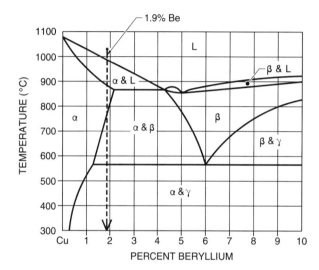

Figure 18-9. Low-beryllium coppers and high-beryllium coppers are the two groups of beryllium coppers.

Water quenching is the usual method of cooling from the solution-annealing temperature. Complex sections are quenched more slowly in oil or in forced air to prevent cracking.

The aging temperature for precipitation hardening is determined by the required properties desired and the amount of prior cold work. The aging temperature ranges between 315°C and 370°C (600°F and 700°F), ±5°C (10°F). See Figure 18-10.

Wrought Beryllium Coppers. Wrought beryllium coppers may be produced with a wide range of mechanical properties by introducing cold work either before or after aging. Wrought beryllium coppers are supplied as rod, tube, plate, strip, wire, and extruded shapes.

C17600, 50 alloy (.3% Be) and C17500, 10 alloy (.5% Be) are wrought low-beryllium coppers. These beryllium coppers are used in lower-strength applications than the high-beryllium coppers.

AGING TREATMENTS FOR STRIP BERYLLIUM COPPERS

Initial Condition	Time (hr)	Treatment Temperature (°C)	Final Temper	Tensile Strength (ksi)	Yield Strength[a] (ksi)	Elongation (%)	Hardness
C17000							
Annealed[b]	—	None	TB00	60 to 78	25 to 35	35 to 60	45 to 78 HRB
¼ hard	—	None	TD01	75 to 88	45 to 75	10 to 35	68 to 90 HRB
½ hard	—	None	TD02	85 to 100	65 to 90	5 to 25	88 to 96 HRB
Hard	—	None	TD04	100 to 120	80 to 110	2 to 8	96 to 102 HRB
Annealed	3	315	TF00	150 to 180	120 to 160	4 to 10	33 to 39 HRC
Annealed	3	345	TF00	160 to 185	125 to 165	4 to 10	34 to 40 HRC
¼ hard	2	315	TH01	160 to 185	125 to 165	3 to 6	34 to 40 HRC
¼ hard	3	330	TH01	170 to 195	130 to 170	3 to 6	36 to 41 HRC
½ hard	2	315	TH02	170 to 195	130 to 170	2 to 5	36 to 41 HRC
½ hard	2	330	TH02	180 to 200	140 to 180	2 to 5	38 to 42 HRC
Hard	2	315	TH04	180 to 200	140 to 180	2 to 5	38 to 42 HRC
Hard	2	330	TH04	185 to 205	155 to 195	2 to 5	39 to 43 HRC
C17200							
Annealed	—	None	TB00	60 to 78	28 to 36	35 to 60	45 to 78 HRB
¼ hard	—	None	TD01	75 to 88	60 to 80	10 to 35	68 to 90 HRB
½ hard	—	None	TD02	85 to 100	75 to 90	5 to 25	88 to 96 HRB
Hard	—	None	TD04	100 to 120	90 to 112	2 to 8	96 to 102 HRB
Annealed	3	315	TF00	165 to 190	140 to 175	4 to 10	35 to 40 HRC
Annealed	½	370	TF00	160 to 190	130 to 175	3 to 10	34 to 40 HRC
¼ hard	2	315	TH01	175 to 200	150 to 185	3 to 6	37 to 42 HRC
¼ hard	¼	370	TH01	170 to 200	140 to 185	2 to 6	36 to 42 HRC
½ hard	2	315	TH02	185 to 210	160 to 195	2 to 5	39 to 44 HRC
½ hard	¼	370	TH02	180 to 210	150 to 195	2 to 5	38 to 44 HRC
Hard	2	315	TH04	190 to 215	165 to 205	1 to 4	40 to 45 HRC
Hard	¼	370	TH04	185 to 215	160 to 205	1 to 4	39 to 45 HRC
C17500							
Annealed	—	None	TB00	35 to 55	20 to 30	20 to 35	20 to 43 HRB
Hard	—	None	TD04	75 to 85	55 to 80	5 to 10	78 to 88 HRB
Annealed[b]	3	480	TF00	100 to 110	80 to 100	8 to 12	92 to 100 HRB
Annealed[b]	3	455	TF00	105 to 120	80 to 105	8 to 12	93 to 100 HRB
Hard	2	480	TH04	110 to 125	100 to 120	5 to 8	95 to 103 HRB
Hard	2	455	TH04	115 to 138	105 to 125	5 to 8	97 to 104 HRB

[a] at .2% offset
[b] All annealing of these alloys is solution treating.

Figure 18-10. The selected aging temperature for beryllium coppers depends on the desired properties and the amount of prior cold work.

C17000, 165 alloy (1.7% Be) and C17200, 25 alloy (1.9% Be) are wrought high-beryllium coppers. These beryllium coppers develop extremely high strength during the aging process.

Cast Beryllium Coppers. Cast beryllium coppers must be solution treated a minimum of 3 hours to homogenize their microstructure. C82000 and C82200 (both .5% Be) are cast low-beryllium coppers used in low-strength applications. C82400 (1.7% Be) and C82500 (2% Be) are cast high-beryllium coppers used for high strength applications.

Brasses

Brasses are wrought alloys of copper and zinc. The zinc content may vary from 5% Zn to 50% Zn. Some wrought brasses may contain additions of tin and other elements.

Brasses are the most popular and least expensive of the copper alloys. They display a wide range of mechanical properties, are easy to work, have pleasing color, and exhibit good corrosion resistance. Cold working generally improves the strength of the brass. Brasses that are extendedly cold worked may be softened by annealing.

The color of brass varies with its copper content. At least 90% Cu is required to resemble the color of copper. Between 90% Cu and 80% Cu a brass exhibits a reddish color. At 65% Cu the color of a brass is yellow. Brasses consist of three groups, which are alpha and beta brasses, tin brasses, and leaded brasses.

Alpha and Beta Brasses. Alpha brasses are wrought, single-phase alloys of copper and zinc containing >64% Cu and <36% Zn. Beta brasses are wrought, two-phase (duplex) alloys of copper and zinc containing <64% Cu and >36% Zn.

Although copper and zinc have different crystal structures, which are FCC for copper and CPH for zinc, their atomic size difference is low (4%). As a result, copper and zinc exhibit extensive solid solubility. The maximum solubility of zinc in copper is approximately 38% at an elevated temperature. The maximum solubility of zinc in copper falls to 35% at room temperature.

On the copper-zinc phase diagram, compositions between point F and point G solidify as alpha solid solution, usually cored. Compositions between point G and point H begin by forming alpha solid solution and finally form some beta solid solution of composition at point H by the peritectic reaction. When cooling to room temperature, the amount of the beta constituent decreases according to the solubility lines GM and HN. Compositions between point H and point J form alpha, which at 905°C (1660°F) reacts with the liquid of the composition at point J to form beta. When cooling, alpha is precipitated from the beta when solubility line HN is passed, forming a Widmanstätten structure. Compositions between point J and point K solidify as beta. Alloys containing <46.6% Zn precipitate some alpha, while alloys containing >50% Zn precipitate gamma solid solution when cooling to room temperature.

Between 450°C and 470°C (845°F and 880°F), the beta phase transforms to a low-temperature modification referred to as beta prime (β'). This transformation is caused by the arrangement of the zinc atoms in the crystal lattice changing from random to ordered.

An increase in zinc increases strength by solid solution hardening and also increases elongation. Maximum strength is obtained at approximately 45% Zn, after which it falls rapidly due to the formation of the gamma phase. Maximum elongation occurs at 30% Zn, at which point the beta phase forms and causes a dramatic drop in elongation.

The single-phase microstructure of alpha brass makes it easy to hot or cold work. The beta phase in beta brass is much harder than alpha and tolerates only a small amount of cold work. Beta brass begins to soften suddenly at 470°C (880°F), the temperature of the order-disorder change. At 800°C (1470°F), beta brass is much easier to work than alpha brass.

Alpha brasses are suitable for mechanical working operations, such as cold rolling into thin sheets, drawing into wire or tube, and pressing. Beta brasses are suitable for hot extrusion and hot stamping.

C21000, gilding metal, is an alpha brass with 5% Zn. This brass is stronger than copper and is used for coins, medals, tokens, fuse caps, primers, emblems, plaques, and as a base for articles to be gold plated or highly polished.

C22000, commercial bronze, is an alpha brass with 10% Zn. This brass has excellent cold-working and hot-working properties and is used for costume jewelry, compacts, lipstick cases, marine hardware, forgings, rivets, and screws.

C23000, red brass, is an alpha brass with 15% Zn. This brass is used for electrical conduit, screw shells, sockets, hardware, condenser and heat exchanger tubes, plumbing pipe, nameplates, tags, and radiator cores.

C24000, low brass, is an alpha brass with 20% Zn. This brass is used for ornamental metalwork, medallions, thermostat bellows, musical instruments, flexible hose, and other deep drawn articles.

C26000, cartridge brass, is an alpha brass with 30% Zn. This brass is widely used for automotive radiator cores, tanks, headlight reflectors, flashlight shells, lamp fixtures, electric socket shells, eyelets, fasteners, grommets, rivets, springs, plumbing accessories, and ammunition components, such as cartridge cases.

C27000, yellow brass, is an alpha brass with 35% Zn. This brass has similar applications to those of the cartridge brass.

C28000, Muntz metal, is a beta brass with 40% Zn. Muntz metal is the most widely used beta brass because of its high-strength and excellent hot-working properties. Muntz metal is used for condenser heads, perforated metal, architectural applications, valve stems, brazing rods, and condenser tubes.

Tin Brasses. Tin brasses are alpha and beta brasses containing small amounts of tin, and occasionally other elements, to improve strength and corrosion resistance. Additions of tin significantly increase the resistance of brass to dezincification. *Dezincification* is a form of corrosion in which the zinc constituent is preferentially removed. Arsenic, antimony, or phosphorus is additionally incorporated to further improve the resistance.

C44300, arsenical admiralty metal, is an alpha tin brass containing 30% Zn, 1% Sn, and .15% As. Arsenical admiralty metal is used for condenser, distiller tubes, heat exchanger tubes, ferrules, strainers, and tubesheets.

C46400, uninhibited naval brass (Tobin bronze), is a beta tin brass containing 40% Zn and approximately 1% Sn. C46500, arsenical naval brass, contains an additional .15% As. The naval brasses are used for condenser tubesheets, welding rod, marine hardware, propeller shafts, valve stems, nuts, bolts, and fittings.

C68700, arsenical aluminum brass is a beta tin brass containing a low percentage of aluminum to improve resistance to erosion-corrosion in high-velocity salt waters, plus a small amount of arsenic. Arsenical aluminum brass is used for condenser tubing plate, heat exchanger tubing, and distiller tubing.

Leaded Brasses. Leaded brasses are alpha, beta, or tin brasses that contain from .5% Pb to 4.0% Pb to improve machinability. The lead is present in the microstructure as uniformly dispersed particles. See Figure 18-11. Mechanical properties, such as ductility, are not affected unless the lead content is extremely high.

Many of the common brasses and tin brasses have leaded equivalents (alloys with similar base com-

position, but with lead added). Leaded brasses are used for valve components, hardware, nuts, screws, and components that are machined at high rates, such as automatic screw machine components. Leaded brasses are often supplied as extruded or drawn rod.

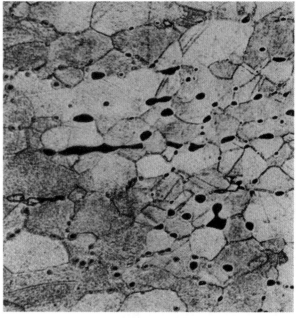

American Society for Metals

Figure 18-11. The microstructure of leaded beta brass shows a dispersion of lead particles.

C31400, leaded commercial bronze, is a leaded alpha brass with 10% Zn. Leaded commercial bronze is used for screw machine components, pickling racks and fixtures, electrical plug-type connectors, and builders' hardware.

C33500 (low-leaded brass), C34000 (medium-leaded brass), C34200 (high-leaded brass), C35600 (extra-high leaded brass), and C36000 (free-cutting brass) are leaded alpha brasses with 35% Zn. The difference between the brasses is the amount of lead they contain.

C37700, forging brass, is a leaded beta brass containing 38% Zn and 2% Pb. Forging brass is the most forgeable copper alloy and is used for all types of forgings and pressings.

C36500, leaded Muntz metal, and C37000, free-cutting Muntz metal, are beta brasses with 40% Zn and increasing alloying amounts of lead. Leaded Muntz metal is used for condenser tube plates, and free-cutting Muntz metal is used for automatic screw machine components.

Casting Brasses

Casting brasses are alpha or beta brasses containing specific alloying additions to improve their castability and strength beyond that of regular copper-zinc binary alloys. The alloying additions consist of combinations of tin, lead, iron, manganese, aluminum, and nickel. Casting brasses can be poured into complex shapes having very low levels of porosity and good mechanical properties. Casting brasses consist of red and semi-red brasses, yellow brasses, and manganese bronzes. When properly cleaned, red, semi-red, and yellow casting brasses can be plated with nickel. Nickel plating improves corrosion resistance.

Red and Semi-red Brasses. Red and semi-red brasses are general purpose casting alloys, which are usually leaded, and are the easiest to cast. C83600, leaded red brass, and C83800, hydraulic bronze, are the most popular alloys. Leaded red brass contains 85% Cu, 5% Sn, 5% Pb, and 5% Zn. Hydraulic bronze contains 83% Cu, 4% Sn, 6% Pb, and 7% Zn. Red and semi-red brasses have moderate strength, ductility, and corrosion resistance, and possess good machinability. These brasses are used for plumbing goods and fire fighting equipment.

C84400 and C84800 are leaded semi-red brasses. These leaded semi-red brasses have higher zinc and lead contents. C84400 contains 81% Cu, 3% Sn, 7% Pb, and 9% Zn. C84800 contains 76% Cu, 2.5% Sn, 6.5% Pb, and 15% Zn. C84400 and C84800 are used for plumbing goods.

Yellow Brasses. Yellow brasses are cheaper and stronger than the red and semi-red brasses. Yellow brasses are usually leaded and include C85200 and C85400. C85200 contains 72% Cu, 1% Sn, 3% Pb, and 24% Zn. C85400 contains 67% Cu, 1% Sn, 3% Pb, and 29% Zn. Yellow brasses are used in die castings for plumbing goods and in accessories for low-pressure valves and fittings.

Manganese Bronzes. Manganese bronzes are high-strength beta brasses containing 55% Cu to 60% Cu, 38% Zn to 42% Zn, 0% Sn to 1.5% Sn, 0% Fe to 2% Fe, 0% Al to 1.5% Al, and up to 3.5% Mn. Manganese bronzes have tensile strengths in the range 60 ksi to 110 ksi (415 MPa to 760 MPa) and possess good corrosion resistance. Lead decreases tensile strength and elongation, and should not exceed .1% Pb in the high-strength versions. Tin is added to the lower-strength alloys to increase resistance to dezincification, but is limited to .1% Sn in high-strength alloys to prevent reduction in strength and ductility. In order of increasing tensile strength, the manganese bronzes include C86400 (60 ksi, or 415 MPa), C86500 (65 ksi, or 450 MPa), C86100 (90 ksi, or 620 MPa), C86200 (90 ksi, or 620 MPa) and C86300 (110 ksi, or 760 MPa).

Many of the elements added to manganese bronzes affect the microstructure in the same way as increasing the zinc content. The addition of nickel has the opposite effect. The following numbers express the approximate zinc replacement capacity (Zn_{rc}) of the various elements:

	Si	Al	Sn	Mg	Fe	Mn	Ni
Zn_{rc}	10	6	2	2	.9	.5	–1.2 (1.2Cu)

One percent silicon has an effect similar to that of 10% Zn, while 1% Ni has an effect similar to that of 1.2% Cu. To figure the zinc equivalent (*ZnE*) of a manganese bronze, the percent zinc is added to the sums of the percentages of each alloying element and their zinc replacement capacity. The following formula is used to figure the equivalent zinc percentage (E_{Zn}):

$$E_{Zn} = \frac{ZnE}{(\% \ Cu + ZnE)}$$

where

E_{Zn} = equivalent zinc percentage

ZnE = zinc equivalent

%Cu = percent copper

Example: Figuring equivalent zinc percentage

What is the equivalent zinc percentage of a manganese bronze with 62% Cu, 34% Zn, 2% Al, and 2% Mn?

Figuring zinc equivalent

$$ZnE = 34\% \text{ Zn} + (2\% \text{ Al} \times 6) + (2\% \text{ Mn} \times .5)$$

$$ZnE = 34 + 12 + 1$$

$$ZnE = \mathbf{47}$$

Figuring equivalent zinc percentage

$$E_{Zn} = \frac{ZnE}{(\%Cu + ZnE)}$$

$$E_{Zn} = \frac{47}{(62 + 47)}$$

$$E_{Zn} = \frac{47}{109}$$

$$E_{Zn} = \mathbf{43.1\%}$$

From the actual zinc content of 34% Zn, the alloy would appear to be an alpha brass. The other alloying elements create the equivalent zinc percentage of 43.1%, which brings the alloy into the two-phase alpha-plus-beta region.

Manganese bronzes are used in applications that require high levels of strength, hardness, and corrosion resistance. See Figure 18-12. Manganese bronzes are use for propellers and fittings on ships, bushings and bearings, and gear components. Manganese bronzes are sometimes hot worked to further improve strength.

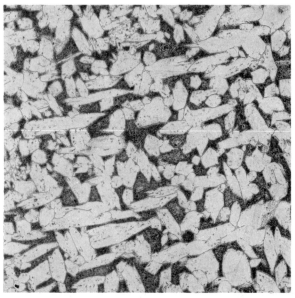

American Society for Metals

Figure 18-12. Manganese bronzes exhibit a duplex microstructure.

Tin Bronzes

Tin bronzes are wrought and cast alloys of copper and tin. Tin bronzes, which are also referred to as phosphor bronzes, contain from 1.25% Sn to 10.00% Sn, plus lead, zinc, nickel, and phosphorus. The phosphorus is added as a deoxidizer during casting, and it improves soundness and cleanliness.

Effects of Alloying Elements. Tin is the most important alloying element in copper. It exerts a significant strengthening effect on copper. The phosphorus addition varies from a few hundred parts per million to 1% P. The amount of phosphorus over and above the amount required for deoxidation exists in solid solution and increases hardness and strength at the expense of ductility and electrical conductivity. Up to 5% Zn is added to improve casting properties, as much as 25% Pb is added to improve machinability and wear resistance, and small amounts of nickel are added to disperse the lead and refine grain size.

On the copper-tin phase diagram, the solubility of tin in copper is 13.5% at 798°C (1468°F) and approximately 1% at room temperature. A peritectic reaction occurs at 798°C (1468°F), leading to the formation of a beta solid solution. When cooled to 586°C (1086°F), beta changes to gamma. At 521°C (970°F), gamma transforms to a eutectoid mixture of alpha and delta. The delta constituent is essentially the compound $Cu_{31}Sn_8$. This compound is pale blue, hard, and brittle.

Equilibrium is not easily achieved, even at 600°C (1110°F), and the duplex alpha-plus-beta structure is only attained after prolonged holding in the temperature range 280°C to 350°C (540°F to 660°F). Cast tin bronzes are far from equilibrium and have a microstructure consisting of alpha-plus-delta eutectoid. The wide separation of the liquids and solidus curves leads to a large composition gradient during solidification, and pronounced coring occurs in tin bronzes. See Figure 18-13.

Tin bronzes have high strength, good toughness, high corrosion resistance, and low coefficient of friction. This makes tin bronzes suitable for bearings operating under high loads. Tin bronzes are preferred over red brasses and semi-red brasses in instances where greater corrosion resistance, leak tightness, and strength at higher operating temperatures are required.

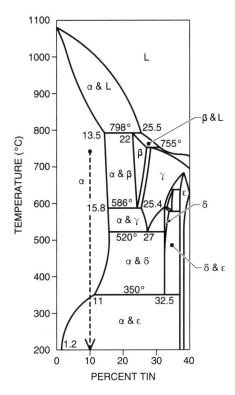

Figure 18-13. The copper end of the copper-tin phase diagram shows a wide temperature range between the liquidus and the solidus, making these alloys highly susceptible to coring.

Wrought Tin Bronzes. Wrought tin bronzes contain 2.5% Sn to 8.5% Sn and .10% P to .35% P, with significantly less lead than their cast equivalents. Wrought tin bronzes are used for applications requiring wear resistance and springiness. Atomic diffusion takes place so slowly that the delta phase actually appears in alloys containing 7% Sn. Alloys with greater amounts of tin can only be worked with difficulty, unless they are homogenized for a prolonged time at 760°C (1400°F) to produce a uniform solid solution.

C50500, phosphor bronze 1.25% E, has 1.25% Sn and exhibits excellent cold formability and good hot formability. Phosphor bronze 1.25% E is used for electrical contacts and flexible hoses.

C51000, phosphor bronze 5% A, has 5% Sn, excellent cold formability, and poor hot formability. Phosphor bronze 5% A is used for bridge bearing

plates, beater bars, bellows, clips, fasteners, bushings, springs, perforated sheets, textile machinery, and welding rods.

C52100, phosphor bronze 8% C, has 8% Sn and has good cold formability and poor hot formability. Phosphor bronze 8% C is used in more severe service conditions than phosphor bronze 5% A.

C52400, phosphor bronze 10% D, has 10% Sn and has good cold formability and poor hot formability. Phosphor bronze 10% D is used for heavy bars and plates under severe compression, such as bridge expansion plates and fittings. Phosphor bronze 10% D is also used for articles requiring extra spring quality and fatigue resistance.

C54400, free-cutting phosphor bronze, has 4% Sn, 4% Pb, and 4% Zn. Free-cutting phosphor bronze has good cold formability but should not be hot formed. Free-cutting phosphor bronze is used for bearings, bushings, gears, pinions, screw machine products, and valve components.

Cast Tin Bronzes. In addition to tin and phosphorus, cast tin bronzes contain zinc and lead. Zinc improves castability and pressure tightness. Lead is added to improve machinability. Cast tin bronzes are identified by the nominal percentages of their principal alloying elements in the order copper-tin-lead-zinc. For example, 88-10-0-2 bronze contains 88% Cu, 10% Sn, 0% Pb, and 2% Zn.

Instead of the wrought equivalents, cast tin bronzes are used where a softer metal is required for operations with slow to moderate speeds, and where the loads do not exceed 800 psi (5.5 MPa). Very high lead contents, from 10% Pb to 25% Pb, are specified where moderate loads are encountered under conditions of poor or nonexistent lubrication, such as in mining or cement plant equipment.

C90300, 88-8-0-4 tin bronze (G bronze), is used for bearings, bushings, pump impellers, piston rings, seal rings, and steam fittings. C90500, 88-10-0-2 tin bronze (gun metal), is used for similar applications to C90300.

C90700, gear bronze, contains 89% Cu and 11% Sn. Gear bronze is used for worm wheels and gears. It is also used for bearings operating under heavy loads at slow speeds.

C92200, 88-6-1.5-4.5 leaded tin bronze (M valve bronze), is used for valve components, flanges and fittings, oil pumps, gears, bushings, bearings, and backings for babbitt-lined bearings. It is also used for low-pressure components up to 3 ksi (20 MPa) and 290°C (550°F).

C92300, 87-8-1-4 leaded tin bronze, is a strong general-utility structural bronze that is used in severe conditions. It is used in valves, expansion joints, and steam pressure castings.

C92600, 87-10-1-2 leaded tin bronze (commercial bronze), is used for high-duty bearings where wear resistance is essential, for example in heavy-pressure bearings and bushings used against hardened steel. Commercial bronze is also used for special high-pressure pipe fittings and pump pistons.

C92700, 88-10-2-0 leaded tin bronze, is used for general purpose bearings, bushings, pump impellers, piston rings, and gears. C93200, 83-7-7-3 high-leaded tin bronze (bearing bronze 660), is a general-utility bearing alloy used in automobile bearings and bushings. C93500, 85-5-9-1 high-leaded tin bronze, is used for small bearings, bushings, and backings for babbitt-lined automotive bearings.

C93700, 80-10-10-0 high-leaded tin bronze, is a general bearing alloy suited to applications where lubrication may be difficult. It is widely used in machine tools, steel mill machinery, and automotive applications such as high-speed and heavy-pressure pumps, impellers, and components requiring corrosion resistance and pressure tightness.

C93800, 78-7-15-0 high-leaded tin bronze (anti-acid metal), is used for locomotive engine casting and general service bearings for moderate pressure, and as a general purpose bearing metal for rod bushings, shoes and wedges, and freight car bearings. It is also used in pump impellers and bodies for acidic mine water.

C94300, 70-5-25-0 high-leaded tin bronze, is used where high loads are encountered under poor or nonexistent lubrication conditions. It is used for bearings under light load and high speed, driving boxes, and railroad bearings.

Aluminum Bronzes

Aluminum bronzes are wrought and cast alloys of copper containing 7.0% Al to 13.5% Al, plus small amounts of manganese, nickel, and iron. Aluminum bronzes are not leaded.

Microstructure. On the copper-aluminum phase diagram, the phases that form at 500°C (930°F) generally do not transform at lower temperatures unless excessively long soaking times are employed. Alloys with <8% Al are single phase. Those with >8% Al initially form beta phase. When slowly cooled, the beta phase transforms by a eutectoid reaction to alpha-plus-gamma$_2$. The slow-cooled microstructure is duplex, consisting of primary alpha and a lamellar mixture of eutectoid alpha-plus-gamma$_2$. See Figure 18-14.

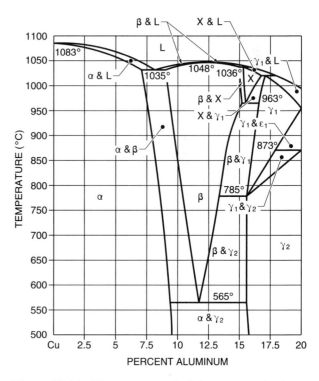

Figure 18-14. The copper end of the copper-aluminum phase diagram indicates that alloys up to approximately 8% Al are single phase.

Duplex aluminum bronzes can be quenched and tempered to improve mechanical properties. When oil or water quenched from a temperature of 815°C to 1010°C (1500°F to 1850°F), the beta phase undergoes a martensite (shear type) reaction to form beta prime. Beta prime is hard and brittle. It is tempered at 595°C to 650°C (1100°F to 1200°F) to improve its ductility and toughness. See Figure 18-15.

Aluminum bronzes have useful strength and excellent corrosion resistance. Their iron content varies from 2% Fe to 4% Fe, which produces a FeAl$_3$ intermetallic compound. This compound accounts for the hardness and wear resistance of aluminum bronzes. The FeAl₃ particles appear in the microstructure as small, round particles, which occasionally provide the surface with a rusty tinge. However, this has no effect on overall corrosion resistance.

Nickel is added to aluminum bronzes to further improve corrosion resistance. The addition of nickel introduces kappa phase into the microstructure. See Figure 18-16. Nickel alters the corrosion resistance of the beta phase, which improves resistance to dealuminification. *Dealuminification* is the preferential corrosion of aluminum that occurs in nickel-free aluminum bronzes.

Wrought Aluminum Bronzes. Wrought aluminum bronzes are usually formed by hot working. Wrought aluminum bronzes may also be formed and strengthened by cold working.

C61400, an aluminum bronze with 7% Al, is a wrought single-phase aluminum bronze. It is used as tubing and plate for condensers and heat exchangers, fasteners, and corrosion-resistant vessels.

C62300 (9% aluminum bronze), C62400 (11% aluminum bronze), and C62500 (13% aluminum bronze) are wrought duplex aluminum bronzes. These bronzes are used for gears, wear plates, cams, bushings, nuts, drift pins, and tie rods.

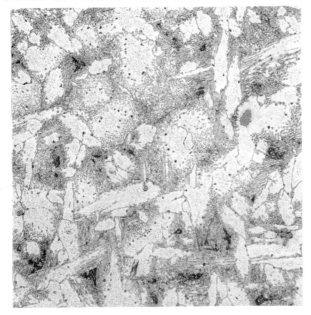

American Society for Metals

Figure 18-16. The microstructure of nickel-aluminum bronzes shows kappa phase and iron rounds.

SLOW COOLING RATE

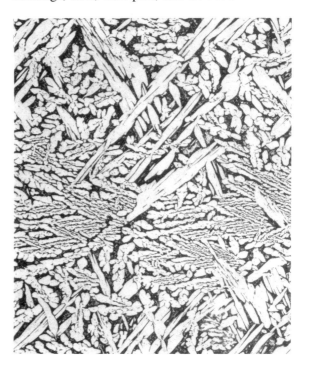

FAST COOLING RATE

American Society for Metals

Figure 18-15. The microstructure of cast duplex aluminum bronzes may be altered by the rate of cooling.

C63000, with 4% Ni and 10% Al, and C63200, with 4% Ni and 9% Al, are wrought nickel-aluminum bronzes. These wrought nickel-aluminum bronzes are used for nuts, bolts, pump components, shafts, and corrosion and spark resistant components for industrial and marine use.

Cast Aluminum Bronzes. Cast aluminum bronzes have superior galling and fatigue resistance compared to the manganese bronzes. Alloys with 8% Al to 9% Al are used for bushings and bearings in light-duty and heavy-duty machinery. Those with 9% Al to 11% Al are used as-cast or heat treated in heavy-duty service. Alloys with >11% Al have high levels of hardness and very low elongation, and are used for guides and aligning plates. Alloys with >13% Al have hardness levels that exceed 300 HB, but they are brittle and suitable for dies and components that are not subjected to impact loads.

C95200 is a cast single-phase aluminum bronze with 9% Al. This aluminum bronze is used for acid-resistant pumps, bearings, bushings, gears, and nonsparking hardware.

C95300, with 10% Al, and C95400, with 11% Al, are cast duplex aluminum bronzes. These aluminum bronzes are used for pump impellers, bearings, gears, and nonsparking hardware.

C95500, with 4% Ni and 11% Al, and C95800, with 4% Ni and 9% Al, are cast nickel-aluminum bronzes. These aluminum bronzes are used for valve guides, valve seals, and ship propellers.

C95700, with 8% Al, 2% Ni, and 12% Mn, is a cast duplex nickel-aluminum bronze. Manganese improves resistance to dealuminification. This bronze is used for propellers, impellers, stator clamp segments, nonsparking hardware, welding rods, valves, pump casings, and marine fittings.

Silicon Bronzes

Silicon bronzes are wrought and cast alloys of copper with 1% Si to 5% Si and additions of manganese, iron, and zinc. Certain silicon bronzes have leaded equivalents. Silicon bronzes have high strength similar to carbon steel, good toughness, and excellent corrosion resistance.

The maximum solubility of silicon in copper is 5.3% at 852°C (1565°F). The eutectoid reaction that

occurs at 554°C (1030°F) is extremely sluggish, so silicon bronzes with <5% Si are generally single phase. See Figure 18-17.

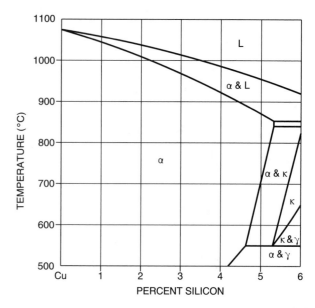

Figure 18-17. The copper-silicon phase diagram indicates that alloys with up to 5% Si are single phase because the eutectoid reaction is very sluggish.

Wrought Silicon Bronzes. Wrought silicon bronzes may be hot worked. The single-phase wrought silicon bronzes may also be cold worked. C65100, low-silicon bronze, is a wrought silicon bronze with 1.5% Si. It is used for aircraft hydraulic pressure lines, screws, bolts, cable clamps, heat exchanger tubes, and welding rod.

C65500, high-silicon bronze, is a wrought silicon bronze with 3% Si. High-silicon bronze is used for similar applications as low-silicon bronze and is used for chemical equipment, screen cloth, and marine propeller shafts.

Cast Silicon Bronzes. Cast silicon bronzes contain additions of zinc or tin to improve fluidity and castability, which provides excellent casting characteristics. C87200, Everdur, is a cast silicon bronze with up to 5% Si and 5% Zn. Everdur is used as a substitute for tin bronze when good corrosion resistance is needed. Everdur is also used in bearings, bells, impellers, pump and valve components, and marine fittings.

C87500 and C87800, Tombasil, are cast silicon bronzes that contain 4% Si and 14% Zn. These cast silicon bronzes are very similar. C87500 and C87800 are used for bearings, rocker arms, valve stems, and small boat propellers.

Copper-Nickels

Copper-nickels (cupronickels) are wrought and cast alloys of copper containing up to 30% Ni, plus minor additions of chromium, tin, beryllium, or iron. Iron is the most important addition in copper-nickels and is added to increase resistance to erosion-corrosion in water. Resistance to erosion-corrosion is the resistance to the effect of velocity or turbulence in a corrosive environment.

Copper-nickels have moderate strength and better corrosion resistance than other copper alloys. Corrosion resistance increases with increasing nickel content. Copper and nickel are soluble in one another, so all copper-nickels are single-phase alloys. Annealing the copper-nickels eliminates the coring that is often present in cast alloys. See Figure 18-18.

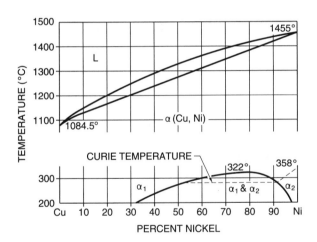

Figure 18-18. The copper-nickel phase diagram indicates that copper and nickel are soluble in one another at all compositions and form a single-phase microstructure.

Wrought Copper-Nickels. Wrought copper-nickels are used in condenser and heat exchanger tubing, tubesheets, water boxes, and ferrules. C70600, C71000, and C71500 are wrought copper-nickels with up to 1% Fe and with 10% Ni, 20% Ni, and 30% Ni, respectively. C70600 is designated 90-10 copper-nickel, C71000 as 80-20 copper-nickel, and C71500 as 70-30 copper-nickel. The 90-10 copper-nickel and 70-30 copper-nickel are commonly used for condensers, condenser tubesheets and tubing, heat exchanger tubing, salt water piping, and ferrules. The 70-30 copper-nickel is used in applications with the severest velocity and temperature conditions.

C72500, tin-bearing copper-nickel, contains 10% Ni and 2% Sn. Tin-bearing copper-nickel is used for relay and switch springs, connectors, lead frames, and brazing filler metals.

C71900, chromium-bearing copper-nickel, contains 30% Ni and 3% Cr. Chromium-bearing copper-nickel is hardened by spinodal decomposition. *Spinodal decomposition* is the growth of compositionally different waves within the crystal structure, without any basic change in crystal structure. Spinodal decomposition is achieved by soaking at 900°C to 1010°C (1650°F to 1850°F) and slow cooling through the range 760°C to 425°C (1400°F to 800°F). Spinodally decomposed C71900 has more than twice the strength of hot-rolled 70-30 copper-nickel. It is used when greater strength and greater hardness are required.

Cast Copper-Nickels. Cast copper-nickels have similar compositions to their wrought alloys. Cast copper-nickels are used for components such as valves, pump bodies, elbows, and flanges that require resistance to salt water corrosion.

C96200, with 10% Ni, and C96400, with 30% Ni, are the cast 90-10 copper-nickel and the cast 70-30 copper-nickel, respectively. These cast copper-nickels are used in sand, centrifugal, and continuous cast components, such as valves, pump bodies, flanges, and elbows for applications requiring resistance to seawater corrosion.

C96600, beryllium copper-nickel, is a cast copper-nickel with 30% Ni and .5% Be. Beryllium copper-nickel is precipitation hardened at 510°C (950°F) to achieve almost twice the strength of a conventional 70-30 copper-nickel. Beryllium copper-nickel is used for high-strength construction components in marine service, such as pressure housings for long unattended submergence, pump bodies, valves, and seawater line fittings.

Nickel-Silvers

Nickel-silvers are wrought and cast alloys of copper, with 5% Zn to 45% Zn and 5% Ni to 30% Ni. Nickel has a strong decolorizing effect on copper-zinc alloys. With >20% Ni, the color is a silver-white and the alloy takes a brilliant polish. Nickel-silvers may also be leaded.

Nickel-silvers have moderate strength and good corrosion resistance. Nickel-silvers provide an excellent base for chromium, nickel, or silver plating.

Wrought Nickel-Silvers. Wrought nickel-silvers are either single phase or duplex. Duplex wrought nickel-silvers are significantly stronger, harder, and less ductile than the single-phase wrought nickel-silvers. Wrought nickel-silvers are easily cold worked, but have poor hot-working characteristics. Wrought nickel-silvers are used for rivets, screws, table flatware, zippers, costume jewelry, nameplates, key blanks, camera parts, and optical equipment.

C75200, nickel-silver 65-18, is a wrought single-phase nickel-silver that contains 65% Cu, 18% Ni, and 17% Zn. C77000, nickel-silver 55-18, is a wrought duplex nickel-silver that contains 55% Cu, 18% Ni, and 27% Zn.

Cast Nickel-Silvers. Cast nickel-silvers can be single phase or duplex. Cast duplex nickel-silvers have a higher zinc content and are stronger but less ductile than cast single-phase nickel-silvers. Cast nickel-silvers are used for hardware fittings, valves and valve trim, sanitary fittings, and parts for musical instruments. C97300, leaded nickel brass, is a cast single-phase leaded nickel-silver that contains 64% Cu, 12% Ni, 20% Zn, 10% Pb, and 2% Sn. C97600, dairy metal, is a cast single-phase leaded nickel-silver that contains 64% Cu, 20% Ni, 8% Zn, 4% Pb, and 4% Sn.

COPPER FABRICATION

Copper and copper alloys are easy to fabricate. Copper and copper alloys are fabricated by forming or machining. Joining by welding poses difficulties in some copper alloys. Copper and copper alloys are easy to braze and solder.

Forming

Copper alloys are formed or shaped by a wide variety of processes that include blanking and piercing, bending, drawing and stretch forming, coining, spinning, and forging. See Figure 18-19. Blanking and piercing is often done using the same tooling that forms and shapes the final component geometry. Most copper alloys are readily blanked and pierced.

FORMING	
Application	**Operation**
Electrical terminals and connectors	Blending, stretch forming, blanking, coining, and drawing
Electronic lead frames	Bending, coining, and blanking
Hollow ware, flatware	Roll forming and blanking
Builder's hardware	Shallow and deep drawing and stretch forming
Heat exchangers	Roll forming, bending, sinking, and blanking
Coinage	Blanking, coining, and embossing
Bellows, flexible hose	Cupping, deep drawing, and bending
Musical instruments	Blanking, drawing, coining, bending, and spinning
Ammunition	Blanking and deep drawing

Figure 18-19. Copper alloys are formed by a wide variety of processes.

The key criterion of bend formability depends on the orientation of the bend direction to the cold rolling direction. Bending at 90° to the rolling direction is optimum with alloys that exhibit strong anisotropy, such as cartridge brass.

Drawing and stretch formability varies with specific copper alloys. For example, low-brass, cartridge brass, yellow brass, and Muntz metal exhibit good drawing and stretch formability. Phosphor bronze 5% A, the nickel-silvers, and some tin brasses exhibit good drawing and stretch formability. Beryllium coppers are sometimes drawn in the solution annealed condition and then precipitation hardened. Grain size is the major property that determines drawing and stretch formability of coppers and single-phase copper alloys. See Figure 18-20.

Coinability depends upon strength and work-hardening rate of copper or copper alloys. Commercially pure coppers, low-zinc brasses, lower-alloy nickel-silvers, and copper-nickels exhibit good coinability.

GRAIN SIZE AND RECOMMENDED APPLICATIONS FOR COPPER ALLOYS		
Average Grain Size mm	in.	Operation and Surface Characteristics
.005 to .015	.0002 to .0006	Shallow forming or stamping. Components will have good strength and very smooth surfaces. Also used for very thin metal.
.010 to .025	.0004 to .0010	Stampings and shallow-drawn components. Components will have high strength and smooth surfaces. General use for metal thinner than .25 mm (.01 in.).
.015 to .030	.0006 to .0012	Shallow-drawn components, stampings, and deep-drawn components that require buffable surfaces. General use for thicknesses under .3 mm (.012 in.).
.020 to .035	.0008 to .0014	This grain size range includes the largest average grain that will produce components essentially free of roughened surfaces. Therefore, it is used for all types of drawn components produced from brass up to .8 mm (.032 in.) thick.
.010 to .040	.0004 to .0016	Begins to show some roughening of the surface when severely stretched. Good deep-drawing quality in .4 mm to .5 mm (.015 in. to .020 in.) thickness range.
.030 to .050	.0012 to .0020	Drawn components from .40 mm to .64 mm (.015 in. to .025 in.) thick brass requiring fairly good surface, or stamped components requiring no polishing or buffing.
.040 to .060	.0016 to .0024	Commonly used for general applications for the deep and shallow drawing of components from brass in .5 mm to 1.0 mm (.02 in. to .04 in.) thicknesses. Moderate roughened surfaces may develop on drawn surfaces.
.050 to .119	.0020 to .0047	Large average grain sizes are used for the deep drawing of difficult shapes or deep-drawning components for gauges 1 mm (.04 in.) and thicker. Drawn components will have rough surfaces except where smoothed by ironing.

Figure 18-20. Grain size is the major property that determines drawability and stretching formability of coppers and single-phase copper alloys.

Spinability is superior in copper alloys with a high plastic strain ratio, high tensile elongation, and low work-hardening rate. Tough pitch copper is the easiest copper alloy to spin. Most brasses, except for Muntz metal, are well suited for spinning. Nickel-silvers with <65% Cu and copper-nickels are also well suited for spinning.

Forging processes, such as closed die forging, upset forging, and ring rolling, are used on copper alloys to produce components for pressure-containing and high-strength applications. The most forgeable copper alloy is forging brass. See Figure 18-21.

Joining

Copper alloys are joined by welding, brazing, or soldering. Weldability varies for different copper alloys, but most are easily brazed or soldered. There are some difficulties that may be encountered when joining copper alloys.

Welding Difficulties. The difficulties that are associated with welding specific copper alloys are related to their characteristic properties. These properties include high thermal conductivity, high coefficient of thermal expansion, hot shortness, and high fluidity.

High thermal conductivity causes heat to be conducted rapidly by the base metal. This conduction may result in lack of fusion unless high heat input is applied or a high-speed process with high heat input, such as gas tungsten arc welding (GTAW) or shielded metal arc welding (SMAW), is used.

High coefficient of thermal expansion may cause distortion or cracking as the metal changes dimensions when cooled. Hot shortness occurs with coring or segregation during solidification and leads to grain boundary separation.

FORGEABILITY OF COPPER ALLOYS

UNS	Nominal Composition	Relative Forgeability[a] (%)
C10200	99.95 min Cu	65
C10400	Cu-.027Ag	65
C11000	99.9 min Cu	65
C11300	Cu-.027Ag + O_2	65
C14500	Cu-.65Te-.008P	65
C18200	Cu-.10Fe-0.90Cr-.10Si-.05Pb	80
C37700	Cu-38Zn-2Pb	100
C46400	Cu-39.2Zn-.8Sn	90
C48200	Cu-38Zn-.8Sn-.7Pb	90
C48500	Cu-37.5Zn-1.8Pb-.7Sn	90
C62300	Cu-10Al-3Fe	75
C63000	Cu-10Al-5Ni-3Fe	75
C63200	Cu-9Al-5Ni-4Fe	70
C64200	Cu-7Al-1.8Si	80
C65500	Cu-3Si	40
C67500	Cu-39Zn-1.4Fe-1Si-.1Mn	80

[a] Ratings are in terms of the most forgeable alloy, forging brass (C37700).

Figure 18-21. Copper alloys may be ranked according to forgeability.

Methods of combatting hot shortness include reducing joint restraint, decreasing the size of the root opening, and increasing the size of the root pass. High fluidity of the molten metal can make out-of-position processes, such as overhead welding, difficult.

The most popular welding processes for copper and copper alloys are gas metal arc welding (GMAW) and GTAW. GMAW is used for welding thicker materials and GTAW is used for the thinner gauge materials. SMAW and oxyfuel gas welding (OFW) are also used.

The American Welding Society specifies welding filler metals for welding applications. Recommended welding filler metals for welding copper alloys may be covered electrodes or bare wire. See Figure 18-22.

Welding Copper Alloys. Commercially pure coppers require high preheating temperatures to counteract high thermal conductivities. Preheating temperatures range from 120°C to 540°C (250°F to 1000°F), depending on the section size. Electrolytic tough pitch copper is difficult to weld because it is susceptible to embrittlement. Electrolytic tough pitch copper should be brazed or soldered. With phosphorus deoxidized copper, silicon-containing electrodes are used to reduce oxygen pick-up. Oxygen-free coppers should be welded as rapidly as possible to avoid oxygen pick-up.

Sulfur, selenium, or tellurium (Te) present in free-machining coppers tends to cause cracking. These alloys are more easily soldered. Anneal-resistant coppers can be welded with less preheating than is required for the commercially pure coppers.

WELDING FILLER METALS FOR FUSION WELDING COPPER ALLOYS

AWS Classification		Common Name	Base Metal
Covered Electrode	Bare Wire		
ECu	ERCu	Copper	Coppers
ECuSi	ERCuSi-A	Silicon bronze	Silicon bronzes and brasses
ECuSn-A and ECuSn-C	ERCuSn-A	Phosphor bronze	Phosphor bronzes and brasses
ECuNi	ERCuNi	Copper-nickel	Copper-nickel alloys
ECuAl-A2	ERCuAl-A2	Aluminum bronze	Aluminum bronzes, brasses, silicon bronzes, and manganese bronzes
ECuAl-B	ERCuAl-A3	Aluminum bronze	Aluminum bronzes
ECuNiAl	ERCuNiAl	—	Nickel-aluminum bronzes
ECuMnNiAl	ERCuMnNiAl	—	Manganese-nickel-aluminum bronzes
	RBCuZn-A	Naval brass	Brasses and copper
	RCuZn-B	Low-fuming brass	Brasses and manganese bronzes
	RCuZn-C	Low-fuming brass	Brasses and manganese bronzes

Figure 18-22. Welding filler metals used for welding copper and copper alloys may be covered electrodes or bare wire.

High-strength coppers must be precipitation hardened after welding to develop full strength.

Beryllium coppers form tenacious oxide films that inhibit wetting and fusion during welding. Joint surfaces of beryllium copper must be absolutely clean. Beryllium coppers are welded in the solution-annealed or overaged condition. Should multiple passes be required, these treatments have greater metallurgical stability. Beryllium coppers are then precipitation hardened to achieve the required mechanical properties. Components in the precipitaion- hardened condition are not normally welded because of cracking susceptibility in the heat-affected zone.

When welding brasses, zinc is subject to vaporization. Vaporization may be minimized by decreasing or eliminating preheating, or by using lower welding currents. High-zinc brasses have lower electrical conductivities and require less preheating than low-zinc brasses. Leaded brasses are not suited for welding because the lead creates porosity and promotes cracking.

Tin bronzes have high thermal conductivities and wide freezing ranges, making them susceptible to hot shortness. To offset these problems, preheating is performed in the range 150°C to 205°C (300°F to 400°F), high currents and high travel speeds are employed, and each pass of weld is peened (mechanically worked by hammering) while hot. Leaded tin bronzes should not be welded.

Aluminum bronzes are susceptible to hot shortness, especially single-phase alloys with <7% Al. Aluminum bronzes with higher additions are weldable with adequate preheating.

Silicon bronzes have relatively low thermal conductivities. Silicon bronzes only require preheating when welding heavy sections.

Copper-nickels have similar thermal and electrical conductivity to carbon steel and are relatively easy to weld. Cleanliness of the copper-nickels before welding is essential. Lead, phosphorus, and sulfur contamination cause intergranular cracking. Preheating is not normally required and the interpass temperature should not exceed 65°C (150°F).

Nickel-silvers are similar to brasses because of their high zinc contents. Nickel-silvers are brazed rather than welded.

Brazing and Soldering. Brazing is widely used for joining copper alloys. Silver alloy and copper-phosphorus welding filler metals are commonly used. See Figure 18-23.

Soldering is used to join many copper alloys except aluminum bronzes, beryllium coppers, silicon bronzes, and manganese bronzes. These copper alloys are difficult to clean properly. Cleanliness is essential to achieve a good joint. Soldering does not soften cold worked alloys. Lead-tin solders are most commonly used solders. Soldering is limited to room temperature applications, such as plumbing. Joint strengths are much lower for soldering than for brazing or welding.

Machining

Equivalent cast and wrought copper alloys have similar machinability. The cast copper alloys are separated into three machinability groups, which are easy, moderate, and difficult. See Figure 18-24.

The easy machinability group includes leaded single-phase alloys. Additions of lead improves machinability of the alloys by facilitating chip breakage and decreasing tool wear.

The moderate machinability group includes alloys with two or more phases. The secondary phases are harder and more brittle than the copper-rich matrix phase and act as internal chip breakers, which reduces chip size and improves overall machinability. Alloys in this group include silicon bronzes, some aluminum bronzes, and high-tin bronzes.

The difficult machinability group consist of two copper alloys. These alloys include the high-strength manganese bronzes and aluminum bronzes with high-iron or high-nickel additions.

COPPER ALLOY PROPERTIES

Because of the various properties that copper alloys exhibit, they are used in many applications. Major properties of importance include high electrical conductivity, high thermal conductivity, wear and galling resistance, and corrosion resistance.

Electrical Conductivity and Thermal Conductivity

Copper is used for its high electrical conductivity and high thermal conductivity more than any other metal. Alloying decreases electrical conductivity, and, to a lesser extent, thermal conductivity. Silver, lead, zinc, and cadmium reduce these conductivities only slightly compared with other elements.

Copper castings are used for their current carrying capacity and as water-cooled parts of melting and refining furnaces. The electrical and thermal conductivity of castings are lower than those of equivalent wrought alloys. To maximize conductivity, the regular deoxidizers used in castings, such as silicon, tin, zinc, aluminum, and phosphorus, cannot be used.

Calcium boride or metallic lithium additions help produce sound castings with high conductivity.

Wear and Galling Resistance

Copper alloys are used for bearings because they offer a combination of strength, corrosion resistance, and either wear resistance or self-lubricity. Several groups of copper alloys used for bearings and wear-resistant applications include tin bronzes, high-leaded tin bronzes, manganese bronzes, aluminum

BRAZING COPPER AND COPPER ALLOYS				
Material	**AWS Classification Brazing Filler Metals**	**AWS Brazing Atmosphere[a]**	**AWS Brazing Flux, No.**	**Remarks**
Coppers	BCuP-2, BCuP-3, BCuP-5, RBCuZn, BAg-1a, BAg-1, BAg-2, BAg-5, BAg-6, and BAg-18	1, 2, or 5	3 or 5	Oxygen-bearing coppers should not be brazed in hydrogen-containing atmospheres.
High coppers	BAg-8 and BAg-1			
Red brasses	BAg-1a, BAg-1, BAg-2, BCuP-5, BCuP-3, BAg-5, BAg-6, and RBCuZn	1, 2, or 5	3A 3, 5 for RBCuZn	
Yellow brasses	BCuP-4, BAg-1a, BAg-1, BAg-5, BAg-6, BCuP-5, and BCP-3	3, 4, or 5	3	Keep brazing cycle short.
Leaded brasses	BAg-1a, BAg-1, BAg-2, BAg-7, BAg-18, and BCuP-5	3, 4, or 5	3	Keep brazing cycle short, and stress relieve before brazing.
Tin brasses	BAg-1a, BAg-1, BAg-2, BAg-5, BAg-6, BCuP-5, BCuP-3, and RBCuZn for low tin	3, 4, or 5	3	
Phosphor bronzes	BAg-1a, BAg-1, BAg-2, BCuP-5, BCuP-3, BAg-5, and BAg-6	1, 2, or 5	3	Stress relieve before brazing.
Silicon bronzes	BAg-1a, BAg-1, and BAg-2	4 or 5	3	Stress relieve before brazing. Abrasive cleaning may be helpful.
Aluminum bronzes	BAg-3, BAg-1a, BAg-1, and BAg-2	4 or 5	4	
Copper-nickel	BAg-1a, BAg-1, BAg-2, BAg-18, BAg-5, BCuP-5, BCuP-3	1, 2, or 5	3	Stress relieve before brazing.
Nickel silvers	BAg-1a, BAg-1, BAg-2, BAg-5, BAg-6, BCuP-5, BCuP-3	3, 4, or 5	3	Stress relieve before brazing, and heat uniformly.

[a] 1 = combusted fuel gas (low hydrogen), room temperature maximum dew point
2 = combusted fuel gas (decarburizing), room temperature maximum dew point
3 = combusted fuel gas, dried, –40°C maximum dew point
4 = combusted fuel gas, dried (carburizing), –40°C maximum dew point

Figure 18-23. Silver alloy and copper-phosphorus filler metals are commonly used to braze copper alloys.

bronzes, and silicon bronzes. These are cast on rolled strips, made into sintered components, or pressed and sintered onto a backing material.

MACHINABILITY OF SELECTED COPPER CASTING ALLOYS		
UNS	Common Name	Machinability Rating %
Easy		
C83600	Leaded red brass	90
C83800	Leaded red brass	90
C84400	Leaded semi-red brass	90
C84800	Leaded semi-red brass	90
C94300	High-leaded tin bronze	90
C85200	Leaded yellow brass	80
C85400	Leaded yellow brass	80
C93700	High-leaded tin bronze	80
C93800	High-leaded tin bronze	80
C93200	High-leaded tin bronze	70
C93500	High-leaded tin bronze	70
C97300	Leaded nickel brass	70
Moderate		
C86400	Leaded high-strength manganese bronze	60
C92200	Leaded tin bronze	60
C92300	Leaded tin bronze	60
C90300	Tin bronze	50
C95600	Silicon-aluminum bronze	50
C95300	Aluminum bronze	35
C86500	High-strength manganese bronze	30
Difficult		
C86300	High-strength manganese bronze	20
C95200	9% aluminum bronze	20
C95400	11% aluminum bronze	20
C95500	Nickel-aluminum bronze	20

Figure 18-24. Cast copper alloys may be divided into three groups according to ease of machining.

Tin bronzes have a high resistance to wear, high hardness, and moderate strength. High-leaded tin bronzes are used when a softer metal that operates at slow to moderate speeds is required. Manganese bronzes have high tensile strength and high shock resistance under motion and high loads. Aluminum bronzes increase in hardness as the aluminum content is increased. Depending on the composition, they are used either as-cast or as-cast and heat treated for applications ranging from light-duty high-speed machinery to heavy-duty service. The hardest and most brittle alloys are used for dies and components not subject to impact loads.

Corrosion Resistance

Copper is a noble metal, but unlike gold and other precious metals, it can be attacked in some corrosive environments. Copper and copper alloys do exhibit excellent corrosion resistance in various forms of water, acids and alkalies, some salt solutions, and organic compounds.

Fresh Water. The freshwater corrosion resistance of copper alloys is generally excellent. The greatest single application of copper tubing is for transporting hot and cold water in buildings.

When steel or aluminum is electrically in contact with copper, they can suffer galvanic corrosion (accelerated corrosion) in the region of contact, such as a weld joint. Galvanic corrosion may lead to premature failure of the joint. One remedy is to electrically insulate the two metals at the junction.

Brasses containing >15% Zn can suffer from dezincification. The zinc is preferentially removed, leaving behind a spongy mass of copper with considerably reduced strength. Tin inhibits dezincification, and additions of arsenic, antimony, or phosphorus further improve resistance to dezincification.

Copper alloys in waters and other environments are subject to erosion-corrosion in fast moving streams. The erosive action of the flowing stream removes protective films on the metal surface, which rapidly increases the rate of corrosion. Erosion-corrosion is characterized by undercut grooves, waves, ruts, gullies, rounded holes, and horseshoe patterns on the metal surface. Erosion-corrosion can be prevented by reducing fluid velocity or alloy upgrading to copper-nickels.

Crevice corrosion (under-deposit attack) may occur under deposits that create a crevice on the metal surface. Such deposits may consist of dirt, shell, vegetation, or rust from corroding steel equipment upstream. A localized oxygen deficient zone is created in the crevice formed by the deposit and the metal. The metal suffers accelerated attack. Crevice corrosion may be reduced by cleaning the surface or maintaining adequate fluid velocity to prevent solids from settling out.

Salt Water. Copper alloys are extensively used for saltwater applications, such as on ships and tidewater power stations. The most resistant alloys (in increasing order) are C44300 to C44500, C68700, C70600, and C71500. C70600 and C71500 are superior in fast moving polluted streams. As a group, copper alloys tend to resist biofouling (deposits) by marine organisms. The most resistant alloys contain at least 85% Cu.

Steam and Steam Condensate. Steam and steam condensate are not usually aggressive toward copper alloys. Steam and steam condensate are aggressive when significant amounts of carbon dioxide, oxygen, or ammonia are present.

Acids and Alkalies. Acids are generally aggressive enough that copper alloys are only considered for dilute applications at low temperatures. Oxidizing acids, such as nitric and chromic, are significantly more aggressive than reducing acids, such as sulfuric and hydrochloric. Copper alloys should not be used for applications involving oxidizing acids.

Copper alloys can generally be used for applications involving alkalies, with the exception of ammonium hydroxide, which slowly dissolves copper alloys. Also, brasses with >15% Zn are at risk of dezincification. Aluminum bronzes can lose aluminum by dealuminification in warm alkaline solu-

tions. Copper may be used in applications involving anhydrous ammonia, but the presence of small amounts of moisture promote rapid corrosion. Stress-corrosion cracking may occur if tensile stresses are present in brasses in moist ammonia and ammonium hydroxide.

Salt Solutions. Copper alloys are used in various applications involving salt solutions. Neutral chlorides, sulfates, and nitrates present no problems. Oxidizing salts, such as ferric chloride, can be aggressively corrosive. Acid salts hydrolyze to produce corresponding dilute acids. Corrosion resistance depends on the type of acid formed and its concentration. Alkaline salts act like alkalies and are generally not very corrosive. Mercury salts and sulfides are corrosive toward copper alloys. Salts of metals, such as silver nitrate, which are more noble than copper, rapidly attack copper by plating out the noble metal on the surface of the copper.

Organic Compounds. Copper alloys are generally used without problems in applications involving organic compounds, such as benzene, gasoline, sugar, or acetic acid. This is true unless the copper alloys are contaminated with aggressive salts, waters, or acids.

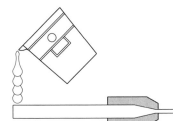

Nickel and Cobalt 19

Nickel is one of the most important nonferrous metals. Nickel is present in many alloys and exerts a significant effect on their properties. Several nickel alloy systems have unique properties, such as very low thermal expansion, that can only be attained through the use of nickel. Cobalt is a major constituent in certain types of wear-resistant, corrosion-resistant, heat-resistant, and magnetic alloys. Compared with the number of applications of nickel, there are relatively few cobalt applications. The principal alloying elements of cobalt are chromium, nickel, and iron.

NICKEL MANUFACTURING

Nickel is extracted from two types of ore to make primary nickel products. These are melted or combined with other materials to produce alloy ingots or powders, which are then made into various finished forms. Nickel alloys are work hardened by various processes and are softened or homogenized by annealing.

Extraction and Production

Nickel-containing ores are concentrated and refined into primary nickel products that contain various percentages of nickel. Extraction and refining is performed on two types of nickel-containing ores, which are sulfide and laterite (silicate) ores, that are found in many parts of the world. The sulfide ores (sometimes associated with iron, copper, and minerals that contain precious metals) are present in hard rock deposits. The laterite ores, by contrast, are clay-like materials that contain no other useful extractable elements.

Sulfide and laterite ores are refined into three types of primary nickel products, which are pure nickel, ferronickel (an alloy containing chiefly iron

and some nickel), and nickel oxide. Primary nickel products are usually remelted to produce nickel alloys. Ferronickel and nickel oxide are used specifically in steelmaking processes for stainless and low-alloy steels. Pure nickel is used in nickel plating, powder metallurgy, battery electrodes, or as an alloying addition in the manufacture of nickel-base alloys. See Figure 19-1. Primary nickel products are shaped into a variety of finished forms.

NICKEL APPLICATIONS	
Application	**Usage**
Stainless and heat-resistant steels	48%
High-nickel alloys	15%
Plating	10%
Other alloy steels	9%
Foundry (ferrous and nonferrous)	8%
Superalloys	5%
Other	5%

Figure 19-1. The largest application of primary nickel products is the manufacturing process of stainless steels.

Production of Finished Forms. The production of finished nickel alloy forms borrows melting and mechanical working practices from high-alloy steelmaking. Strict process control is required. For example, nickel and nickel alloys are extremely susceptible to sulfur embrittlement, so the quality of oil and gas used as heating fuel must meet strict purity standards.

The nickel production sequence consists of melting, hot working, and cold working. Nickel alloys are also produced as powders. Electric melting is the preferred method of melting nickel. It reduces contamination and provides operational flexibility. The purest raw materials are used, and contaminated scrap should be avoided. All scrap is thoroughly degreased before melting. Slag or refractory entrainment during melting is not allowed. The gas content of the melt may be reduced by means of a carbon monoxide boil. Small additions of silicon or aluminum are used to deoxidize the melt. Titanium is employed to fix (combine with) nitrogen. Sulfur is controlled with magnesium or calcium additions.

The molten metal is cast into ingots. For greater purity, the ingots are remelted using vacuum arc or electroslag remelting. These processes are similar to those used in steelmaking to produce high-quality materials for demanding service applications.

The ingots are hot worked into various shapes, such as forgings, plate, bar, and tubing. The temperatures for hot working range from 850°C to 1250°C (1560°F to 2280°F), depending on the alloy and the type of working operation. Careful temperature control is required because the optimum range for hot working most nickel alloys is narrow. Additional heat input may be required if the temperature falls below the specified range, and it must be supplied using low-sulfur fuel. Special lubricants are sometimes necessary during rolling or extruding because the oxide skin, which forms on these alloys at high temperatures, is not slippery enough to overcome frictional forces.

Cold working is used to produce sheet, strip, wire, rod, tube, and other forms. Intermediate annealing may be necessary to soften alloys that have work hardened excessively.

Nickel alloys are also produced as powders that are compacted to make fully dense products by means of continuous rolling, unidirectional pressing, or isostatic pressing. Powders have higher purity than other material forms, and greater chemical compositional precision can be obtained with them.

This compositional precision is particularly important in manufacturing magnetic and controlled expansion alloys.

Working and Heat Treatment

Nickel alloys are strengthened by work (strain) hardening, precipitation (age) hardening, and dispersion hardening (mechanical alloying). Nickel alloys are softened or homogenized by annealing.

Work Hardening. Nickel alloys are hardened by cold working. The rate of work hardening for nickel alloys is greater than for 1020 carbon steel and is usually less than the rate of work hardening for 304 stainless steel. The rate generally increases with the complexity of the alloy, from moderately low for nickel and nickel-copper alloys to moderately high for nickel-chromium alloys and superalloys. Precipitation-hardened alloys have higher work-hardening rates than their solution-hardening counterparts.

Precipitation Hardening. Precipitation hardening is performed on certain nickel alloys to produce significant strengthening. The principal hardening elements are magnesium, aluminum, silicon, and titanium. To produce hardening, the alloy is solution annealed, quenched, and reheated to an intermediate temperature. The alloy is held at the intermediate temperature for a specified time to develop the required mechanical properties. See Figure 19-2.

Nickel alloys are usually precipitation hardened in sealed boxes that are placed in a furnace. The boxes should be loosely packed, yet provide minimum excess space. Small horizontal or vertical furnaces without boxes are also used for the precipitation hardening of nickel alloys.

Dispersion Hardening. Pure nickel and certain nickel-chromium alloys may be strengthened by a uniform dispersion of fine refractory oxides, such as thoria (ThO_2). Oxides are the most beneficial additions to metal because they increase hardness, increase stability at high temperatures, create insolubility in the metal matrix, and are available in fine particulate form. Dispersion-hardened nickel is produced by powder metallurgy.

PRECIPITATION HARDENING FOR SELECTED NICKEL ALLOYS					
Alloy	UNS	Temperature (°C)	Solution Treating Time (hr)	Cooling Method[a]	Age Hardening
Monel K-500	N05500	980	.5 to 1	W	Heat to 595°C, hold 16 hours; furnace cool to 540°C, hold 6 hours; furnace cool to 480°C, hold 8 hours; and air cool
Inconel® 718	N07718	980	1	A	Heat to 720°C, hold 8 hours; furnace cool to 620°C, hold until furnace time for entire age-hardening cycle equals 18 hours; and air cool
Inconel® X-750	N07750	1150	2 to 4	A	Heat to 845°C, hold 24 hours; air cool; re-heat to 750°C, hold 20 hours; and air cool
		980	1	A	Heat to 730°C, hold 8 hours; furnace cool to 620°C, hold until furnace time for entire age-hardening cycle equals 18 hours; and air cool
Hastelloy® X	N06002	1175	1	A	Heat to 760°C, hold 3 hours; air cool; reheat to 595°C, hold 3 hours; air cool

[a] W = water quench and A = air cool

Figure 19-2. Specific nickel alloys are precipitation hardenable.

Annealing. Annealing is a general term that describes several types of heat treatment that soften, stress relieve, homogenize, or improve the corrosion resistance of a metal. Annealing processes for nickel include full annealing, stress relieving, stress equalizing, solution annealing, and stabilizing.

Full annealing is an annealing process that produces a recrystallized grain structure and softens work-hardened alloys. Depending on the alloy composition and the amount of prior work, full annealing usually requires temperatures between 705°C and 1205°C (1300°F and 2200°F). Full annealing is commonly performed by heating the component in a furnace in which the component is protected from oxidation by the products of combustion of the heating fuel.

Stress relieving is an annealing process that is performed to remove stresses in work-hardened, non-precipitation-hardenable alloys and does not recrystallize the grain structure. Depending on the alloy composition and the amount of work hardening, the temperature range of stress relieving is from 425°C to 870°C (800°F to 1600°F).

Stress equalizing is a low-temperature heat treatment that is performed to balance stresses in cold-worked material without an appreciable decrease in the mechanical strength produced by cold working.

Stress equalizing is performed at lower temperatures than stress relieving, depending on the alloy composition. Prolonged stress equalization has no detrimental effects on the nickel alloy or its mechanical properties.

Solution annealing is a high-temperature heat treatment that dissolves precipitates such as carbides and age-hardening compounds. Solution annealing is the first step in precipitation hardening and is also performed to produce a homogeneous metallurgical structure in corrosion-resistant alloys. Solution annealing is performed between 980°C and 1175°C (1800°F and 2150°F).

Stabilizing is an annealing process performed on corrosion-resistant nickel-iron-chromium-molybdenum alloys that contain additions of titanium, niobium, or tantalum to stabilize them. The alloy is heated to a specific value below the solution-annealing temperature to tie up carbon as a precipitate of titanium carbide, niobium carbide, or tantalum carbide. The carbide precipitation prevents sensitization that can reduce the corrosion resistance of the alloy.

Nickel Designation Systems

The American Society for Testing and Materials designations and standards are the most widely used for nickel alloys and cover the range of finished forms. The American Welding Society designations and standards cover brazing and welding filler metals. Aerospace Materials Specifications designations and standards address finished forms for aerospace applications and are generally the most demanding, which is reflected in the stringent quality requirements. Such requirements are not necessarily confined to the aerospace industry.

Unified Numbering System designations for wrought and cast nickel alloys have the prefix uppercase letter N. For example, N02200 is equivalent to commercially pure Nickel 200. There is no other designation system for wrought nickel alloys. The Alloy Castings Institute designations are used for the majority of cast nickel alloys. For example, ACI CZ-100 is the designation for cast commercially pure nickel.

In most standards, the chemical composition for nickel is expressed as nickel plus cobalt. This is because most forms of refined nickel contain approximately .5% Co. Generally, this amount of cobalt does not affect the properties of nickel alloys.

NICKEL ALLOYS

Nickel is one of the most widely used chemical elements. Nickel is incorporated as a major or minor constituent in approximately 3000 alloys. Nickel is also used as a plating material. Over half of all the nickel used is incorporated as an alloying element in steels, chiefly stainless and low-alloy steels.

The main alloying elements for nickel are copper, iron, molybdenum, chromium, and cobalt. Other metallic and nonmetallic elements are added to introduce specific characteristics. The major nickel alloy systems are based on nickel and nickel-chromium. The alloy systems that are based on nickel encompass high-nickel, nickel-copper, nickel-iron, and nickel-molybdenum alloys. These alloys have a wide variety of properties, from corrosion resistance to unique thermal expansion. The nickel-chromium phase diagram provides the basis for many corrosion-resistant and heat-resistant alloys. Corrosion-resistant and heat-resistant alloys encompass nickel-chromium, nickel-chromium-molybdenum, nickel-iron-chromium-molybdenum, nickel-iron-chromium, and superalloys. Many of the heat-resistant alloys and some of the corrosion-resistant alloys that are based on nickel-chromium are referred to as superalloys.

Nickel has an FCC crystal structure and retains a significant range of solid solubility with many of its alloying elements. The nickel-rich solid solution (gamma solid solution) is present in the microstructure of many nickel-base alloys.

High-nickel Alloys

High-nickel alloys are essentially made of pure nickel. High-nickel alloys are comprised of commercially pure nickel and high-purity nickel. See Figure 19-3. They also comprise dispersion-hardened nickel and nickel plating.

Commercially Pure Nickel. Commercially pure nickel contains at least 99.5% Ni. The principal wrought alloys are N02200, Nickel 200 and N02201, Nickel 201. Nickel 200 is a general purpose grade. Many commercially pure nickel alloys are strongly ferromagnetic. Nickel 200 has a restricted carbon content, making it more suitable for cold working and applications above 315°C (600°F). If the carbon content is not restricted, carbon may precipitate as graphite, which leads to embrittlement.

HIGH-NICKEL ALLOYS									
Alloy	**UNS**	**ASTM Specification**	**%Ni**	**%Cu**	**%Fe**	**%C**	**%Mn**	**%Si**	**%Mg**
Nickel 200	N02200	B160, B161, B162, B163, and B336	99.5	.13	.08	.08	.20	.05	—
Nickel 201	N02201		99.6	.13	.01	.01	.20	.05	—
Nickel 205	N02205	F1, F2, F3, and F9	99.5	.10	.02	.02	.20	—	.04
Nickel 270	N02270		99.98	—	.01	.01	—	—	—

Figure 19-3. High-nickel alloys are essentially pure nickel.

The cast version of Nickel 200 is CZ-100. CZ-100 contains approximately .75% C and 1% Si to increase casting fluidity. A higher-carbon, higher-silicon grade is occasionally specified for greater resistance to wear and galling. At room temperature, CZ-100 is slightly magnetic, but magnetism decreases with increased silicon content.

The microstructure of commercially pure nickel is single phase and consists of grains of gamma solid solution. Twinning may be present in cold-worked or annealed material. Carbon is present as finely distributed spheroidal graphite in cast commercially pure nickel.

Wrought and cast commercially pure nickels are used extensively in the chemical and process industries for resistance to the following types of chemical reagents: caustic soda and other alkalis at high temperatures; chlorine and hydrogen chloride at high temperatures; and salts, except for oxidizing chloride salts. These nickels are fabricated into vessels, coils, and pump components. They are also used for transducers, which convert electricity to ultrasonic energy. Compacted nickel powders are used in much of the world's coinage.

High-purity Nickel. High-purity nickel contains at least 99.99% Ni. Wrought high-purity nickels include N02205, Nickel 205, and N02270, Nickel 270. Nickel 205 has controlled amounts of trace elements and is used extensively in the electronics industry. Nickel 270 is suitable for heavy cold-working operations without the need for intermediate annealing.

Dispersion-hardened Nickel. Dispersion-hardened nickel is strengthened by a uniform mixture of inert refractory oxide particles, usually approximately 2% thoria. Dispersion-hardened nickel is made chiefly in sheet form and has good high-temperature strength. Dispersion-hardened nickel is used for combustion system components in advanced gas turbine engines, fixtures for high-temperature tensile testing, and for specialized furnaces and heating elements. Dispersion-hardened nickels must be coated to prevent corrosion at high temperatures.

A typical alloy is thoria dispersion-hardened nickel (TD nickel). TD nickel is mildly radioactive, and specific applications may require licensing.

Nickel Plating. Nickel is applied as a thin layer (plating) on substrate metals as electroplated, electroformed, or electroless nickel. Nickel plating is used extensively as an undercoating for decorative chromium. It is also used as an undercoating where surface build-up may be required, such as in the rebuilding of a worn or undersized component.

Nickel-Copper Alloys

Nickel and copper are completely soluble in one another. Nickel-copper alloys consist of solution-hardening, precipitation-hardening, and free-machining alloys. See Figure 19-4.

Solution-hardening Nickel-Copper Alloys. The base composition of solution-hardening nickel-copper alloys is N04400, alloy 400. Alloy 400 is widely known as Monel 400, or Monel. Alloy 400 contains 66% Ni, 32% Cu, and additions of iron and manganese.

NICKEL-COPPER ALLOYS										
Alloy	UNS	ASTM Specification	%Ni	%Cu	%Fe	%C	%Mn	%Al	%Ti	%S
Alloy 400	N04400	B127, B163, B164, B564, and B336	63	28 to 34	2.5	.30	2.5	—	—	—
Alloy K-500	N05500	—	63	27 to 33	2.0	.25	1.5	.5	2.3 to 3.15	.35 to .85
Alloy R-405	N04405	B164	63	32	2.5	.30	2.0	—	—	.012

Figure 19-4. Nickel-copper alloys may be solution-hardening, precipitation-hardening, or free-machining alloys.

Alloy 400 has better corrosion resistance than pure nickel in reducing environments and copper in oxidizing environments. These features combined with higher strength, good ductility, and good weldability make alloy 400 desirable for many applications in the chemical and process industries and in saltwater services, such as seawater and brackish waters. Alloy 400 is generally the best practical material for hydrofluoric acid service. It is also used in the metallurgical industries for iron and steel pickling equipment.

As copper is added to nickel, the Curie temperature falls steadily and reaches room temperature at approximately 30% Cu. At the composition of alloy 400, the Curie temperature is slightly above ambient temperature. If the alloy is heated, it becomes completely nonmagnetic. This affects the way the material is inspected. For example, eddy current inspection of alloy 400 heat exchanger tubing in the field is usually performed while the equipment is being steam heated, because interpretation of the results is much easier if the material is nonmagnetic.

M-35 is the cast version of alloy 400. M-35 contains a small amount of silicon to increase melt fluidity. The 50-50 nickel-coppers have approximately equal atomic percentages of nickel and copper. These alloys have electrical resistivity that remains constant with increasing temperature. An example is the constantan series of alloys, which contains between 42% Ni and 45% Ni, with the balance of the content being copper. They are used as thermocouple elements in type J, T, and E thermocouples and also for thermocouple compensating leads up to 400°C (750°F).

Precipitation-hardening Nickel-Copper Alloys. Alloy 400 is made precipitation hardenable by the addition of small amounts of aluminum or titanium. Alloy N05500, K-500 (Monel K-500), is such a material. Precipitation hardening is achieved by development of the gamma prime (γ') precipitate, which is very fine and not resolved in the microstructure unless the alloy is overaged.

Free-machining Nickel-Copper Alloys. Free-machining nickel-copper alloys are versions of alloy 400 and K-500 that contain small amounts of sulfur.

The sulfur makes the alloys suitable for automatic screw-machining operations.

N04405, alloy R-405, is a free-machining version of alloy 400. Alloy R-405 is used in screw machine products, water meter components, valve seat inserts, and fasteners for nuclear equipment. N05502, alloy 502, is a free-machining version of alloy K-500. Alloy 502 is used for fasteners, pump and propeller shafts, and valve stems.

Nickel-Iron Alloys

Nickel-iron alloys consist of soft magnet alloys, permanent magnet alloys, low-expansion alloys, and maraging steels. These alloys exhibit a wide variety of unique physical and mechanical properties.

Soft Magnet Alloys. Soft magnet alloys are characterized by high magnetic permeability and little or no retentivity, meaning they lose magnetism if the magnetic field is removed. Nickel-iron alloys with 50% Ni to 78% Ni are among the many families of alloys used to make soft magnet alloys. Soft magnet alloys usually have single-phase microstructures. Soft magnet alloys are used to make magnets for rotors and generators.

Permanent Magnet Alloys. Permanent magnet alloys provide useful and stable magnetic fields and possess high retentivity. Alloys of nickel, cobalt, and aluminum are among the groups of materials used for permanent magnets. The microstructure of the permanent magnet nickel-iron-aluminum alloys are two phase. The microstructure consists of grains of the two isomorphous BCC phases. Nickel-iron-aluminum alloys are used for applications such as telephone receivers, flowmeters, and loudspeakers.

Low-expansion Alloys. Low-expansion alloys have coefficients of expansion that are very low or that match other specific materials. An alloy with 36% Ni, with the balance of the content being iron, has the lowest coefficient of expansion of any metal. This alloy is known commercially as Invar® or Ni-Lo® 36. See Figure 19-5. The low-expansion characteristics of Invar® exist over a narrow temperature range. Low-expansion alloys are used for applications such as moving components that require little or no thermal expansion, for glass-to-metal seals, and components for electronic devices.

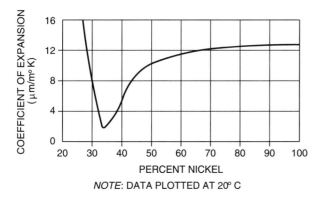

NOTE: DATA PLOTTED AT 20° C

Figure 19-5. An alloy containing 36% Ni, with the balance of the content being iron, has the lowest coefficient of expansion of any metal.

Maraging Steels. Maraging steels are ultrahigh-strength steels that are strengthened by precipitation hardening. Maraging steels contain approximately 18% Ni with additions of cobalt or chromium, and additions of molybdenum, titanium, and aluminum.

The identification system for these steels indicates their yield strength of 200 ksi (1380 MPa). This alloy, grade B (250), and grade C (300) are the most widely used grades. The 18 (350) alloy is an ultrahigh-strength version made in limited quantities for special applications. Cast 18 Ni is used specifically for alloy castings.

The matrix of maraging steels consists of very low-carbon martensite, which makes them extremely strong and tough when precipitation hardened. Maraging steels are also highly weldable. Maraging steels are used in aerospace applications up to 480°C (900°F). At temperatures above 480°C (900°F), the maraging steels begin to overage and weaken. Maraging steels are also used in tooling because of their low dimensional change on precipitation hardening. The tool is machined to size while the alloy is in the annealed condition. The alloy is then precipitation hardened to maximum strength and hardness.

Nickel-Molybdenum Alloys

The nickel-molybdenum phase diagram exhibits 20% solubility for molybdenum at ambient temperature. Solubility rises to almost 40% at the eutectic temperature. See Figure 19-6. Molybdenum improves resistance to reducing acids, such as hydrochloric acid. Nickel-molybdenum alloys containing in excess of 25% Mo develop exceptional resistance to hydrochloric acid.

The principal nickel-molybdenum alloys are B-2 and the B-2 cast equivalents. N10665, alloy B-2 (Hastelloy® B-2), contains 67% Ni and 28% Mo. The cast equivalents of alloy B-2 are ASTM A 494, grades N12M-1 and N12M-2.

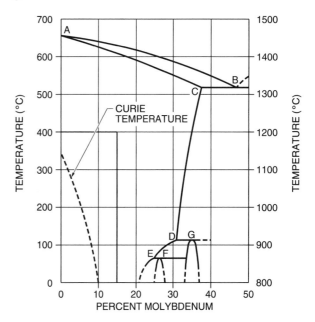

Figure 19-6. Nickel exhibits 20% solubility for molybdenum at ambient temperature, rising to almost 40% at the eutectic temperature.

The chief weakness of alloy B-2 is poor resistance to oxidizing species, even if the species are present in small amounts. For example, the excellent resistance to hydrochloric acid is severely reduced by the entry of air through a pump packing or by the presence of ferric or cupric ions from corrosion products of equipment downstream, all of which are oxidizing species.

Nickel-Chromium Alloys

In the nickel-chromium phase diagram, the solubility of chromium in nickel decreases from 47% at the eutectic temperature to approximately 30% at ambient temperature. Nickel-chromium alloys are generally single-phase alloys with traces of chromium carbides visible either at grain boundaries or within grains, depending on the thermal history. Above approximately 35% Cr, the microstructure becomes two phase. See Figure 19-7.

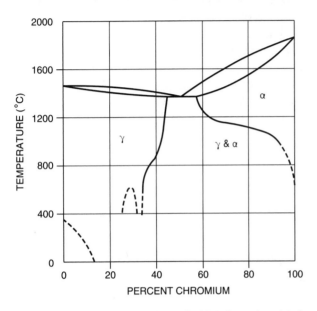

Figure 19-7. Above approximately 35% Cr on the nickel-chromium phase diagram, the microstructure of the nickel-chromium alloys becomes two phase.

The chromium in nickel-chromium alloys provides high-temperature corrosion resistance, and the nickel provides workability. The optimum balance between the two properties is obtained at 20% Cr and 80% Ni. This alloy is drawn into wire and used for electrical heating elements at temperatures up to 1150°C (2100°F).

Wrought and cast alloys containing 50% Ni and 50% Cr are used as heater tube supports in furnaces that burn coal or fuel oil containing appreciable amounts of sulfur. These wrought and cast versions are Inconel® 671 and IN-657®, respectively. The cast version has additions of niobium and restricted levels of silicon and nitrogen.

An alloy with 90% Ni and 10% Cr (Chromel) is used as one half of type K and type E thermocouples. The other halves of these thermocouples are 95% Ni and 5% aluminum alloy (Alumel) for type K and 45% Ni and 55% copper alloy (constantan) for type E.

Nickel-Chromium-Molybdenum Alloys

The addition of chromium and molybdenum to nickel yields alloys with resistance to oxidizing and reducing corrosives. Resistance to oxidizing corrosives is due to the addition of chromium, and resistance to reducing corrosives is due to the addition of molybdenum.

Nickel-chromium-molybdenum alloys contain up to 30% Cr and 16% Mo, plus carbide forming elements and other elements. These alloys are the most corrosion resistant of all the nickel-base alloys. Nickel-chromium-molybdenum alloys also compete favorably with other highly corrosion-resistant alloys, such as tantalum and titanium. Nickel-chromium-molybdenum alloys have the additional advantage of ease of welding over tantalum and titanium. Nickel-chromium-molybdenum alloys are used structurally or as weld overlays or cladding to achieve corrosion resistance at lower cost.

N10276, alloy C-276 (Hastelloy® C-276), has the very best corrosion resistance of the nickel-chromium-molybdenum alloys. Alloy C-276 is sensitized by the precipitation of chromium carbides after short exposure to the temperature range 650°C to 1090°C (1200°F to 2000°F). This sensitization results in a loss of corrosion resistance. Welded alloy C-276 usually does not sensitize if the time of exposure in the sensitizing temperature range is less than approximately 3 minutes. Three minutes is normally acceptable for the welding of most standard thicknesses of equipment and piping. Longer periods of time in the sensitizing temperature range might occur when heavier sections, which require a greater number of weld passes, are welded. In such cases, the alloy will suffer a loss of corrosion resistance.

N06455, alloy C-4 (Hastelloy® C-4), resists sensitizing for considerably longer periods of time than alloy C-276. Alloy C-4 is suitable for fabrication operations such as roll cladding on carbon steel, hot forging, or stress relieving. To achieve greater metallurgical stability, tungsten is omitted, iron is reduced, titanium is added, and the carbon and silicon contents are significantly lowered. Alloy C-4C is the cast version of alloy C-4.

N06022, Hastelloy® C-22, is a modification of alloy C-276. Hastelloy® C-22 contains 13.5% Mo and 22% Cr. The higher chromium content provides better resistance to oxidizing acids.

N06625, alloy 625 (Inconel®625), contains 21.5% Cr and 9% Mo. Alloy 625 also contains 3.5% Nb, plus an addition of tantalum, which provides significant solid solution strengthening and elevated temperature strength. Alloy 625 sensitizes if held at 760°C (1400°F) for more than 1 hour. Alloy CW-6MC is the cast version of alloy 625.

N06617, Inconel® 617, contains 22% Cr, 9% Mo, and 1.2% Al. Inconel® 617 was originally developed as a high-temperature oxidation-resistant alloy for aerospace and turbine applications.

Nickel-Iron-Chromium-Molybdenum Alloys

Nickel-iron-chromium-molybdenum alloys contain large amounts of iron and chromium, and smaller additions of molybdenum. Additions of niobium plus tantalum or additions of titanium are added to certain alloys for stabilization against sensitization. Copper is incorporated to some nickel-iron-chromium-molybdenum alloys to improve corrosion resistance in certain environments. As a group, nickel-iron-chromium-molybdenum alloys are less corrosion resistant than the nickel-chromium-molybdenum types. The nickel-iron-chromium-molybdenum alloys consist of high-nickel and low-nickel alloy groups.

High-nickel Group. The high-nickel group of nickel-iron-chromium-molybdenum alloys is extremely resistant to reducing acids, such as sulfuric and phosphoric acid. The nickel content makes this group extremely resistant to chloride stress cracking. High-nickel group alloys are used for vessels, piping, tubing pumps and valves, and other equipment in chemical processing.

The high-nickel group alloys are supplied in the stabilized condition. The stabilizing carbide precipitates can be redissolved during exposure to a specific temperature range, which might occur during welding or hot working. For example, if the heat input of welding is excessive, as in multipass welding, a small region of the heat affected zone adjacent to the weld experiences temperatures in the range susceptible to carbide dissolution. Carbide dissolution leads to sharply reduced corrosion resistance in these regions. The higher the nickel content of the alloy, the lower the temperature range for carbide dissolution. High-nickel alloys should not be hot worked in the susceptible temperature range to avoid loss of corrosion resistance, and the manufacturer should be contacted for guidance.

High-nickel group alloys include 20Cb-3, CN-7M, 825, G-3, and Hastelloy® G-30. N08020, alloy 20Cb-3 (Carpenter® 20Cb-3), is also known as alloy 20. Alloy 20 contains approximately 34% Ni, 24% Cr, 2.5% Mo, and 3.3% Cu. Alloy 20 is stabilized by niobium and tantalum, and was originally developed for sulfuric acid service, but has good resistance to a wide range of corrosives.

N08007, CN-7M, is a cast material roughly equivalent to alloy 20Cb-3. It contains approximately 28% Ni, 20% Cr, 3% Mo, and 2% Cu. CN-7M is produced under various commercial names, including alloy 20, Durimet® 20, and Worthite® CN-7M, and has similar applications to alloy 20Cb-3.

N08825, alloy 825 (Incoloy® 825), contains approximately 42% Ni, 22% Cr, 3% Mo, and 2.2% Cu. Alloy 825 is stabilized by titanium and exhibits better chloride stress-cracking resistance than alloy 20Cb-3 because of its higher nickel content.

N06985, alloy G-3 (Hastelloy® G-3), is more highly alloyed and contains approximately 44% Ni, 22% Cr, 7% Mo, and 1.9% Cu. Alloy G-3 is stabilized by niobium and tantalum and has better corrosion resistance to reducing environments because of its higher molybdenum content.

N06030, Hastelloy® G-30, contains approximately 43% Ni, 29.5% Cr, 5% Mo, and 1.7% Cu. Hastelloy® G-30 is stabilized with niobium and tantalum. The high chromium content improves resistance to oxidizing environments.

Low-nickel Group. The low-nickel group of nickel-iron-chromium-molybdenum alloys are the superaustenitic stainless steels. These alloys have significantly superior corrosion resistance to regular austenitic stainless steels, such as types 316 and 317. They also exhibit reasonably good chloride stress-cracking resistance. These properties are achieved through rather high levels of molybdenum and nitrogen. The nitrogen counteracts the tendency to form embrittling sigma phase in the absence of high-nickel contents.

The low-nickel group alloys do not require stabilizing elements and do not exhibit the type of sensitizing problems that occur from carbide dissolution in the high-nickel group. Alloys in the low-nickel group include N08367 (AL6X-N®), N08028 (Sanicro® 28), and N08904 (904L). The cast version of N08367 is CK-3MCuN. These alloys are welded with high-alloy welding filler metals to maintain the corrosion resistance of the weld.

Nickel-Iron-Chromium Alloys

In the nickel-iron-chromium phase diagram, the gamma phase field is very wide. It extends to approximately 40% Cr, which is close to the iron corner of the phase diagram. The gamma phase field contains many commercial high-temperature alloys. See Figure 19-8. These alloys are divided into heat-resistant castings and iron-base superalloys.

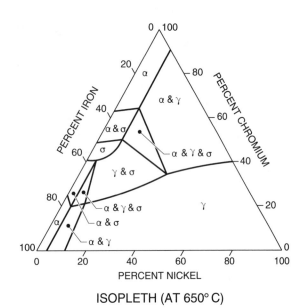

ISOPLETH (AT 650° C)

Figure 19-8. The nickel-iron-chromium phase diagram has a wide gamma solid solution phase field close to the iron side of the diagram, which is the basis for many important heat-resistant alloys.

Heat-resistant Castings.

Heat-resistant castings include a wide range of alloy compositions encompassing stainless steels and nickel-base alloys. They are designated with the prefix uppercase letter H, followed by a letter that designates the approximate alloy content. The higher the letter, the greater the percentage of alloying elements. Nickel-chromium heat-resistant castings contain from 30% Ni to 66%

Ni and 12% Cr to 26% Cr, with the balance of the content being iron.

Superalloys

Superalloys are specifically made for creep resistance in the range 650°C to 1095°C (1200°F to 2000°F). They are produced by casting, rolling, extrusion, forging, and powder processing. These processes are performed under controlled conditions to optimize the metallurgical structure of the alloy. Superalloys are used in aircraft, industrial and marine gas turbines, nuclear reactors, aircraft skins, spacecraft structures, and petrochemical production.

Superalloys consist of three groups. The two superalloy groups that contain appreciable quantities of nickel are the iron-base and nickel-base superalloys. The third group is the cobalt-base superalloys. Superalloys are strengthened chiefly by precipitation hardening. Some alloys rely on solid solution hardening, but the strengthening effect is less pronounced.

Iron-base Superalloys. Iron-base superalloys are wrought alloys containing 9% Ni to 49% Ni, 12% Cr to 22% Cr, 22% Fe to 62% Fe, and additions of other elements. Iron-base superalloys are an extension of austenitic stainless steel technology and are suitable for use up to approximately 760°C (1400°F). See Figure 19-9

Creep resistance is generally obtained by intermetallic compound precipitation in the gamma matrix. The most common type of precipitate is gamma prime (γ'), which is found in S66286 (A-286). Other types of strengthening mechanisms also exist, such as the solid solution hardening that develops in N08800, alloy 800 (Incoloy® 800).

N08810 (alloy 800H) and N08811 (Incoloy® 800HT) are versions of alloy 800 that have carbon contents that are maintained at the high end of the range. These alloys are solution annealed between 1120°C and 1200°C (2050°F and 2200°F) to produce a large grain size. Solution-annealing operations improve high-temperature strength. When field fabrication processes for these alloys include bending or other plastic straining operations, the alloys must be solution annealed before being placed in service. Solution annealing eliminates the plastic strain that can lead to recrystallization and loss of creep resistance during operation.

Iron-base superalloys are used for forgings of turbine wheels; various other components of gas turbines; and sheet metal casings, housings, and exhaust equipment. Iron-base superalloys are also used for piping and manifolds in the refining, power, and petrochemical industries.

Nickel-base Superalloys. Nickel-base superalloys are the most important group of superalloys. Nickel-base superalloys contain various amounts of nickel, chromium, cobalt, iron, molybdenum, tungsten, and niobium. See Figure 19-10. There are over 40 different wrought nickel-base super-

alloys and approximately 30 different cast nickel-base superalloys. Nickel-base superalloys are suitable for service up to 1095°C (2000°F). Nickel-base superalloys are used for aircraft equipment such as manifolds, collector rings, and exhaust valves of reciprocating engines. Nickel-base superalloys are also used in sheet form for combustion liners, tail pipes, and casings of gas turbines and jet engines.

The nickel-chromium-aluminum phase diagram is fundamental to the strengthening mechanism of most nickel-base superalloys. The isopleth at 750°C (1380°F) shows that up to approximately 30% Cr, the gamma solid solution is limited by the precipitation of the isomorphous compound $NiAl_3$, or gamma prime (γ').

IRON-BASE SUPERALLOYS													
Alloy	%Ni	%Cr	%Co	%Mo	%W	%Nb	%Al	%Ti	%Fe	%Mn	%Si	%C	%B
A-286	26.0	15.0	—	1.3	—	—	.20	2.0	54.0	1.30	.50	.05	.015
Discaloy	26.0	13.5	—	2.7	—	—	.10	1.7	54.0	.90	.80	.04	.005
Alloy 901	42.5	12.5	—	5.7	—	—	.20	2.8	36.0	.10	.10	.05	.015
Haynes® 556[a]	20.0	22.0	20.0	3.0	2.5	.1	.30	—	29.0	1.50	.40	.10	—
Incoloy® 800	32.5	21.0	—	—	—	—	.40	.4	46.0	.80	.50	.05	—
Incoloy® 801	32.0	20.5	—	—	—	—	—	1.1	44.5	.80	.50	.05	—
Incoloy® 802	32.5	21.5	—	—	—	—	—	—	46.0	.80	.40	.40	—
Incoloy® 807	40.0	20.5	8.0	.1	5.0	—	.20	.3	25.0	.50	.40	.05	—
Incoloy® 825[b]	42.0	22.0	—	3.0	—	—	.20	.8	28.0	1.00	.50	.05	—
Incoloy® 903	38.0	—	15.0	—	—	3.0	.70	1.4	41.0	—	—	—	—
Incoloy® 907	38.0	—	13.0	—	—	4.7	.03	1.5	42.0	—	.15	—	—
Incoloy® 909	38.0	—	13.0	—	—	4.7	—	1.5	42.0	—	.40	.01	.001
N-155[c]	20.0	21.0	20.0	3.0	2.5	1.0	—	—	30.0	1.50	.50	.15	—
V-57	27.0	14.8	—	1.3	—	—	.30	3.0	52.0	.30	.70	.08	.010
19-9 DL	9.0	19.0	.4	—	1.3	—	—	.3	bal	1.00	.50	.30	—
16-25-6	25.5	16.25	—	6.0	—	—	—	—	bal	2.00	1.00	.10	—
Pyromet® CTX-1	37.7	.1	16.0	.1	—	3.0	1.00	1.7	39.0	—	—	.03	—
Pyromet® CTX-3	38.3	.2	13.6	—	—	4.9	.10	1.6	bal	—	.15	.05	.007
17-14CuMo[d]	14.0	16.0	—	2.5	—	.4	—	.3	62.4	.75	.50	.12	—
20-Cb3[e]	34.0	24.0	—	2.5	—	1.0	—	—	35.0	—	—	.07	—

[a] 2% N, .02% La, and .9% Ta
[b] 2.2% Cu
[c] 15% N
[d] 3% Cu
[e] 3.3% Cu

Figure 19-9. Iron-base superalloys are an extension of austenitic stainless steel technology and are suitable for use up to approximately 760°C (1400°F).

NICKEL-BASE SUPERALLOYS														
Alloy	%Ni	%Cr	%Co	%Mo	%W	%Nb	%Al	%Ti	%Fe	%Mn	%Si	%C	%B	%Zr	%Other
Astroloy	55.0	15.0	17.0	5.3	—	—	4.0	3.5	—	—	—	.06	.030	—	—
Cabot 214	75.0	16.0	—	—	—	—	4.5	—	2.5	—	—	—	—	—	.01 Y
D-979	45.0	15.0	—	4.0	—	—	1.0	3.0	27.0	.30	.20	.05	.010	—	—
Hastelloy® C-22	51.0	22.0	2.5	13.5	4.0	—	—	—	5.5	1.00	.10	.01	—	—	.3 V
Hastelloy® C-276	55.0	15.5	2.5	16.0	3.7	—	—	—	5.5	1.00	.10	.01	—	—	.3 V
Hastelloy® G-30	43.0	29.5	2.0	5.5	2.5	.8	—	—	15.0	1.00	1.00	.03	—	—	1.7 Cu
Hastelloy® S	67.0	15.5	—	14.5	—	—	.3	—	1.0	.50	.40	—	.009	—	.05 La
Hastelloy® X	47.0	22.0	1.5	9.0	.6	—	—	—	18.5	.50	.50	.10	—	—	—
Haynes® 230	57.0	22.0	—	2.0	14.0	—	.3	—	—	.50	.40	.10	—	—	.02 La
Inconel® 587	bal	28.5	20.0	—	—	.7	1.2	2.3	—	—	—	.05	.003	.05	—
Inconel® 597	bal	24.5	20.0	1.5	—	1.0	1.5	3.0	—	—	—	.05	.012	.05	.02 Mg
Inconel® 600	76.0	15.5	—	—	—	—	—	—	8.0	.50	.20	.08	—	—	—
Inconel® 601	60.5	23.0	—	—	—	—	1.4	—	14.1	.50	.20	.05	—	—	—
Inconel® 617	54.0	22.0	12.5	9.0	—	—	1.0	.3	—	—	—	.07	—	—	—
Inconel® 625	61.0	21.5	—	9.0	—	3.5	.2	.2	2.5	.20	.20	.05	—	—	—
Inconel® 706	41.5	16.0	—	—	—	2.9	.2	1.8	40.0	.20	.20	.03	—	—	—
Inconel® 718	52.5	19.0	—	3.0	—	5.1	.5	.9	18.5	.20	.20	.03	—	—	—
Inconel® X-750	73.0	15.5	—	—	—	1.0	.7	2.5	7.0	.50	.20	.04	—	—	—
M-252	55.0	20.0	10.0	10.0	—	—	1.0	2.6	—	.50	.50	.15	.005	—	—
Nimonic® 75	76.0	19.5	—	—	—	—	—	.4	3.0	.30	.30	.10	—	—	—
Nimonic® 80A	76.0	19.5	—	—	—	—	1.4	2.4	—	.30	.30	.06	.003	.06	—
Nimonic® 90	59.0	19.5	16.5	—	—	—	1.5	2.5	—	.30	.30	.07	.003	.06	—
Nimonic® 105	53.0	15.0	20.0	5.0	—	—	4.7	1.2	—	.30	.30	.13	.005	.10	—
Nimonic® 115	60.0	14.3	13.2	—	—	—	4.9	3.7	—	—	—	.15	.160	.04	—
Nimonic® 263	51.0	20.0	20.0	5.9	—	—	.5	2.1	—	.40	.30	.06	.001	.02	—
Nimonic® 942	39.5	12.5	—	6.0	—	—	.6	3.7	37.0	.20	.30	.03	.010	—	—
Nimonic® PE.11	37.5	18.0	—	5.2	—	—	.8	2.3	35.0	.20	.30	.05	.030	.20	—
Nimonic® PE.16	43.0	16.5	1.0	1.1	—	—	1.2	1.2	33.0	.10	.10	.05	.020	—	—
Nimonic® PK.33	56.0	18.5	14.0	7.0	—	—	2.0	2.0	.3	.10	.10	.05	.030	—	—
Pyromet® 860	43.0	12.6	4.0	6.0	—	—	1.25	3.0	30.0	.05	.05	.05	.010	—	—
René® 41	55.0	19.0	11.0	1.0	—	—	1.5	3.1	—	—	—	.09	.005	—	—
René® 95	61.0	14.0	8.0	3.5	3.5	3.5	3.5	2.5	—	—	—	.15	.010	.05	—
Udimet® 400	60.0	17.5	14.0	4.0	—	.5	1.5	2.5	—	—	—	.06	.008	.06	—
Udimet® 500	54.0	18.0	18.5	4.0	—	—	2.9	2.9	—	—	—	.08	.006	.05	—
Udimet® 520	57.0	19.0	12.0	6.0	1.0	—	2.0	3.0	—	—	—	.05	.005	—	—
Udimet® 630	60.0	18.0	—	3.0	3.0	6.5	.5	1.0	18.0	—	—	.03	—	—	—
Udimet® 700	55.0	15.0	17.0	5.0	—	—	4.0	3.5	—	—	—	.06	.030	—	—
Udimet® 710	55.0	18.0	15.0	3.0	1.5	—	2.5	5.0	—	—	—	.07	.020	—	—
Udimet® 720	55.0	17.9	14.7	3.0	1.3	—	2.5	5.0	—	—	—	.03	.033	.03	—
Unitemp AF2-1DA6	60.0	12.0	10.0	2.7	6.5	—	4.0	2.8	—	—	—	.04	.015	.10	1.5 Ta
Waspaloy	58.0	19.5	13.5	4.3	—	—	1.3	3.0	—	—	—	.08	.006	—	—

Figure 19-10. Nickel and cobalt superalloys consist of many wrought and cast alloys that are suitable for services up to temperatures of 1095°C (2000°F).

Gamma prime appears above approximately 5% Al. The aluminum in the gamma prime can be partially replaced by titanium without changing its crystal structure. The gamma prime in this form plays a major role in strengthening the nickel-base superalloys. See Figure 19-11.

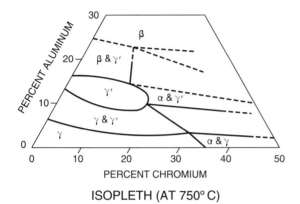

ISOPLETH (AT 750° C)

Figure 19-11. An isothermal section through the nickel-chromium-aluminum phase diagram shows the appearance of the precipitation-hardening gamma prime compound at approximately 5% Al.

Nickel-base superalloys exhibit a variety of microstructures. The microstructure depends on the composition, thermal and mechanical treatment, and service history of the alloy. Within the single-phase gamma matrix, carbide and gamma prime precipitates may be observed. The carbide precipitates are primary titanium carbides and secondary chromium carbides. The gamma prime precipitate is not normally resolvable at low microscopic magnifications unless the alloy is overaged.

N07718, Inconel® 718, is one of the strongest at low temperatures and is the most widely used of all superalloys. Inconel® 718 rapidly loses strength in the range 650°C to 815°C (1200°F to 1500°F).

N07750, Inconel® X-750, is the precipitation-hardening version of Inconel® 600. Inconel® X-750 has three times the yield strength of Inconel® 600 at 540°C (1000°F).

Alternate forms of strengthening occur in a limited number of nickel-base superalloys. They include solution hardening and oxide dispersion hardening, which is found in MA-745.

The basic solution-hardening alloy is N06600, alloy 600 (Inconel® 600), but is often referred to as Inconel®. Alloy 600 contains 76% Ni, 15.5% Cr, and 8% Fe. Alloy 600 has excellent high-tempera-ture oxidation resistance and strength, but generally has lower corrosion resistance than the austenitic stainless steels. CY-40 is the cast equivalent of alloy 600. Alloy X (N06002, or Hastelloy® X) is another alloy that is strengthened by solution hardening. It has useful strength to 815°C (1500°F) and excellent oxidation resistance. Because of these properties, alloy X is used as a lining material to temperatures in excess of 1050°C (1920°F).

Cobalt-base Superalloys. Cobalt-base superalloys have lower creep strengths than the nickel-base superalloys. See Figure 19-12. Cobalt-base superalloys are strengthened by a combination of carbides and solid solution hardeners. Cobalt-base superalloys are used in many applications that are similar to, but less stringent than, the applications of nickel-base superalloys.

Cobalt-base superalloys that are used primarily for applications at temperatures between 650°C and 1150°C (1200°F and 2100°F) include Haynes® 25, Haynes® 188, UMCo-50, and Stellite 6B. Haynes® 25 is the best known cobalt-base superalloy and is widely used for the hot section of gas turbines, components of nuclear reactors, and surgical implants. Haynes® 25 is also used in the cold-worked condition for fasteners and wear plates.

Haynes® 188 is specially designed for sheet metal components, such as combustors and transition ducts in gas turbines. Haynes® 188 has excellent oxidation resistance up to 1095°C (2000°F). Haynes® 188 resists embrittlement during long periods of exposure at service temperatures.

UMCo-50 contains approximately 21% Fe and is not as strong as Haynes® 25. UMCo-50 is not often used in gas turbine applications. UMCo-50 is extensively used in Europe for furnace components and furnace fixtures.

Stellite® 6B has high hot hardness and good resistance to oxidation. The high hot hardness is due to the formation of complex carbides. The good resistance to oxidation is due to the high chromium content. Stellite® 6B is widely used for erosion shields in steam turbines, for wear pads in gas turbines, and for bends in tubing carrying particulate matter at high temperatures and high velocities.

COBALT-BASE SUPERALLOYS														
Alloy	%Ni	%Cr	%Co	%Mo	%W	%Ta	%Nb	%Al	%Fe	%Mn	%Si	%C	%Zr	%Other
Air Resist 213	—	19.0	66.0	—	4.7	6.5	—	3.5	—	—	—	.18	.15	.1 Y
Elgiloy®	15.0	20.0	40.0	7.0	—	—	—	—	16.0	2.0	—	.10	—	.04 Be
Haynes® 188	22.0	22.0	39.2	—	14.0	—	—	—	3.0	—	—	.10	—	—
L-605	10.0	20.0	52.9	—	15.0	—	—	—	—	—	—	.05	—	—
MAR-M 918	20.0	20.0	52.5	—	—	7.5	—	—	—	—	—	.05	.10	—
MP35N®	35.0	20.0	35.0	10.0	—	—	—	—	—	—	—	—	—	—
MP159	25.5	19.0	35.7	7.0	—	—	.6	.2	9.0	—	—	—	—	3.0 Ti
Stellite® 6B	3.0	30.0	52.0	1.5	4.5	—	—	—	3.0	2.0	2.00	1.1	—	—
Haynes® 150	—	28.0	50.5	—	—	—	—	—	20.0	—	.75	—	—	.02 P and .002 S
S-816	20.0	20.0	bal	4.0	4.0	—	4.0	—	3.0	1.2	—	.40	—	—
V-36	20.0	25.0	bal	4.0	—	—	2.3	—	2.4	1.0	—	.32	—	—

Figure 19-12. Cobalt-base superalloys are strengthened by a combination of carbides and solid solution hardeners.

Other Alloy Systems

Nickel is used in several other alloy systems, such as NiTiNOL and nickel-containing brazing and welding filler metals. *NiTiNOL* is an alloy that was developed by the Naval Ordinance Laboratory (NOL) and contains 5% Ni and 45% Ti. NiTiNOL is also referred to as the shape memory metal. It has cubic crystal structure at elevated temperatures, but changes to a hexagonal structure when cooled to approximately 35°C (95°F). If NiTiNOL is formed into a shape at an elevated temperature and then cooled and deformed at ambient temperature, it will recover its original shape when it is reheated between 50°C and 60°C (120°F and 140°F).

Nickel is an important constituent in many families of brazing and welding filler metals. Brazing filler metals consist of nickel alloys with any of the following elements: chromium, boron, silicon, manganese, and gold. Welding filler metals also comprise a variety of alloys. These are specifically designed to match and join the families of nickel alloys, cast irons, and stainless steels.

NICKEL ALLOY FABRICATION

Fabrication processes for nickel alloys include forming, machining, and joining. The chief considerations in fabricating nickel alloys relate to the rate of strain hardening and susceptibility to embrittlement caused by sulfur-containing compounds.

Forming

The principal methods of forming nickel products are forging and cold forming. Most forging operations for nickel alloys are performed in the range 870°C to 1095°C (1600°F to 2000°F). See Figure 19-13. Some nickel-base alloys are extremely strong at high temperature, and the forging press must have sufficient power to avoid the need for frequent reheating, which can lead to coarse grain size. This is an important consideration for superalloys that require controlled and uniform grain size if optimum mechanical properties and service performance are to be obtained. Some nickel alloys have a narrow forging temperature range, and because of their low thermal conductivity, it is necessary to minimize the time of contact between the tool face and the component. This prevents chilling and cracking of the surface. Sulfur contamination must not be allowed during any of the hot-working operation.

Much of the cold forming of nickel alloys is performed by bending, deep drawing, and spinning. Lubrication is extremely important because nickel alloys tend to gall more easily than other metals, such as steel. Tool and die materials must be hard and wear resistant to maintain dimensional accuracy and provide long life. Hard alloy bronzes, chromium cast iron, chromium-plated hardened steel, and tungsten carbide are commonly used die materials. Softer materials, such as pure nickel or the nickel-copper alloys, work harden during cold forming, and intermediate annealing is required to soften them so that the operation can be completed.

FORGING RANGES FOR SELECTED NICKEL ALLOYS		
Alloy	Light Forging Range (°C)	Heavy Forging Range (°C)
Alloy 200	Up to 860	860 to 1225
Alloy 301	860 to 1030	1030 to 1225
Alloy 400	650 to 925	925 to 1175
Alloy K-500	870 to 1030	1030 to 1150
Alloy 600	Up to 650 and 870 to 1030	1030 to 1225
Alloy 625	—	1000 to 1175
Alloy 718	890 to 950	950 to 1120
Alloy 722	975 to 1030	1030 to 1205
Alloy X-750	975 to 1030	1030 to 1205
Alloy 800	Up to 650 and 875 to 1005	1005 to 1205
Alloy 825	Up to 650 and 875 to 1005	1005 to 1175

Figure 19-13. Most forging operations for nickel alloys are performed in the range 870°C to 1095°C (1600°F to 2000°F).

Machining

Machining processes include standard machining operations, electrochemical machining, and electric discharge machining (EDM). Standard machining operations use cutting tools. The major impediment in these operations is the toughness and work-hardening capacity of nickel alloys. The component and tool must be sharp and must have high-quality surface finishes. The tool clearance angle must be no larger than necessary. The supply of cutting fluid must be generous, the feed rate must be high in order to minimize work hardening, and there must be maximum clearance for swarf. *Swarf* is a mixture of grinding chips and fine particles of abrasive. Cutting tool materials are high-speed tool steels or tungsten carbide. More ductile materials, such as nickel-copper alloys, are easier to machine in the cold-worked condition.

Electrochemical machining involves chemical attack of the component. Electrochemical machining is used on any type of nickel alloy, including fully hardened types. The toughness or hardness of the material is not a consideration. Masking procedures are used to retain surfaces and dimensions that are not to be disturbed by the chemical attack.

Electrical discharge machining uses the passage of an electrical discharge between a copper electrode and the component to make a cavity that is a mirror image of the electrode. The metal is removed with sparks produced by the electrical discharge. Electrical discharge machining is used for drilling deep holes or recesses in superalloys.

Joining

Nickel alloys may be joined by welding or brazing. These processes weaken work-hardened or precipitation-hardened alloys in the heat-affected zone of the joint. Work-hardened alloys are not joined by welding or brazing if full strength is required. Precipitation-hardened alloys are re-solution annealed and precipitation hardened after welding. If this is not possible, for example because of dimensional instability, the alloy should, at a minimum, be subjected to the precipitation-hardening step. Before joining, a region approximately 1 in. from both sides of the joint should be mechanically abraded and thoroughly degreased to prevent contamination.

Except for oxyfuel welding, virtually all welding processes may be used to join nickel alloys. Oxyfuel welding has limited applicability and should be used only when other equipment is not available.

Except for pure nickel, the thermal conductivity of nickel-base alloys is rather low. Consequently, the chance of distortion increases during welding because heat tends to be retained rather than rapidly dissipated through the base metal. The thermal expansion coefficients of nickel-base alloys are approximately the same as those for carbon steels and low-alloy steels, except for the controlled expansion series of nickel-iron alloys.

Preheating and interpass temperatures should be kept low to minimize total heat input. Post-weld heat treatment is not a requirement to restore mechanical properties, except for precipitation-hardening alloys. Alloys that sensitize when welded may require post-weld solution annealing in order to restore corrosion-resistant properties.

Nickel alloys are brazed by most conventional processes. The selection of the brazing filler metal depends on the service conditions. The brazing filler metal must not reduce the corrosion resistance of the joint and must withstand any subsequent heat treatment process.

Of the superalloys, the cobalt-base type are the easiest to braze. The process is usually performed in a dry hydrogen atmosphere or in a vacuum using nickel-base brazing filler metal or alloys based on silver, gold, or palladium. Iron-nickel and nickel-base superalloys are successfully brazed in a vacuum using nickel-base, silver-base, or gold-base brazing filler metals. When brazing superalloys, the cycle is carefully controlled to avoid excessive metallurgical reactions between the base metal and the brazing filler metal.

Nickel alloys are rarely soldered because solders tend to have low levels of strength, especially at high temperatures. This makes them unsuitable for many end uses.

NICKEL ALLOY PROPERTIES

The thermal properties of nickel alloys generally affect weldability. Thermal properties of nickel alloys include thermal conductivity and thermal expansion. Nickel alloys possess several outstanding mechanical properties. Mechanical properties of nickel alloys include low-temperature toughness and high-temperature strength. Nickel alloys have excellent corrosion resistance in a wide range of environments, including chemical solutions and high-temperature gases. Other less expensive alloys, such as austenitic stainless steels or copper-nickels, are suitable for many less corrosive services.

Thermal Properties

Except for pure nickel, the thermal conductivity of nickel-base alloys is rather low. This increases the chance of distortion during welding because heat tends to be retained rather than rapidly dissipated through the base metal.

With the exception of the low-expansion nickel-iron alloys, the thermal expansion coefficients of nickel-base alloys are approximately the same as those for carbon and low-alloy steels. Those of aus-

tenitic stainless steels are considerably higher, which introduces problems when joining them to nickel-base alloys.

Mechanical Properties

Nickel-base alloys are extremely tough at low temperatures because of their FCC crystal structure. Nickel additions also have a beneficial effect on carbon and stainless steels. Specific families of nickel-base alloys that have exceptional high-temperature strength begin to be used above 400°C and 590°C (750°F and 1100°F). These temperatures are the approximate upper temperature limits for the application of carbon and chromium-molybdenum steels, respectively.

Corrosion Resistance

Forming or welding operations often have a significant effect on corrosion resistance, and these factors must be recognized when considering an alloy for a particular application. Corrosives may be grouped into waters, inorganic acids, bases, and high-temperature environments.

Waters. Waters include seawater, effluent streams, and boiler feed waters. The principal corrosive agent in waters requiring the use of nickel-base alloys is the chloride ion. Nickel alloys are often over qualified for service in waters. Alloys such as copper-nickels, brasses, and austenitic stainless steels are usually applicable in these situations. In seawater, the low-nickel group of iron-chromium-molybdenum alloys often shows superior resistance because of its high molybdenum content.

Inorganic Acids. Inorganic acids include sulfuric, nitric, and hydrochloric acid. Many nickel alloys are used for applications in these environments. Selection depends on the acid type, temperature, concentration, and impurities present.

High-temperature Corrosives. High-temperature corrosives include sulfur products, hydrogen chloride, hydrogen, and hydrocarbons. Several families of nickel alloys are extremely resistant to high-temperature corrosives. Sulfur and sulfur products are extremely corrosive toward nickel alloys, especially at high temperatures when a low melting point, nickel-sulfur eutectic compound may be formed. This formation can lead to catastrophic failure. Cutting oils and greases, which cause sulfur contamination, must always be removed by thorough degreasing before any operation that involves exposure to elevated temperatures.

Organics. Organics include many types of compounds, such as acetic acid, trichloroethylene, and formaldehyde. Nickel alloys are used for organics only if the organics are too corrosive for less expensive alloys. This occurs frequently at high temperatures, at high concentrations of the organic compounds, or when the alloys are contaminated with water and other impurities.

Bases. Bases include sodium and potassium hydroxide. Nickel 200 and alloy 600 are the two nickel alloys that are the most corrosion resistant to bases. They are especially resistant to sodium hydroxide at high temperatures and high concentrations.

COBALT

Cobalt usually occurs with nickel sulfide ores. The pure metal is used as a target in X-ray tubes and for other special components. Cobalt is used chiefly as an alloying element. Cobalt alloys are used for wear resistance, heat resistance, and corrosion re-

sistance. Some cobalt alloys have specific uses in permanent magnets and as matrix material in cutting tools. The principal alloying elements are chromium, nickel, and iron.

Wear-resistant Alloys

The chief families of wear-resistant cobalt-base alloys are Stellites® and Tribaloys®. Stellites® are wrought powder metallurgy or cast cobalt alloys containing from 25% Cr to 35% Cr, 1% C to 3% C, 4% W to 25% W, plus minor amounts of other alloying elements. See Figure 19-14. Tribaloys® are intermetallic materials made by powder metallurgy.

Stellites®. Although primarily used for wear resistance, Stellites® also possess useful high-temperature strength. Stellites® are high-temperature cobalt alloys that can be classified as cobalt-base superalloys.

Stellites® are considered the most versatile wear-resistant alloys, especially for applications involving combinations of heat, corrosion, and oxidation. Stellites® with higher carbon content are used where high hardness and abrasion resistance are required and impact resistance is not as important. These alloys are excellent when service temperatures exceed 650°C (1200°F), and they resist oxidation up to 980°C (1800°F).

R30006, Stellite® 6, is the most popular cast alloy and is most commonly applied as a hard-facing material. Stellite® 6B is the wrought equivalent of Stellite® 6. Stellite® 6B is used for wear plates, bushings, and shaft sleeves in demanding environments where

STELLITE® MATERIALS										
Alloy	%Cr	%W	%Ni	%Mo	%Si	%Mn	%Fe	%C	%B	%Other
Stellite® 3	31.0	12.5	3.0	—	1.0	1	3	2.4	1	1
Stellite® 6	29.0	4.5	3.0	1.5	1.5	1	3	1.2	1	2
Stellite® 6B	30.0	4.5	3.0	1.5	2.0	2	3	1.1	—	—
Stellite® 12	30.0	8.5	3.0	—	1.0	1	3	1.5	1	3
Stellite® 19	31.0	10.5	3.0	—	1.0	1	3	1.9	1	2
Stellite® 31	25.5	7.5	10.5	—	1.0	1	2	.5	—	2
Stellite® 190	26.0	14.0	3.0	1.0	1.0	1	5	3.1	1	2
Stellite® 98M2	30.0	18.5	3.5	.8	1.0	1	5	2.0	1	2; 4.2 V
Stellite® Star J	32.5	17.5	3.0	—	1.0	1	3	2.5	1	2

Figure 19-14. Stellites® are primarily used for wear resistance, but they also possess useful high-temperature strength.

lubrication is difficult or impossible. Stellite® 12 has more carbon than Stellite® 6 or 6B, which results in higher hardness. Stellite® 6K and Stellite® 12 are used in applications where a keen cutting edge must be maintained. For example, Stellite® 6K is used in knives, and Stellite® 12 is used for cutting timber. The microstructure of the Stellites® consists of hard chromium and tungsten carbide precipitates supported in a cobalt-base alloy matrix.

Tribaloys®. Tribaloy® T-400 has excellent wear resistance, good mechanical properties, and good corrosion resistance. Tribaloy® T-800, which has a higher chromium content than Tribaloy® T-400, has better corrosion resistance and high-temperature oxidation resistance. Tribaloys® are well suited for applications where lubrication is a problem. Tribaloys® are used for piston rings, bushings, cams, seals, pump compressor components, thrust washers, valves, and vanes.

Corrosion-resistant Alloys

The wear-resistant group of cobalt-base alloys also possesses exceptional corrosion resistance. For example, these alloys are used in surgical implants where there is a need for resistance to body fluids. Additionally, there are alloys that develop extremely high strength, coupled with corrosion resistance. These alloys include Elgiloy® and MP35N®. They achieve strength by a combination of cold work and heat treatment. Elgiloy® and MP35N® are used chiefly for corrosion-resistant springs and fasteners.

R30003, Elgiloy®, contains 40% Co, 20% Cr, 15% Ni, 7% Mo, with the balance of the content being iron. To achieve a maximum strength of 325 ksi, Elgiloy® is heat treated at 480°C (900°F) for 5 hours after cold drawing of 80%.

R30035, MP35N®, contains 35% Co, 35% Ni, 20% Cr, and 10% Mo. To achieve a maximum strength of 300 ksi, MP35N is heat treated at 540°C (1000°F) after cold drawing of 53%.

Permanent Magnet Alloys

Cobalt alloys containing iron and nickel are among the groups of materials used for permanent (hard) magnets. Permanent magnets have high resistance to demagnetization. These cobalt alloys are marketed under a wide variety of trade names and designations, such as the Alnico® and Remalloy® series.

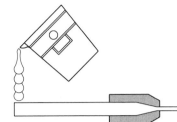

Aluminum 20

Aluminum and aluminum alloys possess a unique combination of properties, making them versatile materials of construction for many applications, from wrapping foil to aerospace components. Like other metals, aluminum is identified by various designation systems. The principal alloying elements in aluminum are copper, manganese, silicon, magnesium, and zinc. Aluminum alloys are wrought or cast and can be strengthened by cold working or by precipitation hardening. Aluminum alloys are formed, machined, and joined using a wide variety of processes. Aluminum is resistant to corrosion in many environments.

ALUMINUM MANUFACTURING

Aluminum is extracted electrolytically from bauxite ore and mechanically formed into wrought products. Aluminum products are also produced by pouring molten aluminum into castings. Aluminum is primarily strengthened by work hardening or by precipitation hardening.

Extraction and Production

Aluminum is made by the electrolysis of aluminum oxide, which is obtained principally from the concentration of bauxite ore. The aluminum oxide is dissolved in electric melting pots (furnaces) in a molten bath of sodium-aluminum fluoride at 940°C to 980°C (1725°F to 1800°F). The pots are made of carbon-lined steel. Under the influence of the electricity, the oxygen in the ore combines with the carbon, leaving pure, molten aluminum at the bottom of the pots. Approximately 9 kilowatt-hours (kWh) of electricity are required to make each pound of aluminum. Because of the expense, aluminum manufacturing plants are located in regions of relatively cheap power.

The molten aluminum is siphoned off and poured into molds to form primary ingots. Alternatively, the molten aluminum is charged directly to a remelt furnace where pure alloying elements or master alloys (concentrated alloys with an aluminum base) are added to produce aluminum alloys. These aluminum alloys are poured into molds to make primary aluminum ingots.

Primary ingots are rolled, extruded, or drawn into wrought mill products, or finished forms. Engineered aluminum products include special extruded shapes, forgings, and stampings that are not available as off-the-shelf mill products.

Aluminum and aluminum alloys are cast to make components when the product shape favors casting. Powder metallurgy (P/M) components are made by compressing powders into shaped dies. The compressed powders are sintered at elevated temperatures to densify and strengthen them. Aluminum and aluminum alloys are cast by all common casting processes, which are die casting, permanent mold casting, sand casting, shell molding, plaster casting, investment casting, and continuous casting.

Die casting is a casting process that consists of using substantial pressure to inject molten metal into the cavity of a metal die. The low melting points of aluminum alloys make them suitable for die casting, which produces a dense, fine-grain surface structure with excellent wear and fatigue properties. Die casting is the most popular casting process for aluminum and is suited to high-volume production.

Permanent mold casting is a casting process that requires a metal mold, which is used repeatedly for producing many castings of the same form. Permanent mold castings are usually larger than die castings. Because of the rapid freezing rate, the castings are sound (free from porosity) and exhibit excellent mechanical properties.

Sand casting is a casting process that consists of pouring molten metal into the cavity of a sand mold. Sand casting is extremely versatile, but it has low dimensional accuracy and produces castings with poor surface finishes.

Shell molding is a casting process that consists of pouring molten metal into the cavity of a mold, which consists of a thin shell of resin-bonded sand. The resulting surface finish and dimensional accuracy are superior to those of sand casting.

Plaster casting is a casting process that consists of pouring molten metal into the cavity of a plaster mold. The high insulating value of the plaster allows thin wall castings to be made with close tolerances, excellent surface finish, and resolution of fine detail.

Investment casting is a casting process that uses a plaster mold cavity that is formed around a wax pattern, which is melted out when the plaster is fired to harden it. The firing is performed before pouring molten metal into the mold cavity. Investment casting is used for small precision components where no subsequent machining is required.

Centrifugal casting is a casting process that consists of pouring molten metal into the cavity of a rotating mold, which can be made of various materials. Centrifugal casting is usually limited to components of circular cross section, such as wheels and rolls.

Continuous casting is a casting process that consists of pouring molten metal into a bottomless water-cooled mold of simple cross section and continuously withdrawing solidified metal from the bottom of the mold. Continuous casting is used in the production of billets for subsequent rolling, extrusion, or forging.

Strengthening

Annealed or cast commercially pure aluminum has one fifth of the tensile strength of structural steel. Aluminum and aluminum alloys are strengthened by solid solution or dispersion hardening, cold working, and precipitation hardening. See Figure 20-1. The strength of commercially pure aluminum may be more than doubled by cold working and further increased by precipitation hardening. This makes the strength-to-weight ratios of hardened aluminum alloys as high as for other structural materials.

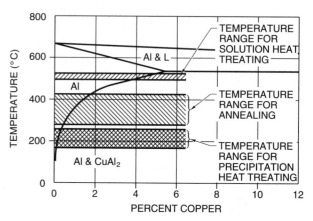

Figure 20-1. Aluminum-copper alloys (2xxx series) may be solution treated for optimum mechanical properties. Precipitation hardening may also be used, but it leads to a significant loss of elongation.

Solid Solution and Dispersion Hardening. Aluminum alloys that are not heat treatable are hardened and strengthened by the addition of alloying elements. Hardening and strengthening occurs by solid solution hardening or by the dispersion of intermetallic compounds.

Cold Working. Aluminum alloys that are not heat treatable are also strengthened by cold working. Cold working is performed below the recrystallization temperature of the particular alloy. Recrystallization temperatures range from 345°C to 425°C (650°F to 800°F). Increasing the percentage of cold work increases strength and hardness. Products hardened by cold working can be restored to a fully soft condition by annealing.

Precipitation Hardening. Precipitation hardening (aging) is a method of strengthening heat-treatable aluminum alloys. Heat-treatable aluminum alloys contain elements that are more soluble at elevated temperatures than at room temperature. When the solid solution is rapidly quenched, a supersaturated condition occurs. As the alloying elements precipitate out of the solution with the passage of time, the strength of the alloy increases. Artificial aging (precipitation hardening performed at elevated temperatures) is used to develop immediate maximum strength.

Annealing. Annealing is performed on heat-treatable and non-heat-treatable alloys to remove the effects of cold work. It is accomplished by heating within 300°C to 450°C (570°F to 840°F). The rate of softening depends on the time at temperature and can vary from several hours at low temperature to seconds at high temperature.

Aluminum Designation Systems

Aluminum and aluminum alloy products are identified by various designation systems, standards, and specifications. Aluminum products are covered by standards produced by the American Society for Testing and Materials, the Society of Automotive Engineers, and the Aerospace Materials Specifications.

Wrought and cast aluminum alloys are identified by various designation systems, which include the Aluminum Association (AA) numbering system and the Unified Numbering System for metals and alloys. See Figure 20-2. Aluminum Association designations for wrought aluminum alloys consists of four digits. The first digit indicates the major alloying element. For example, 2014 is an aluminum-copper alloy and 3003 is an aluminum-manganese alloy, both with specified chemical composition ranges. Aluminum Association designations for cast alloys consist of three numbers, a period, and a fourth number. The first number indicates the major alloying element. For most aluminum alloys, the fourth digit indicates the product form. Zero indicates the product form is a casting, a 1 for a standard ingot, and a 2 for an ingot with a narrower chemical composition range than a standard ingot. For example, 356.0 is an aluminum-silicon-copper-magnesium alloy casting with a specific chemical composition range, while 356.1 and 356.2 are similar alloys, but in ingot form and with slightly different chemical composition ranges.

Temper designations are letters that indicate the final condition of cold-worked (H) or heat-treated (T) material. The temper designation is separated from the alloy designation by a hyphen. For example, 3003-H2 designates a quarter hard aluminum-manganese alloy, and 2014-T4 designates an aluminum-copper alloy that is solution treated, quenched, and allowed to age at room temperature. See Figure 20-3.

The Unified Numbering System for metals and alloys identifies wrought and cast aluminum alloys with the uppercase letter A followed by five numbers that identify a composition range for a specific alloy. For wrought aluminum alloys, the first number is 9 followed by the Aluminum Association number for the alloy. For example, alloy 3003 is equivalent to UNS A93003. For cast aluminum alloys, the first number varies from 0 to 6. For example, alloy 356.0 is equivalent to UNS A03560.

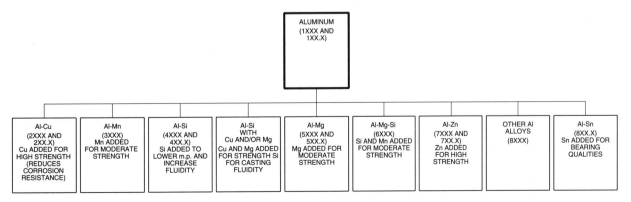

Figure 20-2. Wrought aluminum alloys are designated by various systems, which include the Aluminum Association numbering system.

TEMPER DESIGNATIONS	
Designation	**Condition**
F	As-fabricated
O	Annealed
H1	Strain hardened only
H2	Strain hardened and partially annealed
H3	Strain hardened and thermally stabilized
W	Solution heat treated
T1	Cooled from an elevated-temperature shaping process and naturally aged
T2	Cooled from an elevated-temperature shaping process, cold worked, and naturally aged
T3	Solution heat treated, cold worked, and naturally aged
T4	Solution heat treated and naturally aged
T5	Cooled from an elevated-temperature shaping process and then artificially aged
T6	Solution heat treated and then artificially aged
T7	Solution heat treated and stabilized
T8	Solution heat treated, cold worked, and then artificially aged
T9	Solution heat treated, artificially aged, and then cold worked
T10	Cooled from an elevated-temperature shaping process, cold worked, and then artificially aged

Figure 20-3. Temper designations are used to indicate the cold-worked or heat-treated condition of aluminum alloys.

Wrought Aluminum Alloys. The Aluminum Association designations for the wrought aluminum alloy series is 1xxx for commercially pure aluminum, 2xxx for aluminum-copper, 3xxx for aluminum-manganese, 4xxx for aluminum-silicon, 5xxx for aluminum-magnesium, 6xxx for aluminum-magnesium-silicon, 7xxx for aluminum-zinc, and 8xxx for aluminum-other elements. Chemical compositions and mechanical properties vary for the different wrought aluminum alloys.

The 1xxx series comprises commercially pure aluminum alloys containing >99.00% Al. The 1xxx series alloys have excellent corrosion resistance, high thermal and electrical conductivities, and excellent workability. These weak alloys are moderately strengthened by cold working and are used for chemical equipment, reflectors, heat exchangers,

electrical conductors and capacitors, packaging foil, architectural products, and decorative trim.

The 2xxx series comprises aluminum-copper alloys. The 2xxx series alloys are solution treated to achieve optimum mechanical properties. In some instances, these alloys are precipitation hardened, but this leads to a significant loss of elongation. The 2xxx series alloys are less corrosion resistant than most other aluminum alloys. To improve corrosion resistance, 2xxx series alloys are clad with high-purity aluminum or with an aluminum-magnesium-silicon alloy of the 6xxx series. The resulting alloy combination is an alclad product. An *alclad* is a composite wrought product comprised of an aluminum alloy core having, on one or both surfaces, a metallurgically bonded aluminum or aluminum alloy coating, which is resistant to corrosion. The 2xxx series alloys have high strength-to-weight ratios and are used for truck and aircraft wheels, truck suspension parts, aircraft fuselage, and wing skins. The weldability of the 2xxx series alloys is limited.

The 3xxx series comprises aluminum-manganese alloys. The 3xxx series alloys are strengthened by cold working and have approximately 20% more strength than the 1xxx series. See Figure 20-4. Alloys 3003, 3004, and 3105 are widely used for moderate strength applications requiring good workability. Applications include beverage cans, cooking utensils, heat exchangers, storage tanks, awnings, furniture, highway signs, roofing, siding, and other architectural forms.

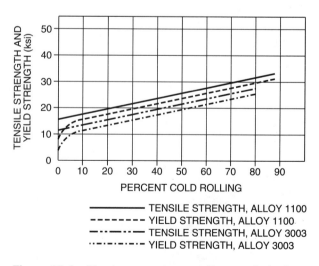

Figure 20-4. Aluminum-manganese (3xxx series) alloys may be heavily strain hardened during cold working.

The 4xxx series comprises aluminum-silicon alloys. Silicon may be added to aluminum in significant quantities (up to 12% Si), which causes lowering of the melting temperature range without embrittlement. See Figure 20-5. Consequently, 4xxx series alloys are often used for welding wire and as brazing filler metal. Although most 4xxx series alloys are not strengthened by heat treatment, when used for welding heat-treatable alloys, they will pick up some of the alloy content of the base metal and respond in a limited way to heat treatment. High-silicon alloys become dark gray to charcoal gray when anodized (specially treated to thicken the surface film) and are used in architectural applications. Alloy 4032, with 12.5% Si, has high wear resistance and is used for forged engine pistons.

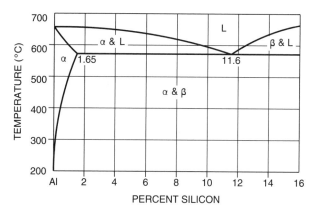

Figure 20-5. Silicon may be added to aluminum in significant quantities up to 12% Si.

The 5xxx series comprises aluminum-magnesium alloys, which are sometimes additionally alloyed with a small amount of manganese. Both elements cause solid solution hardening, although manganese is considerably more effective. The 5xxx series alloys have moderate to high strength, good weldability, and good marine corrosion resistance. To prevent stress-corrosion cracking, the amount of cold working must be limited in alloys with greater than approximately 3.5% Mg and in alloys that operate above 65°C (150°F). The 5xxx series alloys are used for architectural trim, decorative trim, cans and can ends, household appliances, bases for streetlights, boats and ships, cryogenic applications, and crane parts.

The 6xxx series comprises aluminum-silicon-magnesium alloys. The proportions of these elements are adjusted to form the intermetallic compound magnesium silicide (Mg_2Si). See Figure 20-6. The 6xxx series alloys are strengthened by precipitation hardening. These alloys have medium strength (less than the 2xxx or 7xxx alloys), formability, weldability, machinability, and corrosion resistance. The 6xxx series may be formed in the T4 (solution heat-treated) temper, and then precipitation hardened to the T6 temper. The 6xxx series alloys are

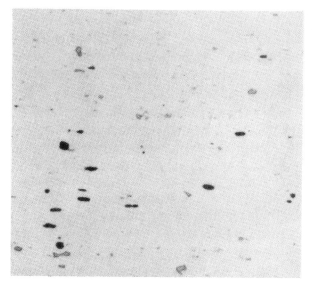

ALLOY 6061-F

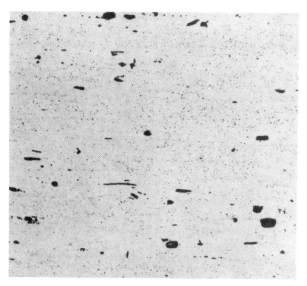

ALLOY 6061-T6

American Society for Metals

Figure 20-6. Alloy 6061-T6 is strengthened by magnesium silicide precipitation.

used for architectural applications, transportation equipment, bridge railings, and welded structures. Alloy 6061 is one of the most versatile of all aluminum alloys.

The 7xxx series comprises aluminum-zinc alloys with 1% Zn to 8% Zn. When coupled with small amounts of magnesium, 7xxx series alloys may be precipitation hardened to very high strength levels. Other elements, such as copper and chromium, are usually added in small quantities. The 7xxx series alloys are used in airframe structures, mobile equipment, and other highly stressed components. Alloy 7075 is one of the strongest alloys available and is used in airframe structures and for other highly stressed components.

The 8xxx series comprises alloy compositions that do not fit into any other series. Examples include alloy 8001, an aluminum-nickel-iron alloy used in atomic energy applications, and alloy 8280, an aluminum-tin-nickel-copper alloy used for bearings.

Cast Aluminum Alloys. The Aluminum Association designations for cast alloys consist of three numbers, a period, and a fourth number. The designations for the cast aluminum alloy series are 1xx.x for commercially pure aluminum, 2xx.x for aluminum-copper, 3xx.x for aluminum-silicon with copper and/or manganese, 4xx.x for aluminum-silicon, 5xx.x for aluminum-magnesium, 6xx.x for the unused series, 7xx.x for aluminum-zinc, and 8xx.x for aluminum-tin. Chemical compositions and mechanical properties vary for the different cast aluminum alloys.

A few casting designations also include a prefix letter, indicating a slight difference in chemical composition from the base alloy or the presence of minor amounts of other elements. Examples of these alloys are A356.0 and B380.0.

The 1xx.x series comprises the commercially pure aluminum alloys and consists of alloys containing 99.00% Al or more. The second and third digits indicate the purity of the alloy, for example, 199.x is 99.99% Al, and 195.x is 99.95% Al. Commercially pure aluminum alloys are seldom used for castings because of hot shortness (tendency to crack from solidification stresses) and high shrinkage during solidification. Commercially pure aluminum alloys are

used only where high conductivity is demanded, such as in squirrel-cage induction motors.

The 2xx.x series comprises aluminum-copper alloys, which also contain quantities of other alloying elements. The 2xx.x series alloys with 4% Cu to 6% Cu may be heat treated to obtain high strength and ductility, but castability of these alloys is poor. Small amounts of silicon are added to improve castability and are widely used in automotive and aerospace applications. Manganese is added to low-copper alloys to improve response to heat treatment. The 2xx.x series alloys with 7% Cu to 8% Cu are not greatly used because of their low strength and poor castability. The 2xx.x series alloys with 9% Cu to 11% Cu are limited to high-temperature wear-resistant applications, such as automotive pistons and cylinder blocks.

The 3xx.x series comprises aluminum-silicon alloys with additions of copper and/or magnesium. Commercially, these are by far the most important casting alloys because of their superior casting characteristics compared with other series. Alloys of silicon, copper, and magnesium are strengthened by precipitation hardening. The most popular of the 3xx.x series alloys is 356 and the higher-purity A356.0, which have excellent casting characteristics, weldability, pressure tightness, and corrosion resistance. These alloys and alloy A357.0 are used for military and aircraft applications. Alloys of silicon, copper, and magnesium containing large amounts of silicon (up to 22% Si) have excellent wear resistance that is developed by a network of primary silicon crystals. Automotive engine blocks are the major use for these alloys.

The 4xx.x series comprises aluminum-silicon alloys. Depending on the silicon content, which varies between 5% Si and 13% Si, 4xx.x series alloys may be hypoeutectic or hypereutectic. Sodium, calcium, or antimony is added to the melt before it is poured to refine the grain structure, which improves the strength and ductility of hypoeutectic alloys. In hypereutectic alloys, the structure is refined through modification of the proeutectic silicon by the addition of phosphorus.

The 5xx.x series comprises aluminum-magnesium alloys. These alloys possess high corrosion and tarnish resistance, especially in marine environments. The 5xx.x series alloys are strengthened by solid solution hardening and are limited by poor castability and weldability.

The 7xx.x series comprises aluminum-zinc alloys. The 7xx.x series alloys usually contain a small quantity of magnesium to improve response to precipitation hardening. The 7xx.x series alloys have high melting points, making them suitable for castings assembled by brazing. The 7xx.x series alloys age naturally at room temperature in a relatively short time, achieving moderately high strength levels. This occurs in the as-cast condition without the need to solution heat treat. These alloys are readily machinable and weldable but exhibit poor castability. The 7xx.x series alloys are used where a good combination of mechanical properties is required without the need for heat treatment. For example, they are used in furniture or very large components where furnace costs would be excessive.

The 8xx.x series comprises aluminum alloys with approximately 6% Sn. The 8xx.x series alloys are developed for bearings and bushings with high load carrying capacity and fatigue strength. Tin provides lubricity. Such alloys have superior corrosion resistance to most bearing materials. Careful control of the casting method is necessary to obtain fine interdendritic spacing, which is required for optimum bearing performance. Foundry practices must also be controlled because the 8xx.x series alloys are susceptible to hot cracking. Alloy 850.0 is commonly used for connecting rods and crankcase bearings for diesel engines.

ALLOYING ELEMENTS

Aluminum alloys are wrought or cast, and both groups contain individual families of alloys that are primarily identified by the principal alloying elements. The principal alloying elements in aluminum are copper, magnesium, silicon, manganese, zinc, and tin. These alloying elements are added to both wrought and cast alloys to strengthen them by solid solution or dispersion hardening, to strengthen them by precipitation hardening, and to improve their castability. Copper increases strength in precipitation-hardened alloys but substantially decreases corrosion resistance. Consequently, wrought alloys that contain copper are frequently clad with pure aluminum to improve corrosion resistance.

Manganese increases strength in work-hardened alloys by dispersion hardening. A combination of magnesium and manganese increases the work-hardening rate of the alloy.

Silicon substantially lowers the melting point of the alloy. Significantly more silicon is added to cast alloys than is added to wrought alloys to increase fluidity, reduce cracking, and improve feeding of the molten metal (minimizes shrinkage porosity). Silicon is an important constituent of welding and brazing filler metals.

Magnesium increases the strength and the work-hardening rate in work-hardened alloys, and it also improves weldability. Special care is required with tempering and operating conditions in alloys with high magnesium contents (>3.5% Mg) to avoid stress-corrosion cracking. The combination of magnesium and silicon improves formability, corrosion resistance, and strength by precipitation hardening.

Zinc is a strengthening agent in precipitation-hardened alloys. Zinc is usually coupled with a small percentage of magnesium and small quantities of other elements, such as copper and chromium.

Tin improves the machinability of aluminum alloys, but its main purpose is to improve bearing qualities. Tin is a component of cast alloys used for bearings and bushings that are subject to high loads and fatigue.

ALUMINUM FABRICATION

Aluminum alloys are easy to form and machine. Aluminum alloys are formed, machined, and joined using a wide variety of processes. Aluminum alloys are joined by welding, brazing, soldering, mechanical fastening, and adhesive bonding.

Forming

Aluminum alloys are hot and cold formed. Hot forming is performed between 260°C and 480°C (500°F and 900°F), depending on the type of alloy and the forming process. Cold forming is performed on non-heat-treatable alloys. Alloys 1100 and 3003 are frequently used for these operations because of their excellent workability and low cost. If somewhat higher strength is required, alloys with magnesium can be used.

The annealed temper (O temper) is the most workable condition for cold forming. Alloys that have been freshly solution treated and quenched (W temper) can be cold formed, which increases their strength when they are aged. Material that has been solution treated at the mill, but not artificially aged (T3, T4, or W tempers) is generally suitable only for mild forming.

Machining

Compared with steel, aluminum alloys are relatively easy to machine. Machinability of aluminum alloys varies widely with composition and temper condition. Machinability varies from poor, for annealed pure metals, to good, for precipitation-hardened or cold-worked metals.

Joining

Welding, brazing, soldering, mechanical fastening, and adhesive bonding have a variety of joining characteristics that make them vastly different. See Figure 20-7. During welding operations, the thin film of aluminum oxide that forms naturally on aluminum increases in thickness with temperature and is a hinderance. The film is relatively thick on heat-treated aluminum alloys. Thick oxide films must be removed mechanically before resistance and fusion welding and must be prevented from re-forming during fusion welding by means of an inert gas shield.

All aluminum alloys can be resistance welded, and most can be fusion welded. Gas metal arc welding (GMAW or MIG) is the most common fusion welding process. Gas tungsten arc welding (GTAW or TIG) is also used for joining aluminum. The alloys that are not heat treatable are the most easily welded. See Appendix. The heat of welding removes the strengthening effect of any cold work in the heat-affected zone of base metal adjacent to the weld.

Precipitation-hardened alloys are less easy to weld. As-welded precipitation-hardened alloys (T4 or T6) have lower strength levels than nonwelded types because the heat-affected zone of base metal adjacent to the weld does not possess optimum strength. To optimize the as-welded properties, postweld solution treating and aging must be performed. There are a variety of welding filler metals for the various aluminum alloys. See Appendix.

Brazing is performed on metals that have very thin cross sections or inaccessible joints. Active (corrosive) fluxes must be used to remove the oxide film during joining. The oxide film must be thoroughly removed or corrosion will result during service. Aluminum brazing filler metals are available as wire, rod, sheet, and cladding on sheet.

JOINING METHODS		
Welding	Resistance	Ultrasonic (no flux)
Fusion	Spot	**Mechanical fastening**
Torch (gas)	Seam	Mechanical clinching
Oxyfuel	Flash	Staking
Arc	Stud	Rivets, hot and cold
Metal arc with flux	High-frequency induction	Nailing
Gas metal arc (GMAW)	**Brazing**	Stapling
Electrogas	Pot	Stitching
Gas tungsten arc (GTAW)	Torch (hand) flux	Screws and bolts
Pressure	Dip-flux pot	**Adhesive bonding**
Explosive	Vacuum furnace (no flux)	Pressure, room temperature
Friction	Furnace (flux)	Weld bonding
Ultrasonic	**Soldering**	Pressure, high temperature
Electron beam	Torch (flux)	
Laser beam	Furnace air (flux)	

Figure 20-7. Aluminum is joined by welding, brazing, soldering, mechanical fastening, and adhesive bonding.

Soldering is similar to brazing, except that the melting points of solders are much lower than brazing filler metals. If a flux is used, it must be removed immediately to avoid corrosion. Solders are made from metals other than aluminum and will cause galvanic corrosion of the aluminum during service. Therefore, the use of solders is limited to noncorrosive situations, except when the assembly (joint area) is coated to protect it from the atmosphere.

Mechanical fastening for aluminum consists of the same techniques used for other metals. Mechanical fastening includes riveting, bolting, nailing, stitching, stapling, and clinching.

Adhesive bonding may be used to improve the overall strength of the joint. Adhesive bonding has several advantages, which include a more evenly distributed load, elimination of stress concentrations, and dampened vibrations. Dissimilar metals and metals of greatly different thicknesses can be joined with adhesive bonding. In addition, adhesive bonding is the only method of joining aluminum to nonmetals. Various adhesives such as elastomers, epoxy resins, and vinyl plastisols can be used to join aluminum alloys.

ALUMINUM PROPERTIES

Desirable properties of aluminum include low density, high specific strength (strength-to-weight ratio), high electrical and thermal conductivity, and good corrosion resistance. The two most useful properties of aluminum alloys are low density and high specific strength. High electrical and thermal conductivity are also very important properties of aluminum alloys. Corrosion resistance is improved by surface treatments, such as anodizing, or by cladding less resistant alloys with pure aluminum. All of these properties are used in many applications.

Density and Specific Strength

The density of aluminum is approximately one third of that of copper and steel. This reduces the cost per pound and increases the specific strength of aluminum. By volume, aluminum is less than half as expensive as copper. The specific strength values of some precipitation-hardened aluminum alloys are as high as many other structural materials. Conse-

quently, aluminum is used in transportation equipment, where weight is an important consideration.

Electrical and Thermal Conductivity

Commercially pure aluminum conducts electricity and heat approximately 60% as well as copper. However, the lower density of aluminum results in approximately a 50% weight savings for equal conductivity when aluminum is substituted for copper. The need for structural support for the aluminum prevents full realization of the economies, but substantial savings are often possible.

Aluminum alloys have substantially greater electrical resistivity (lower conductivity) than pure aluminum. Any form of hardening causes an appreciable increase in resistivity.

The thermal conductivity of aluminum is approximately 50% that of copper and four times that of low-carbon steel. This means that heat must be supplied to aluminum approximately four times as fast as to steel to locally raise the temperature by the same amount. Any form of hardening causes an appreciable decrease in thermal conductivity.

Corrosion Resistance

The thin aluminum oxide film that forms instantaneously on aluminum in air serves as protection against corrosion in many environments, from foods and drinks to industrial chemicals. Contact with some acidic solutions or with moist corrosive materials prevent access of oxygen to the film and result in a breakdown of the film, which leads to severe corrosion. Some alkaline solutions dissolve the film and lead to significant corrosion. Surface treatments help to improve the corrosion resistance of aluminum and aluminum alloys.

The corrosion resistance of aluminum alloys is strongly influenced by the alloy content. Pure aluminum has the best corrosion resistance. The corrosion resistance is reduced by certain alloying elements present in specific families.

Aluminum alloys exhibit various types of corrosion, which include exfoliation, galvanic corrosion, and deposition corrosion. Copper impairs corrosion resistance more than any other alloying element. The 2xxx series (wrought aluminum-copper alloys), the 7xxx series (wrought aluminum-zinc alloys with copper), and the similar cast alloys have the poorest corrosion resistance. Magnesium improves corrosion resistance to alkaline solutions. In 5xxx series wrought alloys with >3% Mg, cold forming must be controlled and the operating temperature must be limited to 65°C (150°F) to avoid susceptibility to stress-corrosion cracking. Silicon, manganese, and zinc have little effect on the corrosion resistance. Surface treatments are used to prevent corrosion of aluminum and aluminum alloys.

Exfoliation.

Exfoliation is a form of corrosion that occurs in certain cold-worked tempers and proceeds along selective subsurface paths parallel to the surface. The layers of noncorroded metal between the parallel paths are split apart by the voluminous (abundant) corrosion products, leading to delamination of the metal.

Exfoliation develops in products that have markedly directional structures, with highly elongated grains that are thin relative to their length and width. Alloys most subject to exfoliation are members of the 2xxx series, 5xxx series, and 7xxx series.

Galvanic Corrosion.

Galvanic corrosion is the selective attack of one metal when it is electrically coupled to another metal or conducting nonmetal in an electrically conductive environment. Galvanic corrosion can cause severe corrosion of the aluminum component of the couple. Electrically insulating the members of the couple from one another diminishes galvanic corrosion.

Metals and alloys that form unfavorable galvanic couples with aluminum are copper, nickel, and tin. Graphite, a conducting nonmetal, has the same unfavorable effect. In certain aggressive environments, such as seawater, other alloys may also cause galvanic corrosion of aluminum. These other alloys include stainless steels, carbon steel, titanium alloys, and lead alloys. Metals that corrode preferentially when coupled galvanically to aluminum include magnesium and zinc.

Deposition Corrosion.

Deposition corrosion is a variation of galvanic corrosion that occurs when metals plate out from solutions on the surface of aluminum. Copper is the most common of these because it is often present in small amounts in waters, such as in plumbing systems composed of copper pipe and fittings. Deposition corrosion occurs by the following reaction:

$$3CuSO_4 + 2Al \rightarrow Al_2(SO_4)_3 + 3Cu\downarrow$$
(solution)   (plates out)

Other metals that plate out include mercury, tin, lead, and iron. Iron in the form of rust, however, does not attack aluminum.

Surface Treatment.

The commonly used surface treatments to prevent the corrosion of aluminum are anodizing and cladding. *Anodizing* is an electrolytic oxidation process in which the protective oxide film is artificially thickened to improve corrosion resistance. An anodized surface has improved resistance to chemicals, rubbing, and the atmosphere. Anodized surfaces may be further improved by sealing them with dichromate solution, followed by a hot water rinse.

Cladding is the bonding together of two and/or three layers of metals to form a composite metal. High-strength aluminum alloys with poor corrosion resistance are frequently clad with pure aluminum to improve corrosion resistance. The combination is called an alclad product.

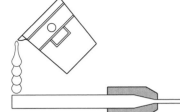

Magnesium 21

Magnesium is the lightest commercial metal. Some magnesium alloys have strength-to-weight ratios comparable with stronger aluminum alloys and low-alloy steels, although their ductility is limited. Magnesium is strengthened by alloying and by precipitation hardening or work hardening. Magnesium alloys are formed at elevated temperatures, and they are relatively easy to machine and join. The corrosion resistance of magnesium alloys is satisfactory in most atmospheres and variable in chemical solutions. Surface treatments are used to improve corrosion resistance of magnesium alloys.

MAGNESIUM MANUFACTURING

Magnesium is extracted from seawater and refined into alloy ingots for structural end uses, sacrificial anode materials, and granules for steelmaking purposes. Magnesium alloys are cast or wrought and are strengthened by precipitation hardening and work hardening. Magnesium alloys are surface treated to combat possible corrosion. The American Society for Testing and Materials designation system for magnesium alloys indicates the approximate chemical composition and mechanically-worked or heat-treated condition.

Extraction and Production

Magnesium is extracted from magnesium chloride in seawater, which contains .13% Mg. It takes approximately 142 gallons of seawater to make one pound of magnesium. The seawater is filtered to remove aquatic life and then mixed with caustic soda or slaked dolomine. *Slaked dolomine* is dolomite rock that has been roasted and mixed with water. These compounds react with the magnesium chloride in the seawater to form a precipitate of magnesium hydroxide.

The slurry of magnesium hydroxide and seawater flows into large ponds where the magnesium hydroxide settles to the bottom. The magnesium hydroxide is raked toward the center of the pond and pumped to a filtering plant. The magnesium hydroxide is then run over drum-like filters to remove as much of the seawater as possible. This produces a thick sludge with the consistency of wet cement.

The magnesium hydroxide is mixed with hydrochloric acid. This produces a solution containing magnesium chloride, the same as the original compound in the seawater, except that the magnesium chloride concentration is increased to 35%. Sulfuric acid is added to the solution to precipitate calcium sulfate from unwanted calcium. After running the solution through more filters and drying it, a free-flowing, granular solid that contains approximately 70% magnesium chloride results.

The magnesium chloride material is fed to electrolytic refining cells consisting of thick-walled steel pots into which vertical graphite electrodes are extended. Current from the electrodes produces molten magnesium metal and chlorine gas. The molten metal floats to the top of the cell and is tapped several times a day into a large track-mounted crucible.

The molten magnesium metal is then poured into a casting furnace. The casting furnace makes magnesium for ingots, anodes, or granules, which have a variety of end uses. For example, alloyed magnesium ingots are used as a basis for sand castings, die castings, ground powders, and products such as forgings, sheet, plate, and extrusions. Unalloyed magnesium ingots are used in the aluminum industry to make aluminum-magnesium alloys. Magnesium anodes are used to protect underground steel tanks and pipelines from corrosion. Magnesium granules are used in the steel industry to reduce the sulfur content of steel.

Wrought Magnesium Alloys. Wrought magnesium alloys are produced by extrusion, forging, and rolling. Extrusion products consist of round rods and various types of bars, tubes, and shapes. Extrusion offers certain unique designs, such as reentrant angles and undercuts. The most important factor influencing the ease of extrusion is the cross-section symmetry of the extrusion. Extrusion dies are inexpensive and dimensions can be held close enough so that final machining is often unnecessary.

Forging develops the best combination of strength characteristics. Forgings are used where light weight coupled with rigidity and high strength is required. Compared with other metals, magnesium is easily forged, which greatly reduces the number of forging operations required to produce the finished product.

Rolling is used to produce sheet and plate. Rolling is applied to a limited number of magnesium alloys that have good formability.

Cast Magnesium Alloys. Cast magnesium alloys are usually produced by sand casting, permanent mold casting, or die casting. The mechanical properties of sand castings and permanent mold castings are similar. Permanent mold casting and die casting are used only when the number of parts justifies the high cost of the casting equipment.

When sand casting in green sand (moist) molds, the moisture in the sand may react with the magnesium to form magnesium oxide, which liberates hydrogen. The magnesium oxide forms blackened areas, or burns, on the casting surface, and the liberated hydrogen may cause porosity. To prevent these problems, inhibitors are mixed with the sand.

Gravity-fed molds for casting magnesium require an extra-high head of molten magnesium so that the pressure is great enough to force gas bubbles out of the casting and encourage the metal to take the mold detail. In designing magnesium castings, extra-large fillets must be provided at all reentrant corners. Stress concentrations caused by sharp fillets lead to hot cracking.

Since the solidification shrinkage of magnesium and aluminum alloys is similar, permanent molds for aluminum and magnesium are interchangeable. The gating system (molten metal inlet design) must be adapted to allow for the sluggish feeding characteristics of magnesium.

With die casting, the rapid solidification caused by contact of molten metal with the cold die produces a casting of dense structure with excellent physical properties, good finish, and close dimensional control. Die castings are not usually heat treated because heat treatment compromises dimensional tolerances.

Strengthening

Many cast and wrought magnesium alloys are strengthened by precipitation hardening. Wrought magnesium alloys may also be strengthened by work hardening. Surface treatments are used to increase the corrosion resistance of magnesium alloys.

Precipitation Hardening. Precipitation hardening is often used to strengthen magnesium alloys. For precipitation hardening to be applicable, the magnesium end of the phase diagram must display a solvus (solid solubility boundary line) which exhibits rapidly decreasing solubility for magnesium with decreasing temperature.

The magnesium end of the phase diagram for the magnesium-aluminum alloy system is the region for precipitation-hardening alloys. See Figure 21-1. The maximum solubility of aluminum in magnesium decreases from 12.1% at 436°C (818°F) to 3.2% at 204°C (400°F). The gamma (γ) phase is the compound $Mg_{17}Al_{12}$, which is responsible for the precipitation-hardening effect.

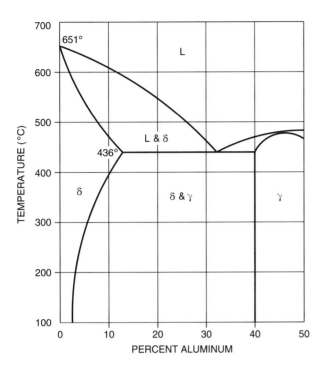

Figure 21-1. Precipitation-hardening magnesium alloys are indicated at the magnesium end of the phase diagram by a rapid decrease in magnesium solubility with a decrease in temperature.

Precipitation hardening is a three-stage process. The stages consist of holding the alloy at elevated temperature to obtain a single-phase solid solution (solution treating), quenching the alloy to obtain a supersaturated solution, and then heating the alloy to a temperature below the solvus (artificial aging), which allows the precipitate to develop within the solid solution.

Protective Surface Treatments. Surface treatments are used to protect magnesium alloys from corrosive or polluted atmospheres. The various types of coatings that are used are divided into dip coatings and anodic coatings. See Figure 21-2.

Dip coatings are thin coatings used primarily for protection during shipment and storage and as primers for subsequent painting. Dip coatings should not be heated above 260°C (500°F). Anodic coatings are thicker and harder than dip coatings. Anodic coatings can be heated to 345°C (650°F) with no reduction in corrosion resistance. Flash (thin) coats are often used as paint bases. Thicker coats without paint give some protection against corrosion in moderately corrosive atmospheres, but sealing the coating by impregnation or painting is usually desirable. *Impregnation* is the process of filling pores of a material with a liquid, such as a lubricant.

Magnesium Designation Systems

The ASTM is the most common designation system for identifying magnesium alloys. The Unified Numbering System for metals and alloys designations for magnesium alloys reference the ASTM system.

The ASTM system for magnesium alloys consists of four parts. The first part indicates the two principal alloying elements. The second part indicates the approximate amounts of the two principal alloying elements. The third part distinguishes between magnesium alloys having the same amounts of the two principal alloying elements. The fourth

		PROTECTIVE SURFACE TREATMENTS			
Common Names	**Treatment**	**Treated Alloys**	**Solutions**	**Temperature (°C)**	**Time (min)**
Chrome pickle, Dow 1	Dip	All alloys	Chromic acid, nitric acid, and water	90 to 100	1 to 15
Dichromate, Dow 7	Dip	All wrought alloys except those containing thorium	Sodium dichromate, calcium or magnesium fluoride, and water	100	30
Dow 17	Anodic	All alloys	Ammonium acid fluoride, sodium dichromate, phosphoric acid, and water	70 to 80	1 to 30
HAE	Anodic	All alloys	Potassium hydroxide, aluminum hydroxide, trisodium phosphate, potassium fluoride, potassium manganate, and water	25 to 30	60 to 90

Figure 21-2. Surface treatments are used to protect magnesium alloys in corrosive atmospheres.

part indicates the temper condition of the alloy and is similar to the codes used for aluminum alloys. See Figure 21-3.

An example of an ASTM designation is the magnesium alloy AZ91C-T6. The *AZ* indicates that aluminum and zinc are the two principal alloying elements. The *91* indicates that the aluminum and zinc are present in percentages of 9% Al and 1% Zn. The *C* indicates that this is the third alloy standardized with 9% Al and 1% Zn as principal alloying elements. The *T6* indicates that the alloy is in the solution-treated and artificially aged condition.

The UNS designation system for magnesium alloys consists of the uppercase letter M followed by five numbers. For example, the alloy ASTM AZ91C is equivalent to UNS M11914. Compositions and mechanical properties of magnesium alloys can be cross-referenced by ASTM and UNS designations.

MAGNESIUM ALLOYS

Magnesium alloys consist of two groups. The first group contains 2% Al to 10% Al, plus minor amounts of manganese, silicon, and zinc. These alloys are widely available at moderate costs. The second group contains various elements, which include manganese, zinc, rare earth elements, and thorium (Th), plus a small amount of zirconium to refine the grain size. These magnesium alloys have better properties at higher temperatures, are more difficult to produce, and are much more expensive.

Magnesium-Aluminum Alloys

Aluminum additions to magnesium provide the greatest strengthening and hardening effects. Alloys with more than 6% Al are precipitation hardenable.

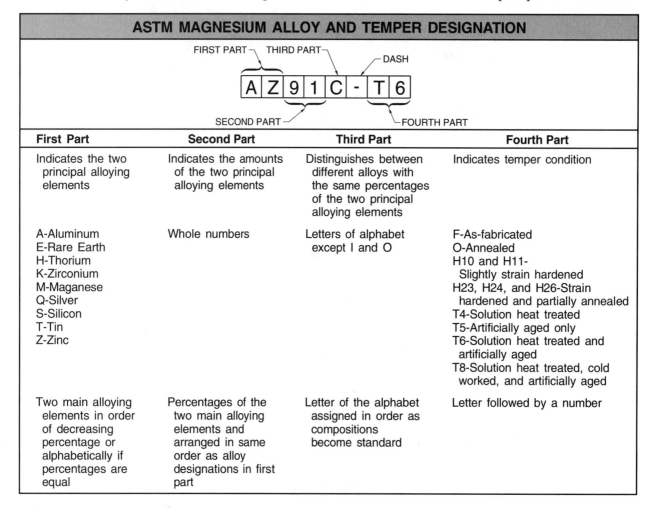

ASTM MAGNESIUM ALLOY AND TEMPER DESIGNATION

First Part	Second Part	Third Part	Fourth Part
Indicates the two principal alloying elements	Indicates the amounts of the two principal alloying elements	Distinguishes between different alloys with the same percentages of the two principal alloying elements	Indicates temper condition

A-Aluminum
E-Rare Earth
H-Thorium
K-Zirconium
M-Maganese
Q-Silver
S-Silicon
T-Tin
Z-Zinc
	Whole numbers	Letters of alphabet except I and O	F-As-fabricated
			O-Annealed
			H10 and H11-
			  Slightly strain hardened
			H23, H24, and H26-Strain
			  hardened and partially annealed
			T4-Solution heat treated
			T5-Artificially aged only
			T6-Solution heat treated and
			  artificially aged
			T8-Solution heat treated, cold
			  worked, and artificially aged

Two main alloying elements in order of decreasing percentage or alphabetically if percentages are equal	Percentages of the two main alloying elements and arranged in same order as alloy designations in first part	Letter of the alphabet assigned in order as compositions become standard	Letter followed by a number

Figure 21-3. The ASTM magnesium alloy and temper designation consists of four parts.

Magnesium-aluminum alloys include those alloyed with manganese (AM), silicon (AS), and zinc (AZ). These alloys are wrought or cast.

AM alloys are magnesium-aluminum alloys with additions of manganese. The manganese is added principally to improve saltwater corrosion resistance of the alloys. AM100A (M10100) is sand cast or permanent mold cast and has a good combination of tensile strength, yield strength, and elongation. AM100A is used for pressure-tight castings, but it has largely been replaced by AZ92A.

AM60A (M10600) is a die cast alloy that has better elongation and toughness than AZ91A or AZ91B alloys. AM60A is used in the as-cast temper (F) for the production of automotive wheels and other components.

AS alloys are magnesium-aluminum alloys with additions of silicon. AS alloys consist of one principal alloy, AS41A (M10410). It is die cast and has good tensile strength, yield strength, elongation, and creep resistance up to 175°C (350°F). AS41A is used in the as-cast temper (F) for products such as crank cases for air-cooled automotive engines.

AZ alloys are magnesium-aluminum alloys with additions of zinc. AZ alloys constitute the largest group of magnesium alloys. Increasing the aluminum content of the AZ alloys raises the yield strength but reduces the ductility for comparable heat treatment conditions. AZ63A (M11630) is sand cast, and although it has maximum toughness, maximum ductility, and moderately high yield strength, it has largely been replaced by AZ91C or AZ81A.

AZ92A (M11920) is sand cast and is used where maximum toughness and ductility, moderately high yield strength, and pressure tightness are required. AZ92A has virtually replaced AM100A. The sand cast microstructure of AZ92A consists of grains of delta (magnesium terminal solid solution) surrounded by a network of delta- plus-gamma eutectic. Gamma is the compound $Mg_{17}Al_{12}$. The delta phase is finer in the permanent mold cast microstructure because of the faster cooling rate that occurs in metal molds. The delta phase dissolves in the solid solution when the alloy is solution treated. Quenching and artificial aging cause the compound to precipitate as fine particles within the grains. See Figure 21-4.

AZ91C (M11914) and AZ81A (M11810) are sand cast or permanent mold cast alloys with good yield strength and ductility up to 120°C (250°F). In most applications, AZ91C and AZ81A have completely replaced AZ63A.

AZ91A (M11910) and AZ91B (M11912) are die cast. AZ91B is permitted a higher copper content than AZ91A. Although this higher copper content

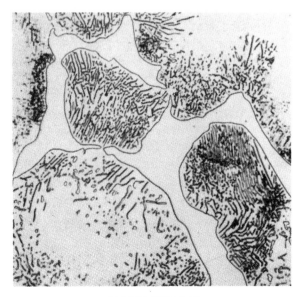

ALLOY AZ92A-F

ALLOY AZ92A-T6

American Society for Metals

Figure 21-4. The quenching and artificial aging of AZ92A cause the compound gamma to precipitate as fine particles within the grains.

lowers corrosion resistance, it allows the alloy to be made from secondary metal (scrap), which reduces the cost. AZ91A and AZ91B are used in the as-cast (F) temper. AZ91A is preferred for pressure-tight applications.

AZ31A (M11310), AZ31B (M11311), and AZ31C (M11312) are widely used wrought materials that are strengthened by work hardening. These alloys offer a good combination of strength, ductility, toughness, formability, and weldability. AZ31B is the most commonly used material for sheet and plate and is available in several tempers. PE is an AZ31-type alloy with a low calcium content. PE is used in photoengraving applications. PE is rolled to make special quality sheet with excellent flatness, corrosion resistance, and etchability.

AZ10A (M11100) is a wrought alloy that is used for low-cost extrusions and has moderate mechanical properties and high elongation. AZ10A is used in the as-extruded (F) temper. Because of the low-aluminum content, AZ10A may be welded without the need for stress relieving.

AZ61A (M11610) is a wrought alloy that has good mechanical properties and is used for general-purpose extrusions and forgings in the as-fabricated (F) temper. AZ61A is used in sheet form for battery applications only.

AZ80A (M11800) is a wrought alloy that has excellent ductility, which is not impaired by precipitation hardening. AZ80A is used for extruded products and press forgings in the as-fabricated (F) and artificially aged (T5) tempers.

AZ21X1 (M11210) is a wrought alloy and is used in the as-extruded (F) temper. AZ21X1 is designed specially for use as impact-extruded battery anodes.

Magnesium-Manganese Alloys

Additions of manganese have little effect upon tensile strength but slightly increase the yield strength of the alloy. The chief magnesium-manganese alloy is M1A (M15100), which is wrought and is used as extrusions and forgings. M1A has moderate mechanical properties and excellent weldability, corrosion resistance, and hot formability. M1A is not strengthened by heat treatment. Applications using M1A are declining because of its relatively low strength compared with other alloys.

Magnesium-Zinc Alloys

Zinc is added to improve the room temperature strength of magnesium alloys. Zinc is usually added in combination with rare earth elements (ZE), thorium (ZH), or zirconium (ZK) to produce precipitation-hardenable alloys. In some cases, zirconium is added to magnesium without any other alloying elements. An example of this type of alloy is K1A (M18010), which is cast. K1A is used in the as-cast (F) temper for its high damping capacity. K1A has slightly better mechanical properties when die cast than when sand cast.

ZE alloys are magnesium-zinc with additions of rare earth elements. The rare earth elements are a series of elements with atomic numbers from 58 to 71. Rare earth elements narrow the freezing range and are added to magnesium-zinc alloys to reduce weld cracking and porosity in castings.

ZE41A (M16410) is sand and permanent mold cast and has medium strength up to 90°C (200°F), good fatigue resistance, and good weldability. ZE41A has improved castability over AZ91C and AZ92A because it is free from microshrinkage during solidification. ZE41A is used in the artificially aged (F5) temper at temperatures up to 160°C (320°F). It is used in applications such as aircraft engines and helicopter and airframe components.

ZE63A (M16630) is sand or permanent mold cast and has high tensile strength and yield strength equivalent to ZK61A. ZE63A is used in the solution-treated and aged (T6) temper. It is especially useful in thin-section castings for applications requiring high mechanical strength and freedom from porosity. Heat treatment in a hydrogen atmosphere is required to develop properties.

ZH alloys are magnesium-zinc alloys with additions of thorium. Thorium is added to magnesium-zinc alloys to provide a significant increase in strength up to temperatures of 370°C (700°F).

ZH62A (M16620) is sand or permanent mold cast and has high yield strength. ZH62A is used in the artificially aged temper (T5) for room temperature applications. See Figure 21-5.

ZK alloys are magnesium-zinc alloys with additions of zirconium. Zirconium is added to magnesium-zinc alloys to refine the coarse-grain, columnar (elongated) as-cast microstructure, which improves mechanical properties.

ZK51A (M16510) is sand or permanent mold cast and has high yield strength, good ductility, and fatigue strength equivalent to the magnesium-aluminum-zinc alloys. ZK51A is used in the artificially aged (T5) temper for small, highly stressed parts.

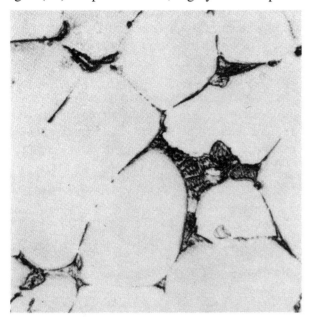

Figure 21-5. The microstructure of magnesium alloy ZH62A-T5 indicates the lamellar eutectic Mg-Th-Zn compound at the grain boundaries of a magnesium-rich solid solution.

ZK61A (M16610) is sand or permanent mold cast, contains slightly more zinc than ZK51A, and is significantly stronger. ZK61A has high yield strength, good ductility, and fatigue strength equivalent to the magnesium-aluminum-zinc alloys. ZK61A is most commonly used in the solution-treated and artificially aged (T6) temper for simple, highly stressed castings of uniform cross section, and also for intricate castings subject to microporosity or cracking during solidification.

ZK21A (M16210) is a wrought, moderate-strength extrusion alloy with good weldability. ZK21A is used in the as-extruded (F) temper. Stress relieving is not required after welding this alloy.

ZK40A (M16400) is a wrought, high yield-strength extrusion alloy. ZK40A is used in the as-extruded (F) and artificially aged (T5) tempers.

ZK60A (M16600) is a wrought alloy that has high strength and good ductility. ZK60A is used for extrusions and press forgings in the artificially aged (T5) temper.

Magnesium-Rare Earth Alloys

Magnesium-rare earth alloys include those with zinc (ZE) and those with zinc and zirconium (EZ). The principal EZ alloy is EZ33A (M12330). EZ33A is sand or permanent mold cast and has excellent pressure tightness and good strength up to 260°C (500°F). EZ33A is more difficult to cast than the AZ alloys because of the formation of dross (nonmetallic scale that floats on the surface of the molten metal). EZ33A is used in the artificially aged (T5) temper. See Figure 21-6.

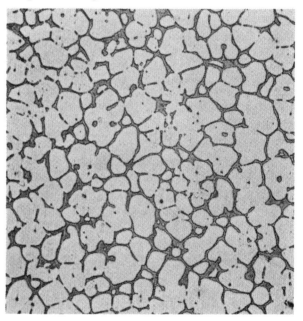

Figure 21-6. The microstructure of sand-cast magnesium alloy EZ33A-T5 consists of a network of massive magnesium-rare earth compounds in a magnesium-rich solid solution.

Magnesium-Thorium Alloys

Thorium additions greatly increase the strength of magnesium alloys at temperatures up to 370°C (700°F) and improve the weldability of alloys containing zinc. Most magnesium-thorium alloys contain 2% Th to 3% Th in combination with zinc (HZ), zirconium (HK), and manganese (HM).

HZ alloys are magnesium-thorium alloys with additions of zinc. The principal magnesium-thorium-zinc alloy is HZ32A (M13320). HZ32A is sand cast and combines moderate strength with optimum properties for medium-term to long-term exposure above 260°C (500°F). HZ32A is used for pressure-tight castings in the artificially aged (T5) temper.

HK alloys are magnesium-thorium alloys with additions of zirconium. The principal magnesium-thorium-zirconium alloy is HK31A (MM13310). HK31A may be sand cast, permanent mold cast, or wrought. HK31A has moderate strength and moderate creep resistance up to 345°C (650°F). Castings are used in the solution-treated and artificially aged (T6) temper. Wrought material has excellent weldability and formability and is rolled to make sheet and plate. Wrought material is used in the annealed (O) temper or in the strain-hardened and partially annealed (H24) temper.

HM alloys are magnesium-thorium alloys with additions of manganese. The principal magnesium-thorium-manganese alloys are HM21A and HM31A. HM21A (M13210) is wrought and is suitable for applications at 345°C (650°F) and above, which makes it superior to HK31A. HM21A is used as rolled sheet and plate, and also as forgings in the solution-annealed, cold-worked, and artificially aged (T8) temper.

HM31A (M13312) is chiefly used for producing extruded products and is suitable for applications up to 345°C (650°F). HM31A is used in the as-extruded (F) temper and artificially aged (T5) temper.

MAGNESIUM FABRICATION

Magnesium alloys have limited ductility and work harden rapidly. Magnesium alloys are formed at elevated temperatures where they are more plastic. Magnesium alloys are easily machined. Joint cleanliness is essential when welding magnesium, and preheating and stress relieving may be necessary.

Forming

Magnesium exhibits a CPH structure. Only three slip systems are available for plastic deformation, which limits ductility. Twinning, which makes more slip systems available, can occur in compression.

Consequently, magnesium exhibits greater ductility in cold-forming operations involving compression, such as rolling, than in operations involving tension, such as bending. In general, cold-forming operations must encompass mild deformations, and generous radii must be used.

Hardened dies are not necessary for most forming operations. The methods and equipment used for forming magnesium are the same as those commonly employed for forming other metals, except for the differences in tooling and technique that are required in elevated-temperature forming operations. Forming at elevated temperature is usually performed in one operation without repeated annealing. This reduces the time involved and eliminates the need for additional equipment that is required for the extra stages of forming.

The maximum forming temperature used depends on the alloy. The maximum time at the forming temperature is limited to prevent adverse effects on mechanical properties. For example, HK31A may be held for 15 minutes at 345°C (650°F), but for only 3 minutes at 400°C (750°F). See Figure 21-7.

FORMING TEMPERATURES		
Alloy	**Temperature (°C)**	**Time**
Sheet		
AZ13B-O	290	1 hr
AZ31B-H24	165	1 hr
HK31A-H24	345	15 min
	370	5 min
	400	3 min
Extrusions		
AZ61A-F	290	1 hr
AZ31B-F	290	1 hr
M1A-F	370	1 hr
AZ80A-F	290	½ hr
AZ80A-T5	195	1 hr
ZK60A-F	290	½ hr
ZK60A-T5	205	½ hr

Figure 21-7. The maximum forming temperature and time permitted for magnesium alloys varies with the composition of the alloy.

Machining

Magnesium is easier to machine than other structural metals. The power required to remove a given amount of magnesium is lower than that required for any other commonly machined metal. All magnesium alloys have similar machinability.

Magnesium can be machined at extremely high speeds and with deep cuts. The excellent free-cutting characteristics indicate that machining chips are well broken. Close dimensional tolerances of ±.1 mm (.004 in.) can be attained. An extremely fine finish may be achieved, so grinding and polishing are often unnecessary for a smoothly finished surface.

The ignition of chips or turnings is a consideration when magnesium is machined. The localized heat evolution produced by the contact of the magnesium with the machining tool may be sufficient to cause the magnesium to ignite. Chips must be heated near the melting point before ignition can occur. Ignition susceptibility is a function of chip size. Roughing cuts and medium-finishing cuts produce coarse chips which cannot readily ignite during machining. Fine-finishing cuts, however, produce fine chips that can be ignited by a spark. To reduce ignition susceptibility, all cutting tools are kept sharp and ground with adequate relief and clearance angles; heavy feeds are used to produce thick chips; mineral oil coolants are used where possible; fine cuts are avoided where coolants are not used; chips are not allowed to accumulate on machines or clothing;

chips are stored in clean, clearly labeled, covered metal cans; and an adequate supply of magnesium fire extinguishers are kept within easy reach of the machine operators.

Joining

Magnesium is readily joined by resistance spot welding, gas metal arc welding, gas tungsten arc welding. Adhesive joining and riveting are also used. Cleanliness is absolutely essential for sound weld joints. Any surface contamination inhibits wetting and fusion. Chemical and mechanical cleaning are used to remove surface contamination. The selection of the proper welding filler metal is of great importance when joining magnesium. See Figure 21-8.

Resistance spot welds in magnesium alloys have good static strength. Resistance spot welds have lower fatigue strength than either adhesive bonded or riveted joints.

For gas metal arc welding and gas tungsten arc welding, welding filler metals with a wider freezing range than the base metal are selected. This provides good wettability and minimizes cracking in the base metal.

Preheating is determined by the product section thickness and the degree of restraint in the joint.

WELDING FILLER METALS									
Alloys	**ER AZ61A**	**ER AZ92A**	**ER EZ33A**	**Base Metal**	**Alloys**	**ER AZ92A**	**ER EZ33A**	**ER AZ101A**	**Base Metal**
Wrought					**Cast**				
AZ10A	X	X			AM100A	X		X	X
AZ31B	X	X			AZ63A	X		X	X
AZ61A	X	X			AZ81A	X		X	X
AZ80A	X	X			AZ91C	X		X	X
ZK21A	X	X			AZ92A	X		X	X
HK31A			X		EK41A		X		X
HM21A			X		EZ33A		X		X
HM31A			X		HK31A		X		X
M1A				X	HZ32A		X		X
					K1A		X		X
					QH21A				X
					ZE41A		X		X
					ZH62A		X		X
					ZK51A		X		X
					ZK61A		X		X

Figure 21-8. Welding filler metals for welding magnesium alloys have a wider freezing range than the base metals they join to provide good wettability and minimize base metal cracking.

Restraint is a measure of the rigidity of the joint. Thick sections may not require preheating unless the restraint is high. See Figure 21-9. High restraint indicates that the joint is not capable of localized deformation to accommodate the thermal stresses that result from welding. Thin sections and highly restrained joints generally require preheating to avoid weld cracking, such as with high-zinc alloys.

Arc welds in magnesium-aluminum-zinc alloys and any alloys with >1% Al are subject to stress-corrosion cracking. These alloys must be stress relieved after welding. The components are placed in a jig or clamping plate, heated to the specified stress-relieving temperature, held for the required time, and cooled in still air. See Figure 21-10.

Adhesive bonding produces joints with good fatigue strength and allows thinner materials to be joined. During adhesive bonding, the adhesive fills the space in the joint and acts as an insulator between the dissimilar metals. This minimizes the opportunity for galvanic corrosion.

Riveting procedures are similar to those used for other metals. Only aluminum rivets should be used if galvanic corrosion is to be minimized.

STRESS RELIEVING FOR ARC WELDS		
Alloy	Temperature (°C)	Time (min)
Sheet		
AZ31B-H24	150	60
AZ31B-O	260	15
Extrusions		
AZ31B-F	260	15
AZ61A-F	260	15
AZ80A-F	260	15
AZ80A-T5	205	60
Castings		
AZ63A	260	60
AZ81A	260	60
AZ91C	260	60
AZ92A	260	60

Figure 21-10. Specific magnesium alloys are stress relieved after welding to avoid stress-corrosion cracking.

RECOMMENDED WELD PREHEATING FOR MAGNESIUM ALLOYS				
Alloy	Temper Before Welding	Final Temper	Weld Preheating	Weld Postheating
AZ63A	T4	T4	Heavy and unrestrained sections: none or local; thin and restrained sections: 175°C to 380°C max	½ hr at 390°C
	T4 or T6	T6		½ hr at 390°C + 5 hr at 220°C
	T5	T5	Heavy and unrestrained sections: none or local; thin and restrained sections: none to 260°C (1½ hours max at 260°C)	5 hr at 220°C
AZ81A	T4	T4	Heavy and unrestrained sections: none or local; thin and restrained sections: 175°C to 400°C max	½ hr at 415°C
AZ91C	T4	T4		½ hr at 415°C
	T4 or T6	T6		½ hr at 415°C + either 4 hr at 215°C or 16 hr at 170°C
AZ92A	T4	T4		½ hr at 410°C
	T4 or T6	T6		½ hr at 410°C + either 4 hr at 260°C or 5 hr at 220°C
AM100A	T6	T6		½ hr at 415°C + 5 hr at 220°C
EK41A	T4 or T6	T6	None to 260°C (1½ hr max at 260°C)	16 hr at 205°C
	T5	T5		16 hr at 205°C
EZ33A	F or T5	T5		5 hr at 215°C; 2 hr at 345°C + 5 hr at 215°C
HK31A	T4 or T6	T6	None to 260°C	16 hr at 205°C; 1 hr at 315°C + 16 hr at 205°C
HZ32A	F or T5	T5		16 hr at 315°C
K1A	F	F	None	None
ZE41A	F or T5	T5	None to 315°C	2 hr at 330°C; 2 hr at 330°C + 16 hr at 175°C
ZH62A	F or T5	T5		16 hr at 250°C; 2 hr at 330°C + 16 hr at 175°C
ZK51A	F or T5	T5		16 hr at 175°C; 2 hr at 330°C + 16 hr at 175°C
ZK61A	F or T5	T5		48 hr at 150°C
	T4 or T6	T6		2-5 hr at 500°C + 48 hr at 130°C

Figure 21-9. Preheating is determined by the product section thickness and the degree of restraint in the joint.

MAGNESIUM PROPERTIES

The most important characteristics of magnesium and magnesium alloys are the low density and the corresponding high strength-to-weight ratio. Corrosion resistance of magnesium is achieved by the formation of a passive surface film, which may be destroyed in certain atmospheres and in some solutions. Atmospheric corrosion resistance is improved by surface treatments. Galvanic corrosion may occur when magnesium is coupled with certain other metals.

Density

Magnesium, with a density of approximately two-thirds that of aluminum and approximately one-fourth that of steel, is the lightest commercial metal. The strength-to-weight ratio of precipitation-hardened magnesium alloys is comparable to that of the stronger alloys of aluminum and alloy steels. See Figure 21-11.

Magnesium alloys are also used where a thick, light form is desired. For example, magnesium alloys are used for complex castings, such as housings in aircrafts, and for components in rapidly rotating or reciprocating machines.

Corrosion Resistance

The corrosion resistance of magnesium is good in most atmospheres and variable in chemical solutions. In clean atmospheres, the alloy surface develops a gray protective surface film consisting of a mixture of magnesium carbonate, magnesium sulfate, and magnesium hydroxide. In industrial, marine, and humid areas, the natural protective film breaks down and corrosion is faster. Over a period of years, corrosion will lead to powdering of the surface. The corrosion of magnesium and magnesium alloys is slow compared to the rusting of steel in the same atmosphere.

In stagnant water, magnesium rapidly forms a passive surface film that prevents further corrosion. If water is constantly replenished by circulation, the corrosion rate rises because the film is continuously dissolved away.

Severe corrosion may occur in salt solutions containing heavy metal ions, such as copper, iron, or nickel ions. Chloride-containing salt solutions are also corrosive because they break down the surface

STRENGTH-TO-WEIGHT RATIOS FOR SELECTED ALLOYS			
Alloy	Density (g/cm³)	Yield Strength (psi)	Ratio (psi:g/cm³)
Magnesium alloy			
AZ80A[a]	1.80	40,000	22,000
AM100A[a]	1.81	22,000	12,000
AZ63A[a]	1.84	19,000	10,500
AZ81A[a]	1.80	12,000	6500
HK31A[a]	1.79	15,000	8500
HK31A[b]	1.79	29,000	16,000
M1A[b]	1.76	26,000	14,500
HM21A[b]	1.78	25,000	14,000
Aluminum alloy			
7075-T6	2.80	72,000	25,500
Titanium alloy[c]	4.50	110,000	24,500
Stainless steel			
Type 302[d]	7.90	140,000	17,500
Low-alloy steel	7.90	55,000	7000

[a] cast
[b] wrought
[c] full hard
[d] 30% cold worked

Figure 21-11. Compared with other metals, the strength-to-weight ratio of magnesium alloys is very favorable.

film. Fluorides are not appreciably corrosive because they form insoluble magnesium fluoride, which is protective.

Magnesium is rapidly attacked by most mineral acids, such as sulfuric acid and nitric acid. The two acids that do not attack magnesium are chromic acid and hydrofluoric acid. Magnesium is resistant to dilute alkalis at relatively low temperatures. The corrosion rate increases rapidly at temperatures above 60°C (140°F). Because they rapidly dissolve the surface hydroxide film, hot alkaline cleaners are used for cleaning magnesium before welding.

Galvanic Corrosion. Magnesium can suffer galvanic corrosion when electrically coupled with certain metals in highly conductive environments, such as salt solutions. Preventive measures to avoid galvanic corrosion include the selection or electroplating of the contact metal so that it does not form an unfavorable galvanic couple with magnesium, the use of a suitable surface treatment or insulating gasket to protect the magnesium from electrical contact with the incompatible metal, and the prevention of water accumulation at the dissimilar metal assembly.

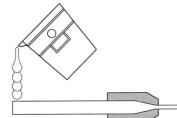

Titanium 22

Titanium is the fourth abundant metal, but the difficulty of its extraction causes an increase in its cost. Titanium alloys are classified as alpha, alpha-beta, or beta alloys. Titanium alloys can be heat treated to develop high strength-to-weight ratios similar to those of high-strength magnesium and aluminum alloys. However, the use of titanium alloys is limited to demanding aerospace, chemical processing, and marine applications because of high manufacturing costs and welding difficulties. Titanium alloys are strengthened by solution hardening or precipitation hardening. Titanium alloys are difficult to form, machine, or join by welding. The two most important attributes of titanium and titanium alloys are low density and excellent corrosion resistance.

TITANIUM MANUFACTURING

Although titanium is the fourth most abundant metal, the difficulty of extracting it from titanium ores imposes severe increases to its cost. The extraction and production of finished forms occurs in four stages. The mechanical properties of titanium are significantly influenced by the processing conditions. Care must be taken during all manufacturing stages to avoid contamination of the titanium, ensure the specified amount of reduction, and maintain the correct working temperature. A limited amount of titanium is produced as castings.

Titanium alloys are designated by their nominal compositions or grade numbers. Titanium alloys are also designated in the Unified Numbering System. Product forms are specified in American Society for Testing and Materials and Aerospace Materials Specifications standards.

Extraction and Production

Titanium is manufactured in four stages. First, titanium ore is reduced to titanium metal (sponge). Second, the sponge (plus reclaimed scrap) is melted into ingots. Third, the ingots are converted to general mill products, or primary products. Fourth, certain primary products are converted into specific shapes or secondary products.

Reduction. Titanium ore is reduced using the Kroll process. The principal titanium ores, rutile and ilmenite, are very abundant. Titanium is the fourth most abundant metal after aluminum, iron, and magnesium. The high cost of production and the complexity of the production process limits the applications of titanium.

In the Kroll process, titanium ore is reduced to sponge. This consists of producing titanium tetrachloride from the ore, purifying it, and reducing the purified titanium tetrachloride with magnesium or sodium. The sodium-reduced sponge is leached with acid to remove the sodium chloride by-product of reduction. Magnesium-reduced sponge is leached, inert gas swept, or vacuum distilled to remove the magnesium chloride by-product of reduction.

Vacuum Melting of Titanium Sponge. Ingots are produced by melting titanium sponge and reclaimed scrap. The materials are vacuum melted twice in an electric furnace, or double consumable-electrode vacuum melted. Alloying elements are also added at this stage. Vacuum melting removes volatile elements and impurities, such as hydrogen, which would be detrimental in the final product.

Primary Fabrication. Primary fabrication consists of mechanical-working operations that convert the ingots into general mill products. General mill products include billet, bar, plate, sheet, and strip.

The ingot is reduced to billets using a press-cogging. This reduction is performed at a temperature that causes grain refinement to occur. The final properties of specific alloys are strongly influenced by the amount of reduction and the temperature range. The billets are rolled into bar, plate, sheet, and strip. Strict temperature control is necessary to achieve the required properties.

Control of the operating parameters during primary fabrication is important because secondary fabrication operations usually have little or no effect on the metallurgical characteristics. As a result, the secondary fabrication operations cannot be used to modify the properties of the final product.

Secondary Fabrication. Secondary fabrication consists of any operations that convert primary products into final products. The most important of these operations are die forging and extrusion.

Die forging is the mechanical reduction of billets under specific temperature conditions to obtain properties that are not achieved in the billets themselves. Extrusion is performed on billets to make rod-like products and tubing.

Casting. Titanium is rarely cast. The mechanical properties of titanium castings are roughly equal to wrought alloys of equivalent chemical composition, except that fatigue strength is usually inferior. Unlike many alloy systems, wrought compositions do not require modification to improve castability.

For corrosion resistant castings, the commercially pure titanium grade 1, grade 2, and grade 3 provide the bulk of the applications. For aerospace and marine applications, Ti-6Al-4V is the dominant alloy.

To produce castings, the alloys are consumable-electrode vacuum arc melted in water-cooled copper crucibles. The alloys are then centrifugally pumped under vacuum into graphite molds.

Titanium Designation Systems

UNS designations for titanium and titanium alloys provide an orderly system of classification. Designations based on nominal composition, such as Ti-6Al-4V, or grade number are also widely used. UNS designations consist of the uppercase letter R (for reactive and refractory metals and alloys) followed by five numbers. Titanium alloys occupy the series R50000 to R59999. For example, titanium grade 2 is R50400. Equivalent cast and wrought alloys have the same UNS designations. See Figure 22-1.

ASTM standards for titanium alloys include plate, sheet, strip, seamless and welded pipe, seamless and welded tubing, bar and billet, castings, and forgings. These standards reference the UNS designation in addition to the traditional grade or compositional designation. AMS standards cover a wide variety of product forms, which include bar, tubing, and welding wire, for many titanium alloys used in the aerospace industry.

TITANIUM ALLOYS

Based on their crystal structures at room temperature, titanium alloys are divided into alpha, alpha-beta, and beta alloys. Alpha alloys are usually subdivided into commercially pure titanium alloys and alpha or near-alpha alloys, which have larger amounts of alloying elements.

Crystal Structure

The crystal structure of pure titanium is CPH (alpha). Above 882°C (1620°F), the titanium changes by allotropic transformation to BCC (beta). The allotropic transformation temperature (beta transus) is affected by the amount of impurities in the titanium or by the alloying elements.

Aluminum additions stabilize the alpha phase and raise the allotropic transformation temperature. In contrast to this, chromium, molybdenum, and vanadium, as well as other elements, stabilize the beta phase and lower the allotropic transformation temperature. With large amounts of beta stabilizers, the beta phase is stable at room temperature and below. When both alpha and beta stabilizing elements are present, some beta phase is present in the alpha matrix at room temperature.

TITANIUM ALLOYS						
Designation	**UNS**	**Nominal Composition**				
		%Al	**%Sn**	**%Zr**	**%Mo**	**Others**
Alpha Alloys						
ASTM grade 1	R50250	—	—	—	—	.18O
ASTM grade 2	R50400	—	—	—	—	.25O
ASTM grade 3	R50550	—	—	—	—	.15O
ASTM grade 4	R50700	—	—	—	—	.40O
ASTM grade 7	R52400	—	—	—	—	.2Pd
Near-Alpha Alloys						
Ti-5Al-2.5Sn grade 6	R54250	5	2.5	—	—	.20O
Ti-5Al-2.5Sn-ELI		5	2.5	—	—	.12O
Ti-8Al-1Mo-1V		8	—	—	1	1V
Ti-6Al-2Sn-4Zr-2Mo		6	2	4	2	—
Ti-6Al-2Nb-1Ta-.8Mo		6	—	—	1	2Nb and 1Ta
Ti-2.25Al-11Sn-5Zr-1Mo		2.25	11	5	1	.2Si
Alpha-Beta Alloys						
Ti-6Al-4V grade 5	R56400	6	—	—	—	4V
Ti-6Al-4V-ELI	R56401	6	—	—	—	4V
Beta Alloys						
Ti-13V-11Cr-3Al	R58010	3	—	—	—	11Cr and 13V
Ti-4.5Sn-6Zr-11.5Mo grade 10	R58030	—	4.5	6	11.5	—

Figure 22-1. Titanium alloys are designated by their UNS numbers and ASTM grade numbers.

Titanium alloys have three basic types of microstructures, which are alpha, alpha-beta, and beta. In alpha-beta alloys, the alpha-beta ratio is important because it influences grain size, workability, toughness, and weldablity.

Alpha Alloys. Commercially pure titanium alloys comprise alpha alloys with extremely low amounts of interstitial elements (nitrogen, oxygen, and carbon). The primary difference between the various grades of commercially pure titanium is the interstitial element content. Another difference is the iron content, which must be severely restricted for specific corrosive services. Alloys with higher purity (less interstitial element content) have lower strength, lower hardness, and a lower alpha-beta transformation temperature. See Figure 22-2.

Grade 1 (R50250), grade 2 (R50400), grade 3 (R50550), and grade 4 (R50700) titanium comprise four commercially pure alloys that have increasing interstitial element contents, which are principally oxygen. Although strength increases with interstitial element content, commercially pure alloys have low to intermediate strength compared with other titanium alloys.

Grade 2 and grade 3 are used in the chemical process industry for vessels, pipe, and tubing. They are also used for marine, aerospace airframe, and engine components. Grade 3 and grade 4 are used in aerospace applications requiring higher strength, such as fire walls, pumps, valves, and piping.

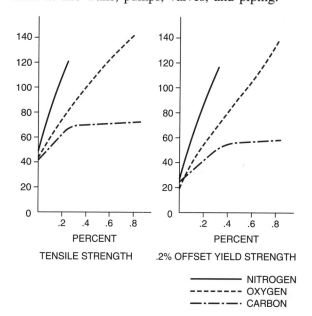

Figure 22-2. Oxygen, nitrogen, and carbon are interstitial elements that cause significant hardening of titanium and must be strictly controlled.

Alpha and near-alpha alloys of titanium principally contain alpha stabilizing elements, therefore a high percentage of the alpha phase is present at room temperature. The principal alpha alloys are based upon additions of aluminum and tin. Alpha alloys have a single-phase structure, are weldable, and have good ductility. They are used in cryogenic and high-temperature applications. Two important alpha alloys are Ti-Pd and Ti-5Al-2.5Sn.

Ti-Pd alloys, which are grade 7 (R52400) and grade 11 (R52250), consist of commercially pure titanium with approximately .15% Pd. Palladium is added to expand the range of corrosion resistance. Grade 11 has a lower interstitial element concentration and better ductility and formability than that of grade 7. Ti-Pd alloys are used in the chemical process industry for applications in which the commonly used grade 2 and grade 3 do not have sufficient corrosion resistance, such as when the medium is mildly reducing (dilute hydrochloric or sulfuric acid).

Ti-5Al-2.5Sn, which is grade 6 (R54250), is an alpha titanium alloy that is available in a standard grade and an extra low interstitial (ELI) grade. The standard grade is used for gas turbine engine casings and rings operating up to 480°C (900°F), structural members in hot spots, and high-temperature chemical processing equipment. The ELI grade is used in pressure vessels for liquified gases and in hardware for service at cryogenic temperatures. The microstructure of grade 6 consists completely of alpha phase. See Figure 22-3.

Alpha-Beta Alloys. Alpha-beta alloys contain percentages of beta stabilizing elements that are high enough to cause the beta phase to be present at room temperature. Alpha-beta alloys can be strengthened by solution treating and aging. The most common alpha-beta alloy is Ti-6Al-4V.

Ti-6Al-4V, which is grade 5 (R56400), is the most widely used titanium alloy. At least a dozen varieties of grade 5 are produced, which include ELI grades that have lower strength and higher ductility than standard grades. Grade 5 is used in both the annealed and aged conditions for aircraft gas turbine disks and blades, airframe structural components, applications requiring strength up to 315°C (600°F), prosthetic implants, and chemical processing equipment.

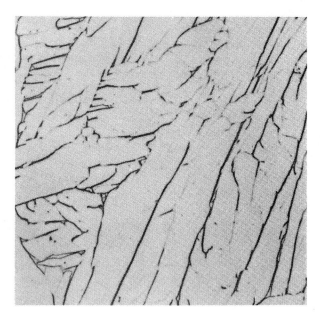

American Society for Metals

Figure 22-3. The microstructure of Ti-5Al-2.5Sn that is hot worked below the alpha transus and annealed above the beta transus consists of coarse, plate-like alpha phase.

Beta Alloys. Beta alloys of titanium contain elements that stabilize the beta phase at room temperature. Beta alloys may be heat treated to extremely high strength levels by precipitation hardening. A principal advantage of beta alloys over alpha alloys is deep hardenability, such as that occurring in very thick forgings. Additionally, beta alloys may be heat treated to very high strength levels from the annealed and formed condition.

Beta alloys have excellent formability in the single-phase condition. When heat treated, beta alloys exhibit high strength-to-weight ratios but have low ductility and fracture toughness. Beta alloys are limited to moderate strength applications up to 315°C (600°F) because of low creep strength at elevated temperature. Beta alloys have exceptional work-hardening characteristics, and are used for fasteners and springs for this reason. Two examples of beta titanium alloys are Ti-13V-11Cr-3Al and Ti-4.5Sn-6Zr-11.5Mo.

Ti-13V-11Cr-3Al (R58010) is formed in the soft, annealed condition and strengthened by precipitation hardening. A typical heat treatment schedule consists of annealing at 845°C to 870°C (1550°F to 1600°F), water quenching or air cooling, followed by aging for 24 hours at 510°C (950°F). Ti-13V-11Cr-3Al is used primarily for aircraft frames and auxiliary applications, which include thick-section forgings, sheet construction, and springs from rod and wire.

Ti-4.5Sn-6Zr-11.5Mo, which is grade 10 (R58030), is easily formed in the annealed condition before precipitation hardening. A typical heat treatment schedule consists of annealing for a short time at 730°C to 760°C (1350°F to 1400°F), water quenching or air cooling, followed by aging for 8 hours at 480°C (900°F) and air cooling. Grade 10 is primarily used for aerospace fasteners and springs.

Strengthening

The type of microstructure (alpha, alpha-beta, and beta) influences the mechanical working or heat treatment employed to strengthen titanium alloys. The two principal methods of strengthening titanium alloys are solid solution hardening and precipitation hardening.

Solid Solution Hardening. Solid solution hardening occurs in alpha alloys and may be of the interstitial or substitutional type. Interstitial solid solution hardening occurs in commercially pure titanium alloys when hydrogen, oxygen, nitrogen, or carbon is added. These elements dissolve interstitially in the crystal lattice and have a significant effect on strength. Increasing these elements decreases toughness and ductility to the point that the material becomes brittle. Consequently, the amounts of these elements are strictly controlled.

Substitutional solid solution hardening occurs in alpha alloys when elements such as aluminum and tin, which exert significant strengthening effects, are added. For example, each percent addition increases tensile strength by 8000 psi (55 MPa) for aluminum and 4000 psi (28 MPa) for tin.

Precipitation Hardening. Precipitation hardening is used to strengthen alpha-beta and beta titanium alloys. These two alloys are solution annealed at a temperature in the two-phase alpha-beta field, quenched, and precipitation hardened at a lower temperature to increase their strength. Precipitation hardening increases the strength of alpha-beta alloys by 30% to 50% over annealed material.

Solution annealing is performed at a temperature that is high in the two-phase alpha-beta field. If the temperature were in the beta phase field, excessive grain growth would occur, and the subsequent formation of alpha would be at the beta grain boundaries. These two factors reduce ductility.

Quenching is performed using water, oil, or another suitable medium. The beta phase present at the solution-treating temperature is either retained on quenching, or partially transformed by a martensitic (shear-type) reaction or by nucleation and growth. The exact mode depends on the specific alloying elements.

The alloy is then aged between 480°C and 650°C (900°F and 1200°F) to precipitate alpha and to produce a fine mixture of alpha or beta in the retained or transformed beta phase. The beta phase present at the solution-annealing temperature may be completely retained during cooling or a portion may transform to alpha. The specific response depends on the alloy composition, solution-annealing temperature, cooling rate, and section size.

During precipitation hardening, fine alpha particles precipitate in the retained or transformed beta phase. This duplex structure is harder and stronger than the annealed structure.

Annealing. Annealing is performed on titanium alloys to improve ductility, dimensional or thermal stability, fracture toughness, and creep resistance. The component is held at the annealing temperature until it is uniformly heated and the desired transformations are completed. The component is then air or furnace cooled to room temperature.

Stress Relieving. Stress relieving is sometimes used to remove residual stresses in a weldment. A *weldment* is an assembly whose component parts are joined by welding. Stress relieving can be beneficial in maintaining dimensions, reducing cracking tendency, and avoiding stress-corrosion cracking in certain alloys. Stress relieving might alter the mechanical properties of the weld zone by aging heat-treatable alloys. This results in the reduction of ductility.

TITANIUM FABRICATION

Compared with copper alloys, carbon steels, and stainless steels, titanium is relatively difficult to form, machine, and join. Rigid quality control is required to perform any of these processes.

Forming

Forming processes for titanium alloys are hampered by a variety of factors. Notch sensitivity may cause cracking and tearing, especially in cold-forming operations. Galling is more severe with titanium alloys than with stainless steels and is most severe in hot-forming operations. Poor shrinkability is sometimes a disadvantage in some flanging operations. Embrittlement may occur from the absorption of hydrogen and other gases during heating. Limited workability and springback restrict cold-forming operations. *Springback* is inaccuracy in shape and dimensions that increases resistance to plastic deformation. The principal methods of forming are hot forming, and cold preforming and hot sizing.

Hot Forming. *Hot forming* is a forming process that increases formability and reduces springback and is used in the production of a great majority of titanium components. Lubricants such as boron nitride are used to reduce galling. Hot forming is usually performed at temperatures between 540°C and 815°C (1000°F and 1500°F). Scaling and embrittlement by oxygen-rich surface layers occurs above 540°C (1000°F). To limit scaling and embrittlement, the maximum time allowed for heating in air is 1 hour at 705°C (1300°F) and 20 minutes at 870°C (1600°F). For any hot-forming operation, scale must be removed after processing.

Cold Preforming and Hot Sizing. Cold preforming and hot sizing are used together to correct springback. *Cold preforming* is the process in which cold-preformed components are clamped into fixtures made to the exact final shape and dimensions of the component. *Hot sizing* is the process in which the fixtures are heated for a time long enough to cause the components to assume the correct shape.

Machining

For successful machining, titanium alloys require rigid quality control. The rigid quality control pertains to setups, slow speeds, heavy feeds, sharp tools, and ample supplies of nonchlorinated coolant.

Joining

Titanium alloys are joined by welding, brazing, adhesive bonding, mechanical fastening, and metallurgical bonding. Titanium alloys are welded by gas shielded arc welding, electron beam welding, laser beam welding, and resistance welding. Welding has the greatest potential for affecting the properties of materials. Stringent surface preparation is needed before welding. In all types of welds, contamination by interstitial impurities, such as oxygen and nitrogen, must be carefully controlled to maintain useful ductility in a weldment.

Oil, fingerprints, grease, paint, and other foreign matter should be removed using a suitable solvent cleaning method. Tap water should not be used to rinse titanium components. Chloride containing solvents can leave residues that cause stress-corrosion cracking at temperatures above 285°C (550°F). Hydrocarbon residues can result in contamination and embrittlement of titanium. Only stainless steel wire brushes should be used to remove residues.

Welding must be performed under inert gas shielding, such as argon shielding, to avoid picking up interstitial elements, particularly oxygen. The argon shield must be maintained on all metal surfaces above a temperature of 540°C (1000°F). This means that both sides of a welded joint, such as a pipe, must be shielded. The metal that is cooling down from the welding temperature must also be protected by a trailing shield. See Figure 22-4.

Commercially pure titanium is usually welded with a welding filler metal one grade below that of the base metal because welding operations lead to slight pick up of interstitial elements. For example, grade 2 titanium (.25% O) is welded with grade 1 titanium welding filler metal (.18% O).

Alpha titanium alloys have good weldability because they possess good ductility. Welding or brazing operations have little effect on the mechanical properties of annealed material.

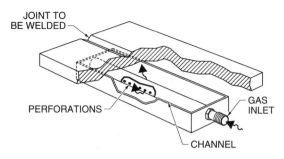

Figure 22-4. When titanium is fusion welded, it must be protected on both sides of the weld to ensure that the hot metal does not pick up interstitial elements as it cools.

Alpha-beta titanium alloys may undergo significant strength, ductility, and toughness changes when welded. Grade 5 has the best weldability of the alpha-beta alloys. It can be welded in either the annealed condition or the solution annealed and partially aged condition. Aging may then be completed during a post-weld stress heat treatment.

Beta titanium alloys are weldable in either the annealed or heat-treated condition. Weld joints have good ductility but relatively low strength as-welded.

Brazing operations have very little effect on the properties of commercially pure titanium and alpha alloys. Beta alloys are unaffected by the brazing cycle when used in the annealed condition. However, if the beta alloys are to be heat treated, the brazing temperature can have an important effect on their properties.

Titanium may also be joined by adhesive bonding and mechanical fastening. If the joint is properly designed, adhesive bonding and mechanical fastening will not affect its properties.

TITANIUM PROPERTIES

Two outstanding properties of titanium are low density and high corrosion resistance. These properties lead to unique applications of titanium in the aerospace and chemical processing industries.

Density

The combination of high strength, stiffness, and good toughness coupled with low density (approximately 4.5 g/cm³) allows for a high strength-to-weight ratio. See Figure 22-5. This high strength-to-weight ratio is used in aerospace applications operating at low to moderately elevated temperatures.

STRENGTH-TO-WEIGHT RATIOS[a] AT VARIOUS TEMPERATURES					
Alloy	25°C	0°C	−100°C	−200°C	−250°C
Ti-5Al-2.5Sn	825	850	1000	1375	1600
Ti-6Al-4V	825	850	975	1325	1625
4340 steel	800	800	900	1000	—
ZK60A-T5 magnesium	600	650	800	850	900
7079-T6 aluminum	600	625	675	750	825
17-7PH stainless steel	675	700	725	825	925

[a] yield strength/1000 × density, or specific strength

Figure 22-5. Titanium alloys have good strength-to-weight ratios at moderate and low temperatures.

Corrosion Resistance

Titanium alloys have outstanding corrosion resistance in most chemicals and waters. The commercially pure grades 1, 2, 3, and 4 are most commonly used for chemical processing and marine applications. The addition of .15% Pd in grade 7 and grade 11 enhances resistance to reducing corrosives.

Titanium alloys are fully resistant to waters (natural, high-purity, body fluids, steam, seawater) at temperatures in excess of 315°C (600°F). If the chloride concentration exceeds 1000 ppm, titanium is susceptible to crevice corrosion at temperatures above 75°C (170°F).

Titanium is resistant to most salt solutions, except for acidic fluorides. Titanium is generally very resistant to alkalis. Higher concentrations of sodium and potassium hydroxide cause measurable corrosion at boiling temperatures. In highly alkaline solutions where the pH is above 12, embrittlement from hydride formation occurs.

Titanium is highly resistant to oxidizing acids over a wide range of concentrations and temperatures. These acids include nitric, chromic, perchloric, and hypochlorous acid. Resistance to reducing acids, such as sulfuric, hydrochloric, hydrobromic, phosphoric, and sulfamic acid is severely limited. Resistance is improved by the presence of oxidizing species, such as cupric or ferric ions. Hydrofluoric acid solutions attack titanium at all concentrations.

Organic compounds such as alcohols, ketones, ethers, aldehydes, and hydrocarbons present no problems for titanium alloys. Traces of moisture, which are normally present in industrial organic streams, are sufficient to maintain the protective oxide film. Totally anhydrous streams may prevent oxide film maintenance and should be avoided. Chlorinated hydrocarbons are only of concern if sufficient hydrolysis occurs, leading to the formation of free hydrochloric acid.

Titanium has excellent resistance to gaseous oxygen and to air up to 430°C (805°F). Above this temperature, excessive oxidation and surface embrittlement occur.

Pyrophoric behavior is the spontaneous ignition of a metal when scratched or struck. Pyrophoric behavior is exhibited in titanium if certain conditions are satisfied. Titanium is extremely reactive if its protective oxide film is stripped or removed for any reason. When the source of energy accompanying the stripping of the film is high enough, titanium may burn in gases such as oxygen or chlorine. For example, it will ignite if dry chlorine is sparged (fed at high velocity) down a titanium tube. Under these conditions, a minimum water content is necessary to maintain the protective film and prevent ignition. See Figure 22-6.

Hydride embrittlement is the loss of ductility caused by corrosion. Titanium hydride is an intermetallic compound having a flake-like shape that develops from reaction with hydrogen. Hydride embrittlement of titanium occurs when hydrogen is released during corrosion. As little as 100 to 200 ppm of hydrogen will severely embrittle titanium.

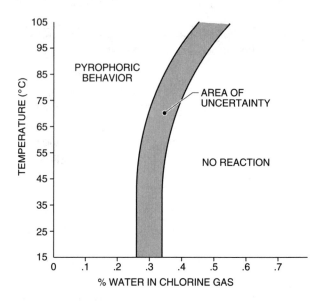

Figure 22-6. Titanium will ignite if exposed to flowing dry chlorine. A minimum amount of water is required to maintain the protective oxide film.

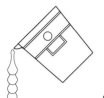

Lead, Tin, and Zinc 23

Lead, tin, and zinc have specific uses as structural metals and as coatings or platings on other metals. Lead is extracted from lead sulfide ore. Lead is dense, malleable, soft, lubricious, highly corrosion resistant, and has a low melting point. Tin is extracted and refined from oxide ores. Tin is a soft, white metal with good corrosion resistance and lubricity. Solders and bearing metals are based on tin and lead alloys. Zinc is extracted and refined from zinc sulfide ore. Zinc is a heavy, bluish-white metal that is used principally for its low cost, corrosion resistance, and alloying properties.

LEAD

Lead is extracted from lead sulfide ore. Lead and lead alloys are made into specific product forms. Lead alloy product forms are used for several specific applications. Lead is also applied as a coating on other metals. It is used for several kinds of high-lead alloys. Type metals are a group of lead-antimony alloys. Lead is dense, malleable, soft, lubricious, highly corrosion resistant, and has a low melting point.

Extraction and Production

Lead is extracted from lead sulfide ore (galena). The recycling of scrap from batteries, sheet, cable, bearings, and solder is also a major source of lead. The sulfide ore is concentrated and then roasted to remove the sulfur constituent and produce a sinter. This is charged into the top of a blast furnace along with suitable fluxes and coke. An impure lead product is tapped from the blast furnace.

The impure lead is then treated in a series of steps to remove the copper, antimony, tin, arsenic, precious metals, zinc, and bismuth. The purified lead is cast into bars, which are remelted to make lead alloys or that are mechanically worked into the specific product forms. These are used in applications such as lead-acid storage batteries, ammunition, cable sheathing, construction, and electrolytic refining and plating.

Applications. Battery grids are the largest single application of lead. Battery grids consist of a series of grid plates made from either cast or wrought calcium-lead or antimonial lead, which are pasted with a mixture of lead oxides and immersed in sulfuric acid.

In ammunition, lead is used for shot and for bullet cores. Shot is alloyed with 1% As and bullet cores are alloyed with up to 2% As.

Lead sheathing is extruded around electrical power cable and communications cable to provide long-term protection against moisture and corrosion damage. When used underground or underwater, lead sheathed cables are protected from mechanical damage by wood, cement, clay, or fiber.

Construction applications of lead include building flashing, shower pans, X-ray and gamma-ray shielding, vibration damping and soundproofing, and chemical process piping and linings.

Lead anodes are used in electrolytic refining and plating processes for metals such as manganese and zinc. Lead anodes are selected chiefly for the high resistance to the sulfuric acid that is used in those electrolytic solutions.

Lead Alloy Coatings

Lead alloy coatings of terne metal are applied to sheet steel by hot dipping. Terne metal is a lead base alloy with 3% Sn to 15% Sn. The addition of tin assists in forming a tight, integral, corrosion-resistant coating on the steel.

The terne coating acts as a lubricant and facilitates the cold forming of the steel. Terne coated steel is used for automobile parts such as gasoline tanks, caskets, file drawer tracks, and small fuel tanks for many gasoline powered components.

Warning: Lead can present a health hazard. Strict procedures must be followed when handling lead or lead alloys, dust, or fumes. Lead must not be used to conduct very soft water for drinking nor should lead come in contact with foods.

High-lead Alloys

High-lead alloys contain <2% of the alloying elements. High-lead alloys include corroding lead, chemical lead, calcium-lead alloys, and antimonial lead alloys. Corroding lead is 99.94% pure and is used for storage batteries, foil, ceramics, collapsible tubes, and seals. Chemical lead contains a minimum of 99.9% Pb and is used principally as a corrosion-resistant material in the chemical industry.

Calcium-lead alloys contain up to .12% Ca and up to 1.5% Sn to increase strength. Calcium-lead is used for high-strength cast parts, such as grids for maintenance-free and standby batteries.

Most antimonial lead alloys contain from 1% Sb to 9% Sb to raise the recrystallization temperature and increase hardness and tensile strength. See Figure 23-1. One percent antimonial lead is used for cable sheathing. Antimonial lead alloys with more antimony are the hard lead alloys. Hard lead alloys are used in applications that require higher strength

than the softer lead alloys can provide, but they exhibit similar corrosion resistance.

	PROPERTIES OF ANTIMONIAL LEAD ALLOYS	
%Sb	Hardness (HB)	Tensile Strength (psi)
0	4.0	2500
1	7.0	3400
2	8.0	4200
3	9.1	4700
4	10.0	5660
5	11.0	6360
6	11.8	6840
7	12.5	7180
8	13.3	7420
9	14.0	7580
10	14.6	7670

Figure 23-1. Antimony increases the hardness and tensile strength of lead.

Type Metals

Type metals are alloys of lead, antimony, and tin that are used extensively in the printing industry for typesetting machines. Small additions of copper are added to some grades to increase hardness for specific applications. The lead base provides low cost, low melting point, and ease of casting. Antimony provides hardness and wear resistance and lowers the casting temperature. Tin increases fluidity, reduces brittleness, and provides a finer structure. Copper is added to further increase hardness of the alloy. Type metals are classified as electrotype, stereotype, Linotype®, monotype, and foundry type. See Figure 23-2.

Electrotype metals are type metals that are used as a backing material for an electroformed copper shell, which carries the impression, and are not required to resist wear. They contain the lowest percentage of alloying additions.

Stereotype metals are type metals that are used directly for printing and must be harder than electrotype metals. Sometimes, for greater wear resistance, stereotype metals are lightly plated with chromium or nickel.

COMPOSITIONS AND PROPERTIES OF TYPE METALS						
Metal	%Pb	%Sn	%Sb	Hardness (HB)[a]	Liquidus (°C)	Solidus (°C)
Electrotype						
General	95.0	2.5	2.5	—	303	246
General	94.0	3.0	3.0	12.4	298	246
Curved plates	93.0	4.0	3.0	12.5	294	245
Stereotype						
Flat plate	80.0	6.0	14.0	23.0	256	239
General	80.5	6.5	13.0	22.0	252	239
Curved plates	77.0	8.0	15.0	25.0	263	239
Linotype®						
Standard	86.0	3.0	11.0	19.0	247	239
Special	84.0	5.0	11.0	22.0	246	239
Ternary eutectic alloy	84.0	4.0	12.0	22.0	239	239
Monotype						
Ordinary	78.0	7.0	15.0	24.0	262	239
Display	75.0	8.0	17.0	27.0	271	239
Case type	64.0	12.0	24.0	33.0	330	239
Rules	75.0	10.0	15.0	26.0	270	239
Foundry type						
Hard (1.5% Cu)	60.5	13.0	25.0	—	—	—
Hard (1.5% Cu)	58.5	20.0	20.0	—	—	—
Hard (2.0% Cu)	61.0	12.0	25.0	—	—	—

[a] 10 mm ball, 250 kg load

Figure 23-2. Type metals are lead-tin-antimony alloys that are used primarily in the printing industry.

Linotype® metals are type metals that are machine die cast for the high-speed composition of newspapers. Linotypes® have compositions close to the ternary eutectic because a low melting point coupled with a short freezing range are of greatest importance in this application. See Figure 23-3.

Monotype metals are type metals in which only one type character is cast at a time so that a rapid cooling rate is possible and a harder alloy with a higher melting range is permissible. Like Linotypes®, monotype metals are also machine die cast.

Foundry type metals are type metals that contain the greatest amount of alloying additions (copper to increase hardness) and are used exclusively to cast type for hand composition. Foundry type metals are used over and over again before being remelted. This service requires the hardest, most wear-resistant alloy practicable.

American Society for Metals

Figure 23-3. Linotypes® have compositions close to the ternary Pb-Sn-Sb eutectic and produce a short freezing range and low melting point.

Properties

The outstanding properties of lead alloys are density, malleability, and corrosion resistance. The less desirable properties, which must be considered in product design, include low strength and high coefficient of thermal expansion.

Density. The high density of lead (approximately 11 g/cm^3) makes it very effective in shielding against X rays. Lead is also an excellent material for attenuating gamma radiation because of its density, coupled with high atomic number and high level of stability. The combination of high density and low stiffness make lead an excellent material for deadening sound and for isolating equipment and structures from mechanical vibrations.

Malleability, Softness, and Lubricity. Malleability, softness, and lubricity are three related properties of lead that make it suitable for many applications. For example, the malleability of lead makes it useful as a caulking agent. Softness and self-lubricity of lead make it useful for bearing alloys or as a coating on sheet metal where it acts as a drawing lubricant.

Corrosion Resistance. Lead is highly resistant to corrosion by the atmosphere, waters, and a wide range of chemicals. Lead develops corrosion resistance through the formation of a protective film of corrosion products. The majority of these products are insoluble lead salts, such as lead sulfate in the case of sulfuric acid, and tend to stifle further attack. In cases where the corrosion product is soluble, such as lead chloride in hydrochloric acid, lead has low corrosion resistance. See Figure 23-4.

Any factor that damages or removes the protective film increases the corrosion rate. For example, high velocity tends to strip the film, exposing unprotected lead and increasing the corrosion rate.

Strength and Coefficient of Thermal Expansion. Lead has low tensile strength and creep strength. In self-supporting designs, commercially pure lead will creep under its own weight and must be supported, for example, with straps. Alloying with

CORROSION RESISTANCE OF LEAD TO SELECTED CORROSIVE AGENTS	
Corrosive Agent	**Resistance[a]**
Acetone	R
Acetylene	R
Acid	
Chromic	R
Citric	M
Hydrochloric	M
Hydrofluoric	R
Sulfurous	R
Tartaric	M
Air	R
Alcohol	
Ethyl	R
Methyl	R
Ammonium hydroxide	R
Ammonium phosphate	R
Benzol	R
Formaldehyde	M
Magnesium chloride	S
Magnesium sulfate	R
Motor fuel	R
Nickel sulfate	R
Oxygen	R
Photographic solutions	G
Sodium carbonate	R
Water	
Chlorinated	R
Sea	R

[a] R = resistant
M = moderate attack
G = generally resistant
S = severe attack

Figure 23-4. There is a wide range in the corrosion resistance of chemical lead.

other metals, notably antimony, strengthens lead for many applications and reduces creep.

The high coefficient of thermal expansion of lead is a disadvantage, especially when coupled with its low strength. For example, in lead roofing, there must be adequate provision for thermal expansion by the use of small pieces and loose locked sheets.

TIN

Tin is extracted and refined from oxide ores. It is a soft, white metal with good corrosion resistance and lubricity. Tin is used chiefly in tinplate, as an alloying addition, and in a limited number of tin-base alloys. Tin is also a major constituent in alloys used for costume jewelry and pewter. The second largest use of tin is for solders, where it is combined with lead or other alloying elements. Tin-lead alloys are also used for bearing materials.

Extraction and Production

Tin is extracted chiefly from cassiterite, which is an oxide ore. Much of it is obtained by dredging and gravel pump mining. Water jets are used to break up the ore, and the material is washed into a sump. The tin-bearing gravel is pumped to an elevated launder where the ore concentrates to approximately 75% by settling out. The concentrate is sent to smelters where it is mixed with anthracite and limestone, and then it is charged into a reverberatory furnace. The charge is heated to 1400°C (2550°F) to reduce the tin oxide to impure tin.

The impure tin is remelted in large cast iron pots. Steam or compressed air containing controlled amounts of other elements is introduced to refine the tin to high levels of purity. Further refining, by liquation (selective dissolution) or electrolytic refining, is often performed to increase the purity level of the tin to 99.99% Sn.

Refined tin is melted and cast into ingots for further processing into products. The most common ingot grade is commercially pure tin (grade A). Commercially pure tin is used in the manufacturing of tinplate.

Tinplate

Tinplate production represents the largest single use of tin. Tinplate is used to make containers and packaging items for foods and beverages, as well as a variety of nonfood items.

Sheet steel is tin-plated on continuous plating lines using acid or alkaline electrolytes. The plating can be produced with an equal amount of tin on the steel surfaces or with a differential coating on each side. The plating thickness varies from .40 μm to 1.5 μm (15 μin. to 60 μin.), and the sheet steel thickness varies from .15 mm to .60 mm (.006 in. to .025 in.).

Commercially pure tin is also used for tin foil and collapsible tubes for food pastes, pharmaceuticals, and artist's paints. Tin sheet is used as a lining for storage tanks and components in applications involving very pure or distilled water. Tin is also used as an alloying element with other metals, such as copper, lead, or antimony.

Tin Alloys

The principal tin alloys are white metal and the pewters. *White metal* is a tin-base casting alloy containing 92% Sn and 8% Sb. White metal is used for costume jewelry.

Pewter is a tin-base alloy containing 90% Sn to 95% Sn and 1% Cu to 3% Cu, with the balance of the content being antimony. A typical pewter is 91Sn-7Sb-2Cu. Pewter is used for coffee and tea services, trays, steins, mugs, candy dishes, bowls, plates, candlesticks, and vases. It can be centrifugally cast to make costume jewelry. Pewter is readily soldered and easy to form by hammering, spinning, and casting.

Tin-Lead Alloys

Tin-lead alloys have two major areas of applications, which are solders and bearing metals. In these alloys, lead and tin are the major components. Other elements that are alloyed with tin include arsenic, antimony, and silver.

Solders. The majority of solders are alloys of tin and lead, sometimes containing other elements such as antimony or silver, which are used to join metals. *Solders* are alloys that melt below 450°C (840°F) and below the lowest melting point of the base metals (metals being joined). Tin is an important constituent in solders because it wets and adheres to many common base metals at temperatures considerably below their melting points. Tin-lead solders have melting points lower than those of tin or lead.

The tin-lead phase diagram indicates that a wide variety of binary solder compositions are possible, depending on the melting range (pasty range) that is required. For example, electronic components require the lowest melting point (eutectic) composition to minimize damage to printed circuit boards. See Figure 23-5.

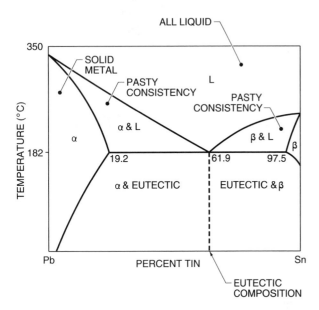

Figure 23-5. Tin is alloyed with lead to produce solders that have lower melting points than either tin or lead.

Solders are divided into tin-lead solders and other solders based on various metals. The tin-lead group includes high-lead, general purpose, and high-tin solders. The other groups of solders include tin-silver, tin-antimony, indium-base, fusible, zinc-containing, cadmium-silver, and precious metal.

High-lead solders contain a minimum of 80% Pb. High-lead solders are used for joining tin-plated containers and automobile radiators.

General-purpose solders vary in compositional contents from 25% Sn and 75% Pb to 50% Sn and 50% Pb. General-purpose solders have lower liquidus temperatures and are used where the improved wetting characteristics make sounder joints. These solders are used for many industrial products and require an inorganic fluxing material. The most widely used general-purpose solders contain 40% Sn to 50% Sn and are used to join copper alloys in electrical connections. These solders were also used extensively in plumbing applications, but the *Federal Safe Drinking Water Act* mandates the use of lead-free solders for plumbing applications.

High-tin solders contain from 50% Sn and 50% Pb to 63% Sn and 37% Pb. High-tin solders have the lowest melting points and narrowest pasty range. They are used in electronic applications or where low soldering temperatures are required. Silver additions are sometimes made to alloys used in electronic applications to reduce dissolution of silver-base coatings.

Tin-silver solders contain from 90% Sn to 96% Sn, with the balance of the content being silver. Tin-silver solders are more expensive than other solders and are used for fine instrument work and in food applications.

Tin-antimony solders contain 95% Sn and 5% Sb. Tin-antinomy solders have excellent strength characteristics and are used where a slightly higher temperature range is required, such as in applications for joining stainless steels where lead contamination must be avoided.

Indium-base solders are alloys of tin, lead, and indium. Indium wets glass, quartz, and many ceramics. This makes the indium-base solders useful in glass-to-metal seals.

Fusible solders are based on bismuth and have additions of lead, tin, and cadmium. Fusible solders do not have good wettability and are used in applications where failure of the solder joint at a predetermined low temperature is required for safety reasons. For example, fusible solders are used for fuse links in sprinkler systems.

Zinc-containing solders are tin-zinc, cadmium-zinc, and zinc-aluminum solders, which are used for joining aluminum. Aluminum is difficult to solder because it forms a tenacious oxide surface film that is difficult to flux (remove). These solders readily wet aluminum and provide joint strength.

Cadmium-silver solders are used where high service temperature is required. For example, a 95% Cd and 5% Ag solder has a solidus temperature of 338°C (640°F).

Precious metal solders are based on gold. These solders are used in the semiconductor industry. Although their cost is high, they have good corrosion resistance, wettability, strength, and compatibility with silicon.

Bearing Metals. Tin-base and lead-base bearing metals are babbitts, or alloys used for lining bearings. Babbitts include compositions ranging from ≥80% Sn and 0% Pb to <5% Sn and 80% Pb.

Antimony and copper are usually present in these alloys. Cadmium, nickel, arsenic, or tellurium is added to these alloys to increase strength. The microstructure of babbitts consists of a soft matrix with hard intermetallic particles. See Figure 23-6. The matrix has the ability to retain a film of lubricant and prevent metal-to-metal contact, even if temporary failure of the lubricant supply occurs.

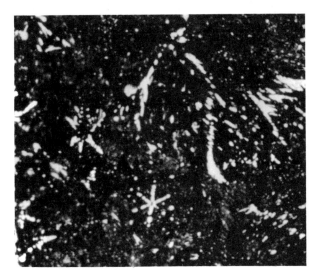

TIN-BASE BABBITT

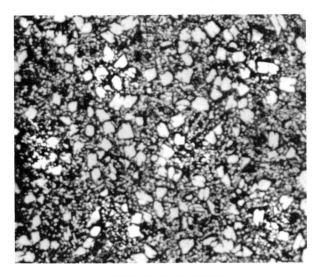

LEAD-BASE BABBITT

American Society for Metals

Figure 23-6. The microstructure of tin-base babbitt exhibits a soft matrix of Sn-Sb eutectic containing hard SnSb cuboids and Cu_6Sn_5 needles. The lead-base babbitt exhibits a soft matrix of Pb-Sn-Sb eutectic containing hard primary Sb-Sn phase.

Babbitts possess excellent conformity and embeddability, which allows foreign particles to be tolerated by burial in the bearing surface and avoid damage to shafts of low hardness. Babbitts are used at high sliding velocities under light loads. Babbitts are centrifugally or gravity cast into steel or bronze bushing shells to improve their load-carrying capacity. Many sleeve bearings are manufactured by casting babbitts onto tin-plated steel strip, and then the material is rolled, blanked, formed, and broached to form half bearings. Babbitt metals are also used to impregnate sintered alloys used for bearings.

In addition to tin-base and lead-base babbitts, there is the third group that is referred to as intermediate lead-tin bearing alloys. All three groups of the bearing alloys are specified by ASTM and SAE.

Tin-base babbitts are alloys of tin with 10% Sb to 15% Sb and 2% Cu to 8% Cu. Lead is restricted between .35% Pb and .50% Pb to avoid the formation of a low melting point lead-tin eutectic, which will reduce bearing strength. The addition of approximately 1% Cd raises bearing fatigue strength and compressive strength by 40%. Tin improves resistance to corrosion caused by the lubricant or by contaminants present in the oil.

Lead-base babbitts are alloys that contain from 75% Pb to 85% Pb, up to 10% Sn, and 12% Sb to 18% Sb. Lead-base babbitts are cheaper than tin-base babbitts, but the lead-base babbitts have lower strength and hardness at elevated temperatures. Lead-base babbitts are prone to segregation of the intermetallic phases during solidification. Cerium, arsenic, and nickel are sometimes added to control segregation.

Intermediate lead-tin bearing alloys consist of 20% Sn to 75% Sn, 10% Pb to 65% Pb, 12% Sb to 15% Sb, and up to 5% Cu. At room temperature, intermediate lead-tin bearing alloys have mechanical properties that match those of the tin-base and lead-base babbitts.

At elevated temperature, mechanical properties decrease rapidly because of the low melting point of the eutectic present in the microstructure. The intermediate lead-tin bearing alloys are used for less exacting applications where enough bearing area is employed to reduce surface loading and surface compression.

ZINC

Zinc is extracted and refined from its sulfide ore. Zinc is a heavy, bluish-white metal that is used principally for its low cost, corrosion resistance, and alloying properties. It is employed primarily as a sacrificial coating or anode material in corrosive applications. A *sacrificial coating* is a coating that reduces the corrosion of a metal by coupling it to another metal that is more active in the service environment. Zinc is also used considerably in pressure die castings. Zinc and zinc alloys are employed as metallic coatings for steel products and also in die castings for household, automotive, and industrial components.

Extraction and Production

Zinc is extracted from the zinc sulfide ore, or zinc blende. The ore is concentrated by flotation and then roasted in the presence of air to yield zinc oxide. The zinc oxide is reduced to zinc by the process of furnace reduction or by electrolytic refining.

In the furnace reduction process, a mixture of zinc oxide and coal is fired, the zinc is vaporized, reduced at high temperature, and then captured by liquefying in a condenser. The purity of the zinc produced by this method is no less than 98%. To obtain 99.99% pure zinc, this product is *fractionally distilled*, a process in which the temperature is controlled to remove elements or contaminants that have different melting points from pure zinc.

In the electrolytic refining process, the zinc is first dissolved in sulfuric acid. The zinc is then redeposited electrolytically to produce a 99.99% pure zinc product.

Refined zinc is extruded into rod and shapes; drawn into rod and wire; and rolled into sheet, strip, and foil. Wrought zinc products are alloyed with small additions of cadmium, lead, or copper. They have specialized uses, such as dry batteries, photoengraving plates, and eyelets.

Zinc Coating

The largest use of zinc is for coating steel, which prevents the steel from corroding. The various types of coatings consist of galvanizing, metallizing, and zinc-rich paints.

Galvanizing. *Galvanizing* is the process of coating a metal with zinc. Galvanizing is the largest single use of zinc. See Figure 23-7. The zinc is applied to the steel either by hot dip galvanizing or electrogalvanizing. In hot dip galvanizing, prepared sheet steel, strip, or wire is continuously passed through a bath of molten zinc.

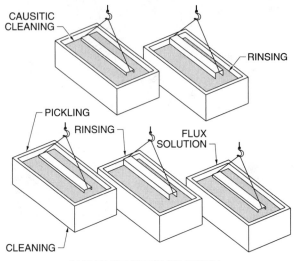

SURFACE PREPARATION

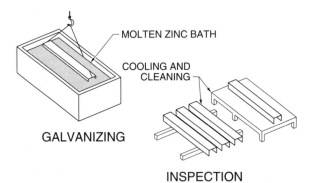

HOT-DIP GALVANIZING PROCESS

Figure 23-7. Hot dip galvanizing consists of immersing steel in molton zinc to coat it.

Individual components to be galvanized are dipped in a bath of molten zinc. Coating thicknesses up to .09 mm (.0035 in.) are obtained. In electrogalvanizing, the zinc is electro-deposited. Coating thicknesses vary from .0040 mm to .0025 mm (.00015 in. to .00010 in.). Electrogalvanized zinc is an excellent base for paint.

Metallizing. *Metallizing* is the formation of a metallic coating by an atomizing spray of molten metal. In zinc metallizing, molten zinc is sprayed on the surface of steel parts to produce a metal coating. Coating thicknesses of .08 mm to .40 mm (.003 in. to .016 in.) are obtained.

Zinc-rich Paints. In zinc-rich paints, metallic zinc powder is suspended in an organic vehicle. For corrosion protection, the zinc must provide continuous electrical coverage. The coating must consist of 80% Zn to 95% Zn.

Sacrificial Coatings. Sacrificial coatings reduce the corrosion of a metal by coupling it to another metal that is more active in the service environment. Zinc is coupled with aluminum to produce a hot dip coating on steel that offers improved corrosion resistance.

Galfan is a zinc alloy with 5% Al. An alloy containing 45% Al and 55% Zn is used for coatings. Both products are applied by hot dipping and contain minor amounts of other alloying elements to improve wettability and/or adhesion during forming.

Zinc Alloys

The major structural use of zinc is in die castings. Another zinc alloy is superplastic zinc. *Superplastic zinc* is a zinc-aluminum eutectic that is formed in the same manner as plastics.

Zinc Die Castings. Zinc die castings are used in several industries, such as in the automobile industry for carburetors, fuel pump bodies, windshield wiper parts, and brake components. In domestic applications, zinc die castings are used for washing machine components, vacuum cleaner parts, and kitchen equipment. In commercial equipment they are used for personal computers, typewriters, recording machines, cash registers, and slicing machines.

Zinc pressure die castings are low cost and have good strength. They can be cast close to dimensional limits, require minimum machining, possess good resistance to surface corrosion, and can be plated

for decorative purposes. They are limited to a maximum operating temperature of 95°C (200°F).

The major zinc die casting alloys are AG40A and AC41A. See Figure 23-8. AG40A, or Zamak-3 (Z33520), is a zinc die casting alloy containing 4% Al and .04% Mg. AG40A has slightly higher ductility and retains its impact strength at elevated temperature better than AC41A.

AC41A, or Zamak-5 (Z35531), is a zinc die casting alloy containing 4% Al, 1% Cu, and .04% Mg. AC41A is slightly harder, stronger, and easier to cast than AG40A.

PROPERTIES OF ZINC DIE CASTING ALLOYS		
Property	**ASTM AG40A**	**ASTM AC41A**
Tensile strength (ksi)	41.0	47.6
% Elongation (in 50 mm)	10.0	7.0
Compressive strength (ksi)	60.0	87.0
Liquidus (°C)	387.0	386.0
Solidus (°C)	381.0	380.0
Shear strength (ksi)	31.0	38.0
Specific heat (J/kg)	420.0	420.0
Thermal conductivity (W/m•k)	113.0	109.0
Thermal expansion (μm/m•k)	27.4	27.4
Density (kg/m^3)	6.6	6.7

Figure 23-8. The major zinc die casting alloys are AG40A and AC41A.

Superplastic Zinc. Superplastic zinc is a eutectoid 78Zn-22Al alloy. This alloy can be formed, like plastics, by vacuum forming, blow molding, and compression forming. The room temperature properties of super-plastic zinc are similar to those of lead.

To achieve a useable engineering condition, the components must be annealed at 340°C (645°F) and slow cooled. Superplastic zinc is used for electronic enclosures, cabinets, and medical and other laboratory instruments.

Corrosion Resistance

Zinc has good corrosion resistance to atmospheres and waters. This corrosion resistance is due to the formation of a protective corrosion product film of zinc hydroxide or zinc carbonate. The corrosion resistance of zinc by itself is further improved by anodizing.

Zinc has good corrosion resistance, but it is widely used sacrificially to prevent the corrosion of steel. The corrosion of steel is prevented through the use of zinc coatings and zinc anodes. Zinc coatings and zinc anodes are coupled electrically to steel and corrode at the expense of the steel. Coatings are widely used for atmospheric exposure, and anodes are used in waters.

In certain water compositions at temperatures above 60°C (140°F), the electrochemical potentials of steel and zinc reverse, meaning that the steel then protects zinc from corrosion. Under these conditions, the steel corrodes at the expense of the zinc. Zinc anodes should not be used to cathodically protect steel above 60°C (140°F).

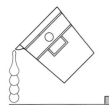

Precious, Refractory, and Specialty Metals 24

The precious metals are silver, gold, and the platinum group, which is composed of platinum, rhodium, ruthenium, palladium, iridium, and osmium. The eight precious metals are generally scarce and expensive. Refractory metals consist of tantalum, niobium (formerly columbium), molybdenum, and tungsten. The four refractory metals exhibit extremely high melting points and high elevated-temperature strength. Beryllium and zirconium are two other commercially important metals with specialized properties, which are conveniently grouped as specialty metals.

PRECIOUS METALS

The eight precious metals are silver, gold, and the platinum group, which is composed of platinum, rhodium, ruthenium, palladium, iridium, and osmium. Precious metals are identified in the UNS by the uppercase letter P followed by five digits. Precious metals are used in a wide variety of industrial and consumer applications that utilize their unique properties, such as high corrosion resistance or low electrical resistivity. See Figure 24-1.

With the exception of gold, precious metals are extracted from ores that contain substantial quantities of other minerals. Precious metals are available in many product forms and are also supplied as cladding or platings on substrate materials. Precious metals are alloyed with one another and with other metals to increase strength, reduce cost, or develop specific properties.

Extraction and Production

Silver is extracted from silver ores, but is principally obtained as a by-product in the extraction of copper, lead, and other metals. Gold is usually found in the pure state mixed with gravel or as veins in rocks.

The pure metal is separated by panning and sluicing or by amalgamation with mercury and later extracted by volatilization of the mercury. Vein deposits are crushed and the gold separated by a cyanide process. Platinum metals are obtained from deposits that are associated with copper or with nickel sulfide ores.

Finished Forms. Finished forms of silver, gold, platinum, rhodium, and palladium consist of rod, wire, sheet, strip, ribbon, tubing, and powder. Precious metals are also clad or deposited on various substrate materials.

Precious metals are drawn into wire as small as 25 μm (.001 in.) diameter. Some platinum-iridium alloys are drawn to diameters of 7.5 μm (.0003 in.). Sheet thicknesses may be as thin as 2.5 μm (.0001 in.). Seamless tubing and welded tubing are manufactured in many sizes. Powders are chemically precipitated or atomized to produce a wide variety of particle sizes.

Deposition and cladding are used to reduce cost by applying a thin layer of the precious metal to a cheaper material substrate with the required structural strength. Product forms include wire, sheet, strip, and formed components, and tubing clad on the inside and outside. The processes used include mechanical or thermal bonding, vacuum coating, electroplating, chemical coating, and fired-on films. Of these, electroplating is the most important process.

Precious metals are also important constituents in many brazing filler metals and solders. For example, gold-nickel alloys are used to braze heat-resistant alloys, and tin-silver solders are used in fine instrument work.

APPLICATIONS OF PRECIOUS METALS	
Metal	**Application**
Silver	Photography, electrical and electronic components, catalysts, brazing filler metals and solders, and dental amalgams
Gold	Investment products, jewelry, and brazing filler metals
Platinum	Thermocouples, catalysts, crucibles, electrical contacts, spark plug electrodes, and anodes
Rhodium	Mirror electroplate and alloying element for platinum and palladium
Ruthenium	Alloying element for platinum and palladium
Palladium	Automatic telephone dial contacts, film resistors, thermocouples, brazing filler metals, and catalysts
Iridium	Alloying element for platinum and crucibles
Osmium	Alloying element for other precious metals, fountain pen nibs, and electrical contacts

Figure 24-1. Industrial applications of the precious metals utilize the unique properties they exhibit.

Silver

Silver is a bright, white metal that is very soft and malleable. Pure silver has the lowest electrical resistivity of any metal and is an excellent conductor of heat. See Figure 24-2. It is the least expensive of the precious metals and has the widest number of uses. The major applications for silver, in order of importance, are photography, electrical and electronic components, and appliance manufacturing.

Silver is resistant to food products, hot concentrated solutions of many organic acids, and hot caustic alkali solutions. At room temperature, silver quickly tarnishes when in contact with sulfur or at-

mospheres that contain sulfur. This results in the formation of a black sulfide film. Silver oxidizes rapidly in air at elevated temperatures. Although resistant to chlorine at room temperature, silver has poor resistance at elevated temperature.

PROPERTIES OF SILVER	
Density (g/cm^3)[a]	10.5
Melting point (°C)	962
Crystal structure	FCC
Tensile strength (ksi)[b]	18.2
Electrical resistivity ($\mu\Omega$•cm)[c]	1.47

[a] at 20°C
[b] annealed
[c] at 0°C

Figure 24-2. Pure silver has the lowest electrical resistivity of any metal and is an excellent conductor of heat.

Silver Alloys. Silver alloys comprise the silver-copper alloys, silver brazing alloys, and dental amalgams. Copper hardens silver and forms a eutectic at 28.1% Cu, which melts at 223°C (435°F). The two most popular silver alloys are sterling silver (92.5% Ag and 7.5% Cu) and coin silver (90% Ag and 10% Cu). Sterling silver is used for flat and hollow tableware and for jewelry. Coin silver, which was formerly used for silver coins, is used for electrical contacts operating under service conditions where commercially pure silver is either too soft or likely to pit.

Silver brazing alloys and silver solders are compositions of silver and various additions of copper, zinc, cadmium, and other metals. They are used as brazing filler metals for copper, nickel, tool steels, stainless steels, carbide tools, and precious metals.

Dental amalgams are alloys of silver, mercury, tin, copper, and zinc that are used for restoring lost tooth structure. They are referred to as silver fillings and are made by mixing particles of a 65% Ag alloy with mercury. The amalgam is then formed in place by packing and carving with instruments before hardening occurs.

Gold

Gold is a bright, yellow metal that is very dense, very malleable, and highly ductile. See Figure 24-3.

Gold is commonly alloyed with copper and silver to increase its hardness and strength, and to minimize cost. Gold is used in jewelry making, visual art, various industrial applications, dental materials, and investment products. It is also a constituent in some brazing filler metals.

Gold for jewelry is described by the karat (K) and color. Pure gold (100% Au) corresponds to 24K and an alloy containing 50% Au corresponds to 12K.

Gold is not attacked by common acids and is not oxidized at any temperature in air. It is attacked by aqua regia (mixture of nitric and hydrochloric acids), nitric-plus-sulfuric acid mixtures, alkali cyanides, and free chlorine above 80°C (175°F).

PROPERTIES OF GOLD	
Density (g/cm^3)^a	19.3
Melting point (°C)	1063
Crystal structure	FCC
Tensile strength (ksi)^b	18
Electrical resistivity (μΩ•cm)^c	2.1

^{a} at 25°C
^{b} annealed
^{c} at 0°C

Figure 24-3. Gold is commonly alloyed with copper and silver to increase hardness, increase strength, and minimize cost.

Gold Alloys. Gold alloys include green, yellow, and red golds, white gold, gold-platinum, and gold-nickel. Green, yellow, and red golds are alloys of gold containing silver and copper, frequently modified with additions of zinc and sometimes modified with nickel. These alloys are used for jewelry and dental applications.

White gold is a group of gold-nickel-copper alloys containing approximately 80% Au. White gold work hardens more rapidly than yellow golds. White gold is used as a substitute for platinum in jewelry.

Gold-platinum is a 70% Au and 30% Pt alloy with exceptional corrosion resistance. It is used for spinnerets in rayon production and as a brazing alloy with a high melting point.

A gold-nickel alloy with 18% Ni is a useful brazing alloy. This alloy is a strong brazing filler metal and is usually applied in a vacuum.

Platinum

Platinum is the least rare and most widely used metal in the platinum group. Platinum is white, ductile and malleable, and retains its brightness at all times. Platinum of the highest purity is used for resistance thermometers and thermocouples.

Platinum is resistant to nitric, hydrofluoric, and sulfuric acid at high temperatures. It is also resistant to perchlorates, persulfates, and hypochlorites. Platinum has reduced resistance to chlorine at elevated temperatures but does not oxidize in air at any temperature. Platinum is alloyed with rhodium, ruthenium, iridium, nickel, tungsten, and cobalt. The majority of these elements have a pronounced hardening effect.

Platinum is one of six metals in the platinum group. Platinum group metals have distinctive properties and similarities. The group is divided into two sets of triplets that correspond to the specific gravity of the metal. Rhodium, ruthenium, and palladium is one set, which has specific gravities of approximately 12 g/cm^3. The second set is platinum, iridium, and osmium, which has specific gravities of approximately 22 g/cm^3.

The platinum group can also be divided into three sets of twins by the crystal structure and properties they exhibit. See Figure 24-4. Ruthenium and osmium have CPH structures, are quite hard, and have limited ductility. Rhodium and iridium have FCC structures, are more ductile than the other platinum group metals, and have the lowest electrical resistivity. Platinum and palladium also have FCC structures, have the lowest annealed hardness, and are easily worked.

Platinum Alloys. Platinum and rhodium are completely miscible in one another. Platinum-rhodium alloys contain from 3.5% Rh to 40% Rh. They are used in platinum/platinum-rhodium (Pt/Pt-Rh) thermocouples, windings for furnaces, crucibles, applications involving molten glass, and as a catalyst in fertilizer production.

Platinum-iridium alloys contain between .4% Ir and 30.0% Ir and have excellent tarnish resistance. These alloys are used for laboratory equipment, jewelry, electrical contacts, and electrodes for aircraft spark plugs.

PROPERTIES OF PLATINUM GROUP METALS					
Metal	**Crystal Structure**	**Electrical Resistivity ($\mu\Omega$•cm)**	**Density (g/cm³)**	**Melting Point (°C)**	**Tensile Strength (ksi)**
Ruthenium	CPH	7.6[a]	12.45	2310	78[c]
Osmium	CPH	9.5[b]	22.58	2700	—
Rhodium	FCC	4.5[b]	12.41	1963	120[b]
Iridium	FCC	8.4[b]	22.65	2447	110[b]
Palladium	FCC	10.8[b]	12.02	1552	21[b]
Platinum	FCC	9.8[a]	21.45	1769	18[b]

[a] at 0°C
[b] at 20°C
[c] hot rolled 55%

Figure 24-4. The platinum group can be divided into three sets of twins by the crystal structures and properties they exhibit.

Platinum-ruthenium alloys contain from 5% Ru to 14% Ru. They are used in applications similar to those of the platinum-iridium alloys. Platinum-ruthenium-rhodium alloys are used for crucibles.

Platinum-nickel alloys contain up to 20% Ni. The nickel increases the elevated-temperature strength of platinum. Platinum-nickel alloys are used for taut band strips for electrical meters.

Platinum-tungsten alloys contain from 2% W to 8% W. Platinum-tungsten alloys are used in a limited number of applications, such as potentiometer wire, in which the alloys exhibit excellent wear resistance and low noise characteristics. Platinum-tungsten alloys are also used for spark plug electrodes.

A platinum-cobalt alloy containing 23.3% Co has unique properties as a permanent magnet alloy where a very short magnet is required. Examples include hearing aid magnets, focusing magnets, electric watch magnets, Pt-Co films for digital magneto-optic recording, and catalytic converter anodes.

Rhodium. Rhodium is the whitest platinum group metal. It is moderately ductile at low temperatures and highly ductile at elevated temperatures. It is used primarily as a hard, non-tarnishing electroplate for searchlight and projector mirrors, and as an alloying addition to platinum and palladium. Rhodium is used as a barrier, or plating under other plated metals.

The corrosion resistance of rhodium is comparable to that of iridium, including resistance to boiling aqua regia, hot hydrochloric acid, and hot chlorine. Rhodium has good oxidation resistance in air at all temperatures.

Ruthenium. Ruthenium is an extremely hard, white metal that is difficult to work at any temperature. It is used mainly as an alloying element to harden platinum and palladium.

Ruthenium resists most acids (including aqua regia) up to 100°C (212°F). It exhibits good resistance to attack by molten lithium, sodium, potassium, copper, silver, and gold in the absence of oxygen. It has poor oxidation resistance in air.

Palladium. Palladium is a white metal with properties similar to those of platinum. It is extremely ductile and can be hot or cold worked. Cold work increases tensile strength from 30 ksi (205 MPa) to approximately 55 ksi (380 MPa) at 50% reduction of area, or 65 ksi (450 MPa) at approximately 75% reduction of area. Palladium has poorer corrosion resistance than platinum, which makes it less resistant to highly oxidizing chemicals. Palladium oxidizes in air in the range 400°C to 800°C (750°F to 1470°F).

Palladium can be electroplated, electroformed, or electrolessly deposited. It is used extensively in automatic telephone dial contacts, film resistors, thermocouples, brazing alloys, selectively permeable membranes, and catalysts.

Palladium is alloyed with silver, copper, gold, and ruthenium. Palladium and silver are completely soluble in one another. Palladium-silver alloys containing between 40% Ag and 99% Ag are used for electric contacts, precision resistance wires, and for brazing Inconel®, stainless steel, and other heat-resistant alloys.

Palladium-silver alloys with additions of copper, gold, zinc, and platinum are age hardenable and have high levels of strength after heat treatment. Palladium-ruthenium alloys are used for jewelry, catalysts, and electrical contacts.

Iridium. Iridium, the densest and most corrosion resistant of all metals, is a white metal with limited malleability at room temperature. It is very difficult to forge or fabricate, even at elevated temperatures, and is not ordinarily used by itself. The major application for iridium is as a hardening agent for platinum. It also is used as a crucible material in crystal growth.

Iridium has an exceptionally high modulus of elasticity and high-temperature strength, comparable with tungsten, up to 1650°C (3000°F). This, coupled with a high melting point, makes iridium useful for crucibles for melting nonmetallic substances at temperatures as high as 2040°C (3700°F).

Iridium is only slightly attacked by boiling aqua regia, hot hydrochloric acid, and hot chlorine, but it is susceptible to attack by molten salts. Iridium oxidizes in air in the range 600°C to 1000°C (1110°F to 1830°F) but reduces directly to the metallic form at higher temperatures.

Osmium. Osmium is a white, very dense, hard metal that is virtually impossible to work at any temperature. It has the highest melting point of all the metals in the platinum group. Osmium is slightly less corrosion resistant than the other platinum group metals and has poor oxidation resistance in air.

Osmium is used as an alloying element to increase the hardness of other precious metals. It is rarely available in the pure form because it forms toxic oxide osmium tetroxide by oxidation in air. Osmium alloys have high corrosion and wear resistance. Osmium-iridium alloys are used for fountain pen nibs, electrical contacts, and instrument pivots.

REFRACTORY METALS

Refractory metals and alloys are characterized by extremely high melting points and high elevated temperature strength. The four refractory metals are tantalum, niobium (formerly columbium), molybdenum, and tungsten. The refractory metals are designated in the UNS by the uppercase letter R followed by five digits.

The maximum temperature at which a metal exhibits useful high-temperature properties is related to its recrystallization temperature. The recrystallization temperature of pure metals is approximately 40% of their melting temperatures and is raised by the addition of alloying elements. See Figure 24-5. Compared with most other metals, refractory metals have very high recrystallization temperatures.

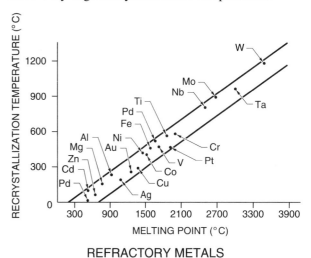

REFRACTORY METALS

Figure 24-5. The recrystallization temperatures of pure metals is approximately 40% of their melting temperatures.

Despite extremely favorable high-temperature mechanical properties, the refractory metals have one significant drawback; they oxidize rapidly in air, even at moderately high temperatures. To take advantage of the mechanical properties at temperatures up to 1370°C (2500°F), some refractory materials are coated with aluminide or silicide compounds. Coatings are applied by various methods, such as dipping, spraying, or pack cementation.

Tantalum

Tantalum is a soft, ductile metal with a melting point of 2996°C (5425°F). It is approximately twice as dense as steel. Unlike many BCC metals, tantalum retains good ductility at very low temperatures and does not exhibit a nil ductility transition temperature.

Tantalum is used in structural applications at service temperatures from 1370°C to 2040°C (2500°F to 3700°F) but requires a protective coating for any exposure to an oxidizing environment above 425°C (800°F). Such coatings are seldom successful. Tantalum is anodized to develop a dense, dielectric oxide film, a property that is utilized in miniature capacitors and is the largest application for tantalum. Tantalum has excellent corrosion resistance and is used in demanding chemical process environments for equipment and components. It is also used as an alloying agent for superalloys.

Manufacture and Production. Tantalum is extracted from ores that contain iron and manganese tantalates, or as a by-product in slag from tin refining. It is purified by fused salt electrolysis or by reduction of tantalum halide (binary compound of tantalum and halogen) with lithium or sodium. The purified tantalum is produced as a powder.

The powdered metal is then consolidated by sintering and vacuum arc or electron beam melting to produce ingots. The ingots are formed into wrought products. To avoid oxidation of the tantalum, protective atmospheres such as helium, argon, or full vacuum are used for all metal processing operations performed above 260°C (500°F). Coatings are used for short term exposure to temperatures from 260°C to 980°C (500°F to 1800°F), but some oxidation does occur.

Fabrication. Tantalum is fabricated by machining, forming, and welding. It is easily embrittled by the interstitial elements (oxygen, nitrogen, hydrogen, and carbon) and must be protected from picking them up during any fabrication operations that involve high temperatures.

The machinability of fully recrystallized, unalloyed tantalum is similar to that of soft copper. Special attention must be paid to tool design, speed, and types of coolants used. The machinability of tantalum alloys is difficult because they are gummy and gall easily. Tantalum and tantalum alloys are easily cold formed using conventional equipment.

Welding is performed by gas metal arc, gas tungsten arc, resistance, electron beam, and laser welding. Fusion welding processes must be performed in vacuum or in high-purity argon or helium atmospheres. Resistance welding is performed in air or under water. Before any welding or heat treatment operation can be performed, the tantalum must be thoroughly degreased and chemically cleaned because it may be embrittled.

Corrosion Resistance. Tantalum is highly resistant to corrosion in many chemical environments, such as most acids, salts, and organic chemicals. It also exhibits good resistance to many corrosive gases and liquid metals.

Tantalum oxidizes rapidly in air above 300°C (570°F). It is attacked by hydrofluoric acid, fuming sulfuric acid that contains free sulfur trioxide, strong alkalis, and salts that hydrolyze to form even trace amounts of hydrofluoric acid.

Tantalum is embrittled by hydrogen when it is electrically coupled to less noble metals in an acid environment or when it is exposed to hydrogen at elevated temperature. Tantalum is attacked by fluorine at room temperature, chlorine above 250°C (480°F), bromine above 300°C (570°F), and iodine at higher temperatures.

Tantalum Alloys. Tantalum is alloyed with the other refractory metals and with zirconium, hafnium (Hf), tungsten, and titanium. This develops strengthening chiefly by solid solution or dispersion hardening. A limited number of alloys are precipitation hardenable.

The other refractory metals, zirconium, and hafnium have extensive solid solubility in tantalum. Tungsten and molybdenum exert the greatest solid solution strengthening effect on tantalum. Niobium is only added in very small amounts for the same reason. Zirconium, hafnium, and titanium strengthen tantalum by dispersion hardening, which is the formation of small, well-dispersed carbides, oxides, and nitrides. Wrought tantalum alloys have good ductility at cryogenic temperatures. See Figure 24-6.

Tantalum alloys include commercially pure tantalum, Ta-2.5W, Ta-10W, T-111, T-222, and Astar 811C. Commercially pure tantalum has severely restricted amounts of the interstitial elements (oxygen, nitrogen, hydrogen, and carbon). Commercially pure tantalum is soft and weak. It is used as a capacitor material for chemical process applications where it is explosion bonded to a stronger structural material. It is also used in medical applications for surgical implants.

Ta-2.5W contains 2.5% W and .5% Nb, and has approximately twice the yield strength of commercially pure tantalum at 200°C (400°F). It is used for chemical process equipment and corrosion-resistant linings.

COMPOSITION OF TANTALUM ALLOYS

Designation	%C	%Hf	%Nb	%Re	%Ta	%W
Ta-2.5W	—	—	0.5	—	97.0	2.5
Ta-10W	—	—	—	—	90.0	10.0
T-111	—	2.0	—	—	90.0	8.0
T-222	.010	2.5	—	—	87.5	10.0
Astar 811C	.025	.7	—	1	90.3	8.0

Figure 24-6. Wrought tantalum alloys have good ductility at cryogenic temperatures.

Ta-10W contains 10% W and is stronger than Ta-2.5W. Ta-10W is the most widely used of these alloys for aerospace components, either bare or coated, at temperatures up to 230°C (450°F). It is also used in the chemical process industry where greater hardness is required than can be obtained with Ta-2.5W.

T-111 contains 2.5% Hf and 8% W. It has greater creep strength than Ta-10W, with comparable fabricability and low-temperature properties. T-111 is more resistant to some liquid metals, such as alkali metals, than Ta-10W, Ta2.5W, and commercially pure tantalum. As a result, T-111 is used for containment of liquid metals.

T-222 contains 10% W, 2.5% Hf, and .01% C. It is stronger than T-111 but has similar fabricability, low-temperature properties, and liquid metal resistance. The applications of T-222 are similar to those of T-111.

Astar 811C contains 8% W, 1% Re, .7% Hf, and .025% C. It is superior to T-111 and T-222 in creep resistance and fabricability. It is intended for high-temperature structural applications and for containment of liquid alkali metals.

Niobium

Niobium, which was formerly referred to as columbium in the United States, is a ductile metal with a melting point of 2468°C (4474°F). Niobium has approximately half the density of tantalum, and its alloys are used extensively in aerospace applications. Niobium is also widely used in nuclear applications for its low-thermal neutron absorption cross section and as an alloying addition to superalloys.

Despite a BCC crystal structure, niobium exhibits excellent toughness at cryogenic temperatures. The mechanical properties of niobium depend highly on its purity. Like tantalum, it is easily embrittled by the interstitial elements. It must be protected from these elements during high-temperature processing operations.

Manufacture and Production. Niobium is extracted from ores containing iron and manganese colombates. Niobium is also present in most ores that contain tantalum and is removed during tantalum refining. Niobium is also purified by exothermic reaction (reduction) with aluminum to form a niobium alloy with 2% Al to 3% Al in the ingot form.

The refined niobium is then prepared as a powder. The powdered metal is refined by electron beam melting and is further purified by electron beam zone refining. Arc melting is an alternative method of production. Niobium alloys are produced by arc melting high-purity niobium in conjunction with alloying elements.

Fabrication. Niobium and niobium alloys are fabricated by machining and welding. The machinability of niobium alloys varies between those of copper and stainless steel. The machinability of niobium is similar to that of tantalum.

Niobium alloys are readily formed at room temperature by all processes, such as rolling or forging. Niobium alloys are broken down above their recrystallization temperature of 1205°C (2200°F) but are often worked at room temperature or below 425°C (795°F). The alloys are then reworked at room temperature or lower.

Welding processes for niobium include electron beam, resistance, and gas tungsten arc welding. The equipment, procedures, and precautions used for titanium are also applicable for niobium. For example, fusion welding for both must be performed in an inert gas atmosphere or vacuum.

Corrosion Resistance. Niobium has good corrosion resistance to many environments because of a naturally formed protective oxide film. Niobium is resistant to strong nitric acid solutions and hydrochloric acid solutions when the hydrochloric acid is under oxidizing conditions. At elevated temperatures, niobium reacts with the halogens, oxygen, nitrogen, carbon, hydrogen, and sulfur. Niobium rapidly oxidizes in the air at temperatures above 400°C (750°F).

Niobium has good resistance to liquid sodium, lithium, and sodium-potassium eutectic. The addition of 1% Zr to niobium increases its resistance to embrittlement that is caused by oxygen absorbed from these liquid metals and compounds.

Niobium Alloys. Niobium is solid solution hardened by tantalum, tungsten, molybdenum, titanium, and zirconium. See Figure 24-7. Commercially pure niobium contains specified low amounts of the interstitial elements. It is produced as commercial grade and reactor grade. Reactor grade contains severely restricted amounts of hafnium. Commercially pure niobium is used in super-conductor, aerospace, and nuclear applications.

Nb-1Zr contains 1% Zr and is most commonly used in the production of sodium vapor lamps as internal supports. Nb-1Zr is also used in nuclear applications because of its low thermal-neutron absorption cross section, superior corrosion resistance to commercially pure niobium, and good resistance to radiation damage. It is used for liquid metal systems operating from 980°C to 1205°C (1800°F to 2200°F). Nb-1Zr is divided into reactor grade and commercial grade.

C-103 contains 10% Hf and 1% Ti. It is used for rocket components requiring moderate strength at temperatures of 1090°C to 1370°C (2000°F to 2500°F). C-103 is also used for thermal shields, piping for chromic acid and other acids, piping for liquid alkali metal containment, and sodium vapor lamp electrodes.

C-129Y and Cb-752 contain 10% W plus additions of hafnium and yttrium or zirconium. They have higher elevated-temperature tensile and creep strength than C-103 but maintain good fabricability. They are used as leading edges, nose caps for hypersonic flight vehicles, and guidance structures for reentry vehicles.

FS-85 and B-66 exhibit high strength at temperatures above 1205°C (2200°F) but remain ductile for fabrication operations. FS-85 contains 28% Ta, 10% W, and 1% Zr. B-66 contains 5% Mo, 5% V, and 1% Zr. FS-85 and B-66 are used as lifting and guidance structures for reentry vehicles, nozzles, and gas turbine components.

Molybdenum

Molybdenum is a ductile metal with a high modulus of elasticity, high thermal conductivity, a low coefficient of thermal expansion, and a melting point of 2610°C (4730°F). Molybdenum has a BCC crystal structure and exhibits a nil ductility transition temperature between –20°C and 95°C (0°F and

COMPOSITION OF NIOBIUM ALLOYS									
Designation	%Hf	%Mo	%Nb	%Ta	%Ti	%W	%V	%Y	%Zr
Nb-1Zr	—	—	99.0	—	—	—	—	—	1.0
B-66	—	5	89.0	—	—	—	5	—	1.0
C-103	—	—	89.0	—	1	—	—	—	—
C-129Y	10	—	80.0	—	—	10	—	.01	—
Cb-752	10	—	87.5	—	—	10	—	—	2.5
FS-85	—	—	61.0	28	—	11	—	—	1.0

Figure 24-7. Niobium is solid solution hardened by tantalum, tungsten, molybdenum, titanium, and zirconium.

200°F). See Figure 24-8. This temperature is influenced by the amount of work hardening, metal purity, grain size, and alloying elements.

The major use of molybdenum is as an alloying element in steels, cast irons, and heat-resistant and corrosion-resistant alloys. Molybdenum additions improve hardenability, toughness, abrasion resistance, corrosion resistance, strength, and creep resistance. Molybdenum alloys are used in several specific applications.

PROPERTIES OF MOLYBDENUM	
Density $(g/cm^3)^a$	10.2
Melting point (°C)	2610
Crystal structure	BCC
Tensile strength $(ksi)^b$	196
Electrical resistivity $(\mu\Omega\cdot cm)^c$	5.2

[a] at 20°C
[b] annealed
[c] at 0°C

Figure 24-8. The major use of molybdenum is as an alloying element in steels, cast irons, heat-resistant alloys, and corrosion-resistant alloys.

Manufacture and Production. Molybdenum is extracted from its sulfide ore. It is purified by successive reaction with oxygen and then hydrogen to produce a powder. The powder is consolidated or vacuum melted to produce billets suitable for further working or refining. Vacuum melted billets generally have lower oxygen and nitrogen contents than those produced from powder billets.

Molybdenum is marketed in many forms, such as ore concentrates, alloys, molybdate salts, and oxides. Molybdenum is chiefly alloyed in small amounts with other metals. There are a few molybdenum alloys with specific end uses, which include cathodes and cathode supports for radar devices, mandrels for winding tungsten filaments, and resistance heating elements in vacuum furnaces.

Molybdenum and molybdenum alloys have low solubility for oxygen, nitrogen, and carbon at room temperature. When cooled from the molten state or from temperatures near the melting point, these interstitial elements are rejected as oxides, nitrides, and carbides. If the impurity content is high enough, a continuous, brittle grain boundary film, which se-

verely limits plastic flow at moderate temperatures, is formed. Warm working below the recrystallization temperature breaks up grain boundary films and produces a fibrous structure that has good strength parallel to the working direction.

Fabrication. Molybdenum is fabricated by machining, forming, and welding. Machining operations are performed using tungsten carbide cutting tools with soluble oil coolants. Electrical discharge machining and electrochemical machining are also performed on molybdenum.

Forming is performed using conventional processes and tools. Fabrication operations must be performed above the nil ductility transition temperature and below the recrystallization temperature. The rate of working and the effects of stress concentration are important because the nil ductility transition temperature is shifted upward by an increasing strain rate.

Welding must be performed in an inert gas atmosphere or in a vacuum to prevent contamination by oxygen. There should be minimum restraint in the weld, and weldments must be stress relieved promptly to prevent cracking in the heat-affected zone.

Corrosion Resistance. Molybdenum is resistant to molten bismuth, lithium, magnesium, potassium, and sodium. Molybdenum is attacked by molten tin, aluminum, iron, nickel, and cobalt. Molten oxidizing salts, such as potassium nitrate, and fused caustic alkalis attack molybdenum.

Molybdenum oxidizes rapidly in air and oxidizing atmospheres above 500°C (930°F) and requires a protective coating above this temperature. Ceramic coatings are used for low-stress applications and metallic coatings, such as the precious metal coatings, for high-stress applications.

Molybdenum Alloys. Molybdenum is alloyed with small amounts of titanium, zirconium, and tungsten to improve high-temperature strength properties. Molybdenum alloys include TZM and Mo-30W.

TZM contains .5% Ti, .1% Zr, and .15% W. It has good high-temperature strength. TZM is used for heat engines, heat exchangers, nuclear reactors, radiation shields, extrusion dies, and boring bars.

Mo-30W contains 30% W and has excellent resistance to corrosive attack by liquid metals, especially liquid zinc. It has a higher melting point than that of TZM or pure molybdenum.

Tungsten

Tungsten is dense, stiff, and brittle. Tungsten has the highest melting point of any metal, 3410°C (6170°F), and a BCC crystal structure. See Figure 24-9. Its major use is for tungsten carbide cutting tools. Another important use of tungsten is as an alloying element in steels, high-speed tool steels, and nickel-cobalt alloys.

PROPERTIES OF TUNGSTEN	
Density (g/cm³)[a]	19.3
Melting point (°C)	3410
Crystal structure	BCC
Tensile strength (ksi)[b]	140
Electrical resistivity (μΩ•cm)[c]	5.3

[a] at 20°C
[b] annealed
[c] at 0°C

Figure 24-9. An important use of tungsten is as an alloying element in steels, high-speed tool steels, and nickel-cobalt alloys.

Manufacture and Production. Tungsten is extracted from minerals in which it is combined with calcium, iron, or manganese. Pure tungsten is obtained by reduction of its oxide with hydrogen. It is compacted, sintered in a hydrogen atmosphere, and worked at successively lower temperatures, which has the effect of lowering the nil ductility transition temperature to below room temperature. This makes the material suitable for structural uses. The high melting point, low vapor pressure, high modulus of elasticity, and basically good electrical conductivity of tungsten provide it with a unique position for lamp filaments.

Fabrication. Tungsten is fabricated by forming, machining, and welding. The extreme notch sensitivity (susceptibility to brittle failure from the presence of surface flaws) of tungsten is an important consideration in any forming operation. Even minute surface flaws must be removed by grinding, oxidizing, or electrolytic polishing. Tungsten is sensitive to the rate of working. The nil ductility transition temperature is shifted upward by increasing strain rate.

Welding requirements for tungsten are similar to those of molybdenum. Welded joints are brittle at room temperature and extremely notch sensitive. The welded surface should be finished smooth and faired (externally blended) gradually into the base metal if possible.

Tungsten Alloys. Recrystallized tungsten exhibits an extremely high nil ductility transition temperature of 205°C (400°F). Alloying elements lower the nil ductility transition and recrystallization temperatures, and strengthen tungsten by dispersion or solution hardening. Tungsten alloys include commercially pure (undoped) and commercially doped tungsten, W-ThO$_2$, W-Mo alloys, and W-Re alloys. A *dopant* is an impurity added, usually in small amounts, to a pure metal to alter its properties.

Unalloyed tungsten is produced as commercially pure and commercially doped (AKS) tungsten. AKS designates that the alloy contains small amounts of aluminum, potassium, and silicon. These additions improve the recrystallization and creep properties, which are especially important when tungsten is used for incandescent lamp filaments.

W-ThO$_2$ alloy is dispersion hardened by 1% Th to 2% Th. The thorium addition enhances thermionic electron emission, which improves the starting characteristics of gas tungsten arc welding electrodes. This is the major application for W-ThO$_2$.

In W-Mo alloys, the molybdenum forms a continuous solid solution with tungsten. W-Mo alloys are used mainly for improved machinability where strength lower than that of tungsten or W-ThO$_2$ can be tolerated.

In W-Re alloys, up to 26% rhenium (Re) is soluble in tungsten. Above 26% Re, the embrittling sigma phase begins to form. W-Re alloys that contain 1.5% Re and 3.0% Re are used to improve resistance to cold fracture in lamp filaments, especially where vibrations are involved. These alloys also contain AKS dopants to improve creep strength in filament wires.

SPECIALTY METALS

All the precious and refractory metals have special, but limited, applications. The precious and refractory metals form two well-defined groups in which beryllium and zirconium do not fit. As a result, beryllium and zirconium are classified as specialty metals for convenience.

Zirconium is more correctly grouped with titanium as a reactive metal. A reactive metal is a metal that is extremely reactive with air or specific chemicals if, for any reason, its naturally formed protective oxide film is removed.

Beryllium

Beryllium is a steel-gray metal with extremely low density. See Figure 24-10. The density of beryllium compares with that of magnesium. This feature, coupled with a high modulus of elasticity, makes it possible to design lightweight, thin members with high stiffness. For example, a beryllium column in pure-buckling applications (application where a load is applied purely along the axis of the column) will have a greater load carrying capacity and will be lower in weight than any other metal of equal size. Beryllium is identified in the UNS by R1 followed by four digits.

PROPERTIES OF BERYLLIUM	
Density (g/cm³)ᵃ	1.85
Melting point (°C)	1283
Crystal structure	CPH
Tensile strength (ksi)ᵇ	55 to 60
Electrical resistivity (μΩ•cm)	4

ᵃ at 20°C
ᵇ annealed

Figure 24-10. The major use of beryllium is as an alloying element in copper and nickel to produce age-hardening alloys for springs, electrical contacts, spot welding electrodes, and nonsparking tools.

The high specific heat of beryllium makes it attractive for lightweight heat sinks, such as aircraft disk brakes and rocket heat shields. For example, a unit mass of beryllium will absorb as much heat as five times the mass of copper and twice the mass of aluminum.

The major use of beryllium is as an alloying element in copper and nickel to produce age-hardening alloys for springs, electrical contacts, spot welding electrodes, and nonsparking tools. Beryllium is added to aluminum and magnesium to produce grain refinement and oxidation resistance. Beryllium is also used unalloyed as a structural material.

Manufacture and Production. Beryllium mill products are made from powder that has been consolidated into a block by hot vacuum pressing. The powder is made by chipping and mechanically pulverizing vacuum cast ingots. The hot vacuum pressed block is then rolled into sheet, bar, rod, or billets. Billets are extruded into other shapes. Cold-worked material may exhibit strong anisotropy. The tensile properties of beryllium vary greatly with the processing method.

Fabrication. Beryllium is fabricated by forming, machining, and welding. Forming is difficult because the CPH crystal structure provides few slip planes, which leads to limited ductility. For this reason, forming is performed at elevated temperatures of 540°C to 815°C (1000°F to 1500°F) using slow speeds and special lubricants. After forming, the beryllium components may require cleaning by wet grit blasting.

Machining of powder metallurgy components is generally easy. Chip formation is similar to that of cast iron. All machining operations produce a damaged surface layer that must be removed by etching a minimum of .05 mm (.002 in.) per surface for critical, highly stressed applications.

Warning: Beryllium and beryllium oxide are toxic. Dust, fumes, and vapor present serious health hazards if inhaled. The ingestion and inhalation of beryllium in any form must be avoided.

Although beryllium is toxic, it does not present a health hazard during forming operations. In forming operations, no fumes or oxide dust is created, and the maximum temperatures used for preheating leads only to the development of a thin, hard oxide film.

Welding, brazing, and soldering of beryllium is hindered by the adherent oxide film that inhibits wetting and flow. Before joining, beryllium components must be adequately cleaned by degreasing and pickling in a solution of 40% nitric acid and 5% hydrofluoric acid, followed by ultrasonic rinsing in deionized water.

Electron beam welding is the most suitable fusion welding process for beryllium. The low heat input provides a narrow heat-affected zone that minimizes loss of ductility. For low-temperature brazing applications, aluminum-silicon brazing filler metals are used, and for high-temperature brazing applications, silver-base alloys are used.

Corrosion Resistance. At room temperature, beryllium forms a thin, protective oxide film that provides excellent corrosion resistance, except in dilute hydrofluoric acid, hydrochloric acid, sulfuric acid, and nitric acid at ambient temperatures and above. Beryllium reacts appreciably with oxygen and nitrogen above 760°C (1400°F). It reacts with fluorine at room temperature and with chlorine, bromine, iodine, hydrogen fluoride, and hydrogen chloride at elevated temperature.

Beryllium Alloys. Unalloyed beryllium is produced in two grades, structural (S) and instrument (I). The structural grade contains significantly less beryllium oxide because increasing amounts of the oxide severely limit formability. The structural grade is used in satellite superstructures, antennae booms, optical support structures, aircraft disk brakes, and rocket nozzles. Applications of instrument grade beryllium include gyroscopes, inertial guidance system components, X-ray tube windows, and precision satellite and airborne optical components.

Zirconium

Zirconium is a silvery-white metal that is similar to titanium in physical characteristics but has a 50% higher density. See Figure 24-11. The major use of zirconium is as zircon (zirconium silicate) sand in foundries. Metallic zirconium is used for fuel element cladding, structures in certain nuclear reactors, crucibles, getters to remove traces of gases such as oxygen and hydrogen from inert gas filled spaces, chemical process equipment, and as a grain refiner for aluminum and magnesium alloy castings. Zirconium alloys are identified by the uppercase letter R in the UNS designation system.

Hafnium always occurs with zirconium in nature, typically in the ratio of one part hafnium to fifty parts zirconium. Hafnium is a strong absorber of thermal neutrons and must be removed from nuclear grades of zirconium. To remove the hafnium, further refining steps are required, which result in an increase in the cost of nuclear grade zirconium.

PROPERTIES OF ZIRCONIUM	
Density (g/cm^3)[a]	6.5
Melting point (°C)	1852
Crystal structure	CPH
Tensile strength (ksi)[b]	55
Electrical resistivity (μΩ•cm)	45

[a] at 20°C
[b] grade 720

Figure 24-11. The major use of zirconium is as zircon sand in foundries.

Manufacture and Production. Zirconium is extracted from zircon sand. The zircon is chlorinated using the Kroll process developed for titanium to produce zirconium sponge. The sponge is crushed, compacted into consumable electrodes, and vacuum melted into ingots. The ingots are usually vacuum melted a second time to further reduce impurities before being mechanically worked into finished forms.

Fabrication. Zirconium is fabricated by forming, machining, and welding. In forming operations, zirconium work hardens very rapidly and reaches maximum strength and hardness after 20% cold working. The work-hardening capacity accounts for a high degree of springback in plastic deformation operations, such as heat exchanger tube rolling. Zirconium is slightly anisotropic in the cold-worked condition and moderately anisotropic in the fully recrystallized condition. The yield strength of zirconium is higher in the transverse direction, and tensile strength is higher in the longitudinal direction.

Recrystallization begins in cold-worked metal at approximately 510°C (950°F). Grain growth is a problem in cold-forming operations, such as straightening and forming, in which a limited amount of cold working is performed (approximately 5% to 8%).

Zirconium is machined using conventional methods that include turning and boring, milling, drilling, wheel grinding, belt grinding, and sawing. Electrical discharge machining is also used for zirconium. The basic requirements for machining zirconium are slow speeds, heavy feeds, and a flood of coolant using a water-soluble oil lubricant. Zirconium has a tendency to gall and work harden, which necessitates higher-than-normal clearance angles on tools to penetrate the previously work-hardened surface and cut a clean, coarse chip.

Zirconium is readily welded with the same processes used for welding titanium. Zirconium has a low coefficient of thermal expansion, which contributes to low distortion during welding. Residual welding stresses are low because of the low modulus of elasticity of zirconium. Welds become embrittled if atmospheric contaminants are encountered during welding. The requirements of absolute joint cleanliness and inert gas purging during welding and cooling are necessary when welding zirconium. Any discoloration of the weld or heat-affected zone is an indication that contamination has occurred and is cause for rejection. When brazing zirconium, joint surfaces must be carefully cleaned to remove oxides and other surface containments. Brazing must be performed immediately afterward in a vacuum or dry atmosphere of argon or helium.

Corrosion Resistance. Zirconium is highly corrosion resistant to many chemicals and waters. A tightly adhering oxide film causes the corrosion resistance. If the oxide film is broken down, zirconium is extremely reactive. In high-temperature water and steam, Zircaloy-2 is superior to commercially pure grades of zirconium. All grades of zirconium have excellent resistance to seawater, brackish water, and polluted water.

Zirconium is resistant to sulfuric acid up to 70% Zr and at the boiling point. It is also resistant to all concentrations of nitric and hydrochloric acid at temperatures above the boiling point. Zirconium is not resistance to hydrofluoric acid.

The presence of oxidizing species in corrosive environments can seriously impair the resistance of zirconium. For example, chlorine, ferric ions, or cupric ions cause rapid corrosion in hydrochloric acid solutions. Acid solutions that contain the fluoride ion are also extremely corrosive. Zirconium may be embrittled by hydride formation when it is coupled with a less noble metal in a reducing acid solution, such as sulfuric acid.

Zirconium forms visible oxide film in air above 205°C (400°F). A loose, white scale develops with prolonged exposure above 815°C (1500°F). Zirconium may burn in air under specific conditions. For example, if the metal surface is in a finely divided form as a result of corrosion and is dealt a sharp blow, ignition may occur. At high temperatures, oxygen and nitrogen are readily soluble in zirconium and cause embrittlement.

Zirconium Alloys. Zirconium is used in the commercially pure form and is also alloyed with elements that stabilize the alpha phase or beta phase. Pure zirconium has a CPH crystal structure (alpha phase) at room temperature. It undergoes allotropic transformation to the BCC crystal structure (beta phase) at 870°C (1600°F). Alpha stabilizers, such as tin and oxygen, raise the alpha-beta transformation temperature (transus). Beta stabilizers, such as boron, chromium, and nickel, lower the alpha-beta transus. Elements such as carbon, phosphorus, and silicon form intermetallic compounds with zirconium.

Most zirconium alloys are dilute alpha alloys and exhibit the alpha phase at room temperature. Zirconium alloys include grade 702, grade 705, Zircaloy-2, and Zircaloy-4. See Figure 24-12.

Commercially pure zirconium, or grade 702 (R60702), has a yield strength of 35 ksi (240 MPa) and a minimum tensile strength of 55 ksi (380 MPa). Grade 702 is used in the chemical process industry for heat exchanger tubing, vessels, and other process equipment components.

Zr-2.5Nb, or grade 705 (R60705), has superior mechanical properties to those of grade 702. Grade 705 contains 2.5% Nb, which is a mild beta stabilizer.

COMPOSITION OF ZIRCONIUM ALLOYS[a]					
ASTM Grade	**%Zr + %Hf min[b]**	**%Sn**	**%Nb**	**%Fe + %Cr**	**$\%O_2$ max**
R60702	99.2	—	—	.2 max	.16
R60704	97.5	1 to 2	—	.2 to .4 max	.18
R60705	95.5	—	2 to 3	.2 max	.18
R60706	95.5	—	2 to 3	.2 max	.16

[a] .0005% H max, .025% N max, and .05% C max
[b] 4.5% Hf max

Figure 24-12. Most zirconium alloys are dilute alpha alloys and consist of alpha phase at room temperature.

Zircaloy-2 (R60802) is a nuclear grade of zirconium, deliberately low in hafnium. Zircaloy-2 contains small amounts of alpha stabilizers, such as tin and iron, and beta stabilizers, such as iron, chromium, and nickel. These alloys are forged in the beta region at 1065°C (1950°F), water quenched, and then hot worked below 840°C (1545°F) in the alpha region. Hot working in this region produces a thin, uniform distribution of intermetallics in a microstructure that is completely alpha. Zircaloy-2 is stronger than commercially pure zirconium.

Zircaloy-4 (R60804) is similar to Zircaloy-2, except it contains no nickel and slightly more iron. Zircaloy-4 is used in similar applications to those of Zircaloy-2. Although Zircaloy-2 and Zircaloy-4 are metallurgically similar to grade 702 and grade 705, they are slightly stronger and less ductile.

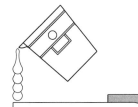

Casting 25

Casting is the process of pouring molten metal into a prepared mold cavity of a desired shape and allowing the metal to solidify. Castings differ from most wrought metal products. Castings are usually made in various finished forms and then fabricated to final shape by machining and joining operations. Consequently, cast metals are fundamentally different from wrought products. The casting practice and casting quality are tailored to produce the optimum product for the service application. The melting practice encompasses methods employed to control the quality of cast metal. Castings are produced by a variety of processes that vary according to cost, production rate, and quality obtained. Melting furnaces provide the source of molten metal for castings.

CASTING PRACTICE

Casting is the process of pouring molten metal into a prepared mold cavity of a desired shape and allowing the metal to solidify. The casting practice consists of technology used to obtain high quality castings. The casting practice is divided into melting practice, casting processes, and melting furnaces.

Melting Practice

The melting practice encompasses methods employed to control the quality of cast metal. The important components of the melting practice are the foundry flowsheet, alloying, refining and purifying, inoculation, and the molds.

Foundry Flowsheet. The *foundry flowsheet* is a diagram that indicates the steps performed to produce castings and recycle scrap material. The foundry flowsheet outlines the production steps, which include the recycling of raw materials and the finishing operations used on the castings. See Figure 25-1.

A foundry produces finished castings to specified quality requirements. The two basic types of foundries are the captive foundry and the independent foundry. The *captive foundry* is a foundry that is usually adjacent to, or connected to, a large manufacturing company, such as an automobile plant. It offers on-time delivery of cast components, such as engine blocks. Captive foundries sometimes operate at a small loss, but speedy delivery is a compensating factor. The *independent foundry* is a foundry that usually performs a wide variety of work and has a large customer base. It is more susceptible to economic downturns because it cannot absorb losses as easily as the captive foundry.

Alloying. Alloying consists of combining various materials (melting stock) to produce a heat of metal suitable for casting. Melting stock is obtained from three principal sources, which are virgin material, foundry returns, and scrap metal.

Virgin material consists of pure metals and specially manufactured alloying additions (master alloys). Master alloys are often used when one of the alloying elements has a low melting point in comparison with the major element, or when the major element oxidizes easily. For example, when making brass, zinc oxide is produced if zinc alone is added to the molten copper. A master alloy of zinc and copper is therefore used to add zinc to the composition. Examples of other master alloys include ferronickel and ferrochromium, which are used to make stainless steels and nickel alloys.

Foundry returns consist of internal scrap metal cast to the required composition but the casting was rejected. Foundry returns also include the excess stock (flash, risers, etc.) removed from castings during finishing operations.

Virgin material and foundry returns provide acceptable melting stock for any type of casting, but the use of external scrap must be carefully controlled. External scrap is not permitted in critical applications because the chemical composition of scrap material is not under the same rigorous control as virgin material and foundry returns. Minor elements in the scrap can have detrimental effects on specific alloys. For example, the presence of small amounts of lead in heat-resistant austenitic castings severely affects creep strength and weldability.

Refining and Purifying. Refining and purifying are techniques that remove atmospheric contaminants from the melt before it is poured. Two of the major problems in melting are the tendency for molten metals to absorb and dissolve gases and the oxidation of metals at elevated temperature. Vacuum furnaces are used to help overcome the problem of gas absorption and oxidation. Other techniques are also used to refine molten metals by removing tramp elements (undesirable elements) before pouring into the mold. These include protective flux coverings, bubbling insoluble gases through the melt, and vacuum degassing. See Figure 25-2.

Protective flux coverings are added to remove undesirable impurities, gases, and oxides. They also reduce surface oxidation by forming protective coatings. For example, the problem of zinc oxidation

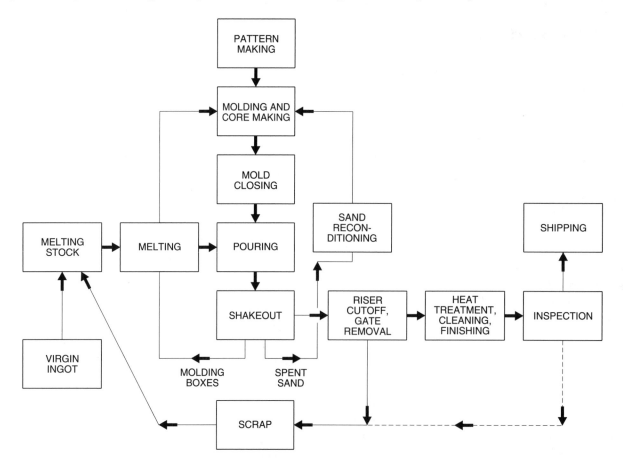

Figure 25-1. The foundry flowsheet indicates the steps performed to produce castings and recycle scrap metal.

in brass melting is reduced by a thin cover of a flux containing glass, soda ash, or borax. *Deoxidizing fluxes* are metal powders that actively combine with the oxygen in molten metal. For example, iron is deoxidized with aluminum, silicon, or manganese. Nickel is deoxidized with calcium or magnesium. Care must be taken to prevent the flux covering from becoming trapped in the molten metal during pouring, which leads to incorporation of the flux into the casting.

Insoluble gases, such as nitrogen, helium, argon, or chlorine, are bubbled through the melt to remove certain impurities and unwanted dissolved gases. The small gas bubbles tend to attach themselves to particles of oxide and float to the top of the melt. Dissolved gases, such as hydrogen, tend to diffuse into insoluble gas bubbles and are carried to the surface of the melt, where they dissipate. Vacuum degassing consists of reducing the pressure over the melt, which causes dissolved gases to be evolved.

Inoculation. Inoculation is a component of the melting process that consists of incorporating additions to the melt to alter the grain size or structure of the cast metal. Inoculants are usually added late in the melting operation and promote the nucleation of solids. The greater the number of solid nuclei in the melt when solidification begins, the finer the grain size of the casting.

For example, when grain refining aluminum alloys, additions of <.2% Ti or .002% B are sufficient to reduce the grain size from 2.5 mm to .1 mm (.100 in. to .005 in.) diameter. Magnesium alloys are usually grain refined with very small additions of zirconium, and magnesium-aluminum alloys can be refined with small additions of carbon.

Gray cast iron is inoculated with magnesium or cesium (Cs), which alters the flake graphite into spheroids, to produce ductile iron. A similar type of treatment is used to improve the ductility of aluminum-silicon alloys containing 8% Si to 10% Si. In this case, minute quantities of strontium (Sr) are added to alter the needle-like Al_3Si_2 precipitate to form a finely divided eutectic structure.

Molds. A *mold* is a hollow shape into which molten metal is poured, allowed to solidify, and removed

in the desired casting shape. A mold is made from a pattern of wood, metal, or wax that has the shape of the required casting. The mold material is formed around the pattern, which is then removed. Molds are made of sand, plaster of paris, ceramic, or metal.

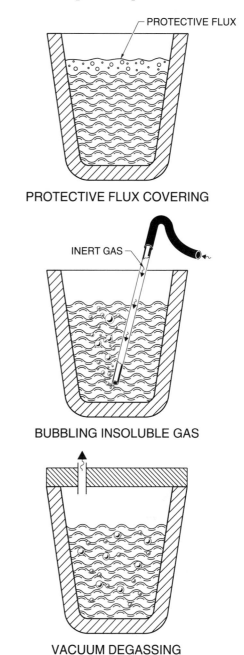

PROTECTIVE FLUX COVERING

BUBBLING INSOLUBLE GAS

VACUUM DEGASSING

Figure 25-2. Methods of refining molten metals include protective flux coverings, bubbling insoluble gases through the melt, and vacuum degassing.

The most widely used molds are made of sand. The cope (top of the mold) and drag (bottom of the mold) are split at the parting line. The mold is split to remove the pattern. The cope and drag are contained in low-carbon steel containers referred to as flasks. See Figure 25-3. The core is another type of mold that is inserted in the mold cavity to allow for hollows. The core is made from its own pattern referred to as a core box. Chaplets are metal spacers inserted between the core and the mold. Chaplets are usually made of the same metals as the casting and fuse with the molten metal when it is poured.

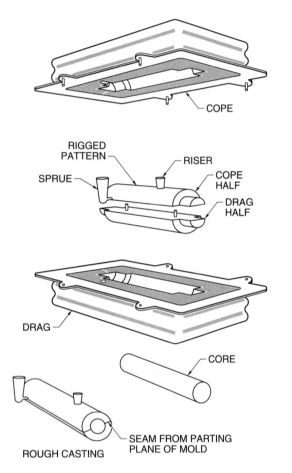

Figure 25-3. A sand mold is complex and consists of several components.

The molten metal feeding system, which supplies molten metal to the casting (mold cavity), is part of the mold and comprises a runner, pouring basin, sprue, and gate. A *runner* is a horizontal channel along which the molten metal flows when it is poured from the furnace. A *pouring basin* is a reservoir placed close to the entrance to the casting, which is designed to provide a head of molten metal

and minimize washing of sand into the casting. A *sprue* is a short channel between the pouring basin and the casting. A *gate* is the entrance to the casting.

Risers, chills, and exothermic feeding aids are incorporated at strategic locations in the mold to encourage the proper feeding of the molten metal to the casting and to ensure a sound product. *Risers* are reservoirs placed on the casting that fill with molten metal and provide a localized head of molten metal. *Chills* are pieces of metal incorporated into the mold surface in order to locally increase the cooling rate of the casting and alter the solidification pattern or metallurgical structure. Exothermic feeding aids consist of aluminum powder, which is incorporated into some riser designs. The aluminum powder exotherms (gives off tremendous heat) when the casting is poured and ensures that there is a head of molten metal to feed the solidifying casting. Contamination of the casting by the exothermic compound must be carefully monitored.

Casting Processes

Castings are produced by a variety of processes that vary according to cost, production rate, and quality obtained. The primary casting processes are sand, centrifugal, investment, permanent mold, and die casting. With the exception of die casting, these processes are performed at atmospheric pressure.

Sand Casting. Sand casting is the most versatile and widely used casting process. In sand casting, sand is packed around a pattern, and when the pattern is removed, a mold cavity in the shape of the product is created. Molten metal is poured into the cavity and allowed to solidify. The solidified metal is a cast replica of the pattern. The sand that forms the mold is friable after the metal is cast, and is broken away for removal of the casting (knockout or shakeout). The various types of sand casting include green sand molding, which is the most widely used, and shell molding.

Green sand molding is sand casting that uses molds made of sand, clay, water, and other materials. The term "green" means that the molded sand remains moist. Green sand is used without any further conditioning. Most ferrous and nonferrous castings are produced in these molds. The molds are prepared, metal is poured, and castings are shaken out in rapid production cycles. Green sand molding is adaptable to automatic machine molding. For example, production rates of 300 molds per hour are common in automated foundries that produce automotive components. Although green sand molding has rapid production cycles, it produces castings with low dimensional accuracy and poor surface finishes.

Shell molding (Croning process) uses sand that is coated with thermosetting resin. The molds or cores are made with metal patterns or core boxes and heated to 150°C to 260°C (300°F to 500°F) to cure and harden the resin. Shell molds are most often used to manufacture cores, which are the most fragile components of the mold assembly. Dimensional accuracy is improved when shell molds are used. Because of environmental problems associated with disposal of thermosetting resins, shell molding is being replaced by other molding processes that use foam patterns and cores.

Centrifugal Casting. Centrifugal casting is performed by pouring molten metal into a rotating cylindrical mold, which generates a centrifugal force that forces the molten metal against the mold wall and forms the desired shape. The two types of centrifugal casting machines are the horizontal type, which rotates about a horizontal axis, and the vertical type, which rotates about a vertical axis.

Centrifugal casting is usually limited to products of circular cross section, such as wheels, rolls, and piping. The mold is either expendable or permanent. Expendable molds are molds that are lined with sand. Permanent molds are molds that are made of steel, copper, or graphite. Generally, the mold is rotated at a speed that will create a centrifugal force ranging from 75 times the force of gravity (g) to 120g. This causes constituents of various densities to separate, with lighter particles, such as slags and nonmetallic impurities, gathering on the inner diameter. These impurities are removed by boring or machining.

Investment Casting. In investment casting, a plaster mold is formed around a wax pattern, which is then melted out, leaving a hollow mold. The

mold is then hardened by firing and filled with molten metal to form a casting. The solidified casting is knocked out, finished, and inspected. See Figure 25-4.

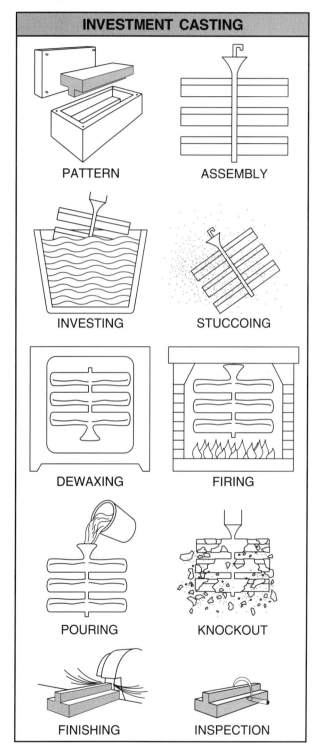

INVESTMENT CASTING

PATTERN

ASSEMBLY

INVESTING

STUCCOING

DEWAXING

FIRING

POURING

KNOCKOUT

FINISHING

INSPECTION

Figure 25-4. Investment casting is used to produce small, precision shapes and is achieved using a wax pattern.

Investment casting is used to produce small, complex components that are difficult, if not impossible, to make by other casting processes or by machining. Examples of components that are investment cast include aircraft engines, air frames and fuel systems, guns and small armaments, hand tools, valves, optical equipment, medical implants, and textile equipment.

Permanent Mold Casting. Permanent mold casting is a casting process that requires a metal mold, which is used repeatedly for producing many castings of the same form. Simple cores are made of metal, but complex cores are sand or plaster.

Permanent mold casting is used for the high-volume production of small, simple shapes that have fairly uniform wall thickness and no undercuts or internal coring. See Figure 25-5. Because of the rapid cooling rate in the mold, permanent mold castings are extremely sound, exhibit excellent mechanical properties, and display excellent surface finishes with good dimensional tolerances.

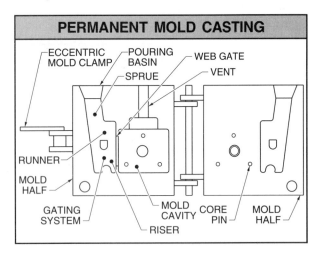

PERMANENT MOLD CASTING

Figure 25-5. Permanent mold casting requires a metal mold that may be reused many times.

Die Casting. Die casting is a casting process that consists of using substantial pressure to inject molten metal into the cavity of a metal mold. After the die has been closed and locked, the molten metal is delivered to a piston pump. The pump plunger is advanced to force the molten metal quickly through the feeding system. While the molten metal enters the die, the air in the die escapes through vents. Enough molten metal is introduced to overflow the die cavities, fill overflow wells, and develop some flash. Pressure is applied as the metal solidifies, and when the casting is solid, the die is opened and the casting is ejected.

With die casting, more complex shapes with thinner walls can be produced than with permanent mold casting. Surface finish is excellent and little finishing is required. Although high production rates are possible, die casting is limited to large repetitive lots. The process is generally limited to alloys that have low melting temperatures, such as those of aluminum and zinc.

Melting Furnaces

Melting furnaces provide the source of molten metal for castings. Melting furnaces include electric arc, induction, reverberatory, crucible, cupola, and vacuum furnaces.

Electric Arc Furnace. The *electric arc furnace* is a melting furnace that has a refractory-lined bowl in which material is melted by electric arcs from three carbon or graphite electrodes mounted in the roof of the furnace. Radiation from the walls of the furnace provide further heat to melt the charge. The electrodes are controlled automatically so that an arc of the proper height can be maintained.

In order to handle bulky charge materials, the molten bath is wider than it is deep. Reactions between the slag and the metal are efficient because of the wide surface area. The charge is fed to the furnace through the roof, and the electrodes are lowered into the charge to maintain the arc. Molten metal is removed by tilting the furnace.

Induction Furnace. The *induction furnace* is a melting furnace that is essentially a transformer, in that a coil of water-cooled copper tubing outside the furnace is the primary winding, and the metal charge inside the furnace acts as the secondary winding. The furnace is charged and an alternating current is passed through the coil. This induces a much heavier current in the charge, causing it to heat up and eventually melt. A second magnetic field is created by the induced current in the charge. Because these two fields are always in opposite directions, they create a mechanical force that causes metal movement and stirring. This stirring action causes excellent alloy and charge absorption into the melt, making it chemically and thermally homogeneous. Induction furnaces produce rapid melting with little metal loss.

The two basic induction furnaces are the coreless type and the channel (core) type. See Figure 25-6. The coreless induction furnace is used more frequently and consists of a water-cooled copper coil that completely surrounds a refractory-lined crucible.

In the channel induction furnace, the coil surrounds a small appendage of the unit, which is called an inductor. The molten metal forms a loop (channel) within the inductor. The volume of molten metal is relatively small compared with the bulk of metal in the furnace, and it must be pumped out of the loop to heat the charge in the furnace. For this reason, the channel furnace must be started with a charge of molten metal in the loop.

Reverberatory (Hearth) Furnace. The *reverberatory furnace* is a melting furnace consisting of a large, refractory-lined hearth that is heated from above by an open flame. The charge is melted by radiation from the hot walls and roof and by convection from the movement of hot gases. Reverberatory furnaces are principally used to remelt large charges or scrap metal and are also as holding furnaces.

Crucible Furnaces. The *crucible furnace* is a melting furnace consisting of a refractory-lined pot that is externally heated by the combustion of a wide variety of fuels. It is extremely versatile and is generally used to handle small charges. See Figure 25-7.

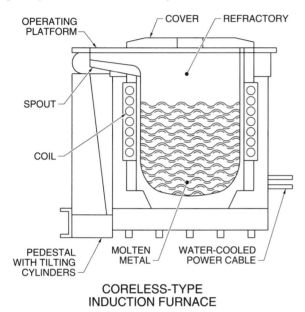

CORELESS-TYPE
INDUCTION FURNACE

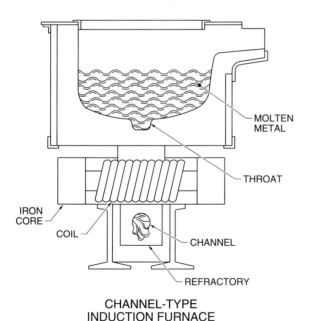

CHANNEL-TYPE
INDUCTION FURNACE

Figure 25-6. The basic types of induction furnace are the coreless and the channel (core) types.

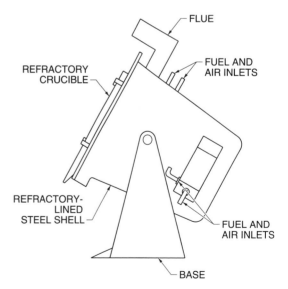

CENTER-AXIS TILTING CRUCIBLE FURNACE

Figure 25-7. The crucible furnace is extremely versatile and is generally used to handle small charges.

Cupola. The *cupola* is a cylindrical vertical shaft furnace that burns metallurgical coke. This effect is intensified by the blowing of air through tuyeres. Alternate layers of metal and replacement coke are charged into the top of the furnace. In its descent, the metal is melted by direct contact with the countercurrent flow of hot gases from combustion of the coke. The molten metal collects at the bottom, where it is discharged for casting by intermittent tapping or continuous flow.

Vacuum Melting Furnace. Vacuum melting furnace processes include vacuum arc remelting (VAR), vacuum induction remelting (VIR), and electron beam remelting (EBR). These process are normally performed for the production of refined ingots and billets. In some cases, vacuum melting furnaces are used to cast shapes in alloys, such as titanium, where no atmospheric contamination can be tolerated.

CASTING QUALITY

The manner in which metal solidifies exerts a profound influence on the quality and properties of castings. Castings are fundamentally different from wrought metals and exhibit characteristic types of defects. Castings are designed to take advantage of the unique characteristics of the process.

Solidification

Factors that affect the solidification of metals are shrinkage and the freezing range. Shrinkage and the freezing range also influence the characteristic types of casting defects.

Shrinkage. *Shrinkage* is a volume reduction in the casting, accompanying the temperature drop from the pouring temperature to room temperature. Three phases of shrinkage occur during solidification. These are liquid shrinkage, solidification shrinkage, and solid shrinkage. See Figure 25-8.

Liquid shrinkage is shrinkage that occurs as the metal cools from the pouring temperature to the liquidus temperature, which is a range of 95°C to

150°C (200°F to 300°F). Liquid shrinkage has no detrimental effect and is of no concern.

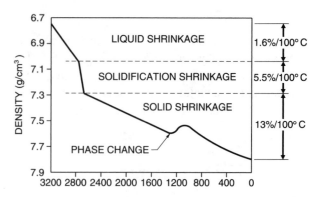

Figure 25-8. Three stages of shrinkage occur during the solidification of a casting.

Solidification shrinkage is shrinkage that occurs as the solidified metal contracts when it cools. Solidification shrinkage causes the most problems during casting. Most metals contract as they pass from the liquid to the solid state. Certain compositions of gray and ductile iron are exceptions to this rule. Solidification shrinkage leads to voids and microporosity unless the casting is adequately fed with molten metal. See Figure 25-9. Risers are incorporated into the mold to provide reservoirs of molten metal to prevent solidification shrinkage and to promote directional and progressive solidification. Exothermic feeding aids are also used to prevent solidification shrinkage.

Solid shrinkage must be compensated for in order to produce a casting of the specified dimensions. To achieve this, the pattern is made slightly larger than the casting dimensions at room temperature. For example, the patternmaker allows and extra ⅛ inch per foot (in./ft) for cast iron and ¼ in./ft for cast steel. Since the amount of solid shrinkage is a function of the specific alloy involved, problems might occur when substituting different alloys for the same pattern.

In casting design, solid shrinkage must also be considered because it is one of the primary causes of warped and cracked castings. These problems result from stresses that are created as the casting cools to room temperature.

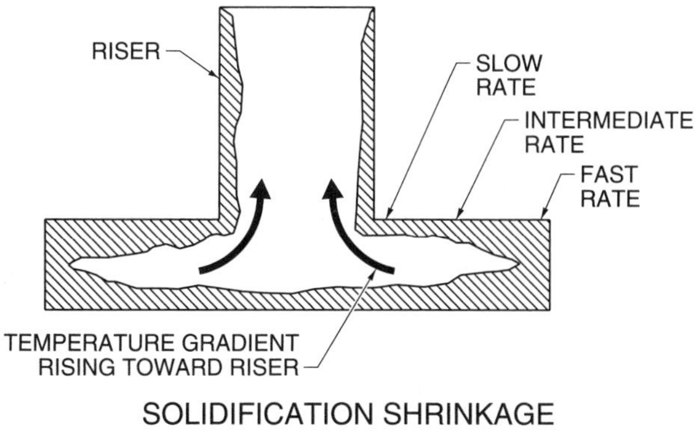

Figure 25-9. Solidification shrinkage leads to voids and microporosity unless the casting is properly fed.

Freezing Range. The *freezing range* is the difference between the solidus and liquidus temperatures. Metals are classified as having a short freezing range or long freezing range. When a metal is poured into a mold, it first freezes by forming a solid skin of equiaxed grains in contact with the mold wall, which is referred to as the chill zone. What happens next depends on whether the metal has a short freezing range or a long freezing range. See Figure 25-10.

Pure metals and eutectic alloys have a short freezing range. They solidify progressively inward, forming columnar grains that grow perpendicular to the mold wall. Short freezing range alloys tend to be pressure-tight because there is less chance of shrinkage voids developing within the grains of solidifying metal. The columnar grains may lead to planes of weakness because impurities tend to be rejected to the solid-liquid interface between solidifying grains.

Alloys with long freezing ranges tend to form solid nuclei within the melt ahead of the solidifying front. These nuclei create an equiaxed grain structure, which begins to form and grow throughout the solidifying mass. This leaves little room for columnar grains. Alloys with long freezing ranges exhibit a mixed grain structure that consists mainly of equiaxed grains. These alloys are characterized by microporosity or inferior pressure tightness, coupled with a uniform distribution of impurities, and less opportunity for planes of weakness to develop.

The cooling rate and temperature gradient in the solidifying casting also help determine the grain size. Molds with high thermal conductivity, such as metal molds, quickly and continuously remove heat from the casting. This results in more rapid cooling and a finer grain size. By contrast, castings poured into sand molds, which have relatively low thermal conductivity, have coarser grain size. The use of chills incorporated into the mold wall encourages finer grain size.

Inoculants greatly increase the number of nuclei available for solidification and encourage an equiaxed grain structure. When inoculants are added to the melt, the chill zone and columnar grains tend to be diminished, whether the alloy has a short or long freezing range.

Casting Defects

Casting defects are discontinuities in castings that exhibit a size, shape, orientation, or location that makes them detrimental to the useful service life of the casting. Some casting defects are remedied by minor repair or refurbishing techniques, such as welding. Other casting defects are cause for rejection of the casting. Casting defects are categorized as metallic projections, cavities, discontinuities, defective surfaces, incomplete castings, incorrect dimensions, and inclusions. See Appendix.

Metallic Projections. Metallic projections include fins (flash), swells, and scabs. *Fins* are excessive amounts of metal created by solidification into the parting line of the mold. Fins are removed by grinding or sandblasting. *Swells* are excessive amounts of metal in the vicinity of gates or beneath the sprue. Scabs are surface slivers caused by splashing and rapid solidification of the metal when it is first poured and strikes the mold wall.

Cavities. *Cavities* are rounded voids with smooth walls that are internal or intersect the surface of the casting. The various types of cavities are blowholes, pinholes, shrinkage cavities, and porosity. Blowholes and *pinholes* are holes formed by gas entrapped during solidification. *Shrinkage cavities* are cavities that have a rougher shape and sometimes penetrate deep into the casting. Shrinkage cavities are caused by lack of proper feeding or nonprogressive solidification. Porosity is pockets of gas inside the metal caused by microshrinkage, such as that ocurring from dendritic shrinkage during metal solidification.

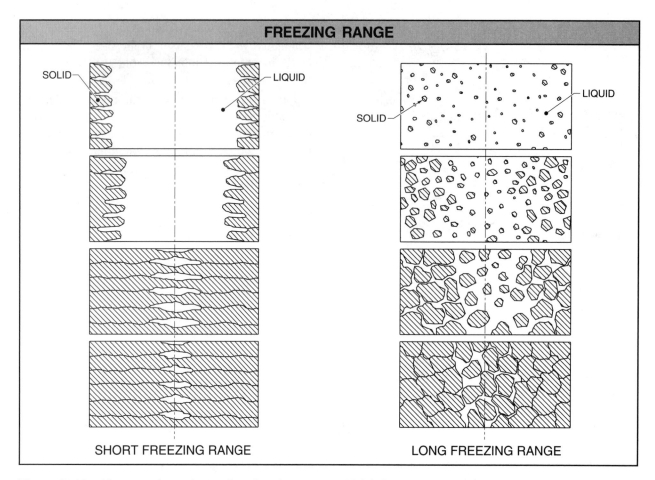

Figure 25-10. Alloys may have short or long freezing ranges, which influences the way they solidify.

Discontinuities. *Discontinuities* are cracks in castings and are caused by hot tearing, hot cracking, and lack of fusion (cold shut). A *hot tear* is a fracture formed during solidification because of hindered contraction. A *hot crack* is a crack formed during cooling after solidification because of internal stresses developed in the casting. Hot cracks are less open than hot tears and usually exhibit less oxidation and decarburization along the crack surface. *Lack of fusion* is a discontinuity caused when two streams of liquid in the solidifying casting meet but fail to unite. Lack of fusion exhibits rounded edges that indicate poor contact between various metal streams during filling of the mold.

Defective Surfaces. *Defective surfaces* are casting surface irregularities that are caused by incipient freezing from too low a casting temperature. Examples of defective surfaces include wrinkles (cold lap), depressions, and adhering sand particles.

Incomplete Castings. *Incomplete castings* are castings with missing portions. Incomplete castings are caused by premature solidification or fracture after solidification.

Incorrect Dimensions. *Incorrect dimensions* are any deviations from the mechanical drawing in terms of dimensions or alignment. Incorrect dimensions cause misfitting components that require extensive rework or cause rejection of the casting.

Inclusions. Inclusions are particles of foreign material in the metal matrix. The particles are usually nonmetallic compounds but may be any substance that is not soluble in the matrix. Slag, dross, and flux inclusions arise from melting slags, products of metal treatment, or fluxes. These are often deep within the casting. Mold or core inclusions come from sand or mold dressings and are usually found close to the surface.

Casting Specifications

Castings are specified differently from wrought materials of similar composition. Casting specifications are influenced by the basic differences between the cast and wrought alloys, design considerations, and inspection requirements.

Cast and Wrought Alloys. Cast alloys are significantly different from their equivalent wrought alloys. The distinctive metallurgical characteristics of castings are acquired during solidification, whereas with wrought materials, they are acquired during mechanical deformation. The primary differences between castings and wrought metals are related to their homogeneity, composition, residual stresses, and anisotropy.

The principal metallurgical difference between castings and wrought materials is that castings lack homogeneity. Castings exhibit coring or segregation. Castings are less homogeneous than wrought alloys because they do not have the benefit of hot work to accelerate the diffusion of the chemical elements to achieve homogenization. Compared with wrought alloys, cast alloys require significantly longer soaking times to achieve homogenization.

Compositional adjustments are also made for other reasons. For example, CF-8 stainless steel is compositionally adjusted to contain a small amount of ferrite in the austenite microstructure to prevent hot tearing. The equivalent wrought alloy, 304 stainless steel, is wholly austenitic in microstructure. The composition of cast alloys are adjusted from their wrought equivalents to improve castability. Cast alloys frequently contain more silicon to improve the fluidity of the molten metal.

Minor elements are sometimes present in castings from the nucleants that are used to refine grain size or alter the grain structure. Solidified castings contain high residual stresses from solid shrinkage, unless they are removed by a stress relief annealing process. Unlike wrought alloys, castings do not exhibit pronounced anisotropy.

Design. The design of castings exerts the most important influence over casting quality, cost, and service life. Factors affecting design are the solidification sequence of the shape, solidification shrinkage, solid shrinkage, and the complexity of the mold.

The solidification sequence of the design must be anticipated to ensure that risers are located in the correct positions and are the last items to solidify. This produces a casting that is free from shrinkage. The greater the solid shrinkage, the greater the problem in feeding the solidifying casting. Sharp corners and reentrant angles should be avoided because they encourage hot tears. See Figure 25-11.

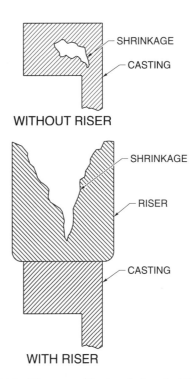

WITHOUT RISER

WITH RISER

Figure 25-11. Risers must be located so that the casting is completely filled and solidified.

In general, castings should be designed for uniform section thickness. This reduces problems with solid shrinkage that leads to excessive residual stresses and cold cracking. To reduce costs, straight mold parting lines should be the goal, and castings should be designed for ease of withdrawal from the mold. See Figure 25-12.

Inspection. In addition to being checked for dimensional accuracy, castings are inspected externally and internally for discontinuities and defects. In certain cases, rather than being rejected, castings are repaired.

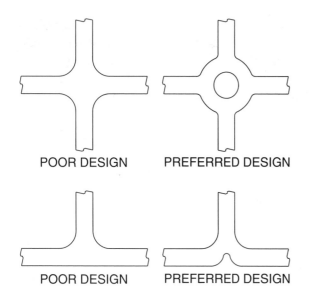

Figure 25-12. Castings must be properly designed to take advantage of their unique characteristics.

Several techniques are used to locate external discontinuities such as shrinkage, improper surface finish, cracks, porosity, and scabs. The inspection techniques include visual examination, liquid penetrant examination (PT), and magnetic particle examination (MT).

Visual examination is done immediately after shakeout and knockout. Visual examination is used to detect major imperfections that can quickly be relayed back to the foundry so that they can be corrected. Specific standards, such as *MSS SP-55 Quality Standard for Steel Castings for Valves, Flanges and Fittings, and other Piping Components*, are used to assist in the examination.

Liquid penetrant examination (PT) is performed principally to detect cracks on casting surfaces. PT is also used to ensure that machined surfaces are sound. If discontinuities are detected by visual examination, liquid penetrant examination will show them to their full extent.

Magnetic particle (MT) examination is used on ferromagnetic material. MT is a highly sensitive technique for indicating cracks at, or just below, the surface in a ferromagnetic material. MT is useful on surfaces that have been cleaned by shot blasting or sand blasting, where the force of the blasting may have made cracks that are too tight for liquid penetrant examination.

Internal inspection techniques include radiographic testing (RT), ultrasonic testing (UT), and leak testing. *Radiographic testing* is an internal inspection technique that is widely used on castings to locate internal shrinkage, porosity, and slag inclusions. Reference radiographs, such as *ASTM E186 Reference Radiographs for Heavy Walled Steel Castings*, are produced to gauge various types of defects in castings. RT is the preferred method of internal inspection.

Ultrasonic testing is an internal inspection technique that is used to detect cracks and measure wall thickness. Internal structural features in castings, such as coarse grain size, porosity, or finely dispersed precipitates, hinder interpretation. Consequently, ultrasonic testing is not commonly used to inspect castings.

Leak testing is an internal inspection technique that is performed on castings that are intended to withstand pressure. The most common is hydrostatic testing, which uses water. A commonly used criterion for hydrostatic testing is one and a half times the design pressure of the casting. Castings that are to be leak tested require fully machined flange faces so that they can be connected to the test flanges of the rig. The extra cost of machining may not be recoverable if the castings fail the hydrostatic test.

Minor defects in castings are often weld repaired. In some cases, impregnation of porosity with sodium silicate is permitted. Most specifications severely limit the amount and size of defects that are repaired. Corrosion-resistant alloys, such as stainless steels, may require a full solution-annealing process followed by quenching, depending on the amount of weld repair that is performed. This process restores optimum corrosion resistance.

Powder Metallurgy, Forming, Machining and Grinding 26

Powder metallurgy (P/M) is the manufacturing of shapes and components from metal powder mixtures. Powder metallurgy offers economical processing of high-volume components close to final dimensions. Forming (metalworking) consists of processes that deform metals using tools or dies to produce shapes or components. Forming is divided into primary and secondary forming processes. Primary forming is mainly used to make primary shapes and secondary forming is used to make finished products. Machining consists of surface removal processes that are used to shape or finish materials.

POWDER METALLURGY

Powder metallurgy (P/M) is a metalworking process for forming precision shapes and components from metal and nonmetal powders or mixtures of the two. Powder metallurgy offers economical processing of high-volume components made exactly to, or close to, final dimensions. When desired, the components can be re-pressed to closer tolerances. Components can be impregnated with oil or plastic, or infiltrated with a material that has a lower melting point. Components can be heat treated, plated, or machined. Production rates range from several hundred to several thousand components per hour.

Process Steps

The process steps of making components by powder metallurgy consists of mixing, compacting, sintering, and secondary operations. Additional steps are employed in specialized cases. See Figure 26-1.

Mixing. Mixing the raw materials is the critical first step of powder metallurgy. Powders consist of elemental metals or alloys and are produced by the reduction of metal oxides, electrolysis, or atomiza-

tion of liquid metals. The powders are mixed together with lubricants or other alloy ingredients to produce a homogeneous blend.

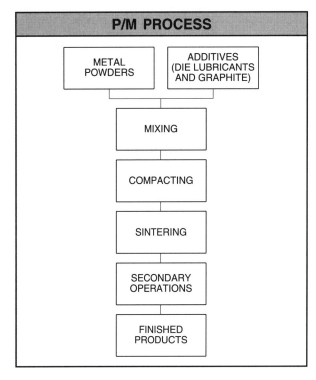

Figure 26-1. Powder metallurgy (P/M) consists of several steps.

Many types of mixers and blenders are used for powder metallurgy. Batch mixers are the most common and include drum, cubic, double cone, twin shell (V), and conical screw (rotating auger) types. The selection of the optimum mixer for a given powder requires careful consideration. Testing must be performed for each case.

Compacting. Compacting consists of automatically feeding a controlled amount of mixed powder into a precision die, after which it is compacted. See Figure 26-2. Compacting is usually performed at room temperature. Pressures range from 10 tons per square inch, or tons/in.^{2}, (138 MPa) to 60 tons/in.^{2} (827 MPa), or more.

Compacting densifies the loose powder into green compact. *Green compact* is an unsintered powder metallurgy compact. The green compact has the size and shape of the finished product when it is ejected from the die. The green compact has enough strength for in-process handling and transportation to the sintering furnace.

Sintering. *Sintering* is the bonding of the particles of the green compact by heating it at a high temperature that is below the melting point of the metal. Representative powder metallurgy sintering temperatures are 1120°C (2050°F) for ferrous alloys, 815°C (1500°F) for bronze alloys, and 1205°C (2200°F) for stainless steels. Typical sintering atmospheres include endothermic gas, exothermic gas, dissociated ammonia, and nitrogen.

Sintering develops metallurgical bonds between powder particles by means of solid-state diffusion. This helps develop the required mechanical properties in the component. Sintering also serves to remove lubricant, reduce oxides, and control the carbon content on the surface and inside the component. In many applications, powder metallurgy components are ready for use after sintering.

Spark sintering is a variation of sintering that is performed with the green compact or powder placed between two punches that also serve as electrodes for conducting a low-voltage, high-amperage current. The powder is heated by the electric current and simultaneously pressed. Spark sintering is mainly used for pressing beryllium powders and titanium alloy powder compacts using graphite molds.

Secondary Operations. Secondary operations are other processes applied to P/M components after sintering to provide or improve specific properties. Components can be re-pressed, impregnated with oil or plastic, or infiltrated with another metal. Components can be finished by machining, tumbling, plating, and heat treating.

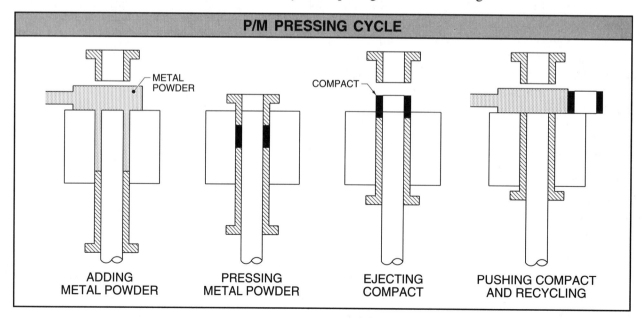

P/M PRESSING CYCLE

METAL POWDER

COMPACT

ADDING METAL POWDER   PRESSING METAL POWDER   EJECTING COMPACT   PUSHING COMPACT AND RECYCLING

Figure 26-2. Compacting consolidates and densifies the component for transportation to the sintering furnace.

Re-pressing is the application of pressure to a previously pressed and sintered P/M component. It is used to increase density, provide greater dimensional accuracy, and improve surface smoothness and hardness. In re-pressing, the sintered P/M component is inserted into a confined die and struck by a punch. The principal reason for re-pressing is to correct distortions that occur during sintering. Re-pressing may be followed by a second sintering operation in order to improve mechanical properties and relieve the stresses induced by the cold work performed during re-pressing.

Oil impregnation of P/M bearings in automobiles is one of the most widely used secondary operations. P/M bearings can hold as much as 10 to 30 times the volume percent of oil. Plastic impregnation is used to make P/M components impervious to liquids or gases, or to make them suitable for plating. Infiltration consists of filling the pores of the P/M component with a molten metal or alloy. The melting point of the infiltrated metal must be much lower than that of the P/M component. The advantages of infiltration are improved mechanical properties, even density variations, increased density, and removal of porosity.

P/M components are finished by operations such as plating, coating, tumbling, burnishing, and coloring. Whenever possible, P/M components are made to final dimensions to eliminate the need for machining. Components with features such as threads, grooves, undercuts, or sideholes require machining operations. The machining characteristics of P/M components are similar to those of cast materials with the same composition.

Special P/M Operations. Special P/M operations include cold isostatic compacting (cold isostatic pressing) and hot isostatic compacting (hot isostatic pressing, or HIP). See Figure 26-3. These special P/M operations apply high pressures simultaneously from all directions on the P/M compact. The metal powder is placed in a flexible mold or container and immersed in a fluid bath or gas within a pressure vessel. The fluid or gas is put under high pressure, which compacts the metal powder. Isostatically compacted components are characterized by their uniform, high density.

Cold isostatic compacting is performed at room temperature. The pressure is applied uniformly to a deformable container that holds the metal powder to be compacted. The technique is especially suitable for manufacturing components that have a large length-to-diameter ratio.

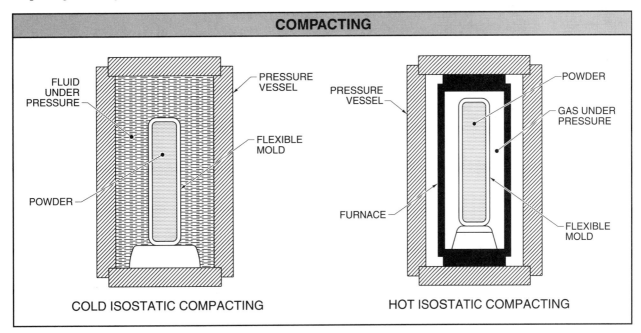

Figure 26-3. Cold isostatic pressing is performed at room temperature with liquid as the pressure medium. Hot isostatic pressing is performed at elevated temperature with gas as the pressure medium.

Hot isostatic compacting is compacting performed using compressed gas at high temperature and pressure. Hot isostatic compacting is capable of temperatures up to 2000°C (3630°F) and pressure up to 30,000 psi (205 MPa). The powder is placed in a flexible, gas-tight capsule that has the shape of the finished component. The capsule is placed in a pressure vessel in which it is heated and subjected to isostatic pressure, which is applied by high-purity argon or helium gas. The component is compressed equally on all sides by the application of the heat and pressure.

P/M Products

Materials that are produced by P/M include iron powders, stainless steels, high-speed tool steels and cemented carbides, superalloys, refractory metals, aluminum alloys, and copper alloys. Iron powders represent the bulk of the industry.

Iron Powders. While iron powder alone is used in some applications, small additions of other powders, such as carbon, copper, or nickel, are added individually or in combination to improve the mechanical properties of the component. Plain carbon steels are made from mixtures of iron and graphite. When the parts of the components are sintered, carburization produces a carbon steel structure with up to .7% C. Increasing the copper content increases strength and hardness. Nickel steels contain from 2% Ni to 8% Ni, with or without copper. Nickel steels have high strength, toughness, and fatigue resistance.

Stainless Steel Powders. Stainless steel powders are sometimes consolidated into billets prior to hot extrusion to form seamless tubing. The advantages of this process over casting and extrusion are a lower inclusion count due to the cleanliness of the powder, closer control of chemical composition, and a more homogeneous structure. Furthermore, the metal is sound and no pores are present. The porosity of conventionally produced stainless steel P/M components may decrease corrosion resistance in specific media.

High-speed Tool Steels and Cemented Carbides. High-speed tool steels and cemented carbides, such as T15, M2, M3, or tungsten carbide, are produced by hot isostatic pressing and are subsequently forged and rolled. The main advantage of P/M over melting and forming is the ability to produce a relatively fine and uniform dispersion of the carbide phase.

Superalloys. Superalloys used for high-performance jet engine rotor disks may be fabricated by P/M. One advantage of P/M is the reduction in the number of processing steps compared with other manufacturing methods.

Refractory Metals. Refractory metals, such as tungsten, tantalum, and molybdenum, are fabricated to wrought intermediate shapes from powder compacts. Frequently, the powders are formed into large green billets by cold isostatic pressing, vacuum sintered at high temperature to close porosity, and then worked by processes such as forging, rolling, or swaging.

Aluminum Alloys. Aluminum alloy P/M is sometimes competitive with many types of aluminum castings, extrusions, and screw-machine products. The corrosion resistance of aluminum P/M products can be improved through the application of chemical conversion coatings or anodizing.

Copper Alloys. Copper alloys produced by P/M are used for applications such as self-lubricating bronze bearings. These bearings are produced from mixtures of elemental copper and tin powders. Most other copper alloy powders are produced from powders made of specific alloys.

PRIMARY FORMING PROCESSES

Primary forming processes are metal deformation methods that are used to make primary shapes. Primary shapes may be subjected to secondary forming processes to make finished products. In some cases, primary forming processes are used to make finished products, such as certain types of forgings.

Primary and secondary forming processes may be performed hot or cold. This depends on whether the operation is performed above or below the recrystallization temperature of the material. Hot forming processes are performed above the recrystallization temperature, and cold forming processes are performed below it. Primary forming processes consist of forging, rolling, and extrusion.

Forging

Forging is the process of working metal to the desired shape by impact or pressure. The process is performed using hammers, upsetters, presses, rolls, and related equipment. The principal advantage of forging is that it orients the grain flow to the contour of the component. This often provides the highest strength in the direction of greatest stress. The higher strength-to-weight ratio achieved results in the use of smaller or lighter components. See Figure 26-4.

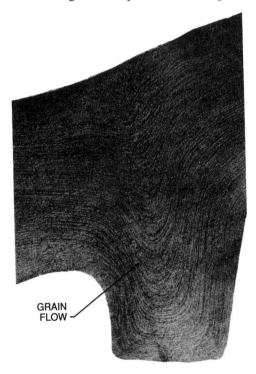

GRAIN
FLOW

Figure 26-4. A macroetched section through a forging indicates that the grain flow follows the contour of the component, which often maximizes strength in the direction of greatest operating stress.

Some disadvantages of forging are the high cost and high residual stress produced. Most forging processes are expensive because of the cost of making dies, so long production runs are usually necessary to reduce costs. The high residual stresses in forgings are often released when they are machined and cause warping when heavy cuts are taken. Forgings should be completely roughed out oversized before any finishing cuts are performed.

Forging processes include open die forging, impression die forging, roll forging, upset forging, ring rolling, rotary (orbital) forging, and radial forging. Forgings may exhibit characteristic defects and are inspected using specific techniques.

Open Die Forging. *Open die forging* is hot mechanical forming between flat or shaped dies in which the metal flow is not completely restricted. The stock is laid on a flat anvil while the flat face of the forging hammer is struck against the stock. Depending upon its size, the stock may be turned by hand using tongs between blows of the hammer. Open die forging allows stock to be rounded or lengthened, and allows thick pieces to be flattened.

Open die forging is also referred to as smith forging, blacksmith forging, hand forging, or flat die forging. The equipment may range from the anvil and hammer to giant hydraulic presses. Most open die forgings are simple geometric shapes, such as disks, rings, or shafts. The process is also used in the steelmaking of ingots or reduced billets. Most forging processes begin with open die forging.

Impression Die Forging. *Impression die forging* is the shaping of hot metal within the cavities or walls of two dies that come together to completely enclose the workpiece. The impression for the forging can be entirely in either die or divided between the top and bottom dies. As the deformation continues, a small amount of metal begins to flow outside the die impression, forming flash. The thin flash cools rapidly, creating a pressure increase inside the workpiece, which assists the flow of metal into the unfilled portion of the impression. See Figure 26-5.

A special type of impression die forging is closed die forging. Closed die forging does not depend on the flash to achieve complete die filling. The material is deformed in a cavity that does not allow excess material to flow outside the impression. In impression die forging, the design and workpiece volume are more critical than in closed die forging.

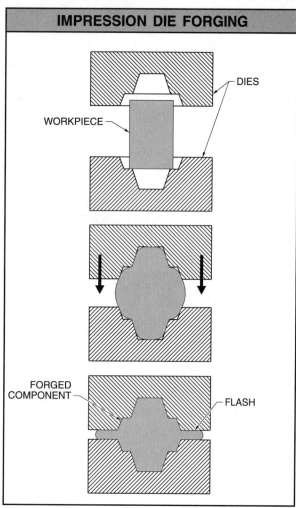

Figure 26-5. In impression die forging, the workpiece is shaped within the cavities of two dies that come together to completely enclose it.

Roll Forging. *Roll forging* is the shaping of stock between two driven rolls that rotate in opposite directions and have one or more matched sets of grooves in the rolls. The stock enters the rolls when the grooves rotate into position. Roll forging is used to produce components with long, tapered, symmetrical sections. Crankshafts and connecting rods are frequently roll forged prior to being finish formed. See Figure 26-6.

Upset Forging. *Upset forging* is a process in which pressure is used to gather a large amount of material

at one end of a bar. It is performed on a horizontal machine into which the bar is inserted between an upsetting head and a fixed anvil plate.

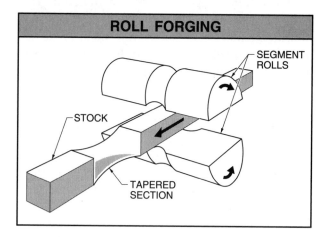

Figure 26-6. Roll forging is used to produce components with long, symmetrical sections, often with a slight taper, by forcing the workpiece between two matched and grooved rolls that rotate in opposite directions.

The upsetting head pushes the bar against an anvil plate, flattening it and developing maximum isotropy. Upset forging can be performed hot or cold. See Figure 26-7.

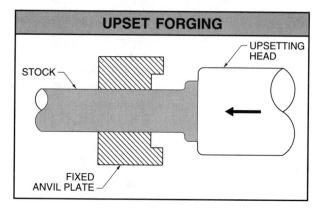

Figure 26-7. Upset forging is used to gather a large amount of material at one end of a bar, such as in the forming of a wrench socket.

Ring Rolling. *Ring rolling* is the shaping of seamless rings by reducing the cross section and increasing the circumference of a heated, donut-shaped blank between two rotating rolls. The ring may be of any desired cross section. Seamless rings are produced in diameters that range from several inches to over 20 ft. Ring rolling is less expensive than closed die forging because there is less material and the ring is much closer to the finished shape. Ring rolling is often used to make flanges. See Figure 26-8.

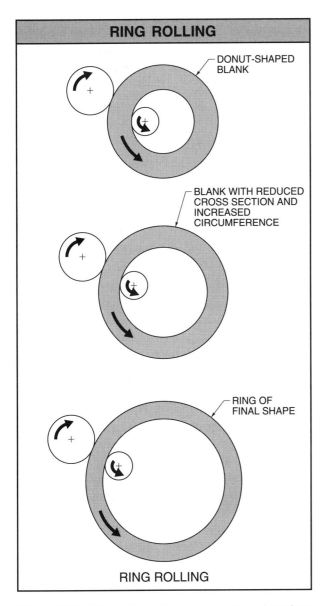

Figure 26-8. Ring rolling is the shaping of seamless rings by reducing the cross section and increasing the circumference of a donut-shaped blank.

Rotary (Orbital) Forging. *Rotary forging* is a process in which the workpiece is subjected to a combined rolling and pressing action between a bottom platen and a swiveling upper die with a conical working face. The cone axis is inclined so that the narrow sector in contact with the workpiece is par-

allel to the lower platen. As the cone rotates about its apex, the contact area also rotates. This causes the workpiece to be compressed by the rolling action. See Figure 26-9.

The primary advantage of rotary forging is that much smaller forging forces are required due to the small area of contact with the workpiece. Rotary forging can be used to form intricate shapes to a high degree of accuracy.

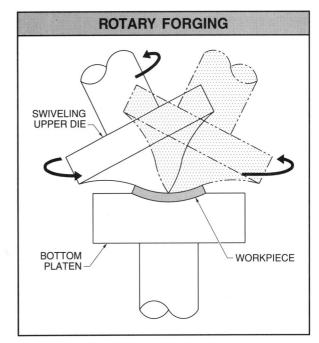

Figure 26-9. Rotary forging requires significantly less force compared with other forging processes.

Radial Forging. *Radial forging* is a process using two or more moving dies or anvils for reducing the cross section of round billets, bars, and tubes. See Figure 26-10. The workpiece is continuously rotated between the blows of the die.

Forging Defects. Forging defects are characteristic flaws that are produced by forging processes. The most common of these are laps, seams, hot tears, bursts, and thermal cracks.

Laps are surface irregularities caused by hot metal folding over and being pressed into the surface. Although the folds are forged into the surface, they are not metallurgically bonded because of the oxide present between the surfaces.

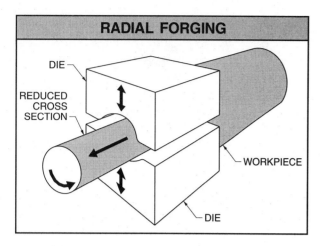

Figure 26-10. Radial forging is used to produce components having a circular cross section.

Seams are surface irregularities that result from a crack, a heavy cluster of nonmetallic inclusions, or a deep lap (a lap that intersects the surface at a large angle). Seams may also be caused by a defect in an ingot surface that has become oxidized and is prevented from welding up during forging.

Hot tears are surface cracks that are caused by rupture of the material during forging from the presence of low melting or brittle phases. Hot tears usually have a ragged appearance.

Bursts are internal ruptures arising from tensile stresses produced by the forging operation. The tensile stresses can be high enough to tear the material apart internally, particularly if the forging temperature is too high.

Thermal cracks are internal or external cracks that occur as a result of nonuniform temperatures in the forging. For example, thermal cracks are present if the forging is heated up too rapidly. The unequal temperatures of the surface and the center of the mass, which results in a difference in thermal expansion, produce tensile stresses near the center. Large section sizes and poor thermal conductivity favor thermal cracking.

Hydrogen Cracking. *Hydrogen cracking* is damage caused by hydrogen present in the alloy. Many of the high-strength steels used in forgings are susceptible to embrittlement by hydrogen. The hydrogen enters the metal from various sources, such as the reaction of water vapor at high temperatures with the liquid metal during melting and pouring.

Hydrogen cracking often occurs in the form of hydrogen flakes, which are internal fissures, parallel to the forging direction. Hydrogen flakes have a bright, shiny appearance.

Inspection of Forgings. Forgings are inspected with destructive and nondestructive techniques. Macroetching is the principal destructive technique. A slice is cut from one end of the forged billet. The slice is ground, macroetched, and examined for grain flow and breakdown of the cast structure.

Nondestructive techniques include liquid penetrant examination, magnetic particle inspection, and ultrasonic testing. Liquid penetrant examination is used for detecting surface discontinuities, such as cracks, which may not be visible to the naked eye. Magnetic particle inspection is used on ferromagnetic materials to detect discontinuities on the surface or slightly below it. Ultrasonic testing is used to locate and measure internal discontinuities.

Rolling

Rolling is the reduction of the cross-sectional area of metal stock, or the general shaping of metal products through the use of metal rolls. Rolling is described by the roll design and the roll configuration. See Figure 26-11. The principal rolling processes are hot rolling and cold rolling. Hot rolling is the most common method of refining the cast structure of ingots and billets to make primary shapes. Cold rolling is most often a secondary forming process that is used to make bar, sheet, and strip. In softer metals, foil is also produced by cold rolling. The main reasons for cold rolling are to obtain a good surface finish and to improve mechanical properties. The machinability of steel improves with cold rolling. For this reason, cold-rolled bar stock is widely used for fast automatic-machining operations.

Rolls are mostly made of chilled or alloyed cast iron or low-alloy steel. Cast iron rolls are extremely hard and low in cost, but their strength is low. Alloy steel rolls are more expensive but have superior strength, rigidity, and toughness. Rolls are refurbished by grinding.

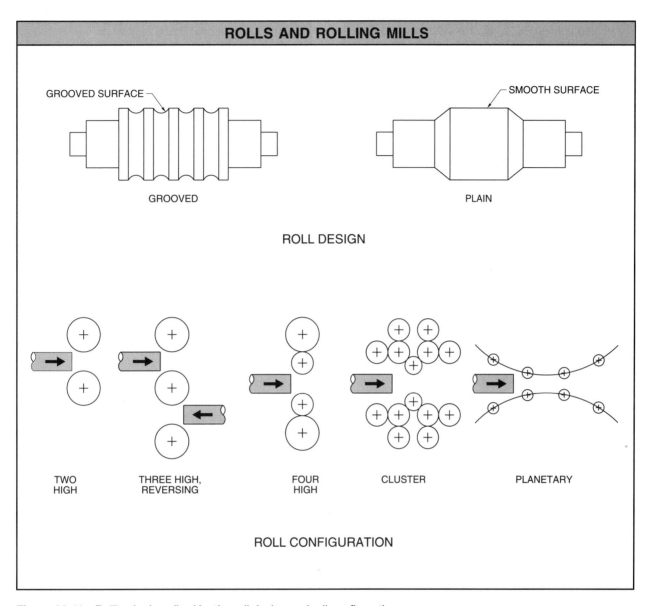

Figure 26-11. Rolling is described by the roll design and roll configuration.

Extrusion

Extrusion is the conversion of ingots or billets into lengths of uniform cross section by forcing the metal to flow plastically through a die by means of a ram. The primary methods of extrusion are forward and backward extrusion. See Figure 26-12.

In forward (direct) extrusion, the die and ram are at opposite ends of the extrusion stock, and the ram travels in the same direction as the extrusion stock. In backward (indirect) extrusion, the die and ram are at the same end of the extrusion stock and the extruded product either flows around the ram or up through the center of a hollow ram.

Some shapes that cannot be rolled because of their geometry, such as those with reentrant angles, are readily extruded. Extrusion is also a less expensive method of producing components in small quantities and may be performed hot or cold, or under impact.

Hot or Cold Extrusion. Hot extrusion uses heated billets and forward extrusion. Hot extrusion is usually performed on horizontal hydraulic presses. Practically all metals can be hot extruded, but extrudability varies with the deformation properties. Examples of alloys that are hot extruded are lead, tin, aluminum, magnesium, zinc, copper, steels, titanium, nickel superalloys, zirconium, beryllium, and molybdenum. Most metal powders can also be hot extruded.

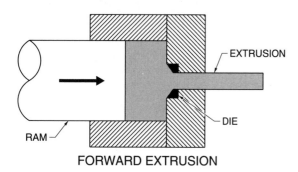

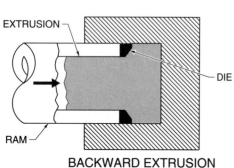

Figure 26-12. The primary methods of extrusion are forward extrusion and backward extrusion.

In cold extrusion, the metal billets are forced to flow by plastic deformation around punches and into or through shape forming dies. Most steels, nonferrous metals, and superalloys can be cold extruded. Cold extrusion is performed forward or backward. Cold extrusion requires less force if performed quickly, particularly with metals that work harden rapidly. Examples of cold-extruded components are cans, fire extinguisher cases, aluminum brackets, projectile shells, wrist pins, and steel gear blanks.

Impact Extrusion. Impact extrusion is a form of cold extrusion that is performed backward, forward, or by a combination of the two processes. See Figure 26-13. Fast ram speeds and short strokes are used in impact extrusion. High production rates are possible because of the impulsive force applied. Impact extrusion is used for nonferrous metals that have low melting points and good ductility. Applications of impact extrusion include the production of collapsible tubular containers (for example, toothpaste and artists' paint tubes), battery cases, cartridge cases, and beverage cans.

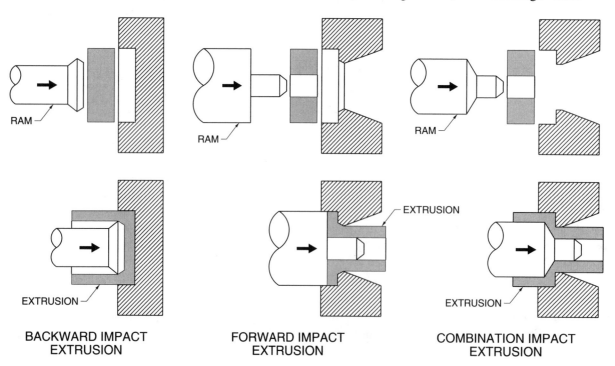

Figure 26-13. Fast ram speeds and short strokes are used in impact extrusion, and, because of the impulsive force applied, high production rates are possible.

SECONDARY FORMING PROCESSES

Secondary forming processes are metalworking processes used to make finished products. Secondary forming processes include tube, pipe, and bar bending; strip, sheet, and plate forming; metal stamping; and squeezing operations.

Tube, Pipe, and Bar Bending

Tube, pipe, and bar bending processes involve stressing the material beyond its yield strength but below its ultimate tensile strength. This is achieved by clamping the workpiece with the tooling while applying a bending force sufficient to cause a permanent set. When the applied force exceeds the yield strength of the material, the outside radius of the workpiece is stretched and the inside radius is compressed. The *neutral axis* is the boundary line between the tensile and compressive stresses in the workpiece. See Figure 26-14.

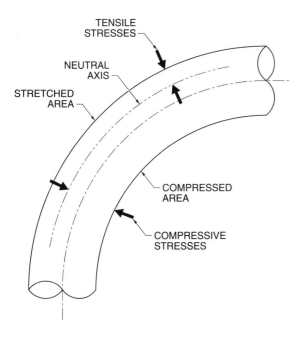

Figure 26-14. The neutral axis is the boundary line between tensile and compressive stresses in bending.

Alloys that work harden rapidly have more tendency toward springback, which can be overcome by slight overbending. Springback is the tendency of a component to return to its original shape after bending, even after forming in the plastic range.

Bending methods include draw bending, compression bending, and press bending. See Figure 26-15.

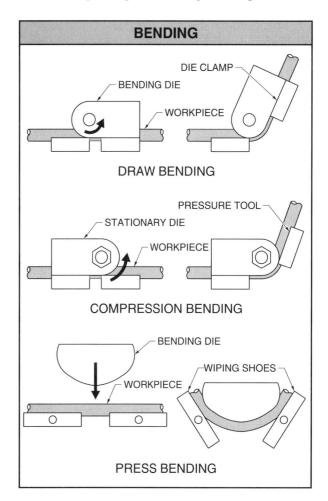

Figure 26-15. Bending methods include draw bending, compression bending, and press bending.

Draw Bending. Draw bending consists of clamping the workpiece against the bending die, locking it into position, rotating the die and clamp, and moving the workpiece through a pressure tool. The pressure tool can be a roller, sliding shoe, or static shoe. A flexible mandrel is often inserted in the pipe or tube to restrain flattening. Draw bending is the most suitable method for tight-radius bends and thin-wall pipe or tube. It is also used to bend extrusions.

Compression Bending. Compression bending consists of clamping the workpiece to a stationary die and wrapping the workpiece around it. It differs from draw bending because the die is stationary.

This method requires much less clamping area than draw bending because the pressure tool forces the material to flow and become wrapped around the die. Compression bending can produce bends in various planes with essentially no straight sections between the bends. Since the flow of metal in compression bending is not as well controlled as in draw bending, some distortion can occur when bending tubing, and mandrels cannot be used.

The main theoretical difference between compression and draw bending is the position of the neutral axis. In draw bending, the neutral axis lies within the inner third of the cross section. The significance of this is that approximately two thirds of the tube or bar cross section is in tension. With compression bending, the neutral axis is in the outer third. This accounts for the severe flattening effect of draw bending and the need for a mandrel.

Press Bending. Press bending is an extension of compression bending that is equivalent to simultaneously achieving two compression bends, each one half of the total desired bend angle. The bending form is mounted on the ram of a press, and two wiping, or pressure, shoes pivot or wipe the material around the bending form. Press bending is used primarily for high-volume production bending of tube, pipe, and bar when the wall thickness and radius are considerably large.

Strip, Sheet, and Plate Forming

Strip, sheet, and plate forming processes include roll bending, roll forming, spinning, and stretch forming. These processes use curving and stretching operations to produce desired shapes.

Roll Bending. *Roll bending* is the curving of plate, sheet, bars, and sections into cylinders or cylindrical segments by means of rolls. During roll bending, the material thickness does not change. Roll bending machines contain two or three rolls that rotate or bend the metal as it passes between them. See Figure 26-16. A special version of the two-roll machine is used to form truncated cones. Roll bending is used to form curved shapes for the fabrication of tanks, pressure vessels, hoppers, bins, appliances, and ordinance components.

Curving of metals in three-roll machines usually requires several passes, with roll adjustments between passes to obtain the desired curvature. Considerable operator skill is required to avoid deviations in roundness.

With two-roll machines, the top steel roll penetrates the bottom urethane rubber roll. Bending is accomplished under very high pressure, with the urethane rubber roll literally wrapping the workpiece around the top steel roll. The workpiece is shaped to the required curvature in one pass, and

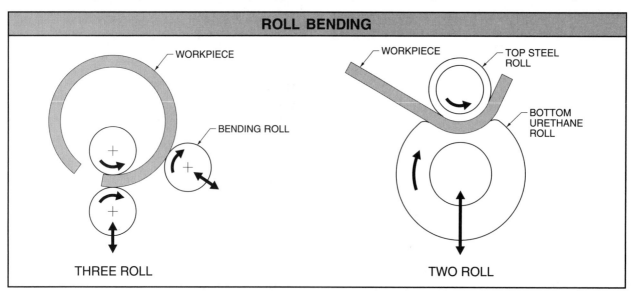

Figure 26-16. Roll bending is used to curve plate, sheet, and bars into cylinders or cylindrical segments using machines with two or three rolls.

consistently repeated accuracy is achieved from component to component.

Roll Forming. *Roll forming* is a continuous process for forming metal from sheet, strip, or coiled stock into shapes of uniform cross section. The material is passed between successive pairs of rolls, which progressively shape it until the desired cross section is produced. During this process, only bending takes place. The material thickness is not changed, except for a slight thinning at the bend radii.

Roll forming is used for the high-volume production of components such as metal window and screen frame members, bicycle wheel rims, furnace jacket rings, garage door trolley rails, metal molding, metal trim, and metal siding. Specific sets of rolls must be made up for each job.

Spinning. *Spinning* is a process of forming disks or tubing into cones, dish shapes, hemispheres, hollow cylinders, and other circular shapes by the combined forces of rotation and pressure. Spinning does not result in any change in thickness. The operation is achieved by forcing the workpiece to conform to a shaped mandrel that rotates concentrically with the component. See Figure 26-17.

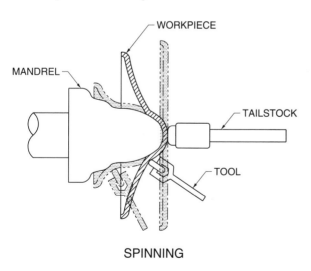

SPINNING

Figure 26-17. Spinning is a process of forcing disks or tubing into cones, disk shapes, hollow cylinders, and other circular shapes by combined forces of rotation and pressure.

Stretch Forming. *Stretch forming* is a process for forming sheet metal by applying tension, or stretch, and then wrapping it around a die of the desired shape. See Figure 26-18. Stretch forming is used most extensively in the aircraft industry to produce components of large radius of curvature, frequently with double curvature.

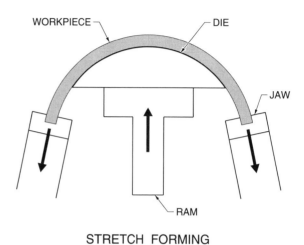

STRETCH FORMING

Figure 26-18. Stretch forming is a process for forming sheet metal by applying tension to it and then wrapping it around a die of the desired shape.

Metal Stamping

The stamping of components from sheet metal consists of a wide variety of operations that cut or shape metal through deformation by shearing, punching, drawing, bending, etc. The workpiece is pressed into a die by a punch to form the desired shape. See Figure 26-19. Production rates are high, and secondary machining operations are not usually necessary. Stampings are made from low-carbon steel, aluminum alloys, copper alloys, and other nonferrous metals. Stamping processes are widely used to make components for all kinds of equipment.

Squeezing Operations

Squeezing operations include cold heading; swaging; sizing, coining, and hobbing; thread rolling; and ironing. Squeezing operations are used to form ductile metals.

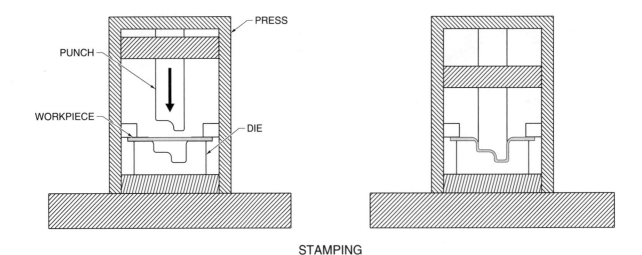

STAMPING

Figure 26-19. In metal stamping operations, sheet metal is formed into the desired shape by the punch and the die.

Cold Heading. *Cold heading* is the forcing of metal to cold flow into enlarged sections by endwise squeezing. See Figure 26-20. It is similar to upset forging, which does much the same type of hot work on larger pieces and harder-to-work materials. Typical cold headed components are tacks, nails, rivets, screws, and bolts under 1 in. in diameter, and a large variety of machine components, such as small gears with stems.

Swaging. *Swaging* is a squeezing process in which the material flows perpendicular to the applied force. Rotary swaging is the main swaging operation. Rotary swaging is a method of reducing the diameter of rods, bars, or tubes. A pair of tapered dies open and shut quickly while the workpiece is fed in rapidly as it rotates. The dies, which are inserted into slots in a spindle, are rotated and forced together repeatedly by the rollers around the periphery as much as several thousand times a minute. See Figure 26-21.

Swaging is usually performed cold so that the metal is work hardened. Rotary swaging is sometimes used to strengthen the end of a length of rod or tubing that is to be cold drawn to a smaller diameter. This prevents the rod or tubing from breaking when gripped and pulled through the dies.

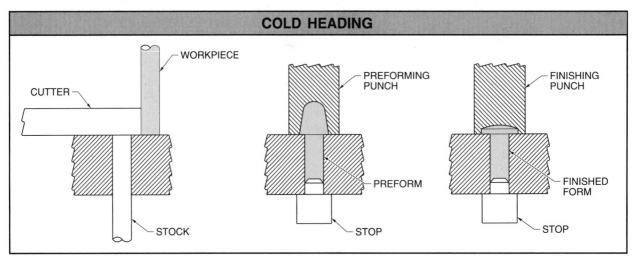

Figure 26-20. Cold heading is used to force metal to cold flow into enlarged sections by endwise squeezing.

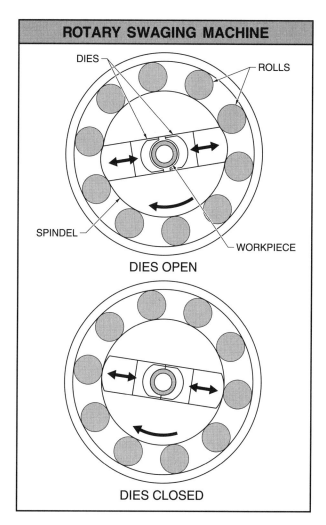

Figure 26-21. In rotary swaging, tapered dies open and shut rapidly, as much as several thousand times a minute, as the workpiece is fed in.

Sizing, Coining, and Hobbing. Sizing, coining, and hobbing are squeezing processes used for finishing. They change metal thickness and configuration by squeezing and working the metal beyond the yield point. See Figure 26-22.

Sizing is a squeezing process that is usually performed in an open die. Sizing is commonly used to sharpen corners on stampings, flatten areas around pierced holes, or develop exact dimensions along a specific axis.

Coining is a closed-die squeezing process in which all surfaces of the component are confined, which results in a well-defined imprint of the die on the component. Coining leads to an accurate reproduction of the die cavity, which is necessary in the production of coins.

Hobbing is a squeezing process used for making molds for the plastics and die castings industry. A punch (hob) is machined from tool steel in the shape of the cavity, heat treated to harden it, and polished. It is pressed into a blank of soft steel to make the mold. This is done slowly and carefully in several stages between annealing operations on the blank. A retaining ring keeps the mold from spreading out of shape. One hob can make several cavities in a series of molds.

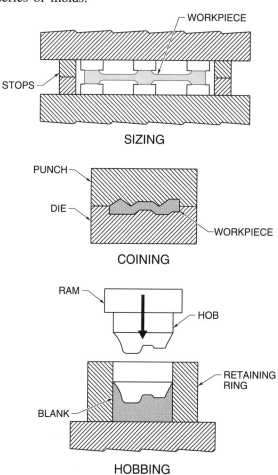

Figure 26-22. Sizing, coining, and hobbing are finishing operations that change metal thickness and configuration by squeezing and working the metal beyond its yield point.

Thread Rolling. *Thread rolling* is the production of threads by rolling the workpiece between two grooved die plates, one of which is in motion, or between rotating grooved circular rolls. See Figure 26-23. The workpiece is depressed to the root (the root diameter of the thread) and raised to fill the crest of the thread. Thread rolling is most commonly used for making screw threads.

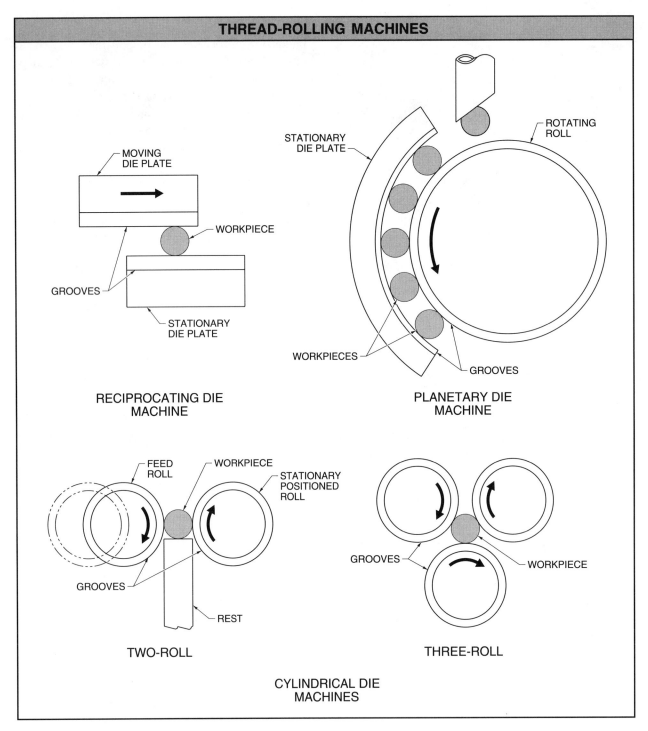

THREAD-ROLLING MACHINES

MOVING
DIE PLATE

WORKPIECE

GROOVES

STATIONARY
DIE PLATE

RECIPROCATING DIE
MACHINE

STATIONARY
DIE PLATE

ROTATING
ROLL

WORKPIECES

GROOVES

PLANETARY DIE
MACHINE

FEED
ROLL

WORKPIECE

STATIONARY
POSITIONED
ROLL

GROOVES

REST

TWO-ROLL

GROOVES

WORKPIECE

THREE-ROLL

CYLINDRICAL DIE
MACHINES

Figure 26-23. Thread rolling is the production of threads by rolling the workpiece between two grooved die plates and is most commonly used for making screw threads.

Ironing. *Ironing* is a process for smoothing and thinning the wall of a shell or cup by forcing the component through a die with a punch. See Figure 26-24. Cold working is severe, and annealing is often necessary between ironing steps.

MACHINING

Machining encompasses procesess that remove material in order to produce components with the required shape and dimensions. Machining includes processes that precisely shape a workpiece by removing some of the workpiece material using various types of tools.

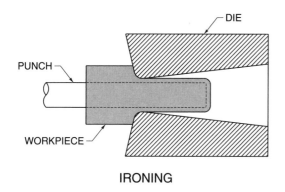

Figure 26-24. Ironing is a process for smoothing and thinning the wall of a shell or cup.

Machinability

In machining, a surface layer of constant thickness (chip) is removed by the relative movement between the cutting tool and the workpiece. Machinability is affected by a number of important criteria. These include tool life, surface finish, size control and sensitivity to changes in speed, and feed or tool angles. The machinability of a material is measured by comparing its performance with another material that is considered to be a standard.

The microstructure of an alloy can affect its machinability, and specific elements may be added to alloys to improve machinability. Such compositions are referred to as free machining or free cutting. They allow high cutting speeds with long tool life, well broken up chips, and smooth, accurate finishes.

Elements that form soft inclusions, such as sulfur, lead, phosphorus, selenium, and tellurium, are added to alloys to improve machinability. Generally, these elements make the alloys susceptible to hot shortness. *Hot shortness* is the tendency for a metal to separate along grain boundaries when stressed or deformed at temperatures near the melting point of the metal. As a result, free-machining alloys should not be welded.

Elements that form hard abrasive compounds decrease machinability. Examples include carbide forming elements in steels and also aluminum and silicon, which form alumina and silica, respectively.

Very soft alloys or alloys that work harden rapidly are also difficult to machine. Most machining processes remove metal by chip formation. A wide variety of tool materials are used for machining.

Chip Formation. In traditional machining processes, the cutting tool is driven through the component to remove chips from it by a shearing process and leave geometrically true surfaces. The cutting tool is characterized by two angles, which are the relief (clearance or α) angle and the rake (γ) angle. See Figure 26-25. A *relief angle* is the angle that provides clearance between the tool and the workpiece. A *rake angle* is the angle between the tool face that contacts the chip and the vertical. An increase in rake angle increases the shear angle between the chip and the tool, decreasing the forces necessary to deform the workpiece but, at the same time, decreasing mass and strength of the tool point.

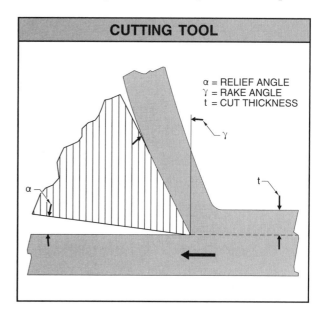

Figure 26-25. The cutting tool is characterized by the rake angle and the relief angle.

Cutting speed is the relative movement between the cutting tool and the workpiece. Cutting fluids are used to cool the tool and workpiece, minimize friction and tool wear, flush away chips, and protect the workpiece from corrosion.

Tool Materials. Tool materials include water-hardening and oil-hardening tool steels, high-speed steels, cemented carbides, ceramics, and diamonds. Water-hardening and oil-hardening tool steel consist of alloys, such as W1 and O6, that are quenched and tempered to high hardness levels. They soften at temperatures above 205°C (400°F) and are limited to slow cutting speeds and light-duty applications.

High-speed steels are used for a wide variety of machining operations. In general, the tungsten grades (T series) are not as tough as the molybdenum grades (M series). Both grades can be hardened to 64 to 66 HRC and resist softening to temperatures of 590°C (1100°F). High-speed steels may be titanium nitride coated by physical vapor deposition to improve their hardness and lubricity.

Cemented carbides are also widely used. They retain high hardness at temperatures above 590°C (1100°F), where the high-speed steels begin to soften. Cemented carbides can be coated with a thin layer of titanium carbide by chemical vapor deposition from the gaseous phase. These coated carbides have significantly improved machining performance, which results from reduced friction, increased surface hardness, and chemical inertness.

Ceramics consist of alumina or alloyed cermets. Alumina is cold pressed and used as an insert for light-duty cutting operations. Alloyed cermets are aluminum oxide-base materials containing various amounts of titanium carbide or other alloying in-

gredients. Alloyed cermets are hot pressed and much tougher than alumina. They are used for heavier duty cutting operations, such as roughing.

Diamonds are used for machining nonferrous alloys and abrasive materials. Man-made diamonds are sintered under very high temperatures and pressures. The man-made diamonds are superior to naturally mined diamonds.

Machining Processes

Machining processes generally hold tolerances down to approximately ±.025 mm (±.001 in.). A surface roughness of 1.6 μm (64 μin) or coarser can be obtained at reasonable cost by general roughing or semifinish machining operations. With finish machining operations, surface roughnesses as fine as .8 μm (32 μin) can generally be obtained. For finer finishes of .05 μm to .8 μm (2 μin to 32 μin), it is necessary to specify grinding processes. Machining processes consist of turning, drilling, trepanning, shaping and planing, milling, broaching, sawing and filing, and electrical and chemical machining.

Turning. *Turning* is the generation of cylindrical forms by removing metal with a single-point cutting tool moving parallel to the axis of rotation of the workpiece. See Figure 26-26. Turning is performed

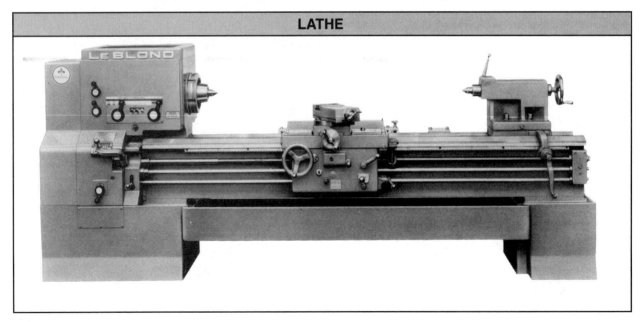

LATHE

Figure 26-26. A lathe is a basic feature of any machine shop and performs a variety of turning functions.

on a lathe. The rated size of a lathe indicates the largest diameter and length of workpiece it will handle. Thus a 14 in. diameter lathe will swing a 14 in. diameter workpiece.

Automatic lathes are used for producing duplicate components in large quantities. Once the automatic lathe is set up, the tools can be applied to the workpiece repeatedly, without having to reset the lathe for each cut.

Drilling. *Drilling* is the process of making holes by using a rotary end-cutting tool with one or more cutting lips. The tool has one or more helical or straight flutes, or tubes, for the ejection of chips and the passage of cutting fluid.

The most common metal-working drill is the twist drill, which is characterized by helical grooves or flutes. The body is the fluted portion of the drill and the point is its cutting end. The drill is held and driven by gripping the shank on the opposite end from the cutting end. Twist drills are made principally of high-speed steel. Carbide-tipped drills are sometimes used for drilling cast iron, high-silicon aluminum alloys, masonry, abrasive materials, and plastic materials with high levels of hardness.

Center drilling is drilling a conical hole in the end of a workpiece. *Core drilling* is enlarging a hole with a chamfer-edged, multiple-flute drill. *Step drilling* is the use of a multiple diameter drill to produce a hole having one or more diameters. *Spade drilling* is drilling with a flat-blade drill tip. *Boring* is the enlarging of a hole with the objective of producing a more accurate hole than by drilling.

Counterboring is enlarging a hole for a limited depth. *Countersinking* (chamfering) is the cutting of an angular opening into the end of a hole. *Reaming* is a hole enlarging process that is used to produce a hole of accurate size and good surface finish, with limited stock removal. *Gun drilling* is drilling using special straight-flute drills with a single lip and cutting fluid at high pressures for deep hole drilling. *Oil hole drilling* (pressurized coolant drilling) is the use of a drill with one or more continuous holes through its shank to permit the passage of a high-pressure cutting fluid, which emerges at the drill point and ejects chips.

Trepanning. *Trepanning* is a machining process for producing a circular hole or groove in solid stock, or for producing a disk, cylinder, or tube from solid stock, by the action of a tool containing one or more cutters (usually single point) revolving around a center. See Figure 26-27. Trepanning is used for producing disks, large shallow through holes, circular grooves, and deep holes.

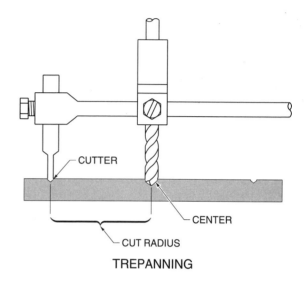

Figure 26-27. Trepanning is a machining process for producing a circular hole or groove in solid stock, or a disk, cylinder, or tube from solid stock.

Shaping and Planing. *Shaping* and *planing* are machining processes that use a reciprocating motion between one or more single-point tools and the workpiece to produce flat or contoured surfaces. There are several types of shaping operations, which may be performed horizontally or vertically. Form shaping consists of the use of a tool that is ground to provide a specific shape. Contour shaping is the shaping of an irregular surface, usually with the aid of a tracing mechanism. Internal shaping is the shaping of internal forms, such as keyways or guides.

Planing is used to produce flat surfaces and is performed on a massive table, fully supported by a heavy bed capable of sustaining heavy loads. The table slides on ways on the bed. Tee slots and holes are provided for bolts, keys, and pins for holding and locating workpieces on the finished table top.

Milling. *Milling* is the removal of material from a workpiece by the use of a rotary tool with teeth that engage the workpiece as it moves past the tool. Many sizes and types of cutters are used for the large variety of work that is done by milling. See Figure 26-28.

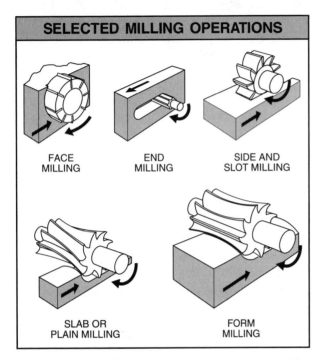

Figure 26-28. Milling is the removal of material from a workpiece by a rotary tool with teeth that engage the workpiece as it moves past the tool.

Face milling is the milling of a surface perpendicular to the axis of the cutter. Peripheral cutting edges remove the bulk of the material while the face cutting edges provide the finish of the surface being generated. *End milling* is milling performed using a tool with cutting edges on its cylindrical surface as well as on its end. *Side* or *slot milling* is milling of the side or slot of a workpiece using a peripheral cutter. *Slab milling* is the milling of a surface parallel to the axis of a helical, multiple-tooth cutter mounted on an arbor. *Form milling* is milling of a contoured surface using a cutter with peripheral teeth, which has a profile that is specially made to match the required contour.

Broaching. *Broaching* is cutting with a bar that has a series of teeth on one face that remove increasing amounts of material from the start to finish of the bar. The bar is pushed or pulled over the surface of the workpiece. See Figure 26-29. Each tooth takes

a thin slice from the surface. The entire cut is made in single or multiple passes of the broach over the workpiece to obtain the required surface contour. Internal or tunnel broaching is broaching of inside surfaces using several broach inserts. Typical internal broaching operations are the sizing of holes and cutting of serrations, straight or helical splines, gun rifling, and keyway shaping.

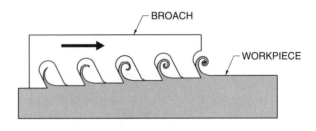

BROACHING

Figure 26-29. Broaching is cutting with a bar that has a series of teeth on one face that is pushed or pulled over the surface of the workpiece.

Sawing and Filing. Sawing and filing are operations that remove metal by the action of many small teeth. Saw teeth act in a narrow line. A saw can sever a sizeable piece of material with a minimal amount of cutting. The teeth of a file act over a wide surface and progress slowly. The cutting edge of a file can be watched and controlled.

Saws and files have been used as hand tools because they require little force and power. There are also machine driven (power) saws and files. Power hack saws, circular saws, and band saws are used to cut off pieces of bar stock, plate, sheet, and other product forms. Reciprocating-filing machines and continuous-filing machines are used for rapid and accurate finishing of plain and irregular surfaces in small quantities.

Electrical and Chemical Machining. Electrical and chemical machining are processes that shape objects or remove material without the use of conventional tooling. These operations include electrical discharge machining (EDM), electrochemical machining (ECM), and chemical milling (CM).

Electrical discharge machining is the removal of metal by rapid spark discharge between two electrodes, one the workpiece and the other the tool, with different polarities. The electrodes are separated by a gap distance of .013 mm to .890 mm (.0005 in. to .0350 in.). This gap is filled with dielectric fluid and metal particles that are melted, partially vaporized, and expelled from the gap. See Figure 26-30. Electrical discharge wire cutting is a form of EDM in which the electrode is a wire. EDM is a means of shaping hard metals and forming deep and complex-shaped holes in electrically conducting materials.

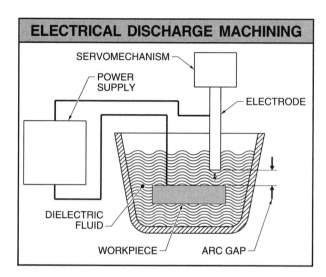

Figure 26-30. Electrical discharge machining is the removal of metal by rapid spark discharge between different polarity electrodes.

Electrochemical machining is controlled metal removal by the corrosion mechanism of anodic dissolution. Direct current passes through a flowing film of electrically conductive solution that separates the workpiece (anode) from the electrode-tool (cathode). The gap between the anode and cathode is a few thousandths of an inch. ECM cuts conductive materials of any hardness and is not confined to holes and cavities. See Figure 26-31.

Chemical milling is the controlled dissolution of material by contact with chemical reagents of varying types and strengths, depending upon the particular alloy being machined. Sometimes electrochemical milling is used. In electrochemical milling, a current is used at low current densities to assist the dissolution process. The workpiece is first cleaned and areas that must not be machined are masked off. The workpiece is then immersed in a tank of chemical solution.

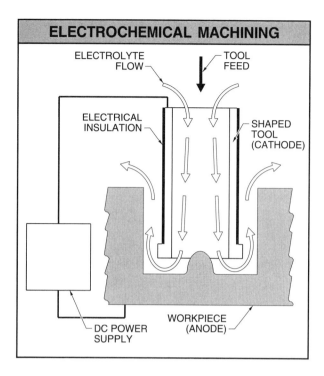

Figure 26-31. Electrochemical machining is the controlled removal of metal by anodic dissolution in an electrolytic cell in which the workpiece is the anode and the tool is the cathode.

For example, caustic soda is used for aluminum. The action usually proceeds at the rate of .025 mm/min (.001 in./min) when the proper concentration of chemical is used. After the required amount of metal is removed, the workpiece is removed, washed, and the masking taken off. Sometimes a two-stage process is used in which the mask is reapplied and trimmed after the first chemical milling cut. See Figure 26-32.

GRINDING

Grinding includes processes that accomplish metal removal mechanically, with hard and brittle grains of an abrasive material. Unlike machining, grinding usually does not remove much stock or remove it at a high rate. Grinding operations include precision grinding and ultrafinishing.

Abrasives

Abrasives are hard substances that are used in various forms as tools for grinding and other surface finishing operations. Abrasives are usually applied to belts or wheels. Abrasives are used to cut materials that are too hard for other tools and to give better finishes and hold closer tolerances than can be economically obtained by other means on most materials. When applied most efficiently, abrasives remove metal by cutting it into chips in the same manner as the other material cutting tools. The chips produced by grinding are generally so small that they must be magnified to be seen.

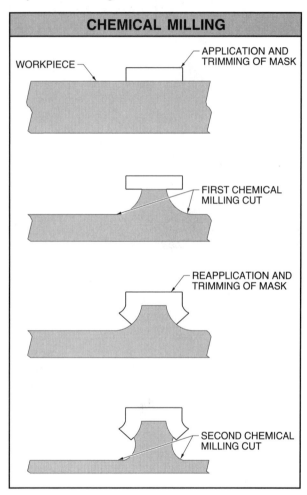

Figure 26-32. Chemical milling is the controlled dissolution of a workpiece surface by contact with chemical reagents varying in type and strength.

Grinding wheels are marked with symbols that designate their properties. See Figure 26-33. The principal abrasive materials are aluminum oxide and silicon carbide. The grain size of the wheel determines the approximate surface roughness that can be obtained on the workpiece. Wheels with finer grain size usually produce surfaces having lower roughness with sacrifice of stock removal capability. Very smooth surfaces require abrasive grain sizes of 220 and finer.

Aluminum oxide is a synthetic material produced in an electric furnace. It has a wide variety of applications. Tough aluminum oxide is used for rough grinding. Semifriable aluminum oxide is a general purpose abrasive. White friable aluminum oxide is used for tool grinding and heat-sensitive materials. Pink friable aluminum oxide is used for alloys that are difficult to grind.

Silicon carbide is also produced in an electric furnace. Silicon carbide is harder than aluminum oxide and is used primarily for grinding nonmetallic, nonferrous, and low-tensile strength materials. Green friable silicon carbide is used for the general purpose grinding of tungsten carbide. Various combinations of silicon carbide and aluminum oxide are also used as abrasives.

Abrasives are held together with bonding material. Bonding materials consist of various substances, such as vitrified clay, resinoid (synthetic, organic, or plastic compounds), rubber, or shellac.

Grinding Processes

Grinding processes encompass a number of procedures, which include surface grinding, cylindrical grinding, centerless grinding, lapping, honing, polishing, buffing, burnishing, power brushing, tumbling, and blasting. *Surface grinding* is the production of a flat surface as a workpiece passes under a grinding wheel.

Cylindrical grinding is the grinding of the outside diameter of a cylindrical workpiece. The workpiece is held between centers and rotated against a grinding wheel.

Centerless grinding is the grinding of cylindrical surfaces without the use of fixed centers to rotate the workpiece. The workpiece is supported and rotated between a grinding wheel, a regulating wheel, and a work guide (rest) blade. Centerless grinding is used to finish internal or external surfaces.

Lapping is a finishing operation that uses fine abrasive grains rolled into a lapping material, such as cast iron, copper, or lead. Lapping is done both by hand and by machine. Lapping provides extremely accurate dimensions, correction of minor shape imperfections, refinement of surface finish, and close fit between joined surfaces. Examples of lapped items are surfaces that must be liquid-tight or gas-tight without the use of gaskets, gauge blocks, and some gear teeth.

Honing is a finishing operation that uses fine-grit abrasive stones to finish round holes and, to a lesser extent, external flat and curved surfaces. Because the abrasive is not free to imbed in the surface, metallic and nonmetallic, as well as hard materials, can be honed. Typical applications of honing are the finishing of automobile engine cylinders, bearings, gun barrels, ring gauges, piston pins, shafts, and flange faces.

Polishing is the removal of metal by the action of abrasive grains carried to the component by a flexible support, generally a wheel or coated abrasive belt. Polishing is performed to produce a smooth finish on surfaces by taking out scratches, tool marks, pits, and other defects. Accuracy in the size and shape of the finished surface is not important. Much of the cost of polishing can be reduced by adequate prior surface preparation.

Buffing generally follows polishing and gives a higher luster to the surface. Buffing consists of two stages, cutting down and coloring. Cutting down removes scratch marks from rough polishing, stretch marks from forming, and die marks or other surface imperfections. Coloring refines the cut-down surface and brings out maximum luster. The workpiece is pressed on cloth wheels or belts on which a fine abrasive in a lubricant binder is sprayed from time to time.

Burnishing is a finishing and sizing operation performed on previously machined or ground surfaces by the displacement, rather than the removal, of minute surface irregularities. Burnishing utilizes smooth point or line contact and fixed or rotating tools.

Power brushing is the application of high-speed, revolving brushes to a workpiece in order to improve the surface appearance, remove sharp edges, burrs, fins, and particles. This tends to blend surface defects and round edges without excessive removal of material. Power brushing materials are wire bristle, hard cord, and tampico (tough fiber wheels). The materials are naturally flexible, able to conform to irregular surfaces, and able to get into otherwise hard to reach places.

Tumbling is a finishing process that consists of loading workpieces into a barrel about 60% full of

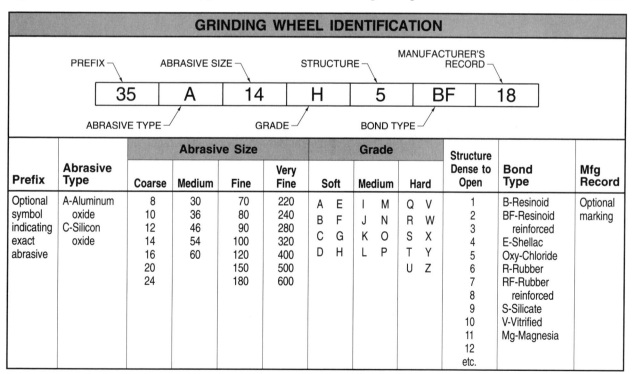

Figure 26-33. Grinding wheels are marked with symbols that designate their properties.

abrasive grains, wood chips, sawdust, natural or artificial stones, cinders, sand, metal slugs, or other scouring agents. Water is usually added, often mixed with an acid, detergent, or rust preventative. The barrel is closed or tilted and rotated at slow speed for approximately 1 to 10 hours. The workpieces and tumbling media slide over one another, producing a scouring action that removes burrs, sharp edges, and fins. This action also polishes and is used on materials before and after plating.

Blasting is a process in which particles moving at a high velocity clean metal surfaces or prepare them for coating. The particles may be metallic shot or grit, or artificial or natural abrasives, such as sand or shells. Shot blasting is also applied under controlled conditions to induce compressive surface stresses in order to increase fatigue resistance. Four common ways of blasting are by compressed air, centrifugal action, high-pressure water, and a mixture of compressed air and water.

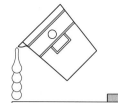

Welding 27

Welding is the most common method of joining metals. It permits great flexibility in design compared with other fabrication techniques. A welder must have knowledge of the metal and its condition, select the proper welding filler metal, and apply the correct welding procedure in order to make a structurally sound joint. Brazing and soldering differ from welding. During brazing and soldering, the brazing filler metal or solder melts below the melting point of the metals being joined and flows into the joint by capillary action. High temperatures and high cooling rates have an effect on the metallurgical structure of welded alloys. Welding processes may introduce characteristic discontinuities into the weld joint area.

WELDING PROCESSES

Welding is the process of joining metals, with or without a welding filler metal, by applying heat and/or pressure. The heat and/or pressure causes fusion or recrystallization across the interface of the metals. Welding differs from other fabrication processes for metals in several important ways. The metal reaches a temperature far higher than in any other fabrication process, and the rate of cooling from the joining temperature is much faster. Contamination of the joint area may have a considerably more serious effect on joint integrity. Also, residual stresses may be significantly higher. Specific alloys react in different ways to the effects of welding. Knowledge of the metal and its condition, selection of the proper welding filler metal, and application of the correct welding procedure are all necessary to make a structurally sound joint. Welding processes are divided into arc, gas, resistance, specialty, and solid-state welding.

Brazing and soldering are different from welding. During brazing and soldering, the brazing filler metal or solder melts below the melting point of the material or materials being joined and then flows into the joint by capillary action. Comprehensive

specifications exist for welding filler metals, brazing filler metals, and solders.

Arc Welding

In arc welding, the heat required to melt the filler metal or fuse the joint is generated by an arc struck between an electrode and the workpiece. The joint area is shielded from the atmosphere until it is cool enough to prevent the absorption of harmful impurities from the atmosphere. Arc welding is the most commonly used method of joining metals. It is also used for rebuilding and overlaying. Rebuilding is used to bring components, such as shafts, back to their original dimensions. Overlaying can also be used to return components to their original dimensions, but the weld metal often has specific properties that reduce wear or corrosion. Arc welding processes include shielded metal arc welding (SMAW), gas metal arc welding (GMAW), gas tungsten arc welding (GTAW), flux cored arc welding (FCAW), and submerged arc welding (SAW). See Figure 27-1.

Shielded Metal Arc Welding (SMAW). Shielded metal arc welding (stick welding, covered electrode

413

welding, or metal arc welding) is most often performed manually with the welder feeding a consumable electrode into the work area. The electrode is coated with a flux, which vaporizes and forms a shielding gas around the molten weld metal. SMAW is most often applied to steels. The process is not usually automated and is frequently used in repair work and in welding large structures.

Gas Metal Arc Welding (GMAW). Gas metal arc welding, or MIG welding, utilizes an arc that is struck between a consumable electrode wire and the workpiece. The work area is shielded by gas that flows through the electrode welding nozzle. Argon is used to shield nonferrous metals, such as aluminum. Carbon dioxide is commonly used to shield steels. GMAW is fast, can be automated, and works

in all positions. The rapid application of welding filler metal makes GMAW the lowest heat-input arc welding process.

Gas Tungsten Arc Welding (GTAW). Gas tungsten arc welding, sometimes referred to as TIG or Heliarc, uses an arc that is struck between a nonconsumable tungsten electrode and the work area. Welding filler metal may be fed in separately. If no welding filler metal is used to make the joint, the weld is an autogenous weld. The work area is shielded by helium or argon gas flowing out of the nozzle that contains the tungsten electrode. GTAW is a relatively slow process, works in all positions, and can be fully automated. Of the arc welding processes, GTAW is capable of having the highest heat input because the welder can "play" with the

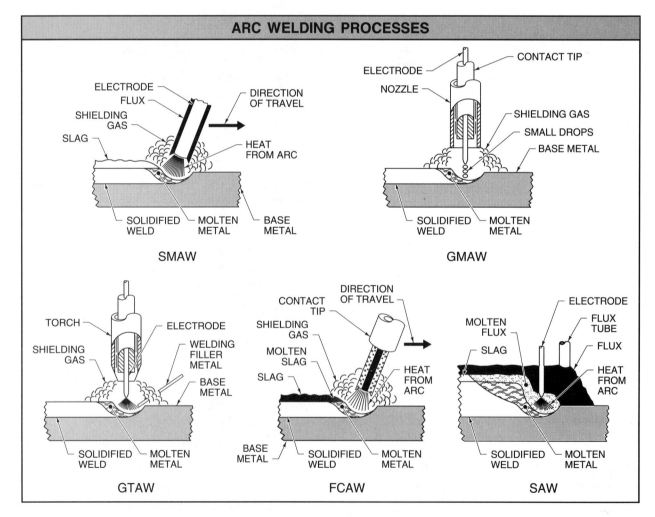

Figure 27-1. Arc welding encompasses several manual and automatic processes.

welding filler metal in the work area, allowing heat to build up.

Flux Cored Arc Welding (FCAW). Flux cored arc welding is a hybrid process with elements of both SMAW and GMAW. The arc is struck between the workpiece and a hollow electrode that contains a flux in the core. Some electrodes are self-shielding, while others require an external shielding gas, usually carbon dioxide. FCAW is used primarily on carbon and low-alloy steels. The process works in all positions, producing a fast, clean weld, and can be automated.

Submerged Arc Welding (SAW). Submerged arc welding uses a mound or line of granular flux that is deposited on the workpiece prior to the consumable electrode. When the arc is struck, the flux melts in the heated area, forming a slag that protects the molten area from the atmosphere. The process is limited to flat, or low-curvature, workpieces. Both flux and filler wire can be fed automatically. SAW is commonly used to join thick metal sections requiring deep penetration.

Oxyfuel Welding (OFW)

In oxyfuel welding processes, various fuel gas combinations are burned in oxygen to produce the necessary heat. The combination of oxygen and acetylene (oxyacetylene) is the most widely used process. See Figure 27-2.

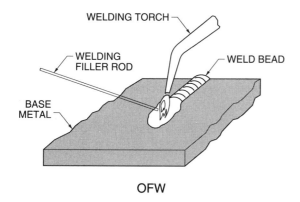

Figure 27-2. Oxyfuel welding uses the burning of various types of fuel gases to produce the heat needed to fuse metal.

Oxyacetylene Welding. Oxyacetylene welding uses the heat of burning oxygen and acetylene. Although the flame temperature is 3540°C (6400°F), the actual temperature due to the diffusion of air into the work area is 3260°C (5900°F). The oxyacetylene flame can be made slightly reducing for welding metals, such as steels, or neutral for heating, brazing, and welding other metals. Welding filler metal can be added if required. With suitable tips and other equipment, oxyacetylene may be used to cut steel.

Resistance Welding (RW)

In resistance welding, heat is generated by passing high-amperage current through the mating workpieces. See Figure 27-3. Resistance welding is used to make localized (spot) or continuous (seam) joints.

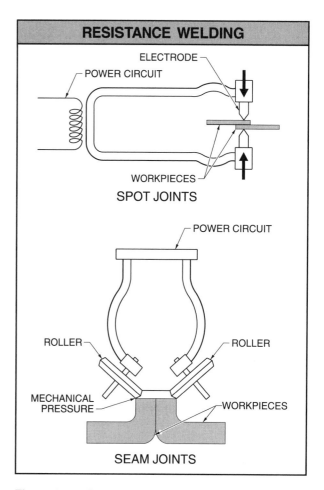

Figure 27-3. Resistance welding generates the heat required to fuse metal by passing high-amperage current through mating workpieces.

No shielding gas is used, except in specific situations for reactive metals, such as titanium. Because of the clamping action required for the weld, access to both sides of the component is necessary.

Resistance welding is commonly used as a mass-production technique requiring special fixtures and automatic handling equipment. The process can be applied to almost all steels, stainless steels, aluminum alloys, and some dissimilar-metal bonds. Stock thicknesses range from .1 mm to 2 cm (.004 in. to .800 in.). There are no limitations on welding position.

Specialty Welding

Specialty welding processes require specialized equipment. Specialty welding process include laser beam welding (LBW), plasma arc welding (PAW), electron beam welding (EBW), and electroslag welding (ESW). See Figure 27-4.

Laser Beam Welding (LBW). Laser beam welding uses a laser to produce a high-intensity beam of coherent light that is focused onto the workpiece by conventional optical elements. It produces a narrow, clean weld with a small heat-affected zone. The beam can be controlled with sufficient accuracy to weld microelectronic elements. LBW is used for joining very thin sections of dissimilar metals in addition to more regular welding. X-ray shielding and vacuum are not required. The beam can be transmitted over long distances and can be easily

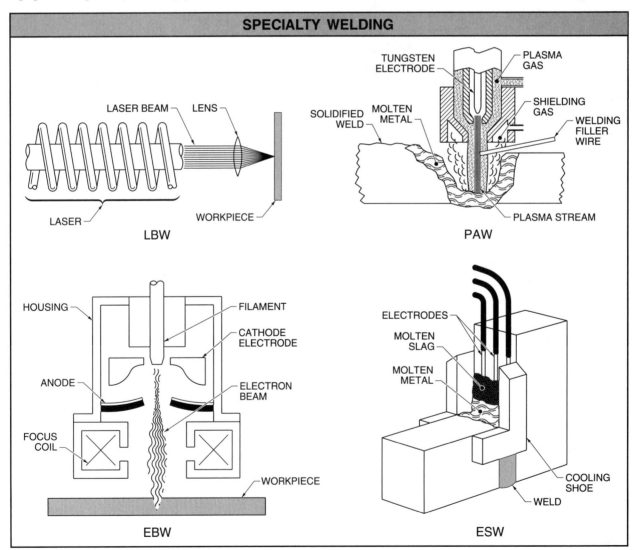

Figure 27-4. Specialty welding processes require specialized equipment.

focused. It is possible to route laser beams by mirrors so that one laser can serve several independent work stations.

Plasma Arc Welding (PAW). Plasma arc welding uses hot ionized gases (plasma) to shield the work area. The plasma is sometimes supplemented with a separate shielding gas. The welding gun is similar to that used with GTAW in that it uses a nonconsumable tungsten electrode. Filler metal, if used, is fed in separately. The weld produced is also similar to GTAW, but the process is faster.

PAW produces a deep, narrow, uniform weld zone and is suitable for refractory metals, low-alloy steels, stainless steels, aluminum, and titanium. It is most frequently used for welds of high quality on high-strength, thin-section material. PAW tolerates great variations in joint alignment.

Electron Beam Welding (EBW). Electron beam welding uses a high-voltage electron beam that is focused on the metal to melt it and produce a narrow, deep joint, without the need for filler metal. The highly localized heating produces welds with significant depth-to-width ratios, usually approximately 25:1. The beam must be generated in a vacuum, but the workpiece may be in air, partial vacuum, or full vacuum, depending on the application. Equipment for the process is relatively costly and X-ray shielding is required. EBW is suitable for mass production and is applicable in all positions. It can be used on a wide variety of materials and is especially suited for joining dissimilar metals, reactive metals, and refractory metals.

Electroslag Welding (ESW). Electroslag welding is similar to submerged arc welding in that the heat is generated by a consumable electrode submerged in a molten flux. It is normally used in the vertical position to join very thick sections up to 90 cm (35.5 in.) with gaps up to 3.8 cm (1.5 in.). Access to all sides of the weld is necessary. Molten weld metal is contained between water-cooled copper shoes. One type of electroslag system has cooling shoes that slide up the joint as welding proceeds, with the electrode wires entering from the side. Another system employs fixed shoes and consumable

tubes that guide the electrode or electrodes into the joint from the top.

Solid-state Welding (SSW)

In solid-state welding, metals are joined below their melting points and various forms of energy are used to produce a metallurgical bond at the joint. Solid-state welding processes include friction (inertia) welding, ultrasonic welding, and explosion welding. See Figure 27-5.

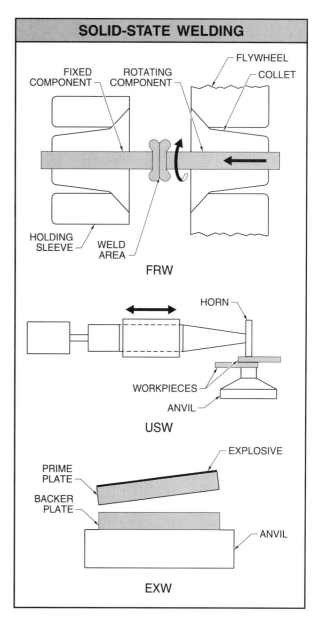

Figure 27-5. Solid-state welding uses various forms of energy to join metals below their melting points.

Friction Welding (FRW). *Friction welding* is a solid-state welding process that uses kinetic energy to produce a weld. The kinetic energy is stored in a rotating flywheel to which one component of the joint is attached. The other part of the joint is fixed. The rotation of the flywheel is accelerated, and at a predetermined speed, driving power is cut and the components are forced together. This converts the kinetic energy of the rotating component into heat, which bonds the two metals.

No flux, filler, or shielding gas is required in friction welding, and no special preparation of the interface surfaces is necessary, unless there is heavy mill scale or a carburized case, which tends to form brittle regions and must be removed by grinding. The major limitation of FRW is that one of components being joined must be axially symmetric.

Ultrasonic Welding (USW). *Ultrasonic welding* is a solid-state welding process that produces a weld by the local application of high-frequency electrical energy (10 kHz to 20 kHz), which is converted to ultrasonic vibrations, as the metal components to be joined are clamped together under pressure. The components are clamped between a movable element (horn) and a fixed anvil. The ultrasonic vibrations are transmitted to the components through the horn. The component next to the horn is rubbed at the ultrasonic frequency against the component that is held stationary on the anvil. No electric energy passes through the components. The shape of the horn is designed to amplify and transmit the vibrations to the work. Both horn and anvil can be made in various configurations to meet specific component shapes.

Explosion Welding (EXW). *Explosion welding,* sometimes referred to as explosion bonding or explosion cladding, is a solid-state welding process that produces a weld by high-velocity impact of the components as a result of controlled detonation. The detonation is used to create a metallurgical bond between two or more dissimilar metals. A layer of explosive is placed in contact with one surface of the metal plate (prime plate), which is maintained at a constant parallel separation with the backer plate. When the explosive is detonated, the prime plate is deflected and accelerated toward the backer plate, producing a bond that has the cross section of a detonation wave.

Many dissimilar metal combinations can be explosively welded. EXW is used extensively to produce clad plate, in which a thin section of expensive, corrosion-resistant metal is bonded to thicker structural steel. These clad plates are fabricated into vessels and other equipment.

Brazing and Soldering

Brazing and soldering are joining processes that use brazing filler metals and solders that melt below the melting temperatures (liquidus) of the metals being joined. The joint clearance is extremely narrow, so that the filler metal flows into the joint area by capillary action. See Figure 27-6. The difference between brazing and soldering is the melting temperatures of the brazing filler metal and the solder. The brazing filler metal has a melting temperature above 450°C (840°F), and solders have a melting temperature below 450°C (840°F). Braze welding is a process in which a brazing filler metal is used to make joints that have large joint clearances that do not fill by capillary action.

Joint Clearance. The clearance between components being brazed or soldered is important. If the joint clearance is too small, it will not allow capillary action to cause the brazing filler metal to flow uniformly throughout the entire joint. If the clearance is too great, filler metal may not flow throughout the joint. In either case, the joint strength will be lower than anticipated. Another factor that influences joint clearance is the length or area of the joint. A smaller joint clearance can be used for smaller lengths and areas or when protective atmospheres are employed rather than fluxes. The joint clearance with fluxes ranges from .025 mm to .630 mm (.001 in. to .025 in.).

Brazing. *Brazing* is a welding process that joins metal components by heating them to the brazing temperature in the presence of a brazing filler metal that melts above 450°C (840°F) and below the melting point of the metals being joined. The brazing filler metal fills the joint area by capillary action. Brazing is usually performed with a torch (torch brazing) or in a furnace (furnace brazing). Other methods of brazing include dip, resistance, induction, and infrared brazing.

Torch brazing is performed in air and is used to join fairly small assemblies made from materials that either do not oxidize at the brazing temperature or can be protected from oxidation with a flux. Torch brazing is the most commonly used brazing process and may be performed manually or automatically. Some brazing filler metals include aluminum-silicon alloys, silver-base alloys, and copper-zinc alloys. A flux is required with these filler metals if a protective atmosphere is not used. Self-fluxing copper-phosphorus alloys are also used.

Furnace brazing is used for larger assemblies and those that can be self-jigged. Although it can be performed in an air furnace with a flux, a protective gas atmosphere is more common. The type of atmosphere required depends upon the materials being brazed and the brazing filler metals used. Components with readily reducible surface oxide films can be brazed in an atmosphere of combusted natural gas or cracked ammonia. Dry hydrogen, a powerful reducing agent, can be used for brazing most stainless steels and many of the nickel, cobalt, and iron-nickel base superalloys. Heat-resistant high-strength alloys that contain appreciable amounts of aluminum or titanium are frequently brazed in a vacuum to prevent the formation of oxides that inhibit wetting and flow of the filler metal. Components made from these alloys can be plated to prevent oxidation during brazing. Plated components can be brazed in a vacuum or in a controlled atmosphere.

Dip brazing is done by immersing the components in a molten bath, which serves as the heating medium as well as the fluxing agent. Uniform heating to the brazing temperature is achieved rapidly. Components are cleaned, assembled, and held together in fixtures during brazing, and are normally preheated before immersion. Residues must be removed after brazing to prevent corrosion. Dip brazing is frequently used to braze aluminum assemblies, such as heat exchangers.

Resistance brazing is used when small areas of a component require brazing and the electrical conductivity of the joint members is high, for example, when brazing electrical contacts to contact holders. Heat is produced by the resistance to the flow of current through the joint members. Resistance welding machines are often used.

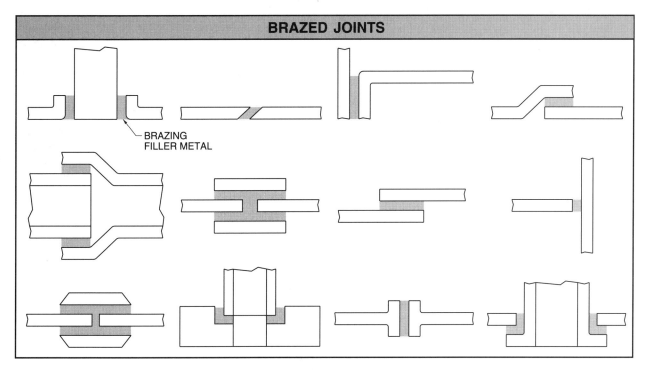

BRAZED JOINTS

BRAZING FILLER METAL

Figure 27-6. The distance (clearance) between the surfaces being joined has a significant effect on the strength of a brazed or soldered joint.

In induction brazing, heat is supplied to the components by inducing an electric current in the metal, for example, by wrapping the component in an AC coil. Induction brazing is used when it is not necessary to heat the entire assembly or when part of the assembly would be adversely affected by the heat. Because the component is heated selectively by the coil, induction brazing reduces distortion. Induction heating is also a very rapid way to bring the joint to the brazing temperature.

Soldering. *Soldering* is a welding process that joins metal components by heating them to the soldering temperature in the presence of a solder that melts below 450°C (840°F) and below the lowest melting point of the metals being joined. The solder flows into the joint by capillary action.

Flame heating is probably the most common way to reach the soldering temperature. Soldering with irons and bits are other methods of reaching the soldering temperature. The usual steps in the soldering process are to clean the joint area, apply the flux, apply the solder, heat, and if necessary, clean the soldered joining. Soldering is frequently used for electrical and electronic applications and may also be used for sealing tin cans and radiator seams.

Braze Welding. *Braze welding* is an arc welding process in which the joint is relatively wide and brazing filler metal is not distributed into it by capillary action. Brass or bronze filler metal is used in braze welding.

WELDING SPECIFICATIONS

Welding specifications are developed to ensure that equipment fabricated by welding is safe to operate at the conditions for which it was designed. Comprehensive specifications exist for welding filler metals, brazing filler metals, and solders. The main specifications are produced by the American Welding Society (AWS). Welding codes govern weld quality for various types of equipment and structures. Welding procedures guide the welder to achieve the required level of quality required by the applicable code.

Welding Filler Metals

Welding filler metals (consumables) are used, or consumed, and become part of the finished joint. Generally, welding filler metals are of different compositions from the base metals they join. For virtually all metals and alloys, modifications in chemical composition will improve their properties in weld metal form. For this reason, welding rods and filler metals have evolved as a separate class of materials, just like as-cast and wrought metals.

Welding filler metals are supplied as wire and are categorized as welding rods or electrodes. If the wire is to be used for welding or brazing and does not conduct electric current, it is a welding rod. If it is used in an electrical circuit, it is a welding electrode.

Welding rods and electrodes are classified by the AWS in a series of specifications that cover specific alloy families. For example, A5.1 covers carbon steel covered arc welding electrodes. See Figure 27-7. See Appendix. The American Society of Mechanical Engineers (ASME) issues similar specifications for boilers and pressure vessels, but uses the prefix uppercase letters SF in their designation system. For example, SF-5.1 covers carbon steel arc welding electrodes.

Welding filler metals are classified in various ways. The primary method of classification is as rod, wire, or electrode. Rod is wire that is cut and straightened. Rod or electrode may be flux covered or bare. Wire is usually bare. Electrodes may be flux cored (tubular), consisting of a metal sheath packed with fluxes and alloying elements. Fluxes, when used separately from filler metals, are also classified. To describe each classification, prefix uppercase letters such as R for rod, E for electrode, RB for rod or wire, and ER for electrode, rod, or wire are used.

Carbon and low-alloy steel covered electrodes are designated by the uppercase letter E followed by four or five numbers. The first two or three numbers indicate the minimum tensile strength in ksi of the as-deposited metal. The second-to-last number indicates the welding position, the number 1 being all-position and the number 2 being flat or horizontal. The last number (from 1 to 8) provides a variety of information. It indicates the recommended power supply, the type of slag, arc characteristics, amount of penetration, and presence of iron powder in the coating. The purpose of iron powder is to permit higher welding currents and to allow larger amounts of the alloy to be transferred from the coating to the weld pool. For example, E6010 is a low-strength covered electrode (60 ksi minimum tensile strength), which may be used in all positions, and is deep penetrating with easily removed slag. E6010 is sometimes used for the root pass in welding carbon steel. E7018 is a higher strength covered electrode (70 ksi minimum tensile strength), which may be used in all positions and is low in hydrogen. E7018 is used for welding carbon steel.

Electrode diameter is determined by the thickness of the steel to be welded. Thicker metals require higher current input to the arc to guarantee that they are melted so that the metals can unite with the deposited metal.

Brazing Filler Metals and Solders

Brazing filler metals and solders are specified by the *AWS A5.8, Brazing Filler Metal.* Brazing filler metals and solders are designated by the prefix uppercase letter B followed by the chemical symbols for the predominant alloying elements.

Welding Codes and Procedures

Welding codes govern the methods by which equipment and structures are fabricated during welding. There are three main welding codes in the United States, which are maintained by the following organizations. The ASME maintains Section IX (Welding Qualifications) of the *Boiler and Pressure Vessel Code.* The American Petroleum Institute (API) maintains *API 1104, Standard for Welding Pipelines and Other Facilities.* The AWS maintains *Standard D1.1, The Structural Welding Code,* which covers items, such as structural steel work, that are not within the mandate of the other codes. These codes contain the required design methods for welded fabrications and the methods that must be followed to qualify welding processes and welders, referred to as welding procedures.

Welding procedures are specific requirements for welding that are described in the various welding codes and that are broken down into the welding procedure specification (WPS), the procedure qualification record (PQR), and the welder performance qualification (WPQ). See Figure 27-8. The first part of qualifying a weld procedure is to prepare a detailed WPS. All information pertaining to joint design, base metal, weld metal, preheating and post-weld heat treatment, shielding gas, purge gas, electrical characteristics, and welding technique are listed. Then, a sample weld is made using the proposed WPS. Next, the actual parameters used to weld the sample are recorded in the PQR. The sample is then cut up, and tensile and bend tests are performed on the weld (plus impact tests, if required). For weld overlay qualifications, a chemical analysis of the overlay is required, as well as bend tests. If the samples tested are deemed acceptable, the WPS is considered qualified.

BRAZING FILLER METAL AND SOLDER SPECIFICATIONS		
AWS Specification	Specification Title	Process[a]
A5.1	Carbon steel covered arc-welding electrodes	SMAW
A5.2	Iron and steel gas welding rods	OAW
A5.3	Aluminum and aluminum alloy arc welding electrodes	SMAW
A5.4	Corrosion-resisting chromium and chromium-nickel steel covered welding electrodes	SMAW
A5.5	Low-alloy steel covered arc welding electrodes	SMAW
A5.6	Copper and copper alloy covered electrodes	SMAW
A5.7	Copper and copper alloy welding rods	OAW and GTAW
A5.8	Brazing filler metal	BR

[a] if GTAW specified, PAW will also apply

Figure 27-7. American Welding Society specifications cover every type of standardized welding filler metal, brazing filler metal, and solder composition.

WELDING PROCEDURES		
Welding Procedure Specification (WPS)	Procedure Qualification Record (PQR)	Welder Performance Qualification (WPQ)
Concerns how to weld	Concerns mechanical properties	Concerns depositing sound weld metal
Is intended to be a guide for the welder	Is intended to provide proof of weldability	Is intended to provide proof of welder's ability
Lists acceptable ranges for all parameters	Lists actual variables used in making test sample	Lists actual variables used
Highly recommended that supporting PQR or PQR's be listed on the WPS	Lists test results	Lists ranges of variables covered
Must be supported by a PQR	Qualifies WPS by welding test coupons, testing specimens and recording weld data, and test results	Lists test results
Must provide detailed description of variables necessary to produce a sound weld as a true guide for the welder and as a good reference for the foreman and inspector		Qualifies by welding test coupons using the parameters of a WPS, testing specimens and recording weld data, test results, and ranges qualified

Figure 27-8. Welding procedures provide all the basic information required to produce a sound joint between two metals.

To fabricate equipment and structures using the qualified weld procedure, it is necessary to qualify the welders who will perform the fabrication. This consists of having the welder perform a weld, per the WPS, in the most difficult position (for example, flat, vertical, or overhead) to be used in the fabrication job. The information as to how the weld was made is entered in the WPQ. The sample weld is tested, either by cutting and running a bend test or by radiographic examination. Once the results of these tests have been deemed acceptable, the welder is qualified to do production welding per the applicable WPS.

Welding Inspection

The welding codes specify what type of inspection is required, how much inspection is required, and what size and type of defects, if any, are acceptable. Most codes in the United States rely heavily on radiography to detect defects. Radiography, or any other single NDT method, is not capable of detecting all defects in a given weld. However, with the margin of safety built into the design section of the codes (including efficiency factors for welds), the resulting equipment has almost always proved capable of operating safely under the conditions for which it was designed.

WELDING METALLURGY

Welding metallurgy is concerned with the effect of high temperatures and fast cooling rates on the metallurgical structure of alloys. Heat is the most important ingredient in welding. The molten weld transfers heat into the base metal. This acts as a heat sink and cools the weld and heat-affected zone of base metal adjacent to the weld very rapidly. The cooling rate in welding is much faster than any other form of quenching, such as water or oil.

The way steels react to the heat input and cooling rate in welding is principally a function of their alloying elements and their susceptibility to embrittlement. The carbon equivalent is used to figure the need for preheating and post-weld heat treatment of the weld.

Welded Joint Regions

As a result of the heat input of welding, a welded joint consists of three metallurgically distinct regions. The regions include the weld bead, the heat-affected zone, and the base metal. See Figure 27-9.

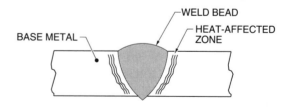

Figure 27-9. A welded joint comprises three metallurgically distinct regions, which are the weld bead, the heat-affected zone, and the base metal.

Weld Bead. A *weld bead* is a weld resulting from the addition of filler metal to a joint. The weld bead exhibits some of the characteristics of a chill-cast structure, such as one made in a metal mold. However, the weld bead is much stronger than a cast metal of similar composition. Little difference exists in the strength of weld metal of the same composition deposited by the different welding processes.

The weld bead includes some melted-out base metal at its boundaries. This leads to dilution or mixing of the weld filler and base metals. When highly corrosion-resistant alloys are joined, the diluted area may be susceptible to accelerated corrosion. To offset the loss of corrosion resistance, the weld bead is often made of a more highly alloyed material for superior corrosion resistance.

When the weld bead has just solidified, it is mechanically weak. The internal stresses developed during welding may cause the weld bead or melted-out region of base metal to be susceptible to hot cracking. Specific alloys are extremely susceptible to hot cracking. Welding processes for such alloys must use low heat input, and joint designs must minimize restraint during cooling. For example, butt welds are favored over fillet welds because butt welds contain less stress.

Heat-affected Zone. The *heat-affected zone* is the narrow region of the base metal adjacent to the weld bead, which is metallurgically altered by the heat of welding. The heat-affected zone is usually the major source of metallurgical problems in welding. It can lead to loss of toughness in low-alloy steels and loss of corrosion resistance in stainless steels and nickel-base alloys.

The width of the heat-affected zone depends on the amount of heat input during welding and increases with the heat input. Generally, the heat-affected zone varies from 1.5 mm to 6.5 mm wide (.06 in. to .25 in.).

Base Metal. The *base metal* is the region of metal that is joined by the welding process and is not metallurgically affected by the heat of welding. This varies with the type of alloy. For example, with carbon steel, the base metal is the part of the joint that has not experienced temperatures of 425°C (800°F) or more.

Effect of Heat

With carbon and low-alloy steels, the rapid cooling rate from the welding temperature is similar to quenching in heat treatment operations. It leads to the formation of hard, brittle martensite in the heat-affected zone.

The higher the carbon or alloy content, the more easily martensite is formed and the harder (more brittle) the martensite is. Since welding produces high residual stresses in the heat-affected zone, this situation may easily cause cracking as the steel cools down. Steels that are susceptible to cracking must be preheated to "cushion" the effects of martensite formation. They are also post-weld heat treated to temper (improve the toughness) any martensite that is formed and additionally stress relive the joint.

The weld bead is less likely to suffer loss of toughness from rapid cooling. However, loss of toughness is a possibility from grain coarsening that occurs at the high temperatures attained during welding. Grain coarsening is rarely a problem because most welds are multipass (consisting of more than one pass of welding filler metal), so each succeeding pass of welding filler metal refines the grain size of the pass below it, improving its toughness. Multipass welding also has a slightly beneficial effect on the toughness of the heat-affected zone.

Carbon Equivalent

The carbon equivalent is a formula based on chemical composition that determines the need to preheat and post-weld heat treat carbon and low-alloy steels during welding. Carbon is the most important element in steels for determining martensite formation and its hardness. Other elements exert a less powerful effect. The carbon equivalent formula weighs the effect of each element in order to combine them with the carbon percentage and to indicate the weldability of the steel. The higher the carbon equivalent, the greater the tendency toward cracking in the heat-affected zone during welding. Therefore, there is a greater the need for preheating and post-weld heat treatment. Plain carbon steels with a carbon equivalent <.4% to .5% are considered readily weldable without the need for preheating or post-weld heat treatment.

The carbon equivalent is found using the following formula:

$$CE = \%C + \%Mn/6 + \%Ni/20 + \%Cr/10 + \%Cu/40 - \%Mo/50 - \%V/10$$

where

CE = carbon equivalent

$\%C$ = percent carbon

$\%Mn$ = percent manganese

$\%Ni$ = percent nickel

$\%Cr$ = percent chromium

$\%Cu$ = percent copper

$\%Mo$ = percent molybdenum

$\%V$ = percent vanadium

Example: Figuring carbon equivalent

What is the carbon equivalent of a 1030 steel with .28% C and .7% Mn?

$$CE = \%C + \%Mn/6 + \%Ni/20 + \%Cr/10 + \%Cu/40 - \%Mo/50 - \%V/10$$

$$CE = .28 + .7/6 + 0/20 + 0/10 + 0/40 - 0/50 - 0/10$$

$$CE = .28 + .12 + 0 + 0 + 0 - 0 - 0$$

$$CE = .40$$

WELDABILITY

The weldability of different metals and alloys varies greatly and is affected by different factors. The weldability of carbon steels is primarily a function of the carbon and manganese contents. The weldability of low-alloy steels is similar to that of medium-carbon steels. The weldability of free-machining steels is poor.

Low-carbon Steels

Low-carbon steels contain up to .3% C and up to 1.2% Mn. Generally, the carbon content does not exceed .15% C. These values (.3% C and 1.2% Mn) represent the compositional limits for ease of welding without cracking. The steels can be welded without the need for preheating or post-weld heat treatment when the joint thickness is <2.5 cm (1 in.) and the weld restraint is not severe. Examples of low-carbon steels include ASTM A36 and AISI 1020.

Medium-carbon Steels

Medium-carbon steels contain .3% C to .6% C and .60% Mn to 1.65% Mn. Examples of these steels are the AISI types 1040 and 1055. These types are extensively used in machinery, for example, in components for tractors, derricks, and pumps. They are often selected for wear resistance rather than strength and are often heat treated to ensure hardness in the required range.

These steels form high-hardness (approximately 63 HRC) martensite in the heat-affected zone, meaning the heat-affected zone is low in toughness and ductility and likely to develop cracks as it cools to room temperature. Additionally, if the weld bead picks up carbon from the base metal, it is likely to display high hardness and a tendency toward brittle failure. To avoid these tendencies, low-hydrogen electrodes are used, plus preheating between 150°C and 260°C (300°F and 500°F) and post-weld heat treatment at 595°C and 680°C (1100°F and 1250°F).

Low-alloy Steels

Examples of low-alloy steels are AISI types 4140 and 4340. They are high-strength materials with yield strengths in excess of 100 ksi (690 MPa) and are used in the quenched and tempered condition. The preferred welding filler metals respond to heat treatment in the same manner as the base metal. Low-alloy steels are welded in the annealed or normalized condition when the microstructure consists of ferrite and pearlite and are heat treated to the desired strength. If the benefit of heat treating to maximum strength after welding is not possible, the specified preheating and post-weld heat treatment must be rigidly maintained.

Type 4140 can be preheated at 345°C (650°F) and post-weld heat treated at 595°C to 675°C (1100°F to 1250°F), followed by slow cooling (under a blanket). Type 4340, which is air hardening, requires the same preheating temperature and should be annealed or normalized and tempered after the welding process.

Free-machining Steels

Free-machining steels contain fairly large amounts of sulfur, phosphorus, or lead to improve their machinability. These steels have poor weldability because they hot crack in the weld bead due to the formation of low-melting point compounds of sulfur and phosphorus at the grain boundaries. The grains may separate from thermal stresses on cooling.

Lead is almost insoluble in steel, forming small globules in the microstructure. The lead can melt during welding and volatilize into the weld fumes, which is of concern on account of its toxicity. Good ventilation during welding is required. Lead may also cause porosity and embrittlement under certain welding conditions.

Normally, free-machining steels should not be welded. If it is absolutely necessary to weld a free-machining steel, low-hydrogen electrodes and low welding currents must be used to limit dilution, porosity, and cracking.

WELDING DISCONTINUITIES AND DEFECTS

Welding processes may introduce characteristic discontinuities into the weld joint area. There are several characteristic discontinuities and defects in welds. A welding discontinuity is not necessarily a defect. Defects are discontinuities of a particular size or orientation that are in excess of the maximum value permitted by the relevant code by which the weld is inspected. The various codes define the size and orientation of discontinuities according to the intended service conditions of the weld.

Discontinuities are invariably caused by improper use of the process employed. Characteristic weld discontinuities include cracking, hydrogen cracking, incomplete penetration, incomplete fusion, porosity, slag inclusions, and undercutting. See Figure 27-10.

Cracking

Cracking is rarely tolerated and must be removed by grinding. Most cracks are caused by shrinkage and run in the longitudinal direction of the weld. Cracks can propagate through several passes of welding filler metal unless removed immediately.

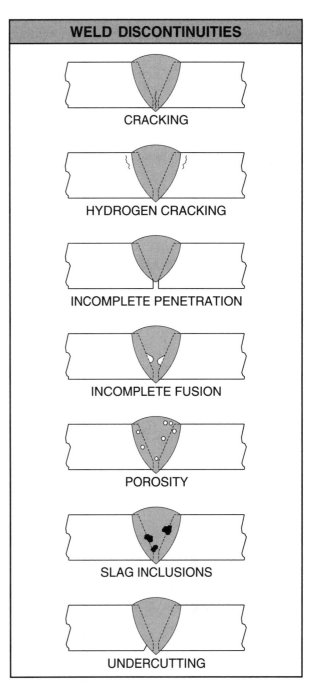

Figure 27-10. Characteristic discontinuities in welds include cracking, hydrogen cracking, incomplete penetration, incomplete fusion, porosity, slag inclusions, and undercuting.

Crack formation is aggravated by welding fixtures that do not permit contraction of the weld during cooling, by narrow joints with large depth-to-width ratios, by poor ductility of the deposited weld metal, or by a high coefficient of thermal expansion coupled with low-heat conductivity in the parent metal.

Hot short metals are more susceptible to cracking. A root pass may crack if insufficient welding filler metal is added.

Hydrogen Cracking

Hydrogen cracking occurs in the heat-affected zone of some steels as hydrogen diffuses into this region when the weld cools. Hydrogen cracking is caused by atomic hydrogen. The sources of atomic hydrogen are organic material, chemically bonded water in the electrode coating, absorbed water in the electrode coating, and moisture on the steel surface at the location of the weld.

Atomic hydrogen forms from any of these sources at the temperature of welding and rapidly diffuses into the molten weld metal. As it solidifies, the hydrogen must escape because solidified metal dissolves considerably less hydrogen. Some hydrogen escapes into the atmosphere. The hydrogen that diffuses into the heat-affected zone causes a problem because the martensite that forms in this region during the rapid cooling is very susceptible to hydrogen embrittlement. The high residual stresses created during the rapid cooling of the weld result in cracking.

Hydrogen cracking usually takes the form of longitudinal fissures a short distance from the fusion line of the weld. It is often referred to as underbead cracking. Underbead cracking is troublesome because it does not necessarily come to the surface, so it cannot be detected by liquid penetrant or magnetic particle testing. Since the medium-carbon or low-alloy steels in which hydrogen cracking occurs are often used for their high strength, such cracking is extremely harmful.

There are several methods of avoiding hydrogen cracking that include using low-hydrogen electrodes, which include baking and storing them in a low-temperature oven; preheating the surface of the steel before welding to remove moisture; and post-weld heat treating immediately to force the hydrogen to escape. Peening immediately after each pass is also beneficial because it induces compressive stresses and offsets the tendency toward cracking.

Incomplete Penetration

Incomplete penetration is caused by welding filler metal that does not penetrate to the furthest part of the joint. It allows the product to lodge in the space during service, leading to contamination if the product being handled begins to degrade. Incomplete penetration may also provide a localized region for corrosion to occur.

Incomplete Fusion

Incomplete fusion occurs when the weld bead does not unite completely with the base metal. It is caused by base metal acting as too high a heat sink or by a welding process that is not operated with sufficient heat input. GMAW processes are prone to incomplete fusion. Incomplete fusion may also occur between weld passes for the same reasons.

Porosity

Porosity is caused by the evolution of gas in the molten weld metal from the presence of rust, dirt, oil, grease, damp electrodes, and/or moisture. Porosity is generally spherical in shape. Cracks do not develop from porosity unless the pores are closely spaced and lined up.

Slag Inclusions

Slag inclusions are caused by careless interpass cleaning. They are small particles of slag that have been trapped in the weld and prevent complete penetration of the welding filler metal. Slag inclusions occur mainly during covered electrode welding processes, such as SMAW.

Undercutting

Undercutting is a notch in the base metal produced during the welding process. The notch is not completely filled by the welding filler metal. Undercutting is usually caused by too hot of an operating temperature. On fillet welds, undercut is usually seen on the vertical leg of a weld.

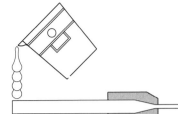

Hard Facing and Surface Treatment 28

Hard facing techniques involve depositing a surface layer on a metal to increase resistance to various types of wear. Hard facing processes consist of welding and thermal spraying. Hard facing prevents wear and galling, gouging, erosion, cavitation erosion, and contact fatigue. Surface treatment comprises methods that are used to either modify the surface of a metal or incorporate a metallic surface coating on it to improve wear resistance and corrosion resistance. Surface treatment processes are divided into conversion coatings, hot dipping, electroplating, and surface modification.

HARD FACING

Hard facing is the application of a coating or cladding to a substrate for the purpose of reducing surface damage. Hard facing processes consist of welding and thermal spraying. Many factors influence the selection of the correct hard facing process.

Surface Damage

Hard facing prevents several types of surface damage. These include adhesive wear and galling, gouging (high-stress abrasion), erosion (low-stress abrasion), cavitation erosion, and contact fatigue. See Figure 28-1.

Adhesive Wear and Galling. Adhesive wear and galling are related phenomena that occur when two metal surfaces come into sliding contact. *Adhesive wear* is the removal of metal from a surface by welding together and subsequent shearing of minute areas of two surfaces that slide across each other under pressure. In advanced stages, adhesive wear leads to galling. *Galling* is a condition whereby excessive friction between surface high spots results in localized welding with subsequent spalling (for-

mation of surface slivers) and further roughening of the rubbing surfaces. This may result in seizure of the components. Examples of components that gall include valve trim, engine camshafts, and threaded connections.

Gouging. *Gouging* (high-stress abrasion) is a severe form of abrasive wear in which the force between an abrasive body and the wearing surface is large enough to macroscopically gouge, groove, or deeply scratch the surface. Gouging is performed by ore grinding mills and back hoe teeth.

Erosion. *Erosion* (low-stress abrasion) is a form of abrasive wear in which the force between an abrasive body and the wearing surface is large enough, but much smaller than those in gouging, to cause the removal of surface material. The forces are much lower and result in the removal of much smaller particles and much less breakdown of the abrasive body. Erosion can occur in moving liquids containing abrasive particles. If the liquid is corrosive, the form of damage is erosion-corrosion. Erosion occurs in coal and ore chutes, slurry pipelines, and fluid catalytic cracking (FCC) units in refineries.

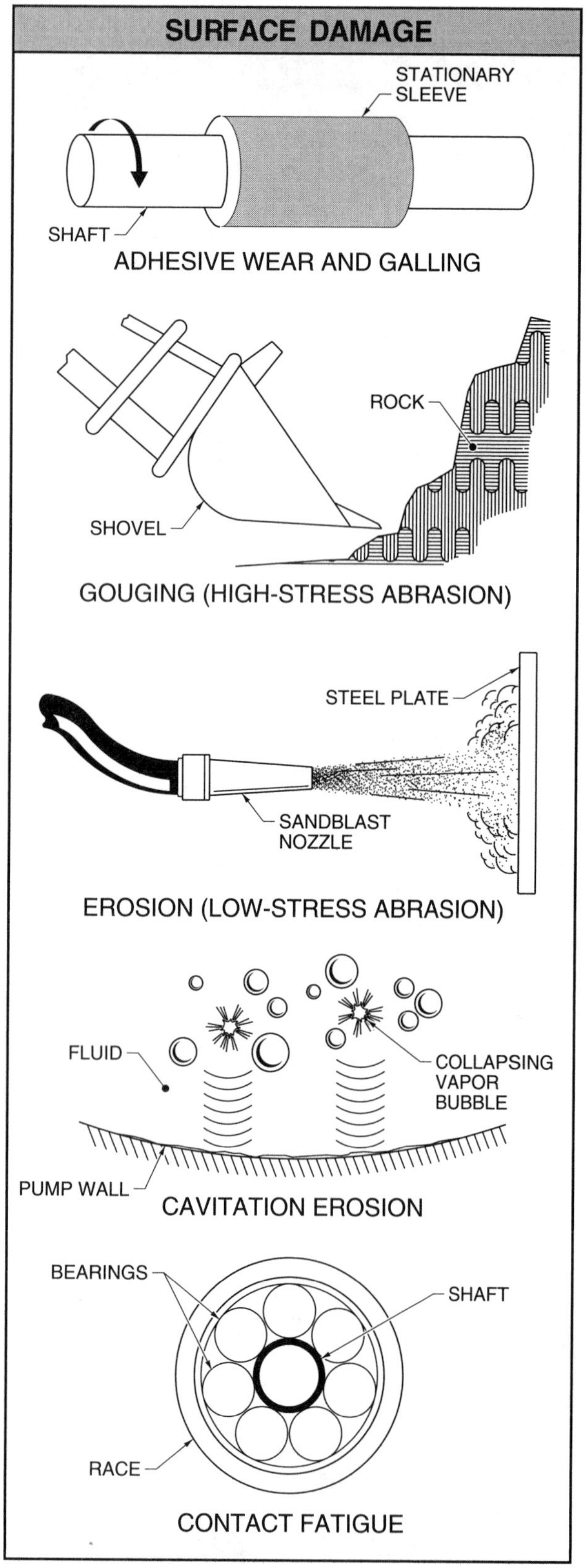

SURFACE DAMAGE

STATIONARY
SLEEVE

SHAFT

ADHESIVE WEAR AND GALLING

ROCK

SHOVEL

GOUGING (HIGH-STRESS ABRASION)

STEEL PLATE

SANDBLAST
NOZZLE

EROSION (LOW-STRESS ABRASION)

FLUID

COLLAPSING
VAPOR
BUBBLE

PUMP WALL   CAVITATION EROSION

BEARINGS

SHAFT

RACE

CONTACT FATIGUE

Figure 28-1. Hard facing is used to combat adhesive wear and galling, gouging (high-stress abrasion), erosion (low-stress abrasion), cavitation erosion, and contact fatigue.

Cavitation Erosion. *Cavitation erosion* is surface damage caused by collapsing vapor bubbles in a flowing liquid. The vapor bubbles form as a result of changes in flow velocity and/or direction, or a reduction in the cross section of the flow passage. An increase in pressure at the nearby location downstream causes the bubbles to collapse. The collapsing bubbles give rise to shock waves or explosions that cause contact stresses on the metal surface. Repetitive explosions lead to spalling and pitting of the surface. Cavitation erosion is common in pumps and engine cylinders.

Contact Fatigue. *Contact fatigue* is the cracking and subsequent break up of a surface that is subjected to alternating stresses, such as those produced under rolling contact or combined rolling and sliding. Contact fatigue is most commonly encountered in rolling element bearings or in gears, where the surface stresses are high due to the concentrated loads that are repeated many times during normal operation.

Weld Overlay

Weld overlay is a form of hard facing that is applied by oxyacetylene or arc welding processes using hard facing welding rods or electrodes. Welding is also used to apply corrosion-resistant alloys as an overlay to the surface of a metal.

Oxyacetylene Welding Overlays. Oxyacetylene welding overlays are widely used on steels where maximum hardness and minimum crack susceptibility are required. They can be applied to most hard facing materials, except for copper alloys.

Although the rate of deposition is not as high as in other processes, oxyacetylene welding has the important advantage of minimizing fusion of the base metal. This minimizes undesirable dilution and loss of hardness of the hard facing alloy. With copper alloys, a greater loss of aluminum or silicon by oxidation occurs during welding compared with arc welding processes, so that the deposit is softer. The absence of steep thermal gradients in oxyacetylene welding reduces cracking or spalling of the overlay because thermal stresses are reduced.

Arc Welding Overlays. Arc welding overlays are applied by gas tungsten arc welding (GTAW), shielded metal arc welding (SMAW), and gas metal arc welding (GMAW). GTAW produces very clean deposits with high rates of deposition. However, the high heat input results in steep thermal gradients, causing dilution and loss of hardness in the overlay coupled with increased cracking susceptibility from high thermal stresses. GTAW is often used where thin overlays are required.

SMAW produces high deposition rates and intermediate dilution. Cleanliness in surface preparation is not as stringent as in other processes. Although some porosity and cracking are present, such discontinuities are usually acceptable in the type of applications for which SMAW is employed. These severe applications, such as earth moving and mining equipment, require thick overlays. It is generally necessary to apply several layers of hard facing to achieve the intended surface hardness.

GMAW is not as widely used for hard facing as the other arc welding processes. Composite filler wire is usually used. This wire, like flux cored electrode, consists of a tubular steel shell in which metallic powders or fine particles of the hard compounds are incorporated. GMAW allows for high deposition rates and low dilution of the hard facing by the base metal.

Hard Facing Alloys. The principal hard facing alloy families are cobalt, nickel or copper alloys, and iron-base alloys with manganese or chromium. See Figure 28-2. With the exception of the copper alloys and manganese steels, wear resistance is developed from the presence of chromium carbide particles. Some alloy compositions have tungsten carbide particles incorporated to further improve wear resistance. Although many hard facing alloys have American Welding Society specifications, others are identified only by commercial designations.

For successful hard facing, a welding procedure must be established. It should specify the welding process, surface preparation requirements, and preheating and post-weld heat treatment temperatures. Additionally, the procedure should indicate any special welding techniques, such as the pattern of hard facing, the method of welding (stringer beads or weave beads), and any post-weld operations such as peening and the method of cooling.

In general, no more than three layers of hard facing alloy are deposited. Excessive numbers of layers are expensive and may introduce problems, such as cracking. For optimum results, it is common practice to prepare the base metal 6 mm ($\frac{1}{4}$ in.) from the final surface. This allows the two passes of hard facing to bring the part to final dimensions. The pass thickness should not exceed 4.5 mm ($\frac{3}{16}$ in.).

The preheating and post-weld heat treatment requirements for hard facing are the same as those for welding the base metal. Similarly, the ease of hard facing specific alloys depends on the weldability of the alloy. See Appendix.

Thermal Spraying

Thermal spraying (*THSP*) is a group of processes in which finely divided metallic or nonmetallic materials are deposited in a molten or semimolten condition to form a coating. The application material may be in the form of powder, ceramic, rod, or wire. The most important aspect of thermal spraying is correct preparation of the component. It must be cleaned and roughened, but sharp corners should be avoided.

Spraying should be performed immediately after the component is cleaned. If the component is not sprayed immediately, it should be protected from the atmosphere by wrapping it in paper containing a vapor-phase corrosion inhibitor. The first pass should be applied as soon as possible and as quickly as possible. Additional coats are applied slowly. It is important to maintain uniformity of temperature throughout the component. The surface of the component does not heat up appreciably during thermal spraying as it does during welding. Distortion during thermal spraying is minimal.

Warning: Thermal spraying requires protective clothing, eye protection, and special ventilation. All safety practices similar to those of welding processes must be observed. Ear protection may also be required.

CHARACTERISTICS OF HARD FACING MATERIALS

Hard Facing Materials	Abrasion Resistance	Oxidation Resistance	Corrosion Resistance	Impact Wear Resistance	Hot Hardness
Tungsten carbide composite AWS RWC – ⅝ through RWC – 40/120[a] for welding rods, and AWS EWC – 12/30 through EWC 40/120 for electrodes	Superior abrasive resistance Varies with volume of undissolved carbides Carbide granule size will affect surface smoothness	Vulnerable to oxidation above 650°C	Not recommended	Very low	Good up to 540°C
Austenitic high-chromium iron AWS – FeCr – A1	Low-stress abrasion resistance is good High-stress grinding abrasion resistance is only medium	Good up to 982°C due to high %Cr	Poor due to Cr in matrix being low Will rust in moist air	Low Cannot tolerate plastic deformation	Decreases slowly up to temperatures of 425°C to 480°C
Cobalt-base alloys AWS – CoCr – A,B,C	Depends upon carbon contents Alloy A is low Alloy C is medium	Good resistance due to high %Cr	Excellent in most environments	Low resistance Decreases as carbon increases Inferior to martensitic steels	Excellent Can be used above 650°C and is immune to tempering
Nickel-base alloys AWS – NiCr – A,B,C	Not recommended for high-stress abrasion Alloy C has medium low-stress abrasion resistance	Good up to 980°C	Very good for atmosphere, steam, salt water, salt spray, and mild acids	Low resistance Decreases as carbon increases	Very good up to 540°C
High-speed steel (Martensitic steels) AWS – Fe5 – A,B,C	Medium	Oxidizes readily because of high molybdenum content Use a reducing atmosphere in heat-treating furnaces	Very low	Medium Will improve with tempering	Good up to 595°C due to slow diffusion of tungsten and molybdenum
Austenitic manganese steel AWS – FeMn – A,B	Medium resistance	Resistance not greater than mild steel	No resistance greater than mild steel	High	None Use at ambient temperatures
Copper-base alloys AWS – CuAl – A2,B,C,D,E CuSi, CuSi – A CuSn, CuSn – A,C,D CuZn – E	Low	Generally good	Generally good	CuZn is very low CuSn and CuAl are low	Good up to 205°C

[a] The numerical designation refers to the mesh size of the tungsten carbide granules.

Figure 28-2. Hard facing materials are identified by their AWS designations or by their commercial names if no AWS designations exist.

Slight preheating of the component may be necessary depending on the alloy content of the component. The four principal thermal spraying processes are flame spraying, spray and fuse, plasma spraying, and high velocity oxyfuel (HVOF) thermal spraying. See Figure 28-3.

Flame Spraying. Flame spraying is a thermal spraying process that uses an oxyfuel gas flame as a source of heat for melting the coating material. Compressed air is used for atomizing and propelling the material to the workpiece. Two variations of flame spraying exist. One uses metal in wire form and is sometimes referred to as metallizing. The other uses materials in powder form. In both versions, the material is fed through a gun and a nozzle and melted in the oxyfuel gas flame. Atomizing, if required, is performed by an air jet that propels the atomized particles to the workpiece.

Flame spraying allows hard, thin coatings to be deposited quickly and uniformly. Deposits range from .25 mm to 2.00 mm (.01 in. to .08 in.) thick.

The coatings are porous and usually brittle. They do not resist excessive mechanical abuse.

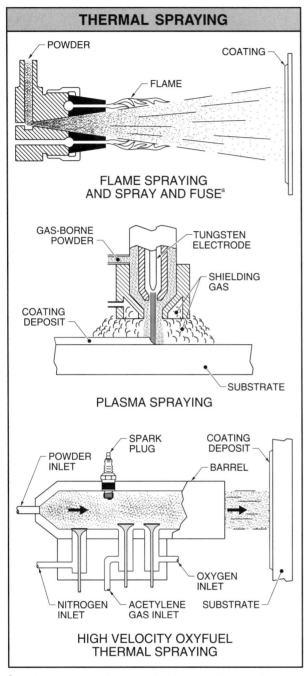

THERMAL SPRAYING

FLAME SPRAYING
AND SPRAY AND FUSE[a]

POWDER
COATING
FLAME

PLASMA SPRAYING

GAS-BORNE POWDER
TUNGSTEN ELECTRODE
SHIELDING GAS
COATING DEPOSIT
SUBSTRATE

HIGH VELOCITY OXYFUEL
THERMAL SPRAYING

SPARK PLUG
COATING DEPOSIT
POWDER INLET
BARREL
OXYGEN INLET
NITROGEN INLET
ACETYLENE GAS INLET
SUBSTRATE

[a] coated component is placed in furnace to fuse coating during spray and fuse process

Figure 28-3. The four principal thermal spraying processes are flame spraying, spray and fuse, plasma spraying, and high velocity oxyfuel thermal spraying.

Spray and Fuse. Spray and fuse is a variation of flame spraying in which the coating is fused after application by using a heating torch or by placing the component in a furnace. Spray and fuse coatings are usually made of nickel or cobalt self-fluxing alloys that contain silicon or boron to depress the melting point between 370°C and 425°C (700°F and 800°F) below the melting point of most steels. Tungsten carbide particles are added for increased wear resistance.

Fusing is performed by gradually increasing the temperature between 1010°C and 1300°C (1850°F and 2370°F), depending on the type of alloy being used. The fusion process permits the fluxing additives of silicon and boron to react with oxide films on the surface and with powder particles. This allows them to wet and interdiffuse with the substrate, resulting in full densification of the coating.

Plasma Spraying. *Plasma spraying* is a thermal spraying process in which a plasma torch is used as a heat source for melting and propelling the surfacing material to the workpiece. The temperature of the plasma arc is so much higher than that of flame spraying that additional material with higher melting points can be sprayed by this process. Most inorganic materials that melt without decomposition can be used.

The material to be sprayed is in a powder form and is carried to the plasma spray gun suspended in a carrier gas. The high-temperature plasma immediately melts the powdered metal and propels it to the surface of the workpiece. Since inert gas and high gas temperatures are used, the mechanical and metallurgical properties of the coatings are generally superior to either type of flame spraying, and bond and tensile strengths are higher.

High Velocity Oxyfuel (HVOF) Thermal Spraying. High velocity oxyfuel thermal spraying is quite different from other thermal spraying processes. In this process, a mixture of oxygen and a combustible gas, such as acetylene, is fed into the barrel of the spray gun with a charge of surfacing powder. The mixture is ignited and the detonation wave accelerates the powder to the workpiece while heating it close to or above its melting point. The cycle is repeated many times a second. The noise level is extremely high, and the process must be performed in a soundproof room.

HVOF thermal spraying is most successful in applying dense, hard, carbide, and oxide coatings to critical areas of precision components. Since the substrate surface is seldom heated above 150°C (300°F), the component can be fabricated and fully heat treated prior to coating.

Process Selection

The selection of hard facing processes depends on consideration of the wear process involved, the rate of wear, the operating temperature of the component, and the corrosiveness of the environment. For example, valve trim is sometimes subjected to a combined wear process of adhesive and low-stress abrasive wear from a combination of the process fluid and contact between mating components. In moderately corrosive environments and at moderately high temperatures, this leads to low wear rates, which can be eliminated by making the trim of quenched and tempered 410 stainless steel.

Another example is a pump shaft operating in process water subjected to a wear process of high-stress abrasive wear from the packing gland or bearing, leading to low wear rates. At moderate temperatures in the mildly corrosive environment, a plasma sprayed coating can be used. Weld overlays are not usually desirable because of the potential for warpage. However, if the contact stress is high enough, only weld overlays will exhibit adequate life.

A final example is a fluid catalytic cracker (FCC) slurry pump subjected to a wear process of low-stress abrasion from the slurry, leading to high wear rates at moderate operating temperatures in a mildly corrosive environment. Two alternative solutions to this are a weld overlay with a thick deposit or a thinner hard facing containing tungsten carbide, which is harder than the abrasive catalyst.

SURFACE TREATMENT

Surface treatment of metals encompasses a wide variety of processes that either modify the surface or apply a coating to it in order to provide resistance to wear, corrosion, or high-temperature corrosion. Surface treatment processes are divided into con-version coatings, hot dipping, electroplating, and surface modification.

Conversion Coatings

Conversion coatings are surface treatments that convert the surface layer of a metal by oxidation into a constituent of a coating, contributing to a strong bond between the metal and the conversion coating. Conversion coatings consist of black oxide, phosphate, chromate, and anodic coatings (anodizing). See Figure 28-4. Each coating has different finishes and is applied to varying substrates for a variety of applications.

Black Oxide Coatings. *Black oxide coatings* are conversion coatings applied to steel, stainless steel, copper, and other metals for decorative or functional purposes. These coatings are usually applied by simple immersion of the component in hot chemical solutions, which are usually highly alkaline and contain oxidizing agents. The oxidizing agents are usually nitrates or nitrites. The oxide for steel is magnetite (Fe_3O_4), and for copper, it is cupric oxide (CuO).

The black oxide coating provides a poor corrosion barrier because it is thin (usually less than 2 microns, or .08 mils) and porous. The coating is frequently sealed with an oil or wax to extend its life and lubricity. Black oxide coatings are applied to firearms, grenades, periscopes, telescopes, and items that require a black surface finish to avoid detection from reflected light.

Phosphate Coatings. *Phosphate coatings* are conversion coatings of zinc or magnesium phosphate that are developed on steels to prepare them for painting. A phosphate solution is used to develop these coatings. It consists of phosphoric acid and soluble phosphates of zinc and magnesium. The phosphoric acid reacts with the metal surface, increasing the alkalinity of the interfacial solution (thin layer of solution in contact with the metal surface). This leads to the formation of metal phosphates in the thin layer. The metal phosphates are insoluble in the interfacial solution and crystallize on the metal surface. Along with substrate metal phosphates that are also formed, a firmly anchored deposit is produced.

The microcavities and capillaries in phosphate coatings provide a large surface area. This greatly improves the adhesion of paint compared with a clean, untreated metal surface. Phosphate coatings are applied by spraying or dipping articles such as nuts, bolts, barrels, and other hardware.

Zinc phosphate coatings are also used in conjunction with a lubricant (usually soap) to facilitate the deformation of steel components during processes such as drawing, stamping, and extrusion. The phosphate holds the lubricant tenaciously by mechanical and chemical forces, preventing the seizure of the deforming metal in the die or tool.

Chromate Coatings. *Chromate coatings* are conversion coatings applied to aluminum, brass, cadmium, copper, and zinc by immersion in a solution containing a dichromate or chromic acid, plus an activator. The activator consists of a nitrate, sulfate, chloride, formate, or fluoride.

An oxidation-reduction reaction occurs on the metal surface with the formation of base metal ions and trivalent chromium ions. There is an accompanying increase in alkalinity of the solution adjacent to the metal surface. This results in the precipitation of a gelatinous film comprised largely of chromic hydroxide and substrate metal compounds, in which soluble chromates are incorporated. After rinsing and drying, this coating quickly hardens. A 24-hour waiting period is usually observed before handling.

CHARACTERISTICS AND APPLICATIONS OF CONVERSION COATINGS			
Primary Finish Characteristics	**Coating**	**Substrate**	**Application**
Corrosion resistance	Anodic-sealed	Al,Mg,Zn	Military equipment, aircraft components, pump impellers
	Chromate	Al,Zn,Cd	Electronic chassis, bicycle parts, carburetors, fuel pumps
	Zinc phosphate plus oil or wax	Fe	Underhood automobile parts, hardware, cartridge clips
Decoration	Anodic-dyed	Al	Tumblers, giftware, kitchenware, jewelry
	Anodic-clear bright	Al	Automobile trim, hollowware, appliances, reflectors
	Anodic-integral color	Al	Outdoor architectural components
	Chromate-clear bright	Zn,Cd,Al	Hand tools, fasteners, wire goods, shelving
	Black oxide	Fe,Cu,Brass	Hand guns, tools, fasteners, lamp components
Wear resistance	Anodic-hard coat	Al	Weapon and missile components, pistons, gears
	Manganese phosphate plus lubricant	Fe	Piston rings, gears, camshafts
Paint or lacquer bonding	Zinc phosphate	Fe	Automobile bodies and frames, washing machines, vending machines
	Iron phosphate	Fe	Metal stampings, file cabinets, coil stock
	Chromate	Al,Zn	Automotive components, house siding, refrigerator shelves
	Anodic	Al,Mg	Aircraft components, brakes
Nonreflective and solar-heat absorbing	Chromate-black or olive drab	Zn,Cd	Camera components, hand grenades
	Black oxide	Fe,Cu, Stainless steel	Periscopes, weapons components
	Zinc or manganese phosphate	Fe	Weapons components
Facilitation of metal deformation	Zinc phosphate plus lubricant	Fe	Tubing, wire, cartridge cases, projectiles
Color coding	Chromate dyed	Zn,Cd,Al	Fasteners, fuse bodies, simulated brass

Figure 28-4. Conversion coatings are surface treatments that modify the surface layer of a metal by oxidation.

Chromate coatings provide increased corrosion resistance not only by providing a barrier to the environment but also as a result of scratching or abrasion, because soluble chromates leached from the chromate film provide protection to the freshly exposed area. The chromate film, which is also applied as the last step after the plating of zinc and cadmium, can provide color, leading to a wide variety of applications.

Anodic Coatings. *Anodic coatings* are conversion coatings applied by the anodic oxidation of certain base metals. Aluminum is the most frequently anodized metal, but the process is also applied to magnesium, titanium, and tantalum.

The anodizing of aluminum is performed by making the component the anode in an electrolyte in which the naturally formed surface oxide film is soluble, such as 10% to 12% sulfuric acid. Direct current is used to develop a superior oxide film that consists of a nonporous barrier layer on the metal surface and a porous outer layer. The porous outer layer may be sealed by hot water treatment to augment corrosion resistance.

Anodized coatings can be dyed to provide many decorative finishes that are both abrasion and corrosion resistant. The degree to which dyes can be absorbed into the coating depends on the thickness and porosity of the coating. The thicker the coating, the darker the color that can be obtained. The construction industry uses dyed anodized aluminum for door and window frames on residential and industrial buildings. See Figure 28-5.

Hot Dipping

Hot dipping is a processes in which a component is dipped into a molten metal to produce a thin coating that enhances corrosion resistance or workability. The principal hot dipping processes are hot dip galvanizing, hot dip aluminizing, hot dip tin, and hot dip lead.

Hot Dip Galvanizing. *Hot dip galvanizing* is a method of applying a coating of zinc on ferrous substrates, primarily steel. The primary purpose of hot dip galvanizing is corrosion protection provided by the zinc. The corrosion protection is a function of the coating thickness and the exposure conditions.

Hot dip galvanizing is performed by dipping cleaned and activated steel components in molten zinc at 445°C to 465°C (830°F to 870°F). At these temperatures, the molten zinc is interlocked with the steel by forming various iron-zinc alloy layers, creating a metallurgical bond. These layers provide protection to the base metal. The thickness of the hot dip galvanized coating is measured in grams of coating per square meter of surface (ounces of coating per square foot of surface).

Most specifications call for a minimum coating thickness in the range of 85 microns to 100 microns (3 mils to 4 mils) or 500 g/m^2 to 600 g/m^2 (1.6 oz/ft^2 to 2.0 oz/ft^2). Since forms such as sheet are usually coated on both sides, the coating weight per unit area on each side is approximately one half the average weight of coating per unit area. *ASTM A123, Specification for Zinc (Hot Galvanized) Coatings on Products Fabricated from Rolled, Pressed and Forged Steel Shapes, Plates, Bars and Strip*, and *ASTM A153, Specification for Zinc Coating (Hot Dip) on Iron and Steel Hardware*, provide coating weight requirements for hot dip galvanized coatings. Items such as nails, reinforcing rolls, anchor bolts, pipe and pipe fittings, refuse containers, and guardrails are typically galvanized.

APPLICATIONS OF ANODIZED ALUMINUM			
Medium Thickness			
μm	in.	Description	Applications
2.5	.0001	Interior, limited abrasion	Housewares, automobile interior components
5.0	.0002	Interior, moderate abrasion	Appliances, nameplates, lawn furniture
7.5	.0003	Automotive exterior	Exterior automobile trim
10.0	.0004	Architectural	Maintained exterior facades, windows
17.5	.0007	Architectural, heavy duty	Unmaintained exterior facades

Figure 28-5. Anodizing is the anodic oxidation of specific base metals to enhance their naturally formed oxide films.

Zinc coatings of any type should never be used on critical steel components that reach temperatures above 260°C (500°F) because zinc may diffuse into grain boundaries and cause embrittlement of the steel. Zinc coatings are not to be used on high-strength steel components or fasteners because the corrosion of the coating may produce hydrogen and lead to cracking of the steel. Zinc coatings can produce bulky corrosion products during exposure to marine or tropical environments and are not to be used where the products may cause binding and prevent functioning of equipment. In specific hot aqueous solutions above 60°C (140°F) the electrochemical potentials of zinc and steel reverse, which means that the steel corrodes preferentially to the zinc and may cause premature corrosion failure of the steel.

Hot Dip Aluminizing. *Hot dip aluminizing* is a method of applying aluminum coatings to sheet steel, strip, or wire in a continuous length and also to fabricated components. Aluminumized coatings considerably outperform galvanized coatings of the same thickness in all environments, except on cut edges and bare areas in nonsaline exposure sites.

Aluminized steel is particularly effective in retarding high-temperature corrosion and oxidation and is used on automobile mufflers, furnace components, and heater tubes. Aluminized steel is sometimes used for chain-link fencing and barbed wire to provide outdoor corrosion resistance. A hot dip 55% Al and 45% Zn alloy has particularly good sacrificial protection on steel and long term corrosion resistance.

Hot Dip Tin Coating. *Hot dip tin coating* is a method of applying a tin coating to a base metal. It is more common for tin coatings to be applied by electroplating.

Hot Dip Lead Coating. *Hot dip lead coating* is a method of applying a lead coating to a base metal. Lead coatings usually contain small amounts of tin to improve bonding. Lead coatings provide effective protection to steel in various natural environments.

Lead alloys containing from 3% Sn to 20% Sn are referred to as terne coatings. Terne coatings are used for roofing materials, gasoline tanks, and stampings that require deep drawing, where the coating acts as a die lubricant. Terne coatings are also applied by electroplating.

Electroplating

Electroplating (electrodeposition) is a process for depositing (plating) a thin layer of metal onto a metallic component that is made the cathode in an electrical circuit and immersed in a solution containing ions of the metal to be plated. In some cases, nonmetals can be electroplated. Electroplating is used to improve corrosion or wear resistance, increase surface hardness, provide an attractive appearance, change electrical characteristics, prepare surfaces for joining, improve dimensional tolerances, or build up work components.

The largest use of electroplating is on steel, but many other substrate materials, such as aluminum, brass, cadmium, chromium, copper, tin, zinc, and plastic, are electroplated. *Electroforming* is a process in which a plating process is used to make a structural form.

Electroplating Principles. Electroplating consists of making the component to be plated the cathode in the plating solution. In a simplified operation, the components are suspended by a hook or wire from a cathode bar. Low-voltage direct current is supplied by a rectifier connected by a cable or buss bar to the anode and cathode components of the electric circuit in the plating tank via anode and cathode bars. These anode and cathode bars are fabricated of copper and are designed to carry the weights of the anodes and cathodes as well as conduct the current that is applied. See Figure 28-6.

The anodes are electrodes that are suspended in the plating solution. They are usually made of the same metal that is being plated. For example, copper anodes are used for copper plating and nickel anodes for nickel plating. During the plating operation the anode dissolves, replenishing the solution of metallic ions that are depleted by metal deposition on the cathode.

Insoluble anodes are used for specific plating operations. The purpose of insoluble anodes is solely to provide current to the electrolyte. Insoluble anodes are used in situations where the electrolyte dissolves the anode at a rate faster than the metal is plated out of solution, or alternatively where the cost of a pure anode made of the metal to be plated is prohibitive. When insoluble anodes are used, the plating bath must be replenished continually to maintain the optimum concentration of ions in solution. Examples of insoluble anodes are carbon for precious metals and lead alloys for chromium.

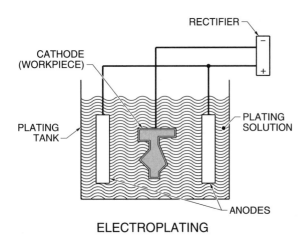

Figure 28-6. An electroplating bath consists of a tank, anodes, cathodes, and rectifier. The tank is filled with the plating solution (electrolyte).

During plating, low-voltage direct current is supplied to the electrolyte by the rectifier. Depending on the process, the voltage is in the range of 6 V to 12 V, with a typical current density of 20 amps to 100 amps per square foot. A cathode area of 10 ft² would require a rectifier capacity of 200 A to 1000 A. In production plants, rectifiers with a capacity of 5000 A to 20,000 A are common.

The plating solution is an electrolyte. The electrolyte is generally an aqueous solution containing dissolved salts of the metal to be plated and other chemicals (wetting agents, leveling agents, and brighteners) as are required.

An electrolyte is a conductive media that allows current to pass from the anode to the cathode. As direct current is applied to soluble anodes, metal is dissolved and forms positive metal ions. The current causes these positive metal ions to migrate toward the cathode, where they are reduced to the metallic state and are deposited on the cathode. See Figure 28-7. Due to the complexity of many elec-

trolytes, other secondary reactions may also occur. The principal reaction is the deposition of metal on the cathode.

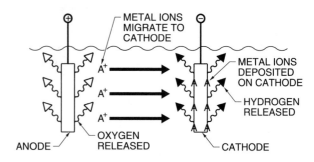

Figure 28-7. During electroplating, positive ions in the plating solution migrate toward the cathode, where they are reduced to the metallic state and are deposited on the cathode.

During electroplating, the current is applied for a predetermined amount of time until the desired amount of metal is deposited on the cathode workpiece. Theoretically, a deposit of one gram equivalent (GE) of the metal (atomic weight divided by the valency, expressed in grams) is obtained for each faraday. A *faraday* is a unit of electricity equal to 96,500 ampere-seconds. See Figure 28-8. For example, for each faraday, 31.8 g of copper is deposited, where 31.8 g is the gram equivalent weight of copper with a valency of two.

AMOUNT OF DEPOSITED METAL[a]			
Metal	**Valence**	**Atomic Weight**	**Grams[b]**
Cadmium	2	112.41	56.21
Chromium	6	52.01	8.67
Copper	1	63.57	63.57
Copper	2	63.57	31.78
Gold	1	197.20	197.20
Nickel	2	58.71	29.36
Silver	1	107.88	107.88
Tin	4	118.70	29.67
Zinc	2	65.38	32.69

[a] at 100% efficiency
[b] grams deposited per faraday

Figure 28-8. The amount of metal deposited in electroplating is determined by its equivalent weight.

The actual weight of metal deposited is given by the following equation:

$$W = \frac{I \times t \times GE}{F} \times \text{Current Efficiency}$$

where

I = current in amps

t = time in seconds

GE = gram equivalent

F = faraday (96,500)

The current is a measure of the percentage of current that actually deposits metal. A current efficiency of 100% is not achieved because part of the current is consumed in parasitic reactions, such as hydrogen evolution at the cathode surface.

Example: Figuring weight of metal deposited

For a current of 20 amps applied for 30 minutes (1800 s) at a current efficiency of 90%, what is the weight of copper deposited from a copper sulfate ($CuSO_4$) bath?

$$W = \frac{I \times t \times GE}{F} \times \text{Current Efficiency}$$

$$W = \frac{20 \times 1800 \times 31.8}{96,500} \times .9$$

$W = $ **10.7 g deposited**

A proper distribution of the plating is achieved when the anodes and cathodes are positioned so that the current flow is uniform across the entire cathode surface. Areas of the component, such as a protruding corner, that are closer to the anode than other areas will receive a higher current density and consequently develop a thicker deposit. Proper positioning and masking of the component in the plating solution will alleviate the problem and produce a deposit of uniform thickness.

Electroplating Procedures. Electroplating consists of several steps that include surface preparation of the workpiece, cleaning, activation, electroplating, rinsing, and drying. Surface preparation is required when defects such as scale, pits, mold marks, grinding lines, tool marks, or scratches must be removed. Polishing, buffing, sand blasting, or grinding operations are used to prepare the surface.

Cleaning removes unwanted soils, greases, and oils that come from fabrication and surface preparation. Cleaning operations include solvent cleaning, vapor degreasing, and emulsion cleaning. An effectively cleaned surface is water-break free, meaning that water does not coalesce into droplets on the surface when allowed to run off.

Activation (pickling) is performed just before plating. It consists of exposing the workpiece to an acid such as sulfuric, hydrochloric, nitric, hydrofluoric, or mixtures of these, which removes oxides and scale from the metal surface. After activation, the workpiece is rinsed with water and placed immediately in the plating solution.

Several methods of handling the workpiece in the plating bath include rack, barrel, basket, brush, and selective plating. With rack plating, a framework (rack) is used to hold the workpiece in a fixed position. Barrel plating is used to plate small workpieces in bulk, such as nuts, screws, washers, and bolts. The workpieces are loaded into a barrel containing many small perforations to allow the flow of electrolyte. Basket plating is also a bulk method in which the workpieces are manually loaded into a basket that is suspended by a hook from the cathode.

Brush plating and selective plating (electrochemical metallizing) are methods of electrodepositing metal without an immersion tank. Brush or selective plating is used to touch up damaged plating or to locally repair worn or scored shafts or other items in the field. See Figure 28-9.

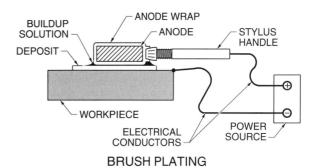

BRUSH PLATING

Figure 28-9. Brush or selective plating is used to touch up damaged plating or to locally repair worn or scored shafts or other items in the field.

In brush plating, the negative lead of the plating rectifier is connected to the workpiece and the positive lead is connected to the brush or swab, which is usually wrapped in an absorbent cloth. The swab is dipped in the plating solution and the area to be plated is swabbed. The procedure is repeated until the area is sufficiently plated. Selective plating is a refinement of brush plating. Higher current densities and more concentrated plating solutions are used for selective plating. The anodes are usually composed of carbon. The plating solution is circulated and cooled. After plating, the chemicals are rinsed off the workpiece with clean water. This is followed by thorough drying to prevent staining or rusting.

Cadmium Plating. Cadmium plating is generally applied to steel substrates and is used primarily as protection against corrosion because it provides a sacrificial coating. Cadmium does not form crust-like corrosion products as zinc does and is superior to zinc in marine environments. Cadmium is the most widely used sacrificial coating for steel in military and aerospace applications. It has excellent solderability and lubricity.

Cadmium is usually applied as a thin coating of <25 microns (1 mil) thick. It is applied from cyanide baths using soluble cadmium anodes. Hydrogen embrittlement of high-strength steels may result when cadmium is electroplated from the commonly used plating baths. Special bath formulations must be used to avoid this problem. Brighteners are added to produce uniformity of deposit and improve throwing power. *Throwing power* is the ability of an electrolyte to deposit metal of a uniform thickness across a cathode surface.

Cadmium plating is produced principally to *Federal Specification QQ-P-416, Plating, Cadmium (Electrodeposited)*. Applications of cadmium plating include aerospace fasteners, landing gear, steel springs, bearing rings and assorted hardware, and electronic components.

Warning: Cadmium is toxic and should not be used in any food-handling applications.

Zinc Plating. Zinc plating is much cheaper than cadmium plating. It is applied to steel to provide sacrificial protection. However, corrosion products of zinc in severe environments tend to form white, crust-like deposits. This does not occur in less severe environments or indoors. Unlike hot dip galvanized coatings, electrodeposited zinc forms an attractive, bright, silvery coating. It is applied as a thin film of 7 microns to 15 microns (.3 mils to .6 mils), although coatings up to 20 microns (.8 mils) are used for more severe exposure conditions. A single dip of clear chromate finish may be applied to the zinc to provide a decorative luster.

Mechanical Plating of Zinc. Zinc is mechanically plated by articles with impact media, such as glass beads, in a liquid slurry containing powdered zinc. This process is not electrochemical and is used to coat high-strength steel components, such as lockwashers, that could be embrittled by electroplating.

Tin Plating. Tin plating is the process of applying tin to sheet steel and copper wire by electroplating. Tin is plated from acid or alkaline solutions using soluble anodes. The position of tin in the galvanic series in seawater indicates that it will not protect steel in seawater or in the atmosphere. In the presence of food acids, such as citric acid and oxalic acid, the electrochemical potentials of tin and steel are reversed, and tin is protective to steel.

Copper wire is tin plated to prevent discoloration from reactions with substances in the insulation covering. Continuous coils of steel strip are electroplated with thin tin deposits of 2.5 microns to 75.0 microns (.1 mils to 3.0 mils) for the production of cans.

The tin plate is usually heat-flow brightened, which consists of immersion in a hot liquid to eliminate porosity. Heat-flow brightening forms an iron-tin compound on the plating surface that improves protection.

Hard Chromium Plating. Hard (engineering, or industrial) chromium plating is plating that is resistant to abrasion, corrosion, heat, friction, and galling and is applied principally for wear resistance from 12.5 microns to 250.0 microns (.5 mils to 10.0 mils) thick. It is not normally considered for chemical resistance unless the base metal has good corrosion resistance, such as a base metal with a corrosion rate of <5 mils per year (mpy). If the base metal has a higher corrosion rate, alternative plating methods are considered.

Hard chromium plating is applied to steels, tools steels, stainless steels, cast irons, nickel-base alloys, and brasses. Plating of other metals is not very common. The hardness of the deposit varies from 600 HV to 1100 HV, depending on the plating conditions.

Standard, nonfluoride proprietary, and high-speed baths are all used for hard chromium plating. These baths contain a chromic acid and sulfate solution. The nonfluoride proprietary and high-speed baths also contain catalysts that allow the use of lower chromic acid concentration. This reduces cost and also allows for a wider temperature and current density range. The catalysts also greatly improve metallurgical properties of the deposit, such as hardness, crack structure, and wearability. One disadvantage of the fluoride-containing bath is the extreme corrosivity toward the components being plated. For this reason, all areas not requiring plating must be masked with either paint, plastic, tape, or wax to keep them from contacting the electrolyte.

Since chromium plating baths have very poor throwing power, conforming anodes must be used on components with intricate shapes. These anodes are placed .5 cm to 2.5 cm (.2 in. to 1.0 in.) from the component and conform exactly to the surface configuration. Both conforming and regular tank anodes are made of a lead alloy, typically composed of 93% Pb and 7% Sn.

Chromium plating magnifies surface imperfections on the workpiece. Very little to no leveling is provided by chromium plating. *Leveling* is the ability of an electrolyte to deposit metal in order to smooth out surface scratches and imperfections by filling them up with deposit. Careful surface finishing of the workpiece prior to plating is mandatory. Hard chromium plating is usually finish ground.

Hard chromium plating tends to form a microcracked structure. A crack-free deposit is obtained by altering the bath conditions, but it is relatively soft (approximately 600 HV), smooth, and silvery rather than bright and shiny. Normal hard chromium (800 HV to 900 HV) is bright and contains 400 cracks to 900 cracks per inch. Very hard deposits (1000 HV) show 100 cracks to 300 cracks per inch. Cracks lead to a reduction in corrosion and fatigue

resistance, but such structures are an advantage in bearing applications. In bearing applications, these cracks tend to fill with oil or lubricant and provide low coefficient of friction.

Chromium plating reduces fatigue strength 60% to 70% compared with the unplated condition. See Figure 28-10. The reduction in fatigue strength is related to the tensile stresses in the plated layer. Baking at 190°C (375°F) for 4 hours removes hydrogen and helps reduce the effect but does not return the workpiece to its preplated fatigue strength.

EFFECT OF CHROMIUM PLATING		
Condition	Fatigue Strength (ksi)	% Loss
As-finished (unplated)	84	—
Chromium plated	38	45
Chromium plated and baked	59	70
Shot peened and chromium plated	81	96
Shot peened, chromium plated, and baked	83	99

Figure 28-10. Chromium plating causes a significant reduction of fatigue strength unless steps are taken to prevent the loss.

Shot peening before plating almost completely offsets the loss in fatigue strength. Shot peening involves cold working the metal surface with a high-velocity stream of steel or chilled iron shot. This produces a cold-worked surface layer that is in compression and offsets the residual tensile stress in the chromium plating. Shot peening must be performed on highly stressed components subject to dynamic loads. Shot peening produces a rougher surface finish on the workpiece than machining or grinding. The application of shot peening and baking on hard chromium plated workpieces is covered in *Military Specification MIL-S-13165* and *Federal Specification QQ-C-320B*.

Hard chromium plating is applied to components such as drawing dies, molds for plastics, mixing equipment, gun barrel interiors, thread guides, hydraulic shafts, piston rings, gear rolls of various types, punch and die sets, gauges, and diesel-engine cylinder liners. Hard chromium plating is usually applied directly over the base metal unless significant buildup is required for component restoration from wear or mismachining. In these cases, an undercoat of nickel is first applied to the required thickness.

Decorative Chromium Plating. Decorative chromium plating is an extremely thin chromium plating that usually does not exceed 1.25 microns (.05 mils) and that is applied to components that require a bright and aesthetically pleasing appearance. It is applied over a nickel or nickel-plus-copper plated underlayer, making the total plating thickness approximately 12.5 microns (.5 mils) thick. The underlying layers of copper and nickel protect the steel substrate from corrosion because, as with hard chromium, decorative chromium contains microcracks.

The current efficiency of decorative chromium plating baths is low, approximately 15% to 20%. The metal is plated at conventional speeds by the use of high current densities. Catalysts are added to the bath to promote deposition at higher speeds, greatly increasing the throwing power. Despite this, chromium baths generally have very poor throwing power and require proper racking techniques and careful control of optimum bath composition. Decorative chromium is used on automobile bumpers and trim, household appliances, furniture, and many other articles that require a bright, attractive appearance.

Copper Plating. Copper plating is copper applied to steel substrates and is from 8 microns to 50 microns (.3 mils to 2.0 mils) thick. It is primarily used as an undercoat for nickel or chromium. It is also used for antiqued components, where it is intentionally oxidized to obtain a desired color. Copper is plated primarily from cyanide and pyrophosphate alkali baths, and sulfate and fluoborate acid baths.

Copper plating is used as a plating barrier in carburizing operations in which only selective areas are to be carburized. On zinc-base die castings, copper plating is used as the initial coating to cover the pores in the casting.

Nickel Plating. Nickel plating is capable of exhibiting a wide variety of properties, depending on the composition of the plating bath and the operating conditions. It is used extensively as an undercoat for decorative chromium. Nickel plating is also used

as an undercoat for hard chromium where surface build up is required, such as in the rebuilding of components. Although the chromium top layer is microcracked, the nickel undercoat is continuous and creates a highly corrosion-resistant system.

Numerous baths are used for nickel plating. See Figure 28-11. The two most widely used baths are the Watts bath and the nickel sulfamate bath. The Watts bath is the oldest and most widely used bath. This is a mixture of nickel sulfate and nickel chloride, with roughly five parts nickel to one part chloride. Approximately half a part of boric acid is added for buffering purposes. Wetting, brightening, and levelling agents are also added to the bath. The wetting agent prevents pitting of the deposit caused by the evolution of hydrogen gas and bath contamination. Brightening agents are used to refine the grain structure and promote brightness. Leveling agents are used to improve the microthrowing power, promoting fast brightening with minimal cost.

The nickel sulfamate bath is more concentrated than the Watts bath, and higher deposition rates are obtained. One of the biggest advantages of this bath is the low residual stress it produces in deposits. When heavy coatings are to be applied, or fatigue is a concern, the plating rate coupled with the low residual stress make nickel sulfamate the choice over the Watts bath.

Soluble anodes are used for nickel plating. These anodes, in the shape of buttons, are referred to as SD nickel and contain a dispersion of sulfur to increase their reactivity. The anodes are contained in perforated baskets.

A double-layer nickel deposit is much more effective than a single layer of the same total thickness in preventing substrate corrosion. The first layer is deposited from a typical Watts bath with organic semibright additives and usually accounts for three quarters of the total nickel thickness. The second layer is bright nickel and accounts for one quarter of the nickel thickness.

The surface layer of bright nickel is anodic to the semibright and corrodes sacrificially. Penetration is slowed at the semibright interface, and subsequent corrosion proceeds laterally into the bright surface layers, reducing the penetration rate. The bright layer thus provides sacrificial protection to the inner layer, which preserves its integrity as a barrier. Double-layer nickel deposits are produced to various thicknesses depending on the service application.

Electroforming. Electroforming is the physical formation of articles using electrodeposition. Nickel and copper are most often used for electroforming and are usually deposited from acid baths to provide low-stress deposits. Intricate components are easily formed, and extremely fine detail can be produced.

Electroformed deposits may be applied to reusable, permanent mandrels (molds) or to mandrels that are dissolved or melted out after the electroform is completed. Typical applications of electroforming are phonograph record stampers, pens, parabolic reflectors, radar wave guides, whisker guides for electric razors, precision screens and filters, miniature bellows, electrotype printing plates, and hypodermic needles.

Electroless Nickel Plating. Electroless nickel plating is nickel plating achieved without an electric current. Nickel is chemically reduced to the metallic state and deposited onto the workpiece from a solution containing nickel ions. This is achieved by the use of a reducing agent, such as sodium hypophosphite. Plating rates are very slow compared with electroplated nickel, and very few applications over 125 microns (5 mils) thick are economically feasible. Additionally, chemical costs are high for electroless nickel plating.

Unlike electroplated nickel, electroless nickel deposits contain an appreciable amount of phosphorous. This accounts for superior corrosion resistance, high hardness, and relatively low ferromagnetic properties compared with electroplated nickel. The primary advantage of electroless nickel is that it is not dependent on current density or throwing power. An electroless nickel coating is uniform in thickness and composition without the limitations that apply to electroplating. Complex components can be

NICKEL PLATING BATHS			
Properties	**Watts Bath**	**Sulfamate Bath**	**Fluoborate Bath**
Composition			
Nickel sulfate	30 to 55 oz/gal.	—	—
Nickel chloride	4 to 8 oz/gal.	0 to 4 oz/gal.	0 to 2 oz/gal.
Nickel sulfamate	—	35 to 60 oz/gal.	—
Nickel fluoborate	—	—	30 to 40 oz/gal.
Total nickel as metal	7.7 to 14.2 oz/gal.	8.2 to 15 oz/gal.	7.6 to 10.5 oz/gal.
Boric acid	4 to 6 oz/gal.	4 to 6 oz/gal.	2 to 4 oz/gal.
Operating conditions			
pH	15 to 5.2	3 to 5	2.5 to 4
Temperature	46°C to 71°C	38°C to 60°C	38°C to 71°C
Current density	10 to 100 A/ft^2	25 to 300 A/ft^2	25 to 300 A/ft^2
Mechanical properties of deposits			
Tensile strength	50 to 100 ksi (345 to 690 MPa)	55 to 155 ksi (380 to 1070 MPa)	55 to 120 ksi (380 to 830 MPa)
Vickers hardness	100 to 250 HV	130 to 600 HV	125 to 300 HV
Elongation in 50 mm (2 in.)	10% to 35%	3% to 30%	5% to 30%
Stress	15 to 30 ksi (105 to 205 MPa)	.5 to 16 ksi (3 to 110 MPa)	13 to 30 ksi (90 to 205 MPa)

Figure 28-11. Nickel plating is capable of exhibiting a wide variety of properties depending on the composition of the plating bath and the operating conditions.

plated to close tolerances. The bath for electroless nickel plating is very stable.

The hypophosphite-reduced nickel coatings are one of the few metallic glasses that are used as an engineering material. Depending on the formulation of the plating solution, commercial coatings may contain from 6% P to 12% P dissolved in the nickel. As plated, these coatings are amorphous, containing no crystal structure or grain boundaries.

As electroless nickel is heated above approximately 230°C (445°F), changes start to occur. Particles of nickel phosphide form within the material. At 315°C (600°F), the deposit begins to crystallize and lose its amorphous character.

With continuing time at these temperatures, the nickel phosphide begins to agglomerate and a two-phase alloy forms. With coatings containing over 8% P, the matrix is nickel phosphide, whereas at lower nickel contents, the matrix is pure nickel. The net effect of the increase in phosphorus is the increase in hardness and wear resistance at the expense of corrosion resistance and ductility. The best corrosion resistance is obtained in the as-plated condition, and alloys containing more than 10% P have better corrosion resistance than lower phosphorus alloys.

Heating above 230°C (445°F) also causes a shrinkage of approximately 4% to 5%, which increases the tensile strength of the coating. As deposited, electroless nickel coatings have a hardness of 500 HV to 600 HV. Heat treatment can increase hardness to over 1000 HV.

Electroless nickel coatings cause a reduction in fatigue strength of 10% to 50% in alloy steels. Coatings lower in phosphorus contain the highest residual stress and cause the greatest decrease in fatigue strength. Thicker deposits also have a greater effect in reducing fatigue strength.

Although much less hydrogen is produced by electroless nickel than by electrolytic plating, the amount is sufficient to crack high-strength steels. Steels with hardnesses of greater than 40 HRC should be baked after plating to drive out hydrogen. Baking is performed at 190°C (375°F) for a minimum of 4 hours, the same conditions as for chromium and nickel electroplating.

Electroless nickel is used on dies for molding and extruding plastics, valves, air compressors, pumps, and other components where a corrosion-resistant and good wearing surface is required. Heat-treated electroless nickel is not as effective as hard chromium plating for severe wear applications. For optimum hardness, the component is heat treated at 175°C (350°F).

Two ASTM standards provide extremely useful information on electroless nickel plating. *ASTM B733, Standard Guide for Autocatalytic Nickel-Phosphorus Coatings on Metals* defines five service conditions, specifies minimum coating thicknesses for each, and provides a guide to coating thicknesses to meet various types of corrosion and wear conditions. *ASTM B656, Standard Guide for Autocatalytic Nickel-Phosphorus Deposition on Metals for Engineering Use* contains useful information on pretreatments, plating bath control and maintenance, as well as possible post treatments.

Precious Metal Plating. All the precious metals, which include silver, gold, rhodium, osmium, platinum, and palladium, can be plated. Thicknesses range from .15 microns to 125.00 microns (.006 mils to 5.000 mils). They are plated from acid cyanide or neutral baths and exhibit excellent chemical resistance and good physical properties. The largest application for precious metal plating is in the electronics industry, where they are used to reduce contact resistance and increase corrosion resistance.

Plating On Plastics. Products made of plated plastics are visually indistinguishable from products made of plated metals. Plastics that are often plated include acrylonitrile-butadiene-styrene (ABS), polysulfone, and polypropylene. Plated plastic components include knobs, handles, automobile grills and headlight bezels, interior automobile parts, plumbing fixtures, cosmetic containers, camera equipment, marine hardware, household appliances, and office furniture.

A thin electroless nickel or copper deposit is first applied to provide a conductive base for electroplating. A bright acid-sulfate copper deposit is then applied to produce a leveling and brightening effect. A bright nickel deposit is next applied to develop an additional luster and to serve as a suitable base for the final decorative deposit, which is usually a chromium deposit.

Surface Modification

Surface modification processes alter the surface layers of a metal to provide desirable properties, such as wear resistance, corrosion resistance, and high-temperature oxidation resistance. Surface modification processes include aluminum pack diffusion, ion implantation, physical vapor deposition (PVD), and chemical vapor deposition (CVD). Case hardening heat treatment processes, such as carburizing and nitriding, are other types of surface modification processes.

Aluminum Pack Diffusion. *Aluminum pack diffusion* is a surface modification process used for enriching the surface layers of iron, nickel, cobalt, and copper alloys. Because of the high temperatures involved, the aluminum becomes alloyed with the surface layers of the substrate metal. Unlike hot dip aluminizing, a pure aluminum overlay does not form.

Precleaned components are packed in a retort (vessel) that contains powdered aluminum or a powdered ferroalloy, a ceramic phase to prevent agglomeration of the metallic components, and a volatile halide that acts as a chemical transfer medium for the aluminum. The retort is heated between 815°C and 980°C (1500°F and 1800°F) for 6 to 24 hours, after which the components are removed and heated in air to diffuse the aluminum. Coating depths from 25 microns to 1000 microns (1 mil to 40 mils) are obtained through this process.

Ion Implantation. *Ion implantation* is the modification of the physical or chemical properties of the immediate surface of a metal by embedding appropriate atoms into it from a beam of ionized particles. Penetrations of .1 microns to .2 microns (.004 mils to .008 mils) can be produced. The concentration of the implanted species increases to a peak value below the surface and then falls off with increasing depth below the surface. If plotted, the concentration of the implanted species forms a skewed bell curve.

Ion implantation is performed in a vacuum using an ion accelerator. See Figure 28-12. The ion accelerator provides a source of ions and accelerates them as a high-energy beam toward the target to be implanted. As the ion beam enters the target, it undergoes collisions with free electrons in the target and atom nuclei on lattice sites. Each collision reduces the energy of the ions, eventually causing them to stop moving. The higher the beam energy, the greater the depth of penetration.

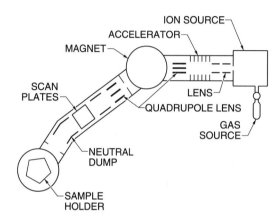

Figure 28-12. Ion implantation is performed in a vacuum using an ion accelerator.

Practically any element in the periodic table can be applied by ion implantation. The major applications of ion implantation are in the electronics industry where it is a useful method of doping semiconductors to produce controllable, homogeneous concentrations of doping agents.

Apart from the cost, some disadvantages of ion implantation include the extremely shallow surface layer that is produced and the fact that it is a line-of-sight process. Only regions directly in line with the beam can be modified.

Advantages of ion implantation include the ambient application temperature, which minimizes diffusion reactions and allows concentrations of the implanted species to far exceed equilibrium solubility limits with the base metal. Dimensional stability is excellent.

A principal metallurgical application of ion implantation is the use of nitrogen to enhance wear resistance of steels and cemented carbides. Some examples include improvement of rolling contact fatigue resistance on 52100 low-alloy steel and 440C stainless steel roller bearings, improved wear resistance of D2 punches and dies and M2 taps for phenolic resin, and improved wear resistance of cobalt bonded tungsten carbide swaging dies. These applications are generally limited to service temperatures below 500°C (930°F). Above this temperature, the nitrogen begins to diffuse from the surface, reducing hardness and wear resistance.

Physical Vapor Deposition (PVD). *Physical vapor deposition* is a surface modification process in which a substrate is coated by a streaming vapor of specific atoms or molecules in a high-vacuum environment. Three methods of physical vapor deposition, which are related to the method of generating the vapor of atoms or molecules, are sputtering, evaporation, and ion plating.

In sputtering, the atoms are transported from the target to the substrate by means of bombardment of the target by gas ions that have been accelerated at high voltage. Atoms from the target are ejected and move across the vacuum to be deposited on the substrate by mechanical impact. See Figure 28-13.

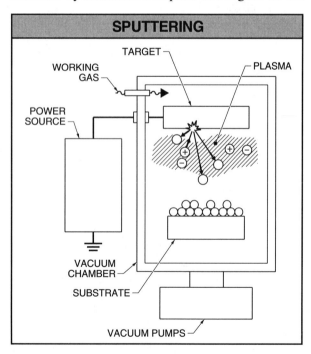

Figure 28-13. In sputtering, the atoms are transported from the target to the substrate by means of bombardment of the target by gas ions that have been accelerated at high voltage.

The chamber is initially evacuated, back filled with argon, and the target made negative by the application of 500 V to 5000 V of direct current. A low-pressure glow discharge plasma is created around the target cathode, creating positively charged argon ions that are accelerated toward the target. The momentum transfer, due to the impact of the argon ions, is enough to eject target atoms that travel to the substrate and coat it.

In evaporation, the ions or molecules are obtained by vaporizing a source. Compared with other PVD methods, the relatively low energy required for the process means that adhesion of the deposited element to the substrate is poorer. Additionally, the temperature of the substrate is not increased by the deposition, so there is little or no diffusion to help adhesion. For these reasons, care is needed to remove all impurities from the chamber and produce a perfectly clean substrate. See Figure 28-14.

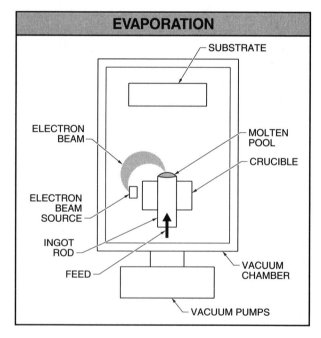

Figure 28-14. In evaporation, the ions or molecules are obtained by vaporizing the source.

In ion plating, the atoms are also generated by thermal evaporation of an appropriate source, such as a wire, which is heated during the process to provide the required ions. Specific differences between ion plating and evaporation exist. The chamber is first evacuated and then back filled with argon in ion plating. A potential is applied across the system with the substrate acting as the cathode. This produces a cleaning effect on the substrate that prepares it for ion plating. After cleaning, the coating source is energized and the coating material is vaporized in order to develop a coating on the surface of the target. Dense coatings that posess excellent adhesion are achieved through ion plating. See Figure 28-15.

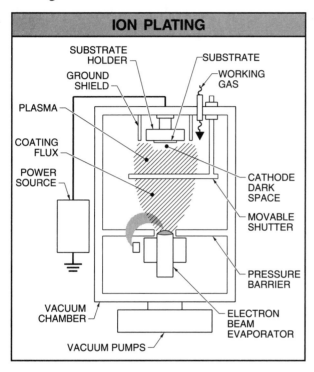

Figure 28-15. In ion plating, dense coatings may be achieved.

By introducing gases such as nitrogen, methane, or oxygen, sputtering and ion plating can be used to produce metal nitride, carbide, or oxide ceramic coatings. The coatings formed by PVD processes are produced by nucleation and growth. This involves the absorption of incident atoms (adatoms) on the substrate surface. The adatoms then diffuse on the substrate surface to preferred sites within the substrate, such as vacancies, to form growing clusters of nuclei.

A highly successful application of PVD is the titanium nitride coating of high-speed steels for machining and cutting and of other tool steels for hot and cold working molds and dies. These coatings are from 1 micron to 6 microns (.04 mils to .20 mils) thick and have a uniform gold color. They are extremely hard, approximately 2000 HV (compared with typical tool steel hardnesses of 800 HV), and have good lubricity.

As with ion implantation, a limitation of PVD is that it is a line-of-sight process. To partially offset this disadvantage, the component may be rotated to provide uniform coverage.

Chemical Vapor Deposition (CVD). *Chemical vapor deposition* is a surface modification process in which all the reactants are gases. A chemical reaction takes place in the vapor phase adjacent to, or on, a metal object, depositing the reaction products on the object. A by-product material is concurrently liberated as a gas and is removed from the processing chamber along with the excess reactant gas.

Chemical vapor deposition is used to apply a wide variety of coatings to metal surfaces for a variety of electronic, corrosion-resistant, oxidation-resistant, heat-resistant, and machining applications. Unlike ion implantation and PVD, CVD is not a line-of-sight process and has good ability to cover complex shapes.

> **Warning:** Potential hazards exist in CVD processes that are associated with handling some of the chemicals used at elevated temperatures. All safety practices related to handling chemicals at elevated temperatures must be observed.

Most CVD process require temperatures of 800°C (1470°F) or higher so that CVD coated steels require preheating to strengthen them after coating. Commonly applied CVD coatings include chromium (chromizing) and titanium carbide.

Chromizing is chromium enrichment of the surface of steel and other alloys by a pack cementation process. The pack cementation process is similar to aluminum pack diffusion. The process is strictly CVD because coating occurs by thermal decomposition of a gaseous chromium carrier. The component is packed in a mixture of elemental chromium, an inert filler, and an activator, such as ammonium iodide. The pack is heated between 1000°C and 1150°C (1830°F and 2100°F). Chromium diffuses into the surface layers of the workpiece at appreciable rates by the formation of chromous iodide in the vapor.

When it is applied to steel, chromizing converts the surface layer into a heat-resistant or corrosion-resistant stainless steel. If the steel contains appreciable amounts of carbon (above .6% C), chromium carbides precipitate and cause an increase in wear resistance. In this mode, chromizing is used to increase wear resistance of drop forging dies, tools, hydraulic rams, pistons, and pump shafts.

Titanium carbide coatings are applied by the hydrogen reduction of titanium tetrachloride in the presence of methane, or some other hydrocarbon, at a temperature of 900°C to 1010°C (1650°F to 1850°F). The process is usually performed at a reduced pressure of .1 atmospheres to improve uniformity of deposition. Titanium carbide coatings are applied to tools made of cemented carbide and tool steels, such as D2, to improve the coefficient of friction between the tool and the workpiece.

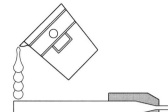

Nondestructive Testing 29

Nondestructive testing (NDT) is any type of testing performed on an object that leaves it unchanged. It is a basic tool of industry for determining and maintaining product quality. NDT is also referred to as nondestructive evaluation (NDE) or nondestructive inspection (NDI). Thirty-five to 40 nondestructive testing techniques exist. In many cases, several techniques are employed to characterize flaws or imperfections that may lead to premature failure. Areas of applicability of nondestructive tests overlap one another. The level of sensitivity required in specific testing techniques is determined by the service and by controlling codes or standards. The six most commonly used techniques are penetrant testing (PT), magnetic-particle testing (MT), ultrasonic testing (UT), radiographic testing (RT), eddy-current testing (ET), and acoustic-emission testing (AE).

NONDESTRUCTIVE TESTING

Nondestructive testing (NDT) is any type of testing performed on an object that leaves it unchanged. NDT is sometimes called nondestructive evaluation (NDE) or nondestructive inspection (NDI). Nondestructive testing involves the cooperative efforts of design and materials engineers, NDT specialists, and NDT technicians. The factors that must be considered when selecting an NDT technique include the type of material under test, method of processing to finished form, method of fabrication, component size and shape (geometry), types of flaws or imperfections that significantly affect product quality or service life, mode of operation, sensitivity and resolution required in the detection techniques, and cost effectiveness of the technique.

The design and materials engineers list the orientation, geometry, and types of significant flaws or imperfections that are present in a component. These may result from the inherent material characteristics, method of processing and fabrication, or operating conditions in service. For example, surface defects are more significant in components that operate under fatigue stresses because fatigue failures usually initiate at the surface.

The NDT specialist establishes the particular NDT techniques that most easily and economically detect the flaws or imperfections to the required level of sensitivity. Selection of NDT techniques and test parameters is based on experience with the sensitivity of the various techniques in detecting specific flaws or imperfections. Many NDT standards have been developed to suit different product forms and service categories. For example, the cast steel valve and fitting industry has established specific requirements for surface finish in the Manufacturers Standardization Society standard *MSS SP-55*.

The NDT technician performs the applicable tests using the specified parameters. Technicians may be qualified and certified in different techniques according to regulations established by the American Society for Nondestructive Testing (ASNT).

Thirty-five to 40 nondestructive testing techniques exist. Though each testing technique has particular strengths, the areas of applicability of nondestructive tests overlap one another. More than one testing technique may be required to define a specific flaw or imperfection with the required degree of confidence.

The six major nondestructive testing techniques are penetrant testing, magnetic-particle testing, ultrasonic testing, radiographic testing, eddy-current testing, and acoustic-emission testing. Penetrant and magnetic-particle testing are primarily used for detecting surface flaws or imperfections. Ultrasonic and radiographic testing are used chiefly to examine the interior of components. Eddy-current testing is predominantly a surface or near-to-surface inspection tool. Acoustic-emission testing is used to find the location of significant flaws or imperfections in components, which must then be identified and sized by another technique.

Virtually every basic principle of physics has been used as an NDT source to obtain information on the condition of materials and components. The five common elements of most NDT systems are an energy source, a modification of the energy source, a detection system, an indication and recording system, and an interpretation system. See Figure 29-1.

The energy source provides a specific probing medium that is used to inspect the component being tested. Typical energy sources include ultrasonic sound and electromagnetic radiation. The modification of the energy source occurs as a result of its interaction with the flaws or imperfections in the component being tested. The detection system measures the modification of the energy source and provides a signal that indicates the size, orientation, and nature of the flaws or imperfections. The indication and recording system collects and displays the signals obtained from the detection system. The interpretation system interprets the data and allows a decision to be made based on the measurement.

Penetrant Testing

Penetrant testing (PT) is a nondestructive testing technique for detecting discontinuities, such as cracks and porosity, on the surface of a nonporous solid. PT is used with excellent success on nonporous metals, ceramics, glasses, and polymers.

Testing Method. The surface to be tested must be dry and free of any contaminants. The penetrant is a light, oil-like liquid. See Figure 29-2. It is applied to the surface by brushing, dipping, or spraying. The penetrant is drawn into cracks or surface discontinuities by strong capillary action. It is given a dwell time (time to seep in) for the specific test. At the end of the dwell time, the portion of penetrant remaining on the surface is removed by wiping or washing. The removal of the excess penetrant leaves penetrant within the cracks or surface discontinuities. The component is then treated with dry powder or a suspension of powder in a liquid, referred to as developer. The developer acts like a sponge or blotting paper, drawing the penetrant from the cracks or surface discontinuities and enlarging the size of the area of the penetrant indication.

For the inspection process to be completed, the penetrant must be easily observed in the developer. One method used to enhance observation is dying the penetrant red and using a white developer. Another method (fluorescent liquid penetrant testing) is the use of a fluorescent penetrant, which is visible under ultraviolet (black) light.

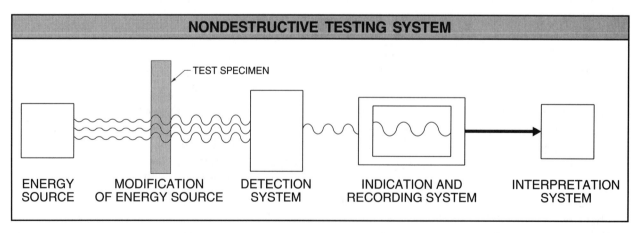

Figure 29-1. All nondestructive testing systems consist of an energy source, a modification of the energy source, a detection system, an indication and recording system, and an interpretation system.

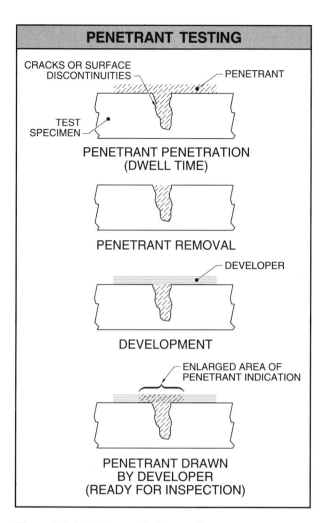

PENETRANT TESTING

CRACKS OR SURFACE
DISCONTINUITIES — PENETRANT

TEST
SPECIMEN

**PENETRANT PENETRATION
(DWELL TIME)**

PENETRANT REMOVAL

DEVELOPER

DEVELOPMENT

ENLARGED AREA OF
PENETRANT INDICATION

**PENETRANT DRAWN
BY DEVELOPER
(READY FOR INSPECTION)**

Figure 29-2. PT is used with excellent success on nonporous metals, ceramics, glasses, and polymers.

Testing Applications. PT is used to detect surface flaws and imperfections on forgings, castings, welds, and other components. It is most often used in conjunction with other nondestructive testing techniques. PT is typically rapid and convenient. PT kits for field and laboratory applications consist of three cans, which contain solvent for preparing the surface and removing excess penetrant, penetrant, and developer.

Although little operator training or skill is required, the specified PT procedure must be followed. Most importantly, the correct dwell time must be used and the method of removing excess solvent must be performed correctly. For example, to maximize the detection of tight cracks or cracks filled with corrosion products, it is necessary to use long dwell times and to carefully remove excess penetrant by swabbing.

Warning: The use of penetrant testing solvents in confined spaces may present a health hazard. The regulations on the use of these chemicals must be strictly followed.

Magnetic-particle Testing

Magnetic-particle testing (MT) is a nondestructive testing technique for detecting surface and subsurface imperfections and flaws in ferromagnetic materials, such as iron and steel. MT has two advantages over PT. MT detects much finer imperfections and flaws, such as fatigue cracks, which may be too tight to draw in liquid penetrant. MT can also detect imperfections and flaws that are not open to the surface up to $\frac{1}{16}$ in. below the surface. Unlike PT, MT cannot be used on nonferromagnetic materials.

Testing Method. In MT, the component being inspected is magnetized. The component is then covered with fine iron powder. The iron powder is preferentially attracted to surface imperfections and flaws where the magnetic field leaks out to the surface. The leaks form small north and south poles that attract the iron powder. See Figure 29-3. For discontinuities further below the surface, a smaller and smaller portion of the magnetic field is deflected to the surface, so less and less iron powder is attracted. If the discontinuity is far below the surface, no leakage of field is obtained and no iron particles are attracted.

Direct current, alternating current, and rectified alternating current are used for magnetizing purposes. Direct current is more sensitive than alternating current for detecting flaws and imperfections that are not open to the surface. Rectified alternating current is between direct current and alternating current in sensitivity.

The two methods of applying the iron powder are the wet method and the dry method. In the wet method, the iron particles are suspended in an oil or water vehicle. The iron particles are visible as either red, gray, or black, or fluorescent in order to maximize contrast. Wet fluorescent magnetic-particle testing is very common because of its sensitivity. In wet fluorescent magnetic-particle testing, viewing must be performed under ultraviolet or black light. In the dry method, the iron particles are transported by air to the component. After MT, the component must be demagnetized so that it does not pick up stray particles of iron or rust in service.

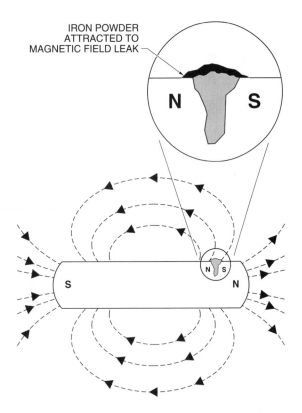

Figure 29-3. In MT, the magnetic powder accumulates at the leak of the magnetic field.

Testing Applications. The applications of MT depend on the technique of magnetizing the component. The two magnetizing techniques used are direct field and indirect field. In direct field magnetic-particle testing, the magnetic field is produced in the component by passing electric current through it. In indirect field magnetic-particle inspection, the magnetic field is produced by passing electric current through a conductor that is in close proximity to the component.

The direct field is obtained by electrically contacting the component or area of interest to the terminals of probes or end plates. To obtain the indirect field, a permanent coil or flexible cable is wrapped around the component, or a hand held yoke is placed in contact with the component. When using the hand-held yoke, care must be taken not to accidentally discharge sparks into the component. The sparks may cause hard zones from martensite formation and result in cracking during service.

In magnetic-particle testing, the detection of cracks depends on the direction of the magnetic field. Cracks are most easily seen when the magnetic field is perpendicular to them. Therefore, it is always recommended to magnetize the component in two directions at 90° to one another. This is accomplished by using circular magnetism and longitudinal magnetism. See Figure 29-4.

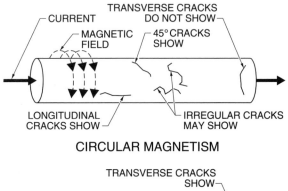

CIRCULAR MAGNETISM

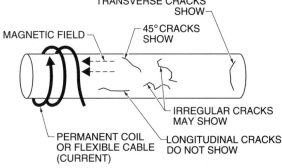

LONGITUDINAL MAGNETISM

Figure 29-4. Cracks show best when oriented 90° to the magnetic field.

Ultrasonic Testing

Ultrasonic testing (*UT*) is a nondestructive testing technique that uses electrically produced high-frequency sound waves to penetrate solids and detect flaws and imperfections inside components. Specific UT techniques can also detect flaws and imperfections that are near the surface. The basic UT techniques are straight-beam UT, angle-beam UT, and immersion UT.

Testing Method. The high-frequency sound waves used in UT are produced by a transducer that contains a piezoelectric crystal, which is capable of converting electrical energy into sound energy and vice versa. Frequencies between 500 kHz and 25 MHz are used. The ultrasonic vibrations emitting from the component are received by the transducer, which converts them into a signal that can be amplified, displayed, and analyzed.

Smooth surfaces are more suitable for UT, especially where the transducer must contact the component. For proper transmission of the sound waves, no air gap must exist between the component and the transducer. An air gap destroys the sound waves. Glycerin is usually applied to the transducer to maintain close contact. When feasible, the components are placed in a tank of water, oil, or glycerin to eliminate this problem.

UT is used to locate such flaws and imperfections as internal bursts, cracks, inclusions, seams, laps, slivers, laminations, or porosity in forgings, castings, wrought mill products, and welds. UT is also used extensively to measure wall thickness in equipment that is not accessible from both sides, such as pressure vessels and piping.

Straight-beam UT. Straight-beam UT is the most widely used ultrasonic technique. Straight-beam UT is used to measure the thickness of materials and also the location and size of flaws and imperfections in materials. The most common form of straight-beam UT is pulse-echo. In this technique, the high-frequency sound is transmitted and received by the same transducer. See Figure 29-5.

When the sound waves contact the surface of the material, a small portion of the waves is reflected back to the transducer. The sound waves that reach the back side of the material are also reflected back. If there is an internal flaw or imperfection between the front and back surfaces in the path of the sound waves, sound waves are also reflected back to the transducer. These reflections are displayed on an oscilloscope as a measure of elapsed time. Calibration of the signals received from a similar material of known thickness allows the thickness of an unknown material to be determined. The depth and size of imperfections can be estimated, providing that they are good reflectors

of sound. The signal is digitized to automatically record the wall thickness of equipment such as pressure vessels and piping.

In straight-beam UT, there is a surface-layer dead zone, which is caused by reverberations of the sound beam as it enters the component. Any flaws or imperfections in the dead zone, which extends approximately $\frac{1}{16}$ in. deep, may not be detected.

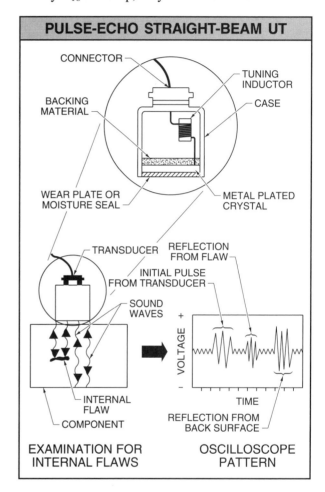

Figure 29-5. Pulse-echo straight-beam UT uses one transducer to send and receive the signal.

Angle-beam UT. Angle-beam UT employs a transducer that transmits the ultrasonic signal at a specific angle into the surface of the material. Angle-beam UT is especially suited to the inspection of welds because the weld surface does not have to be ground flush. Serious flaws in welds, such as cracks or incomplete fusion, usually extend longitudinally into the weld and give especially clear signals when the sound beam strikes the welded joint at the appropriate angle.

Angle-beam UT requires considerably more operator skill and experience than straight-beam UT. In addition to weld inspection, angle-beam UT is used in the detection and location of seams, laps, slivers, and cracks at or near the surfaces of rolled products. See Figure 29-6.

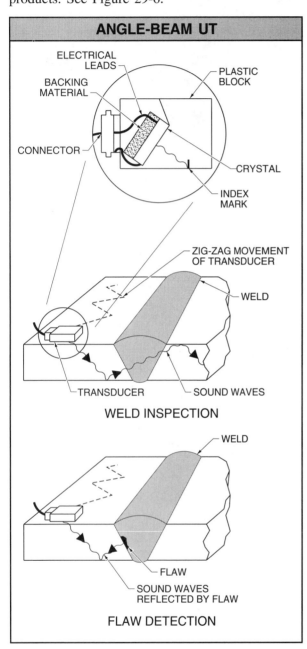

Figure 29-6. Angle-beam UT is used extensively to detect weld defects.

Immersion UT. Immersion UT involves placing the component and transducer in a tank of water. In this technique, there is no contact between the transducer and the component. The ultrasonic signal is focused on the component. This, coupled with the space maintained between the transducer and component, causes a significant reduction in the front surface dead zone. See Figure 29-7. Therefore, immersion testing is more capable than pulse-echo UT in detecting imperfections much closer to the front surface. Immersion UT is routinely used for checking the internal cleanliness of plates and billets, and for inspecting complex geometries, such as forgings.

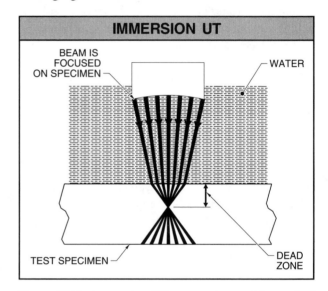

Figure 29-7. Immersion UT uses a focused beam that helps eliminate the front surface dead zone in the test specimen.

Radiographic Testing

Radiographic testing (RT) is a nondestructive testing technique in which electromagnetic radiation (X rays and gamma rays) is used to determine the interior soundness of solids. X rays and gamma rays are the same in nature, except for the way they are produced. X rays are generated electronically when high-velocity electrons give up energy upon striking a target anode, such as tungsten, and gamma rays are produced as characteristic radiation during the decay of radioactive isotopes.

Testing Method. Both X rays and gamma rays used in RT have useful characteristics. They penetrate matter; they are preferentially absorbed depending on the material density, contour, thickness, and the presence of flaws or imperfections; they travel in straight lines; and they produce effects on photographic film.

A radiograph is a shadow picture of the component obtained on photographic film. The radiation darkens (shadows) the film so that regions of lower density, such as voids, which readily permit penetration, appear dark on the negative as compared with regions of higher density, such as material, which absorb more radiation. The radiographic film is usually fixed to one side of the component and the source is placed approximately 2 ft or 3 ft on the opposite side of the component. See Figure 29-8. The divergence of the radiation causes the shadows of the flaws to appear larger than the actual flaw.

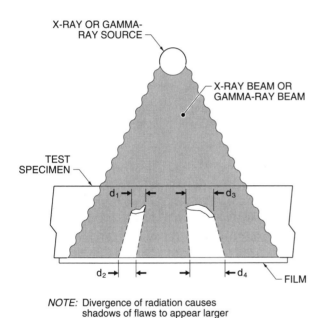

NOTE: Divergence of radiation causes shadows of flaws to appear larger than actual flaw (d₂ > d₁, and d₄ > d₃).

Figure 29-8. Radiography permits a permanent record of the internal condition of materials to be obtained.

Lead letters are placed on the radiograph to identify it. A penetrameter, or image quality indicator, is placed on the source side of the component to indicate contrast and resolution in the image obtained. Penetrameters are pieces of metal made of the same material as that being radiographed. The penetrameter thickness establishes the contrast sensitivity of the radiograph, and usually varies between 1% and 4%. To establish the resolution, three holes of different diameters are drilled in the penetrameter. The first hole is equal to the penetrameter thickness (T), the second hole is equal to twice the penetrameter thickness (2T), and the third hole is equal to four times the penetrameter thickness (4T). The sensitivity reading of the radiograph is then given by the percent contrast sensitivity and the smallest penetrameter hole resolvable.

For a 2% penetrameter, the required sensitivity and resolution is usually 2-2T. This means that the radiograph quality should be capable of distinguishing density differences of 2% of metal thickness and resolving features of similar size. A 2-1T radiograph would have more resolution than required, and a 2-4T would have less.

The exposure time required to produce a suitable radiograph depends on several factors, including the density and thickness of the material, the type of radiation employed, and the speed of the film. The thicker or more dense the material, the greater the exposure time required.

Because radiation can be scattered as well as absorbed, the radiograph can be fogged by scattered radiation that provides no information. A thin lead screen is placed in front of the film to absorb this scattered radiation and prevent fogging of the film. A thicker lead screen is placed behind the film to absorb backscatter coming from behind the film. The film is loaded in cassettes in a dark room to provide a convenient way to handle it on the job.

RT is primarily used to inspect welds and castings for internal defects, such as porosity, shrinkage, and slag inclusions. Compared with other methods of NDT, RT is relatively expensive, slow, and often entails delay in final product acceptance while the film is developed and analyzed. The advantage of RT is that it provides a direct and permanent record of the quality of a product.

X-ray Radiography. X-ray radiography is a nondestructive testing technique in which an X-ray generator is used as the source of radiation. X-ray radiography is performed by bombarding a target anode, such as tungsten, with a stream of rapidly moving electrons.

X-ray generators produce high-energy sources. These sources can be used to inspect a 24 in. piece of steel. Portable X-ray units used in the field are limited in energy output and are not commonly used.

Gamma-ray Radiography. Gamma-ray radiography is a nondestructive testing technique in which an artificially produced radioactive isotope is used as the source of radiation. The two most common isotopes are cobalt 60 and iridium 192. The specific activity of a radioactive species decreases with time. The time factor is measured in terms of the half-life of the radioactive species. The half-life is the period of time over which the specific activity drops to one-half of the specified (nameplate) value. For example, the half-life of cobalt 60 is 5.3 years, in which time a source with a specified activity of 30 curies drops to 15 curies. For iridium 192, the half-life is 75 days. Therefore, it is necessary to know how long a particular source has been in use in order to calculate the required exposure time.

Gamma-ray radiography is widely used in the field for examining welds. The maximum section thickness of steel that can be inspected with iridium 192 is 3 in. The maximum with cobalt 60 is 9 in.

Warning: Strict personnel safeguards must be in place when radiographic testing is used. These safeguards include mandatory controls on human exposure to radiation through the use of proper shielding, a reliable monitoring program for personnel involved with testing, and controls on the proximity of ancillary personnel to the radiation source.

Eddy-current Testing

Eddy-current testing (ET) is an electromagnetic nondestructive testing technique for measuring physical and mechanical parameters of metals. ET also measures surface or near-to-surface imperfections in electrically conducting materials.

Testing Method. In ET, the test probe (a coil), which carries an alternating current, is placed near the test specimen. See Figure 29-9. The alternating magnetic field generated in the test probe is a result of the alternating current. The alternating magnetic field generated emanates much like radio waves into the space around the test probe. Upon encountering the test specimen, the alternating magnetic field induces swirling electric currents (eddy currents) in

the test specimen. *Loading* is the interaction between the test probe and the induced eddy currents, where the induced eddy currents produce a reactive magnetic field that opposes or reduces the magnetic field in the test probe that produced the currents.

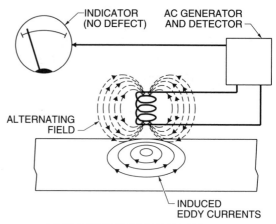

TEST SPECIMEN WITH NO FLAW

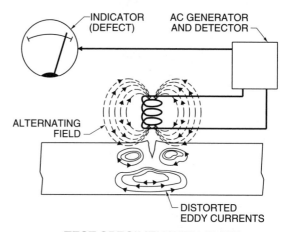

TEST SPECIMEN WITH FLAW

Figure 29-9. The eddy-current test probe induces eddy currents, in the test specimen, which may be used to sort materials.

Impedance is the electrical property that changes with the loading. Impedance is detected by measurement of current through, or voltage across, the test probe. This measurement is displayed on a cathode-ray tube, printed on a chart, or electronically stored.

The shape of the test probe is critically dependent on the purpose of the test and the shape of the test specimen. When used to detect flaws or imperfections, it is essential that the flow of eddy currents be as nearly perpendicular to the orientation of the flaw or imperfection as possible.

Eddy-current testing instruments contain an oscillator, some method of detecting the impedance changes in the test probe, and a method for displaying the results. The most common eddy-current signal detection system is a simple bridge circuit. See Figure 29-10. When the bridge is balanced against a standard test specimen, no current flows through the amplifier for display purposes. When the test probe is placed close to a test piece containing a specific flaw or defect, the coil impedance changes, and current flows through the amplifier, displaying a signal.

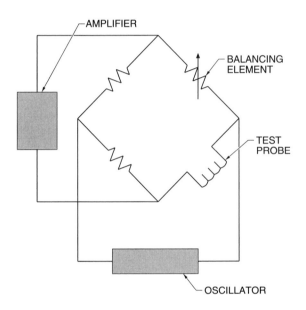

ET INSTRUMENTATION

Figure 29-10. The most common eddy-current detection system is a simple bridge circuit.

Testing Applications. ET is ideal for determining coating thickness, material thickness (for thin components), defect detection in plate and tubing, and alloy sorting. ET is used for automatic, in-line flaw detection without contact at high production rates, such as welded tubing as it leaves the production mill. ET is widely used for in-plant inspection and quality control.

The versatility of ET in responding to alloy composition differences, hardness, and heat treatment, and the presence of defects, sometimes makes it difficult to separate the parameter of concern (important responses) from responses to all other vari-

ables. Standardization of the test procedures, test material, and the effects of the flaws or imperfections being analyzed are important.

Acoustic-emission Testing

Acoustic-emission testing (AE) is a nondestructive testing technique that is used to detect spontaneously generated elastic waves created by localized movement of a material under stress, such as a pressure vessel or structural beam. AE differs from the other nondestructive testing techniques because it provides only the location of possible flaws and imperfections in components. When the flaws and imperfections are located, they must be identified and sized using another nondestructive testing technique.

Testing Method. Acoustic emissions are sonic pulses that radiate in all directions from a growing flaw or imperfection. Piezoelectric transducers act as microphones and record the presence of the elastic waves and their time of arrival. By knowing the velocity of sound in the material, the locations of defects are pinpointed. See Figure 29-11.

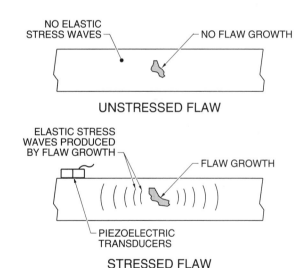

Figure 29-11. Acoustic-emission testing records the sound emitted from growing flaws or imperfections in a material.

To perform AE, the piezoelectric transducers are placed on the component at strategic points. The component is then stressed in order to cause any flaws or imperfections to grow slightly and emit sonic waves. For example, in a pressure vessel, this is achieved by a hydrostatic test that causes yielding. The transducers locate the exact position of the defects. These must then be identified and sized by another technique, such as ultrasonic testing.

Testing Application. AE is used to monitor the entire condition of structures and vessels using a fast, one-step process. The test is usually repeated at regular time intervals to keep track of the integrity of the component. The test itself requires only limited access to the piece of equipment. AE may also be used to limit the maximum pressure during the testing of containment systems if a large flaw or imperfection that may enlarge catastrophically and cause fracture is present. For example, if the sum of emissions recorded rises significantly during the application of a load, the test must be terminated immediately to avoid component failure. The larger the flaw, the smaller the load required to cause failure. See Figure 29-12.

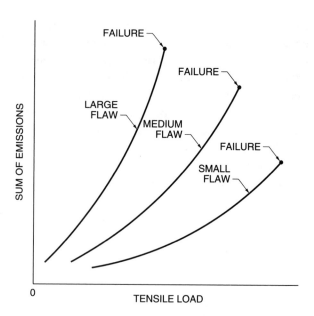

Figure 29-12. The larger the flaw in a component, the lower the permissible load applied during an acoustic-emission test.

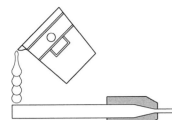

Effect of Temperature on Metals 30

The mechanical properties of metals are profoundly influenced by the temperature range in which they operate. At high temperatures, the most important properties are resistance to creep, rupture, and high-temperature oxidation. High-temperature corrosion resistance is often a limiting factor in the selection of high-temperature alloys. High-temperature corrosion testing is used to assess the comparative behaviors of high-temperature alloys. At low temperatures, the most important property is notch toughness. Low temperatures are divided into the two categories of low temperature and cryogenic temperature.

METALS AT HIGH TEMPERATURE

The most important mechanical property in metals operating at high temperature is creep resistance. The metallurgical requirements for creep-resistant alloys are often significantly different from those required for alloys that operate at ambient or moderately elevated temperatures. High-temperature mechanical testing is performed to determine such properties. High-temperature corrosion resistance is often a limiting factor in the selection of high-temperature alloys. High-temperature corrosion testing is used to assess the comparative behaviors of high-temperature alloys.

High-temperature Mechanical Property Tests

Poor correlation exists between the results of mechanical tests used to develop property data at room temperature and high temperatures. High-temperature mechanical property tests include creep, stress-rupture, and high-temperature tensile tests.

Creep Testing. Creep is time-dependent deformation that occurs after the application of a load to a solid. Metals and alloys undergo creep at elevated temperatures. Tensile strength and yield strength is used to calculate design stresses at approximate temperatures up to 425°C (800°F) for carbon steels, to 480°C (900°F) for chromium-molybdenum steels, and to 540°C (1000°F) for austenitic stainless steels. Above these temperatures, design stresses are usually based on special high-temperature tests, such as creep or stress-rupture tests.

Secondary stresses that do not occur from direct loading are also important in high-temperature service. Secondary stresses include stresses from thermal gradients caused by nonuniform heating and cooling and stresses caused by the effects of thermal expansion.

Creep testing is a form of tensile testing that is performed at a constant load and temperature. The test specimen is held at a constant temperature in an electric resistance furnace and subjected to a static tensile load. The load causes the test specimen to elongate gradually. The elongation is measured periodically. The total elapsed time of the test may vary from a few hours to tens of thousands of hours. Precise control of furnace temperature, specimen alignment, and applied stress are mandatory and essential in obtaining satisfactory results.

The method of conducting creep tests is covered by *ASTM E139, Recommended Practice for Conducting Creep, Creep-Rupture and Stress-Rupture Tests of Metallic Materials*. The total creep, or percent elongation, is plotted against time for the entire duration of the test.

The creep curve consists of three principal regions or stages. See Figure 30-1. When the load is first applied, there is an instantaneous elastic elongation. In the first stage, primary creep, the test specimen elongates gradually with decreasing creep rate. In the second stage, secondary (steady state) creep, the creep rate becomes essentially constant for a period of time. The slope of the curve in the secondary stage is the minimum creep rate. Finally, if the test time is long enough, the third stage of tertiary creep occurs in which the creep rate increases, eventually leading to the fracture of the test specimen.

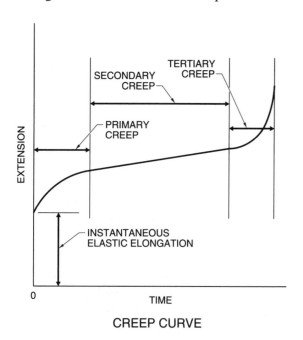

CREEP CURVE

Figure 30-1. The creep curve consists of three principal regions or stages.

Creep cracks originate as tiny cavities in the metal, which grow by the diffusion of voids toward the crack locations. The tiny cracks link together to form larger cracks, eventually leading to failure.

To obtain creep data, a series of tests are run under different stresses at a single temperature. Each series of tests is repeated at different temperatures. The creep rate during secondary creep is measured and plotted against the applied stress. Using log-log coordinates, the points normally exhibit a straight line. See Figure 30-2.

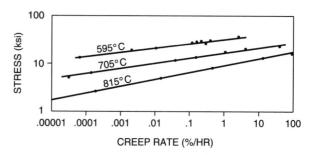

Figure 30-2. The creep rate during secondary creep is measured and plotted against the applied stress and usually exhibits a straight line.

The two most commonly used criteria for creep strength are as follows:

1. The stress needed to produce a minimum creep rate of .00001 percent per hour (%/hr), which is equivalent to 1% in 10,000 hours (a little over a year). This criterion is sometimes used for components, such as pyrolysis tubing, where a moderate amount of creep can be tolerated.

2. The stress needed to produce a creep rate of .000001 %/hr, which is equivalent to 1% in 100,000 hours (a little over 11 years). This is the criterion used for moving components where very little creep can be tolerated.

Creep tests are seldom performed for the lengths of time corresponding to the intended service life. The results are based on the conclusions from short term tests to long service times.

Stress-rupture Testing. Stress-rupture testing is similar to creep testing, except that the loads and, consequently, the creep rates are higher. Also, stress-rupture testing is performed to failure of the metal. The stress-rupture test apparatus is similar to that used for creep testing, but less sensitive instrumentation is used to measure the strain.

In recording stress-rupture data, the stress is plotted against rupture time on log-log paper. This usually produces a straight line for each test temperature, often with a change in slope at some specific time. See Figure 30-3. Because of this change in slope, it is very desirable to continue stress-rupture tests for a considerable time period.

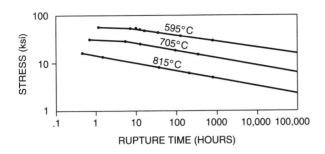

Figure 30-3. Stress-rupture curves are plots of the stress and rupture time at various test temperatures.

Rupture values are usually recorded as the stress to cause rupture in 100, 1000, 10,000, and 100,000 hours. The 100,000 hour rupture strength is one of the fundamental properties used in establishing design stresses.

High-temperature Tensile Testing. High-temperature tensile testing is short-term tensile testing performed in the creep range at strain rates that are available in regular tensile testing machines. The duration of the testing is usually only a few minutes and the effect of time at temperature is not measured. Elastic properties at elevated temperature are not realistic since their value depends on the time between load applications and their accuracy depends on the sensitivity of the extensometer.

High-temperature tensile testing fails to predict what will happen over prolonged time periods and has few useful applications. It provides an estimate of the static load-carrying capacity under short-term tensile loading and a comparative measurement of the ductility that can be expected.

Creep-resistant Alloys

The creep resistance of alloys is greatly affected by small variations in their microstructures and processing histories. Grain size is an important factor because coarse-grain materials perform better at elevated temperatures. This is in contrast to alloys that operate at ambient or low temperature, where fine grain size is desirable for toughness.

The presence of solute atoms, even in minor amounts, tends to increase creep resistance by interfering with the movement of dislocations. Finely dispersed second phase particles are even more beneficial in this respect. Creep-resistant alloys include carbon steels, chromium-molybdenum steels, chromium-molybdenum-vanadium steels, stainless steel, and superalloys.

Carbon Steels. Carbon steels are widely used for condenser tubes, heat exchangers, boilers, superheaters, and stills. Depending on the operating stresses, the maximum service temperature varies from 400°C to 480°C (750°F to 900°F). Above approximately 480°C (900°F), the pearlite in carbon steel begins to transform to graphite, which leads to embrittlement.

Chromium-Molybdenum Steels. Chromium-molybdenum (chrome-moly) steels contain .5% Cr to 9.0% Cr and either .5% Mo or 1.0% Mo. The chromium content improves scaling resistance. Molybdenum increases strength at elevated temperature and provides resistance to graphitization by forming molybdenum carbides. The carbon content is normally <.2% C to maintain weldability. Chrome-moly steels are widely used for piping and vessels operating at temperatures up to 540°C (1000°F) in the petroleum refining industry and in steam power generation. Chrome-moly steels are available in many product forms. See Figure 30-4.

The two most popular chrome-moly steels are those composed of 1.25% Cr and .5% Mo, and 2.25% Cr and 1.0% Mo. All chrome-moly steels are air hardening and require preheating and post-weld heat treatment when welded. They are welded with matching welding filler metals. See Figure 30-5.

Chromium-Molybdenum-Vanadium Steels. Chromium-molybdenum-vanadium steels contain up to .5% C, 5% Cr, 1% Mo, and .85% V. The vanadium addition raises the softening temperature of the steel. These steels are used for bolts, steam turbine rotors, and other components operating at temperatures up to 540°C (1000°F). These steels are usually not welded. Examples of chromium-molybdenum-vanadium steels include UNS K14675, K23015, K22770, T20811, and T20813. T20811 and T20813 are more commonly known as H11 and H13 hot work tool steels.

SPECIFICATIONS FOR CHROMIUM-MOLYBDENUM STEEL PRODUCTS					
Type	Forgings	Tubes	Pipe	Castings	Plate
.5Cr-.5Mo	A182-F2	A213-T2	A335-P2 A369-FP2 A426-CP2	—	A387-Gr2
1Cr-.5Mo	A182-F12 A336-F12	A213-T12	A335-P12 A369-FP12 A426-CP12	—	A387-Gr12
1.25Cr-.5Mo	A182-F11 A336-F11/F11A A541-C15	A199-T11 A200-T11 A213-T11	A335-P11 A369-FP11 A426-CP11	A217-WC6 A356-Gr6 A389-C23	A387-Gr11
2Cr-.5Mo	—	A199-T3b A200-T3b A213-T3b	A369-FP3b	—	—
2.25Cr-1Mo	A182-F22/F22a A336-F22/F22A A541-C16/6A	A199-T22 A200-T22 A213-T22	A335-P22 A369-FP22 A426-CP22	A217-WC9 A356-Gr10 A643-GrC	A387-Gr22 A542
3Cr-1Mo	A182-F21 A336-F21/F21A	A199-T21 A200-T21 A213-T21	A335-P21 A369-FP21 A426-CP21	—	A387-Gr21
5Cr-.5Mo	A182-F5/F5a A336-F5/F5A A473-501/502	A199-T5 A200-T5 A213-T5	A335-P5 A369-FP5 A426-CP5	A217-C5	A387-Gr5
5Cr-.5MoSi	—	A132-T5b	A335-P5b A426-CP5b	—	—
5Cr-.5MoTi	—	A213-T5c	A335-P5c		—
7Cr-.5Mo	A182-F7 A473-501A	A199-T7 A200-T7 A213-T7	A335-P7 A369-FP7 A426-CP7		A387-Gr7
9Cr-1Mo	A182-F9 A336-F9 A473-501B	A199-T9 A200-T9 A213-T9	A335-P9 A369-FP9 A426-CP9	A217-C12	A387-Gr9

Figure 30-4. Chrome-moly steels are widely used for elevated temperature applications up to 540°C (1000°F).

RECOMMENDED MINIMUM PREHEATING TEMPERATURES[a]			
Steel[b]	Thickness (in.)		
	Up to 0.5	0.5 to 1.0	Over 1.0
.5Cr-.5Mo 1Cr-.5Mo 1.25Cr-.5Mo	100°F	200°F	300°F
2Cr-.5Mo 2.25Cr-1Mo	150	200	300
3Cr-1Mo	250	300	400
5Cr-.5Mo 7Cr-.5Mo 9Cr-1Mo	400	400	500

[a] for welding chromium-molybdenum steels with low-hydrogen covered electrodes
[b] maximum carbon content of 0.15%

Figure 30-5. All chrome-moly steels require preheat and post-weld heat treatment when welded.

Wrought Stainless Steels. Wrought stainless steels used for high-temperature applications include martensitic, ferritic, and austenitic stainless steels. The high chromium contents of the stainless steels provide the steels with excellent scaling resistance.

Martensitic stainless steels that are used in high-temperature applications include type 410 (UNS S41000) and type 422 (UNS S42200). Type 410 has the lowest strength of the martensitics and is used for bolts, steam valves, pump shafts, and other components up to 540°C (1000°F). Type 422 has the highest strength of the martensitics and is used in similar applications up to 650°C (1200°F). Martensitic stainless steels must be tempered approximately 55°C (100°F) above the maximum operating temperature to prevent softening in service.

Ferritic stainless steels have the best scaling resistance because they have high chromium contents. The structural applications of ferritic stainless steels are limited to temperatures below 370°C (700°F) because of the precipitation of alpha prime (470°C or 885°F embrittlement), which seriously impairs room temperature ductility. Ferritic stainless steels also suffer from sigma-phase precipitation and related embrittlement on continuous exposure to temperatures in the range 650°C to 870°C (1200°F to 1600°F). Where structural strength is not a consideration, type 430 (UNS S43000) is used up to 815°C (1500°F) and type 446 (UNS S44600) up to 1095°C (2000°F).

The austenitic stainless steels exhibit the best creep strength of all of the stainless steels. They also exhibit excellent scaling resistance. These properties, coupled with ease of fabrication, make them widely used at temperatures up to 870°C (1600°F). See Figure 30-6. A major problem with austenitic stainless steels is the high coefficient of expansion they exhibit. This must be compensated for when designing austenitic stainless steel components.

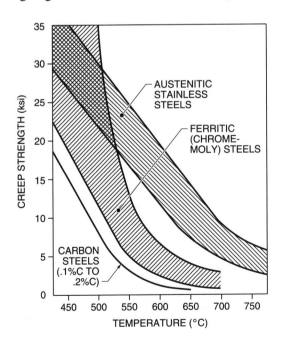

Figure 30-6. Austenitic stainless steels exhibit the best creep strength of all stainless steels.

Type 304 (UNS S30400) is used up to 815°C (1500°F). Type 316 (UNS S31600) has the highest stress-rupture capability of all 300 series alloys. Type 316 is susceptible to rapid oxidation in still air at elevated temperatures. This occurs because of its molybdenum content, which leads to the formation of a volatile oxide film under stagnant conditions. Type 321 (UNS S32100) and type 347 (S34700) are similar to type 304, except that type 347 contains niobium and type 321 contains titanium. Niobium and titanium additions are carbide forming elements that minimize sensitization (chromium carbide formation) on prolonged exposure to high temperatures. Type 309 (UNS S30900) and type 310 (UNS S31000) have inferior strength to type 316, but have the best scaling resistance of the 300 series.

Low-carbon grades must not be substituted for the regular carbon grades of austenitic stainless steels in high-temperature service. Examples of low-carbon grades are type 304L (UNS S30403) or type 316L (UNS S31603). Carbon improves the creep strength of austenitic stainless steels. Alloys with the suffix H, such as 304H, 316H, and 321H, have controlled carbon contents at the upper end of the specification range. These alloys are given a high-temperature annealing operation to ensure a coarse grain size. They have superior creep strength over the regular carbon grades.

Welding filler metals for austenitic stainless steels that are used in high-temperature service must be completely austenitic. If welding filler metals containing ferrite are used, the ferrite may transform to embrittling sigma phase during prolonged exposure to temperatures in the range of 650°C to 870°C (1200°F to 1600°F).

Heat-resistant Castings. Heat-resistant castings are used for demanding service applications above 650°C (1200°F). Heat-resistant castings include a wide range of alloy compositions encompassing stainless steels and nickel-base alloys. Heat-resistant castings are designated by the uppercase letter H followed by a letter that indicates the approximate alloy content. The higher the letter, the greater the percentage of alloying elements. Heat-resistant nickel-chromium castings contain from 30% Ni to 66% Ni, 12% Cr to 26% Cr, with the balance of the content being iron.

Superalloys. Superalloys are iron-base, nickel-base, and cobalt-base alloys that have good creep properties when used in the temperature range of 650°C to 1095°C (1200°F to 2000°F). Many superalloys contain additions of titanium and aluminum that precipitate at high temperatures and increase creep strength.

Iron-base superalloys are an extension of austenitic stainless steel technology and are suitable for use up to 1100°C (1400°F). Nickel-base superalloys are the most important group of superalloys. Over 40 wrought and 30 cast nickel-base superalloys exist. These nickel-base superalloys are used up to 1095°C (2000°F). Cobalt-base superalloys are used primarily in applications at temperatures between 650°C and 1150°C (1200°F and 2100°F).

High-temperature Corrosion

High-temperature corrosion resistance is an important requirement for alloys that operate at elevated temperatures. The forms of high-temperature corrosion include oxidation, carburization, sulfidation, chlorination, and hydrogen attack.

Oxidation. Oxidation occurs in air and steam-containing environments by the process of diffusion of oxygen inward and alloying elements outward. This results in the formation of a thin oxide film that develops into a thicker scale. Scaling occurs when oxidation becomes excessive.

In carbon steels, the amount of oxidation in air is negligible below approximately 535°C (1000°F). Above this temperature, the rate of oxidation increases rapidly. This temperature coincides with the formation of wustite (FeO). Wustite contains many ionic defects that permit the transportation of reactants across the oxide layer, which increases the rate of oxidation.

The most important alloying element for increasing oxidation resistance is chromium. Chromium forms a tightly adherent layer of chromium-rich oxide on the surface of the metal. This layer retards the inward diffusion of oxygen. The maximum amount added to carbon steels is 9% Cr. In stainless steels, additions as high as 29% Cr are made to some grades. Stainless steels with the high chro-

mium contents, such as type 309, type 310, and type 446, have oxidation resistance up to 1095°C (2000°F). See Figure 30-7. Other elements, such as silicon and aluminum, also improve oxidation resistance, particularly when added to a steel that contains chromium.

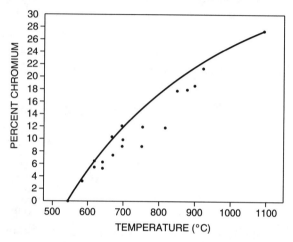

EFFECTS OF CHROMIUM CONTENT ON OXIDATION TEMPERATURES

Figure 30-7. Chromium is the most important alloying element for increasing oxidation resistance of steels and stainless steels.

Carburization. *Carburization* is a form of high-temperature corrosion that occurs in carbonaceous (carbon containing) environments where the carbon in the environment diffuses into the base metal and forms carbides. Unlike oxidation, scale formation does not occur during carburization. With time, carburization reduces ductility and creep resistance. Carburization occurs in steam-methane reforming furnaces for ammonia synthesis and in ethylene pyrolysis. Nickel and silicon are added in order to reduce carburization.

Metal dusting is a form of high-temperature corrosion associated with carburization. Metal dusting occurs in gas compositions rich in carbon monoxide rather than carbon dioxide. Whole grains of metal are dislodged and the surface becomes sugary to touch. Metal dusting occurs in the range of 500°C to 800°C (930°F to 1470°F). Steam, ammonia, or sulfur additions to the gas stream reduce susceptibility to metal dusting.

Green rot is corrosion that occurs in part-carburizing and part-oxidizing environments. It has the appearance of a green fracture surface when chromium oxide forms during oxidizing conditions following a precarburizing exposure at 900°C to 980°C (1650°F to 1800°F). Such conditions occur during startup when there is incomplete combustion leading to carburizing conditions.

Sulfidation. *Sulfidation* is a form of high-temperature corrosion that occurs in sulfur-containing environments, such as hydrogen sulfide and sulfur dioxide. Sulfidation occurs more rapidly than oxidation because sulfide scales contain more defects and have lower melting temperatures. These environments can be generated during the combustion of fossil fuels.

Alloy steels and stainless steels are usually the first choice for sulfidation resistance. The chromium steels with 9% Cr to 12% Cr are used for organic sulfides and hydrogen sulfide. Types 310 stainless steel, alloy 800, and RA 330 are also used for sulfidation resistance.

Chlorination. *Chlorination* is a form of high-temperature corrosion that occurs in environments containing chlorine or hydrogen chloride. The mechanism of corrosion is the same as for oxidation and sulfidation, but protective scales do not generally form because the corrosion products are volatile and vaporize. Chlorination occurs in some extractive metallurgical operations and also in waste incineration.

Nickel-base alloys are used for chlorination resistance. Aluminum additions, like those which are added in alloy 214, are beneficial.

Hydrogen Attack. *Hydrogen attack* is a form of high-temperature embrittlement that occurs in environments containing hydrogen, which is caused by diffusion of hydrogen into the steel where it reacts with the carbides to form methane gas, creating zones of weakness within the alloy. Chrome-moly steels have resistance to hydrogen attack up to specific temperatures and pressures, depending on their compositions. See Figure 30-8. Austenitic stainless steels are immune to hydrogen attack.

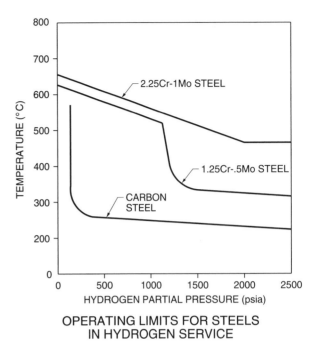

OPERATING LIMITS FOR STEELS
IN HYDROGEN SERVICE

Figure 30-8. Chrome-moly steels have greater resistance to hydrogen attack than carbon steels.

High-temperature Corrosion Testing

High-temperature corrosion testing is performed in simulated laboratory environments and in operating equipment to assess the comparative behavior of high-temperature alloys. High-temperature corrosion testing is performed in the laboratory (in-house) or in the field.

Laboratory testing is performed in a tube furnace. The coupons are usually small cylinders or sheets contained in ceramic boat-shaped crucibles. They are inserted in the furnace tube and exposed to high-temperature corrosion. High-temperature corrosion is calculated as the weight change in milligrams per square centimeter (mg/cm^2) of surface area and plotted against time. Depending on the test environment, coupons may gain or lose weight during testing.

After the last exposure period is completed, the coupons are weighed and sectioned transversely across their midpoints. These sections are mounted, polished, and examined for the depth of the corrosion attack. See Figure 30-9. During examination of the depth of the corrosion attack, the metal loss (M_L) and the maximum corrosion attack (A_{max}) is figured. *Metal loss* is the loss of metal due to massive oxides and sulfides. *Maximum corrosion attack* is the loss of metal due to all forms of oxidation and to sulfides. The following formula is used to figure the metal loss and the maximum corrosion attack:

$$M_L = D - D_1$$

where

M_L = metal loss

D = original diameter of the metal (measured with a micrometer)

D_1 = diameter of the structurally useful metal (measured at 100X)

and

$$A_{max} = D - D_2$$

where

A_{max} = maximum corrosion attack

D = original diameter of the metal (measured with a micrometer)

D_2 = diameter of metal unaffected by oxides and sulfides (measured at 100X)

Example: Figuring metal loss and maximum corrosion attack

What is the metal loss and the maximum corrosion attack in a metal component with an original diameter of 20 mm, a diameter of structurally useful metal of 17.5 mm, and a diameter of unaffected metal of 15 mm?

Figuring metal loss

$$M_L = D - D_1$$
$$M_L = 20 - 17.5$$
$$M_L = \textbf{2.5 mm}$$

Figuring maximum corrosion attack

$$A_{max} = D - D_2$$
$$A_{max} = 20 - 15$$
$$A_{max} = \textbf{5 mm}$$

Standard procedures have been developed to assess the amount of attack caused by certain types of high-temperature corrosion. For example, carburization test evaluation is described in *ASTM G79, Practice for the Evaluation of Samples Exposed to Carburization Environments*.

Field testing in operating equipment uses coupons that are flat bar or sheet. The coupons are sturdily bolted on a test rack made of a heat-resistant alloy, such as type 310 stainless steel. The test rack must be welded into the equipment so that it does not become detached in service. After exposure, when the equipment is down, the test rack is removed from the equipment by burning and carefully dismantled so that coupons may be examined.

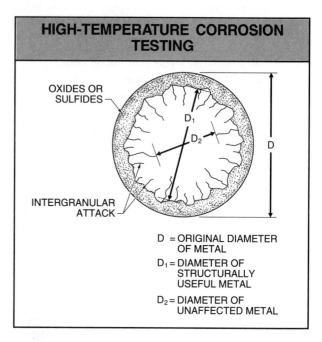

Figure 30-9. In high-temperature corrosion testing, coupons are sectioned and metallographically examined to determine the depth of the attack.

METALS AT LOW TEMPERATURE

Low temperatures can exert a significant influence on the mechanical properties of metals. Low temperatures are divided into the two categories of low temperature and cryogenic temperature. The low-temperature category includes temperatures to –100°C (–150°F). The cryogenic temperature category includes temperatures to absolute zero (–273°C, or –459°F). Specific alloys are used for low-temperature and cryogenic-temperature service.

Effect on Mechanical Properties

As the temperature is decreased below room temperature, the hardness and yield strength of all metals and alloys increase. With a few exceptions, the ultimate tensile strength and modulus of elasticity of all metals and alloys also increases. See Figure 30-10. The most important mechanical property change that is affected by decreasing temperature is toughness. Some metals become extremely brittle at low temperatures.

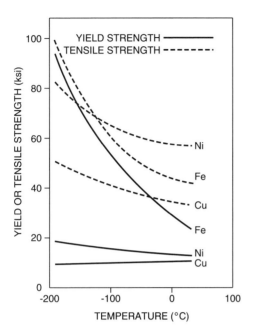

Figure 30-10. As the temperature is decreased below room temperature, the yield strengths of all metals and alloys increase, and with few exceptions, so does the ultimate tensile strength.

Low temperatures or cryogenic temperatures also effect the nil ductility transition (NDT) temperature. The NDT temperature is the temperature at which the impact behavior of a metal changes from ductile to brittle in the presence of a stress raiser. The NDT temperature is an important parameter in defining the serviceability limits of certain families of alloys at low temperatures or cryogenic temperatures. In carbon steels, this temperature is influenced principally by the carbon content and grain size of the steel.

Nil Ductility Transition Temperature. The nil ductility transition temperature is also referred to as the ductile-to-brittle transition temperature (DBTT). Carbon steels show a sharp transition from ductile to brittle at a specific temperature. Austenitic stainless steels do not exhibit an NDT temperature and remain ductile at cryogenic temperatures.

The usual method of obtaining the NDT temperature in carbon and low-alloy steels is the Charpy V-notch impact test. The test is performed at decreasing temperatures and the impact energy absorbed is plotted against the temperature. With many steels, there is a temperature range over which the impact energy absorbed values drop sharply. See Figure 30-11. The maximum temperature at which this occurs is the NDT temperature.

The lower the NDT temperature, the more-resistant the steel is to the embrittling effect of stress concentrations, high loading rates, and low temperature. Because there is no absolute NDT temperature value for a material, the NDT temperature determined by the Charpy impact test does not indicate the specific temperature at which brittle fracture occurs in items such as structural components or threaded fasteners. The test does provide a method of rating the notch toughness of steels at low temperatures.

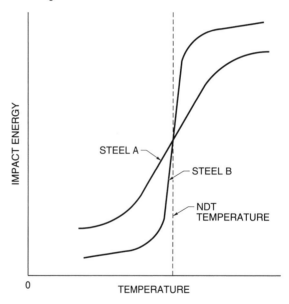

Figure 30-11. The NDT temperature is the temperature at which the impact behavior of a metal changes from ductile to brittle in the presence of a stress raiser.

The principal mechanical factors that influence the NDT temperature are the shape of the notch and the strain rate. The sharper the notch, the higher the NDT temperature. Increasing the strain rate also raises the NDT temperature. For example, the difference between the transition temperature in a normal tensile test and a notched bar impact test may be more than 100°C (180°F).

Alloys for Low-temperature and Cryogenic Temperature Service

Low-temperature applications include service in cold climates and the storage and handling of some liquified gases, such as propane, anhydrous ammonia, carbon disulfide, and ethane. Cryogenic temperature includes applications such as the storage and handling of liquified gases such as nitrogen, argon, hydrogen, and helium. The major structural alloys for use in low-temperature and cryogenic temperature services are carbon steels, nickel steels, austenitic stainless steels, and aluminum alloys.

Carbon Steels. The notch toughness of carbon steels decreases with decreasing temperatures. This is of critical importance in the selection of carbon steels for low-temperature service.

Factors that affect the NDT temperature of carbon steels include carbon content and grain size. Carbon has the greatest effect on notch toughness. Lowering the carbon content lowers the NDT temperature. For maximum toughness, the carbon content should be kept as low as possible and consistent with strength requirements. See Figure 30-12.

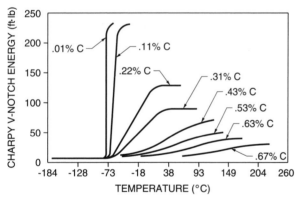

EFFECTS OF CARBON ON
TRANSITION CURVE

Figure 30-12. Carbon content should be kept as low as possible and consistent with strength requirements.

Reducing the grain size is also beneficial to the toughness level of carbon steels. For a particular type of steel and toughness level, fine-grain material has a lower NDT temperature than coarse-grain material. Killed steels are superior to other steels because the aluminum addition helps refine the grain

size. Normalizing has a similar effect and is an important requirement in specifications.

Carbon steels are generally limited to approximately –45°C (–50°F), depending on the section thickness. A commonly used structural steel is ASTM A516, which is aluminum killed and normalized. ASTM A516 is available in four grades with increasing carbon content and strength.

Nickel Steels. Nickel steels are low-carbon steels that contain from 2% Ni to 9% Ni for service at low temperatures and cryogenic temperatures. The nickel additions significantly lower the NDT temperature. Nickel steels are ferritic alloys used for applications involving exposure to temperatures from 0°C to –195°C (32°F to –320°F). Applications include storage tanks for liquefied hydrocarbon gases and machinery designed for use in cold climates. See Figure 30-13.

EFFECT OF NICKEL ON NDT TEMPERATURE	
%Ni	**Temperature NDT (°C)[a]**
0	– 45
2.25	– 68
3.5	–101
5	–170
9	–195

[a] based on Charpy impact test for 1 in. plate

Figure 30-13. Increases in nickel content significantly lower the NDT temperature.

The 2.25% Ni steel is used extensively for applications to –60°C (–75°F). The 3.5% Ni steel is used for the storage of liquefied gases to –100°C (–150°F). The 5% Ni and 8% Ni steels are suitable down to –170°C (–275°F), and the 9% Ni steel is suitable down to –195°C (–320°F). To maximize toughness at low temperatures and cryogenic temperatures, the nickel steels are used in the normalized, quenched, and tempered condition.

When welding nickel steels, preheating up to 150°C (300°F) may be required, depending on the degree of restraint of the joint. Post-weld heat treatment up to 635°C (1175°F) may also be necessary in heavier sections. Matching welding filler metals are used for low-nickel alloys (up to 3.5% Ni). Higher nickel contents require a nickel alloy welding filler metal such as Inconel® 82 or Inconel® 182.

Austenitic Stainless Steels. Austenitic stainless steels are FCC and do not exhibit a low NDT temperature. Austenitic stainless steels are excellent materials over the entire range of cryogenic applications. They exhibit excellent ductility, notch toughness, and corrosion resistance. Austenitic stainless steels remain tough at the boiling points of many gases, such as hydrogen and helium. See Figure 30-14.

BOILING POINT OF GASES AND RECOMMENDED STEELS FOR THOSE APPLICATIONS			
	Temperature[a]		
Gas	**°C**	**°F**	**Recommended Steel[b]**
Butane	− 0.6	30.9	ASTM A 333 (grades 1, 6, and 7)
Sulfur dioxide	− 10.0	14.0	ASTM A 334 (grades 1, 6, and 7)
Isobutane	− 10.2	13.6	ASTM A 516[c]
Methyl chloride	− 23.7	− 10.7	ASTM A 537[d]
Fluorocarbon refrigerant$_{12}$	− 30.0	− 22.0	ASTM A 662 (grade A)
Ammonia	− 33.3	− 27.9	ASTM A 734 (type A)
Fluorocarbon refrigerant$_{22}$	− 10.6	− 41.0	ASTM A 736 (classes 2 and 3) and
Ketene	− 41.0	− 41.8	ASTM A 782
Propane	− 42.3	− 44.1	ASTM A 517
Propylene	− 47.0	− 52.6	ASTM A 203
Hydrogen sulfide	− 59.6	− 75.3	ASTM A 203 (3½% Ni steel)
Carbon dioxide	− 78.5	−109.3	ASTM A 645 (5% Ni steel)
Acetylene	− 84.0	−119.2	ASTM A 333 (grade 3)
Ethane	− 83.3	−126.9	ASTM A 334 (grade 3)
Nitrous oxide	− 89.5	−129.1	Stainless steels (AISI 300 series)
Ethylene	−103.8	−154.8	
Xenon	−109.1	−164.4	
Ozone	−111.9	−169.4	ASTM A 645 (5% Ni steel)
Krypton	−151.8	−241.2	ASTM A 553 (grades A and B[e])
Methane	−161.4	−258.5	ASTM A 353 (9% Ni steel)
Oxygen	−183.0	−297.4	ASTM A 333 (grade 8)
Argon	−185.7	−302.3	ASTM A 334 (grade 8)
Fluorine	−187.0	−304.6	Stainless steels (AISI 300 series)
Carbon monoxide	−192.0	−313.6	
Nitrogen	−195.8	−320.4	
Neon	−245.9	−410.6	ASTM A 213 (AISI type 304)
Tritium	−248.0	−414.4	ASTM A 240 (AISI type 304)
Deuterium	−249.5	−417.1	ASTM A 269 (AISI type 304)
Hydrogen	−252.7	−422.9	ASTM A 312 (austenitic
Helium	−269.9	−453.8	stainless steels)[f]

[a] approximate boiling point at 1 atm
[b] Steels listed intended as general guide to steel selection. Steels listed in each grouping may not be suitable for all gases listed in corresponding gas groupings.
[c] normalized
[d] with modifications
[e] grade B to −170°C (−275°F)
[f] types 304, 304L, 304H, 304N, 309, 310, 316, 316L, 316H, 316N, 317, 321, 321H, 347, 347H, 348, 348H, XM15, XM19, and XM29

Figure 30-14. Austenitic stainless steels remain tough at the boiling points of many gases.

Ferritic and martensitic stainless steels are not recommended for cryogenic use. These two stainless steels are BCC and exhibit a relatively high NDT temperature.

Aluminum Alloys. Aluminum alloys are FCC and retain good ductility at cryogenic temperatures. Cast and wrought alloys are used for applications such as pumps, heat exchangers in gas liquefaction plants, tanks for the transport of liquefied natural gas, and liquid gas fuel tanks in aerospace applications. Aluminum alloys that are used for these applications include both solid solution hardening and precipitation-hardening types, such as 2014-T6, 3003, 5456, 6061-T6, and 7075-T6.

Corrosion 31

Corrosion is an electrochemical reaction between a metal and an electrolyte (corrosive environment). Electrochemical principles are used to explain corrosion. Corrosive environments are electrolytes that contain positive and negative ions. The anode and cathode reactions are the basic corrosion reactions. The corrosiveness of an environment is influenced by its characteristics and the exposure conditions present. Most forms of corrosion are localized. Corrosion prevention methods include the selection of resistant materials, proper design, electrochemical control, and modification of the environment.

CORROSION

Corrosion is the electrochemical dissolution of a metal in a corrosive environment or electrolyte. An *electrolyte* is a liquid, most often a solution, that conducts an electrical current. During the corrosion of a metal, two distinct reactions occur simultaneously on the metal surface. Electrochemical principals are used to explain corrosion. Corrosive environments are diversified, and each environment has distinctive characteristics. The corrosiveness of an environment may be increased by the exposure conditions (external factors) present.

Corrosion Reactions

Corrosive environments are electrolytes that contain positive and negative ions. The anode and cathode reactions are the two most important components of corrosion reactions. The anode and cathode reactions are the basic corrosion reactions.

Anode and Cathode Reactions. In the anode reaction, metal atoms are converted into positively charged ions and electrons, leading to dissolution of the metal. In the cathode reaction, specific positively charged ions in the electrolyte consume the electrons. See Figure 31-1.

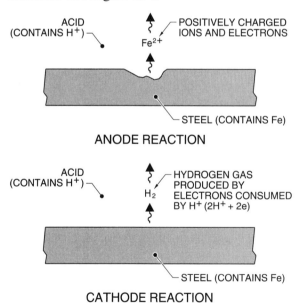

Figure 31-1. Corrosion consists of anode and cathode reactions that take place at different locations on a metal surface.

The positive charge passing from the metal into the electrolyte at anode locations on the metal surface is balanced by an equal amount of negative charge passing from the electrolyte to the metal at cathode locations. In electrical terms, when corrosion occurs, a current flows from the anode to the cathode. Since the anode and cathode currents are equal and opposite, no electric shock is experienced from a corroding system.

When the corrosion is general, the anode and cathode sites are switching continuously, causing relatively uniform metal loss over the surface. When corrosion is localized, such as in crevice corrosion or pitting, the anode and cathode regions are located on different parts of the metal surface. When corrosion is galvanic, the anode and cathode are two different metals. See Figure 31-2.

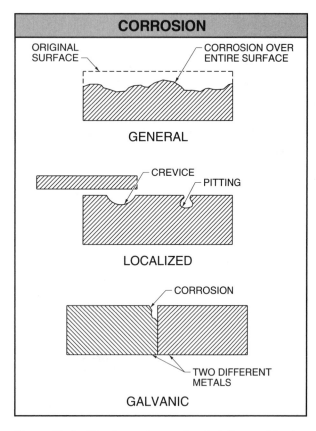

Figure 31-2. The type of corrosion is influenced by the location of the anode reaction.

Cathode reactions accept the electrons created by the anode reaction. Unlike anode reactions that are all of one type, cathode reactions depend on the nature of the electrolyte. The reaction is different in different types of electrolytes, such as aerated salt solution, aerated sulfuric acid, and deaerated sulfuric acid. In these three conditions, the anode reaction consists of the steel corroding at anode sites to form iron ions and electrons. This is shown in the following equation:

$$Fe \rightarrow Fe^{2+} + 2e$$

where

e = electrons in electrolyte

The balancing cathode reactions are:

in aerated salt solution

$$\tfrac{1}{2}\,O_2 + H_2O + 2e \rightarrow 2OH^-$$

and in aerated sulfuric acid

$$O_2 + 4H^+ + 4e \rightarrow 2H_2O$$

and in deaerated sulfuric acid

$$2H^+ + 2e \leftrightarrow H_2\uparrow$$

Many forms of cathode reactions often control the corrosion rate. The anode and cathode reaction products frequently combine chemically to form corrosion products, such as rust. If these corrosion products are not removed, they usually stifle further reaction and reduce the corrosion rate.

Electrochemical Principles

Electrochemical principles are used to explain corrosion by various engineering concepts. The engineering concepts include electrochemical potential, polarization, corrosion potential (E_{corr}), and anode and cathode polarization curves.

When the anode and cathode are electrically connected, their half-cell potentials polarize to the corrosion potential. The point where the anode and cathode polarization curves intersect determines the corrosion rate. Electrochemical corrosion testing consists of plotting the variation in corrosion-current density (corrosion rate) with potential.

Electrochemical Potential. The anode and cathode develop an electrical potential, or their half-cell potentials. The anode reaction is the oxidation reaction and the cathode reaction is the reduction reaction. The tendency for corrosion to occur in an environment is given by the difference between the half-cell (open circuit) potentials of the anode and cathode.

Reduction-oxidation (redox) potentials is often used to describe the calculated half-cell potentials.

Half-cell potentials are measured by comparison with a set standard. By convention, the chosen standard is the standard hydrogen electrode (SHE), which is the zero against which practical reference potential systems are based. The two commonly used standards in engineering applications are the copper/copper sulfate ($Cu/CuSO_4$) electrode and the saturated calomel electrode (SCE). The copper/copper sulfate $Cu/CuSO_4$ electrode is extensively used for underground cathodic protection studies. The $Cu/CuSO_4$ electrode is approximately 340 mV more positive than the SHE. The SCE is extensively used for laboratory electrochemical studies. The SCE is approximately 240 mV more positive than the SHE.

The net potential (cell potential) of a corroding system is figured using the following equation:

$$E = E_c - E_a$$

where

E = net potential

E_c = cathode half-cell potential

E_a = anode half-cell potential

If the difference is positive, the corrosion reaction is thermodynamically spontaneous, which means that corrosion is possible. See Figure 31-3.

HALF-CELL POTENTIALS	
Reaction	**Volts**
$O_2 + 4H^+ + 4e \rightarrow 2H_2O$	+1.23
$Fe^{3+} + e \rightarrow Fe^{2+}$	+ .77
$O_2 + 2H_2O + 4e \rightarrow 4OH^-$	+ .40
$Cu \rightarrow Cu^{2+} + 2e$	+ .34
$2H^+ + 2e \leftrightarrow H_2$	.00
$Fe \rightarrow Fe^{2+} + 2e$	− .44
$Zn \rightarrow Zn^{2+} + 2e$	− .76

Figure 31-3. Anodes and cathodes develop half-cell, or redox, potentials, which may be measured with reference to the standard hydrogen electrode (SHE).

With copper in deaerated and aerated hydrochloric acid, the relevant half-cell potentials under standard conditions of temperature, pressure, and acid concentration are as follows:

For the anode reaction in both deaerated and aerated hydrochloric acid

$$Cu \rightarrow Cu^{2+} + 2e, \text{ where } E_a = + .34 \text{ V}$$

and for the cathode reaction in deaerated acid

$$2H^+ + 2e \leftrightarrow H_2, \text{ where } E_c = 0 \text{ V}$$

and for the cathode reaction in aerated acid

$$O_2 + 4H^+ + 4e \rightarrow 2H_2O, \text{ where } E_c = + 1.23 \text{ V}$$

The net potentials for the two corrosion reactions described are as follows:

In deaerated acid

$$E = E_c - E_a$$
$$E = 0 - .34$$
$$E = -.34 \text{ V}$$

and in aerated acid

$$E = E_c - E_a$$
$$E = 1.23 - .34$$
$$E = +.89 \text{ V}$$

Using the rule for corrosion being thermodynamically spontaneous, under standard conditions corrosion is not possible in the deaerated acid, but it is possible in the aerated acid. This is also found to be true in practice.

Half-cell potentials change with the concentration of the ions in solution and the temperature. If the conditions are not standard, the half-cell potentials must be calculated for the actual conditions.

Polarization. Polarization is the shift of the half-cell potentials toward each other as a result of corrosion. Polarization indicates behavior when the anode and cathode are coupled (short circuited), which occurs during corrosion. Polarization determines the corrosion-current density, which is a measure of the corrosion rate. The magnitude of the anode and cathode half-cell potential shifts are the anode overpotential and cathode overpotential. See Figure 31-4.

When zinc corrodes in a reducing acid, such as hydrochloric acid, the two half-cell reactions are as follows:

Anode reaction

$$Zn \leftrightarrow Zn^{2+} + 2e$$

and

Cathode reaction

$$2H^+ + 2e \leftrightarrow H_2$$

Before polarization, there is an exchange current situation at the anode and cathode regions, which is expressed as exchange current density (exchange current per unit area). The exchange current density at the anode and cathode regions is equal and opposite. Before polarization and at the anode, the rate of zinc ions entering the solution is equal to the rate of zinc ions reverting to zinc atoms (zinc metal). At the cathode, the rate of hydrogen ions converting to hydrogen atoms equals the rate of hydrogen atoms reverting to hydrogen ions. This is a dynamic equilibrium, therefore no corrosion takes place.

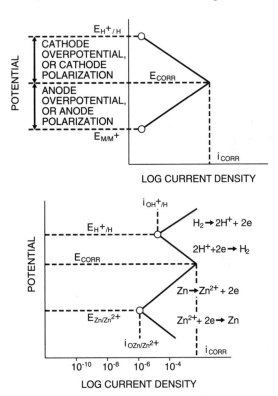

Figure 31-4. When corrosion occurs, the anode and cathode are short circuited and their half-cell potentials are displaced toward each other.

When the anode and cathode are coupled, their half-cell potentials no longer correspond to the original half-cell values, and they are displaced toward each other (polarized). The polarization results in the corrosion of zinc, which forms zinc ions by the anode reaction. The hydrogen atoms combine to evolve hydrogen molecules by the cathode reaction. The anode and cathode current densities are equal and opposite at the corrosion potential.

Corrosion Potential (E_{corr}). The *corrosion potential* is the potential exhibited by a metal in an electrolyte under steady-state conditions (chemical equilibrium). The corrosion potential is measured with the aid of a reference electrode using a high-impedance voltmeter. See Figure 31-5.

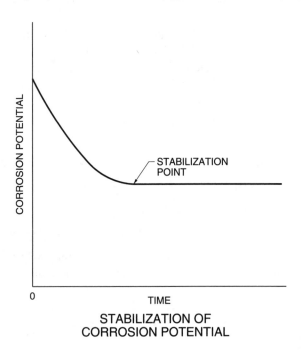

STABILIZATION OF
CORROSION POTENTIAL

Figure 31-5. After it has stabilized, the corrosion potential is measured by means of a high-impedance voltmeter.

It takes time for the corrosion potential to reach a steady value after initial exposure of the metal to the electrolyte. Before starting an electrochemical corrosion test, it is mandatory to allow the corrosion potential of the test specimen to stabilize to a steady corrosion potential.

Anode Polarization Curves. Anode polarization curves indicate the polarization behavior of metal anodes, which are classified as active and active-passive. An active metal displays a polarization curve in which current density increases with increasing polarization. An active-passive metal shows the same increase initially. In both cases, this region corresponds to corrosion and to the formation of soluble corrosion products, so the corrosion rate increases with increasing potential.

With the active-passive metal, however, the formation of a protective or passive film becomes thermodynamically possible at a specific potential. The corrosion rate falls abruptly to a very low value and remains there when this potential is exceeded. At some higher potential, the corrosion rate again begins to rise, corresponding to the dissolution or breakdown of the passive film. See Figure 31-6.

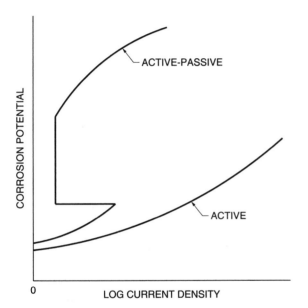

Figure 31-6. Metal anodes display two types of polarization curves, which are active and active-passive.

Both types of anode behavior are plotted against log current density. A logarithmic (log) scale is used to plot current density in order to facilitate the study of corrosive behavior in the passive region, where the corrosion rate is extremely low.

Active-passive metals can exhibit a very low corrosion rate over a wide potential range (range of corrosive conditions). Some metals, such as titanium, show active-passive behavior in many corrosive environments. Other metals, such as steel, display active-passive behavior in a limited number of corrosive environments. For a very low corrosion rate, a metal must be in its passive state.

Cathode Polarization Curves. Cathode polarization curves indicate the polarization behavior of cathodes. Cathode polarization curves exhibit various shapes, which are illustrated by the three common cathode reactions of hydrogen evolution, oxygen reduction, and ferric ion reduction. See Figure 31-7. During hydrogen evolution, the cathode polarization curve slopes downward linearly, indicating a chemical reduction reaction. Hydrogen evolution is expressed by the following equation:

$$2H^+ + 2e \leftrightarrow H_2\uparrow$$

Oxygen reduction in an acid solution is expressed by the following equation:

$$O_2 + 4H^+ + 4e \rightarrow 2H_2$$

Oxygen reduction in a neutral and alkaline solution is expressed by the following equation:

$$\tfrac{1}{2} O_2 + H_2O + 2e \rightarrow 2OH^-$$

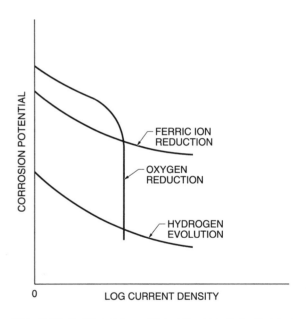

Figure 31-7. The shape of the cathode polarization curve depends on whether an oxygen diffusion or chemical reduction reaction influences depolarization.

An *overvoltage* is the difference between the half-cell potential and the potential resulting from polarization. At low overvoltage, the cathodic polarization curve slopes downward linearly, as with hydrogen evolution. At higher overvoltage, the curve begins to drop vertically because once the oxygen molecules close to the corroding surface have been used up, additional oxygen must diffuse in from the bulk of the solution for corrosion to continue. Under stagnant conditions, a limiting current density for oxygen reduction is rapidly reached. Increasing the solution velocity or the temperature causes depolarization. Depolarization is a decrease in the slope of the polarization curve. Depolarization often increases corrosion rates by causing the anode and cathode polarization curves to intersect at higher current densities (corrosion rates). The increases in solution velocity and temperature increase the rate of oxygen diffusion to the surface.

Ferric ion reduction is expressed by the following equation:

$$Fe^{3+} + e \rightarrow Fe^{2+}$$

Ferric ion reduction is similar to hydrogen reduction in that it is a chemical reduction reaction and not influenced by diffusion of species in the electrolyte to the cathode. Ferric ions often exert a powerful effect on corrosion by causing the anode and cathode polarization curves to intersect at high corrosion rates.

Electrochemical Corrosion Testing. In electrochemical corrosion testing, polarization curves are created and used to study corrosion and predict corrosion behavior. Electrochemical corrosion testing is performed by immersing the test specimen in the electrolyte and systematically displacing the potential from the corrosion potential. The most common technique is potentiodynamic polarization, in which the potential is continuously displaced and the resulting current density is plotted as a function of the potential.

A potentiodynamic polarization test is conducted in a laboratory corrosion cell consisting of a reference electrode, working electrode (test specimen), and counter electrode. See Figure 31-8. These are connected to a potentiostat, which is used to alter the potential and measure the resulting current. The resulting potential/current density curve is automatically plotted.

A potentiodynamic polarization curve is developed by sweeping first in the anode direction and then reversing the potential scan and sweeping in the cathode direction. The potentiodynamic polarization curve that is plotted represents displacement from the equilibrium conditions at the corrosion potential, where the anode and cathode reactions are equal and opposite.

The current density at any specific potential on this combined polarization curve represents a summation of the anode and cathode polarization behavior at that potential. From the shape of the poteniodynamic polarization curve, it is possible to predict the corrosion behavior, such as susceptibility to localized corrosion, of the alloy in the electrolyte at the test temperature.

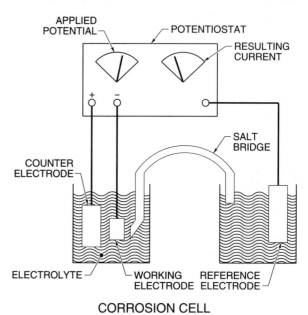

Figure 31-8. Electrochemical corrosion testing consists of plotting the variation between the corrosion current density and the potential using a potentiostat.

Corrosive Environments

Corrosive environments include acids, alkalis, salt solutions, seawater, and soil. Corrosive environments have characteristics that influence their corrosiveness. These characteristics are pH, aggressive species, and oxidizing species.

pH. The *pH* is the hydrogen-ion activity and denotes the acidity or basicity of a solution. It can be an important indicator of corrosiveness. See Figure 31-9. Solutions of pH 0 to 7 are acidic, and solutions of pH 7 to 14 are alkaline. The lower the pH, the greater the amount of free acid present and the more corrosive the solution. As a rule of thumb, the minimum pH for carbon steel applications, without excessive corrosion, is 5. Metals such as zinc, aluminum, and lead are susceptible to corrosion in highly alkaline (high pH) solutions.

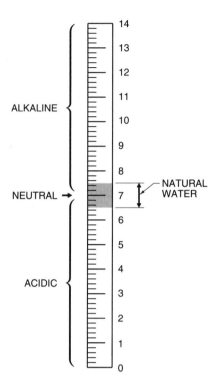

Figure 31-9. The pH scale can be an important indicator of corrosiveness.

Aggressive Species. Aggressive species are corrosive materials that can penetrate and break down protective surface films or modify the shape of the cathodic polarization curve. The chloride ion is the most commonly encountered aggressive species. The chloride ion increases susceptibility to general corrosion, pitting, crevice corrosion, and stress-corrosion cracking in many alloys. When coupled with acids and oxidizing agents, the chloride ion is extremely corrosive.

Sulfur compounds, such as polythionic acids and thiosulfates, also increase the corrosiveness of many environments. Sulfur compounds lead to general corrosion, pitting, and stress-corrosion cracking.

The severity of the corrosion problems caused by the chloride ion and sulfur compounds is chiefly due to their almost universal presence. The chloride ion is present in most waters, and sulfur is found in lubricants and in the air.

Oxidizing Species. Oxidizing species exhibit mild to severe effects on corrosiveness. Oxygen can be relatively mild, while ferric or cupric ions are almost always severe. Oxidizing species depolarize corrosion reactions.

Alloys such as alloy 400, nickel 200, and alloy B-2 often lose their excellent corrosion resistance when oxidizing species are present. Mild aeration may be sufficient to cause severe corrosion. Zirconium is extremely susceptible to accelerated corrosion when the chloride ion is present with an oxidizing species, such as the ferric ion.

Exposure Conditions

Exposure conditions, when superimposed on characteristics of the environment, may increase corrosiveness of the environment. The characteristics of the environment and exposure conditions influence the cathode reaction. Exposure conditions that influence the corrosiveness of an environment are temperature and aeration.

Temperature. Temperature is an extremely important factor and may effect the corrosion rate in either direction. Operating below design temperature can be as harmful as operating above it. Increasing the temperature usually increases the corrosion rate by a factor that depends on the controlling mechanism for the cathode reaction.

If the cathode reaction is controlled by the diffusion of oxygen to the metal surface, the corrosion rate approximately doubles for every 30°C (55°F) rise in temperature. If the cathode reaction is controlled by a chemical reaction, for example by the reduction of hydrogen ions to hydrogen atoms, the corrosion rate increases more rapidly with temperature, and may double for each 10°C (20°F) rise.

The solubility of oxygen in aqueous solutions decreases as temperature increases. This causes the effect of the decreasing solubility of oxygen to compete with the effect of temperature on the corrosion rate. For example, with carbon steel, the effect of decreasing oxygen solubility eventually overcomes the effect of increasing temperature on the corrosion rate. As a result, the corrosion rate of carbon steel in water increases up to approximately

80°C to 90°C (175°F to 195°F) and then begins to fall. See Figure 31-10.

When considering the effect of temperature on corrosion, it is the surface temperature of the metal and not the actual temperature of the corrosive environment that is important. For example, in selecting materials for steam-heated exchanger tubes, corrosion testing at the operating temperature of the process environment may indicate that 304 stainless steel is an acceptable material of construction. To be absolutely sure, the 304 stainless steel must be evaluated in the process environment at the actual temperature of the tubes, which is higher than operating temperature of the steel.

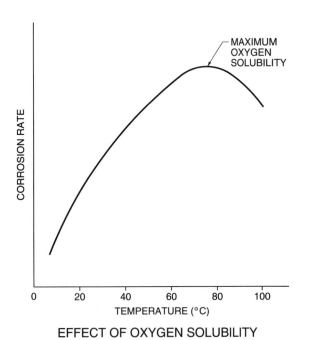

Figure 31-10. The corrosion rate of steel versus temperature rises to a maximum value and then falls.

A decrease in temperature of the metal or the environment may also increase corrosion. This may result from the condensation of a corrosive species. Severe corrosion may take place under these conditions if the condensation occurs at high temperatures, leading to localized boiling of the corrosives on the surface of the metal. This type of attack is dew point corrosion and sometimes occurs in flue gas lines.

Temperature cycling can also lead to corrosion problems by causing corrosive species to concentrate on the metal surface. This is what leads to the chloride stress-corrosion cracking of austenitic stainless steels under wet thermal insulation. Water enters the insulation and evaporates, leaving behind the chloride ion. Over a prolonged period of time and with increasing numbers of temperature cycles, a very low chloride concentration in the water can build up to the threshold level that causes chloride stress-corrosion cracking.

Aeration. Aeration is generally related to the flow rate (velocity). Aeration may cause a significant change in the rate of corrosion, depending on how the conditions affect the stability of the protective surface films or corrosion products.

The greater the velocity, the greater the amount of oxygen that can be brought to the surface of the metal. This, in turn, depolarizes the corrosion reaction. With active metals, depolarization increases the corrosion rate, but with passive metals, the corrosion rate may decrease.

When there is no velocity (stagnant conditions), oxygen diffusion is greatly hindered and solids tend to settle out. This leads to oxygen concentration cells between bare and covered metal surfaces, which increases susceptibility to localized corrosion under the solid deposits.

CORROSION PREVENTION

Three methods can be used to prevent corrosion. They include selecting the optimum materials and design, using electrochemical control techniques, and modifying the environment to make it less corrosive.

Materials Selection and Design

The materials engineer predicts the performance of candidate alloys in corrosive environments through the accumulation of experience, data in handbooks, and corrosion testing programs. The objective is to predict the general corrosion rate and susceptibility to localized attack of candidate materials. With this information, material selection can be optimized.

The scope of selection is not limited to metals. Coatings, linings, polymers, and ceramics also come under close scrutiny.

Design principals that are used to avoid corrosion include prevention of crevices by using continuous welds, avoiding locations where corrosives can concentrate, and preventing excessive velocity on metal surfaces. See Figure 31-11. Coupon corrosion testing is frequently used to select materials and avoid corrosion.

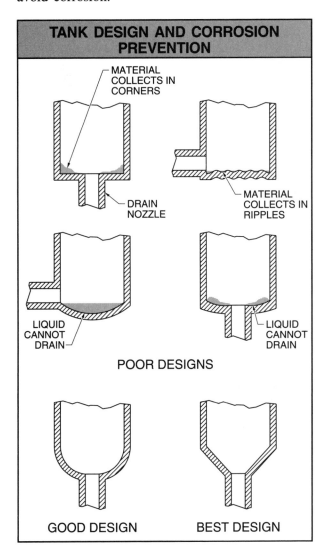

TANK DESIGN AND CORROSION PREVENTION

MATERIAL COLLECTS IN CORNERS

DRAIN NOZZLE

MATERIAL COLLECTS IN RIPPLES

LIQUID CANNOT DRAIN

LIQUID CANNOT DRAIN

POOR DESIGNS

GOOD DESIGN

BEST DESIGN

Figure 31-11. Corrosion is minimized by good design.

Coupon Corrosion Testing. Coupon corrosion testing consists of exposing representative samples (coupons) of the metals, or other test materials, to the corrosive environment. This may be performed in the field in the actual corrosive environment or in a laboratory using a simulation of the environment that the material is exposed to. See Figure 31-12. The corrosion rate, type, and extent of the coupon must be recorded. The most commonly used corrosion rate units are mils per year (mpy), where 1 mil equals .001 in., and millimeters per year (mmpy). The conversion factor between millimeters per year and mils per year is approximately 1 mmpy = 40 mpy. The corrosion rate on the coupons is figured by applying the following formula:

$$C = \frac{3450 \times 1000 \times W}{D \times A \times T}$$

where

C = corrosion rate (in mpy)

W = weight loss (in g)

D = coupon density (in g/cm^3)

A = exposed surface area (in cm^2)

T = exposure time (in hr)

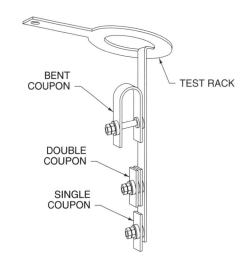

BENT COUPON

TEST RACK

DOUBLE COUPON

SINGLE COUPON

Figure 31-12. Coupon corrosion testing consists of exposing samples (coupons) of various materials to the corrosive environment for sufficient time to obtain a ranking of their performance.

Example: Figuring corrosion rate

Figure the corrosion rate of an aluminum coupon that has a density of 2.7 g/cm^3, a weight of 30 g before exposure, an exposed surface area of 150 cm^2, an exposure time of 240 hr, and a weight of 24 g after exposure.

$$C = \frac{3450 \times 1000 \times W}{D \times A \times T}$$

$$C = \frac{3450 \times 1000 \times (30 - 24)}{2.7 \times 150 \times 240}$$

$$C = \frac{20,700,000}{97,200}$$

$$C = \textbf{212.96 mpy}$$

Electrochemical Control Techniques

Electrochemical control techniques are used to modify the electrochemical characteristics of the environment by means of an externally applied current. This modification allows the use of cheaper, less corrosion-resistant materials of construction. If the electrochemical control parameters are not maintained within the defined range in which the material is resistant, rapid corrosion may occur.

To apply electrochemical control, the electrochemical characteristics of the environment must be characterized. This is usually done by means of electrochemical testing and potential monitoring of the structure and resistivity of the environment. The major types of electrochemical control are cathodic protection and anodic protection.

Cathodic Protection. Cathodic protection is the most widely used form of electrochemical control. It is used chiefly to protect carbon steel from corrosion in soil and seawater. Cathodic protection consists of passing a current into a corroding metal at a rate equal to or greater than the current that was flowing out. This may be accomplished by connecting a large enough DC source to the corroding metal and to an electrode that becomes the anode. For complete protection, the anode potential must be equal to or less than the half-cell potential of the corroding structure.

Cathodic protection is not necessary for every soil system. The need for it depends on the resistivity of the soil and various other factors. Generally, cathodic protection is considered necessary if the resistivity of the soil is <5000 Ωcm.

The greater the corrosion rate, the greater the current required to arrest it. For this reason, coatings are always applied in conjunction with cathodic protection to reduce the amount of bare metal exposed and, thus, reduce the current requirement. The cathodic protection anode can be an inert material, such as silicon iron powered by an AC rectifier, or a sacrificial metal, such as zinc or magnesium. In a soil application, the inert material is usually buried in backfill. The backfill contains coke breeze or gypsum, which helps improve current distribution. See Figure 31-13.

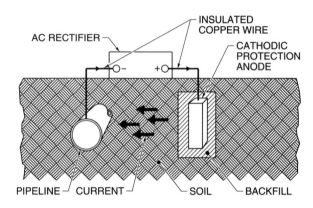

Figure 31-13. Current cathodic protection is extensively used to protect underground pipelines.

Anodic Protection. Anodic protection is used primarily to achieve passivity in stainless steels and carbon steels in applications involving concentrated sulfuric acid. An impressed current is used to raise the potential of the equipment in the passive region. Once the potential has been shifted over the nose of the anode polarization curve, the current density required to maintain protection is very low. The benefits of anodic protection are increased equipment life and lower iron contamination.

Process Control and Inhibitor Additions

Process control and inhibitor additions are the two means of modifying the environment to reduce corrosion. *Process control* is the manipulation of process or operating parameters to reduce corrosion. Process control techniques include reducing the operating temperature, preventing the build-up of aggressive species (particularly chloride, by bleeding off recirculation loops), bleeding in air where aeration is advantageous in maintaining passivity, eliminating air intrusion where aeration increases corrosion, reducing velocity and turbulence where erosion-corrosion is a problem, and increasing velocity where deposit attack is a problem (particularly if aggressive species are present).

Inhibitor additions are additions of chemical compounds to the environment that cause a reduction in corrosion rate. Adding inhibitors to process environments or recirculating water systems is sometimes an economical way of controlling corrosion. For example, refinery overhead streams contain acids, such as hydrochloric, in the aqueous phase. The addition of neutralizing agents and film-forming inhibitors helps reduce the corrosion rate of carbon steel equipment and piping.

One of the largest groups of inhibitors are the barrier layer types. These interfere with the corrosion reaction by forming a barrier on the metal surface. The barrier layer inhibitors are divided into two major groups, the absorbed layer and the oxidizing layer.

One example of an absorbed layer inhibitor is the cathodic inhibitor. A cathodic inhibitor moves the potential into the cathode direction, which reduces the corrosion rate. These inhibitors are used in acid solutions, such as hydrochloric acid chemical cleaning operations.

Oxidizing inhibitors (passivators) are used in neutral solutions and shift the potential in the anode direction into the passive region. Examples of oxidizing inhibitors are nitrites on carbon steel, which are frequently used in cooling water service. See Figure 31-14.

FORMS OF CORROSION

The forms of corrosion may be divided into general corrosion, which is one specific type, and localized corrosion, which are all the others. The forms of localized corrosion are erosion-corrosion, crevice corrosion, pitting, stress accelerated corrosion, stress-corrosion cracking, embrittlement, galvanic corrosion, and microstructural corrosion.

Rarely is one specific form of corrosion totally responsible for failure. Several forms may play a part in the initiation, propagation, and final failure. In many cases, the final failure is from mechanical overload because the component can no longer support the load.

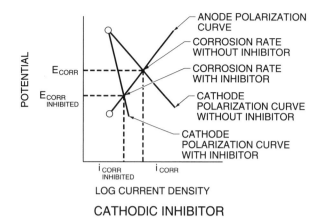

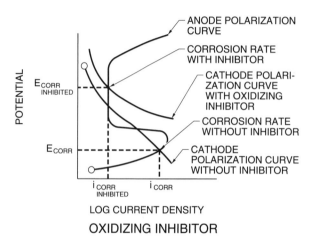

Figure 31-14. Barrier layer inhibitors consist of cathodic and oxidizing (anodic) types.

For example, the failures of hardened alloy steel relief-valve springs in corrosive plant atmospheres often occur when initiation site or general corrosion first takes place, causing hydrogen to be evolved as a result of the corrosion reaction. The hydrogen leads to a crack, which slowly propagates into the metal. When the spring cross section is too small to support the load on it, the spring fails by mechanical brittle fracture. In this failure, general corrosion, hydrogen-assisted cracking, and brittle fracture all make key contributions. See Figure 31-15.

General Corrosion

General corrosion is relatively uniform in appearance. The rate of general corrosion is expressed in various units. A corrosion allowance may be incorporated into equipment and components to allow for general corrosion. The most common type of general corrosion is the rusting of steel.

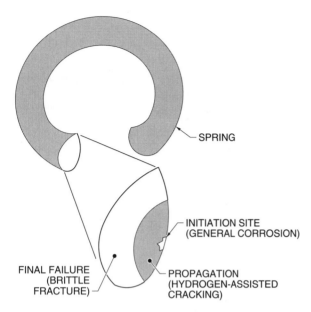

Figure 31-15. More than one form of corrosion may participate in the failure of a component.

General Corrosion Rate. General corrosion proceeds in a relatively uniform manner over the entire surface of a metal. The material becomes thinner as it corrodes until its thickness is reduced to the point where failure occurs, or the part begins to develop perforations that result in leakage or collapse. See Figure 31-16.

A generally corroded surface is usually roughened, pitted, or pock marked. Unless the surface is uneven, the remaining wall thickness can be measured by ultrasonic testing if the probe is placed on the noncorroded side.

Corrosion Allowance. A corrosion allowance is an extra thickness of metal added to the design thickness of a component so that it will have useful life in a corrosive environment. The corrosion allowance is calculated from the anticipated corrosion rate and the required life of the equipment. For example, a carbon steel vessel with a corrosion allowance of .25 in. has an anticipated life of 10 years at a maximum general corrosion rate of 25 mpy and 20 years at 12 mpy.

Rusting of Steel. The rusting of steel is the most common form of general corrosion. *Rusting* is the reaction of water containing dissolved oxygen with steel to form a series of corrosion products leading to a mixture of various iron oxides (geothite). Rusting occurs when the relative humidity of the atmosphere is above a specific, critical value of 63%. Rusting does not take place in dry atmospheres. The worst situation for rusting is when the steel surface cycles through wet and dry cycles, which, for example, occurs between day and night.

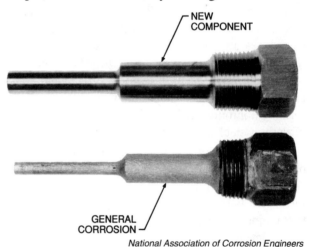

National Association of Corrosion Engineers

Figure 31-16. General corrosion of a component proceeds until failure occurs by overload of the thinned wall, or a leak develops from perforation due to excessive thinning of the wall.

The rate of rusting is dramatically increased by two common atmospheric contaminants, sulfur dioxide and the chloride ion. Sulfur dioxide is a pollutant in industrial and urban atmospheres and exerts two important effects. First, sulfur dioxide lowers the critical relative humidity of the atmosphere, making condensation of moisture more likely to occur. Second, sulfur dioxide forms some iron sulfate in combination with the steel, which catalyzes the general rust formation reaction.

Chloride ions are present in seacoast atmospheres and in the vicinity of highways where deicing salts are used. Chlorides reduce the protective qualities of the iron oxide rust coating so that the rate of rusting does not slow down as the rust film builds up.

Painting is the most common method of preventing the rusting of steel. The paint forms a physical barrier to moisture. Two methods of preventing rusting during equipment storage are to use a nitrogen blanket (nitrogen atmosphere) or vapor phase inhibitor. The nitrogen blanket prevents the access of oxygen, an essential component of the rust reaction. Vapor phase inhibitors form a barrier on the surface and prevent the rust reaction from taking place. If paint is applied on a previously rusted surface that has not been properly cleaned, such as with white-metal blasting, the rust reaction will continue to propagate under the paint causing it to degrade.

Erosion-Corrosion

Erosion-corrosion occurs on metals under turbulent flow conditions created by high fluid velocity. Turbulent flow helps remove protective corrosion products from the surface, accelerating the corrosion rate. If the cathodic reaction is oxygen reduction, turbulent flow or increased velocity also causes depolarization, which increases the corrosion rate of active metals, such as carbon steel or copper. Because increasing velocity helps to maintain their protective passive film, active-passive metals, such as stainless steels, usually benefit from cathodic depolarization. See Figure 31-17.

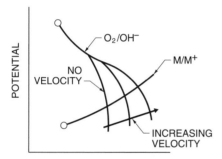

ACTIVE METAL

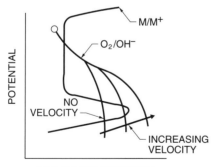

ACTIVE-PASSIVE METAL

Figure 31-17. Erosion-corrosion occurs under agitated conditions, such as high flow rate, leading to depolarization, often coupled with the removal of protective corrosion products. With active-passive metals, these conditions may help maintain the passive film.

Erosion-corrosion is usually localized and is assisted by factors that encourage turbulent flow such as excessive weld penetration in pipelines. Specific forms of erosion-corrosion include liquid impingement erosion, wire drawing, and cavitation.

Liquid Impingement Erosion. Liquid impingement erosion occurs chiefly in components that are subjected to wet steam at high velocity, such as low-pressure turbine blades, low-temperature steam piping, and condenser tubes subjected to direct impingement by wet steam. See Figure 31-18.

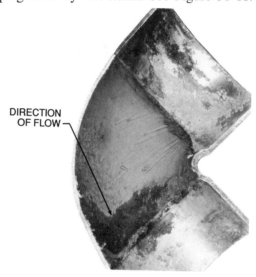

Figure 31-18. Liquid impingement erosion exhibits a characteristic horseshoe pattern, with U's pointing in the direction of the flow.

Damage occurs at locations where the direction of flow changes, for example, at elbows or U bends. Large radius bends or baffles reduce susceptibility to attack, but the use of erosion-resistant materials, such as austenitic stainless steel, is more effective.

Wire Drawing. Wire drawing occurs in valve seats and mechanical seals subject to leakage of high-pressure fluids or steam. Wire drawing appears as channels in the interface, indicating the passage of the high-pressure environment. Materials that exhibit a combination of high hardness and high corrosion resistance, such as Stellite® 6, are used where wire drawing is a potential problem.

Cavitation. Cavitation is surface damage caused by collapsing vapor bubbles in a flowing liquid. See Figure 31-19. The vapor bubbles form as a result of changes in flow velocity or flow direction, or a reduction in the cross section of the flow passage. An increase in pressure at a nearby location downstream will collapse the bubbles. The imploding effect of the collapsing bubbles give rise to shock waves that cause contact stresses on the metal surface. Repetitive implosions lead to spalling (flaking) and pitting of the surface. A cavitated surface has a rough, cratered appearance.

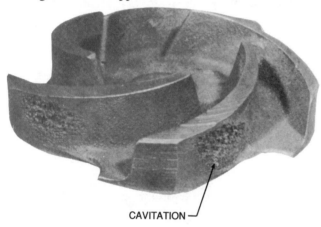

Figure 31-19. Cavitation is caused by imploding vapor bubbles that set up shock waves on the surface of the component.

Crevice Corrosion

Crevice corrosion is localized corrosion that occurs at mating (faying) or closely fitting surfaces where easy access to the bulk corrosive environment is hindered. See Figure 31-20. Crevice corrosion may occur at mating surfaces, assemblies of metals and nonmetals, and deposits on the surfaces of metals (deposit attack). Microbiologically influenced corrosion is a form of crevice corrosion.

The geometry of a crevice prevents access of the bulk corrosive environment, particularly oxygen. The oxygen deprived region within the crevice corrodes preferentially. Because the crevice region retains the corrosive environment, crevice corrosion often continues when the area outside the crevice dries out.

Active-passive alloys, such as stainless steels and aluminum, are extremely susceptible to crevice corrosion because their protective passive films are maintained by aeration or oxidizing species, which become used up inside a crevice. Additionally, aggressive species, such as chlorides, can concentrate inside the crevice and increase the rate of attack.

National Association of Corrosion Engineers

Figure 31-20. Crevice corrosion is caused by an oxygen concentration cell, which is formed between exposed and shielded portions on a metal surface.

Deposit Attack. Deposit attack is a form of crevice corrosion that occurs under foreign objects such as dirt, pieces of shell, vegetation, rust, or scales. These materials locally screen the metal surface leading to oxygen-deficient regions that are preferentially corroded. Deposit attack is controlled by keeping surfaces clean, for example, by increasing solution velocity. Condensers and heat exchanger tubing are periodically cleaned to prevent accumulation of deposits, which may lead to corrosion and also reduce heat transfer.

Microbiologically Influenced Corrosion. *Microbiologically influenced corrosion* is a form of crevice corrosion that occurs in natural waters under deposits that harbor microorganisms. Failure occurs by severe localized attack under the deposits. Microbiologically influenced corrosion primarily affects carbon and stainless steels.

With the exception of microorganisms that produce hydrogen sulfide, microorganisms are generally not corrosive unless they settle on the metal surface. The settlement of microorganisms is assisted by the presence of tubercle or slime forming bacteria, and also by roughness of the metal surface.

Two common microorganisms are the sulfate-reducing types and the iron-oxidizing types. These are extremely corrosive toward carbon steel and stainless steel, respectively. Biocide additions are effective in killing free-swimming or floating microorganisms in contaminated waters. However, biocide additions are ineffective in eliminating microorganisms that are physically attached to the surface because it is difficult for the biocide to penetrate the tubercle or slime that harbors the microorganism.

The use of microbiological or bacteria-contaminated water in equipment must be limited to short time exposures, such as for hydrotesting. Longer exposures may result in serious corrosion. It is important to drain and dry stainless steel equipment after hydrotesting to prevent attack from occurring.

Pitting

Pitting is localized corrosion that has the appearance of cavities (pits) and occurs on freely exposed surfaces. Pitting usually occurs on active-passive metals in environments containing aggressive ions, such as chlorides. Pitting initiates at defects in the passive film and may lead to premature failure by penetration of the wall thickness.

Pitting and crevice corrosion are sometimes confused with each other. The difference between pitting and crevice corrosion is that pitting occurs on an unshielded metal surface, making it less predictable than crevice corrosion.

Pit growth is encouraged by concentration of the corrosive in the pit and the unfavorable anode-cathode ratio. The interior of the pits tends to become more acidic compared with the bulk environment outside of the pits. Since the pit is a cavity, it is difficult for the solution to leave the pit. Additionally, an unfavorable ratio exists between the large passive cathode area surrounding the pits and the small active anode area of the pits. These factors make pitting a self sustaining (autocatalytic) process. Two common forms of pitting are chloride pitting and oxygen pitting.

Chloride Pitting. Chloride pitting occurs on active-passive alloys in chloride containing environments. The chloride ion is the most common cause of pitting. Aluminum alloys and stainless steels are par-

ticularly susceptible to chloride pitting, especially if the environment is stagnant, hot, or contains oxidizing ions. Chloride pits are usually spherical and undercut the surface, meaning that their actual volume is greater than the surface appearance would indicate. See Figure 31-21.

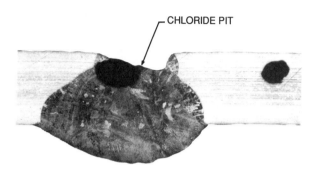

National Association of Corrosion Engineers

Figure 31-21. Chloride pitting of stainless steels consists of spherical cavities that undercut the surface so that the extent of damage is worse than it appears.

Alloying with molybdenum improves the resistance of austenitic stainless steels to pitting in neutral chloride solutions. Austenitic stainless steels, in order of increasing molybdenum and increasing pitting resistance, are type 304 (0%Mo), type 316 (2.5% Mo), type 317 (3.5% Mo), 904L (4.5% Mo), and AL6X-N (6.5% Mo). These alloys do not resist highly oxidizing chloride solutions such as ferric chloride. Titanium and alloy C-276 are often suitable for these environments.

Oxygen Pitting. Oxygen pitting occurs on carbon steel in boiler feedwater or steam condensate that has not been adequately deaerated. Oxygen pitting begins at weak points in the protective iron oxide film. It has the appearance of large, jagged pits. If the attack is actively continuing, the pits are filled with black iron oxide powder. If inactive, the powder is red iron oxide. See Figure 31-22.

Stress-accelerated Corrosion

Stress-accelerated corrosion is localized corrosion that occurs in certain environments at places on a metal surface that have experienced plastic strain from stresses of yield-point magnitude. Stress-accelerated corrosion may occur adjacent to welds or in heavily cold-worked areas. Two common forms of stress-accelerated corrosion are corrosion fatigue and fretting.

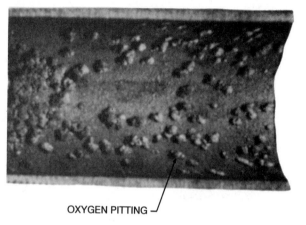

OXYGEN PITTING

National Association of Corrosion Engineers

Figure 31-22. Oxygen pitting of carbon steel consists of large jagged pits.

Corrosion Fatigue. Corrosion fatigue is the acceleration of mechanical fatigue failure from cyclic stresses due to the action of a corrosive environment. Corrosion fatigue depresses or eliminates the threshold stress (fatigue limit) for fatigue failure, meaning that cyclically loaded structures in corrosive environments may eventually fail, even when lightly loaded. See Figure 31-23.

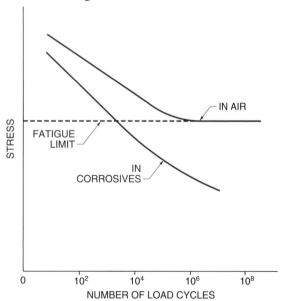

Figure 31-23. Corrosion fatigue eliminates the threshold stress (endurance limit) for fatigue failure.

Fretting. *Fretting* is localized surface damage produced by vibration that occurs at close fitting, highly loaded surfaces. Fretting may occur at press-fitted components, splines, keyways, and bearings. Fretting is caused by localized welding of minute regions of the two faces that are in contact. With continuous oscillatory movement, the welded particles tear away from the opposing surfaces, react chemically with the atmosphere, and form a debris of powder in the joint. Fretting is identified in steel components by pits filled with dark brown iron oxide. See Figure 31-24. Fretting ruins bearings, destroys dimensions, and reduces fatigue strength by creating surface stress concentrations.

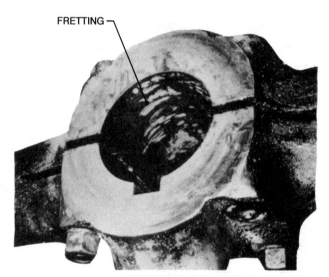

FRETTING

National Association of Corrosion Engineers

Figure 31-24. Fretting in steel components appears as jagged pits filled with dark brown powder.

The best way of preventing fretting is to remove the source of vibration by tighter clamping or more rigid mounting. Other methods include raising the hardness of mating surfaces, inserting rubber gaskets in joints to absorb motion, lubricating with a dry medium, and sealing the area to exclude the atmosphere. Surface treatments, such as case hardening, shot peening, or hard (engineering) chromium plating, lessen susceptibility to fatigue crack propagation resulting from fretting on shafts.

Stress-corrosion Cracking

Stress-corrosion cracking is crack formation that occurs in some alloys in specific corrosive environments under the combined action of a tensile stress and corrosion. The stress-corrosion cracks are usually extremely branched and tightly packed with corrosion products. This makes stress-corrosion cracking almost impossible to repair by grinding or by gouging out the cracks followed by rewelding. The heat created by these operations causes the cracks to rapidly propagate in many directions.

Certain corrosives are responsible for stress-corrosion cracking in specific alloys. See Figure 31-25. The concentration and temperature of the corrosive must be above a threshold value in order for cracking to occur. In certain cases, stress-corrosion cracking occurs within a specific range of electro-chemical potential.

Three common forms of stress-corrosion cracking are chloride cracking of austenitic stainless steels, caustic cracking of carbon steels, and ammonia cracking of brass. Various remedial methods are used to prevent stress-corrosion cracking.

Chloride Cracking of Austenitic Stainless Steels. Chloride cracking of austenitic stainless steels (300 series) is the most common form of stress-corrosion cracking. Chloride cracking occurs at temperatures

STRESS-CORROSION CRACKING SUSCEPTIBILITY									
Environment	**Aluminum Alloys**	**Copper Alloys**	**Nickel Alloys**	**Titanium Alloys**	**Zirconium Alloys**	**Carbon Steels**	**Stainless Steels**		
							Austenitic	**Duplex**	**Martensitic**
Amines (aqueous)		✓				✓			
Ammonia (anhydrous)						✓			
Ammonia (aqueous)		✓							
Bromine				✓					
Carbonates (aqueous)						✓			
Carbon monoxide, carbon dioxide, and water mixture						✓			
Chlorides (aqueous)	✓		✓		✓		✓	✓	
Chloride (concentrated; boiling)			✓				✓	✓	
Chlorides (dry; hot)			✓	✓					
Chlorinated solvents				✓	✓				
Cyanides (aqueous; acidified)						✓			
Fluorides (aqueous)			✓						
Hydrochloric acid									
Hydrofluoric acid			✓						
Hydroxides (aqueous)						✓	✓	✓	✓
Hydroxides (concentrated; hot)			✓				✓	✓	✓
Methanol plus halides				✓	✓				
Nitrates (aqueous)		✓				✓			✓
Nitric acid (concentrated)				✓					
Nitric acid (fuming)				✓					
Nitrites (aqueous)		✓							
Nitrogen tetroxide				✓					
Polythionic acids			✓				✓		
Steam		✓							
Sulfides plus chlorides (aqueous)							✓	✓	✓
Sulfurous acid							✓		
Water (high-purity and hot)	✓		✓						

Figure 31-25. Certain corrosive environments cause stress-corrosion cracking in specific alloys.

above 60°C (140°F) where the chloride can concentrate on the surface of austenitic stainless steels. Chloride cracking often occurs in waters. Waters usually contain small amounts of chlorides of the order of parts per million. With cyclical wet-dry conditions on the hot surfaces, the chloride ions concentrate to levels at which cracking takes place. Small amounts of oxygen are also necessary for chloride cracking. Oxygen is usually in plentiful supply in most waters unless deliberately removed.

An example of the mechanism of chloride cracking is in water-cooled vertical heat exchangers. See Figure 31-26. The tubes at the upper tubesheet contain residual stresses from tube rolling and are exposed to splash from the cooling water. If the temperature is hot enough, the space between the tube and tubesheet is an ideal location for chloride ions in the cooling water to progressively concentrate and result in chloride cracking.

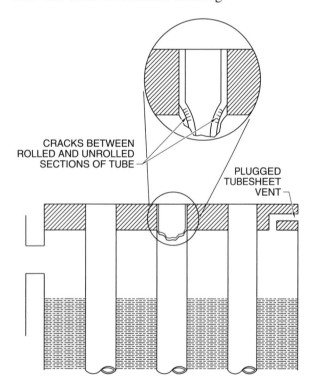

Figure 31-26. The concentration of chloride ions at the top of a vertical heat exchanger tube with a tubesheet made of austenitic stainless steel can lead to chloride cracking.

The austenitic stainless steels are extremely susceptible to chloride cracking. Duplex stainless steels, ferritic stainless steels, and high-nickel austenitic alloys have much greater resistance.

Caustic Cracking of Carbon Steels. Caustic cracking of carbon steels is a form of stress-corrosion cracking that occurs on carbon steels in environments containing the hydroxyl ion. A common case of caustic cracking is where a steam valve is allowed to leak onto valve bonnet bolting made of carbon steel. The steam condenses to water on the bolting and then evaporates, allowing residual hydroxyl ions in the water treatment chemicals in the steam to concentrate. The bolting that is under tensile stress from loading can fail catastrophically.

Carbon steel tanks and piping in caustic service may require stress relieving, depending on the temperature and concentration of the caustic. When the combined caustic concentration and temperature are too severe for stress-relieved carbon steel, nickel or nickel alloys are substituted. See Figure 31-27.

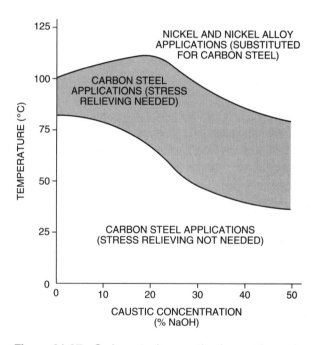

Figure 31-27. Carbon steels operating in caustic service above certain temperatures and concentrations must be stress relieved.

Ammonia Cracking of Brass. Ammonia cracking of brass is a form of stress-corrosion cracking that occurs in alloys of copper (brasses) containing large amounts of zinc, usually more than 15% Zn, in environments that contain ammonia or ammonium hydroxide. Both oxygen and moisture must be present. Residual stresses from cold-forming operations are enough to cause cracking in susceptible alloys.

Stress-corrosion Cracking Prevention. Stress-corrosion cracking prevention is achieved by several techniques, which are used in combination or alone. These techniques include stress relieving, peening, cathodic protection, changing the alloy, and design.

Stress relief and peening are methods of prevention that attempt to remove the residual tensile stress necessary for stress-corrosion cracking to occur. Stress relief is not always successful because it may not completely remove all tensile stresses. In certain metal/environment combinations, even a small amount of residual tensile stress can lead to cracking. Peening the surface to induce a residual compressive stress is often extremely effective, providing that the corrosion rate is sufficiently low to prevent the compressive layer from being corroded away to reveal a layer that is in tension.

Cathodic protection lowers the potential of the alloy from the range where stress-corrosion cracking occurs. Cathodic protection may cause hydrogen to be evolved on the surface of the metal and must not be used on materials that are susceptible to hydrogen embrittlement.

An obvious method of prevention is to substitute an alloy that is not susceptible to cracking in the subject environment. Economics and fabricability often dictate the use of resistant alloys because they are invariably more expensive or have different fabrication characteristics.

Using a design that prevents the access of a corrosive or its concentration on the surface is an effective method of prevention. For example, surface coatings provide a physical barrier to access of the corrosive, and flooding the surface prevents concentration of the corrosive species responsible for cracking.

Embrittlement

Embrittlement is the loss of ductility in a metal caused by corrosion. Two forms of embrittlement are caused by hydrogen uptake in metals. These are hydrogen-assisted cracking and hydride embrittlement. The third, which is caused by liquid metals, is liquid metal embrittlement.

Hydrogen-assisted Cracking. Hydrogen-assisted cracking is a form of embrittlement that occurs in high-strength steels from hydrogen atoms created at the surface of the metal by corrosion. The hydrogen diffuses into the metal and interferes with its ability to yield elastically under stress, increasing the tendency toward brittle fracture.

When a cathodic reaction produces hydrogen atoms on the surface, the hydrogen atoms may or may not combine with each other. If they combine, hydrogen molecules that evolve from the surface are produced. If they do not combine, the hydrogen atoms may diffuse into the metal and cause hydrogen-assisted cracking. Hydrogen-assisted cracking is a potential problem at service temperatures between –20°C (–5°F) and 120°C (250°F). At lower temperatures, the rate of hydrogen diffusion is extremely low. Higher temperatures encourage hydrogen to diffuse out of the metal rather than enter it.

Certain species, such as poisons, are extremely harmful because they prevent the combination of hydrogen atoms to hydrogen molecules. Poisons include sulfides, such as hydrogen sulfide. Sulfide stress cracking is a form of hydrogen-assisted cracking that is widespread in the oil and gas production industry and is caused by the presence of hydrogen sulfide. Susceptibility toward hydrogen-assisted cracking increases with the hardness of the steel. In environments containing poisons, carbon and low-alloy steels must be restricted to a maximum hardness of 22 HRC.

Hydride Embrittlement. Hydride embrittlement is the embrittlement of titanium, zirconium, and tantalum by the formation of hydrides. Hydrides are intermetallic compounds having the shape of flakes that develop from the reaction of these metals with hydrogen. See Figure 31-28. Hydrides significantly reduce strength and ductility. Hydride formation progresses from the surface inward.

Titanium, zirconium, and tantalum are extremely susceptible to hydride embrittlement. As little as 100 ppm of hydrogen will severely embrittle tantalum and 100 to 200 ppm will embrittle titanium or zirconium. The hydrogen can be supplied by a galvanic couple. Since most metals are anodic to titanium, zirconium, or tantalum and corrode when coupled to them, these metals are extremely susceptible to embrittlement when they are part of a galvanic couple. Surface iron contamination can also provide hydrogen by this mechanism. Titanium equipment should never be cleaned with a steel wire brush because it leaves traces of iron on the surface. A stainless steel brush is acceptable.

American Society for Metals

Figure 31-28. Hydride embrittlement occurs in alloys such as titanium by inward diffusion of hydrogen to form brittle metal hydrides.

Liquid Metal Embrittlement. *Liquid metal embrittlement* is the intergranular penetration of one metal by a liquid, or molten, metal. Examples of liquid metal embrittlement include molten zinc embrittlement of stainless steels and mercury embrittlement of alloy 400 or aluminum. Galvanized steel must not be welded to stainless steel because the molten zinc created by the heat of welding will embrittle the stainless steel. Galvanizing must be completely removed by grit blasting a minimum of 13 mm (½ in.) from the joint to ensure that it will not melt during welding and cause embrittlement.

Galvanic Corrosion

Galvanic corrosion is the acceleration of corrosion that occurs when one metal is electrically coupled to a more noble metal or a conducting nonmetal in an electrolyte. While the galvanic corrosion rate (galvanic effect) is increased on the anodic member of the couple, it is decreased on the cathodic member. The factors that influence the magnitude of the galvanic effect are the potential difference between the metals, the polarization behavior of the couple, the conductivity of the environment, and the relative surface areas of the components.

The potential difference between the less noble (anode) and more noble (cathode) components indicates the tendency for galvanic corrosion to occur. See Figure 31-29. As a guide to predicting galvanic corrosion, galvanic series that list the electrochemical potentials of metals and some conducting nonmetals, such as graphite, have been developed for specific corrosive environments. The potential difference between the two components may be obtained by subtracting the less noble potential from the more noble potential. The greater the value, the greater the tendency toward galvanic corrosion.

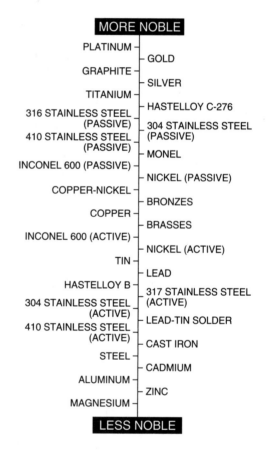

Figure 31-29. The potential difference between the less noble (anode) and more noble (cathode) components indicates the tendency for galvanic corrosion to occur.

The polarization behavior of the components of the galvanic couple is the key consideration in determining the magnitude of the galvanic corrosion effect. For example, if the cathode is not easily depolarized, the anode potential shift is minimized and the corrosion of the anode is relatively slight. For the same potential difference between two metals and an easily depolarized cathode, the corrosion of the anode is relatively great. Metals such as titanium, zirconium, and tantalum are not easily depolarized because of their passive films. Although they are very noble in any galvanic couple, they do not cause significant corrosion of other less noble metals. Easily depolarized alloys, that occupy noble positions in the galvanic series are nickel-chromium alloys, such as alloys 600, 625, and C-276, plus nickel and copper. These metals cause severe corrosion of less noble metals, such as aluminum or steel.

Active-passive alloys can occupy two positions in the galvanic series. When passive, they become noble if the passive film is removed or damaged. The active-passive behavior of alloys such as stainless steels makes the prediction of their galvanic effects extremely difficult. When tantalum, titanium, or zirconium form the noble component of a galvanic couple, care must be taken to prevent hydrogen from forming on them. Hydrogen formation leads to hydrogen embrittlement.

The electrical conductivity of the environment determines the extent of galvanic corrosion. The severest corrosion always occurs on the less noble component immediately adjacent to the more noble component. In low-conductivity solutions, this effect is concentrated in a narrow region adjacent to the cathode. In high-conductivity solutions, the band of attack is wider.

The relative surface areas of the anode and cathode are important factors because the total current flow on the anode and cathode is equal. Consequently, the smaller the anode area, the greater the current density or corrosion on the anode area. A large cathode/small anode couple combination may result in pronounced galvanic corrosion if the conditions for galvanic corrosion are favorable. The reverse is true for a large anode/small cathode couple. Anodic components should never be painted or coated, because holidays (defects) in the coating will expose small areas of metal that may suffer accelerated attack. Cathodes, on the other hand may be coated because coating reduces the total surface area

exposed and lowers the current density on the anode. For example, when a painted carbon steel storage tank containing stainless steel internals is coated, the stainless steel must also be coated.

Microstructural Corrosion

Microstructural corrosion is localized corrosion that selectively follows the microstructural constituents of alloys. The microstructures of alloys are not homogeneous and can display selective corrosion behavior, particularly when altered by heat treatment or mechanical-working operations. Some of the most common forms of microstructural corrosion are dealloying, intergranular corrosion, knifeline attack, end-grain attack, and exfoliation.

Dealloying. *Dealloying* is the preferential dissolution of one microstructural constituent of an alloy, leaving a porous, weakened mass that retains the original shape of the component. The most common form of dealloying is graphitic corrosion and dezincification. Other copper alloys may occasionally, but rarely, suffer dealloying, such as destannification (copper-tin), denickelification (copper-nickel), and dealuminification (copper-aluminum).

Graphitic corrosion is the dealloying of gray cast iron, in which the metal constituent is selectively attacked, leaving behind a semicontinuous network of graphite flakes. This structure will fail with the slightest mechanical shock. Graphitic corrosion is most prevalent in underground water pipes. The cast iron is usually attacked from the inside and the outside of the pipe. For critical services, such as fire-main pipe, gray cast iron should be replaced by ductile iron, which is much less susceptible to graphitic corrosion because the graphite constituent is not continuous. External coating coupled with cathodic protection also helps reduce corrosion from the soil side of the pipe when gray cast iron is used for water-main service.

Dezincification is dealloying that occurs in duplex brasses in potable, cooling, and weakly acidic waters. The zinc constituent of the brass is selectively removed, leaving behind a porous spongy mass of copper. Dezincification exhibits two forms, which are plug type (localized) and layer type (generalized). Plug type dezincification is more damaging because it can lead to wall penetration in a relatively short time. The remedy for dezincification is to use inhibited alloys that contain small amounts of elements such as antimony, arsenic, or tin.

Intergranular Corrosion. Intergranular corrosion is microstructural corrosion that occurs along the grain boundaries of austenitic and ferritic stainless steels and chromium-containing nickel alloys. The heat of fabricating or processing operations can cause the chromium and carbon in stainless steels and chromium-containing nickel alloys to combine, forming chromium carbide. This tends to occur preferentially along the grain boundaries. The grain boundaries become depleted in chromium and susceptible to corrosion in specific environments.

Chromium carbide (sensitization) occurs in austenitic stainless steels in the temperature range of 480°C to 815°C (900°F to 1500°F) and most rapidly at approximately 650°C (1200°F). See Figure 31-30. Severe intergranular corrosion may cause entire grains of metal to drop out (grain dropping). In susceptible alloys, intergranular corrosion may occur in the heat affected zones of welds. These regions, which are a slight distance from the fusion line of the weld, experience the optimum tempera-

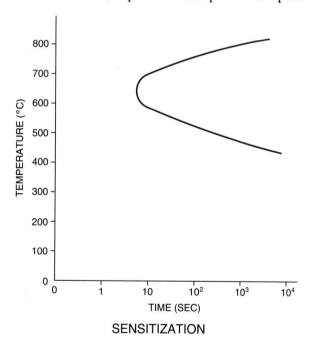

Figure 31-30. Sensitization of austenitic stainless steels is a function of time and temperature.

ture range for chromium carbide precipitation. This phenomenon is referred to as weld decay.

Intergranular corrosion may be prevented by using material in the solution annealed condition. Solution annealing redissolves chromium carbides. Alternatively, low-carbon or stabilized grades of stainless steels, which do not sensitize as readily, may be used. The low-carbon grades will resist sensitization during most welding and fabrication operations. Stabilized grades contain titanium or niobium, which combines preferentially with carbon to form titanium or niobium carbides when given a stabilizing heat treatment. Intergranular corrosion is most likely to occur in environments such as nitric acid, sulfuric acid, hydrofluoric acid, and mixtures of these acids with organics.

Knifeline Attack. *Knifeline attack* is an extremely narrow band of corrosive attack that may occur in stabilized stainless steels and some nickel alloys. Alloys that are stabilized with titanium or niobium are normally resistant to sensitization because the titanium or niobium preferentially forms carbides in the sensitizing temperature range. The titanium or niobium carbides can redissolve at very high temperatures of approximately 1260°C (2300°F). These temperatures may occur in a very narrow band of base metal adjacent to the weld fusion line. Chromium is then able to combine with the free carbon to form chromium carbides. As a result, the narrow band is susceptible to localized corrosion. See Figure 31-31.

End-grain Attack. *End-grain attack* is localized corrosion in a plane parallel to the direction of working. When an alloy is mechanically worked, the grains and nonmetallic inclusions become elongated. The edges of a wrought material, such as the end of a pipe or the edges of a plate perpendicular to the working direction, have a high density of end-exposed nonmetallic inclusions. In specific environments, a corrosion cell is set up between the inclusions and the alloy matrix, leading to deep pitting adjacent to the inclusions. The remedy for end-grain attack is to butter susceptible edges with a weld to seal them off. *Buttering* is surfacing in which one or more layers of welds are deposited on the edge of a metal.

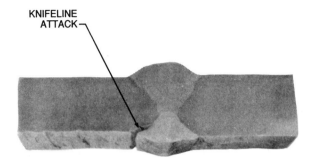

KNIFELINE
ATTACK

National Association of Corrosion Engineers

Figure 31-31. Knifeline attack occurs in a sharply defined zone of base metal adjacent to the fusion line of the weld.

Exfoliation. Exfoliation is a form of corrosion that occurs in certain cold-worked tempers and proceeds along selective subsurface paths parallel to the surface. Corrosion products force the metal away from the body of material, giving rise to a layered appearance, much like the pages of a book. See Figure 31-32.

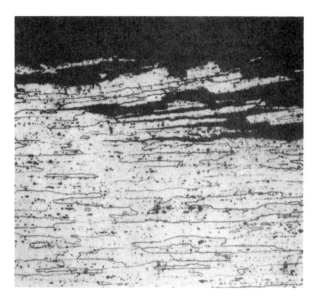

American Society for Metals

Figure 31-32. Exfoliation primarily affects aluminum alloys and causes flaking and swelling.

Appendix

A

DENSITY OF METALS					
Metal	**Density (g/cm^3)**	**Metal**	**Density (g/cm^3)**	**Metal**	**Density (g/cm^3)**
Pure Metals		**Pure Metals**		**Rare Earth Metals**	
Antimony	6.62	Ruthenium	12.2	Lanthanum	5.97[c]
Beryllium	1.848	Selenium	4.79	Lutetium	9.85[d]
Bismuth	9.80	Silicon	2.33	Neodymium	7.00[b]
Cadmium	8.65	Silver	10.49	Neodymium	6.80[c]
Calcium	1.55	Sodium	0.97	Praseodymium	6.77[b]
Cesium	1.903	Tantalum	16.6	Praseodymium	6.64[c]
Chromium	7.19	Thallium	11.85	Samarium	7.49[e]
Cobalt	8.85	Thorium	11.72	Scandium	2.99[d]
Gallium	5.907	Tungsten	19.3	Terbium	8.25[d]
Germanium	5.323	Uranium	19.07	Thulium	9.31[d]
Hafnium	13.1	Vanadium	6.1	Ytterbium	6.96[a]
Indium	7.31	Zirconium	6.5	Yttrium	4.47[d]
Iridium	22.5	**Rare Earth Metals**		**Other Metals**	
Lithium	0.534	Cerium	8.23[a]	Aluminum	2.6989[f]
Manganese	7.43	Cerium	6.66[b]	Copper	8.96
Mercury	13.546	Cerium	6.77[c]	Iron	7.874
Molybdenum	10.22	Dysprosium	8.55[d]	Gold	19.32
Niobium	8.57	Erbium	9.15[d]	Lead	11.34
Osmium	22.583	Europium	5.245[c]	Magnesium	1.738[g]
Plutonium	19.84	Galdolinium	7.86[d]	Nickel	8.9
Potassium	0.86	Holmium	6.79[d]	Platinum	21.45
Rhenium	21.04	Lanthanum	6.19[b]	Tin	7.3
Rhodium	12.44	Lanthanum	6.18[a]	Zinc	7.133

[a] face-center cubic structure
[b] hexagonal structure
[c] body-center cubic structure
[d] close-packed hexagonal structure
[e] rhombohedral structure
[f] 99.996% pure
[g] 99.8% pure

		\multicolumn{2}{c}{HARDNESS CONVERSIONS}					

| Brinell Indentation Dia (mm) | Brinell Hardness Number[a] | Rockwell Hardness Number[b] | | Rockwell Superficial Hardness Number | | | Shore Scleroscope Hardness | Approximate Tensile Strength (ksi) |
		B-Scale[c]	C-Scale[d]	15-N Scale 15-kg Load	30-N Scale 30-kg Load	45-N Scale 45-kg Load		
2.95	429	—	45.7	83.4	64.6	49.9	61	217
3.00	415	—	44.5	82.8	63.5	48.4	59	210
3.05	401	—	43.1	82.0	62.3	46.9	58	202
3.10	388	—	41.8	81.4	61.1	45.3	56	195
3.15	375	—	40.4	80.6	59.9	43.6	54	188
3.20	363	—	39.1	80.0	58.7	42.0	52	182
3.25	352	**110.0**	37.9	79.3	57.6	40.5	51	176
3.30	341	**109.0**	36.9	78.6	56.4	39.1	50	170
3.35	331	**108.5**	35.5	78.0	55.4	37.8	48	166
3.40	321	**108.0**	34.3	77.3	54.3	36.4	47	160
3.45	311	**107.5**	33.1	76.7	53.3	34.4	46	155
3.50	302	**107.0**	32.1	76.1	52.2	33.8	45	150
3.55	293	**106.0**	30.9	75.5	51.2	32.4	43	145
3.60	285	**105.5**	29.9	75.0	50.3	31.2	—	141
3.65	277	**104.5**	28.8	74.4	49.3	29.9	41	137
3.70	269	**104.0**	27.6	73.7	48.3	28.5	40	133
3.75	262	**103.0**	26.6	73.1	47.3	27.3	39	129
3.80	255	**102.0**	25.4	72.5	46.2	26.0	38	126
3.85	248	**101.0**	24.2	71.7	45.1	24.5	37	122
3.90	241	100.0	22.8	70.9	43.9	22.8	36	118
3.95	235	99.0	21.7	70.3	42.9	21.5	35	115
4.00	229	98.2	20.5	69.7	41.9	20.1	34	111
4.05	223	97.3	**18.8**	—	—	—	—	—
4.10	217	96.4	**17.5**	—	—	—	33	105
4.15	212	95.5	**16.0**	—	—	—	—	102
4.20	207	94.6	**15.2**	—	—	—	32	100
4.25	201	93.8	**13.8**	—	—	—	31	98
4.30	197	92.8	**12.7**	—	—	—	30	95
4.35	192	91.9	**11.5**	—	—	—	29	93
4.40	187	90.7	**10.0**	—	—	—	—	90
4.45	183	90.0	**9.0**	—	—	—	28	89
4.50	179	89.0	**8.0**	—	—	—	27	87
4.55	174	87.8	**6.4**	—	—	—	—	85
4.60	170	86.8	**5.4**	—	—	—	26	83
4.65	167	86.0	**4.4**	—	—	—	—	81
4.70	163	85.0	**3.3**	—	—	—	25	79
4.80	156	82.9	**0.9**	—	—	—	—	76
4.90	149	80.8	—	—	—	—	23	73
5.00	143	78.7	—	—	—	—	22	71
5.10	137	76.4	—	—	—	—	21	67
5.20	131	74.0	—	—	—	—	—	65
5.30	126	72.0	—	—	—	—	20	63
5.40	121	69.8	—	—	—	—	19	60
5.50	116	67.6	—	—	—	—	18	58
5.60	111	65.7	—	—	—	—	15	56

[a] 10 mm standard or tungsten carbide ball; 3000 kg load
[b] bold entries are beyond normal range; for information only
[c] 1/16" diameter steel ball; 100 kg load
[d] Brale penetrator; 150 kg load

SELECTED MEMBERS OF ANSI

AA	Aluminum Association
AAMA	Architectural Aluminum Manufacturers Association
AAMI	Association for the Advancement of Medical Instrumentation
AATCC	American Association of Textile Chemists and Colorists
ACI	American Concrete Institute
ADA	American Dental Association
AFBMA	Anti-Friction Bearing Manufacturers Association, Inc.
AGMA	American Gear Manufacturers Association
AHA	American Hardboard Association
AHAM	Association of Home Appliance Manufacturers
AIMM	Association for Information and Image Management (formerly National Micrographics Association)
AITC	American Institute of Timber Construction
ANS	American Nuclear Society
API	American Petroleum Institute
ARI	Air-Conditioning and Refrigeration Institute
ASAE	American Society of Agricultural Engineers
ASHRAE	American Society of Heating, Refrigerating and Air-Conditioning Engineers
ASLE	American Society of Lubricating Engineers
ASME	American Society of Mechanical Engineers
ASQC	American Society for Quality Control
ASSE	American Society of Sanitary Engineers
ASTM	American Society for Testing and Materials
AWS	American Welding Society
AWWA	American Water Works Association
BHMA	Builders Hardware Manufacturers Association
BIFMA	Business and Institutional Furniture Manufacturers Association
CABO	Council of American Building Officials
CAPPA	Crusher and Portable Plant Association
CEMA	Conveyor Equipment Manufacturers Association
CGA	Compressed Gas Association
DCDMA	Diamond Core Drill Manufacturers Association
EIA	Electronic Industries Association
FCI	Fluid Controls Institute
FIPS	Federal Information Processing Standards Publication
HIMA	Health Industry Manufacturers Association
HPMA	Hardwood Plywood Manufacturers Association
HPSSC	Health Physics Society Standards Committee
ICEA	Insulated Cable Engineers Association
IEEE	Institute of Electrical and Electronics Engineers
IES	Illuminating Engineering Society
IIAR	International Institute of Ammonia Refrigeration
IME	Institute of Makers of Explosives
IPC	Institute of Printed Circuits
ISA	Instrument Society of America
ISANTA	International Staple, Nail and Tool Association
ISDSI	Insulated Steel Door Systems Institute
ISO	International Organization for Standardization
ITE	Institute of Traffic Engineers
MPTA	Mechanical Power Transmission Association
MSS	Manufacturers Standardization Society
NAAMM	National Association of Architectural Metal Manufacturers
NB	National Board of Boiler and Pressure Vessel Inspectors
NBS	National Bureau of Standards
NCCLS	National Committee for Clinical Laboratory Standards
NEMA	National Electrical Manufacturers Association
NFPA	National Fire Protection Association
NMA	National Micrographics Association (now Association for Information and Image Management)
NWMA	National Woodwork Manufacturers Association
PTI	Power Tool Institute, Inc
RMA	Rubber Manufacturers Association
RVIA	Recreational Vehicle Industry Association
SAAMI	Sporting Arms and Ammunition Manufacturers Institute
SAE	Society of Automotive Engineers
SAMA	Scientific Apparatus Makers Association
SDI	Steel Door Institute
SI	Salt Institute
SMA	Screen Manufacturers Association
SPR	Simplified Practice Recommendation
TAPPI	Technical Association of the Pulp and Paper Industry
UL	Underwriters Laboratories
VRCI	Variable Resistive Components Institute
Vol Prod Std	Voluntary Product Standard

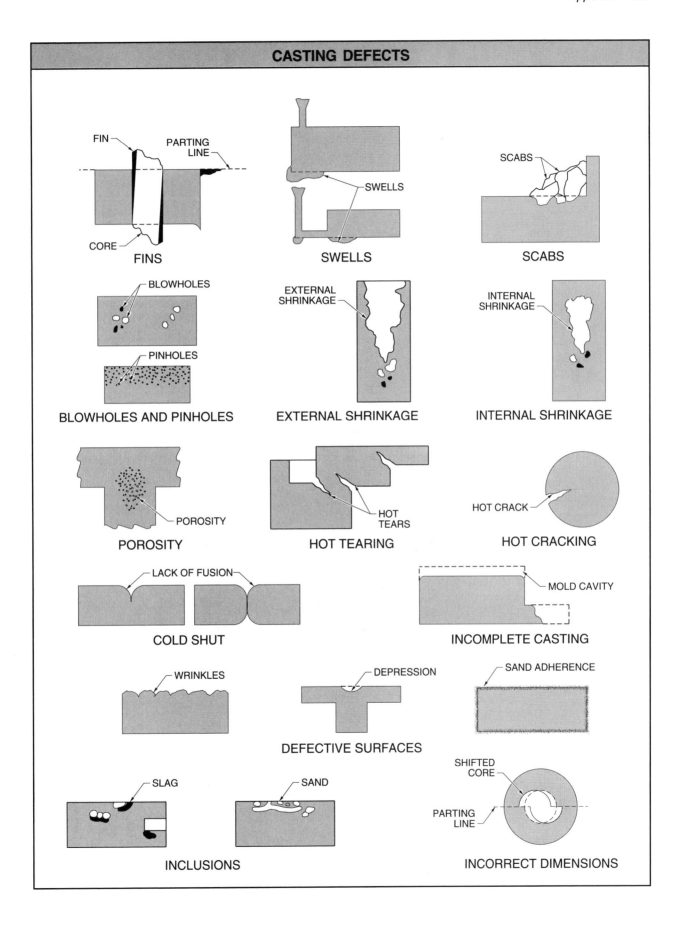

ASTM COPPER AND COPPER ALLOY TEMPER DESIGNATIONS

Designation	Temper Name or Condition	Designation	Temper Name or Condition	Designation	Temper Name or Condition
Cold-worked tempers		O20	Hot forged and annealed	TM02	½ HM
H00	⅛ hard	O25	Hot rolled and annealed	TM04	HM
H01	¼ hard	O30	Hot extruded and annealed	TN06	XHM
H02	½ hard	O31	Extruded and precipitation heat treated	TM08	XHMS
H03	¾ hard	O40	Hot pierced and annealed	**Quench-hardened tempers**	
H04	Hard	O50	Light annealed	TQ00	Quenched hardened
H06	Extra hard	O60	Soft annealed	TQ50	Quenched hardened and temper annealed
H08	Spring	O61	Annealed		
H10	Extra spring	O65	Drawing annealed	TQ55	Quenched hardened and temper annealed, cold drawn and stress relieved
H12	Special spring	O68	Deep-drawing annealed		
H13	Ultra spring	O70	Dead-soft annealed		
H14	Super spring	O80	Annealed to temper, ⅛ hard	TQ75	Interrupted quench hardened
H50	Extruded and drawn	O81	Annealed to temper, ¼ hard	**Precipitation-hardened, cold-worked, and thermal-stress-relieved tempers**	
H52	Pierced and drawn	O82	Annealed to temper, ½ hard		
H55	Light drawn; light cold rolled	OS005	Average grain size, .005 mm	TR01	TL01 and stress relieved
H58	Drawn general purpose	OS010	Average grain size, .010 mm	TR02	TL02 and stress relieved
H60	Cold heading; forming	OS015	Average grain size, .015 mm	TR04	TL04 and stress relieved
H63	Rivet	OS025	Average grain size, .025 mm	**Solution-treated and spinodal-heat-treated temper**	
H64	Screw	OS035	Average grain size, .035 mm		
H66	Bolt	OS050	Average grain size, .050 mm	TX00	Spinodal hardened
H70	Bending	OS060	Average grain size, .060 mm	**Tempers of welded tubing**	
H80	Hard drawn	OS070	Average grain size, .070 mm	WH00	Welded and drawn to ⅛ hard
H85	Medium-hard-drawn electrical wire	OS100	Average grain size, .100 mm	WH01	Welded and drawn to ¼ hard
H86	Hard-drawn electrical wire	OS150	Average grain size, .150 mm	WH02	Welded and drawn to ½ hard
H90	As-finned	OS200	Average grain size, .200 mm	WH03	Welded and drawn to ¾ hard
Cold-worked and stress-relieved tempers		**Solution-treated temper**		WH04	Welded and drawn to full hard
HR01	H01 and stress relieved	TB00	Solution heat treated	WH06	Welded and drawn to extra hard
HR02	H02 and stress relieved	**Solution-treated and cold-worked tempers**		WM00	As-welded from H00 (⅛-hard) strip
HR04	H04 and stress relieved	TD00	TB00 cold worked to ⅛ hard		
HR08	H08 and stress relieved	TD01	TB00 cold worked to ¼ hard	WM01	As-welded from H01 (¼-hard) strip
HR10	H10 and stress relieved	TD02	TB00 cold worked to ½ hard		
HR20	As-finned	TD03	TB00 cold worked to ¾ hard	WM02	As-welded from H02 (½-hard) strip
HR50	Drawn and stress relieved	TD04	TB00 cold worked to full hard		
Cold-rolled and order-strengthened tempers		**Solution-treated and precipitation-hardened temper**		WM03	As-welded from H03 (¾-hard) strip
HT04	H04 and order heat treated	TF00	TB00 and precipitation hardened	WM04	As-welded from H04 (full-hard) strip
HT08	H08 and order heat treated	**Cold-worked and precipitation-hardened tempers**			
As-manufactured tempers				WM06	As-welded from H06 (extra-hard) strip
M01	As-sand cast	TH01	TD01 and precipitation hardened		
M02	As-centrifugal cast	TH02	TD02 and precipitation hardened	WM08	As-welded from H08 (spring) strip
M03	As-plaster cast	TH03	TD03 and precipitation hardened	WM10	As-welded from H10 (extra-spring) strip
M04	As-pressure die cast	TH04	TD04 and precipitation hardened		
M05	As-permanent mold cast	**Precipitation-hardened and cold-worked tempers**		WM15	WM50 and stress relieved
M06	As-investment cast			WM20	WM00 and stress relieved
M07	As-continuous cast	TL00	TF00 cold worked to ⅛ hard	WM21	WM01 and stress relieved
M10	As-hot forged and air cooled	TL01	TF00 cold worked to ¼ hard	WM22	WM02 and stress relieved
M11	As-forged and quenched	TL02	TF00 cold worked to ½ hard	WM50	As welded from annealed strip
M20	As-hot rolled	TL04	TF00 cold worked to full hard	WO50	Welded and light annnealed
M30	As-hot extruded	TL08	TF00 cold worked to spring	WR00	WM00; drawn and stress relieved
M40	As-hot pierced	TL10	TF00 cold worked to extra spring	WR02	WM02; drawn and stress relieved
M45	As-hot pierced and rerolled	**Mill-hardened tempers**		WR03	WM03; drawn and stress relieved
Annealed tempers		TM00	AM	WR04	WM04; drawn and stress relieved
O10	Cast and annealed (homogenized)	TM01	¼ HM	WR06	WR06; drawn and stress relieved
O11	As-cast and precipitation heat treated				

			EFFECTS OF ELEMENTS ON TOOL STEELS	
Element	**Symbol**	**Content**	**When Used With**	**Effect**
Carbon	C	.30% to 2.35%		Most important increasing hardness and wear resistance. Maximum hardness generally possible with carbon content of 0.6 to 0.8%. Wear resistance may be increased with additional carbon because more carbides are produced in microstructure.
Chromium	Cr	.2% to 14%		Increases depth of hardness but to a lesser degree than manganese. Contributes to increased wear resistance (not hardness) and toughness. Low and medium Cr steels do not hold size as accurately as Mn. Increased temperature required for hardening.
		12% to 14%		High-chrome steels have good wear resistance and size accuracy when hardened. Provides mild corrosion resistance in hardened steels.
Cobalt	Co	5% to 12%		Primarily used in high-speed steels. Contributes to red hardness. Increases tendency for decarburization. Decreases toughness.
Niobium	Nb	0% to .1%		Increases wear resistance. Decreases tendency toward decarburization. Raises maximum allowable hardening temperature.
Iron	Fe	60% to 98%		Predominant element in any steel and not usually mentioned in analysis.
Manganese	Mn	.15% to 3%		Increases depth of hardness. Lowers temperature for which steels have to be heated to harden. Effective in deoxidizing steels in final melt stages when small quantities are added.
Molybdenum	Mo	.15% to 10%		Increases depth of hardness more effectively than tungsten. Increases red hardness. Increases wear resistance. Causes decarburization in forging and heat treating.
			Si	Improves toughness.
Nickel	Ni	.29% to .3%		Adds to toughness, wear resistance, and depth of hardness slightly. Lowers hardening temperature a small amount. Increases annealing difficulties in high-alloy steels.
Phosphorous	P	.015% to .05%		Harmful impurity.
Sulfur	S	.015% to .05%		Harmful impurity. In some air-hardening tool steels, sulfur improves machinability (free machining).
Silicon	Si	.15% to 2%		Increases chance of decarburization.
			Mn, Mo, or Cr	Improves strength and toughness dramatically.
		Above .3%		Improves hardenability.
Tungsten	W	.5% to 20%	C	Wear resistance enhanced.
		12% to 20%	C and Cr	Improves red hardness.

continued

continued

EFFECTS OF ELEMENTS ON TOOL STEELS

Element	Symbol	Content	When Used With	Effect
Vanadium	V	.15% to 5%	Mo, Cr, and W	Increases toughness. Increases red hardness. Inhibits growth of grains in heat-treated steel. With medium to high vanadium content, wear resistance is improved in hardened steel because of formation of vanadium carbides. Raises maximum allowable hardening temperatures, reducing possibility of damage because of moderate overheating.
Aluminum Titanium Zirconium	Al Ti Zr	0% to .1%		Aid in deoxidizing the steel during melt stages. Help control grain size.
Nitrogen Copper	N Cu	.03% to .2%		Generally not purposely added to tool steels.

LIST OF MACROETCHANTS

Alloy Family	Common Name for Etchant	Composition	Procedure[a]
Carbon and low alloy steels	Hydrochloric-water	50 ml HCl (conc) 50 ml H_2O	Immerse specimen in solution heated to 70°C to 80°C (160°F to 180°F) for 15 to 30 minutes. Desmut by vigorous scrubbing with vegetable fiber brush under running water.
Tool steels	Picral	ASTM E407 No. 76	—
Stainless steels	Marble's	10 g $CuSO_4$ 50 ml HCl 50 ml water	Immerse or swab. Made more active by adding a few drops of H_2SO_4 just before use. Desmut by dipping in warm 20% HNO_3 to give a bright finish.
Al	Caustic-Water	10 g NaOH 100 ml H_2O	Immerse sample 5 to 15 minutes in solution heated to 60°C to 70°C (140°F to 160°F). Rinse in water, and remove smut in strong HNO_3 solution. Rinse and repeat etching if necessary.
Cu	Nitric-Water	10 ml HNO_3 90 ml H_2O	Immerse specimen in solution at room temperature for a few minutes. Rinse in water and dry.

[a] See numerical list of Etchants

LIST OF MICROETCHANTS[a]		
Alloy Family	**Common Name for Etchant**	**ASTM E407 No.[b]**
Carbon and low alloy steels	Nital or Picral	74a, 76
Tool steels	Nital	74a
Cast irons	Nital	74a
Austenitic stainless steels	Oxalic	13b
Precipitation hardening stainless steels	Fry's	79
Ferritic and martensitic stainless steels	Viella's	80
Heat resistant castings	Glyceregia	87
Ni, Ni-Cu, and Ni-Fe	Acetic-nitric-water	134
Ni-Mo	Chromic-HCl	143
Ni-Cr-Mo (Alloy C-276)	Oxalic	13c
Ni-Cr-Mo (all other)	Hydrochloric-methanol	23
Ni-Fe-Cr-Mo (Alloy 20 Cb-3)	Hydrochloric-nitric	88
Ni-Fe-Cr-Mo (all other)	Hydrochloric-copper sulfate	25
Ni and Fe base superalloys	Kalling's or Glyceregia	94, 87
W, Mo	Murakami's	98c
Ta, Cb	Sulfuric-HF-peroxide	163
Ti	Kroll's	192
Zr	Nitric-HF-hydrochloric	66
Al	Keller's	3
Mg	Acetic-glycol	119
Pb	Acetic-nitric-glycerin	113
Sn, Sn-Pb	Nital	74d
Zinc	Chromic-sodium sulfate	200
Cu alloys	Ammonium hydroxide-peroxide	44
Cu-Zn	Phosphoric-water	8b

[a] Exercise extreme caution in handling all chemicals, especially HF. Follow safety precautions described in ASTM E407.
[b] See numerical list of Etchants

NUMERICAL LIST OF ETCHANTS

Etchant	Composition	Procedure
3	2 ml HF 3 ml HCl 5 ml HNO$_3$ 190 ml water	(a) Immerse 10 to 20 sec. Wash in stream of warm water. Reveals general structure. (b) Dilute with four parts water—color constituents—mix fresh.
8	10 ml H$_3$PO$_4$ 90 ml water	(b) Electrolytic at 1 V to 8 V for 5 to 10 sec.
13	10 g oxalic acid 100 ml water	Electrolytic at 6 V. (b) 1 min (c) 2 to 3 sec
23	5 ml HCl 95 ml ethanol (95%) or methanol (95%)	Electrolytic at 6 V for 10 to 20 sec.
44	50 ml NH$_4$OH 20 to 50 ml H$_2$O$_2$ (3%) 0 to 50 ml water	Use fresh. Peroxide content varies directly with copper content of alloy to be etched. Swab or immerse up to 1 min. Film on etched aluminum bronze removed by No. 82.
66	30 ml HF 15 ml HNO$_3$ 30 ml HCl	Swab 3 to 10 sec or immerse to 2 min.
74	1 to 5 ml HNO$_3$ 100 ml ethanol (95%) or methanol (95%)	Etching rate is increased, selectivity decreased with increased percentage of HNO$_3$. (a) Immerse a few seconds to a minute. (d) Swab or immerse several minutes.
75	5 g picric acid 8 g CuCl$_2$ 20 ml HCl 6 ml HNO$_3$ 200 ml ethanol (95%) or methanol (95%)	Immerse 1 to 2 sec at a time and immediately rinse with methanol. Repeat as often as necessary. Long immersion times result in copper deposition on surface.
76	10 g picric acid 10 ml ethanol (95%) or methanol (95%)	Composition given will saturate the solution with picric acid. Immerse a few seconds to a minute or more.
79	40 ml HCl 5 g CuCl$_2$ 30 ml water 25 ml ethanol (95%) or methanol (95%)	Swab a few seconds to a minute.
80	5 ml HCl 1 g picric acid 100 ml ethanol (95%) or methanol (95%)	Swab or immerse a few seconds to 15 min. Reaction may be accelerated by adding a few drops of 3% H$_2$O$_2$.
82	5 g FeCl$_3$ 5 drops HCl 100 ml water	Immerse 5 to 10 sec.
87	10 ml HNO$_3$ 20 to 50 ml HCl 30 ml glycerin	**Warning:** Nitrogen dioxide gas given off. Use hood. Mix HCl and glycerin thoroughly before adding HNO$_3$. Do not store. Discard before solution attains a dark orange color. Swab or immerse a few seconds to a few minutes. Higher percentage of HCl minimizes pitting. A hot water rinse just prior to etching may be used to activate the reaction. Sometimes a few passes on the final polishing wheel is also necessary to remove a passive surface.

continued

continued

Etchant	Composition	Procedure
88	10 ml HNO$_3$ 20 ml HCl 30 ml water	**Warning:** Nitrogen dioxide gas given off. Use hood. Discard before solution attains a dark orange color. Immerse a few seconds to a minute. Much stronger reaction than No. 87.
94	2 g CuCl$_2$ 40 ml HCl 40 to 80 ml ethanol (95%) or methanol (95%)	Submerged swabbing for a few seconds to several minutes.
98	10 g K$_3$Fe(CN)$_6$ 10 g KOH or NaOH 100 ml water	**Warning:** Extremely poisonous hydrogen cyanide given off. Use hood. Poisonous by ingestion as well as contact. To discard, neutralize (or turn basic) with ammonia and flush down acid drain with water. Use fresh. (c) Swab 5 to 60 sec. Immersion will produce a stain etch. Follow with water rinse, alcohol rinse, dry.
113	15 ml acetic acid 15 ml HNO$_3$ 60 ml glycerin	Use fresh solution at 80°C (175°F).
119	1 ml HNO$_3$ 20 ml acetic acid 60 ml diethylene glycol 20 ml water	Swab 1 to 3 sec for F and T6, 10 sec for T4 and 0 temper.
134	70 ml H$_3$PO$_4$ 30 ml water	Electrolytic for 5 V to 10 V for 5 to 60 sec. Polishes at high currents.
143	0.01 to 1 g CrO$_3$ 100 ml HCl	Allow solution to age a few minutes before using. Swab or immerse a few seconds to a few minutes.
163	30 ml H$_2$SO$_4$ 30 ml HF 3 to 5 drops H$_2$O$_2$ (30%) 30 ml water	Immerse 5 to 60 sec. Use this solution for alternate etch and polishing.
192	1 to 3 ml HF 2 to 6 ml HNO$_3$ 100 ml water	Swab 3 to 10 sec or immerse 10 to 30 sec. HF attacks and HNO$_3$ brightens the surface of titanium. Make concentration changes on this basis.
200	A– 40 g CrO$_3$ 3 g NaSO$_4$ 200 ml water B– 40 g CrO$_3$ 200 ml water	Immerse in Solution A with gentle agitation for several seconds. Rinse in Solution B.

RECOMMENDED PREHEATING FOR SELECTED METALS

Group	Designation	Carbon Content (%)	Preheating Temperature
Plain carbon steels	Plain carbon steel	Below .20	Up to 95°C
	Plain carbon steel	.20 to .30	95°C to 150°C
	Plain carbon steel	.30 to .45	150°C to 260°C
	Plain carbon steel	.45 to .80	260°C to 425°C
Carbon-moly steels	Carbon-moly steel	.10 to .20	150°C to 260°C
	Carbon-moly steel	.20 to .30	205°C to 315°C
	Carbon-moly steel	.30 to .35	260°C to 425°C
Manganese steels	Silicon structural steel	.35	150°C to 260°C
	Medium-manganese steel	.20 to .25	150°C to 260°C
	SAE T1330 steel	.30	205°C to 315°C
	SAE T1340 steel	.40	260°C to 425°C
	SAE T1350 steel	.50	315°C to 480°C
	12% Mn steel	1.25	Usually not required
High-tensile steels (see also steels below)	Manganese-moly steel	.20	150°C to 260°C
	Manten steel	.30 max	205°C to 315°C
	Armco high-tensile steel	.12 max	Up to 95°C
	Mayari R steel	.12 max	Up to 150°C
	Nax high-tensile steel	.15 to .25	Up to 150°C
	Cromansil steel	.14 max	150°C to 205°C
	Corten steel	.12 max	95°C to 205°C
	Yoloy steel	.05 to .35	95°C to 315°C
	Jalten steel	.35 max	205°C to 315°C
	Double-strength 1 steel	.12 max	150°C to 315°C
	Double-strength 1A steel	.30 max	205°C to 370°C
	Otiscoloy steel	.12 max	95°C to 205°C
	A.W. Dyn-El steel	.11 to .14	Up to 150°C
	Cr-Cu-Ni steel	.12 max	95°C to 205°C
	Cr-Mn steel	.40	205°C to 315°C
	Hi-steel	.12 max	95°C to 260°C
Nickel steels	SAE 2015 steel	.10 to .20	Up to 150°C
	SAE 2115 steel	.10 to .20	95°C to 150°C
	2½% nickel steel	.10 to .20	95°C to 205°C
	SAE 2315 steel	.15	95°C to 260°C
	SAE 2320 steel	.20	95°C to 260°C
	SAE 2330 steel	.30	150°C to 315°C
	SAE 2340 steel	.40	205°C to 370°C
Medium nickel-chromium steels	SAE 3115 steel	.15	95°C to 205°C
	SAE 3125 steel	.25	150°C to 260°C
	SAE 3130 steel	.30	205°C to 370°C
	SAE 3140 steel	.40	260°C to 425°C
	SAE 3150 steel	.50	315°C to 480°C
	SAE 3215 steel	.15	150°C to 260°C
	SAE 3230 steel	.30	260°C to 370°C
	SAE 3240 steel	.40	370°C to 540°C
	SAE 3250 steel	.50	315°C to 595°C
	SAE 3315 steel	.15	260°C to 370°C
	SAE 3325 steel	.25	480°C to 595°C
	SAE 3435 steel	.35	480°C to 595°C
	SAE 3450 steel	.50	480°C to 595°C
Moly-bearing chromium and chromium steels	SAE 4140 steel	.40	315°C to 425°C
	SAE 4340 steel	.40	370°C to 480°C
	SAE 4615 steel	.15	205°C to 315°C
	SAE 4630 steel	.30	260°C to 370°C
	SAE 4640 steel	.40	315°C to 425°C
	SAE 4820 steel	.20	315°C to 425°C

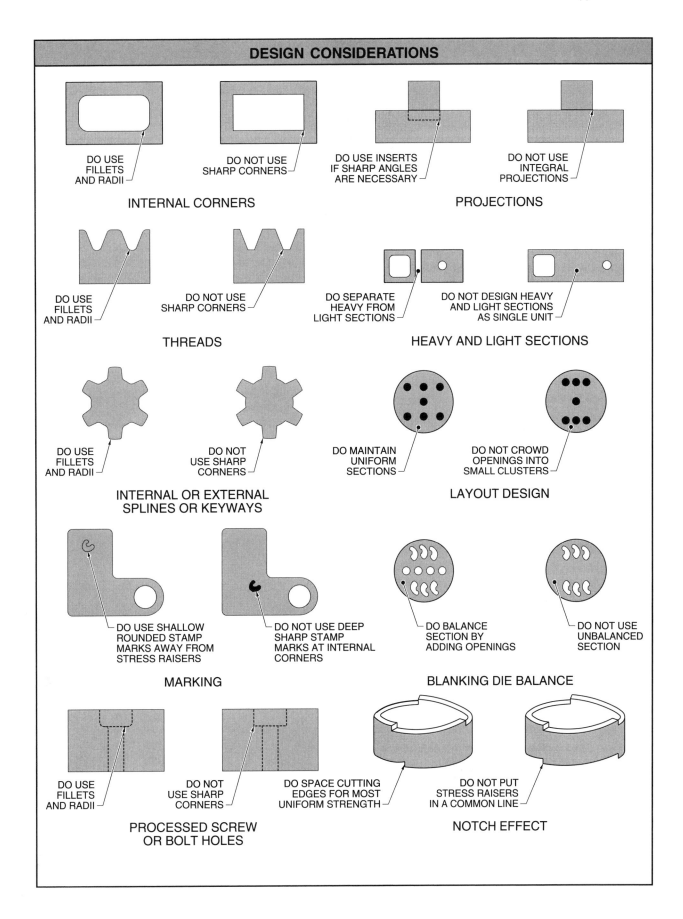

WELDING FILLER METALS FOR ALUMINUM ALLOYS								
Base Metal	**319.0, 333.0, 354.0, 355.0, and C355.0**	**356.0, A356.0, A357.0, 359.0, 413.0, and 433.0**	**514.0, A514.0, and B514.0**	**7005, 7039, 7046, 7146, 710.0, and 712.0**	**6070**	**6005, 6061, 6063, 6101, 6151, 6201, 6351, and 6951**	**5456**	**5454**
1060 and 1350	ER4145[c,i]	ER4043[i,f]	ER4043[e,i]	ER4043[i]	ER4043[i]	ER4043[i]	ER5356[c]	ER4043[e,i]
1100, 3003, and Alclad 3003	ER4145[c,i]	ER4043[i,f]	ER4043[e,i]	ER4043[i]	ER4043[i]	ER4043[i]	ER5356[c]	ER4043[e,i]
2014, 2024, and 2036	ER4145[g]	ER4145	—	—	ER4145	ER4145	—	—
2219	ER4145[g,c,i]	ER4145[c,i]	ER4043[i]	ER4043[i]	ER4043[f,i]	ER4043[f,i]	ER4043	ER4043[i]
3004 and Alclad 3004	ER4043[i]	ER4043[i,f]	ER5654[b]	ER5356[e]	ER4043[e]	ER4043[b]	ER5356[e]	ER5654[b]
5005 and 5050	ER4043[i]	ER4043[i]	ER5654[b]	ER5356[e]	ER4043[e]	ER4043[b]	ER5356[e]	ER5654[b]
5052 and 5652[a]	ER4043[i]	ER4043[b,i]	ER5654[b]	ER5356[e]	ER5356[b,c]	ER5356[b,c]	ER5356[b]	ER5654[b]
5083	—	ER5356[c,e,i]	ER5356[e]	ER5183[e]	ER5356[e]	ER5356[e]	ER5183[e]	ER5356[e]
5086	—	ER5356[c,e,i]	ER5356[e]	ER5356[e]	ER5356[e]	ER5356[e]	ER5356[e]	ER5356[b]
5154 and 5254[a]	—	ER4043[b,i]	ER5654[b]	ER5356[b]	ER5356[b,c]	ER5356[b,c]	ER5356[b]	ER5654[b]
5454	ER4043[i]	ER4043[b,i]	ER5654[b]	ER5356[b]	ER5356[b,c]	ER5356[b,c]	ER5356[b]	ER5554[c,e]
5456	—	ER4043[b,i]	ER5356[e]	ER5556[e]	ER5356[e]	ER5356[e]	ER5556[e]	
6005, 6061, 6063, 6101, 6151, 6201, 6351, and 6951	ER4145[c,i]	ER5356[c,e,i]	ER5356[b,c]	ER5356[b,c,i]	ER4043[b,i]	ER4043[b,i]		
6070	ER4145[c,i]	ER4043[e,i]	ER5356[c,e]	ER5356[c,e,i]	ER4043[e,i]			
7005, 7039, 7046, 7146, 710.0, and 712.0	ER4043[i]	ER4043[b,i]	ER5356[b]	ER5039[e]				
514.0, A514.0, and B514.0	—	ER4043[b,i]	ER5654[b,d]					
356.0, A356.0, A357.0, 359.0, 413.0, and 443.0	ER4145[c,i]	ER4043[d,i]						
319.0, 333.0, 354.0, 355.0, and C355.0	ER4145[c,d,i]							

[a] Base metal alloys 5254 and 5652 are sometimes used for hydrogen peroxide service. ER5654 filler metal is used for welding both alloys for service at 150°F and below.
[b] ER5183, ER5356, ER5554, ER5556, and ER5654 may be used. In some cases, they provide: (1) improved color match after anodizing treatment, (2) highest weld ductility, and (3) higher weld strength. ER5554 is suitable for elevated temperature service.
[c] ER4043 may be used for some applications.
[d] Welding filler metal with the same analysis as the base metal is sometimes used.
[e] ER5183, ER5356, or ER5556 may be used.
[f] ER4145 may be used for some applications.
[g] ER2319 may be used for some applications.
[i] ER4047 may be used for some applications.
[j] ER1100 may be used for some applications.

continued

continued

	WELDING FILLER METALS FOR ALUMINUM ALLOYS								
Base Metal	**5154 and 5254**[a]	**5086**	**5083**	**5052 and 5652**[a]	**5005 and 5050**	**3004 and Alc. 3004**	**2219**	**2014, 2024, and 2036**	**1100, 3003, and Alc. 3003**
1060 and 1350	ER4043[e,i]	ER5356[c]	ER5356[c]	ER4043[i]	ER1100[c]	ER4043	ER4145	ER4145	ER1100[c]
1100, 3003, and Alclad 3003	ER4043[e,i]	ER5356[c]	ER5356[c]	ER4043[e,i]	ER4043[e]	ER4043[e]	ER4145	ER4145	ER1100[c]
2014, 2024, and 2036	—	—	—	—	—	—	ER4145[g]	ER4145[g]	
2219	ER4043[i]	ER4043	ER4043	ER4043[i]	ER4043	ER4043	ER2319[c,f,i]		
3004 and Alclad 3004	ER5654[b]	ER5356[e]	ER5356[e]	ER4043[e,i]	ER4043[e]	ER4043[e]			
5005 and 5050	ER5654[b]	ER5356[e]	ER5356[e]	ER4043[e,i]	ER4043[d,e]				
5052 and 5652[a]	ER5654[b]	ER5356[e]	ER5356[e]	ER5654[a,b,c]					
5083	ER5356[e]	ER5356[e]	ER5183[e]						
5086	ER5356[b]	ER5356[e]							
5154 and 5254[a]	ER5654[a,b]								

[a] Base metal alloys 5254 and 5652 are sometimes used for hydrogen peroxide service. ER5654 filler metal is used for welding both alloys for service at 150°F and below.
[b] ER5183, ER5356, ER5554, ER5556, and ER5654 may be used. In some cases, they provide: (1) improved color match after anodizing treatment, (2) highest weld ductility, and (3) higher weld strength. ER5554 is suitable for elevated temperature service.
[c] ER4043 may be used for some applications.
[d] Welding filler metal with the same analysis as the base metal is sometimes used.
[e] ER5183, ER5356, or ER5556 may be used.
[f] ER4145 may be used for some applications.
[g] ER2319 may be used for some applications.
[i] ER4047 may be used for some applications.
[j] ER1100 may be used for some applications.

	RELATIVE EASE OF JOINING				
	Non-Heat-Treatable Wrought Aluminum Alloys				
	Process[a,b,c]				
Alloy	GMAW GTAW	RSW RSEW	B	S	Melting Range, °F
1060	A	B	A	A	1195 to 1215
1100	A	A	A	A	1190 to 1215
1350	A	B	A	A	1195 to 1215
2014	C	A	X	C	950 to 1180
2017	C	A	X	C	955 to 1185
2024	C	A	X	C	935 to 1180
2036	C	A	X		1030 to 1200
2218	C	A	X	C	940 to 1175
2219	A	A	X	C	1010 to 1190
2618	C	A	X	C	1040 to 1185
3003	A	A	A	A	1190 to 1210
3004	A	A	B	B	1165 to 1205
3105	A	A	B	B	1175 to 1210
5005	A	A	B	B	1170 to 1210
5050	A	A	B	B	1155 to 1205
5052					
5652	A	A	C	C	1125 to 1200
5083	A	A	X	X	1095 to 1180
5086	A	A	X	X	1085 to 1185
5154					
5254	A	A	X	X	1100 to 1190
5182	A	B	X	X	1070 to 1180
5252	A	A	C	C	1125 to 1200
5454	A	A	X	X	1115 to 1195
5456	A	A	X	X	1055 to 1180
5457	A	A	B	B	1165 to 1210
5557	A	A	A	A	1180 to 1215
5657	A	A	B	B	1175 to 1210
6009	A	A	A	B	1040 to 1200
6010	A	A	A	B	1085 to 1200
6061	A	A	A	B	1080 to 1200
6063	A	A	A	B	1140 to 1210
6070	A	A	C	B	1050 to 1200
6101	A	A	A	A	1150 to 1210
6201	A	A	A	B	1125 to 1210
6951	A	A	A	A	1140 to 1210
7005	A	A	B	B	1125 to 1195
7039	A	A	C	B	1070 to 1180
7072	A	A	A	A	1185 to 1215
7075	C	A	X	C	890 to 1175
7178	C	A	X	C	890 to 1165

[a] GMAW = gas metal arc welding
GTAW = gas tungsten arc welding
RSW = resistance spot welding
RSEW = resistance seam welding
B = brazing
S = soldering

[b] Rating, based on the temper most readily joined:
A = readily joined by process
B = joinable by process for most applications
C = joining by process difficult
X = joining by process not recommended

[c] All alloys can be joined by ultrasonic welding or adhesive bonding.
Some alloys can be ultrasonic or abrasive soldered.

RELATIVE EASE OF JOINING

Non-Heat-Treatable Cast Aluminum Alloys

Alloy	Process[a,b,c]				Melting Range, °F
	GMAW GTAW	RSW RSEW	B	S	
Sand Castings					
208.0	B	B	X	C	970 to 1160
222.0	C	C	X	X	965 to 1155
242.0	C	C	X	X	990 to 1175
295.0	C	C	X	C	970 to 1190
319.0	B	B	X	X	960 to 1120
355.0	B	B	X	X	1015 to 1150
356.0	A	A	C	C	1035 to 1135
B443.0	A	A	A	C	1065 to 1170
514.0	A	B	X	X	1110 to 1185
B514.0	B	B	X	X	1090 to 1170
520.0	B	C	X	X	840 to 1120
A712.0	B	C	B	B	1105 to 1195
D712.0	B	C	A	B	1135 to 1200
Permanent Mold Castings					
213.0	B	C	X	C	965 to 1160
222.0	C	C	X	X	965 to 1155
242.0	C	C	X	X	990 to 1175
A332.0	B	B	X	X	1000 to 1050
F332.0	B	C	X	X	970 to 1080
330.0	B	B	X	X	960 to 1085
354.0	B	B	X	X	1000 to 1105
355.0	B	B	X	X	1015 to 1150
C355.0	B	B	X	X	1015 to 1150
356.0	A	A	C	C	1035 to 1135
A356.0	A	A	C	C	1035 to 1135
A357.0	A	B	B	C	1035 to 1135
359.0	B	B	B	C	1045 to 1115
B443.0	A	A	A	C	1065 to 1170
A514.0	A	B	X	X	1075 to 1180
Die Castings					
360.0	B	C	X	X	1035 to 1105
380.0	B	C	X	C	1000 to 1100
413.0	B	C	X	X	1065 to 1080
518.0	B	C	X	X	995 to 1150

[a] GMAW = gas metal arc welding
GTAW = gas tungsten arc welding
RSW = resistance spot welding
RSEW = resistance seam welding
B = brazing
S = soldering

[b] Rating, based on the temper most readily joined:
A = readily joined by process
B = joinable by process for most applications
C = joining by process difficult
X = joining by process not recommended

[c] All alloys can be joined by ultrasonic welding or adhesive bonding.
Some alloys can be ultrasonic or abrasive soldered.

	BRAZING FILLER METAL AND SOLDER SPECIFICATIONS	
AWS Specification	**Specification Title**	**Process[a]**
A5.1	Carbon steel covered arc-welding electrodes	SMAW
A5.2	Iron and steel gas welding rods	OWA
A5.3	Aluminum and aluminum alloy arc welding electrodes	SMAW
A5.4	Corrosion-resisting chromium and chromium-nickel steel covered welding electrodes	SMAW
A5.5	Low-alloy steel covered arc welding electrodes	SMAW
A5.6	Copper and copper alloy covered electrodes	SMAW
A5.7	Copper and copper alloy welding rods	OAW, GTAW, and PAW
A5.8	Brazing filler metal	BR
A5.9	Corrosion-resisting chromium and chromium-nickel steel bare and composite metal cored and standard arc welding electrodes and rods	GTAW, GMAW, SAW, and PAW
A5.10	Aluminum and aluminum alloy welding rods and bare electrodes	OAW, GTAW, GMAW, and PAW
A5.11	Nickel and nickel alloy covered welding electrodes	SMAW
A5.12	Tungsten arc welding electrodes	GTAW and PAW
A5.13	Surfacing welding rods and electrodes	OAW, GTAW, and CAW
A5.14	Nickel and nickel alloy bare welding rods and electrodes	OAW, GTAW, GMAW, SAW, and PAW
A5.15	Welding rods and covered electrodes for welding cast iron	OAW, SMAW, CAW
A5.16	Titanium and titanium alloy bare welding rods and electrodes	GTAW, GMAW, and PAW
A5.17	Bare carbon steel electrodes and fluxes for submerged-arc welding	SAW
A5.18	Carbon steel filler metals for gas shielded arc welding	GTAW, GMAW, and PAW
A5.19	Magnesium alloy welding rods and bare electrodes	OAW, GTAW, GMAW, and PAW
A5.20	Carbon steel electrodes for flux cored arc welding	FCAW
A5.21	Composite surfacing welding rods and electrodes	OAW, SMAW, and GTAW
A5.22	Flux cored corrosion-resisting chromium and chromium-nickel steel electrodes	FCAW
A5.23	Bare low-alloy steel electrodes and fluxes for submerged arc welding	SAW
A5.24	Zirconium and zirconium alloy bare welding rods and electrodes	GTAW, GMAW, and PAW
A5.25	Consumables used for electro-slag welding of carbon and high strength low alloy steels	ES
A5.26	Consumables used for electrogas welding of carbon and high strength low-alloy steels	GMAW and FCAW
A5.27	Copper and copper alloy gas welding rods	OAW
A5.28	Low-alloy steel filler metals for gas shielded arc welding	GTAW, GMAW, and PAW

[a] if GTAW specified, PAW will also apply

CHEMICAL ELEMENTS

Name	Symbol	Atomic Weight[a]	Atomic Number	Name	Symbol	Atomic Weight[a]	Atomic Number
Actinium	Ac	[227]	89	Neon	Ne	20.183	10
Aluminum	Al	26.9815	13	Neptunium	Np	[237]	93
Americium	Am	[243]	95	Nickel	Ni	58.71	28
Antimony	Sb	121.75	51	Niobium	Nb	92.906	41
Argon	Ar	39.948	18	Nitrogen	N	14.0067	7
Arsenic	As	74.9216	33	Nobelium	No	[255]	102
Astatine	At	[210]	85	Osmium	Os	190.2	76
Barium	Ba	137.34	56	Oxygen	O	15.9994	8
Berkelium	Bk	[247]	97	Palladium	Pd	106.4	46
Beryllium	Be	9.0122	4	Phosphorus	P	30.9738	15
Bismuth	Bi	208.980	83	Platinum	Pt	195.09	78
Boron	B	10.811	5	Plutonium	Pu	[244]	94
Bromine	Br	79.909	35	Polonium	Po	[210]	84
Cadmium	Cd	112.40	48	Potassium	K	39.102	19
Calcium	Ca	40.08	20	Praseodymium	Pr	140.907	59
Californium	Cf	[251]	98	Promethium	Pm	[145]	61
Carbon	C	12.01115	6	Protactinium	Pa	[231]	91
Cerium	Ce	140.12	58	Radium	Ra	[226]	88
Cesium	Cs	132.905	55	Radon	Rn	[222]	86
Chlorine	Cl	35.453	17	Rhenium	Re	186.2	75
Chromium	Cr	51.996	24	Rhodium	Rh	102.905	45
Cobalt	Co	58.9332	27	Rubidium	Rb	85.47	37
Copper	Cu	63.54	29	Ruthenium	Ru	101.07	44
Curium	Cm	[247]	96	Samarium	Sm	150.35	62
Dysprosium	Dy	162.50	66	Scandium	Sc	44.956	21
Einsteinium	Es	[254]	99	Selenium	Se	78.96	34
Erbium	Er	167.26	68	Silicon	Si	28.086	14
Europium	Eu	151.96	63	Silver	Ag	107.870	47
Fermium	Fm	[257]	100	Sodium	Na	22.9898	11
Fluorine	F	18.9984	9	Strontium	Sr	87.62	38
Francium	Fr	[223]	87	Sulfur	S	32.064	16
Gadolinium	Gd	157.25	64	Tantalum	Ta	180.948	73
Gallium	Ga	69.72	31	Technetium	Tc	[97]	43
Germanium	Ge	72.59	32	Tellurium	Te	127.60	52
Gold	Au	196.967	79	Terbium	Tb	158.924	65
Hafnium	Hf	178.49	72	Thallium	Tl	204.37	81
Helium	He	4.0026	2	Thorium	Th	232.038	90
Holmium	Ho	164.930	67	Thulium	Tm	168.934	69
Hydrogen	H	1.00797	1	Tin	Sn	118.69	50
Indium	In	114.82	49	Titanium	Ti	47.90	22
Iodine	I	126.9044	53	Tungsten	W	183.85	74
Iridium	Ir	192.2	77	Unnilennium	Une	[266]	109
Iron	Fe	55.847	26	Unnilhexium	Unh	[263]	106
Krypton	Kr	83.80	36	Unniloctium	Uno	[265]	108
Lanthanum	La	138.91	57	Unnilpentium	Unp	[262]	105
Lawrencium	Lr	[256]	103	Unnilquadium	Unq	[261]	104
Lead	Pb	207.19	82	Unnilseptium	Uns	[262]	107
Lithium	Li	6.939	3	Uranium	U	238.03	92
Lutetium	Lu	174.97	71	Vanadium	V	50.942	23
Magnesium	Mg	24.312	12	Xenon	Xe	131.30	54
Manganese	Mn	54.9380	25	Ytterbium	Yb	173.04	70
Mendelevium	Md	[258]	101	Yttrium	Y	88.905	39
Mercury	Hg	200.59	80	Zinc	Zn	65.37	30
Molybdenum	Mo	95.94	42	Zirconium	Zr	91.22	40
Neodymium	Nd	144.24	60				

[a] a number in brackets indicates the mass number of the most stable isotope

PERIODIC TABLE OF THE ELEMENTS

Column headings: I · II · III · IV · V · VI · VII · 0

Group I

Symbol	Z	Name	At. wt.	Shells
H	1	Hydrogen	1.0079	1
Li	3	Lithium	6.941	2-1
Na	11	Sodium	22.98977	2-8-1
K	19	Potassium	39.098	2-8-8-1
Rb	37	Rubidium	85.4678	-18-8-1
Cs	55	Cesium	132.9054	-18-8-1
Fr	87	Francium	223	-18-8-1

Group II

Symbol	Z	Name	At. wt.	Shells
Be	4	Beryllium	9.01218	2-2
Mg	12	Magnesium	24.305	2-8-2
Ca	20	Calcium	40.08	2-8-8-2
Sr	38	Strontium	87.62	-18-8-2
Ba	56	Barium	137.34	-18-8-2
Ra	88	Radium	226.0254	-18-8-2

Shell column labels on far left: 2; 2-8; 2-8-8; -18-8; -18-8; -18-8; -32-18-8; -18-8; -8; -2

Transition elements

Symbol	Z	Name	At. wt.	Shells
Sc	21	Scandium	44.9559	-8-9-2
Y	39	Yttrium	88.9059	-18-9-2
(57-71)				
(89-103)				
Ti	22	Titanium	47.90	-8-10-2
Zr	40	Zirconium	91.22	-18-10-2
Hf	72	Hafnium	178.49	-32-10-2
Rf-Ku	104	Rutherfordium; Kurchatovium	261	-32-10-2
V	23	Vanadium	50.9414	-8-11-2
Nb	41	Niobium	92.9064	-18-12-1
Ta	73	Tantalum	180.9479	-32-11-2
Ha	105	Hahnium	262	-32-11-2
Cr	24	Chromium	51.996	-8-13-1
Mo	42	Molybdenum	95.94	-18-13-1
W	74	Tungsten	183.85	-32-12-2
	106		263	-32-12-2
Mn	25	Manganese	54.9380	-8-13-2
Tc	43	Technetium	97	-18-13-2
Re	75	Rhenium	186.207	-32-13-2
Fe	26	Iron	55.847	-8-14-2
Ru	44	Ruthenium	101.07	-18-15-1
Os	76	Osmium	190.2	-32-14-2
Co	27	Cobalt	58.9332	-8-15-2
Rh	45	Rhodium	102.9055	-18-16-1
Ir	77	Iridium	192.22	-32-15-2
Ni	28	Nickel	58.70	-8-16-2
Pd	46	Palladium	106.4	-18-18-0
Pt	78	Platinum	195.09	-32-17-1
Cu	29	Copper	63.546	-8-18-1
Ag	47	Silver	107.868	-18-18-1
Au	79	Gold	196.9665	-32-18-1
Zn	30	Zinc	65.38	-8-18-2
Cd	48	Cadmium	112.40	-18-18-2
Hg	80	Mercury	200.59	-32-18-2

Group III

Symbol	Z	Name	At. wt.	Shells
B	5	Boron	10.81	2-3
Al	13	Aluminum	26.98154	2-8-3
Ga	31	Gallium	69.72	-8-18-3
In	49	Indium	114.82	-18-18-3
Tl	81	Thallium	204.37	-32-18-3

Group IV

Symbol	Z	Name	At. wt.	Shells
C	6	Carbon	12.011	2-4
Si	14	Silicon	28.086	2-8-4
Ge	32	Germanium	72.59	-8-18-4
Sn	50	Tin	118.69	-18-18-4
Pb	82	Lead	207.2	-32-18-4

Group V

Symbol	Z	Name	At. wt.	Shells
N	7	Nitrogen	14.0067	2-5
P	15	Phosphorus	30.97376	2-8-5
As	33	Arsenic	74.9216	-8-18-5
Sb	51	Antimony	121.75	-18-18-5
Bi	83	Bismuth	208.9804	-32-18-5

Group VI

Symbol	Z	Name	At. wt.	Shells
O	8	Oxygen	15.9994	2-6
S	16	Sulfur	32.06	2-8-6
Se	34	Selenium	78.96	-8-18-6
Te	52	Tellurium	127.60	-18-18-6
Po	84	Polonium	209	-32-18-6

Group VII

Symbol	Z	Name	At. wt.	Shells
F	9	Fluorine	18.99840	2-7
Cl	17	Chlorine	35.453	2-8-7
Br	35	Bromine	79.904	-8-18-7
I	53	Iodine	126.9045	-18-18-7
At	85	Astatine	210	-32-18-7

Group 0

Symbol	Z	Name	At. wt.	Shells
He	2	Helium	4.00260	2
Ne	10	Neon	20.179	2-8
Ar	18	Argon	39.948	2-8-8
Kr	36	Krypton	83.80	-8-18-8
Xe	54	Xenon	131.30	-18-18-8
Rn	86	Radon	222	-32-18-8

Lanthanide series (57–71)

Symbol	Z	Name	At. wt.	Shells
La	57	Lanthanum	138.9055	-18-9-2
Ce	58	Cerium	140.12	-19-9-2
Pr	59	Praseodymium	140.9077	-21-8-2
Nd	60	Neodymium	144.24	-22-8-2
Pm	61	Promethium	145	-23-8-2
Sm	62	Samarium	150.4	-24-8-2
Eu	63	Europium	151.96	-25-8-2
Gd	64	Gadolinium	157.25	-25-9-2
Tb	65	Terbium	158.9254	-27-8-2
Dy	66	Dysprosium	162.50	-28-8-2
Ho	67	Holmium	164.9304	-29-8-2
Er	68	Erbium	167.26	-30-8-2
Tm	69	Thulium	168.9342	-31-8-2
Yb	70	Ytterbium	173.04	-32-8-2
Lu	71	Lutetium	174.97	-32-9-2

Actinide series (89–103)

Symbol	Z	Name	At. wt.	Shells
Ac	89	Actinium	227	-18-9-2
Th	90	Thorium	232.0381	-18-10-2
Pa	91	Protactinium	231.0359	-20-9-2
U	92	Uranium	238.029	-21-9-2
Np	93	Neptunium	237.0482	-22-9-2
Pu	94	Plutonium	244	-24-8-2
Am	95	Americium	243	-25-8-2
Cm	96	Curium	247	-25-9-2
Bk	97	Berkelium	247	-27-8-2
Cf	98	Californium	251	-28-8-2
Es	99	Einsteinium	254	-29-8-2
Fm	100	Fermium	257	-30-8-2
Md	101	Mendelevium	258	-31-8-2
No	102	Nobelium	255	-32-8-2
Lr	103	Lawrencium	260	-32-9-2

Glossary

Note: Terms in this glossary are defined as they relate to metallurgy.

A

absolute zero: Theoretical condition, which is the common base on the Kelvin and Rankine scales, at which no heat is present. See *Kelvin* and *Rankine*.

acid cleaning: Cleaning method that uses special corrosive chemicals to strip extremely stubborn deposits on a metal sample.

acoustic-emission testing: Nondestructive testing technique that is used to detect spontaneously generated elastic waves created by localized movement of a material under stress, such as a pressure vessel or structural beam. See *nondestructive testing*.

adhesive wear: Removal of metal from a surface by welding together and subsequent shearing of minute areas of two surfaces that slide across each other under pressure.

AISI-SAE designation system: Designation system for steel products that consists of a four digit numerical classification that is partially descriptive of the composition.

alclad: Composite wrought product comprised of an aluminum alloy core having on one or both surfaces a metallurgically bonded aluminum or aluminum alloy coating, which is resistant to corrosion.

allotropic metal: Metal that exhibits more than one unit cell structure. See *unit cell*.

allowable stress data: Maximum allowable stress on a component under specific operating conditions. See *stress*.

alloy: Material that has metallic properties and is composed of two or more chemical elements, of which one is a metal.

alloy iron: Cast iron that contains one or more intentionally added alloying elements, up to a total of 30%, in order to enhance specific properties.

alloy steel: Steel that contains specified quantities of alloying elements other than carbon and the common amounts of manganese, copper, silicon, sulfur, and phosphorous.

alloy system: Method of describing all the alloys that can be formed (all possible proportions) by the individual chemical elements of an alloy.

aluminum pack diffusion: Surface modification process used for enriching the surface layers of iron, nickel, cobalt, and copper alloys.

amorphous solid: Solid that does not exhibit a crystal structure. See *crystal structure*.

anisotropy: Variance of mechanical properties of a component in relation to the direction in which they are measured.

annealing: Heat treatment process that is used to soften a metal and consists of heating a component to a suitable temperature, holding at temperature, and cooling at a suitable rate. See *heat treatment process*.

annealing twins: Twin bands formed in the grain structure during recrystallization of certain cold-worked face-centered cubic metals. See *recrystallization*.

anodic coating: Conversion coating applied by the anodic oxidation of certain base metals.

anodizing: Electrolytic oxidation process in which the protective oxide film is artificially thickened to improve corrosion resistance.

arrest lines: Imprints of the temporary positions of the fracture front during the growth of a progressive fracture.

artifact: False indication that does not correspond to the true microstructure and is usually introduced during the various stages of sample preparation. See *microstructure*.

atom: Smallest building block of matter that can exist alone or in combination and cannot be divided without changing its basic character.

atomic bond: Force that holds atoms together.

atomic bonding: Process that occurs when atoms are bonded to each other, or held together by a force of attraction, to form physically visible solid materials.

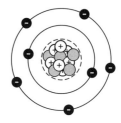

atom

atomic diameter: Closest distance two atoms approach each other in the solid state. See *atom*.

atomic number: Number of protons in the nucleus of a chemical element's atom. See *proton*, *nucleus*, and *atom*.

atomic structure: Organization of atoms and their basic parts that comprise the chemical elements.

atomic weight: Sum of the protons and neutrons in a chemical element's atom. See *proton*, *neutron*, and *atom*.

austempering: Interrupted quenching technique that consists of austenitizing followed by quenching in a medium maintained in the bainite transformation temperature range for the steel. See *quenching* and *austenitizing*.

austenite: Gamma solid solution of one or more elements in face-centered cubic iron.

austenite transformation product: Transformation product produced when steel is slowly cooled from the austenitizing temperature and includes the formation of ferrite, pearlite, or cementite. See *ferrite, pearlite,* and *cementite*.

austenitizing: Process of heating a steel to a temperature above the upper critical temperature to produce a microstructure of austenite.

autogenous weld: A weld in which no welding filler metal is used to make the joint.

autographic pyrometer: See *recording pyrometer*.

automatic polishing: Polishing process that establishes a complex motion for the mount relative to the rotation of the polishing wheel. See *mount*.

B

backlighting: Lighting method that uses a diffused light source to eliminate or soften shadow detail.

backscattered electrons: Electrons from the electron beam reflected back after interaction with the sample surface. See *electron*.

bainite: Austenite transformation product that is formed under continuous cooling or isothermal transformation conditions, which are between those that produce pearlitic or martensitic structures. See *austenite transformation product*.

bar: Elongated hot-worked or cold-worked steel product that is relatively thick and narrow. See *hot working* and *cold working*.

base metal: Region of metal that is joined by the welding process and is not metallurgically affected by the heat of welding.

billet: Square-shaped, semifinished form that is less than 20 cm × 20 cm (8 in. × 8 in.).

bimetallic coil pyrometer: Temperature-measuring instrument with two strips of metal (one with a high coefficient of thermal expansion and one with a low coefficient of thermal expansion) bonded together and wound into a helical coil. See *pyrometer*.

bimetallic coil pyrometer

binary phase diagram: Equilibrium or constitutional diagrams that indicate the phases present in binary alloy systems. See *alloy system*.

black oxide coating: Conversion coating applied to steel, stainless steel, copper, and other metals for decorative or functional purposes. See *conversion coating*.

blasting: Process in which particles moving at a high velocity clean metal surfaces or prepare them for coating.

bloom: 1. Slight haze that appears and is evidence of the first appearance of a microstructure during the etching process. 2. Square-shaped, semifinished form that is greater than 20 cm × 20 cm (8 in. × 8 in.). See *etching*.

blowhole: Hole formed by gas entrapped during solidification.

blue brittleness: Brittleness that occurs in some carbon and alloy steels after being tempered between 200°C and 370°C (392°F and 698°F).

boring: Enlarging a hole with the objective of producing a more accurate hole than by drilling.

braze welding: Method of welding that uses a welding filler metal with a liquidus above 450°C (840°F) and below the solidus of the base metal. The filler metal does not fill the joint by capillary action. See *liquidus* and *solidus*.

brazing: Welding process that joins metal components by heating them to the brazing temperature in the presence of a brazing filler metal that melts above 450°C (840°F) and below the melting point of the metals being joined. The brazing filler metal fills the joint area by capillary action.

build-up lighting: Lighting method that combines (adding or deleting one or more) light sources to achieve the desired lighting effect.

Burgers vector (b): Displacement of the material above the slip plane relative to the material below. See *slip*.

burning: Permanent damage to a metal caused by austenitizing too close to the melting temperature. See *austenitizing*.

burnishing: Finishing and sizing operation performed on previously machined or ground surfaces by the displacement, rather than the removal, of minute surface irregularities.

bursts: Internal fissures in ingots caused by internal tensile stresses resulting from heating and cooling. See *fissure*.

buttering: Surfacing in which one or more layers of welds are deposited on the edge of a metal.

C

capped steel: Steel similar in surface condition to rimmed steel, but intermediate between rimmed and semikilled steel in other properties. It is produced when the rimmed steel process is allowed to begin, but is then terminated after approximately 1 minute by sealing the mold with a cast iron cap. See *rimmed steel*.

capped steel

Canadian Standards: Standards administered by the Canadian Government Standardization Board (CGSB). These standards are identified by the letters CSA followed by a number. See *standard*.

captive foundry: Foundry that is usually adjacent to or connected to a large manufacturing company, such as an automobile plant.

carbide stabilizer: Alloying elements that promote cementite or alloy carbide formation. See *cementite*.

carbon diffusion: Spontaneous movement of carbon atoms within a material.

carbon equivalent: Factor that indicates the total relationship of the percentage of carbon, silicon, and phosphorus in the chemical composition.

carbonitriding: Case-hardening process for carbon steels and alloy steels that consists of holding them above the upper critical temperature to simultaneously absorb carbon and nitrogen. See *case hardening*.

carbon potential: Percentage of carbon in steel that will neither be carburized nor decarburized by gas. See *carburizing* and *decarburizing*.

carbon steel: Alloy of iron with carbon, manganese, and silicon, specifically containing up to 1.6% Mn and .6% Si, plus smaller amounts of sulfur and phosphorus.

carburization: Form of high-temperature corrosion that occurs in carbonaceous environments where the carbon in the environment diffuses into the base metal and forms carbides.

carburizing: Case-hardening process for low-carbon steels that uses an environment with sufficient carbon potential and a temperature above the upper critical temperature. See *case hardening* and *carbon potential*.

case hardening: Group of heat treatment processes that develop a thin, hard surface layer on a component and leave the core relatively soft, strong, and tough.

casting: Process of pouring molten metal into a prepared mold cavity of a desired shape and allowing the metal to solidify.

casting defect: Discontinuity in a casting that exhibits a size, shape, orientation, or location that makes it detrimental to the useful service life of the casting.

cathodic cleaning: Cleaning method that uses the flow of electrical current to strip adherent deposits on a metal sample.

cavitation: Surface damage caused by collapsing vapor bubbles in a flowing liquid.

cavity: Rounded void with smooth walls that is internal or intersects the surface of a casting.

cementation: Introduction of one or more elements to the outer portion of a metal object by diffusion at high temperature. See *diffusion*.

cementite: Compound of iron and carbon referred to as iron carbide, or Fe_3C.

center drilling: Drilling a conical hole in the end of a workpiece. See *drilling*.

centerless grinding: Grinding of cylindrical surfaces without the use of fixed centers to rotate the workpiece.

centrifugal casting: Casting process that consists of pouring molten metal into the cavity of a rotating mold, which can be made of various materials.

ceramic engineering: See *ceramics*.

ceramics: Study of the development and production of any product or related product made from a nonmetallic mineral by firing at high temperatures.

certificate of analysis: Notarized statement of chemical analysis that is legally binding.

certificate of heat analysis: Statement of the chemical analysis and weight percent of the chemical elements present in an ingot or a billet.

certificate of product analysis: Statement of the chemical analysis in weight percent of the end product (plate, wire, forgings, tubing, etc.) manufactured from an ingot or a billet.

checker chamber: In a furnace, a chamber containing a complex arrangement of heat-absorbent bricks used for heating air that moves through the chamber.

chemical element: Basic substance consisting of atoms of one type that, alone or combined with other chemical elements, constitute all matter. See *atom*.

chemical milling: Controlled dissolution of material by contact with chemical reagents of varying types and strengths, depending upon the particular alloy being machined. See *alloy*.

chemical polishing: Polishing technique where chemicals are used to dissolve rough peaks on a mount. See *mount*.

chemical property: Characteristic response of a material in chemical environments.

chemical vapor deposition (CVD): Surface modification process in which all the reactants are gases. A chemical reaction takes place in the vapor phase adjacent to, or on, a metal object, depositing the reaction products on the object.

chevron patterns: V-shaped marks that have apexes pointing toward the origin of a fracture.

chevron patterns

chill: A piece of metal incorporated into a mold surface in order to locally increase the cooling rate of the casting and alter the solidification pattern or metallurgical structure.

chilled iron: Areas of a casting that solidify more slowly and have a readily machinable gray iron microstructure that contains graphite. See *casting*.

chlorination: Form of high-temperature corrosion that occurs in environments containing chlorine or hydrogen chloride.

chromizing: Chromium enrichment of the surface of steel and other metals or alloys by a pack cementation process. See *cementation*.

cladding: Bonding together of two and/or three layers of metals to form a composite metal.

code: Standard or set of applicable regulations that a jurisdictional body has lawfully adopted.

coefficient of linear expansion: The increase in unit length per degree temperature rise.

coefficient of thermal expansion: The increase in a dimension of a metal per unit dimension per unit degree rise in temperature.

cold drawing: Pulling steel through a series of tapered dies in order to reduce the cross section to the required size.

cold extrusion: Converting the ingot to uniform lengths and shapes by forcing the steel plastically through a die orifice.

cold heading: Forcing of metal to cold flow into enlarged sections by endwise squeezing.

cold isostatic compacting: Process for forming a powder metallurgy compact by applying pressure equally from all directions to a metal powder in a sealed mold. See *powder metallurgy*.

cold preforming: Process in which cold-preformed components are clamped into fixtures made to the exact final shape and dimensions of the component. It is used with hot forming to correct springback. See *springback*.

cold rolling: Passing of unheated and previously hot-rolled bars, sheet, or strip through rollers at a distance from each other of less than the size of the steel entering them.

cold shut: See *lack of fusion*.

cold working: Plastic deformation processes performed below the recrystallization temperature, which leads to work hardening. See *recrystallization*.

cold work tool steel: Steel that has an alloy composition designed to provide moderate-to-high hardenability and good dimensional stability during heat treatment.

color coding: Identification method consisting of colored stripes painted on one end of the product prior to storage at a metal service center or user's warehouse.

composite engineering: See *composites*.

composites: Study of the applicability of combinations of engineering materials.

composition ratio: Chromium equivalent divided by nickel equivalent.

compression: Force that occurs when a load is applied axially on a test specimen in a compressive manner.

compression

contact fatigue: Cracking and subsequent break up of a surface that is subjected to alternating stresses, such as those produced under rolling contact or combined rolling and sliding.

continuous casting: Casting process that consists of pouring molten metal into a bottomless, water-cooled mold of simple cross section and continuously withdrawing solidified metal from the bottom of the mold.

continuous cooling transformation diagram: Plot of temperature against log time that indicates the austenite transformation products for specific steels when continuously cooled. See *austenite transformation product*.

controlling pyrometer: Temperature-measuring instrument with temperature-sensing elements and electrical circuitry that maintains a preset temperature in a furnace or any operation in which control is required. See *pyrometer*.

conversion coating: Surface treatment that converts the surface layer of a metal by oxidation into a constituent of a coating, contributing to a strong bond between the metal and the conversion coating.

cooling rate: Amount of degrees a metal cools per unit of time.

core drilling: Enlarging a hole with a chamfer-edged, multiple-flute drill.

coring: Condition of variable composition between the center and the surface of a unit of microstructure.

corrosion inhibitor: Chemical compound that, when added to a corrosive chemical, prevents attack of the metal, but allows dissolution of scales and deposits to continue.

corrosion potential: Potential exhibited by a metal in an electrolyte under steady-state conditions. See *electrolyte*.

counterboring: Enlarging a hole for a limited depth.

countersinking (chamfering): Cutting an angular opening into the end of a hole.

coupon: Specimen that is weighed and measured before cleaning and is made of the same metal as the failure specimen.

covalent bonding: Atomic bonding that occurs when the valency electrons are shared between like atoms so that each partner achieves a stable electron configuration. See *valency electron*.

creep resistance: Resistance to stress-induced strain. See *stress* and *strain*.

crevice corrosion: Localized corrosion that occurs at mating or closely fitting surfaces where easy access to the bulk corrosive environment is hindered.

critical cooling rate: Slowest continuous cooling rate that produces 100% martensite in a given steel. See *cooling rate*.

critical resolved shear stress: Resolved shear stress required to cause slip in a designated slip direction on a given slip plane and that is induced by a tensile or compressive force on the crystal. See *slip*.

critical temperature: Temperature in any specific steel composition at which the austenite phase change begins or is completed (for a specific rate of heating or cooling). See *austenite*.

crucible furnace: Melting furnace consisting of a refractory-lined pot that is externally heated by the combustion of a wide variety of fuels.

crystal rotation: Movement of individual crystals that are under an applied force, so that the crystallographic planes move into the most favorable orientation for slip. See *slip*.

crystal structure: Arrangement of atoms in a metal. See *physical metallurgy*.

cupronickel: Copper-nickel alloy that contains between 90% Cu and 30% Cu (the remaining balance is nickel). Cup-

ronickels are used extensively as tubing for heat exchangers and condensers.

Curie temperature: Temperature of magnetic transformation above which a metal is nonmagnetic, and below which it is magnetic.

cutting speed: Relative movement between a cutting tool and a workpiece.

cycle: Each complete application of stress. Also referred to as cyclic stress. See *stress*.

cylindrical grinding: Grinding of the outside diameter of a cylindrical workpiece.

D

damping capacity: Rate at which a material dissipates energy of vibration.

damping capacity test: Dynamic mechanical test that measures the decrease in amplitude of the torsional vibrations of a twisted cylindrical bar.

dealloying: Preferential dissolution of one microstructural constituent of an alloy, leaving a porous, weakened mass that retains the original shape of the component.

dealuminification: Preferential corrosion of aluminum that occurs in nickel-free aluminum bronzes.

decarburizing: Annealing process in which carbon is removed from steel by diffusion. See *annealing* and *diffusion*.

defective surface: Casting surface irregularity that is caused by incipient freezing from too low a casting temperature.

dendritic growth: Solid, three-dimensional growth that occurs in the melt in the form of shooting spikes. See *melt*.

density: Mass per unit volume of a material.

deoxidation: Removal of oxygen from molten metal.

deoxidizing flux: Metal powder that actively combines with the oxygen in molten metal.

dendritic growth

depolarization: Decrease in the polarization of an electron. Shown by a decrease in the slope of the polarization curve. See *polarization*.

deposition corrosion: Variation of galvanic corrosion that occurs when metals plate out from solutions on the surface of aluminum. See *galvanic corrosion*.

depth of field: Total depth that an image can be maintained in focus.

destructive examination: Examination that requires a test specimen to be cut, machined, broken, drilled, melted, or dissolved.

dew point: Temperature at which gas is saturated with water vapor.

dezincification: Form of corrosion in which the zinc constituent is preferentially removed.

diamagnetic metal: Metal that has low negative values of magnetic susceptibility and is weakly repelled by a magnetic field.

diamond rouge: Polishing powder used in polishing metal.

die casting: Casting process that consists of using substantial pressure to inject molten metal into the cavity of a metal mold.

diffused light: Lighting source that uses a semi-opaque screen (such as ground glass) to diffuse the light source, reduce glare, and soften harsh details.

diffusion: 1. Spontaneous movement of atoms or molecules to new locations within a material. 2. Spreading of a constituent in a gas, liquid, or solid, which tends to produce a uniform composition in a component.

dip coatings: Thin coatings used primarily for protection during shipment and storage and as primers for subsequent painting.

dipole: Atom that has positive and negative centers of charge that are slightly separated.

direct-reduced iron: Iron product made by using methods other than blast furnace reduction.

direct reduction: The process in which direct-reduced iron is produced.

discontinuities: Cracks in castings, which are caused by hot tearing, hot cracking, and lack of fusion (cold shut). See *casting*.

dislocation: Linear imperfection in the structure of a crystal.

dopant: Impurity added, usually in small amounts, to a pure metal to alter its properties.

draft: Single drawing step.

drilling: Process of making holes with a rotary end-cutting tool with one or more cutting lips.

ductile dimples: Mass of half cups that make up the morphology of a ductile fracture surface.

ductility: The ability of a material to deform plastically without fracturing.

E

eddy-current testing: Electromagnetic nondestructive testing technique for measuring physical and mechanical parameters of metals. See *nondestructive testing*.

edge dislocation: Row of mismatched atoms along a straight edge that are formed by extra partial planes of atoms within the body of a crystal.

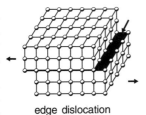

edge dislocation

elastic limit: Maximum stress to which a material is subjected without any permanent strain remaining after stress is completely removed. See *stress*.

electrical conductivity: Rate at which electrons move through atoms causing current to flow. See *electron* and *atom*.

electrical discharge machining: Removal of metal by rapid spark discharge between two electrodes, one being the workpiece and the other being the tool, with different polarities.

electrical resistivity: Electrical resistance of a unit volume of a material.

electric-arc furnace: Melting furnace that has a refractory-lined bowl in which

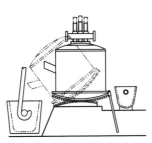

electric-arc furnace

material is melted by electric arcs from three carbon or graphite electrodes mounted in the roof of the furnace.

electrochemical machining: Controlled metal removal by the corrosion mechanism of anodic dissolution.

electroforming: Process in which a plating process is used to make a structural form.

electroless plating: Process in which metal ions in a dilute aqueous solution are deposited on a substrate by means of chemical reduction. See *ion*.

electrolyte: Liquid, most often a solution, that will conduct an electrical current.

electrolytic polishing: Polishing technique where rough peaks on a mount are dissolved by the passing of current over the mount through an electrolytic solution.

electron: In an atom, a particle with a negative electrical charge that is equal in magnitude to the protons.

electron probe microanalysis: Process used in conjunction with a scanning electron microscope to determine the chemical composition of a specimen.

electroplating: Process for depositing a thin layer of metal onto a metallic component that is made the cathode in an electrical circuit and immersed in a solution containing ions of the metal to be plated.

electrotype metal: Type metal that is used as a backing material for an electroformed copper shell, which carries an impression, and is not required to resist wear. See *type metal*.

embrittlement: Loss of ductility in a metal caused by corrosion.

emissivity: Measurement of the extent to which a surface deviates from the ideal radiative surface, which is a perfect emitter and absorber of thermal radiation.

enamelling iron: Sheet product made by decarburizing sheet steel. See *decarburizing*.

end-grain attack: Localized corrosion in a plane transverse to the direction of working.

end-quench hardenability curve: Plot of hardness readings versus the distance from the quenched end of a test specimen.

end milling: Milling performed with a tool having cutting edges on its cylindrical surface as well as on its end.

endothermic heat: Gaseous atmosphere made by passing a mixture of fuel gas (usually propane) and air over an externally heated catalyst.

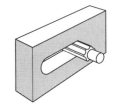

end milling

energy dispersive X-ray analysis (EDXA): Quantitative method of metal identification and chemical analysis of the surface area of a sample scanned by the electron beam.

erosion: Form of abrasive wear in which the force between an abrasive body and the wearing surface is large enough, but much smaller than those in gouging, to cause the removal of surface material. See *gouging*.

etching: Controlled selective attack on a metal surface for the revealing of microstructural detail. See *microstructure*.

eutectic reaction: Isothermal transformation in which a liquid transforms into two solid phases. See *phase*.

eutectoid reaction: Isothermal transformation of a higher temperature solid phase into two new solid phases. See *phase*.

eutectoid steel: Steel that has a carbon content of approximately .8% C.

excessive carbide dissolution: Process in which too much carbon dissolves in the steel.

exfoliation: Form of corrosion that occurs in certain cold-worked tempers and proceeds along selective subsurface paths parallel to the surface.

exothermic gas: Gaseous atmosphere made by passing a partially burned gas-air mixture over a catalyst that is heated by the heat from partial combustion.

explosion bonding: See *explosion welding*.

explosion cladding: See *explosion welding*.

explosion welding (EXW): Solid-state welding process that produces a weld by high-velocity impact of the components as a result of controlled detonation. Also referred to as explosion bonding or explosion cladding.

extractive metallurgy: Study of the extraction and purification of metals from their ores.

extrusion: Conversion of ingots or billets into lengths of uniform cross section by forcing the metal to flow plastically through a die by means of a ram. See *ingot* and *billet*.

F

face milling: Milling of a surface perpendicular to the axis of a cutter.

failure analysis report: Document that describes the cause(s) of failure and offers recommendations about preventing future failures.

faraday: Unit of electricity that is equal to 96,500 ampere-seconds.

fatigue: Failure of a material or component under alternating stresses, which has a maximum value less than the static tensile strength of the material.

face milling

fatigue strength: Stress at which a material fails by fatigue after a specific number of cycles.

ferrite: Alpha solid solution of one or more elements in body-centered cubic iron. When applied to carbon steels, it is an interstitial solid solution of carbon in alpha iron.

ferrite number: Arbitrary, standardized value indicating the ferrite content of an austenitic stainless steel casting or weld metal.

ferroalloys: Specific alloys of iron and other chemical elements that are used to introduce alloying elements into steel.

ferromagnetic metal: Metal that has extremely high and variable magnetic susceptibility and is strongly attracted to a magnetic field.

ferrous metallurgy: Study of alloys that have iron as the major alloying element. See *alloy* and *nonferrous metallurgy*.

fiber structure: Elongation of inclusions and impurities in the direction of cold work.

file hardness test: Hardness test for metals that uses a file to rub against the surface of the metal and results in degree of bite, which indicates hardness.

fill lighting: Lighting method that uses a small region of brighter light to increase detail on a dark area of a subject.

final polishing: Polishing in which a .3 micron-size to .05 micron-size alumina slurry in water is applied to a medium-nap cloth.

fins: Excessive amounts of metal created by solidification into the parting line of a mold.

fissure: Narrow opening or crack.

flaking: Internal cracks or bursts usually occurring during cooling down from rolling or forging.

flame hardening: Case-hardening process in which an intense flame is used to heat the surface layer of a workpiece to a temperature above the upper critical temperature. See *case hardening*.

flashlight: Lighting source that provides a pulse of very intense light.

flexure: In mechanical testing, a force that causes the bending of a test specimen.

foreign national standards: Standards developed by standard setting organizations in many countries, often in coordination with government agencies. See *standard*.

forging: Process of working metal to the desired shape by impact or pressure.

flame hardening

form milling: Milling of a contoured surface using a cutter with peripheral teeth having a profile that is specially made to match the required contour.

foundry flowsheet: Diagram that indicates the steps performed to produce castings and recycle scrap material.

foundry mark: Identification marking embossed on the exterior of a casting.

foundry type metal: Metal that contains the greatest amount of alloying additions and is used exclusively to cast type for hand composition. See *type metal*.

fractional distillation: Process in which the temperature is controlled to remove elements or contaminants that have different melting points from pure zinc.

foundry mark

fractography: Descriptive study of fracture surfaces utilizing photographs, sketches, and text.

fracture surface morphology: Texture, or topography, of the fracture surface.

fracture surface orientation: Angular relationship between a fracture and the direction of applied stress.

freezing range: Difference between the solidus and liquidus temperatures of a metal. See *solidus* and *liquidus*.

fretting: Localized surface damage at close fitting, highly loaded surfaces, produced by vibration.

full annealing: Heat treatment in which a steel component is held in the austenitizing temperature range and then cooled inside a furnace. See *austenitizing*.

G

galling: Condition whereby excessive friction between surface high spots results in localized welding with subsequent spalling (formation of surface slivers) and further roughening of the rubbing surfaces.

galvanic corrosion: Selective attack of one metal when it is electrically coupled to another metal or conducting nonmetal in an electrically conductive environment.

galvanizing: Process of coating a metal with zinc.

gas carburizing: Case-hardening technique in which a workpiece is placed in a furnace containing a gaseous carburizing environment. See *carburizing*.

gas nitriding: Nitriding technique that is performed with ammonia (NH_3). See *nitriding*.

gas or vapor pressure pyrometer: Temperature-measuring instrument with a bulb filled with a volatile liquid that is exposed to the metal to be measured. See *pyrometer*.

gas stirring: Refining technique that consists of bubbling gases, such as nitrogen or argon, through the ladle.

gate: Entrance to the casting.

gouging: Severe form of abrasive wear in which the force between an abrasive body and wearing surface is enough to macroscopically gouge, groove, or deeply scratch the surface.

grain boundary: Interface between grains that affects the behavior of metals.

grain growth: Undesirable increase in grain size that occurs as a result of heating the steel to elevated temperatures above the upper critical temperature. See *critical temperature*.

grain refinement: A method used to induce finer-than-normal grain size and may be achieved by thermally cycling a steel through the critical temperature range. See *critical temperature*.

grain boundary

grain size: Measure of the average dimensions of representative grains in a test specimen.

graphitic corrosion: Dealloying of gray cast iron, in which the metal constituent is selectively attacked, leaving behind semicontinuous network of graphite flakes. See *dealloying*.

graphitization: Transformation of pearlite slowly to graphite, which results in rendering the steel brittle.

graphitizer: Alloying elements that promote the formation of graphite.

green compact: Unsintered powder metallurgy compact. See *sintering* and *powder metallurgy*.

green rot: Corrosion that occurs in part-carburizing and part-oxidizing environments.

green sand molding: Sand casting that uses molds made of sand, clay, water, and other materials. The term "green" means that the molded sand remains moist.

growth: In cast irons, a permanent increase in volume as a result of prolonged exposure to elevated temperatures or repeated cycles of heating and cooling.

gun drilling: Drilling using special straight-flute drills with a single lip and cutting fluid at high pressures for deep hole drilling.

H

hardenability: Measure of the depth of hardening obtained when a metal is quenched. See *quenching*.

hardenability band: Bands that define the boundaries for the minimum and maximum end-quench hardenability curves for standard steels. See *end-quench hardenability curve*.

hard facing: Depositing of a surface filler metal to increase resistance properties.

heat-affected zone: Narrow region of the base metal adjacent to the weld bead, which is metallurgically altered by the heat of welding. See *weld bead*.

heat capacity: Ratio between the heat applied to a material and the temperature rise in the material.

heat checking: Cracking of metal due to alternating heating and cooling of the extreme surface of a metal.

heat colors: See *temper colors*.

heat sink: Metal or nonmetal that has high thermal conductivity and may be used to rapidly conduct heat away from a location. See *thermal conductivity*.

heat treatment process: Heating and cooling of metals in order to obtain desired properties.

heavy press forging: Plastically deforming metals between dies in presses at temperatures high enough to avoid strain hardening. See *strain hardening*.

high-speed tool steel: Steel that is typically employed for high-speed cutting operations.

hobbing: The squeezing process used for making molds for the plastics and die castings industry.

hot crack: Crack formed during cooling after solidification because of internal stresses developed in the casting.

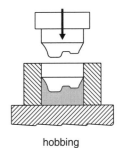

hobbing

hot dip aluminizing: Method of applying aluminum coatings to sheet steel, strip, or wire in continuous lengths and also to fabricated components.

hot dip galvanizing: Method of applying a coating of zinc onto ferrous substrates, primarily steel. See *substrate*.

hot dip lead coating: Method of applying a lead coating to a base metal.

hot dip tin coating: Method of applying a tin coating to a base metal.

hot dipping: Processes in which a component is dipped into a molten metal to produce a thin coating that enhances corrosion resistance or workability.

hot extrusion: Converting an ingot to uniform lengths and shapes by forcing the heated metal plastically through a die orifice.

hot forming: Forming process that increases formability and reduces springback and is used in the production of a great majority of titanium components. See *springback*.

hot isostatic compacting: Process for simultaneously heating and forming a powder metallurgy compact by applying pressure equally from all directions to a metal powder in a sealed mold at a temperature high enough for sintering to occur. See *powder metallurgy* and *sintering*.

hot rolling: Reduction of steel ingot size by rollers rotating in opposite directions and spaced at a distance less than the steel entering them. See *ingot*.

hot sizing: Process in which the fixtures are heated for a time long enough to cause the components to assume the correct shape. It is used with cold preforming to correct springback. See *springback*.

hot tear: Fracture formed during solidification because of hindered contraction.

hot top: Refractory-lined container placed on top of an ingot mold that absorbs heat less rapidly than an ingot mold and, therefore, maintains a reservoir of molten steel.

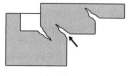

hot tear

hot working: Controlled mechanical operations performed above the recrystallization temperature for the purpose of shaping a product.

hot work tool steel: Steel that has been developed to withstand combinations of heat, pressure, and abrasion associated with manufacturing operations performed at high temperatures from 480°C to 760°C (900°F to 1400°F).

hydride embrittlement: Loss of ductility caused by corrosion. See *ductility*.

hydrogen attack: Form of high-temperature embrittlement that occurs in environments containing hydrogen, which is caused by diffusion of hydrogen into the steel where it reacts with the carbides to form methane gas, creating zones of weakness within the alloy. See *diffusion*.

hydrogen cracking: Damage caused by hydrogen present in an alloy. See *alloy*.

hypereutectoid steel: Steel with a carbon content that lies to the right of the eutectoid point on the iron-carbon diagram. It contains more than approximately .8% C.

hypoeutectoid steel: Steel with a carbon content that lies to the left of the eutectoid point on the iron-carbon diagram. It contains less than approximately .8% C.

I

ideal critical diameter: Largest diameter of any specific steel bar that is hardened to 50% martensite by a perfect quench. See *martensite* and *perfect quench*.

impedance: In eddy-current testing, the electrical property that changes with the loading. See *eddy-current testing* and *loading*.

impregnation: Process of filling pores of a material with a liquid, such as a lubricant.

impression die forging: Shaping of hot metal within the cavities or walls of two dies that come together to completely enclose it.

inclusion: 1. Small nonmetallic compound that is produced from metal production processes. 2. Particles of foreign material in the metal matrix. Also referred to as *structural anomalies*.

inclusion shape control: Refining technique that is used to alter the morphology of nonmetallic inclusions in steel. See *morphology*.

incomplete casting: Casting with missing portions resulting from premature solidification or from fracture after solidification.

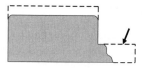

incomplete casting

incorrect dimensions: Any deviations from a mechanical drawing in terms of dimensions or alignment.

independent foundry: Foundry that usually performs a wide variety of work and has a large customer base.

induction furnace: A melting furnace that is essentially a transformer, in that a coil of water-cooled copper tubing outside the furnace acts as the primary winding, and the metal charge inside the furnace acts as the secondary winding.

induction hardening: Case-hardening process in which the surface layer of the workpiece is heated by electromagnetic induction. See *case hardening*.

ingot: A casting of simple shape, suitable for working or remelting.

ingot iron: Extremely pure type of iron that is soft and can be easily magnetized and demagnetized.

inhibitor additions: Additions of chemical compounds to the environment that cause a corrosion rate reduction.

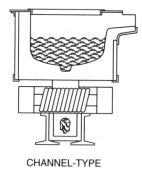

CHANNEL-TYPE

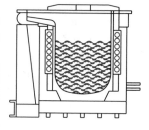

CORELESS-TYPE

induction furnace

insufficient carbide dissolution: Incomplete austenitizing that occurs when the temperature is too low. See *austenitizing*.

integrated steelworks: Large and complex operations consisting of all the necessary production units to manufacture a wide range of semifinished forms in carbon steels and alloy steels.

intergranular corrosion: Corrosion of a metal along its grain boundaries. See *grain boundary*.

intergranular separation: Fracture that propagates along the grain boundaries of a metal.

intergranular separation

intermediate phase: Structure usually formed between dissimilar chemical elements, often at a fixed composition or over a narrow composition range. Intermediate phases usually have covalent or ionic bonding and are nonmetallic in nature. See *covalent bonding* and *ionic bonding*.

intermetallic compound: Chemical compound that is formed between metallic chemical elements that are nonmetallic in behavior.

interrupted quenching techniques: Stepwise quenching processes that develop specific microstructures in steels and minimize distortion and cracking.

interstitial atom: Smaller atom of another element occupying a space between the atoms of a metal in a lattice. See *atom*.

interstitial atom

interstitial compound: Chemical compound formed between metals, such as titanium, tungsten, tantalum, or iron, and nonmetals, such as hydrogen, oxygen, carbon, boron, and nitrogen. The component chemical elements of interstitial compounds bear a fixed ratio to one another.

interstitial solid solution: Solid solution formed when the interstitial atoms (solute atoms) fit into the interstices (spaces) of the solvent metal crystal structure. See *interstitial atom*.

investment casting: Casting process that uses a plaster mold cavity that is formed around a wax pattern, which is melted out when the plaster is baked to harden it.

ion: Atom with a positive or negative electrical charge. See *atom*.

ionic bonding: Bonding that occurs when the valency electrons are exchanged between unlike atoms so that each atom gains or loses electrons to achieve a stable electron configuration. See *valency electron*.

ion implantation: Modification of the physical or chemical properties of the immediate surface of a metal by embedding appropriate atoms into it from a beam of ionized particles. See *atom*.

ion nitriding: Nitriding technique performed in a vacuum where high-voltage electrical energy forms a plasma through which nitrogen ions accelerate to impinge on the workpiece. See *nitriding*.

ironing: Process for smoothing and thinning the wall of a shell or cup by forcing a component through a die with a punch.

ironmaking: First stage in the manufacture of steel.

isopleth: Vertical sections through a space diagram that simplify ternary and more complex diagrams by indicating (freezing) the composition of one or more alloying components.

isotherm: Section through a phase diagram that depicts all phases in equilibrium for an alloy composition at one temperature. See *phase diagram*, *phase*, and *alloy*.

isothermal transformation diagram: Plot of temperature against log time that indicates austenite transformation products for specific steels under isothermal conditions.

isotope: Version of the same chemical element with the same atomic number but different atomic weight. See *atomic number* and *atomic weight*.

isotropy: Uniform properties in all directions.

J

Jominy end-quench test: Laboratory procedure for determining the hardenability of steels. See *hardenability*.

K

Kelvin: Temperature scale on which the unit of measure is equal to the Celsius degree and on which absolute zero is 0° (–273.16°C).

killed steel: Steel that is completely deoxidized by a deoxidizing agent. See *deoxidation*.

knifeline attack: Extremely narrow band of corrosive attack that may occur in stabilized stainless steels and some nickel-base alloys.

killed steel

Kroll process: Process for production of metallic titanium by reduction of titanium chloride with a more active metal, such as magnesium. It yields titanium in the form of powder.

L

lack of fusion: Discontinuity caused when two streams of liquid in a solidifying casting meet but fail to unite. Also referred to as *cold shut*.

ladle: Refractory-lined receptacle that is used for transferring and pouring molten steel.

ladle refining: Process that reduces detrimental internal characteristics in steel products, such as dissolved gases, sulfur, or nonmetallic inclusions.

lap: Surface irregularity caused by hot metal folding over and being pressed into the surface.

lapping: Surface finishing operation that uses fine abrasive grains rolled into a lapping material, such as cast iron, copper, or lead.

leak testing: Internal inspection technique that is performed on castings that are intended to withstand pressure.

leveling: Ability of an electrolyte to deposit metal in order to smooth out surface scratches and imperfections by filling them up with deposit. See *electrolyte*.

line defect: Defect associated with planes of atoms in the lattice. See *dislocation* and *stacking fault*.

Linotype® metal: Type metal that is machine die cast and used for the high-speed composition of newspapers. See *type metal*.

liquid carburizing: Case-hardening technique in which the workpiece is held in a molten salt bath to introduce carbon and sometimes nitrogen into the surface. See *carburizing* and *case hardening*.

liquid expansion pyrometer: Temperature-measuring instrument with a bulb containing a liquid, such as mercury or alcohol, that is exposed to the metal to be measured; bulb expands or contracts with temperature change. See *pyrometer*.

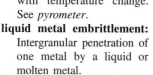

liquid expansion pyrometer

liquid metal embrittlement: Intergranular penetration of one metal by a liquid or molten metal.

liquid nitriding: Nitriding technique performed in a molten salt bath that contains a mixture of sodium or potassium chloride and sodium or potassium cyanide or cyanate. See *nitriding*.

liquid shrinkage: Shrinkage that occurs as the metal cools from the pouring temperature to the liquidus temperature, which is a range of 95°C to 150°C (200°F to 300°F).

liquidus: Locus (connection) of points on a phase diagram representing the temperatures at which each alloy in the system begins to solidify during cooling or completes melting during heating. See *alloy*.

loading: 1. Interaction between a probe and induced eddy currents, where the induced eddy currents produce a reactive magnetic field that opposes or reduces the magnetic field in the test probe that produced the currents. 2. Application of external forces. See *eddy-current testing*.

M

macrograph: Sketch of the etched surface of a specimen illustrating the key features.

macroscopic examination: Examination of a fracture surface performed with the naked eye, magnifying glass, or light microscope.

magnetic-particle testing: Nondestructive testing technique for detecting surface and subsurface imperfections and flaws in ferromagnetic materials, such as iron and steel. See *nondestructive testing*.

magnetic permeability: Measurement of the ease with which a ferromagnetic metal can be magnetized and demagnetized. See *ferromagnetic metal*.

magnetic susceptibility: Intensity of magnetism produced in a material when it is placed in a magnetic field.

magnetostriction: Dimensional change that a ferromagnetic metal exhibits when magnetized. See *ferromagnetic metal*.

main lighting: Primary lighting method that has a light source at an angle to the subject that is within a 40° to 60° range.

maintaining processes: Case-hardening processes that maintain the surface composition of steel, which consist of heating the component very rapidly so that only the surface layer is austenitized. See *austenitizing*.

martempering: Interrupted quenching technique that consists of quenching an austenitized component into molten salt or hot oil close to the M_s temperature, holding until temperature equalization has taken place, and cooling in air to a temperature below the M_f to complete the transformation to martensite. See *austenitizing* and *martensite*.

martensite: Austenite transformation product that forms when austenite is cooled rapidly, producing no time for carbon to diffuse and form pearlitic products. See *austenite transformation product* and *pearlitic product*.

materials engineering: Study of the evaluation of the characteristic properties of all materials.

materials sciences: Scientific and technological aspects of materials that are used to make engineering materials.

maximum corrosion attack: Loss of metal due to all forms of oxidation and to sulfides.

mechanical metallurgy: Study of the techniques and mechanical forces (factors) that shape or make finished forms of metal.

mechanical property: Characteristic dimensional change in response to an applied external or internal mechanical force.

mechanical twinning: Movement of planes of atoms in a lattice so that the two parts are mirror images of each other across the twinning plane. See *atom*.

mechanical twinning

melt: Molten metal.

melting point: Temperature at which a material passes from a solid to a liquid state.

merchant mills: Small steel manufacturing operations that produce a limited range of carbon steel finished forms.

metal: 1. Pure chemical element. Also referred to as *pure metal*. 2. Compound of two or more chemical elements of which one is a pure metal. Also referred to as *alloy*.

metallic bonding: Type of atomic bonding that occurs in a solid metal when the valency electrons leave individual atoms and are shared between all atoms in a free electron cloud. See *valency electron*.

metallizing: Formation of a metallic coating by an atomizing spray of molten metal.

metallograph: Metallurgical microscope that is equipped with a camera facility for photographing microstructures. See *microstructure*.

metallographic examination: Study of the microscopic features of material surfaces that have been specially prepared by cutting, grinding, polishing, and etching.

metallographic mount: See *mount*.

metallography: Technique for studying the grain structure (microstructure) of a metal.

metallurgical mount: See *mount*.

metal loss: Loss of metal due to massive oxides and sulfides.

metallurgical engineering: See *metallurgy*.

metallurgy: Science of metals and alloys devoted to the study of engineering materials. Also referred to as *metallurgical engineering*.

microbiologically influenced corrosion: Form of crevice corrosion that occurs in natural waters under deposits that harbor microorganisms.

microcrack: Small crack that is the result of localized hardening that develops in the metal from the fatigue stress.

microhardness testing: Testing technique used for measuring the hardness of the constituents in the microstructure of a metal.

microscopic examination: Examinations performed with a scanning electron microscope.

microstructure: Microscopic arrangement of the components, or phases, within a metal. See *phase*.

microstructural corrosion: Localized corrosion that selectively follows the microstructural constituents of alloys.

microvoid: Tiny cavity formed as the metal separates in areas with inclusions.

microvoid coalescence: Mechanism of ductile fractures that permits the fracture to zigzag across the metal surface.

milling: Removal of material from a workpiece by the use of a rotary tool with teeth that engage the workpiece as it moves past the tool.

modulus of elasticity: Ratio of stress-strain in the region below the proportional limit on the stress-strain curve. See *stress* and *strain*.

mold: Hollow shape into which molten metal is poured, allowed to solidify, and removed in the desired casting shape.

mold steel: Steel that contains chromium and nickel as the principal alloying elements. See *alloy*.

Monel 400 (CuNi alloy 400): Alloy widely used for corrosion-resistant applications.

monotype metal: Type metal in which only one type character is cast at a time so that a rapid cooling rate is possible and a harder alloy with a higher melting range is permissible. See *type metal*.

morphology: Shape of an inclusion.

mottled iron: Mixed structure of iron carbides and graphite in the boundary zone between the white and gray regions in chilled iron. See *chilled iron*.

mount: Device that is used for holding a specimen during metallographic preparation. Also referred to as *metallurgical mount* and *metallographic mount*. See *metallographic examination*.

N

napped cloth: Woven cloth containing fiber.

necking down: Reduction of the cross-sectional area of the test specimen when a load is applied.

neutral axis: Boundary line between the tensile and compressive stresses in a workpiece.

neutron: Atomic particle with a neutral electrical charge.

necking down

nil ductility transition (NDT) temperature: Temperature at which the impact behavior of a metal changes from ductile to brittle in the presence of a stress raiser. See *stress raiser*.

NiTiNOL: Alloy that was developed by the Naval Ordinance Laboratory (NOL) and contains 5% Ni and 45% Ti. Also referred to as *shape memory metal*.

nitriding: Subcritical case-hardening process that introduces nitrogen into the surface of a steel. See *case hardening*.

nitrocarburizing: Subcritical case-hardening process that consists of diffusing carbon and nitrogen into the surface of a workpiece for three hours at a temperature below 680°C (1255°F). See *case hardening*.

nondestructive testing: Any type of testing performed on an object that leaves it unchanged.

nonferrous metallurgy: Study of pure metals and alloy systems that do not contain iron as the major alloying element. See *alloy* and *ferrous metallurgy*.

normalizing: Commonly used heat treatment that decreases pearlite interlamellar spacing and refines grain size. The steel component is held in the austenitizing temperature range and cooled in still air. See *pearlite*.

notched bar impact test: Test that measures the force (produced by a dynamic load) needed to break a small machine-notched test specimen.

O

oil hole drilling: Use of a drill with one or more continuous holes through its shank to permit the passage of a high-pressure cutting fluid that emerges at the drill point and ejects chips.

open die forging: Hot mechanical forming that occurs between flat or shaped dies in which the metal flow is not completely restricted.

optical emission spectroscopy (OES): Metal identification method that uses light emitted from an unknown metal surface when it is arced (caused to spark) by an electric current.

optical pyrometer: Temperature-measuring instrument with two major components (telescope and control box) that compares the intensity of light emitted from the hot source with the intensity of light emitted from a standard source, such as a lamp filament. See *pyrometer*.

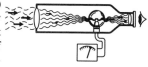

optical pyrometer

order-disorder transformation: Reversible phase change between two solid solutions that have the same unit cell structure. See *phase* and *unit cell*.

orthorhombic crystal structure: Crystal structure with crystals that have equal axes that are at right angles to each other.

overaging: Heat treating a metal for a longer time or at higher temperature, resulting in reduced hardness and strength.

overvoltage: Difference between the half-cell potential and the potential resulting from polarization.

oxidation: Corrosion reaction in which the corroded metal forms an oxide.

P

pack carburizing: Case-hardening technique in which carbon monoxide derived from a solid carbon containing compound decomposes at the metal surface into nascent (newborn) carbon and carbon dioxide. See *case hardening.*

paramagnetic metal: Metal that has low values of magnetic susceptibility and are weakly attracted to a magnetic field.

passive surface film: Film on the outer surface layer of a metal that has superior corrosion resistance.

pearlitic products: Austenite transformation products that consist of ferrite, pearlite, and cementite, which are formed when the cooling rate is slow enough to allow carbon to diffuse out of the austenite. See *austenite transformation product.*

pearlite: Lamellar (layered) aggregate of ferrite and cementite formed from the eutectoid decomposition of austenite during slow cooling. See *ferrite* and *cementite.*

pearlite interlamellar spacing: Distance between the ferrite and cementite lamellae. See *ferrite* and *cementite.*

penetrant testing: Nondestructive testing technique for detecting discontinuities, such as cracks and porosity, on the surface of a nonporous solid. See *nondestructive testing.*

perfect quench: Theoretical quench in which the surface of the bar cools instantaneously from the austenitizing temperature to the temperature of the quenching medium. See *quenching media.*

peritectic reaction: Isothermal reaction in which a solid phase reacts with the liquid from which it is solidifying to yield a second solid phase.

permanent mold casting: Casting process that requires a metal mold, which is used repeatedly for producing many castings of the same form.

pewter: Tin-base alloy that contains 90% Sn to 95% Sn and 1% Cu to 3% Cu, with the balance of the content being antimony.

pH: Hydrogen-ion activity that denotes the acidity or basicity of a solution.

pH scale: Scale from 0 to 14 that indicates whether a solution is acidic, neutral, or alkaline. A solution with a value of 0 to 7 is acidic, 7 is neutral, and 7 to 14 is alkaline.

phase: Any structure that is physically or crystallographically distinct and usually visible under a metallurgical microscope.

phase diagram: Graphic representation of the phases present in an alloy system at various temperatures and percentages of the alloying chemical elements.

ALKALINE

NEUTRAL →

ACIDIC

14
13
12
11
10
9
8
7
6
5
4
3
2
1
0

pH scale

photomacrograph: Photograph of etched surfaces of metals.

photomicrograph: Photograph of microstructures of metals. See *microstructure.*

physical metallurgy: Study of the effect of structure on the various properties of metals. See *crystal structure* and *microstructure.*

physical property: Characteristic response of a material to various forms of energy such as heat, light, electricity, and magnetism.

physical vapor deposition (PVD): Surface modification process in which a substrate is coated by a streaming vapor of specific atoms or molecules in a high-vacuum environment. See *substrate.*

pickling: Removal of surface oxides from metals by chemical or electrochemical reaction.

piercing: Cutting of holes into a metal.

pig iron: Metallic product obtained by the reduction of iron ores in a blast furnace. An impure form of iron.

pipe: 1. Central shrinkage cavity located in the upper portion of an ingot. It is formed during solidification of an ingot. 2. Hollow product of circular cross section having a continuous perimeter.

pitting: Localized corrosion that has the appearance of cavities (pits) and occurs on freely exposed surfaces.

planing: Machining process that uses a reciprocating motion between one or more single-point tools and a workpiece to produce flat or contoured surfaces.

plasma spraying: Thermal spraying process in which a plasma torch is used as a source of heat for melting and propelling the surfacing material to the workpiece. See *thermal spraying.*

plaster casting: Casting process that consists of pouring molten metal into the cavity of a plaster mold.

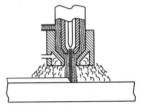

plasma spraying

plastic deformation: Alteration of shape that remains permanent after removal of the applied load that caused the alteration.

plastic strain: Strain that remains permanent after the stress is removed. See *strain* and *stress.*

plating: Forming an adherent layer of metal or alloy on an object.

point defect: Defect associated with discrete points in the crystal lattice. They include vacancies and interstitial atoms. See *vacancy* and *interstitial atom.*

point of failure: Point where a fracture occurs in a metallurgical test specimen.

polarization: The shift of the half-cell potentials toward each other as a result of corrosion. It indicates behavior when the anode and cathode are coupled.

polarizer: Device into which normal light passes and from which polarized light emerges.

polishing: Removal of metal by the action of abrasive grains carried to the component by a flexible support, generally a wheel or coated abrasive belt.

polycrystalline metals: Metals that have the orientation of the crystallographic planes that vary from grain to grain.

polygonization: Reduction in dislocation density caused by the formation of minute subgrains (about 1 micron across) within the stressed grains. See *dislocation.*

polymer engineering: See *polymers.*

polymers: Study of the development and production of synthetic organic materials.

porosity: Pockets of gas inside a metal that may be macroscopic or microscopic. See *macroscopic examination* and *microscopic examination.*

pouring basin: Reservoir placed close to the entrance to the casting, which is designed to provide a head of molten metal and minimize washing of sand into the casting.

powder metallurgy (PM): Metalworking process for forming precision components and shapes from metal and nonmetal powders, or mixtures of the two.

power brushing: Application of high-speed, revolving brushes to a workpiece in order to improve the surface appearance, remove sharp edges, burrs, fins, and particles.

precipitation (age) hardening: Delayed precipitation reaction consisting of the precipitation of finely dispersed particles of a second phase in a supersaturated solid solution, or one containing a second phase in excess of its solubility limits. See *precipitation reaction.*

precipitation reaction: Formation of a second phase within the grains of an original phase.

primary forming processes: Metal deformation methods that are used to make primary shapes.

process annealing: Annealing process performed below the lower critical temperature, which is designed to restore ductility to cold-worked steel products. See *annealing.*

process control: Manipulation of process or operating parameters to reduce corrosion.

proeutectoid phase: Phase that forms between the solvus and eutectoid temperatures. See *phase.*

progressive fractures: Fractures that grow in stages and have rest periods when no growth occurs (the component is idle or out of service).

property: Measurable or observable attribute of a material that is of a physical, mechanical, or chemical nature. See *physical property, mechanical property,* and *chemical property.*

proportional limit: Maximum stress at which strain is directly proportional to stress. See *stress* and *strain.*

proton: In an atom, a particle with a positive electrical charge. See *atom.*

pyrometer: Instrument (beyond the range of mercurial thermometers) that measures temperatures by the increase of electrical resistance in a metal, generation of electrical current of a thermocouple, or by the increase in intensity of light radiated by an incandescent body.

pyrophoric behavior: Spontaneous ignition of a metal when scratched or struck.

Q

quantitative metallography: Use of metallography to measure specific aspects of microstructures, such as grain size and density of nonmetallic inclusions. See *metallography.*

quench cracking: Fracture of a steel during quenching. See *quenching.*

quenching: Rapid cooling of a heated metal.

quenching media: Materials that are used to quench metal.

quench welding: Joining technique where a small length (2 in. to 3 in.) of metal is welded and then quenched with a wet rag. See *quenching.*

R

rachet marks: High and low points on a fracture surface. They are a sign of multiple fracture origins.

radial forging: Process using two or more moving dies or anvils for reducing the cross section of round billets, bars, and tubes.

radial lines: Markings on the fracture surface that point back toward the origin of the fracture.

radiation pyrometer: Temperature-measuring instrument that measures, at a convenient distance from a hot source, the radiant energy emitted from the hot source. See *pyrometer.*

radial lines

radiographic testing: Internal inspection technique that is widely used on castings to locate internal shrinkage, porosity, and slag inclusions. See *shrinkage, porosity,* and *inclusions.*

rake angle: In a cutting tool, the angle between the tool face that contacts the chip and the vertical.

Rankine: Absolute temperature scale on which the unit of measure is equal to a Fahrenheit degree and on which the freezing point of water is 491.69° and the boiling point is 671.69°.

reaming: Hole enlarging process that is used to produce a hole of accurate size and good surface finish, with limited stock removal.

recommended practice: Set of instructions for performing one or more operations or functions, other than identification, measurement, or evaluation of material.

recording pyrometer: Temperature-measuring instrument that automatically measures temperature and records the information on a chart or stores it in a computer. See *pyrometer.*

recovery: Reduction of residual stresses in a cold-worked, fabricated, or cast component, which is effected by holding the component at an elevated temperature. See *residual stress.*

recrystallization: Formation of a new, strain-free grain structure from that existing in a cold-worked metal. It is usually accomplished by heating.

reducing acid: Acid that reduces (dissolves) the passive surface film of a stainless steel. See *passive surface film.*

reduction: Removal of the chemically combined oxygen in iron ore.

reflected light: Lighting source that bounces light off a white card, wall, or ceiling.

relief angle: In a cutting tool, the angle that provides clearance between the tool and the workpiece.

replication: Technique used to make impressions of some or all of the fracture face.

replication cleaning: Cleaning method that uses replication tape to strip adherent deposits on a metal specimen. See *replication tape.*

relief angle

replication tape: Acetate film that is solvent-softened then pressed and allowed to harden to produce a replica of a surface.

re-pressing: Appling pressure to previously pressed and sintered powder metallurgy component. See *powder metallurgy.*

residual element: Additional chemical element in carbon steel that is introduced by the steelmaking process and is not detrimental if present in suitably low amounts. See *chemical element*.

residual stresses: Stresses that remain within a metal as a result of plastic deformation. See *stress* and *plastic deformation*.

resistance pyrometer: Temperature-measuring instrument that uses the change in electrical resistance of a metal, such as copper or nickel, with the change in temperature. See *pyrometer*.

restraint: Measure of the rigidity of a joint.

retained austenite: Austenite that has survived a heat treatment cycle in which it would have been expected to transform to other products.

reverberatory furnace: Melting furnace consisting of a large, refractory-lined hearth that is heated from above by an open flame.

rimmed steel: Steel that has little or no deoxidation. See *deoxidation*.

ring rolling: Shaping of seamless rings by reducing the cross section and increasing the circumference of a heated, donut-shaped blank between rotating wheels.

riser: A reservoir placed on a casting that fills with molten metal and provides a localized head of molten metal.

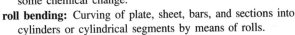

rimmed steel

roasting: Heating to effect some chemical change.

roll bending: Curving of plate, sheet, bars, and sections into cylinders or cylindrical segments by means of rolls.

roll forging: Shaping of stock between two driven rolls that rotate in opposite directions and have one or more matched sets of grooves in the rolls.

roll forming: Continuous process for forming metal from sheet, strip, or coiled stock into shapes of uniform cross section.

rolling: Reduction of the cross-sectional area of metal stock, or the general shaping of metal products through the use of metal rolls.

rotary forging: Process in which a workpiece is subjected to a combined rolling and pressing action between a bottom platen and a swiveling upper die with a conical working face.

rough polishing: Polishing process that is performed on a series of rotating wheels covered with a low-nap cloth. Successively finer grades of diamond rouge are applied to each wheel, usually starting at 45 micron size. See *napped cloth* and *diamond rouge*.

round: The semifinished form that is a circular section of any diameter.

runner: Horizontal channel along which molten metal flows when it is poured from a furnace.

rusting: Reaction of water containing dissolved oxygen with steel to form a series of corrosion products leading to a mixture of various iron oxides (geothite).

S

sacrificial coating: Coating that reduces the corrosion of a metal by coupling it to another metal that is more active in the service environment.

sand casting: Casting process that consists of pouring molten metal into the cavity of a sand mold.

scabs: In ingot casting, surface slivers caused by splashing and rapid solidification of metal when it is first poured and strikes the mold wall.

scaling: Loss of metal from a metal surface by the formation of a scale (oxide layer).

Schaeffler diagram: Geographical representation of the phases present by plotting nickel equivalent against chromium equivalent.

scleroscope hardness test: Hardness test that uses a test specimen that is freely supported horizontally and a glass tube that contains a diamond-pointed hammer positioned vertically over the specimen.

screw dislocation: Corresponding distorted lattice adjacent to the axis of a spiral structure in a crystal.

seam: Surface irregularity that results from a crack, a heavy cluster of nonmetallic inclusions, a deep lap, or a defect in the ingot surface that has become oxidized and is prevented from welding up during forging. See *inclusion* and *lap*.

screw dislocation

secondary electrons: Electrons emitted from the sample surface as a result of the electron beam interaction. See *electron*.

secondary hardening: Formation of alloy carbides that occurs in the temperature range 500°C to 650°C (930°F to 1200°F).

segregation: Any concentration of alloying chemical elements in a specific region of a metallic object. See *alloy* and *chemical element*.

semiconductor: Material that has a resistance to current flow that falls between the low resistance of metals (conductors) and the high resistance of nonmetals (insulators).

semifinished forms: One of four basic shapes of which all finished steel products are produced, which include blooms, billets, slabs, and rounds. See *bloom, billet, slab,* and *round*.

semikilled steel: Steel in which a deoxidizer, such as aluminum or silicon, is added to the molten metal to suppress the rimming action. See *deoxidation* and *rimmed steel*.

sensitization: Precipitation of chromium carbide in austenitic stainless steels. See *precipitation reaction*.

shape memory metal: See *NiTiNOL*.

shaping: Machining process that uses a reciprocating motion between one or more single-point tools and a workpiece to produce flat or contoured surfaces.

semikilled steel

shear stress (shear): Stress caused by two equal and parallel forces acting upon an object from opposite directions.

shell mold casting: Casting process that consists of pouring molten metal into the cavity of a mold, which consists of a thin shell of resin-bonded sand.

shock-resisting tool steel: Steel that has a low carbon content (.4% C to .6% C) and contains manganese, silicon, tungsten, and molybdenum, which offers a combination of high strength, high toughness, high ductility, and low-to-medium wear resistance.

shrinkage: Volume reduction in a casting, accompanying the temperature drop from the pouring temperature to room temperature.

shrinkage cavity: Cavity that has a rough shape and sometimes penetrates deep into a casting. They are caused by lack of proper feeding or nonprogressive solidification.

sintering: Bonding of the particles of the green compact by heating it at a high temperature below the melting point of the metal. See *green compact*.

sizing: Squeezing process usually performed in an open die and commonly used to sharpen corners on stampings, flatten areas around pierced holes, or develop exact dimensions along a specific axis.

slab: Rectangular-shaped, semifinished form that has a width-to-thickness ratio of 2:1 or greater.

slab milling: Milling of a surface parallel to the axis of a helical, multiple-tooth cutter mounted on an arbor.

slack quenching: Incomplete hardening of a steel caused by quenching from the austenitizing temperature at a rate slower than the critical cooling rate. See *critical cooling rate*.

slab milling

slaked dolomine: Dolomite rock that has been roasted and mixed with water. See *roasting*.

slip: Process of plastic deformation in which one part of a metal crystal (grain) undergoes a shear displacement relative to another. See *plastic deformation* and *crystal structure*.

slip bands: Groups of closely spaced parallel slip displacements, which appear as single lines when observed under the optical microscope. See *slip*.

slip systems: Combination of slip planes and directions in the crystal lattice in which plastic deformation by slip occurs most favorably. See *slip* and *plastic deformation*.

solder: Alloy that melts below 450°C (840°F) and below the lowest melting point of the base metals. See *base metal*.

special-purpose tool steel: Steel that contains small amounts of chromium, vanadium, nickel, and molybdenum and is used in applications requiring good strength, good toughness, scratch-resistant surfaces, and gall resistance.

S-N curve: Record of the amplitude of the cyclic stress (Σ or S) plotted against the number of cycles to failure (N). See *cycle*.

soldering: Welding process that joins metal components by heating them to the soldering temperature in the presence of a solder that melts below 450°C (840°F) and below the lowest melting point of the metals being joined.

solidification (solid) shrinkage: Shrinkage that occurs as solidified metal contracts when it cools.

solid solution: Product formed when the base metal (solvent metal) incorporates atoms of the other metal (solute metal) into its crystal structure. See *atom* and *crystal structure*.

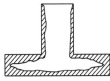

solidification shrinkage

solidus: Locus of points on a phase diagram representing the temperatures at which each alloy in the system completes solidification during cooling or begins to melt during heating. See *phase diagram*.

solution annealing: High-temperature heat treatment that dissolves precipitates such as carbides and age-hardening compounds. See *heat-treatment processes*.

solvus: Temperatures at which various compositions of solid phases coexist with other solid phases.

space lattice: Regular array of points produced by lines connected through the points.

spade drilling: Drilling with a flat-blade drill tip.

spark testing: Method of identifying metals, such as carbon steels, low-alloy steels, and tool steels, by visual examination of the spark stream (spark pattern).

specification: Statement of technical and commercial requirements that a product must meet.

specific gravity: Ratio of the density of a material to the density of water. See *density*.

specific heat: Ratio of the heat capacity of a material to the heat capacity of water. See *heat capacity*.

spheroidizing: Annealing process that produces a spheroidal (globular) form of carbide in a steel product. See *annealing*.

spinning: Process of forming disks or tubing into cones, dish shapes, hemispheres, hollow cylinders, and other circular shapes by the combined forces of rotation and pressure.

spinodal decomposition: Growth of compositionally different waves within the crystal structure, without any basic change in crystal structure. See *crystal structure*.

spotlight: Intense lighting source obtained from a single bulb in a reflector.

springback: Inaccuracy in shape and dimensions that results in an increase of resistance to plastic deformation. See *plastic deformation*.

sprue: Short channel between a pouring basin and a casting. See *pouring basin* and *casting*.

stabilizing: Annealing process performed on corrosion-resistant nickel-chromium-molybdenum-iron alloys that contain additions of titanium, niobium, or tantalum to stabilize them.

stacking fault: Two-dimensional deviation from the normal stacking sequence of atoms in a crystal.

staged heating: Heating at a controlled rate to a set temperature, holding the component until the temperature equalizes throughout the section, and then continuing the heating at a higher rate to the austenitizing temperature. See *austenitizing*.

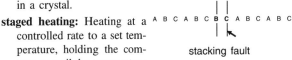

stacking fault

standard: Document that, by agreement, serves as a model in the measurement of a property or the establishment of a procedure.

static tensile strength: Tensile strength of a material before it is subjected to stresses. See *tensile strength* and *stress*.

steelmaking: The process of refining pig iron, directly reduced iron, and scrap steel into carbon steel and various alloys of steel.

stencil marking: Identification method that uses printed markings to describe alloy type, relevant standard designations, and dimension of a product form.

step drilling: Use of a multiple diameter drill to produce a hole having one or more diameters.

stereotype metal: Type metals that are used directly for printing and must be harder than electrotype. See *type metal*.

strain: Accompanying change in dimensions when a load induces stress in a material. See *stress*.

strain hardening: Increase in strength and hardness of a metal and the corresponding decrease in ductility due to plastic deformation by cold working. See *cold working*.

stress: Force per unit area.

stress-corrosion cracking: Crack formation in an alloy exposed to a specific corrosive, often intensified by the presence of tensile stresses. See *stress*.

stress equalizing: Low-temperature heat treatment that is performed to balance stresses in cold-worked material without appreciably decreasing the mechanical strength produced by cold working. See *cold working*.

stress raiser: Change in contour or discontinuity in structure that results in localized stresses. See *stress*.

stress relieving: Heating a metal to a suitable temperature, holding it long enough to reduce residual stresses, and cooling it slowly enough to minimize the development of new residual stresses. See *residual stress*.

stretch forming: Process of forming sheet metal by applying tension, or stretch, and then wrapping it around a die of the desired shape.

striations: Minute steps that indicate the growth of a fatigue crack.

stringers: Elongated configuration of foreign material aligned in the direction of working.

structural anomalies: See *inclusions*.

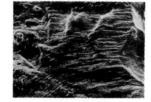

striations

structural shapes: Hot-rolled flanged shapes having at least one dimension of the cross section 3 in. or greater.

substitutional solid solution: Solid solution formed when the solute metal atoms are substituted for the solvent metal atoms in the crystal structure. See *crystal structure*.

substrate: Layer of metal underlying a coating.

sulfidation: Form of high-temperature corrosion that occurs in sulfur containing environments, such as hydrogen sulfide and sulfur dioxide.

superalloy: High-strength, often complex alloy having resistance to high temperature. See *alloy*.

superconductivity: Complete reduction of electrical resistance in a material at temperatures near absolute zero ($-273°C$ or $-460°F$).

superplastic zinc: Zinc-aluminum eutectic that is formed in the same manner as plastics.

surface grinding: Production of a flat surface as a workpiece passes under a grinding wheel.

swaging: Squeezing processes in which the material flows perpendicular to the applied force.

swarf: Mixture of grinding chips and fine particles of abrasive.

swell: Excessive amount of metal in the vicinity of gates or beneath the sprue. See *gate* and *sprue*.

T

taper sectioning: Mounting technique that increases the magnification available when examining a specimen.

technical society: Organization comprised of groups of engineers and scientists that are united by a common professional interest.

temper colors: Indications of the thickness of an oxide film that forms when steel is heated in air. They provide an approximate visual method for estimating the temperature of steel when it is heat treated.

temperature-indicating crayon: Material that melts when a temperature range has been met and is used for monitoring the temperature of a surface that must meet some specified minimum or maximum value.

temper designations: Letters that indicate the final condition of cold-worked (H) or heat-treated (T) material.

tempered martensite embrittlement: Formation of embrittling cementite from the decomposition of retained austenite during the second stage of tempering.

temper embrittlement: Embrittlement that occurs mostly in low-alloy steels that are tempered and then slowly cooled between $575°C$ and $375°C$ ($1067°F$ and $677°F$).

tempering: Heating of a quenched steel in a furnace to a specified temperature below the lower critical temperature and then allowing it to cool at any desired rate.

tensile strength: Ratio of maximum load to original cross-sectional area. Also referred to as *ultimate strength*.

tension: Force that occurs when the load (stress) is applied axially (parallel to the axis) on the test specimen in a stretching manner. See *stress*.

tension

tentative standard: Standard that is issued on a trial basis. See *standard*.

terminal solid solution: Any solid phase of limited compositional range based on one of the components of the alloy system, pure metal, intermetallic compound, or interstitial compound. See *alloy system, metal, intermetallic compound,* and *interstitial compound*.

ternary phase diagram: Equilibrium or constitutional diagram that indicates the phases present in ternary alloy systems, or alloys consisting of three components (chemical elements). See *alloy*.

ternary solid solution alloy: Alloy that solidifies in a manner similar to binary solid solutions. See *alloy*.

terne: An alloy of lead containing 3% Sn to 15% Sn.

test method: Set of instructions for the identification, measurement, or evaluation of the properties of a material.

thread rolling: Production of threads by rolling a workpiece between two grooved die plates, one of which is in motion, or between rotating grooved circular rolls.

thermal capacity: Amount of thermal energy required to raise a unit mass by one degree. See *thermal energy*.

thermal conductivity: Rate at which thermal energy (heat) flows through a material. See *thermal energy*.

thermal crack: External or internal crack that occurs as a result of nonuniform temperatures in the forging.

thermal energy: Energy produced by heat.

thermal spraying (THSP): Group of processes in which finely divided metallic or nonmetallic materials are deposited in a molten or semimolten condition to form a coating.

thermocouple: Device with two dissimilar metal wires that measure potential at the hot junction.

thermoelectric pyrometer: Temperature-measuring instrument with a thermocouple and a millivolt-meter that uses thermoelectric potential to measure temperature. See *pyrometer*.

thermopile: Small group of thermocouples contained within a pyrometer. See *thermocouple*.

thermowell: Protective sheath used to protect the thermocouple from the corrosive effects of a high-temperature environment and also from mechanical abuse. See *thermocouple*.

throwing power: The ability of an electrolyte to deposit metal of a uniform thickness across the surface of a cathode. See *electrolyte*.

tie line: Horizontal line drawn on a phase diagram that represents constant temperature and that connects the compositions of a pair of coexisting phases. See *phase*.

tool steel: Class of steel, generally with high-carbon and high-alloy content. They are characterized by high hardness and wear resistance, often accompanied by toughness and resistance to elevated temperature softening.

torr: Unit of measure equal to $\frac{1}{760}$ of atmospheric pressure.

torsion: In mechanical testing, an internal resisting force in which shear stresses occur by twisting of a test specimen. See *shear stress*.

torsion

toughness: Ability of a metal to absorb energy (high strain rates) and deform plastically before fracturing.

toughness testing: Test used to assess the resistance of metals to brittle fracture propagation (spreading) in the presence of stress raisers.

trade association: Organization representing the producers of metals or specific types of products.

transgranular cleavage: Fracture that propagates within the grains along specific crystallographic planes.

transgranular cleavage

trepanning: Machining process for producing a circular hole or groove in solid stock, or for producing a disk, cylinder, or tube from solid stock, by the action of a tool containing one or more cutters (usually single point) revolving around a center.

tropometer: Device used to measure degree of twist.

tubing: Hollow steel product of round, square, or other cross section having a continuous perimeter.

tubular shapes: Steel products that are hollow.

tumbling: Finishing process that consists of loading workpieces into a barrel approximately 60% full of abrasive grains, wood chips, sawdust, natural or artificial stones, cinders, sand, metal slugs, or other scouring agents.

turning: Generation of cylindrical forms by removing metal with a single-point cutting tool moving parallel to the axis of rotation of the workpiece.

type metals: Alloys of lead, antimony, and tin used extensively in the printing industry for typesetting machines.

U

ultimate strength: See *tensile strength*.

ultrasonic cleaning: Cleaning method that uses an organic solvent to remove lightly adhering grease, deposits, or liquid penetrant residues.

ultrasonic testing: Internal inspection technique that is used to detect cracks and measure wall thickness.

Unified Numbering System (UNS): Designation system, created jointly by American Standards for Testing and Materials and Society of Automotive Engineers, that unifies all families of metals and alloys. This system of identification consists of an uppercase letter followed by five numbers.

unit cell: Smallest arrangement of atoms that repeats itself through the space lattice.

United States Government Military Standards (MIL Standards): Department of Defense standards that cover the specifications of many materials used by the armed forces, but are not to be restricted to them.

upset forging: Process in which pressure is used to gather a large amount of material at one end of a bar.

V

vacancy: Unoccupied lattice point.

vacuum degassing: Refining technique where molten steel is subjected to low pressure to help remove hydrogen and oxygen.

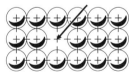

vacancy

valency electron: Electron that occupies the outer shell of an atom. See *electron* and *atom*.

Van der Waals bonding: Bonding that occurs when there is no exchanging or sharing of electrons. See *electron*.

W

weld bead: Weld resulting from the addition of filler metal to a joint.

welding electrode: Component of the welding circuit through which current is conducted and that terminates at the arc, molten conductive slag, or base metal. See *base metal*.

welding rod: Form of welding filler metal, usually manufactured in straight lengths, that does not conduct the welding current.

weldment: Assembly whose component parts are joined by welding.

weld overlay: Form of hard facing that is applied by oxyacetylene or arc welding processes using hard facing welding rods or electrodes. See *hard facing*.

wettability: Ability of a liquid to adhere to a solid.

white metal: Tin-base casting alloy containing 92% Sn and 8% Sb.

wire: Steel which is thin, flexible, continuous, and usually circular in cross section.

wire rod: Steel product rolled from billet in a rod mill and is used primarily in the manufacture of wire.

wrought iron: Pure iron containing silicate slag elongated in the hot-working direction. See *hot working*.

X

X-ray map: Image of the distribution of the characteristic X rays emitted for a specific element on a selected area of the sample surface.

X-ray mapping: Qualitative method of displaying the results of energy dispersive X-ray analysis. See *energy dispersive X-ray analysis*.

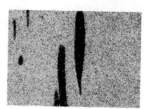

X-ray map

Y

yield point: Point at which strain occurs without an increase in stress. See *strain* and *stress*.

Index

A

Abrasion-resistant iron, 247–248

Abrasives, 409–10, 412

Absolute temperature scale, 15–16

Absolute zero, 13, 15, 464

Acid cleaning, 127

Active-passive metals, 472–473, 481, 483

Adhesive wear, 427

Aeration, 475–476

Age hardening. See Precipitation hardening

Aggressive species, 474–475, 478–479, 482

Alclad products, 326, 332

Allotropic metals, 52, 89, 181–182, 184, 346, 375

Allotropy, 89

Allowable stress data, 23, 35

Alloy, 5

Alloy Castings Institute (ACI), 153, 254, 308, 427

Alloy iron, 241, 247

Alloy steels, 178–79, 191, 199, 202–04

Alloy system, 73

Alloying, 247, 377

Alloying elements. See Master alloys
of steel, 206, 292

Alloys
crystal structure of, 52
ternary, 73, 75, 92–93

Alpha alloys, 347

Alpha brasses, 289–90

Alpha-beta alloys, 348

Aluminizing. See Hot dipping

Aluminum, 323

Aluminum alloys, 153, 310, 312, 323–26, 379, 392, 401, 407, 416, 466–68, 483

Aluminum Association (AA), 153, 325

Aluminum bronzes, 156, 281–282, 284, 294–96, 301–04

Aluminum pack diffusion, 443–45

American Iron and Steel Institute (AISI), 152, 156, 175–79, 201, 229–30, 254, 261, 424

American National Standards Institute (ANSI), 18, 156–57

American Petroleum Institute (API), 153, 180, 421

American Society of Mechanical Engineers (ASME), 35, 152, 154, 180, 254, 420–21

American Society for Testing and Materials (ASTM), 25, 31, 33, 37, 39–41, 43, 45, 69–72, 77, 113, 128, 134, 136, 138–40, 151–52, 154–55, 156–58, 177, 179–80, 188, 200–01, 243, 246, 254, 284, 311, 335–36, 346, 359, 388, 424, 434, 442, 458, 464, 466

American Water Works Association (AWWA), 153

American Welding Society (AWS), 140, 154, 254, 420–21

Ammonia cracking, 485, 487

Amorphous solids, 51

Analytical balance, 142–43

Angstrom, 55, 130

Anistropy, 106

Anneal-resistant coppers, 286, 300

Annealing, 14, 19, 90, 106, 108–13, 161, 171–72, 182, 195, 210, 218–20, 235, 238, 245, 259–60, 263, 265, 266–67, 272, 275–76, 282, 284, 285–87, 289, 297, 305–07, 309, 314–15, 319, 324–25, 340, 348–49, 387–88, 403–04, 461, 490

Annealing twins, 112

Anode polarization curves, 472, 478

Anode reactions, 469

Anodic coating, 434

Anodic protection, 478

Anodizing, 327, 331–32, 362, 368, 392, 432, 434

Arc welding, 413–15, 417, 429

Arc welding overlays, 429

Arrest lines, 120–22

Artifacts, 57, 59, 62, 67–68, 147

As-quenched, 201, 213, 215

Atom, 2

Atomic bonding, 10, 47, 49–51, 85, 100

Atomic diameter, 49

Atomic number, 47

Atomic percent, 75

Atomic structure, 47

Atomic weight, 47–48

Austempering, 193, 217

Austenite, 181–97, 207–17, 221, 235, 242, 250, 263, 387
transformation diagrams, 194
transformation products, 191–93, 195, 197

Austenitic stainless steels, 11, 45, 112, 152, 255, 261, 263–78, 313–14, 317, 320, 457–67, 476, 481, 483, 485–86, 490

Austenitizing, 191, 194, 197–98, 200, 229, 233–36, 259

Autographic pyrometer. See Recording pyrometer

B

Babbitts, 294, 358–59

Backlighting, 129

Backscattered electrons, 130, 132

Backward extrusion, 397

Bainite, 191, 193, 196, 198, 202, 207, 216–17

Bar, 171

Base metal, 423

Basic oxygen furnace, 162

Bauxite ore, 323

Beachmarks. See Arrest lines

Bearing metals, 357–58

Bending, 399

Benefication, 1

Beryllium, 286–88, 297–98, 301, 373–74

Beryllium coppers, 282, 284, 286–88, 297–98, 301

Beta alloys, 348

Beta brasses, 90, 289–91

Billets, 166

Bimetallic coil pyrometer, 16

Binary phase diagrams, 78

Black oxide coating, 432

Blast furnace reduction, 159